ABOUT ISLAND PRESS

Island Press, a nonprofit organization, publishes, markets, and distributes the most advanced thinking on the conservation of our natural resources—books about soil, land, water, forests, wildlife, and hazardous and toxic wastes. These books are practical tools used by public officials, business and industry leaders, natural resource managers, and concerned citizens working to solve both local and global resource problems.

Founded in 1978, Island Press reorganized in 1984 to meet the increasing demand for substantive books on all resource-related issues. Island Press publishes and distributes under its own imprint and offers these services to other nonprofit organizations.

Support for Island Press is provided by Apple Computers, Inc., Mary Reynolds Babcock Foundation, Geraldine R. Dodge Foundation, The Educational Foundation of America, The Charles Engelhard Foundation, The Ford Foundation, Glen Eagles Foundation, The George Gund Foundation, William and Flora Hewlett Foundation, The Joyce Foundation, The J. M. Kaplan Fund, The John D. and Catherine T. MacArthur Foundation, The Andrew W. Mellon Foundation, The Joyce Mertz-Gilmore Foundation, The New-Land Foundation, The Jessie Smith Noyes Foundation, The J. N. Pew, Jr. Charitable Trust, Alida Rockefeller, The Rockefeller Brothers Fund, The Florence and John Schumann Foundation, The Tides Foundation, and individual donors.

WETLAND CREATION AND RESTORATION

WETLAND CREATION AND RESTORATION

THE STATUS OF THE SCIENCE

Edited by

JON A. KUSLER

and

MARY E. KENTULA

Foreword by Senator George J. Mitchell

ISLAND PRESS

Washington, D.C. □ *Covelo, California*

Library of Congress Cataloging-in-Publication Data

Wetland creation and restoration : the status of the science / edited by Jon A. Kusler and Mary E. Kentula.
p. cm.
Includes bibliographical references.
ISBN 1-55963-045-0 (alk. paper).—ISBN 1-55963-044-2 (pbk. : alk. paper)
1. Wetland ecology. 2. Restoration ecology. 3. Wetland conservation.
I. Kusler, Jon A. II. Kentula, Mary E.
QH541.5.M3W46 1990
333.91′8—dc20 90-4053
CIP

Printed on recycled, acid-free paper

Manufactured in the United States of America
10 9 8 7 6 5 4 3

This project was originally funded by the United States Environmental Protection Agency (EPA) and conducted at EPA's Research Laboratory in Corvallis, Oregon, through Contract 68-C8-0006 to NSI Technology Services Corp. and Contract CR-814298-01-0 to the Association of State Wetland Managers.

CONTENTS

Part 2 Perspectives

FOREWORD

The wetland ecosystems of our nation are some of our most biologically productive areas. They are libraries of nature that contain volumes of priceless genetic information.

Unfortunately, much of America's wetlands heritage has been destroyed. Less than half of the 215 million acres present in the 48 conterminous states at the time of European settlement now remain. From the mid-1950s to the mid-1970s, 9 million acres of wetlands were drained, filled, and cleared. More troubling is that the destruction of these aquatic areas continues today at a rate of 450,000 acres per year—an area 12 times the size of the District of Columbia.

For most of this century, wetlands have been viewed narrowly either as wastelands or as areas providing little benefits beyond support of waterfowl populations.

Over the past decade, however, there has been a growing appreciation that wetlands are essential not only to waterfowl, but also to protect fisheries, shellfisheries, drinking water supplies, and flood-prone areas. The link between maintenance of wetland ecosystems and preservation of biological diversity has become increasingly clear as more and more wetland-dependent animals and plants require protection under the Endangered Species Act.

As the American people have become more aware of the pace at which wetlands are being destroyed and the importance of preserving these ecosystems, they have recognized that the nation's past wetland policies have produced unacceptably high economic and environmental costs. This growing public concern, in turn, has prompted governments at the federal, state, and local levels to take steps to stem and reverse losses of these aquatic areas through increased acquisition, restoration and enhancement, tighter regulation, removal of incentives that encourage wetlands destruction and creation of incentives to encourage wetlands protection by the private sector.

During the 1980s, Congress enacted legislation to implement some form of nearly every one of these approaches to wetlands protection. The 1982 Coastal Barrier Resources Act, the 1985 Food Security Act, and the 1986 Tax Reform Act all removed incentives that had encouraged wetlands destruction. Through the Conservation Reserve Program, the Food Security Act also has begun to provide incentives for farmers to protect wetlands. The Emergency Wetlands Resources Act of 1986 doubled the amount of guaranteed funding for federal wetlands acquisition.

More recently, efforts by Congress and federal agencies to maintain the acreage and values of the nation's wetlands have expanded toward encouraging federal, state, and private restoration of degraded or previously functioning wetlands.

The 1985 Food Security Act increased opportunities for wetlands restoration on Farmers Home Administration inventory lands and on Conservation Reserve Program lands. As a result, the U.S. Fish and Wildlife Service restored almost 55,000 acres of wetlands on these agricultural lands from 1987 through 1989. These wetland restoration opportunities are continuing to increase. Approximately 90,000 acres are targeted for restoration by the service in 1990 and 1991. Legislation before the U.S. Senate, which has been proposed for inclusion in a new farm bill, would provide financial incentives to farmers to restore and protect millions of acres of wetlands.

The North American Wetlands Conservation Act enacted in 1989 will greatly expand restoration of wetlands over the next 15 years by providing $25 million annually in federal matching funds for cooperative efforts with states and private organizations for this purpose. This new law encourages partnerships among public agencies and private interests to restore, protect, and enhance wetland ecosystems for waterfowl and other migratory birds. Restoration, protection, and enhancement of nearly 2 million acres of wetland habitats in the United States and almost 4 million acres in Canada are needed to rebuild the continent's waterfowl populations to the levels of a decade ago.

The 101st Congress is considering legislation to fund, at unprecedented levels, the restoration and creation of coastal wetlands in Louisiana. Each year, from the early 1950s to the late 1970s, about 50 square miles, or 32,000 acres, of coastal wetlands were lost in that state. These losses account for over two-thirds of the national coastal wetlands loss during that period. Wetlands restoration and creation offer the only real hope of regaining past losses and offsetting future losses.

Wetlands restoration and creation are receiving greater attention as means of meeting the Clean Water Act's goal to restore and maintain the nation's wetlands and other waters. Under Section 404 of that act, restoration, enhancement, or creation of wetlands is required, where practicable, to offset any unavoidable adverse impacts that can not otherwise be minimized.

Questions about how well restored or created wetlands compensate for destruction of existing, naturally functioning wetlands assume increased significance as the United States moves toward implementing a goal of no overall net loss of wetlands. Because there will always be some wetlands loss that can not be avoided, this goal clearly envisions offsetting unavoidable destruction with compensatory efforts to restore, enhance, and create wetlands.

Decisions about how much reliance and emphasis to place on wetlands restoration or creation under Section 404 of the Clean Water Act, the North American Wetlands Conservation Act, farm bill programs, and future efforts to achieve no net wetlands loss must be based on sound science. Wetlands policy under these statutes and in future legislation must recognize the limits of scientific knowledge about restoration and creation and set priorities for future research.

This publication, which resulted from research initiated by the Environmental Protection Agency on wetlands restoration and creation, makes significant contributions toward these information needs. It examines the adequacy of the science base, the success of restoration and creation efforts, and the gaps in scientific knowledge and provides recommendations to wetland managers.

In 1936, the conservationist Ding Darling told natural resource professionals that "whatever we may have been doing is not wildlife conservation, since we continue to have less instead of more." Unfortunately, the same could be said over half a century later with respect to wetlands conservation. As our understanding and experience with wetlands restoration and creation increase, these techniques will offer an important means of reversing current trends in the destruction of the nation's wetlands so that someday we may have more instead of less.

George J. Mitchell

PREFACE TO THE ISLAND PRESS EDITION

In recent years, federal, state, and local wetland protection efforts have expanded greatly—due in large measure to increased interest in compensating for wetland losses by creating new wetlands or restoring degraded ones. Research and dialogue show wide agreement on several key points with regard to restoration and creation, but many doubts remain concerning technical certainties as well as proper utilization of creation or restoration to compensate for impacts affecting wetlands.

This book reports the current and scientific knowledge on wetland creation and restoration. The effort gathered scientists who represent much of the expertise concerning wetland creation and restoration in the United States. The authors of the papers attempted to infer broad similarities and general "truths," both positive and negative, with respect to potential success of wetland creation and restoration projects. In addition to information routinely collected in research efforts, they attempted to draw upon material that has not been reported in the peer-reviewed literature. Through the cooperation of many who contributed information, in many cases data from personal files, previously unpublished material is now available.

The content of this work shows clearly the enthusiasm, personal conviction, and professionalism of the experts, both named and unnamed, throughout the United States who participated. It reflects the diverse points of view of the authors, and is *not* a policy document, nor is it intended to define, express, or endorse any federal policy toward issues of wetland mitigation. The timeliness of the topic and the broad scope of *Wetland Creation and Restoration* should make this publication both useful and stimulating to further thought. We thank Island Press for helping to make it available to a wide audience.

INTRODUCTION

BACKGROUND

The U.S. Fish and Wildlife Service estimated that 30-40% of the original wetlands in the United States have been lost and that destruction continues at 300-400,000 acres per year (Tiner 1984). In the last decade, interest has increased in wetland restoration and creation at all levels of government, in the scientific community, and in the private sector. Restoration and creation have been advocated to:

-- reduce the impacts of activities in or near wetlands,
-- compensate for additional losses,
-- restore or replace wetlands already degraded or destroyed, and
-- serve various new functions such as wastewater treatment, aquaculture, and waterfowl habitat.

The U.S. Environmental Protection Agency in January 1986 adopted a Wetlands Research Plan (Zedler and Kentula 1986) to assist the Agency in implementing its responsibilities to protect the nation's wetlands resource. Agency personnel surveyed in the planning process agreed that there was a pressing need to determine how well created and restored wetlands compensate for losses permitted under Section 404 of the Clean Water Act. The research proposed was designed to improve methods of creating, restoring, and enhancing wetlands and wetland functions; to provide guidance for the design of effective projects; and to develop methods for evaluating the potential and actual success of projects.

This status report is the first major publication resulting from the research initiated on wetland creation and restoration. Conceived as a mechanism for identifying the adequacy of the available information, this status report will help set priorities for the research program and provide Agency personnel with an analytical framework for making 404 permit decisions based on the status of the science of wetland creation and restoration.

Concern about the status of the wetland resource and interest in enhancing it through wetland creation and restoration continues to be strong in the U.S. Numerous meetings and symposia have been held to discuss wetlands issues. Recently, at the request of the EPA, the Conservation Foundation convened the National Wetlands Policy Forum to address major policy concerns. The goal was to develop sound, broadly supported guidance on how federal, state, and local wetlands policy could be improved. In its final report (The Conservation Foundation 1988) the Forum specifically recommended that:

> "the nation establish a national wetlands protection policy to achieve no overall net loss of the nation's remaining wetlands base, as defined by acreage and function, and to restore and create wetlands, where feasible, to increase the quality and quantity of the nation's wetland resource base".

The Forum went on to emphasize that the goal of no net loss does not imply that individual wetlands will be untouchable. Therefore, a substantial increase in efforts to restore and create wetlands is inherent to attaining the Forum's objective. These recommendations attest to the timeliness of the research prescribed in the EPA Wetlands Research Plan (Zedler and Kentula 1986).

This status report is not the first attempt to gather information on wetland creation and restoration (Table 1). Previous works are cited throughout this report. The purpose of this endeavor was to build upon previous work, not to duplicate it. An effort was made to capture information not published elsewhere and incorporate it with published literature to produce a unique resource.

HOW THE REPORT WAS PREPARED

A meeting was held in February 1987 to discuss a draft plan for this status report. The objectives were to insure that appropriate topics were covered and presented in a format that would be useful to Agency 404 personnel. At the meeting the scientists actively involved in wetland creation and restoration, who ultimately became authors for the chapters in the report, together with representatives of the EPA Regions and the Office of Wetlands Protection critiqued a proposed outline and a list of potential topics. Key questions considered were: What information about wetland creation and restoration is needed? For what wetland types in what parts of the country is there sufficient information about creation and restoration to form a unit for a regional review? What specific information should be presented in each of the regional reviews and how should it be organized? The recommendations were incorporated into the final plan. The authors were then commissioned to prepare individual chapters in the report. Once prepared in draft form, the papers underwent an extensive peer review process.

The authors of the chapters in this volume were selected because of their expertise in particular areas of wetland science or their active involvement in a specific aspect of wetland creation and restoration. An effort was made to separate the

Table 1. The best known of the compilations of information on wetland creation and restoration in the U.S.

PROCEEDINGS OF THE ANNUAL CONFERENCE ON WETLANDS RESTORATION AND CREATION

15 years of proceedings sponsored by
the Hillsborough Community College, Tampa, Florida

THE WORK OF THE DREDGED MATERIAL PROGRAM OF THE U.S. ARMY CORPS OF ENGINEERS

See reports published, such as:

Saucier, R.T., C.C. Calhoun, Jr., R.M. Engler, T.R. Patin, and H.K. Smith. 1978. Executive overview and detailed summary: dredged material research program. Tech. Rep. DS-78-212. U.S. Army Engineers Waterways Exp. Station, Vicksburg, Mississippi.

Newling, C.J. and M.C. Landin. 1985. Long-term monitoring of habitat development at upland and wetland dredge material disposal sites, 1974-1982. Tech. Rep. D-85-5. U.S. Army Engineers Waterway Exp. Station, Vicksburg, Mississippi.

CREATION AND RESTORATION OF COASTAL PLANT COMMUNITIES

1982. Lewis, R.R. (Ed.).
CRC Press, Inc., Boca Raton, Florida.

WETLAND RESTORATION AND ENHANCEMENT IN CALIFORNIA

1982. Josselyn, M. (Ed.). Rep. T-CSGCP-007.
Tiburon Centr. Environ. Studies, Tiburon, California.

WETLAND CREATION AND RESTORATION IN THE UNITED STATES FROM 1970 TO 1985: AN ANNOTATED BIBLIOGRAPHY

1986. Wolf, R.B., L.C. Lee and R.R. Sharitz.
Wetlands 6(1): 1-88.

MITIGATING FRESHWATER WETLAND ALTERATIONS IN THE GLACIATED NORTHEASTERN UNITED STATES: AN ASSESSMENT OF THE SCIENCE BASE.

1987. Larson, J.S. and C. Niell (Eds.). Publ. 87-1.
Environ. Inst., Univ. Mass., Amherst, Massachusetts.

WETLAND FUNCTIONS, REHABILITATION, AND CREATION IN THE PACIFIC NORTHWEST: THE STATE OF OUR UNDERSTANDING

1987. Strickland, R. (Ed.). Publ. 86-14.
Wash. State Dept. Ecol., Olympia, Washington.

PROCEEDINGS OF THE NATIONAL WETLAND SYMPOSIUM: MITIGATION OF IMPACTS AND LOSSES

1988. Kusler, J.A., M.L. Quammen and G. Brooks (Eds.).
Association of State Wetland Managers, Berne, New York.

PROCEEDINGS OF A CONFERENCE: INCREASING OUR WETLAND RESOURCES

1988. Zelanzny, J. and J.S. Feierabend.
Nat. Wildl. Fed., Washington, D.C.

science perspectives presented in this status report from the policy views of agencies involved in wetland management. Therefore, primarily scientists who are not associated with a government agency were commissioned to prepare papers. Authors were also requested to avoid policy judgments on key topics fraught with policy implications, such as onsite/offsite mitigation, in-kind/out-of-kind mitigation, and mitigation banks.

It was also recognized that the success of this project depended on the participation of the many other experts in the field. Attempts were made to involve them through presentations and discussions at meetings of wetland scientists, the information gathering process, and the review procedure.

Meetings of authors were held to assess their progress at various stages of the project. To get early feedback, authors also presented outlines of their papers in special sessions of previously scheduled meetings of wetland scientists, such as the annual meeting of the Society of Wetland Scientists. EPA personnel were encouraged to attend and to provide input. Authors of regional reviews of inland and coastal wetlands were later assembled in separate meetings to discuss their first drafts and to identify common issues. In September 1987, some of the authors presented their draft papers at the Association of State Wetland Manager's National Symposium: Hydrology.

The National Wetland Technical Council was invited to assist in the evaluation of research needs. A meeting of the Council was held to discuss the research issues presented in the draft manuscripts. The Council's recommendations are reported in the statement of research needs which constitutes the final chapter of Volume I, prepared by Council members Dr. Joy B. Zedler and Dr. Milton W. Weller.

WHAT THE REPORT IS AND IS NOT

This report is a preliminary evaluation of the status of the science of wetland creation and restoration in the United States. It is, by no means, the final word. It intentionally avoids a variety of key issues which were deemed more policy than scientific in nature.

This status report is composed of two volumes. The first volume is a series of regional reviews. Each review summarizes wetland creation and restoration experiences in broadly defined wetland "regions" (e.g., Pacific coastal wetlands, wooded wetlands of the Southeast). The authors were asked to summarize the available information, identify what has and has not been learned, and recommend research priorities. Their primary task was to synthesize and evaluate information from as many sources as possible, including personal experience.

The second volume is a series of theme papers, covering a wide range of topics of general application to wetland creation and restoration (hydrology, management techniques, planning).

The amount and quality of information available to the authors was uneven by region and topic, so the papers vary in length and level of detail. This is particularly apparent in the regional reviews. The most quantitative and best documented information was available for Atlantic coastal wetlands, consequently, the reports on these systems heavily cite the juried literature. Conversely, information on the creation and restoration of inland freshwater wetlands was spotty, at best, so the authors drew more heavily on personal experience.

Much was learned from this effort to document the status of the science despite the information gaps; the key conclusions are presented in the Executive Summary. Throughout the preparation of this report, authors and informed contributors continually affirmed that the creation and restoration of wetlands is a complex and often difficult task. This, in turn, pointed to the need for setting clear, ecologically sound goals for projects and developing quantitative methods for determining if they have been met. To validate the goal setting process, wetland science must progress and the role of wetlands in the landscape must be understood. Only then can one truly evaluate which ecological functions of naturally occurring wetlands are provided by created and restored wetlands.

ACKNOWLEDGEMENTS

We appreciate the contribution made by many individuals during the preparation of this status report. Personnel from the EPA Office of Wetlands Protection and Regions responded generously with their time when we needed advice and reviews. Dr. Eric M. Preston, the EPA Project Officer, was supportive of the effort and provided valuable advice.

We especially want to thank those who contributed information, in many cases data from personal files. With these contributions previously unpublished material is now available. A host of wetland scientists and managers reviewed the manuscripts, providing valuable counsel that led to improved papers. Dr. Mary Landin of the U.S. Army Corps of Engineers' Waterways Experiment Station, Dr. Charles Seqelquist of the U.S. Fish and Wildlife Service's National Ecology Center, and Dr. Mary Watzin of the U.S. Fish and Wildlife Service's National Wetlands Research Center coordinated reviews of all the manuscripts from their respective labs.

The manuscripts were copy edited by Gail Brooks and word-processed by Joyce G. Caron and Carol DeYoung. Members of the EPA Wetlands Research Team responded, sometimes at a moment's notice, with help on a variety of tasks. Arthur D. Sherman balanced numerous jobs, while tracking the progress of the individual papers. Without his assistance many tasks in this project could not have been completed. Stephanie Gwin, Frances Morris, Jean Sifneos, and Donna Frostholm performed a myriad of chores to prepare the manuscripts for printing. Their attention to detail was greatly appreciated.

Jon and I want to personally thank the authors of this document. They responded to our requests with enthusiasm and a sense of humor, despite the short deadlines that we set. Most of all, we appreciate their dedication to this project and their commitment to the advancement of wetland science.

Mary E. Kentula and Jon A. Kusler

LITERATURE CITED

The Conservation Foundation. 1988. Protecting America's Wetlands: An Action Agenda; The Final Report of the National Wetlands Policy Forum. Washington, D.C.

Tiner, R.W., Jr. 1984. Wetlands of the United States: Current Status and Recent Trends. U.S. Fish and Wildlife Service, National Wetland Inventory. Washington, D.C.

Zedler, J.B. and M.E. Kentula. 1986. Wetland Research Plan. EPA/600/3-86/009, Environmental Research Lab., U.S. Environmental Protection Agency, Corvallis, Oregon. Nat. Tech. Infor. Serv. Accession No. PB86 158 656/AS.

EXECUTIVE SUMMARY

Jon A. Kusler and Mary E. Kentula

INTRODUCTION

This executive summary is divided into three principal sections: (1) conclusions concerning the adequacy of our scientific understanding concerning wetland restoration and creation; (2) recommendations for filling the gaps in scientific knowledge; and (3) recommendations for wetland managers with regard to restoration and creation based upon the status of our scientific understanding.

The following general conclusions are offered with regard to the adequacy of our scientific understanding and the success of restoration and creation projects in meeting particular project goals.

ADEQUACY OF THE SCIENCE BASE

1. Practical experience and the available science base on restoration and creation are limited for most types and vary regionally.

Experience in wetland restoration and creation varies with region and wetland type, as does the evaluation and reporting of such experience in the scientific literature. Hundreds and perhaps thousands of coastal and estuarine mitigation projects have been constructed along the Eastern seaboard. These projects have been subject to a fair amount of follow-up monitoring and have been quite widely reported in the literature. Fewer projects have been implemented on the Gulf and Pacific coasts and, correspondingly, there is a smaller literature base.

In general, much less is known about restoring or creating inland wetlands. However, two types of inland wetland projects have been quite common: impoundments to create waterfowl and wildlife marshes, and creation of marshes on dredged spoil areas along major rivers. Despite the number of these impoundment projects and a relatively large literature base dealing with waterfowl production and other related topics, only a modest portion of the literature critically examines these efforts. A modest literature base is available on wetlands created on dredged spoil. The best known research is that of the U.S. Army Corps of Engineers Dredged Materials Program.

2. Most wetland restoration and creation projects do not have specified goals, complicating efforts to evaluate "success".

Project goals have rarely been specified, even in cases where wetlands have been intentionally restored or created. This has complicated efforts to evaluate "success". Lacking such goals, success has commonly been interpreted as the establishment of vegetation that covers a percentage of the site and exists for a defined period of time (e.g., 2-3 years). Such measures of success, however, do not indicate that a project is functioning properly nor that it will persist over time. Often these criteria have some relationship to the characteristics of natural wetlands of the same type in the region, but this relationship is limited. In the rare cases where project goals have been formulated and follow-up studies conducted, there have been situations where failure to meet specific goals has occurred although there was partial or total revegetation of the site.

Ideally, success should be measured as the degree to which the functional replacement of natural systems has been achieved. This is much more difficult to assess and cannot be routinely quantitatively determined. The ability to estimate success of future projects will be fostered through establishing specific goals that can be targeted in an evaluation.

3. Monitoring of wetland restoration and creation projects has been uncommon.

Despite thousands of instances in which wetlands have been intentionally or unintentionally restored or created in the United States, in the last 50 years there has been very little short term monitoring and even less long term monitoring of sites. Monitoring of sites and comparisons with naturally occurring wetlands over time would provide a variety of information including rates of revegetation, repopulation by animal species, and redevelopment of soil profiles, patterns of succession, and evidence of persistence.

SUCCESS OF RESTORATION AND CREATION

1. Restoration or creation of a wetland that "totally duplicates" a naturally-occurring wetland is impossible; however, some systems may be approximated and individual wetland functions may be restored or created.

Total duplication of natural wetlands is impossible due to the complexity and variation in natural as well as created or restored systems and the subtle relationships of hydrology, soils, vegetation, animal life, and nutrients which may have developed over thousands of years in natural systems. Nevertheless, experience to date suggests that some types of wetlands can be approximated and certain wetland functions can be restored, created, or enhanced in particular contexts. It is often possible to restore or create a wetland with vegetation resembling that of a naturally-occurring wetland. This does not mean, however, that it will have habitat or other values equaling those of a natural wetland nor that such a wetland will be a persistent, i.e., long term, feature in the landscape, as are many natural wetlands.

2. Partial project failures are common.

For certain types of wetlands, total failures have been common (e.g., seagrasses, certain forested wetlands). Although the reasons for partial or total failures differ, common problems include:

* lack of basic scientific knowledge;
* lack of staff expertise in design, and lack of project supervision during implementation phases;
* improper site conditions (e.g., water supply, hydroperiod, water depth, water velocity, salinity, wave action, substrate, nutrient concentration, light availability, sedimentation rate, improper grades (slopes);
* invasion by exotic species;
* grazing by geese, muskrats, other animals;
* destruction of vegetation or the substrate by floods, erosion, fires, other catastrophic events;
* failure of projects to be carried out as planned;
* failure to protect projects from on-site and off-site impacts such as sediments, toxics, off-road vehicles, groundwater pumping, etc.; and
* failure to adequately maintain water levels.

3. Success varies with the type of wetland and target functions including the requirements of target species.

A relatively high degree of success has been achieved with revegetation of coastal, estuarine, and freshwater marshes because elevations are less critical than for forested or shrub wetlands, native seed stocks are often present, and natural revegetation often occurs. Marsh vegetation also quickly reaches maturity in comparison with shrub or forest vegetation. However, some types of marshes, such as those dominated by Spartina patens, have been difficult to restore or recreate due to sensitive elevation requirements.

Much less success has been achieved to date with seagrasses and forested wetlands. The reasons for lack of success for seagrasses are not altogether clear, although use of a site where seagrasses have previously grown seems to improve the chances for establishing the plants. Lack of success for forested wetlands is due, at least in part, to their sensitive long term hydrologic requirements. Such systems also reach maturity slowly.

Although certain types of wetland vegetation may be restored or created, there have been few studies concerning the use of restored or created wetlands by particular animal species. Restoration or creation of habitat for ecologically sensitive animal or plant species is particularly difficult.

4. The ability to restore or create particular wetland functions varies by function.

The ability to restore or create particular wetland functions is influenced by (1) the amount of basic scientific knowledge available concerning the wetland function; (2) the ease and cost of restoring or creating certain characteristics (e.g., topography may be created with relative ease, while creation of infiltration capacity is difficult); and (3) varying probabilities that structural characteristics will give rise to specific functions. For example (note this is meant to be illustrative only):

* Flood storage and flood conveyance functions can be quantitatively assessed and restored or created with some certainty by applying the results of hydrologic studies. Topography is the critical parameter and this is probably the easiest parameter to restore or recreate.
* Waterfowl production functions may be assessed or created with fair confidence in some contexts, due to the large amount of experience, scientific knowledge, and information on marsh design, and marshes are, relatively speaking, easily restored or recreated.
* Wetland aesthetics may or may not be difficult

to restore or create, depending on the wetland type and the site conditions. Visual characteristics are, in general, much easier to restore than subtle ecological functions.

* Some fisheries functions may be assessed and restored or created. However, the ability to restore or create fisheries habitat will depend on the species and the site conditions.

* Some food chain functions may be assessed, restored, or created. Other more subtle functions are difficult due to the lack of basic scientific knowledge and experience.

* Certain pollution control functions (e.g., sediment trapping) may be relatively easy to assess and create. However, others (e.g., immobilization of toxic metals) may be difficult to create, particularly in the long term because of uncertainties concerning the long term fate of pollutants in wetlands and their impact on the wetland system.

* Groundwater recharge and discharge functions are difficult to assess and create. One confounding factor is that soil permeability may change in a creation or restoration context (e.g., a sandy substrate may quickly become impermeable due to deposition of organics).

* Heritage or archaeological functions (e.g., a shell midden located in a marsh) are impossible to restore or create since they depend upon history for their value.

5. Long term success may be quite different from short term success.

Revegetation of a restored or created wetland over a short period of time (e.g., one year) is no guarantee that the area will continue to function over time. Unanticipated fluctuations in hydrology are a particularly serious problem for efforts to restore or create wetland types (e.g., forested wetlands) with very sensitive elevation or hydroperiod requirements. Droughts or floods may destroy or change the targeted species composition of projects.

Hydrologic fluctuations also occur in natural wetlands. But hydrologic minima and maxima as well as "normal" conditions exist within tolerable ranges at particular locations, otherwise the natural wetland types would not exist. Natural wetlands have been tried and tested by natural processes and are, in many instances, "survivors".

Long term damage to or destruction of restored or created systems may be due to many other factors in addition to unanticipated hydrologic changes. Common threats include pollution, erosion and wave damage, off-road vehicle traffic, and grazing. Excessive sediment is a serious problem for many restored or created wetlands located in urban areas with high rates of erosion and sedimentation. Unlike many natural wetlands, restored or created wetlands also often lack erosional equilibrium (in a geomorphologic sense) with their watersheds.

6. Long term success depends upon the ability to assess, recreate, and manipulate hydrology.

The success of a project depends to a considerable extent, upon the ease with which the hydrology can be determined and established, the availability of appropriate seeds and plant stocks, the rate of growth of key species, the water level manipulation potential built into the project, and other factors. To date, the least success has been achieved for wetlands for which it is very difficult to restore or create the proper hydrology. In general, the ease with which a project can be constructed and the probability of its success are:

* Greatest overall for estuarine marshes due to (1) the relative ease of determining proper hydrology; (2) the experience and literature base available on restoration and creation; (3) the relatively small number of wetland plant species that must be dealt with; (4) the general availability of seeds and plant stocks; and (5) the ease of establishing many of the plant species. However, it is difficult or impossible to restore or create certain estuarine wetland types due to narrow tidal range or salinity tolerances, e.g., high marshes dominated by Spartina patens on the East Coast. The same is true of estuarine wetlands in regions or areas with unique local conditions, e.g., the hypersaline soils common in southern California salt marshes.

* Second greatest for coastal marshes for the same reasons as those given for estuarine wetlands. However, high wave energies and tidal ranges of the open coast reduce the probability of success.

* Third greatest for freshwater marshes along lakes, rivers, and streams. The surface water elevations can often be determined from stream or lake gauging records. There is a fair amount of literature and experience in restoring and managing these systems. However, vegetation types are often more complex than those of coastal and estuarine systems. Problems with exotic species are common. Determination and restoration or creation of hydrology (including flood levels) and hydrology/sediment relationships are more difficult. This is frequently compounded by altered hydrology and sedimentation patterns due to dams and water extractions.

* Fourth greatest for isolated marshes supplied predominantly by surface water. There is limited experience and literature on restoring or creating such wetlands except for waterfowl production where water levels are manipulated on a continuing basis. Determination and restoration of hydrology is very difficult unless mechanisms are available for actively managing the water supply. Depending on the wetland type, plant assemblages can also be complex.

* Fifth greatest for forested wetlands along lakes, rivers, and streams. Determination and restoration or creation of hydrology is very difficult due to narrow ranges of tolerance. Water regimes may be evaluated with the use of records for adjacent waters, but such records are often not sensitive enough. There is also limited literature or experience in restoring such systems. Vegetation is diverse; both the understory and canopy communities may need to be established. Moreover, it may take many years for a mature forest to develop.

* Sixth greatest for isolated freshwater wetlands (ranging from marshes to forested wetlands) supplied predominantly with ground water. Determining and creating the hydrology is very difficult. There is limited experience and literature except on some prairie pothole wetlands.

7. Success often depends upon the long term ability to manage, protect, and manipulate wetlands and adjacent buffer areas.

Restored or created wetlands are often in need of "mid course corrections" and management over time. Original design specifications may be insufficient to achieve project goals. Created or restored wetlands are also particularly susceptible to invasion by exotic species, sedimentation, pollution, and other impacts due to their location in urban settings and the inherent instability of many of their systems. Careful monitoring of systems after their original establishment and the ability to make mid course corrections and, in some instances, to actively manage the systems, are often critical to long term success.

Efforts to create or enhance waterfowl habitat by wildlife agencies and private organizations through the use of dikes, small dams, and other water control structures have been quite successful due, in large measure, to the ability to control and alter the hydrologic regime over time. Water levels may be changed if original water elevations prove incorrect for planned revegetation. Drawdown and flooding may be used to control exotics and vegetation successional sequences.

However, most wetland restoration or creation efforts proposed by private and public developers do not involve water control measures. In addition, few developers are willing to accept long term responsibility for managing systems. Water level manipulation capability and long term management capability are also insufficient, in themselves, without long term assurances that the system will be managed to achieve particular wetland functional goals. For example, water level manipulation and long term management capability exist for most flood control, stormwater, and water supply reservoirs. But wetlands along the margins of these reservoirs are often destroyed by fluctuations in water levels dictated by the primary management goals.

Restored or created wetlands should be designed as self sustaining or self managing systems unless a project sponsor (such as a wildlife agency or duck club) clearly has the incentive and capability for long term management to optimize wetland values.

The management needs of restored or created wetlands are not limited to water level manipulation. Common management needs for both wetlands subject to water level manipulation and those not subject to such manipulation include:

-- Replanting, regrading, and other mid-course corrections.

-- Establishment of buffers to protect wetlands from sediment, excessive nutrients, pesticides, foot traffic, or other impacts from adjacent lands.

-- Establishment (in some instances) of fences and barriers to restrict foot traffic, off-road vehicles, and grazing animals in wetlands.

-- Adoption of point and non-point source pollution controls for streams, drainage ditches, and runoff flowing into wetlands.

-- Control of exotics by burning, mechanical removal, herbicides, or other measures.

-- Periodic dredging of certain portions of wetlands subject to high rates of sedimentation (e.g., stormwater facilities).

8. Success depends upon expertise in project design and upon careful project supervision.

Hydrologic and biological as well as botanical and engineering expertise are needed in the design of many projects. In addition, the involvement of experts with prior experience in wetland restoration or creation is highly desirable. This is particularly true where a wetland with multiple functions is to be created from an upland site. Less expertise may be needed where restoration is to occur, the original hydrology is intact, and nearby natural seed stocks exist.

Careful project supervision is also needed to insure implementation of project design. It is not enough to design a project and turn it over to traditional construction personnel. For example, bulldozer operators often need guidance with regard to critical elevation requirements, drainage, and the spreading of stockpiled soil. Plantings must be shaded from the sun and kept moist until they are placed in the ground.

9. "Cook book" approaches for wetland restoration or creation will likely be only partially successful.

Too little is known from a scientific perspective about wetland restoration to provide rigid, "cook book" guidance. The interdependence of a large number of site-specific factors also warrants against too rigid an approach. For example, in a salt marsh, maxima and minima in hydrologic conditions for particular plant species may depend not only on elevation but on salinity, wave action, light, nutrients, and other factors. Often the best model is a nearby wetland of similar type.

Although "cook book" prescribing rigid design criteria are not desirable, guidance documents suggesting ranges of conditions conducive to success are possible. Requirements for wetland creation that incorporate such general criteria, combined with incentives and flexibility to allow for experimentation offer an increased probability of success as well as a contribution to the information base.

FILLING THE GAPS IN SCIENTIFIC KNOWLEDGE

A variety of measures are needed to fill the gaps in our scientific knowledge. Authors list specific research needs in their papers. The National Wetland Technical Council provides an overview of research needs in the final paper of Volume I of this report (see Zedler and Weller).

The full range of topics needing further research is impressive and perhaps intimidating, given the limited funds available for wetland research. Cost effective measures will be needed to fill the gaps, relying, to the extent possible, upon cooperative sources of funding and innovative strategies. For example, the private and public development sector may be able to provide a portion of the needed research through the monitoring of various restoration and creation projects. Research in wetland restoration and creation may also take place cost-effectively as part of broader lake restoration, strip mine restoration, river restoration, reforestation, Superfund clean up, or post natural disaster (flood, fire, landslide) recovery efforts. Some of the measures needed to fill the gaps include:

1. Systematic monitoring of restoration or creation projects.

Given the high cost of demonstration projects, the greatest potential for filling gaps in scientific knowledge may lie with careful monitoring of selected types of new restoration or creation projects.Standardized methods for project evaluation and project monitoring are needed to facilitate determination of "success" and comparisons between systems and approaches. A regional and national database on projects should be created.

Monitoring should involve:

-- Careful baseline studies on the original wetland systems before they are degraded or destroyed,

-- Monitoring of selected features of the new or restored systems at periodic intervals (e.g., six months, a year, three years, five years, ten years, twenty-five years, etc.) to determine characteristics of the restored or created wetlands (vegetation types, vegetation growth rates, fauna, etc.), functions of the wetlands and persistence of the wetlands. The precise features needing monitoring and the level of monitoring detail will differ, depending upon the type of wetland and specific research need.

Monitoring of new projects can be made a condition of project approval, although, equitably and practically, there must be limits to the prior or post-construction studies and to the duration of the post-construction monitoring period. Project sponsors may be required to carry out monitoring to insure project success over a specified period of time, but they may balk at more basic research responsibilities. Cooperative projects between project sponsors and academic institutions and non profit or government research organizations may reduce the burden on project sponsors while improving the quality of long term monitoring. Such cooperative projects may also involve comparisons between restored or created wetlands and natural wetlands in the region.

After the fact monitoring of restoration and creation projects already in existence may also provide invaluable information, although many such projects lack detailed baseline information concerning the original wetland or the specifics of the restoration or creation effort (e.g., size, substrate, planting, etc.).

2. Demonstration projects.

Wetland demonstration projects established by universities, research laboratories, or agencies to test various restoration or creation approaches offer the greatest "control" and have the greatest potential for answering some research questions. The National Wetland Technical Council (see Zedler and Weller, this volume) recommends the establishment of a series of such demonstration projects on a regional basis. However, such projects will likely be expensive to establish and monitor. Funds may be generated by making such projects multi-objective like the riverine wetland demonstration projects on the Des Plaines River north of Chicago established by Wetland Research. This demonstration project also provides a regional park and wetland education area.

3. Traditional scientific research.

More traditional scientific research is needed on a wide variety of specific topics. Many of the topics relate to basic issues in wetland science, not simply wetland restoration or creation. Some of this research needs to be conducted on natural as well as altered or created systems. The research could involve laboratory experiments, traditional field research, the monitoring of restoration or creation projects, and the establishment of demonstration projects.

Particularly critical topics include:

1. The hydrologic needs and requirements of various plants and animals, minima water depths, hydroperiod, velocity, dissolved nutrients, and the role of large scale but infrequent hydrologic events such as floods and long term fluctuations in water levels.

2. The importance of substrate to flora, fauna, and various wetland functions such as removal of toxics.

3. Characteristics of rates of natural revegetation in contrast with various types of plantings.

4. A comparison of the functions of natural versus restored or created wetlands with special emphasis upon habitat value for a broad range of species, food chain support, and water quality protection and enhancement functions.

5. An evaluation of the stability and persistence of restored or created systems in various contexts and in comparison with natural systems.

6. An evaluation of the impact of sediment, nutrients, toxic runoff, pedestrian use, use by off-road vehicles, grazing, and other impacts upon restored or created wetlands and their functions in various contexts. Further investigation of management alternatives to reduce or compensate for such impacts is also needed.

7. Landscape level comparisons of natural and restored or constructed systems from a broad range of perspectives (see Zedler and Weller, this volume).

Further research into wetland restoration and creation will help provide the scientific know-how for restoring systems which are already degraded as well as for reducing future impacts. It will, more broadly, test the limits of knowledge of wetland ecosystems and how they function. The result will be the production of invaluable, broadly applicable, scientific information. Without such knowledge, the restoration and creation of wetlands in many contexts will continue to be largely a matter of trial and error.

4. Continued synthesis of existing scientific knowledge.

Additional specific guidance documents based upon existing information could be prepared for the restoration and creation of specific types of wetlands and specific functions. The present synthesis, like previous efforts, has been limited by time, funding, and geographical scope. Moreover, pertinent information is constantly being generated.

Such synthesis efforts might productively draw upon the "grey" wetland literature, such as permit files and the records of wildlife refuge managers and nonprofit land management organizations. They might also productively draw (where applicable) upon the larger body of scientific literature with information of potential interest to specific aspects of restoration or creation. These include scientific reports and studies pertaining to restoration of lakes, restoration of streams, restoration of strip-mined areas, Superfund clean up efforts, and restoration of other ecological systems such as prairies. Studies of natural response and recovery processes for systems impacted by floods, volcanoes, fires, and other natural processes should also be consulted.

Synthesis efforts should focus not only upon the creation or restoration of systems but their subsequent maintenance and management. Particularly good candidates for such syntheses (because of the large number of restoration or creation efforts now being attempted) include:

-- Wetlands created to serve as stormwater detention areas,

-- Wetlands along the margins of flood control and water supply reservoirs and other impoundments designed to provide habitat, control erosion, protect water quality, etc.,

-- Wetlands designed to serve as primary, secondary, or tertiary treatment facilities.

RECOMMENDATIONS FOR WETLAND MANAGERS

There are many policy questions and mixed policy-science questions which the wetland regulator must address in evaluating permits proposing wetland restoration and creation such as prior site analysis requirements (e.g., alternative site analysis); acceptable levels of degradation for the original wetlands; desired levels of compensation (e.g., acreage ratios); types of compensation (e.g., in-kind, out-of-kind), and location of compensation (on-site, off-site). These questions were not addressed by the present study which focused only upon the adequacy of the scientific base for wetland restoration and creation. However, based upon this science review, recommendations may be made which have broad scale applicability to restoration or creation efforts wherever they may occur:

1. **Wetland restoration and creation proposals must be viewed with great care, particularly where promises are made to "restore" or "recreate" a natural system in exchange for a permit to destroy or degrade an existing more or less natural system.**

Experience to date indicates that too little is known about restoration and creation and there are too many variables to predict "success" for restoration or creation in many contexts. There have been too few projects with too little monitoring, and, there is too limited a literature base. This does not mean, however, that wetlands with characteristics approximating certain natural wetlands or with specific functions resembling those of the natural wetlands cannot, in some instances, be restored or created. Enough is known to suggest key factors or considerations in restoration and creation. And there is a considerable body of experience pertaining to certain types of wetlands (e.g., marshes for water fowl production) in certain contexts.

2. **Multidisciplinary expertise in planning and careful project supervision at all project phases is needed.**

Experience to date suggests that project success will depend, to a considerable extent, upon the care with which plans are prepared and implemented and the expertise of the project staff. Restoration and creation projects require slightly different types of inputs at each phase:

-- Project design. Wetland restoration or creation without hydrologic design will fail. This does not mean that a hydrologist must be involved in every project but that hydrology must be carefully considered. Careful documentation of elevations and other hydrologic characteristics of naturally occurring systems, including either the original unaltered system or nearby systems, can be a helpful guide. Individuals with hydrologic as well as botanical and biological expertise are needed in project design. A soils expert may also be needed (depending upon the project).

-- Project implementation. Careful supervision of bulldozer operators and other implementation personnel by someone with a complete understanding of the critical parameters for the project such as grade, drainage, soil, and planting needs is critical.

-- Post project monitoring and mid-course corrections. Botanical and biological expertise is often needed for project monitoring and to design mid-course corrections.

A project applicant should provide information concerning the qualifications of project staff at each phase of project design and implementation, such as degree qualifications, work experience, etc.

3. **Clear, site-specific project goals should be established.**

Because no wetland can be restored or duplicated exactly, it is important that the applicant establish site-specific goals for a restoration and creation project related to existing and proposed wetland characteristics and functions. These goals should be used to assist design, monitoring, and follow-up as well as to act as a benchmark for success. These goals can, depending upon the circumstances, relate to the size of the area to be restored, the functions to be restored, the type of vegetation, the density of vegetation, vegetation growth rates, target fauna species, intended management activities and other parameters.

4. **A relatively detailed plan concerning all phases of a project should be prepared in advance to help the regulatory agency evaluate the probability of success for that type of wetland, at that site, meeting specific goals.**

Generalized project information indicating that a project applicant will "create a wetland" at a particular site provides no real basis for determining the probable success of a project.

Although needed information will differ, depending upon the type of wetland and area, at a minimum, a plan is needed:

-- setting forth clear project goals and measures for determining project success,

-- indicating the boundaries of the proposed restoration or creation area,

-- indicating proposed elevations,

-- indicating sources of water supply and connection to existing waters and uplands,

-- indicating proposed soils and probable sedimentation characteristics,

-- indicating proposed plant materials,

-- indicating whether exotics are, or may be, present and, if so, what is to be done to control them,

-- indicating the methods and timing for plantings (if replanting is to take place),

-- setting forth a monitoring program, and

-- setting forth proposed mid-course correction and project management capability.

The amount of formality and detail needed for a restoration plan may depend upon the size of the project, its location, the type of wetland, and other factors.

5. Site-specific studies should be carried out for the original system prior to wetland alteration.

Due to complexities in natural systems, the lack of an extensive scientific base, and difficulties in formulating standards for restoration or creation at a site, a careful inventory of wetland characteristics (size, hydroperiod, soil type, vegetation types and densities, fauna) should take place prior to wetland destruction to determine wetland values and functions, act as a guide for restoration at the site or creation at an analogous site, and form the comparative basis for determining the success of the restoration or creation project.

6. Careful attention to wetland hydrology is needed in design.

Although the basic design needs for "successful" (i.e., meeting specified goals) wetland restoration or creation will differ by type of wetland and area, wetland hydrology is the key (although not necessarily sufficient in itself) to long term functioning systems. Relevant hydrologic factors include: water depths (minima, maxima, norms), velocity, hydroperiod, salinity, nutrient levels, sedimentation rates, levels of toxics and other chemicals, etc.

7. Wetlands should, in general, be designed to be self-sustaining systems and "persistent" features in the landscape.

To the extent possible, restoration and creation projects should be self-sustaining without the need for continued water level manipulation or other management over the life of the project unless such management is an intentional feature of the project (e.g., a wildlife refuge for waterfowl production) and a government agency or other responsible body with long term maintenance powers will have responsibility for the project.

Restoration or creation projects attempting to replace natural wetlands or designed to serve long term objectives should also include design features insuring the long term existence of such projects. To be persistent, wetlands must not be located in areas where natural or man-made processes such as wave action, excessive sedimentation, toxics, or changes in water supply will destroy them. However, many must also undergo periodic major stresses such as fires, floods, and icing over which interrupt the vegetational sequences that occur in most natural wetlands. Such stress must be of a magnitude sufficient to interrupt successional sequences but not great enough to destroy the wetland.

8. Wetland design should consider relationships of the wetland to the watershed water sources, other wetlands in the watershed, and adjacent upland and deep water habitat.

Although cost may prevent broad scale analyses for every restoration and creation project, an analysis of a proposed restoration or creation in a broader hydrologic and ecological context is needed, particularly where "in-kind" goals are not to be applied, where the existing wetland is already degraded, where specific habitat or other values dependent upon the broader context are to be created, or where expected urbanization or other alterations in the watershed or on adjacent lands may threaten the wetland to be created or restored.

9. Buffers, barriers, and other protective measures are often needed.

Protective measures are needed for many restored or created wetlands which may be threatened by excessive sedimentation, water pollution, diversion of water supply, foot traffic, off-road vehicles, and exotic species. Such measures are particularly needed in urban or urbanizing areas with intensive development pressures. Measures may include buffers, fences or other barriers, and sediment basins.

10. Restoration should be favored over creation.

In general, wetland restoration at the site of an existing but damaged or destroyed wetland will have a greater chance of success in terms of recreating the full range of prior wetland functions and long term persistence than wetland creation at a non-wetland site. This is due to the fact that preexisting hydrologic conditions are often more or less intact, seedstock for wetland plants are often available, and fauna may reestablish themselves from adjacent areas.

11. The capability for monitoring and mid-course corrections is needed.

Due to the lack of basic scientific knowledge, lack of experience in restoring and creating many types of wetlands, and the possibility that any effort will fail to meet one or more goals, restoration and creation projects should be approached as "experiments". The possibility of mid-course corrections should be reflected in project design in the event that the project fails to meet one or more specified goals. Such corrections may involve replanting, regrading, alterations in hydrology, control of exotics, or other measures.

12. The capability for long term management is needed for some types of systems.

In some instances, long term management capability is critical to the continued functioning of a system. such management may include water level manipulation, control of exotics, controlled burns, predator control, and periodic sediment removal.

13 Risks inherent in restoration and creation and the probability of success for restoring or creating particular wetland types and functions should be reflected in standards and criteria for projects and project design.

Risks and probability of success should be reflected in the stringency of design requirements, area ratios (e.g., 1:1.5, 1:2) and standards for possible mid-course corrections for projects. Where restoration or creation is very risky or the possibility of project failure may have serious consequences (e.g., destruction of endangered species), successful completion of the restoration or creation project prior to damage or destruction to the original wetland is needed.

14. Restoration for artificial or already altered systems requires special treatment.

Restoration and creation efforts for wetlands already in an altered condition raise special issues and special problems. In restoring an altered wetland, an historical analysis suggesting natural conditions and functions may provide better guidance for restoration than simple documentation and replication of the status quo. A regional analysis of wetland functions and values and "needs" is also desirable.

PART 1

REGIONAL REVIEW

WETLAND MITIGATION ALONG THE PACIFIC COAST OF THE UNITED STATES

Michael Josselyn
Romberg Tiburon Center for Environmental Studies
San Francisco State University

Joy Zedler and Theodore Griswold
Pacific Estuarine Research Laboratory
San Diego State University

ABSTRACT. Mitigation to compensate for coastal wetland losses has taken place under federal and state permit policies for over 15 years. As a result, a substantial data base has developed in the scientific and governmental literature on which to base recommendations for improvement in mitigation practice. The purpose of this chapter is to review the status of wetland mitigation along the Pacific coast based on the available literature and more recent evaluations.

An important distinction must be made when evaluating the effectiveness of mitigation in off-setting wetland losses. Many projects have failed due to lack of compliance with permit requirements, e.g., have never been implemented or were completed without regard to permit specifications. On the other hand, mitigation effectiveness for those projects which have been completed is more difficult to assess. Functional success involves evaluation not only based on objectives; but on our knowledge of wetland hydrology and ecology, fields of science which have only recently received significant attention. Given the rarity of Pacific coastal wetlands and the substantial losses which occurred prior to the Clean Water Act, mitigation must be considered only after avoidance measures are thoroughly considered.

Mitigation for Pacific coastal wetlands is not a "cookbook" exercise. The concept that wetlands are simple ecosystems that can be re-created with little forethought must be rejected. Hydrologic characteristics of a mitigation site are especially important as they structure the possible wetland habitats that can be created. Within the Pacific coastal zone, four general hydrological types occur:

- Wetlands associated with small coastal rivers or lagoons, often subject to sandbar closure,
- Wetlands associated with major estuaries and coastal embayments,
- Wetlands associated with rivers, and
- Non-tidal wetlands such as vernal pools.

Within each of these broad categories, specific opportunities and constraints must be considered prior to approving mitigation proposals. In-kind habitat replacement will not be feasible if mitigation is proposed across types, given their significant differences. Most importantly, watershed management must also be considered within each hydrologic type as an important criteria in evaluating the potential success of a mitigation proposal.

Permit applications must include information on project goals and habitat objectives. Goals and objectives must be specific and stated within a time frame that can be monitored. In addition, a number of elements need to be included within mitigation proposals:

- Description of existing conditions including information on site history, topography, hydrology, sedimentation, soil types, presence of existing wetlands and wildlife, and adjacent land uses.
- Description of proposed hydrological conditions as related to the specific requirements of the wetland vegetation and habitat desired.
- Means by which mitigation site constraints such as subsidence, excessive sedimentation, and poor substrate are to be ameliorated.

- Planting procedures, especially within tidal sites with poor soils or limited seed recruitment. If planting is not required, the period of time after implementation during which full plant establishment is expected should be determined and justified in light of the habitat lost.

- Determination of appropriate buffers that provide protection to the wetland.

- Enforceable procedures to provide construction project oversight by qualified engineers, hydrologists.

- Monitoring programs to allow enforcement of permit requirements and provide further information on the effectiveness of mitigation projects as a means to increase wetland resources rather than to simply offset losses.

Outside of site specific review, resource agencies must assess the long-term implications of individual permit approvals. Re-appraisal should be based on detailed analysis contained within accessible database files. Such appraisal can provide important information on regional trends and means by which mitigation can be re-directed to better serve fish and wildlife needs.

INTRODUCTION

The distribution and functioning of Pacific Coast wetland systems has been described in numerous reviews (Macdonald 1977, 1986, Zedler 1982, Zedler and Nordby 1986, Josselyn 1983, Seliskar and Gallagher 1983, Simenstad 1983, Phillips 1984, Strickland 1986). Several symposia, workshops, and guidebooks have been published on the status of wetland restoration within the region (Josselyn 1982, Harvey et al. 1982, Hamilton 1984, Josselyn and Buchholz 1984, Zedler 1984, Strickland 1986, Josselyn 1988b). A baseline for determining the distribution of wetland types as well as trends in habitat losses and gains is also available (Boule' et al. 1983, Marcus 1982, Handley and Quammen in press). With this background of information, why then are scientists and agency personnel debating whether wetland restoration and creation are effective management tools for mitigating wetland habitat degradation and loss on the Pacific coast?

Race (1985) brought the controversy to the forefront with her article concerning the lack of success of mitigation projects in San Francisco Bay. She argued that many wetland mitigation sites had failure due to lack of establishment by either natural or planted wetland vegetation or problems in creating appropriate elevations for marsh vegetation. She also noted problems in determining project adherence to mitigation requirements due to poor permit descriptions. Her analysis included experimental plantings, dredge material disposal sites, and sites restored by dike breaching. Harvey and Josselyn (1986) countered that Race had overemphasized experimental plantings in her analysis and had ignored the fact that significant wetland habitats had been created despite the lack of specific planning objectives. Nevertheless, Race's arguments and the experience with mitigation in other regions of the country have continued to fuel the debate (O'Donnell 1988).

Further analyses of wetland mitigation success in San Francisco Bay were conducted by Eliot (1985) and Bay Conservation and Development Commission (BCDC) (1988). They found inconsistencies between completed projects and stated goals, most of which could be attributed to lack of enforcement or poor planning and implementation. Kentula (1986) observed similar problems in evaluating wetland mitigation in the Pacific Northwest and criticized the lack of quantitative data necessary to evaluate project effectiveness. Quammen (1986) best described the evaluation problem by distinguishing criteria for: compliance (how well permit and regulatory obligations were met) and function (how well the created wetland functions replace those of natural wetlands) success. Previous reviews have failed to separate these; and what may be a failure in one evaluation can be successful in another.

This chapter focuses on the functional success or effectiveness of wetland mitigation projects within the coastal zone of the western United States (exclusive of Alaska and Hawaii). Though we make recommendations on implementation and monitoring activities that can improve compliance success, our primary attention is centered on the technical and scientific criteria used to plan and implement a wetland mitigation project. This book is written for those reviewing Section 404 permits in which wetland mitigation is proposed. We assume that all steps necessary to avoid or minimize the requirement for habitat replacement have been taken prior to consideration of mitigation.

WETLAND TYPES ALONG THE PACIFIC COAST

The extent of wetlands within the coastal counties of California, Oregon, and Washington is estimated at 67,100 hectares (166,000 acres), or 1.4 percent of the nation's total coastal wetland acreage (Table 1). If the inland counties around San Francisco Bay are included in the total, the amount of wetland extent is doubled to 134,000 hectares (332,000 acres). Two-thirds of the Pacific coast's salt marsh and tidal mudflats are located in San Francisco Bay. The greatest amount of forested wetland is found along the north coast in Washington.

This chapter focuses primarily on emergent wetlands (marshes) as that term is used by the U.S. Fish and Wildlife Service Wetland Classification System (Cowardin et al. 1979). We recognize, however, that most marshes are a complex of unconsolidated bottom, open water, unconsolidated shore, and emergent vegetation. Furthermore, we emphasize estuarine and palustrine habitats within the coastal zone, though some habitats such as vernal pools extend far inland. We have not considered riparian or wooded wetlands.

Coastal wetlands can be further distinguished by the type of estuarine system with which they are associated. The hydrologic characteristics of these systems greatly influence the functioning of the emergent wetland and dictate the types of restoration and creation feasible.

WETLANDS ASSOCIATED WITH SMALL COASTAL RIVERS OR LAGOONS SUBJECT TO SANDBAR CLOSURE

Most wetlands along the Pacific coast are associated with relatively small coastal streams and rivers. Of 85 named wetland systems for the Pacific coast (summarized in Macdonald 1986), at least 50 would be considered within this category. Though some are open to tidal action throughout the year via either natural or man-made channels, others are subject to sandbar closure during some or all of the year. The emergent wetlands are clearly subject to annual and interannual variation in salinity and hydrologic regimes. Frequently, these systems exhibit sharp haloclines where the denser oceanic water entering through the sandbar occurs on the bottom of the lagoon while freshwater remains on the surface. These systems are also subject to significant sedimentation when the sandbar is present.

Examples of lagoonal wetlands system include: Netarts Bay, and Salmon River, Oregon; and Albion River, Pescadero Marsh, San Elijo Lagoon, Estero Americano, and Tijuana Estuary in California.

WETLANDS ASSOCIATED WITH MAJOR ESTUARIES AND COASTAL EMBAYMENTS NOT SUBJECT TO SAND BAR CLOSURE

These wetlands cover the largest acreage along the Pacific Coast, especially those associated with San Francisco Bay. These wetlands have been created by the deposition of sediment and subsequent inundation of low lying lands as sea level has risen. Estuarine wetlands are subject to seasonal salinity changes related to freshwater inflow, but the variation is usually not as great as that experienced in coastal lagoons. Estuarine wetland soils exhibit both vertical and horizontal gradients in salinity due to tidal fluctuations and freshwater flow. Saline tidal marshes are found near the estuary mouth and freshwater tidal marshes at the head. Too much sedimentation can be a problem in these wetland systems, particularly in small estuaries downstream from lumbering and agricultural land uses.

Examples of estuarine wetlands include: Grays Harbor, in Washington, Willapa Bay, Humboldt Bay, Tomales Bay, San Francisco Bay, and San Diego Bay, in California.

WETLANDS ASSOCIATED WITH RIVERS ENTERING AT THE COASTAL ZONE

A few large rivers discharge directly into the Pacific Ocean with no enclosed basin in which freshwater and salt water mix appreciably. Tidal action may cause vertical zonation of the vegetation, but the primary factor influencing marsh composition is freshwater flow. Due to the relatively rapid river flow, sedimentation and erosion modify sites, causing shifts between emergent and mudflat communities.

Examples of this type of wetland system are the Columbia River and Eel River.

VERNAL POOLS

Vernal pools are seasonally wet habitats which have hard pan soils that retain water in shallow depressions. Plant species of varying tolerance to the depth and duration of flooding germinate within the pool. Most species are annual and have short life histories; the seeds provide the mechanism for survival during the long dry period. Invertebrate species are also adapted to the ephemeral nature of the pools and survive the dry period by such mechanisms as cyst formation. Vernal pools within the coastal zone are more common in the southern portion of the west coast, though many have been lost to development.

Table 1. Wetland area in hectares for the coastal counties of Washington, Oregon, and California (from Alexander et al. 1986). San Francisco Bay data from National Wetlands Inventory maps prepared by US Fish and Wildlife and includes tidal and diked marsh. Dash (--) indicates no data available.

STATE	SALT MARSH	FRESH MARSH	TIDAL FLATS	WOODED	TOTAL
California	8741	1780	5423	1376	17320
SF Bay	41253	60	25876	140	67329
Oregon	7608	2549	10198	--	20355
Washington	9591	7123	890	11817	29421
Total	67193	11512	42387	13333	134425

While more prevalent inland, examples of vernal pool habitats can be found on coastal terraces near Santa Barbara and San Diego.

MAN-MADE WETLANDS

A number of man-made wetland types play an important role in regional wildlife habitats. For example, many of the tidal wetlands around San Francisco and San Diego Bays have been converted to salt ponds. These impoundments currently support a number of wading birds such as black-necked stilts, avocets, and herons, which may not have been as common historically as they are now. In Suisun Marsh within northern San Francisco Bay, the tidal wetlands have been diked and managed as seasonal impoundments to attract waterfowl. Water management is designed to encourage specific plants to attract waterfowl and to discourage disease and vector problems.

Throughout the west coast, dikes and levees have converted many tidal wetlands to seasonal wetlands due to the impoundment of rainwater during the winter. These habitats include sites that support wetland species throughout the year as well as fields that are farmed for a significant portion of the year. In many instances, they provide important habitat for migratory waterfowl during the winter and spring months.

Finally, wetlands designed to provide treatment for point or non-point wastewater sources are increasing in number throughout the west coast. The Arcata Wetland Complex in Humboldt Bay and the Mountain View wetland in San Francisco Bay provide wildlife habitat while reducing nutrient and suspended sediment loads (Demgen 1981). More recently, wetlands have been designed to treat urban runoff (Meiorin 1986). Though designed to replicate natural wetlands, these sites must be highly managed to control water quality and potential health problems.

SUMMARY

Each of these wetland types has a different set of functional values that must be considered when attempting to manage or restore them (Table 2). Small, freshwater/brackish water ponds have a broad range of functions, whereas vernal pools provide only a few functions. Diked wetlands do not provide significant fishery habitat, but are more important as flood storage basins than tidal marshes. In addition, there are different management issues associated with each. For example, land uses within the watershed must be considered when attempting to restore wetlands in coastal rivers and lagoon systems, as they are subject to high siltation. On the other hand, estuarine wetlands are less susceptible to siltation, but can be impacted by pollutants entering the estuary.

DISTINGUISHING FEATURES OF PACIFIC COASTAL WETLANDS

The volume and timing of freshwater inflow is the most important natural variable affecting the distribution and functioning of coastal and estuarine wetlands along the Pacific coast of the

Table 2. Significant wetland functions associated with various West coast wetland types and major management problems to be considered in their restoration or enhancement. Wetland functions from Adamus (1983) include: (SA) shoreline anchoring, (ST) sediment trapping, (NR) nutrient retention, (FCS) food chain support, (FH) fishery habitat, (WH) wildlife habitat, (AR) active recreation, (PA) passive recreation, (GWD) ground water discharge, (GWR) ground water recharge, (FS) flood storage. Most functions are applicable to all wetland types; this table focuses on those most important in the management and restoration of these habitats.

WETLAND TYPE	IMPORTANT WETLAND FUNCTIONS											MANAGEMENT ISSUES ASSOCIATED WITH RESTORATION
	SA	ST	NR	FCS	FH	WH	AR	PR	GWD	GWR	FS	
Wetlands associated with low flow rivers and lagoons		X		X	X	X	X	X		X	X	Fluctuating freshwater inflow, erosion in watershed with deposition in wetland, seasonally variable mouth closure.
Major estuarine wetlands	X		X	X	X	X	X	X				High density land use adjacent to site; pollutants in estuary.
Coastal riverine wetlands		X			X	X	X	X	X			Watershed activities affect downstream functioning, possible catastrophic change in wetland distribution due to floods.
Vernal pools				X		X						Small size, need for seed source, subject to seasonal and interannual variability in water availability.
Man-made wetlands												
Diked, managed wetlands			X	X		X	X	X		X	X	Subsidence of land surfaces due to compaction, management of water control structures to affect habitat modification.
Unmanaged seasonal wetlands		X	X	X		X		X		X	X	High adjacent land use, seasonal and interannual variability depending upon rainfall.
Freshwater/brackish ponded wetlands		X	X	X	X	X	X	X	X	X	X	Watershed management and non-point source pollutants.
Wastewater wetlands			X	X	X		X				X	Water quality maintenance, possible.

United States (Onuf and Zedler 1987, Macdonald 1986). Freshwater inflow affects water and soil salinity, sediment and nutrient transport, and, consequently, plant and animal distributions and productivity. Other factors usually considered in explaining the biogeography of wetlands are of lesser significance along the Pacific coast. For example, annual temperature differences between Washington and southern California are moderated by the maritime influence of the Pacific Ocean; altitude is not important as all coastal marshes are near sea-level; and soil types are usually similar as a consequence of recent sedimentation and limited peat accumulation.

Along the Pacific coast, from 30 to 50 degrees North latitude, Macdonald (1977) described four climatic zones, all of which are based on available water (the difference between precipitation and evaporation) (Table 3). There is a net annual excess of water above 40 degrees and a net deficit below that latitude. The coastal wetland vegetation ranges from dominance by sedges and tules in the north to cordgrass and succulents in the south.

The frequency and duration of freshwater inflows markedly affect the structure and ecological function of coastal systems. There are several scales of variability in streamflow, with definite seasonal pulses (most streams in southern California do not flow at all during the long summer drought), occasional major floods, and rare catastrophic floods that completely alter wetland morphometry (e.g., 40% reduction in the low-tide volume of Mugu Lagoon during a recent flood, (Onuf 1987)). The impacts are large because the downstream sediment traps (coastal wetlands) are relatively small. In addition to changing morphometry, major floods can eliminate fish and invertebrate populations, most of which have limited tolerance to low-salinity water, especially when seawater is rapidly replaced by fresh water.

No less important are the seasonal pulses of freshwater, which control the establishment of many wetland plants. Most coastal halophytes are tolerant of hypersalinity, but their seeds may not exhibit high germination percentages in salt water, and their seedlings may not survive in marine conditions. The "low-salinity gap" that accompanies occasional winter floods is responsible for germination and establishment of many species in the arid southwest (Zedler and Beare 1986). At the same time, artificial prolongation of winter freshwater inflows (from post-flood discharges from reservoirs or year-round release of treated wastewater) can allow native salt marsh vegetation to be invaded by, or even replaced by, weedy hydrophytes (e.g., Typha and Scirpus species). Restoration of coastal wetlands is especially difficult where the natural hydrology cannot be recovered because permanent structures (dams) regulate the timing of flows and other human needs (wastewater treatment, importation of water) dictate volumes of flow.

Coastal geography also influences the role of runoff on wetland systems. Unlike the broad, gently sloping coastal plain of the Atlantic and Gulf coasts of the United States, the Pacific coast is characterized by mountain ranges and steep cliffs punctuated by small rivers flowing to the ocean. Most of these rivers have mean annual flows of less than 100 cubic meters per second with peak flows typically occurring between November and April. However, storm events, or less frequently, geologic episodes (vulcanism and earthquakes) can cause catastrophic floods that rush down steep slopes and narrow river valleys. Such events affect coastal wetlands for decades after the floods.

The size of coastal wetlands is limited by the steep topography and lack of protection by barrier beaches. Significant exceptions to this rule exist, such as the estuarine systems of Puget Sound and the broad tidal marsh plains within Humboldt and San Francisco Bays. Boule' et al. (1983), in a review of the coastal wetlands of Oregon and Washington, determined that within the emergent vegetated wetland category, wetlands greater than 100 acres comprised 75% of the total acreage within estuaries, whereas 90% of the palustrine (includes riparian) wetlands were less than 100 acres, with 45% less than 10 acres.

Coupled with size is the isolation of most wetland systems along the Pacific coast. In the Pacific Northwest, the steep coastal topography separates the adjacent watersheds. In areas of low landforms, urbanization has isolated wetlands. In the Los Angeles, San Diego, San Francisco Bay and Seattle-Tacoma areas, population growth has been extremely rapid, exceeding an 85% increase between 1950 and 1980. In Oregon and in less urbanized counties, population density and growth rates have been less than 60% during the same period. Their population densities are an order of magnitude smaller than urbanized areas (Figure 1). Urban development consumes wetland habitat and isolates the remnant wetlands, making it difficult for wildlife species to move between wetlands. Kunz et al. (1988) observed in Washington that the highest number of wetland impacts coincided with the counties of highest population and proximity to water bodies. In California, isolation and loss of wetland habitats are the major reasons for the comparatively high number of rare and endangered species in coastal California.

Given the loss of wetlands within the Pacific coastal zone (70% along the California coast (Marcus 1982), the relative isolation of wetlands, and the frequency of catastrophic events, the remaining systems gain importance for their rarity, rather than their abundance (Onuf et al. 1978). These wetlands provide critical habitat for species during specific life history phases, e.g., larval stage, breeding, nesting, and wintering. One cannot

Table 3. Environmental factors affecting wetland distribution and function along the pacific coast of North America.

LOCATION	LATITUDE (°N)	RAINFALL (cm)	EVAPORATION (cm)	NET WATER BUDGET	TEMPERATURE RANGE (°C)	CLIMATIC ZONE
SEATTLE	47.36	90	30	60	14	Mesothermal
EUREKA	40.45	100	40	60	6	Humid meso thermal
SAN FRANCISCO	37.45	50	120	<70>	7	Semi-arid
SAN DIEGO	32.53	25	160	<135>	8	Arid

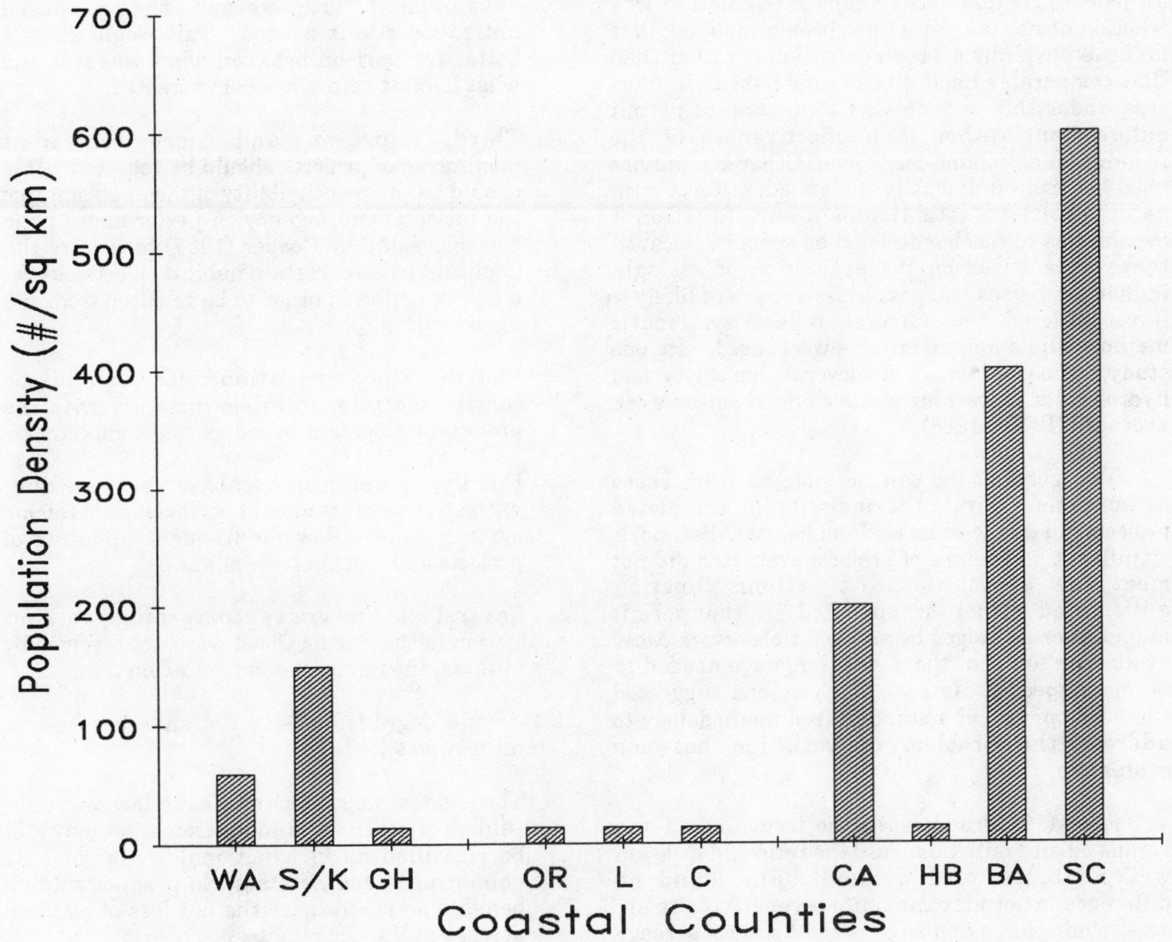

Figure 1. Population density for coastal counties along the Pacific coast. Densities for all coastal counties in each state precede selected counties within those states. S/K refers to the Snohomish/King County area around Seattle/Tacoma, GH refers to Gray's Harbor, L refers to Lincoln County, C refers to Coos County, HB refers to area surrounding Humboldt Bay, BA refers to all counties directly bordering on San Francisco Bay (excluding Suisun Marsh), and SC refers to all counties between Santa Barbara and San Diego.

measure value simply by determining productivity or contribution to the marine or estuarine food chains, but must consider that many species have been extirpated or severely reduced in number due to lack of wetland habitat during certain periods of their life history.

STATUS OF WETLAND MITIGATION ALONG THE PACIFIC COAST

Before evaluating the ecological aspects of wetland mitigation, a review of the extent and type of projects which have occurred is necessary. It is not possible to list all the projects which have been undertaken as records are inconsistent and incomplete. Instead, we have relied on compilations by other authors, recognizing that overlaps and omissions have occurred (Table 4).

Unfortunately, the evaluation of completed projects is inconsistent and such projects are difficult to compare. The simplest evaluation was whether or not the project has been completed; that is "have physical acts been carried out" rather than "has comparable habitat been established...". Success under this criteria was a function of permit enforcement rather than effectiveness of the restoration or enhancement plan. Other evaluations used general (and usually undefined) criteria such as vegetation establishment or functional capabilities to reach a decision on success. Because these were based on the evaluation of a single individual, biases and past experience were likely to have affected the outcome unless systematic methods and standard criteria were used. In one study, the opinions of several biologists and hydrologists concerning a site's effectiveness were averaged (BCDC 1988).

Two conclusions can be gleaned from these evaluations. First, the majority of completed projects did create some wetland habitat. Second, a significant percentage of projects evaluated did not meet the expectations of wetland function anticipated either as specified in the permit language or as judged by an expert observer. Most evaluators felt that the permit language needed to be more specific. Interestingly, no one suggested the development of a standardized methodology to address the problem of variation between evaluators.

Eliot (1985) addressed the issues of off-site versus on-site mitigation and the ratio of developed wetland to restored wetland. She found no difference in permit compliance between off-site and on-site mitigation and no consistent ratio of acreage restored to acreage filled. Kunz et al. (1988) found that within Washington between 1980 and 1986, mitigation negotiations had resulted in a net loss of wetland acreage: from 62 hectares of functional wetland to 40 hectares of created/restored wetland. Such continued net losses in wetland acreages have stimulated some agencies to increase the ratio of wetland restored to wetland filled (BCDC 1988).

Each of the studies provided a number of recommendations for future projects:

- The most common recommendation was to improve the permit to include more detail on the habitat to be created with specific design objectives and features.

- Second, a more precise habitat evaluation of the development site versus the proposed mitigation site is needed. This would allow a better comparison between what was lost and what habitat values are being created.

- Third, improved and more consistent monitoring of projects should be required. This should be the responsibility of the applicant, not the jurisdictional agency. A performance bond was suggested by Cooper (1987) to require the applicant to meet certain habitat objectives over a 5-year period in order to be released from the monetary bond.

- Fourth, the mitigation site should be constructed prior to or concurrently with the project development to reduce non-compliance.

- Finally, a uniform database for recording mitigation projects should be developed. Such a system would allow for frequent updating of projects and tracking compliance.

Several controversies concerning wetland mitigation in the Pacific Coast were not resolved by the authors, though most commented on them.

1. Should degraded offsite wetlands be used as mitigation sites?

   These sites may appear to have low value for wildlife or other wetland functions, yet may still be classified as jurisdictional wetlands. By enhancing these habitats, do greater wildlife benefits accrue despite the net loss of wetland acreage at the permit sites?

2. How far should wetland mitigation sites be from the permit site?

   Do fish and wildlife which utilized the developed site move readily to the distant site? For many migratory species, movement to

Table 4. Reports of wetland restoration/enhancement projects along the West Coast of the contiguous United States. Habitat type refers to general USFWS classification of project considered. Review refers to source(s) used to locate sites or provide criteria for evaluation, e.g., objectives stated in permit application. Method refers to general procedures to make evaluations; office refers to examination of literature or personal communication and field refers to site visits by author(s).

AUTHOR	LOCATION REVIEWED	HABITAT TYPE STANDARD	NO. SITES	REVIEW	METHOD	EVALUATION	COMMENTS
Harvey et al (1982)	California (San Francisco Bay)	Estuarine	9	Actual site work	Field	No evaluation though remarks given for ind. sites.	Dealt with restoration from diked wetlands.
Josselyn and Buchholz (1982)	California	Estuarine/Palustrine/ Lacustrine	33	Permit file	Office	No evaluation though remarks given for sites.	Included experimental plantings, References to 5 previous listings.
McCreary and Robin (1984)	California	Estuarine/Palustrine	30[1]	Coastal Cons. records	Office	No evaluation	Description of Conservancy projects; enhancement and restor projects.
Eliot (1985)	California (San Francisco Bay)	Estuarine	58	Corps/ BCDC permits	Office with 15 field visits	Complete: 31% In progress: 44% Incomplete[2]: 24%	Of 58 projects, only 2 claimed as "successful".
Race (1985)	California (San Francisco Bay)	Estuarine	11	BCDC permit and the "successful establishment" of wet. veg.	Field visits	Success of all projects reviewed is "debatable".	Included experimental plantings.
Kentula (1986)	Washington Oregon	Estuarine/Palustrine/ Lacustrine/Riverine	77	Oregon State Lands and US Fish and Wildlife.	Office	No evaluation	General review, only a few specific projects described.
Cooper (1987)	Washington	Estuarine/Palustrine	22[3]	USFWS admin. records	Office	No evaluation	Reviewed various components needed for good restoration plan.
Zentner (1988)	California	Estuarine	63	Adamus Tech.	Field visits	Success: 65% Partial: 25% Failure: 10%	Federal/state projects had lowest % success.

Table 4. (Continued)

AUTHOR	LOCATION REVIEWED	HABITAT TYPE STANDARD	NO. SITES	REVIEW	METHOD	EVALUATION	COMMENTS
BCDC (1988)	California (San Francisco Bay)	Estuarine	18	BCDC permit	Field visits	Successful: 33% Partial: 39% No decision: 6% Failure: 11% In Progress: 11%	
Schafer (pers. comm.) WSDOT[4]	Washington	Palustrine	9	WSDOT files	Field	Completed projects; no evaluation	All 0.5 - 2.0 acre replacement projects.
Pritchard (pers. comm.) Seattle Aquarium[5]	Washington	Estuarine	5	Adopt-a-Beach	Field	Completed or in progress	Information on status is provided.

[1] Of 30 projects reviewed, 11 partially or fully implemented; remainder are pending or planned.

[2] Incomplete by date specified in permit.

[3] Of 22 projects reviewed, 14 were implemented; 8 planned.

[4] Project Development Office, WSDOT, Olympia, WA 98504.

[5] Adopt-a-Beach, Seattle Aquarium, Pier 59, Waterfront Park, Seattle, WA 98101.

another location may be feasible, but for less mobile species or species dependent upon specific environmental factors in the region of the developed site, the mitigation site may not be suitable.

3. Should mitigation provide for in-kind habitat losses?

   Most authors argue that agency staff must be prepared to manage for regional habitat goals, not on a project-by-project basis. Regional habitat goals would not necessarily call for in-kind habitat replacement.

4. How long is appropriate for wetland succession to occur prior to determining whether the site is successful?

   Most authors state that 3 to 4 years is needed for a site to reach a stage where it can be evaluated in terms of meeting the project objectives. On some sites with unusual conditions, e.g., non-native soils, poor conditions for plant establishment, or where subsidence has occurred, this period may be as long as 7 to 20 years. There is little information on what constitutes a comparable wetland in terms of species diversity and abundance.

5. Should natural events such as floods, siltation, or sea-level changes be considered in determining whether or not a site is successful?

   In other words, if natural events modify a site's ability to achieve the desired objectives, should these events be controlled or the site reconstructed?

FUNCTIONAL DESIGN OF RESTORATION PROJECTS

A number of reports have outlined the necessary planning procedures for the implementation of wetland restoration projects along the Pacific coast (Sorensen 1982, Zedler 1984, Josselyn and Buchholz 1984, Kunz et al. 1988, Boule' 1988). In addition, specific engineering criteria have also been developed, especially for wetlands involving dredge material disposal (U.S. Army Corps of Engineers 1983 a,b). Each has recognized the necessity of establishing specific project objectives and consideration of hydrologic, biologic, engineering, and cost constraints and opportunities in the design process. The purpose of this section is to review the experience of wetland restoration along the Pacific Coast and to make specific recommendations that should be considered when evaluating Section 404 mitigation proposals.

REGIONAL PERSPECTIVE

"Rarity" is the term which distinguishes the wetland habitats of the Pacific coastal plain. The scarcity of continuous wetland habitats along the coast makes regional planning critical to the viability of fish and wildlife species associated with wetlands. When proposals are presented to eliminate habitats in one part of a biogeographic region and to consolidate or replace those habitats in another part of the region, consideration must be given to the ability of species to accommodate such shifts.

While migratory birds are able to move great distances, resident species with restricted ranges are unlikely to become re-established in the new locations, especially where development separates the various wetland habitats. At present, there have been no successful transplants of wetland fauna to new restoration sites, although it has been attempted with the salt marsh harvest mouse in San Francisco Bay and with rare beetles in the Tijuana Estuary. Most project proponents assume that the fauna will move or colonize a new wetland area over a period of time. This assumption is likely to result in a loss of species diversity.

The practice of giving mitigation credits for converting one wetland habitat type (e.g., shallow water) to another (e.g., deep channels) should be carefully reviewed for impacts on regional wetland resources. This is particularly true in southern California where entire lagoonal systems are being modified as mitigation for port development. Such projects may completely eliminate the use of a wetland by particular species which must seek suitable habitat in more distant locations.

The perceived importance of commercial and recreational fisheries over non-game wildlife resources may also drive mitigation planning. In Washington, Kunz et al. (1988) found that most mitigation projects were directed towards anadromous fish habitat regardless of the type of wetland lost, resulting in a gradual loss in the diversity of wetland habitats. For the six year period analyzed, an average of 2.1 habitats were lost/project and only 1.4 habitat types/project were restored through mitigation. This may be an appropriate planning strategy if part of an overall regional plan; however, it is more likely the inadvertent result of individual permit decisions.

Regional planning is difficult to implement as most projects are reviewed individually rather than as a whole. The U.S. Fish and Wildlife Service

National Wetland Inventory can be utilized to reduce this problem. A comprehensive inventory is being completed for San Francisco Bay which will provide information on the acreage and average size of various wetland types (Peters and Bohn 1987). The NWI maps and data can be used to establish the general amount and diversity of various wetland types in the proposed project area and the impact of loss or change of a particular wetland habitat. At the very least, permit applications should discuss the significance of the proposed habitat modification relative to the quantity of various wetland types in the region. In addition, agency staff need to evaluate individual projects on the basis of cumulative impact over time and should periodically review the regional wetland habitat changes that have resulted from permit approvals.

CONDITIONS AT THE MITIGATION SITE

Existing conditions represent the starting point for all project planning. Mitigation proposals should review and evaluate the hydrologic, edaphic, and biologic conditions for both the development and mitigation sites (Sorensen 1982). Increasingly, studies have shown that even degraded wetland sites may have important wetland values to rare and endangered plants and animals, migratory birds, and hydrologic functions in the region. Sites presumed to have little value may provide vital refuge for species during storm events or support rare and endangered species due to lower interspecific competition with other species within these marginal habitats. As a result, one cannot assume that the loss of a degraded wetland can be appropriately mitigated by either the creation or restoration of habitat at another site.

Habitat evaluation techniques such as the U.S. Fish and Wildlife Service's Habitat Evaluation Procedure (HEP) and the joint Federal Highway Administration and U.S. Army Corps of Engineers Wetland Evaluation Technique (WET, Version 2) can be used to assess both the development and mitigation site to determine their relative wetland values. Only a few training classes are offered each year and therefore the lack of familiarity with these methodologies limits their application. In addition, the absence of uniform habitat suitability models in coastal wetlands limits the application of HEP. Ideally, WET and HEP offer the ability to assess wetlands in the absence of long-term data bases. Otherwise, the evaluation should include year-long data on habitat use, especially for sites that support migratory birds or seasonal fish nursery areas. The likelihood that a single year's data represents the regional "norm" should also be indicated by comparison of rainfall, stream flow, and temperature data for measurement years and the period of record.

As an alternative to in-depth site evaluation, most regulators must rely on descriptive site evaluations. The information that should, at a minimum, be included is listed in Table 5. A site containing landfill material or habitat for rare species will require more research to address those concerns. In addition, the type of wetland proposed for the mitigation site may also require more specific data. For instance, a mitigation site receiving urban run-off will require further evaluation to determine the quality of the water entering the site.

The final product produced from this evaluation should be a series of maps illustrating the existing topography, hydrologic features, vegetation, and presence of any unique attributes which need to be preserved or enhanced in the mitigation plan.

CRITICAL ASPECTS TO A SUCCESSFUL MITIGATION PLAN

The important aspects of a mitigation plan that are specifically applicable to the Pacific West Coast are dependent on the type of wetland restored or created. However, there is no "cookbook" solution and the issues discussed below should be considered the minimum information necessary to evaluate a permit application.

Wetlands Associated with Low Flow Rivers and Lagoons

These systems are dominated by seasonal hydrographic events which affect salinity, sand bar closure, and sedimentation. Therefore, long-term hydrographic records are important in evaluating the feasibility of mitigation plans. These may be developed from available data or through the use of computer models (Zedler et al. 1984). A number of features need to be determined: frequency of various flow events, duration, and flow rates, as has been done for the Tijuana Estuary (Williams and Swanson 1987). Examination of aerial photographs and some knowledge of off-shore currents may be required to predict the frequency and duration of sandbar closure as well. Given the dynamic nature of these systems, mitigation plans within these wetlands should not attempt to create wetland types different than those that existed historically.

Hydrographic information is necessary in order to predict likely vegetation patterns in the system. Generally, pickleweed (Salicornia virginica) is tolerant of high salinity and will persist in systems that are closed either periodically or permanently by an outer sandbar (Onuf and Zedler 1987). Cordgrass (Spartina foliosa), on the other hand, requires some freshwater inflow in spring and regular tidal fluctuations. Extremely high flows may completely shift the balance within the emergent plant community to one of fresh and

Table 5. Existing conditions which should be included in the evaluation of proposed mitigation sites.

FEATURE	DESCRIPTION	PRODUCT
Site history	Past uses of site including past functioning as wetland.	Description, map, or photographs illustrating historic uses.
Topography	Surface topography including elevations of levees, drainage channels, ponds, islands.	Topographic map with 1 foot contour intervals, 1" to 100 or 200 scale.
Water control structures	Location of culverts tide gates, pumps, and outlets.	Elevations for all structures, size and type of structure, current operational status.
Hydrology	Description of hydrologic conditions affecting the site.	Water budget for site including inflow, precipitation, evaporation, and outflow; tidal range, history of sand-bar closure.
Flood events	Current potential for flooding by high flows, extreme tides, and storms. Adequacy of any external or internal levees.	Evaluation of current flood control protection using appropriate runoff models.
Sediment Budget	Analysis of sediment inflow, outflow, and retention.	Evaluation of historic sedimentation rates and projected due to watershed development.
Edaphic characteristics	Description of existing soils with analysis of suitability for supporting wetland plants.	Presence of hydric and non-hydric soils, salinity, toxic compounds in filled areas.
Existing wetland characteristics	Determination of COE jurisdictional wetlands, if any.	Boundary map illustrating wetland extent on mitigation site.
Existing vegetation	Description of existing habitat with analysis of degraded areas and any habitat with current high value to wildlife.	Vegetation map with list of dominant species and location of non-native nuisance species and species of regional concern.
Existing wildlife	Description of wildlife using the site, indicating those species which may be displaced by mitigation activity.	Listing of wildlife known to use site, especially species of special concern (incl. rare and endangered).
Adjacent site conditions	Analysis of wildlife habitat adjacent to site, indicating those species likely to benefit or be impacted by the mitigation.	Map showing site in reference to surrounding habitats, preferably NWI maps. List of species benefited or impacted.

brackish water species (Zedler 1983). Closure of the sandbar may lead to a decline in all species due to long-term inundation or hypersaline conditions; however, pickleweed is most likely to persist (Ferren 1985).

Wetlands Associated with Major Estuaries

Most sites available for the mitigation of estuarine wetlands loss are within highly urbanized areas (San Diego Bay, San Francisco Bay, Humboldt Bay, and Puget Sound). High density land uses adjacent to the mitigation sites present serious constraints to the development of wildlife habitat. In addition, pollution from urban runoff, adjacent landfills, and within the bay water itself may affect the suitability of certain sites.

Wetland mitigation projects within these areas frequently have involved restoration of former tidal wetlands that had been diked. Most diked areas have undergone some degree of subsidence due to a variety of factors, including oxidation of organic matter, drying and compaction of the clay and peat soils, erosion of soil during farming activities, overdrafting of ground water, or lack of sediment sources sufficient to maintain elevations relative to rising sea-level. In some cases, diked areas that have received fill have also been used for mitigation. In these cases, the underlying marsh surface has been lowered by the compression of marsh clays and peats. Because all the fill material needs to be removed prior to restoration, the end result is a former marsh surface below its natural level.

The most common problem in restoring diked or filled wetlands is the re-establishment of elevations necessary for marsh plants. This can be done through placement of dredged material, on-site grading, control of water levels, or allowing natural sedimentation (Harvey et al. 1982, Josselyn 1988b). The selection of the most appropriate method depends on economic feasibility and the time allowed for establishment of an emergent marsh. For example, grading and construction of water control structures can be expensive; however, marsh planting can begin immediately following construction. Allowing sedimentation and accretion to create suitable elevations for emergent vegetation is less costly, but may delay marsh establishment for 10 to 20 years. However, in the interim, the site will provide wetland habitat values associated with open water. Use of dredge material can be effective, especially when the economics of other disposal methods is prohibitive. The site can then be planted almost immediately.

Where man-made fill is to be removed, it is important that all the fill material be excavated even if this requires over-excavation and backfilling. At a site in Humboldt Bay, wood debris was left after excavation to create intertidal elevations. When tidal action was restored, the wood debris floated to the surface and was deposited throughout the newly created wetland. Five years following restoration, decomposition of the wood debris continued to cause water quality problems and methane gas production (Josselyn 1988a). Similar problems can occur in sites subject to previous industrial processes. Despite five years of daily tidal action, a former salt crystallizer that was restored to tidal action in San Francisco Bay has failed to re-vegetate due to the presence of thick layers of salt that were not excavated (Josselyn pers. obs.).

Dredged material placement and grading of formerly submerged muds can lead to acidic soils or "cat-clays", especially in the creation of high marsh and islands (Josselyn and Buchholz, 1984). Liming is recommended as a treatment for this problem (Clar et al. 1983) but has had mixed results in San Francisco Bay (Josselyn and Buchholz 1984, Newling and Landin 1985). The acidic soils may last for 10 years or more and, therefore, it is best to prevent such conditions from developing. If islands or high marsh areas are planned, it is preferable to use upland soils that are low in sulfate.

Within estuaries, tidal ranges may either increase or decrease as one moves from the ocean to the head of an estuary. Consequently, plant distribution will shift in relation to the individual species' tolerance to inundation. In addition, some species such as tules and cattails are less tolerant of submergence as salinity increases (Atwater et al. 1979). As a result, their vertical distribution range is reduced towards the mouth of the estuary. Mitigation plans should include information on typical distributional profiles for nearby habitats to assure appropriate elevations.

Coastal Riverine Wetlands

These wetland types are most vulnerable to activities within the watershed. Lumbering, mining, and urbanization have generally increased rates of sedimentation and the movement of logs, debris, and pollutants through these wetlands. As a result, the tidal prism has decreased within these systems due to filling at the river mouth and deposition of material to form levees along the edges of the river as it enters the coastal plain. The levees isolate the natural riparian wetlands along the edge of the river by blocking flood and tidal flows. Furthermore, sedimentation may result in the loss of tidal flats and eelgrass beds.

The two primary means of enhancing or restoring these systems are (1) watershed management to reduce sedimentation or log deposition downstream and (2) removal of levees along river edges and restoration of stream-bank vegetation. Without effective watershed management, downstream restoration will fail. Watershed management requires that a detailed analysis be conducted which considers climate, precipitation rates, vegetation types, slope stability and aspect, soil type and composition, water discharge, and land use. Unfortunately, the required coordination between agencies responsible for watershed management and coastal resources is a significant problem. In California, the State Coastal Conservancy has provided a unique model to provide the coordination and funding for a number of watershed management programs in Tomales Bay, Morro Bay, Los Penasquitos Lagoon, and others.

Vernal Pools

Vernal pool restoration and creation projects are subject to unique problems. As isolated wetlands within upland habitats, the shallow pools are especially susceptible to off-road vehicle damage. Thus, sites must be fenced. Their special hydrology, i.e., short hydroperiods, is difficult to create. Soils must have the necessary clay layer that will retain water during the wet season. Detailed topographic mapping before construction and contouring during construction are required. Not only must the pool depth and cross-sectional profile be correct, but the local drainage area must not be too large or too small. Placement within a large catchment will increase water depth and prolong inundation. In one vernal pool preserve, leakage of an irrigation pipe artificially prolonged the hydroperiod and cattails and other marsh species invaded the pools within one growing season (P. Zedler, pers. comm.). Eradication measures are required, because the cattails will persist even after the pool's normal hydroperiod is restored. Construction of pools within coastal scrub areas also damages the surrounding vegetation which may require restoration to provide a natural transitional landscape.

Most of the biological information on vernal pools describes the vascular plants. The plankton, food webs, and nutrient dynamics are also critical in assessing pool functions before destruction or after restoration or creation. In Southern California, where pool destruction has endangered certain plants, we recommend that no future permits for the destruction of existing pools be granted until restoration/creation efforts can demonstrate that natural functions can be replaced. To insure that vernal pool biodiversity is maintainable, artificial pools should be monitored during environmental extremes, including catastrophic drought.

General Considerations Applicable to All Sites

Replacement acreage ratios--

The loss of wetlands due to development pressures has been substantial. Permit reviews have indicated that often the loss of acreage has not been fully compensated within the mitigation project (Eliot 1985, Kunz et al. 1988). In addition, considerable delays between the wetland loss at the development site and wetland establishment at the mitigation site result in net decline in habitat availability to fish and wildlife (BCDC 1988). Therefore, in no case should mitigation be permitted on less than a one-to-one replacement and in most projects, especially those anticipating delay in implementation, the replacement ratio should be greater.

Timing of mitigation projects--

Construction of mitigation projects should be timed to reduce impacts to existing fish and wildlife and should precede the permitted wetland development.

Most mitigation sites are either located within or adjacent to existing wetlands. The protection of the existing fish and wildlife must be considered in selecting the period of time during which construction activity is allowable. For example, migratory waterfowl primarily utilize wetlands during the spring months (March through April) and in the fall (September through November). In Washington and Oregon, the birds may be present earlier during the fall migration and some may overwinter in southern California wetlands. Most juvenile fish are found within wetlands during the spring and summer months, and construction in sites that are utilized by juvenile salmon smolt should be restricted during the spring. The critical periods (e.g., nesting) in the life stages of rare and endangered species must be considered. Restrictions for protection of wildlife must be balanced with the feasibility of construction during certain seasons.

It is not known how long mitigation sites will take to develop wetland habitat values comparable to natural wetlands (see examples cited in Kentula 1986). Differences in construction techniques, variation in plant establishment, and differences in the colonization of new sites by animals makes it difficult to predict the rate of succession in restoration sites. In western tidal marshes, plant invasion generally proceeds from the higher to lower elevations of the site (Niesen and Josselyn 1981). Thus, high marsh sites are likely to be rapidly colonized by vegetation within a few years whereas lower portions are more slowly revegetated, unless artificial propagation is undertaken. Areas above tidal action invariably require initial plantings and irrigation.

Plant establishment in tidal marshes usually takes two to four years, though it can be accelerated by planting. If natural seed sources are not readily available, plant establishment may take five to seven years. This is particularly true for Pacific cordgrass. Freshwater and brackish marshes are colonized by cattails and tules within one year, whereas woody species such as willows and alders establish quickly, but maturity isn't reached until three to five years.

Because restoration sites may not become vegetated in the first year or two, construction of the mitigation project site should begin prior to the permitted development. Given the frequent delays in construction projects, it is not unreasonable to request mitigation to begin several years prior to the proposed development project (see Zedler in press). Kunz et al. (1988) found that out of 26 projects permitted in Washington State, only two habitat compensation projects were completed before the project impacts began. The average time lag for compensation was 1-3 growing seasons.

Construction oversight and inspection--

Few engineering or construction firms have specific experience in wetland creation and restoration. Working drawings and bid specifications must be prepared by a civil engineer and only experienced contractors in wetland restoration should be hired. However, there are several stages during the development of working drawings and actual site construction that should also include the involvement of a wetland biologist and/or hydrologist (Table 6). As with any complicated construction project, problems are likely to occur and judgments from professional wetland scientists will be needed. Regulatory personnel do not have the time to make such decisions. Therefore, the permit itself must require the applicant to assure on-site review by knowledgeable personnel.

Revegetation of mitigation sites--

There is often debate over the necessity to revegetate mitigation sites. Most agree that given the correct hydrologic regime, wetland plants will eventually become established. The question is one of timing and desired composition. How long is an acceptable period of time before emergent vegetation is established? Race (1985) considered a number of restoration projects in the San Francisco Bay area failures due to the lack of coverage by vegetation; but many of these sites became fully vegetated after 3 to 4 years of natural recolonization (Josselyn pers. obs.).

There are a number of reasons to require re-vegetation by artificial means:

1. A specific wetland plant community may be necessary to support certain fish and wildlife species. Where a rare habitat type will be destroyed by the permitted development, replacement habitat should be provided in advance.

2. Salvaging wetland plants from the development site may be desirable to preserve genetic diversity. In addition, some species, especially woody plants associated with riparian areas, are valuable due to their age. It might be desirable to transplant these species to the mitigation site rather than propagate from seedlings. However, they must often be cut back severely to become established and the high cost of transplanting mature plants limits the number that can be transplanted. For annual and herbaceous species, transferring soil from the impacted wetland to the mitigation site may be sufficient to re-establish a mixed plant community.

3. Invasive and/or exotic species may be better controlled if a revegetation program is initiated early in the restoration project. For example, native high marsh and transition zone species will have difficulty colonizing if more aggressive exotic species become established first. The planting of native species may give the desirable plants a competitive edge over the non-native species.

4. Isolated wetlands may not receive sufficient water-borne seeds or native species nearby may not produce large numbers of viable seeds. For example, in the southern portion of San Francisco Bay, Pacific cordgrass does not produce seeds during dry years when water salinities are high; thus, its invasion there is much slower than in the northern portion of the bay where water salinities are regularly reduced by freshwater inflow in the spring.

West coast re-vegetation techniques have been reviewed by Knutson (1976) and Zedler (1984). The technique selected depends upon the availability of plant material, the area to be planted, the soil conditions, the need to protect plants from grazers and other plant competitors, and other site-specific factors. Regardless of the technique used, the following general guidelines are recommended:

1. Use native species from the local region.

2. Do not allow commercial "cultivars" as a replacement for native species.

3. If transplanting from nearby sites is necessary, require monitoring of the donor site to assure that it is not decimated by the extraction of plant material.

4. The time of planting for high marsh and transition zone species should occur during the fall, prior to the rainy season, unless irrigation is used.

5. Soil conditions on the site should be monitored prior to planting to assure that salinity, moisture, and pH are appropriate for the planned species.

Table 6. Stages in engineering and construction that should be inspected and approved by a wetland biologist and/or hydrologist. It is assumed that a civil engineer, licensed contractor, and appropriate construction supervisory staff are also involved.

ACTIVITY	INDIVIDUAL	APPROVAL
Working drawings	Biologist	Approval of overall plans, elevations, and planting specifications, if any.
	Hydrologist	Location, elevation, and type of water control structures.
Bid documents	Biologist	Timing of construction, type of equipment, if specified.
Start work meeting	Biologist	Location of all construction activity, including staging areas and disposal sites.
On-site supervision	Biologist Hydrologist	On-call to review any changes to the plan due to unforeseen problems.
Acceptance of work	Biologist Hydrologist	Approval of elevations, work modifications, site clean-up.

Use of buffers between mitigation sites and upland development--

The use of buffers between wetlands and upland development is a controversial topic. Buffers are usually upland areas that are not within areas regulated by federal and state agencies. Landowners consider such uplands as developable property. However, resource agencies usually require buffers as part of approved land-use plans. As a result, the nature, size, and use of the buffer zone often becomes a subject of intense debate fueled by the lack of scientific information on the effectiveness of various buffer configurations.

Ecologically, a buffer is a transition zone between one type of habitat and another. Often, these habitat edges support a more diverse flora and fauna than either of the two adjoining habitats (Jordan and Shisler 1988). Buffers are also important in reducing excess nutrient and sediment loading to wetlands. On the other hand, land-use planners view buffers as transitions from higher density uses on adjoining lands. Under such a definition, the buffer itself may have a land-use such as public access, recreation, green-belt, or even parking lots or roads. Thus, the use of buffers in land-use planning is to reduce connectivity between the upland and wetland rather than encourage wildlife utilization.

Width is the most frequently cited criterion for buffers. Generally, thirty meters is recommended. However, the selection of any distance is subjective; the width should depend on the use of the upland area and the sensitivity of the species planned for the mitigation site. Ecologists usually do not have the information necessary to determine the most appropriate distance for individual situations. Based on field study, White (1986) recommended that 30m buffers would reduce impacts of human intrusion on Belding's savannah sparrows, an endangered species in Southern California. Josselyn et al. (1988) determined that water birds varied in their response to disturbance depending upon acclimation and distance, but usually reacted within 25 to 45m.

Other features of buffer zones include fencing to reduce human and pet intrusion, vegetation and berms to reduce noise and visual intrusion, and open water to separate pedestrians from wetlands. Of these, vegetation, especially native shrubs and trees, may be the most effective, as it also provides wildlife habitat.

In reviewing mitigation plans, the planned land-use adjacent to the mitigation site must be considered. Habitat for highly sensitive species cannot be planned next to industrial parks unless elaborate buffers are provided. In addition, upland development will introduce unwanted predators on wetland wildlife such as Norway rats, feral cats, and unleashed dogs. Though predator control plans have been proposed for some mitigation projects, no plan has been implemented nor shown to be effective. High density upland uses will limit the opportunities for on-site mitigation. It is then appropriate either to scale down the development proposal or consider off-site mitigation.

MONITORING PROGRAMS

Two levels of monitoring are necessary. The first level is one of enforcement. Is the project being completed in compliance with the permit specifications? This level of monitoring can be accomplished through annual site visits by regulatory personnel or, at a minimum, submission of site photography by the permittee. A requirement in the issued permit for an annual report on the mitigation project during the first 3-10 years may ensure regular review. The length of time required should be dependent upon the complexity of and experience with the type of habitat proposed.

The second level of monitoring involves evaluating the effectiveness of the restoration project. A variety of monitoring programs have been suggested (Zedler et al. 1983, Zedler 1984, Josselyn and Buchholz 1984), and the need for monitoring has been expressed by many agencies. Unfortunately, few mitigation projects have required detailed monitoring; and of those that have, the results have not been used to effect changes. In a review of 404 permits in Washington State, Kunz et al. (1988) found that only 31% of the permits incorporated monitoring study requirements.

Typically, mitigation site monitoring is of short duration (one year) and only requires data collection on vegetation coverage. If monitoring is to be an effective tool to evaluate the success of achieving mitigation goals, it should be redirected towards understanding the functions of the mitigation site as opposed to its appearance (Shisler and Charette 1984). Josselyn and Buchholz (1984) outlined the types of monitoring required for various habitat functions.

Thom et al. (1987) reported on a monitoring program directed towards understanding the functioning of a restoration project on the Puyallup River estuary in Washington that was designed for several target species. Portions of the restoration were designed to support juvenile salmonids, waterfowl, shorebirds, raptors, and small mammals. A design was then selected that includeda sedge marsh, a cattail marsh, unvegetated mudflats, and channels. The monitoring program was directed to sampling selected physical, chemical, and biological parameters to evaluate the successful establishment of vegetation and the utilization by the target species.

Several wetland sites along the Pacific Coast have been examined at least five years after creation (Josselyn and Buchholz 1984, Newling and Landin 1985, Faber 1986, Josselyn 1988a, Josselyn et al. 1987, Landin et al. 1987, BCDC 1988). These studies have generally observed that, given appropriate hydrologic conditions, functioning wetland habitat in terms of both vegetation colonization and animal utilization evolved over time. The restoration sites did not meet all the objectives stated but provided functions not anticipated. Thus, monitoring and evaluation of success or failure should be flexible and based on ecologic function rather than preconceived notions of how a wetland should look. For example, in Humboldt Bay, a high marsh habitat was planned for the Bracut mitigation bank. Poor substrate conditions have restricted plant establishment by typical high marsh species, but have allowed the extensive colonization by a rare plant species, Humboldt Bay owl's clover (Josselyn 1988a). On the basis of the permit objectives, this project would be considered a failure, yet it provides a very effective habitat for a rare species. Nevertheless, one must carefully evaluate whether in-kind habitat replacement is still being provided if the mitigation site varies too much from the stated objectives.

Monitoring is often accepted as a passive activity with little enforcement. Agency personnel must review monitoring reports and request peer-review to assure unbiased reporting (Zedler 1988). Action should be taken to correct problems at the mitigation site. Kunz et al. (1988) recommended that contingency plans be included as part of the permit if the mitigation should fail. They and Cooper (1987) suggested a performance bond as a means to assure funding for corrective measures.

INFORMATION GAPS AND RESEARCH NEEDS

Successful restoration is based on past experience and predictions from experimental research. With this in mind, future projects should necessarily include in their design at least some degree of experimentation to ensure site-specific success, while contributing to our overall understanding of how best to restore a full, healthy wetland. Because many Pacific coastal wetlands are small and isolated, they are often managed for a relatively few target resources (i.e., endangered species habitat). Information is therefore needed regarding the establishment and maintenance of particular species as well as the general ecosystem. Important research needs that may be incorporated into a plan include (but are not limited to) plant/hydrologic relationships, wildlife habitat utilization, and nutrient removal functions of wetlands (Weller et al. 1988; LaRoe 1988).

Significant gaps exist in the understanding of the effect of various hydroperiods on west coast plants. Despite this, hydrologic modifications (dike breaching, tidal restoration, etc.) are often major components of restoration plans, and improper hydrology often leads to colonization of mitigation areas by non-target species. Period and frequency of soil saturation affect different plant species in different ways, though hydrologic limits are not well defined. Salicornia virginica growth, for example, is inhibited but not entirely eliminated by prolonged inundation. Whether this inhibition allows for the establishment of other species (i.e., Spartina) is not yet clear, nor is it clear how long Salicornia can withstand complete inundation.

In a similar way, there are significant gaps regarding the effect of salinity variation on plant growth and seedling establishment. Widely variable salinities may decrease diversity in wetlands (Zedler 1982), however experimental evidence is needed before salinity effects on individual species can be adequately described. While it is generally accepted that the proper design of the physical ecosystem component (topography, soils properties, hydraulic circulation, sediment budget, etc.) is crucial to a successful mitigation plan, these areas have received much less research than biological aspects of coastal marshes. The extensive body of research on estuarine and fluvial processes must be extended to wetland processes.

The goal of most mitigation projects extends to the restoration of an entire ecosystem. A workable, self perpetuating system that supports the desired (target) wildlife species is often desired but seldom realized, particularly on sites of less than 5 acres. One of the reasons for this is simply an incomplete understanding of the systems being restored. The physical factors and even outward appearance (vascular plants) of an area may seem right for a given species to colonize, however several more subtle factors may deter colonization. Prey species must be available for the target species, extending the need for information to the lower trophic levels of the food chain. Adjacent land uses may affect the desirability of the restored area for wildlife use. The area may just be too small for the desired species, or perhaps misplaced among surrounding habitats. To gain specific knowledge of these factors and how they affect the desired wildlife, broad range experimentation can be incorporated into future project plans, increasing the success of those mitigation projects and the ones to follow.

With ever-increasing human population in the Pacific coast region, wetlands have become important for their capacity to remove nutrients from sewage effluent and urban runoff. In some projects, water quality improvement is an incorporated goal in the mitigation agreement. How this function affects the other functions of a restored wetland is still unclear however, and presents another need for further research. The effect of pollutants on the food chain is likely to have a direct influence on the health of the wildlife that colonizes the area, yet information regarding these effects is scarce. Controlled experimentation incorporated into project plans would greatly assist in the understanding of the impact of these factors.

ACKNOWLEDGEMENTS

The authors wish to acknowledge the review and guidance provided by Jon Kusler and Mary Kentula in the preparation of this chapter. In addition, anonymous reviews by over 10 individuals greatly assisted in giving a larger perspective to the text. The following individuals supplied reports or preprints of material that were important to our review: Mary Burg, Washington State Department of Ecology; Ken Brunner, U.S. Army Corps of Engineers; James Schaefer, Washington State Department of Transportation; Kenneth Raedeke, University of Washington; Ken Pritchard, Adopt-a-Beach Program; Ron Thom, University of Washington; Francesca Demgen, Demgen Aquatics; Jeff Haltiner, Philip Williams and Associates, and John Zentner, Zentner and Zentner. Finally, students at San Francisco State University who assisted in the preparation of the manuscript are: Michael Nelson, Molly Martindale, and John Callaway.

LITERATURE CITED

Alexander, C.E., M.A. Broutman, and D.W. Field. 1986. An Inventory of Coastal Wetlands of the USA. National Oceanic and Atmospheric Administration, Washington, D.C.

Adamus, P.R., and L.T. Stockwell. 1983. A Method for Wetland Functional Assessment. Federal Highway Administration, U.S. Department of Transportation, FHWS-IP-82-23.

Atwater, B.F., S.G. Conard, J.N. Dowden, C.W. Hedel, R.L. MacDonald, and W. Savage. 1979. History, landforms, and vegetation of the estuary's tidal marshes, p. 265-285. In T.J. Conomos, (Ed.), San Francisco Bay: The Urbanized Estuary. Pacific Division, Amer. Assoc. Advance. Sci., San Francisco, California.

Bay Conservation and Development Commission. 1988. Mitigation: An Analysis of Tideland Restoration Projects in San Francisco Bay. Bay Conservation and Development Commission, San Francisco, California.

Boule', M.C. 1988. Wetland creation and enhancement in the Pacific Northwest, p. 130-136. In J. Zelazny and J.S. Feierabend (Eds.), Increasing Our Wetland Resource: Proceedings of A Conference. National Wildlife Federation. Washington, D.C.

Boule', M.C., N. Olmsted, and T. Miller. 1983. Inventory of Wetland Resources and Evaluation of Wetland Management in Western Washington. Washington State Department of Ecology, Olympia.

Camp, Dresser and McKee, Inc. 1980. Bracut Marsh Restoration Project. Prepared for the California State Coastal Conservancy. Oakland, California.

Clar, M., R. Kort, H. Hopkins, N. Page, and C. Lee. 1983. Restoration Techniques for Problem Soils. EL-83-1, Final Report. U.S. Army Corps of Engineers, Waterways Experiment Station, Washington, D.C.

Cooper, J.W. 1987. An overview of estuarine habitat mitigation projects in Washington State. <u>Northwest Environ. Jour.</u> 3(1): 113-127.

Cowardin, L.M., V. Carter, F.C. Golet, and E.T. LaRoe. 1979. Classification of Wetlands and Deep-Water Habitats of the United States. U.S. Fish and Wildlife Service, Biological Services Program, Washington, D.C. FWS/OBS-79/31.

Cuneo, K. 1978. Hayward Shoreline Marsh Restoration Project. Prepared for East Bay Regional Park District by Madrone Associates. Oakland, California.

Demgen, F. 1981. Enhancing California's Wetland Resource Using Treated Effluent. Prepared for California State Coastal Conservancy by Demgen Aquatic Biology. Vallejo, California.

Eliot, W. 1985. Implementing Mitigation Policies in San Francisco Bay: A Critique. Prepared for California State Coastal Conservancy. Oakland, California.

Faber, P. 1986. Report on the Current Status of the Muzzi Marsh. Prepared for Golden Gate Bridge Highway and Transportation District. San Francisco, California.

Ferren, W. 1985. Carpenteria Salt Marsh: Environment, History, and Botanical Resources of a Southern California Estuary. Publ. Number 4. Department of Biological Sciences, University of California, Santa Barbara.

Hamilton, S.F. 1984. Estuarine Mitigation. The Oregon Process. Oregon Division of State Lands, Salem, Oregon.

Handley, L.R. and M.L. Quammen. In press. Wetland Trend Analysis for Lower San Francisco Bay. <u>Coastal Zone '87 Supplemental Volume</u>. Coastal Zone Civil Engineers, New York.

Harvey, T.H., P. Williams, and J. Haltiner. 1982. Guidelines for Enhancement and Restoration of Diked Historic Baylands. Prepared for San Francisco Bay Conservation and Development Commission. San Francisco, California.

Harvey, T.H. and M.N. Josselyn. 1986. Wetland restoration and mitigation policies: comment. <u>Environ. Management</u> 10(5): 567-569.

Jordan, R.A. and J.K. Shisler. 1988. Research needs for the use of buffer zones for wetland protection, p. 433-435. In J.A. Kusler, M.L. Quammen, and G. Brooks (Eds.), Mitigation of Impacts and Losses. Association of State Wetland Managers Technical Report No. 3. Berne, New York.

Josselyn, M. (Ed.). 1982. Wetland Restoration and Enhancement in California. Proceedings of A Conference, February, 1982, California State University, Hayward. California Sea Grant Program Report No. T-CSGCP-007. Tiburon Center for Environmental Studies, Tiburon, California.

Josselyn, M. 1983. The Ecology of San Francisco Bay Tidal Marshes: A Community Profile. U.S. Fish and Wildlife Service. FWS/OBS-83/23. Washington, D.C.

Josselyn, M. 1988a. Bracut Wetland Mitigation Bank, Biological Monitoring: 1987, Final Report. Prepared for California State Coastal Conservancy. Oakland, California.

Josselyn, M. 1988b. Effectiveness of coastal wetland restoration: California, p. 246-251. In J.A. Kusler, M.L. Quammen, and G. Brooks (Eds.), Mitigation of Impacts and Losses. Association of State Wetland Managers Technical Report No. 3. Berne, New York.

Josselyn, M.N. and J. Buchholz 1982. Summary of past wetland restoration projects in California, p. 1-10. In M.N. Josselyn (Ed.), Wetland Restoration and Enhancement in California. California Sea Grant College Program. Report #T-CSGCP-007. LaJolla, California.

Josselyn, M. and J. Buchholz. 1984. Marsh Restoration in San Francisco Bay: A Guide to Design and Planning. Technical Report No. 3. Tiburon Center for Environmental Studies, San Francisco State University, California.

Josselyn, M., J. Duffield and M. Quammen. 1987. Evaluation of habitat use in natural and restored tidal marshes in San Francisco Bay, California, p. 3085-3095. In Coastal Zone '87. American Society of Civil Engineers, New York.

Josselyn, M., M. Martindale, J. Duffield. 1988. Public Access and Wetlands: A Study of the Impact of Recreational Use on Wetlands. Prepared for California State Coastal Conservancy. Oakland, California.

Kentula, M. 1986. Wetland creation and rehabilitation in the Pacific Northwest, p. 119-150. In R. Strickland, (Ed.), Wetland Functions, Rehabilitation, and Creation in the Pacific Northwest: The State of Our Understanding. Washington State Department of Ecology, Olympia, Washington.

Knutson, P.L. 1976. Dredge Spoil Disposal Study: San Francisco Bay and Estuary: Appendix K. Marsh development. U.S. Army Corps of Engineers, San Francisco, California.

Kunz, K., M. Rylko, and E. Somers. 1988. An assessment of wetland mitigation practices in Washington state. Nat. Wetlands Newsletter 10(3): 2-4.

Landin, M.C., C.J. Newling and E.J. Clairain, Jr. 1987. Miller Sands Island: A dredged material wetland in the Columbia River. In Proceedings of conference, Eighth Annual Meeting of the Society of Wetland Scientists, May, 26-29, 1987. Seattle, Washington.

La Roe, E.T. 1988. Mitigation research needs: summary, p. 431-432. In J.A. Kusler, M.L. Quammen, and G. Brooks (Eds.), Mitigation of Impacts and Losses. Association of State Wetland Managers Technical Report No. 3. Berne, New York.

Macdonald, K.B. 1977. Plant and animal communities of Pacific North American salt marshes, p. 167-191. In V.J. Chapman, (Ed.), Ecosystems of the World. Volume 1. West Coast Ecosystems. Elsevier Scientific Publications Co., Amsterdam.

Macdonald, K.B., (Ed.). 1986. Pacific Coast Wetlands Function and Values: Some Regional Comparisons. Symposium Proceedings, Western Society of Naturalists, 63rd Annual Meeting, Long Beach, December 1982. Prepared by K.B. Macdonald & Assoc., Inc. San Diego, California.

Marcus, M.L. 1982. Wetlands Policy Assessment: California Case Study. Prepared for Office of Technology Assessment, Washington, D.C. by ESA/Madrone Consultants. Novato, California.

Meiorin, E.C. 1986. Demonstration of Urban Stormwater Treatment at Coyote Hills Marsh: Final Report. Association of Bay Area Governments, Oakland, California.

Newling, C.J. and M.C. Landin. 1985. Long-Term Monitoring of Habitat Development at Upland and Wetland Development Sites, 1974-1982. Technical Report D-85-5. U.S. Army Corps of Engineers, Waterways Experimental Station, Vicksburg, Massachusetts.

Niesen, T. and M. Josselyn. 1981. The Hayward Regional Shoreline Marsh Restoration: Biological Succession During the First Year Following Dike Removal. Tiburon Center for Environmental Studies Technical Report No. 1. San Francisco State University.

O'Donnell, A. 1988. The policy implications of wetlands creation, p. 141-144. In J. Zelazny and J.S. Feierabend (Eds.), Increasing Our Wetland Resource: Proceedings of A Conference. National Wildlife Federation. Washington, D.C.

Onuf, C.P. 1987. The Ecology of Mugu Lagoon: An Estuarine Profile. U.S. Fish and Wildlife Service. Biol. Rep. 85. Washington, D.C.

Onuf, C.P., M.L. Quammen, G.O. Shaffer, C.H. Peterson, J.W. Chapman, J. Cermak, and R.W. Holmes. 1978. An analysis of the values of Central and Southern California coastal wetlands, p. 189-199. In P.W. Greeson, J.R. Clark, and J.E. Clark (Eds.), Wetlands Functions and Values: The State of Our Understanding. American Water Resources Association, Minneapolis, Minnesota.

Onuf, C. and J.B. Zedler. 1987. Pattern and process in arid-region salt marshes, p. 570-581. In D.D. Hook, W.H. McKee, Jr., H.K. Smith, J. Gregory, V.G. Burrell, Jr., M.R. DeVoe, R.E. Sojka, S. Gilbert, R. Banks, L.H. Stolzy, C. Brooks, T.D. Matthew, and T.H. Shear (Eds.), The Ecology and Management of Wetlands. Timber Press, Portland, Oregon.

Peters, D., and J.T. Bohn. 1987. Wetland Mapping: San Francisco Bay/Delta Area, California. Coastal Zone '87 Supplemental Volume. Coastal Zone Civil Engineers, New York.

Phillips, R.C. 1984. The Ecology of Eelgrass Meadows in the Pacific Northwest: A Community Profile. U.S. Fish and Wildlife Service. FWS/OBS-84/24.

Quammen, M.L. 1986. Summary of conference and information needs for mitigation in wetlands, p. 151-158. In R. Strickland (Ed.), Wetland Functions, Rehabilitation, and Creation in the Pacific Northwest: The State of Our Understanding. Washington State Department of Ecology, Olympia, Washington.

Race, M.S. 1985. Critique of present wetlands mitigation policies in the United States based on an analysis of past restoration projects in San Francisco Bay. Environ. Management 9: 71-82.

San Francisco Bay Conservation and Development Commission. 1988. Mitigation: An Analysis of Tideland Restoration Projects in San Francisco Bay. Staff Report.

Seliskar, D.M. and J.L. Gallagher. 1983. The Ecology of Tidal Marshes of the Pacific Northwest Coast: A Community Profile. U.S. Fish and Wildlife Service. FWS/OBS-82/3.

Shisler, J.K. and D.J. Charette. 1984. Evaluation of Artificial Salt Marshes in New Jersey. New Jersey Agricultural Station Publ. No. P-40502-01-84. Rutgers University, New Brunswick, New Jersey.

Simenstad, C.A. 1983. The Ecology of Estuarine Channels of the Pacific Northwest Coast: A Community Profile. U.S. Fish and Wildlife Service, Office of Biological Services. FWS/OBS-83/05. Washington, D.C.

Sorensen, J. 1982. Towards an overall strategy in

designing wetland restorations, p. 85-95. In M. Josselyn (Ed.), Wetland Restoration and Enhancement in California. California Sea Grant Program Report No. T-CSGCP-007. Tiburon Center for Environmental Studies, Tiburon, California.

Strickland, R., (Ed.). 1986. Wetland Functions, Rehabilitation and Creation in the Pacific Northwest: The State of Our Understanding. Proceedings of conference, April 30 - May 2, 1986, Port Townsend, Washington. Publ. No. 86-14. Washington State Department of Ecology, Olympia.

Swift, K. 1988. Salt marsh restoration: assessing a southern California example. M.S. Thesis, San Diego State University, San Diego, California.

Thom, R.M., R.G. Albright, and E.O. Salo. et al. 1985. Tidal Wetland Design at Lincoln Street on the Puyallup River. Unpublished report. Fisheries Research Institute University of Washington, Seattle.

Thom, R.M., C.A. Simenstad, and E.O. Salo. 1987. The Lincoln Street Wetland System in the Puyallup River Estuary, Washington. Phase 1 Report: Construction and Initial Monitoring, July, 1985 to December, 1986. FRI-UW-8706. Fisheries Research Institute, University of Washington, Seattle.

United States Army Corps of Engineers. 1983a. Engineering and Design: Dredging and Dredged Material Disposal. EM1110-2-5025. U.S. Army Corps of Engineers, Washington, D.C.

United States Army Corps of Engineers. 1983b. Engineering and Design: Beneficial Uses of Dredged Material. EM1110-2-5026. U.S. Army Corps of Engineers, Washington, D.C.

United States Fish and Wildlife Service. 1988. Biological Opinion 1-1-78-F-14-R2: The Combined Sweetwater River Flood Control and Highway Project (Project), San Diego County, California. March 30, 1988. Portland, Oregon.

Weller, M.W., J.B. Zedler, and J.H. Satner. 1988. Research needs for better mitigation: future directions, p. 428-430. In J.A. Kusler, M.L. Quammen, and G. Brooks (Eds.), Mitigation of Impacts and Losses. Association of State Wetland Managers Technical Report No. 3. Berne, New York.

White, A. 1986. Effects of habitat type and human disturbance on an endangered wetland bird: Beldings Savannah Sparrow. M.S. Thesis, San Diego State University, San Diego, California.

Williams, P.B. and M. Swanson. 1987. Tijuana Estuary Enhancement Hydrologic Analysis. Prepared for San Diego State University. San Diego, California.

Philip Williams and Associates. 1981. Final Marsh Management Plan for the Village Shopping Center, Corte Madera: Phase Two, Preliminary Design. Report No. 134. Philip Williams and Associates, San Francisco, California.

Zedler, J.B. 1982. The Ecology of Southern California Coastal Salt Marshes: A Community Profile. U.S. Fish and Wildlife Service, Office of Biological Services. FWS/OBS-81/54. Washington, D.C.

Zedler, J.B. 1983. Freshwater impacts in normally hypersaline marshes. Estuaries 6:306-346.

Zedler, J.B. 1984. Salt Marsh Restoration: A Guidebook for Southern California. California Sea Grant College Program. Report No. 7-CSGCP-009. La Jolla, California.

Zedler, J.B. 1988. Salt marsh restoration: lessons from California, p. 123-238. In J. Cairns (Ed.), Management for Rehabilitation and Enhancement of Ecosystems. CRC Press, Boca Raton, Florida.

Zedler, J.B. and P.A. Beare. 1986. Temporal variability of salt marsh vegetation: the role of low salinity gaps and environmental stress, p. 295-306. In D. Wolfe (Ed.), Estuarine Variability. Academic Press, New York.

Zedler, J.B., J. Covin, C. Nordby, P. Williams, and J. Bolland. 1986. Catastrophic events reveal the dynamic nature of salt-marsh vegetation in Southern California. Estuaries 9(1):75-80.

Zedler, J.B., M. Josselyn, and C.P. Onuf. 1983. Restoration techniques, research, and monitoring: vegetation, p. 63-72. In Josselyn, M. (Ed.), Wetland Restoration and Enhancement in California. California Sea Grant Program Report No. T-CSGCP-007. Tiburon Center for Environmental Studies, Tiburon, California.

Zedler, J.B., W.P. Magdych, and San Diego Association of Governments. 1984. Freshwater Release and Southern California Coastal Wetlands: Management Plan for the Beneficial Use of Treated Wastewater in the Tijuana River and San Diego River Estuaries. San Diego Association of Governments, San Diego, California.

Zedler, J.B. and C.S. Nordby. 1986. The Ecology of Tijuana Estuary, California: An Estuarine Profile. U.S. Fish and Wildlife Service and California Sea Grant Program. Biological Report 85(7.5).

Zedler, P.H. and C. Black. (submitted). Species preservation in artificially constructed habitats: Preliminary evaluation based on a case study of vernal pools at Del Mar Mesa, San Diego County. Proceedings, Urban Wetlands Conference, Association of State Wetland Managers, Berne, New York.

Zentner, J.J. 1988. Wetland restoration success in coastal California, p. 216-219. In J. Zelazny and J.S. Feierabend (Eds.), Increasing Our Wetland Resource: Proceedings of A Conference. National Wildlife Federation. Washington, D.C.

APPENDIX I: SUGGESTED READING

Bay Conservation and Development Commission. 1988. Mitigation: An Analysis of Tideland Restoration Projects in San Francisco Bay. Bay Conservation and Development Commission, San Francisco, California.

Field studies of success of eighteen mitigation projects required by BCDC. Includes analysis of effectiveness of mitigation system.

Bay Institute of San Francisco. 1987. Citizen's Report on the Diked Historic Baylands of San Francisco Bay. L. Treais (Ed.). The Bay Institute of San Francisco, Sausalito, California.

This report describes the types and extent of proposals which effect the 80 square miles of wetlands along the San Francisco Bay shoreline by reviewing the development and restoration proposals of sixteen diked historic bayland sites around the bay.

Bierly, K.F. 1987. Mitigation Bank Handbook: Procedures and Policies for Oregon. Division of State Lands, Salem, Oregon.

This handbook outlines the concept of mitigation banking under Oregon state law. It describes management and ecological terms related to mitigation, the process of estuarine mitigation, and specifics of mitigation banking, such as, how to establish and operate a mitigation bank, and possible sites for mitigation banks in Oregon.

Blomberg, G. 1987. Development and mitigation in the Pacific Northwest. Northwest Env. Jour. 3(1): 63-91.

Briefly reviews legislative framework for mitigation, federal and state agencies involved in mitigation, and coastal zone management programs in Washington and Oregon. Evaluates federal, state, and local mitigation requirements.

Boule', M.C., N. Olmsted, and T. Miller. 1983. Inventory of Wetland Resources and Evaluation of Wetland Management in Western Washington. Washington State Department of Ecology, Olympia.

This report presents a comprehensive inventory of wetlands and discusses the trends in development of wetlands in the last 100 years in western Washington State. It also evaluates the effectiveness of Washington State's Shoreline Management Act and recommends improvements to the program.

Chan, E., G. Silverman, and T. Bursztynsky. 1982. San Francisco Bay Area Regional Wetlands Plan for Urban Runoff Treatment. Volume 1: Plan and Amendments to the Environmental Management Plan. Association of Bay Area Governments, Berkeley, California.

This document presents the results of ABAG's 1981-82 water quality planning program including such topics as wetlands in relation to water quality, the environmental function of wetlands in the Bay Area, state and local policies governing wetland development and guidelines for wetland creation and enhancement.

Cooper, J.W. 1987. An overview of estuarine habitat mitigation projects in Washington State. Northwest Environ. Jour. 3(1): 113-127.

Analyzes mitigation projects that have been designed to offset or compensate for estuarine losses in Washington State in the previous four years. The projects were designed to either: 1) create new replacement habitat, or 2) rehabilitate and upgrade existing wetlands.

Demgen, F. 1981. Enhancing California's Wetland Resource Using Treated Effluent. Prepared for California State Coastal Conservancy by Demgen Aquatic Biology, Vallejo, California.

Describes existing projects in California and the agencies with permitting authority over wastewater wetland projects. Also contains a list and brief discussion of coastal and estuarine dischargers in California.

Dennis, N.B. and M.L. Marcus. 1984. Status and Trends of California Wetlands. Prepared for the California Assembly Resources Subcommittee on Status and Trends by ESA/Madrone Assoc, Novato, California.

This report documents the importance of California's wetlands, reviews their present status and predicts the future of the remaining wetlands.

Eliot, W. 1985. Implementing Mitigation Policies in San Francisco Bay: A Critique. Prepared for California State Coastal Conservancy. Oakland, California.

This report evaluates the effectiveness of state and federal mitigation policies for wetland creation and restoration. Most of the 58 mitigation projects examined were unsuccessful; recommendations were made for improvements in mitigation policy.

Faber, P.M. 1982. Common Wetland Plants of Coastal California: A Field Guide for the Layman. Pickleweed Press, Mill Valley, California.

A field guide for brackish, freshwater and salt marsh plants using photocopy reproductions of the plants for easy identification.

Ferren, W. 1985. Carpenteria Salt Marsh: Environment, History, and Botanical Resources of a Southern California Estuary. Publ. Number 4. Department of Biological Sciences, University of California, Santa Barbara.

Inventory and evaluation of the botanical resources of a southern California estuary in the context of its environment and history.

Hamilton, S.F. 1984. Estuarine Mitigation: The Oregon Process. Oregon Division of State Lands, Salem.

The official administrative rules for mitigation in Oregon's estuaries are presented in this publication. The state's estuarine ecosystems are described in detail along with a system for assigning "relative value" to different habitat types for the purpose of providing full compensatory mitigation.

Harvey, T.H., P. Williams, and J. Haltiner. 1982. Guidelines for Enhancement and Restoration of Diked Historic Baylands. San Francisco Bay Conservation and Development Commission, San Francisco, California.

This technical report provides guidelines for selecting sites and designing restoration and enhancement projects for wetlands.

Horak, G.C. 1985. Summaries of Selected Mitigation Evaluation Studies. U.S. Fish and Wildlife Service, Washington, D.C. Report No. WELUT-86/W03.

Presents brief summaries of past mitigation evaluation studies followed by a critique on the effectiveness of the studies and recommendations for improvement.

Horak, G.C. 1985. Bibliography and Selected Characteristics of Mitigation Evaluation Studies. U.S. Fish and Wildlife Service, Washington, D.C. Report No. WELUT-86/W02.

A listing of all U.S. mitigation evaluation reports organized alphabetically by author; includes a chart listing the type of project and resources effected by the action.

Josselyn, M. (Ed.). 1982. Wetland Restoration and Enhancement in California. Proceedings of A Conference, February, 1982, California State University, Hayward. California Sea Grant Program Report No. T-CSGCP-007. Tiburon Center for Environmental Studies, Tiburon, California.

Conference proceedings which discuss selected aspects of wetland ecology including hydrology, sedimentation and salt marsh fauna in addition to reviewing potential wetland restoration sites in California, regulations governing wetland restoration and design strategies for restoration projects.

Josselyn, M. and J. Buchholz. 1984. Marsh Restoration in San Francisco Bay: A Guide to Design and Planning. Technical Report No. 3. Tiburon Center for Environmental Studies, San Francisco State University.

Evaluates the success of past marsh restoration projects in Marin County, California and reviews several topics important to wetland design including erosion and sedimentation, vegetation and wildlife habitat.

King County Dept. of Planning and Community Development. 1986. Viability of Freshwater Wetlands for Urban Surface Water Management and Nonpoint Pollution Control: An Annotated Bibliography. Washington State Department of Ecology, Olympia. Report No. 87-7B.

Most listings refer to the use of wetlands for the treatment of secondary sewage effluent. Entries primarily from conference and symposia proceedings, research reports, government publications, and scientific journals.

Kunz, K., M. Rylko, and E. Somers. 1988. An assessment of wetland mitigation practices in Washington State. Nat. Wetlands Newsletter 10(3):2-4.

Examines types of projects for which mitigation is required, their location, and impacted wetland acreage; characteristics of the mitigation projects such as habitat type, habitat loss and gain, and habitat functions lost and gained; the characteristics of the mitigation process such as permit specifications and implementation time.

Lewis, J.C. and E.W. Bunce (Eds.). 1980. Rehabilitation and Creation of Selected Coastal Habitats: Proceedings of a Workshop. U.S. Fish and Wildlife Service, Biological Services Program, Washington, D.C. FWS/OBS-80/27.

Collection of papers containing information which may be applied in the rehabilitation and creation of coastal habitats.

Macdonald, K.B. 1977. Plant and animal communities of Pacific North American salt marshes, p. 167-191. In Ecosystems of the world. Volume 1. West coast ecosystems. Elsevier Scientific Publications Co. Amsterdam.

This paper summarizes the present knowledge of salt-marsh ecosystems developed along the west coast of North America; emphasis is placed on vascular plants most characteristic of the salt marshes.

Macdonald, K.B., (Ed.). 1986. Pacific Coast Wetlands Function and Values: Some Regional Comparisons. Symposium Proceedings, Western Society of Naturalists, 63rd Annual Meeting, Long Beach, December 1982. Prepared by K.B. Macdonald & Assoc., Inc., San Diego, California.

Papers treating the ecological function and potential values of Pacific Coast wetlands. Full texts of invited papers and abstracts of contributed papers.

Marcus, M.L. 1982. Wetlands Policy Assessment: California Case Study. Prepared for Office of Technology Assessment, Washington, D.C. by ESA/Madrone Consultants, Novato, California.

This report traces the history of wetland losses in California, reviews regulations governing wetland management and evaluates the effectiveness of the Corps of Engineers 404/10 program.

Phillips, R.C. 1984. The Ecology of Eelgrass Meadows in the Pacific Northwest: A Community Profile. U.S. Fish and Wildlife Service. FWS/OBS-84/24.

This report describes the physiographic setting of the eelgrass community, the distribution of the grass beds, autecology of the eelgrass in terms of growth and reproductive strategies and physiological requirements and functions.

Roelle, J.E. 1986. Mitigation Evaluation: Results of a User Needs Survey. National Ecology Center, U.S. Fish and Wildlife Service, Ft. Collins, Colorado.

Results of a survey of U.S. Fish and Wildlife Service's Ecological Services offices on the importance of mitigation evaluation activities and on the utility of three different data bases.

Seliskar, D.M. and J.L. Gallagher. 1983. The Ecology of Tidal Marshes of the Pacific Northwest Coast: A Community Profile. U.S. Fish and Wildlife Service. FWS/OBS-82/32.

This report describes the structure and ecological functions of tidal marshes of the Pacific Northwest coast. It includes such topics as the physical and chemical marsh environment, marsh distribution, biotic communities, ecological interactions and management of tidal marshes.

Smith, S.E., (Ed.). 1983. A Mitigation Plan for the Columbia River Estuary. Columbia River Estuary Study Taskforce (CREST). Astoria, Oregon.

This report evaluates the resources and habitats of the Columbia River Estuary on an estuary-wide basis. It provides a detailed, step-by-step description of the permit application/mitigation planning process in order to expedite this process.

Strickland, R., (Ed.). 1986. Wetland Functions, Rehabilitation and Creation in the Pacific Northwest: The State of Our Understanding. Proceedings of conference, April 30-May 2, 1986, Port Townsend, Washington. Publ. No. 86-14. Washington State Department of Ecology, Olympia.

A technical review of wetland functions including hydrology and sedimentation, water quality, nutrient cycling, primary production and wildlife use, written for wetland managers and policy makers. Transcripts of conference working groups discuss successes and current limitations in wetland rehabilitation.

Warrick, S.F. and E.D. Wilcox. 1981. Big River: The Natural History of an Endangered Northern California Estuary.Environmental Field Publication No. 6. University of California, Santa Cruz.

This publication documents the geography and natural history of the Big River Estuary, located near Mendocino, California. By addressing such topics as the geology of the estuary, vegetation, and aquatic and terrestrial life this study describes the negative effects of sedimentation due to logging in the watershed.

Weinmann, F., M. Boule', K. Brunner, J. Malek and V. Yoshino, 1984. Wetland plants of the Pacific Northwest. U.S. Army Corps of Engineers, Seattle District.

Fifty-nine species of wetland plants are described and illustrated with color photographs. Definitions and a general introduction to wetlands are also provided.

Wiedemann, A.M. 1984. The Ecology of Pacific Northwest Coastal Sand Dunes: A Community Profile. U.S. Fish and Wildlife Service. FWS/OBS-84/04.

An ecological description of Pacific Northwest coastal sand dunes.

Zedler, J.B. 1984. Salt Marsh Restoration: A Guidebook for Southern California. California Sea Grant College Program. La Jolla, California.

This book offers technical advice for all stages of the restoration and enhancement of disturbed salt marshes, based on a six year study of the Tijuana Estuary.

Zedler, J.B., J. Covin, C. Nordby, P. Williams, and J. Bolland. 1986. Catastrophic events reveal the dynamic nature of salt-marsh vegetation in Southern California. Estuaries 9(1): 75-80.

Reports on a sixteen year study of the effect of hydrological disturbances including flooding, dry-season streamflow, and drought, on cordgrass (Spartina foliosa) distribution in the Tijuana Estuary.

Zedler, J.B., W.P. Magdych, and San Diego Association of Governments. 1984. Freshwater Release and Southern California Coastal Wetlands: Management Plan for the Beneficial Use of Treated Wastewater in the Tijuana River and San Diego River Estuaries. San Diego Association of Governments, San Diego, California.

Presents five recommendations based on findings of previous technical studies. Identifies various agencies involved in administration and regulation in the estuaries and indicates those which will be responsible for implementation of each of the five recommendations.

Zedler, J.B. and C.S. Nordby. 1986. The Ecology of Tijuana Estuary, California: An Estuarine Profile. U.S. Fish and Wildlife Service and California Sea Grant Program. Biological Report 85(7.5).

This report discusses the diverse ecological communities of the Tijuana Estuary, analyzes data on the vegetation, algae, invertebrates, fishes and birds found in the estuary, their ecological interrelationships and relationships of the biota with the physical environment.

APPENDIX II: SELECTED PROJECT PROFILES

VERNAL POOL PROJECT IN SAN DIEGO COUNTY

Project Profile: Mitigation for destruction of vernal pools by creating artificial pools.

Wetland Type: Vernal pool (palustrine, emergent wetland, nonpersistent).

Location: Del Mar Mesa, just west of Interstate Highway 15 near Rancho Penasquitos, San Diego County, California.

Size: Forty pools ranging in size from 15.3 m^2 to 385.7 m^2 in area. The total pool surface area created was 1,854 m^2.

Goals of project:

To create artificial vernal pool ecosystems that would replace those lost during highway construction at another site.

The objectives were:

1. Providing the correct hydroperiods by controlling pool depth and substrate type;
2. Introducing vascular plant species, with special attention given to the endangered plant, mesa mint (Pogogyne abramsii);
3. Predicting long-term success from studies of plant population dynamics through a five-year study, and understanding reasons for success or failure of population establishment and maintenance.

Implementation involved:

1. Creating 40 depressions approximately 10 km north of the highway construction site in a 2-km-long area that contained natural vernal pools. About one-third of the pools were interspersed among natural pools (within one soil type), and the remaining two-thirds were located in two clusters away from natural pools (on a different soil type);
2. Collecting seed and soil from natural pools near the mitigation site and from the site that was later covered by Highway 52;
3. Seeding the artificial depressions; and
4. Monitoring hydroperiods and vascular plant growth to assess success.

Judgement of success:

It is not possible to judge the long-term success of the ecosystem creation program, because it has been less than two years since construction in fall 1986. Native plants have become established in most of the artificial vernal pools, and most support the target endangered plant. However, the pools differ from natural pools in their basic appearance. The peripheral vegetation is sparse, and different soil characteristics (crusting) may be responsible. Rainfall has been highly variable, making it difficult to predict what the long-term success will be.

Monitoring of the total ecosystem was not required in the mitigation agreement. Thus, it is not possible to say whether the native aquatic community of algae, invertebrates, frogs, and birds has been created.

Significance:

This mitigation program is significant in two respects: (1) it is an attempt to create an entire ecosystem and its endangered plant populations, and (2) it is linked to a university research program that has focused on the ecology of mesa mint for the past 10 years.

A substantial monitoring/research program has been approved in concept. The research program is designed to identify conditions that lead to successful establishment of the native plants (especially the mesa mint), as well as to explain reasons for any failure. The field work involves detailed measurements of water levels in selected pools, intensive surveys of mesa mint densities and distributions of all vascular plant species across the pools, and monitoring the phenology and reproduction of mesa mint.

The research project has helped determine measures necessary for successful habitat creation. By using an experimental design with planted and unplanted pools, the necessity of providing seed has been determined. Of the pools left unplanted, none supported mesa mint at the end of the second growing season. Only a few other native plants have become established in unplanted pools. Pools provided with seed have developed native plant species, and these will be followed to determine whether the populations increase, remain stable, or decrease through time. Thus, the long-term success can be predicted from longer-term monitoring, and association of population changes with pool characteristics (e.g., surface and groundwater levels) will help to explain reasons for success or failure.

Reports:

Reports are prepared annually for the California Department of Transportation (Caltrans); Zedler, P.H. and C. Black (submitted).

Contacts:

Monitoring and research: Dr. Paul H. Zedler, Professor of Biology, San Diego State University, San Diego, CA 92182-0057.

Project management: John Rieger, Caltrans, 2829 Juan Street, Old Town, PO Box 81406, San Diego, CA 92138.

SWEETWATER/PARADISE MARSH MITIGATION IN SAN DIEGO COUNTY

Project Profile: Mitigation for highway construction through salt marsh by modifying existing salt marsh and converting upland to tidal salt marsh.

Wetland type: Tidal salt marsh, mud flat, and freshwater wetlands.

Location: Sweetwater/Paradise Creek marsh, west of I-5 at State Route 54 interchange, San Diego County, Calif.

Size: Ten hectares of disturbed high salt marsh changed to tidal channels and low marsh plus approx. 7 hectares of uplands to be converted to tidal channels and intertidal marsh.

Goals of project:

To create nesting habitat for the Light-Footed Clapper Rail (Rallus longirostris levipes) and foraging area for the California Least Tern (Sterna albifrons browni) by enhancing the wetland area adjacent to the highway and flood control channel. To establish a viable population of salt marsh bird's beak (Cordylanthus maritimus ssp. maritimus).

The specific objectives of Phase 1 of the restoration project were to:

1. Create appropriate elevations for low, middle, and high salt marsh habitat;
2. Revegetate graded sites with the appropriate salt marsh plants;
3. Improve tidal influence by creating tidal channels in salt marsh segments;
4. Increase habitat area for prey species of the clapper rail (crabs and other invertebrates) and least tern (fish) by creating mudflat and increasing deep channel area;
5. Salvage native salt marsh plants from impacted areas for propagation and use in later stages of restoration;
6. Convey approximately 300 acres of marshland and environmentally sensitive upland in perpetuity to public ownership for the preservation of endangered species prior to the initiation of Phase 2 construction.

Project implementation involved:

1. Grading existing (degraded) salt marsh to create the appropriate topography for low and middle salt marsh in the form of eight islands;
2. Creating deep channels surrounding the restored marsh area to prevent intrusion by pests, and creating smaller tidal channels to facilitate tidal flushing within the marsh, enhance plant growth and, therefore, rail habitat;
3. Opening both areas to tidal flushing (north area completed Sept. 1984, south area in Oct. 1984);
4. Planting the islands and side banks with low-marsh vegetation in the form of transplants (collected from a nearby source);
5. Creating a tidal Spartina nursery and an irrigated middle-marsh plant nursery near the restoration site for salvaging plants removed from the project area, to be used in subsequent and remedial plantings; and
6. Monitoring the success of the planting, remedial planting as needed, and fencing areas that show signs of grazing;

Judgement of success:

The project has had success in establishing cordgrass, although the area with plants is less than the area planted. Present and future habitat value to the clapper rail is uncertain due to several problems incurred during the restoration process:

1. Grading plans were not followed in the formation of the islands and channels. This has resulted in a reversal of planned water flows in some areas, increasing erosion along many of the creek banks, and significantly altering the graded wetlands. Some channels are filling in rapidly and are fully exposed at medium-low tide, reducing their effectiveness as barriers to intrusion by predators and humans.
2. In the northern half of the restoration site, cordgrass plantings had high mortality. In some places this was due to changing hydrology and erosion of the planted area. A tidal gate was inadvertently installed, and it impounded water in the northern half of the restoration site until its removal, in stages, in 1987 and 1988. The elevation of best cordgrass transplant success was about 0.15 m higher where water impounded than in the fully tidal area downstream. In general, cordgrass mortality could not be explained by soil characteristics (salinity, organic matter content, soil texture), but a successful experimental replanting or "ecoassay" suggested that poor techniques, rather than site factors, were responsible for high mortality of the initial transplants (Swift 1988).
3. Remedial planting has not occurred. A tidal gate to the upstream restoration area was inadvertently installed (only the frame was slated for construction in phase 1 of the project, with the remaining structure to be completed after the flood control channel was completed). This has resulted in the artificial impounding of seawater in the upstream area and significant alteration of the hydrologic conditions for cordgrass. Though measured and planted at the same absolute elevations as downstream plantings, upstream transplants appear to be a full foot higher in the marsh. This is largely due to sluggish tides caused by a culvert and a tide gate that separates the two areas. (This problem has been recognized by the permittee and the tide gate is slowly being disassembled). Relative tidal elevation (including hydrological influences), therefore, may differ considerably from absolute elevation, and should be considered in the planting process.
4. Planting densities in some areas did not match permit specifications (planting centers increased to about 15cm) because of an inadequate amount of transplanting stock. Establishment time and habitat value both suffer because of this, and it is an avoidable problem.
5. During the excavation and dredging operations, a former dump site was uncovered, yielding large amounts of broken glass and debris that complicated the transplanting process. Toxic concentrations of lead were also found, resulting in the need to remove substrata to an off-site disposal.
6. Nursery areas differed in their ability to propagate cordgrass, with elevation of excavation the most important limiting factor. One cordgrass nursery has filled in and is the largest monospecific stand in the

restoration area. The other cordgrass nursery (which has sparse cordgrass) gets little tidal flushing and is apparently at a higher relative elevation. The success of salvaging middle- and high-marsh plants is difficult to evaluate at this point.

7. Significant items of trash, including a 10-m long catamaran hull that repeatedly flattened large areas of planted cordgrass (Swift 1988), enter the site as floating debris and smother transplants. This hinders the establishment process.

In addition, the hydrology of the restored area will change significantly during later phases of the project when tidal flows will also come through the flood control channel between the two sites. The rate of this change will be crucial to the final success of the project due to the inability of many marsh plants to adjust to rapid changes in hydrology.

In August 1988, the mitigation site will become part of a large (approx. 120 hectare) Refuge, managed by the U.S. Fish and Wildlife Service, as ordered by the U.S. District Court in settlement of a lengthy lawsuit.

Significance:

The most significant aspect of this project is the establishment of functional goals for the constructed and modified wetlands and the requirement of high standards for achieving goals. Four concepts of function, maintenance of species diversity, food chain support, and ecosystem resilience, were adopted in the process of setting mitigation objectives and in assessing success. The project sponsors are required to manage the area until these objectives are achieved. Rather than setting a specific time limit for compliance, sponsors must monitor the site to show that the goals have been reached and maintained for two consecutive years. These requirements are the strongest yet required for a coastal wetland mitigation program, and may well set the tone for all future projects.

This site was also chosen to flag several different problems that can develop in the course of mitigating wetland loss. These include:

1. **Legal battles over land transfer--timing is important and may delay the project.** In some cases, the transfer of lands is the key to a mitigation proposal's acceptance. When the transfer agreement is changed or comes under legal scrutiny after the development/construction project is underway (as is the case at Sweetwater marsh), the entire mitigation proposal must be re-evaluated--a costly and time consuming job. This can be avoided by requiring the land transfer to be completed prior to the commencement of construction or development.

2. **Excessive demands on mitigation lands for urban open space uses.** The City of Chula Vista tried to change the initial mitigation agreement by imposing 7 easements for road crossings, utilities, extensive public access to wetlands, public marina construction, and for local management rights to supersede federal mitigation goals. The Sierra Club sued to enforce the original agreement. After three years of litigation, a plan that was substantially more protective of natural resources was mandated by the court as part of the settlement agreement. The result was an overall reduction in the amount of land that could be developed.

3. **History of the site (e.g., former dump sites) and contingency plans.** Preliminary research required by the project should include a brief chronology of the uses of the site. Many areas in or near wetlands have previously been used as solid waste dump sites, and excavation and grading operations will need to deal with proper disposal of potential hazardous wastes. If the history of the property is known, then proper cleanup measures should be included in the mitigation plan. If the historical information is not available, then contingency plans should be included in the plan.

4. **Failure to follow plans (grading, tide gate, etc.).** By not following the grading plans specified in the mitigation agreement, the actual topography and hydrology of the restored marsh differs considerably from that which was proposed. The premature completion of the tidegate represents a potentially damaging influence on the upstream marsh by impounding water and slowing drainage. These conditions are also likely to change the structure of the marsh away from what was proposed. In either case, if the mitigation plan was followed or consultation (prior to change) was required by the plan, the adverse effects could have been avoided.

5. **Poor monitoring.** Since the completion of the revegetation program, there has been little quantitative information gathered on use of the restored area by birds or prey species or on vegetation establishment. Because the mitigation plan did not require an explicit monitoring program (of both plant establishment and subsequent use by birds), the determination of success is completely subjective. In all projects that set specific goals, monitoring programs must be required and enforced if they are to be accurately judged for success. In recognition of the need for improved monitoring, the new compensation measures require extensive monitoring for approximately 5 years.

6. **Factors other than elevation to consider in planting cordgrass and other marsh plants.** A project-wide planting scheme based on absolute elevations would not be advised in most projects, because there may be very different physical characteristics at the same absolute elevation. Two areas in the project site, for example, may be at the same absolute elevation, but one is upstream and the other downstream. These two areas may be at totally different tidal elevations due to muted hydrology. This demonstrates that when creating a revegetation/restoration program, site-specific characteristics (e.g., hydrology) should be incorporated into the plan.

Reports:

Reports are prepared annually for the CaliforniaDepartment of Transportation (Caltrans); Swift, K. (1988); U.S. Fish and Wildlife Service (1988).

Contacts:

<u>Project management:</u> John Rieger, Caltrans, 2829 Juan Street, Old Town, PO Box 81406, San Diego, CA 92138.

AGUA HEDIONDA CREEK AND LAGOON, SAN DIEGO COUNTY

Project Profile: Mitigation for road construction in a wetland, dredging and discharging fill material for construction of a desiltation system, and damage from utility pole movement.

Wetland type: Tidal salt marsh, brackish ponds, freshwater marsh, riparian woodland, and upland transition

Location: Agua Hedionda Creek and Lagoon at Hidden Valley Road, Carlsbad, San Diego County, California.

Size: 5.6 hectares of road fill, 7.5 hectares wetland enhancement, and over 70 hectares of land title transfer.

Goals of project:

To create and enhance wetland habitats lost or impacted by road construction within the lagoon wetland boundaries.

The objectives of this project were to:

1. Expand tidal wetlands with the provision of additional salt marsh habitat;
2. Increase tidal circulation and expand tidal prism of the lagoon;
3. Increase quality and diversity of salt marsh food chain;
4. Enhance lagoon's water quality and salinity levels by increasing tidal circulation;
5. Enhance functional capacity of the overall lagoon ecosystem;
6. Create new wetland habitat from existing upland;
7. Establish a controlled interface (restricting access) along the lagoon;
8. Assure success through specified monitoring program;
9. Promote public appreciation of lagoon ecosystem; and
10. Acquire wetlands for wildlife preservation.

Implementation involved:

1. Constructing openings in the "fingers" area along dikes adjacent to upland area to facilitate tidal flushing and prevent intrusion by humans or potential predators;
2. Lowering and recontouring the ends of three peninsular fill areas to create one hectare of additional salt marsh habitat (to increase tidal action to new and existing salt marsh and mud flats, and provide protected open water habitat for shorebirds);
3. Revegetating the recontoured (newly intertidal) area with the proper salt marsh species at specific densities;
4. Revegetating the transitional area with plant species suitable for wetland/upland interface (as opposed to solely upland) to create a dense landscape buffer zone and natural barrier to human intrusion;
5. Extending and widening existing tidal channels at east end of lagoon to provide increased intertidal habitat, increasing tidal influence to the salt marsh, and providing additional bird nesting habitat;
6. Revegetating channel banks with _Salicornia_ salvaged from a nearby area;
7. Removing sediments, fill, and debris at the mouth of a storm drain and extending freshwater wetlands to a broader area (5.5 acres). Revegetating this area with representative freshwater marsh and willow riparian species;
8. Lowering of an open field (3 hectares) adjacent to the project to expand freshwater wetlands habitat. Revegetating this area with representative marsh and riparian woodland species;
9. Supplying sufficient water to wetland extension areas to insure not only survival of the plants, but natural recruitment as well;
10. Creating a 0.8 hectare "bird nesting island" with fill material and sediment excavated from the storm drain wetland extension;
11. Erecting a temporary fence along the north side of Cannon Road to restrict off road vehicle use in the flood plain during construction;
12. Creating two (0.1 hectare) small brackish pond habitat areas to provide additional wildlife use site away from the main lagoon area in an area not normally vegetated with wetland plants (amended);
13. Monitoring progress of implementation and any need for remedial action in annual reports to be sent to the Army Corps of Engineers, U.S. Fish and Wildlife Service, the Environmental Protection Agency, and California Department of Fish and Game for three years. If after the 2nd year, less than 80% survival rate is evident, plants shall be replaced to ensure 80% survival;
14. Creating an interpretive center (kiosk) explaining the historical and biological resources of the lagoon; and
15. Transferring title of over 70 hectares of low lying land from the landowner to the Wildlife Conservation Board of the State of California.

Judgement of success:

This project involved the restoration of several different types of wetlands throughout the lagoon, none of which can currently be considered successful. The overall impact of the mitigation on the lagoon ecosystem cannot be considered beneficial. Nearly two years after initial planting of the revegetation areas, nearly all of the plantings have failed to produce their specified habitat values, and "enhanced functional capacity of [the] overall lagoon" is not apparent.

Lowering and recontouring did not increase tidal salt

marsh habitat. Though the surrounding dikes have been breached and tidal channels exposed, most of the area that was revegetated lies above tidal influence. As a result, initial revegetation plantings have either died or failed to spread due to inadequate circulation, animal grazing, human disturbance, and weed invasion.

Tidal channels at the east end of the lagoon have been widened and banks have been planted with a single row of Salicornia virginica that is spreading very slowly. Revegetating the wetland/upland transition area with the establishment of a controlled interface (proposed dense landscape buffer zone) to restrict access to the lagoon is not evident in most areas. Temporary fences along the north side of Cannon road have fallen into disrepair and no longer exclude off-road vehicle activity in the flood plain or marsh areas. Uncontrolled off-road vehicle activity has disturbed substantial portions of mud flat and salt marsh habitat and hindered restoration efforts.

Freshwater wetland expansion efforts have been only slightly more successful. Sediments and debris removed from a storm drain are still piled next to the channel, presumably awaiting use later in the project. Initial non-irrigated riparian (willow) woodland plantings have completely failed, and unused planting material from initial revegetation efforts is scattered throughout the area. Renewed efforts at revegetation with irrigation (drip lines and sprinkler systems) have just begun in two expansion areas that are slated for freshwater marsh and riparian habitat. Channel clearing and recontouring in these areas were supposed to supply sufficient water to ensure survival of plants and natural recruitment. In both cases, plantings are currently irrigation dependent and sufficient water supply for natural recruitment is not evident.

The "bird nesting island" does not appear to have been created at all, however this is not necessarily a detriment. The site proposed for the island (which was to have been constructed with fill from the aforementioned storm drain) is now Salicornia marsh, and construction of the island at this time would result in a net loss of wetland habitat.

Excavation of the two brackish ponds revealed a deeper water table than was suspected, negating the possibility of a brackish water habitat away from the main lagoon area. One pond is now tidal and unvegetated, and the other appears to have been filled in (or never created).

Interpretive center (kiosk) explaining the historical and biological resources of the lagoon has not yet been constructed and there is no evidence of an ongoing monitoring program. The agencies that were supposed to receive annual progress reports have not received any data or preliminary reports as of July 1987 (Nancy Gilbert, USFWS, pers. comm.).

Finally, the transfer of over 70 hectares of low lying land from the landowner to public ownership (State of California) has not yet occurred.

Significance:

This project was chosen for the variety of modifications and improvements promised. Its lack of success serve to identify the following "red flags".

1. **The importance of designing proper topographic and hydrologic (inundation and salinity) conditions for wetlands.** Wetland vegetation, by definition, is governed by the dominant local hydrologic conditions. Without the proper hydrologic conditions, the revegetation process will not produce the desired habitat conditions.

2. **The importance of restricting access to wetlands.** From construction/revegetation phases until vegetated buffers have fully filled in, it is imperative that access be limited to wetlands. Inadequate public awareness and the degraded appearance of wetlands during revegetation periods often seems to invite off road vehicle activity that can permanently disrupt existing wetland areas as well as newly restored ones.

3. **The need to create and enforce explicit monitoring programs.** In this case, the project calls for 80% survival after two years with remedial plantings to replace those plants short of an 80% survival rate (at year 2). Initial plantings experienced closer to a 20% survival rate, and remedial plantings have just recently been initiated (as the second anniversary of the plantings nears). The survival of these plantings is to be monitored for one year.

4. **The need to require thorough assessment of successful compliance with project objectives.** This includes not only specific objectives such as planting schemes and grading plans, but also overall objectives for the ecosystem. If the project states that a habitat of some form is to be created, then all aspects of the restoration should be examined to assess whether the project indeed has created habitat, or is merely a group of plantings.

Reports:

Progress reports required by the project permit are as yet unavailable.

Contacts:

Implementation and monitoring: Pacific Southwest Biological Services Inc., PO Box 985, National City, CA 92050.

Project management: Wayne Callaghan, c/o Cal Communities, 38 Red Hawk, Irvine, CA 92714.

Nancy Gilbert, U.S. Fish and Wildlife Service, 24000 Avila Road, Laguna Niguel, CA 92656.

HAYWARD REGIONAL SHORELINE, ALAMEDA COUNTY

Project Profile: Mitigation for access road construction to new bridge crossing; loss of salt pond, emergent marsh, and mudflat habitat.

Wetland type: Tidal salt marsh, mudflat, and island.

Location: Alameda County at West Winton Blvd. on eastern shore of south San Francisco Bay.

Size: 80 hectares.

Goals of project:

To restore tidal action to former salt crystallizers and recreate tidal emergent salt marsh.

Specific objectives in plan included:

1. Creation of extensive area to be colonized by cordgrass (Spartina foliosa);
2. Excavation of channels and deep basins that would retain water during low tide to serve as waterbird habitat;
3. Creation of islands for nesting habitat; and
4. Provision of public access pathways and educational signs.

Project implementation involved:

1. Excavation and grading of site prior to reintroduction of tidal action to create basins, channels, and islands;
2. Construction of levees and bridges for public access; and
3. Breaching of levees in two locations to allow tidal flow, with a sill left to retain water within restoration site during low tide.

Judgement of success:

A number of studies have been completed on this site. Niesen and Josselyn (1981) reported on studies completed the during first year following re-introduction of tidal action. Salt and boron levels dropped dramatically in surface layers of former crystallizer to concentrations similar to those observed in natural marshes. However, lower layers of soil retained high salt concentrations in those areas formerly used in the final process of salt production. Island areas developed cat clay problems with extremely low pH values. Fish and invertebrate species colonized rapidly and appeared at similar levels as nearby mudflat areas. Extensive bird use was noted throughout the site.

Race (1985) noted that vegetation had not become established as expected after two years following dike breaching. Most vegetation appeared to colonize at debris line within wetland. A number of experimental plantings of cordgrass were conducted on the site and appeared to have stimulated establishment so that by 1987, cordgrass was distributed throughout the site, though coverage was still less than 5% overall (San Francisco Bay Conservation and Development Commission 1988) Denser coverage by cordgrass was noted closer to tidal breach and few plants were observed in the former crystallizer region. Erosion of the shoreline, both within and outside the site is occurring.

Josselyn et al. (1987) noted that while bird use is high, it is generally less than observed at a nearby natural marsh. The channels and open basin that were excavated have silted in with sediment. The "nesting islands" were not observed to support any substantial bird utilization.

Significance:

This project is an example of a site which provides wetland functions, but in a different manner than planned. There is high bird and fish use of the mudflats. However, revegetation has been very slow. Some "red flags" which should be considered are:

1. Planting of cordgrass appears to be essential in areas of poor seed sources as in south San Francisco Bay. Project managers place less importance on planting as there is a general attitude of letting "nature take its course". This may be appropriate in some situations, but in regions of slow regrowth, may result in reduced habitat quality for desired species while plants are becoming established.
2. Siltation and erosion were not considered in the planning of the site. The creation of specific habitat features needs to consider the realistic expectations of longevity. Channels and basins sizes should be considered in relation to similar features in natural marshes, not as a feature which can be designed to suit human perceptions of appropriate habitat distribution.
3. Soil conditions must be anticipated as limitations to plant growth. Salt layers in former crystallizers appear to still limit plant establishment. These areas should have been disced or plowed to allow break up of the salt layers. As for islands, development of cat clays has been described by many authors and should be anticipated in any excavation and deposition of former bay muds at elevations above mean high water (MHW).
4. Islands are not natural features of wetlands and their size and configuration need to be compared to actual habitats used by waterbirds.

Reports:

Cuneo (1978), Niesen and Josselyn (1981), Josselyn et al. (1987).

Contacts:

Peter Koos, East Bay Regional Park District, Skyline Blvd, Oakland, CA.

SHOREBIRD MARSH, MARIN COUNTY

Project Profile: Mitigation for fill within seasonal wetland for construction of regional shopping center. Wetland created for purposes of consolidation of wetland acreage and flood storage.

Wetland type: Seasonally tidal saltmarsh with open water, emergent vegetation, and islands

Location: Corte Madera, California.

Size: 14 hectares.

Goals of project:

To provide for flood control during winter months and to create tidal saltmarsh during summer months.

Specific objectives in plan included:

1. Creation of extensive area to be planted with cordgrass (Spartina foliosa);

2. Creation of islands and channel habitat for fish and wildlife use;

3. Provision of water control structure to pump water out of site during flood periods and to limit maximum tidal height during summer months to +0.3 m NGVD, approximately 1 m lower than normal MHHW level;

4. Creation of linear basin to provide treatment of urban runoff from shopping center and nearby streets; and

5. Provision of public pathways and vegetated buffers.

Project implementation involved:

1. Excavation and grading of site to create desired topography;

2. Construction of pump station and water control structure to provide for flood control and dampened tidal action;

3. Excavation of outflow channel to Bay; and

4. Planting of marsh and buffer vegetation.

Judgement of success:

The site was to be managed for flood control from October 15 to April 15 of each year and then operated as controlled tidal marsh during remainder of year. The town employees (Public Works) were responsible for management. During first year following completion, no tidal action was introduced to site as the Town did not have a copy of the management plan. During second year of operation, a number of problems confounded the operation of the site as a tidal marsh:

1. Flap gates protecting local businesses leaked allowing high water to back up into drainage ditches;

2. The electronic system used to control the water control structure was defective and difficult to use;

3. The outfall channel had silted up and retarded flow from the water control structure; and

4. Town employees were still not familiar with the purposes of operation of the site.

A local citizen's committee took responsibility for oversight of the marsh operation and a consulting firm (Wetlands Research Associates, Inc.) was hired to implement a biological monitoring program and develop a planting plan. A number of steps were taken to alleviate the problems noted above, including replacement of the water control operating system, repair of the flap gates, and dredging of the outfall channel and outer mudflat. The local citizen's committee educated the responsible Town employees on the need to properly manage the wetland system.

Several experimental plantings were conducted by the biological consultant, however, the success of the plantings was limited by lack of summer tidal action. By the third year, tidal action was implemented during the summer and full planting of marsh vegetation is planned. No recruitment of marsh vegetation has been noted and there has been a substantial die-back of brackish water vegetation that had previously grown on the site.

Buffer vegetation has been planted. Coyote Brush (Baccharis pilularis) and marsh gumplant (Grindelia humilis) have been successful when planted prior to winter rains. Otherwise, irrigation is required for successful upland plant establishment.

The effectiveness of a portion of the marsh designed for urban runoff pollutant control has never been tested and the weir system proposed to increase residence time in this portion of the marsh never used. Several proposals have been made to develop a small peninsula that extends into the marsh, though none have been implemented. Finally, drainage problems for a portion of the surrounding business has necessitated the construction of a holding basin with a pump station that will discharge into the marsh. Since this drainage area includes mostly automobile dealers and gas stations, it is likely to increase pollutant discharge to the marsh.

A large number of waterfowl utilize the marsh during the migratory season. Herons feed extensively within the marsh during the summer. The Marin Audubon Society has kept records of the bird species utilizing the wetland. Several low islands receive use as a gull roosting area. The higher, steep sloped islands receive little bird use during daylight hours.

Significance:

This project is an example of a site which is designed to serve multiple purposes: mitigation for fill within a wetland, flood storage, and urban runoff pollutant control. It points out several problems inherent in complex projects.

1. The purpose regarded as having the greatest economic advantage to the community will receive highest priority. The town is primarily interested in the flood storage function and is eager to lower the water level in the marsh as soon as possible prior to the rainy season and maintain it at low levels as long as possible during the spring.

2. Most communities do not have the staff to manage complex systems, especially those which do not have a direct impact on the well-being of local residents. Therefore, the urban runoff pollutant control portion of the marsh has never been managed as planned due to the difficulty of setting up the system and the lack of knowledge of its function. Staff are not inclined to "calibrate" the tidal action within the marsh for the sole benefit of wildlife.

3. Local citizen groups and an effective biological consultant can be instrumental in stimulating the proper implementation of a marsh design. This interest must be sustained, however, so that problems are handled over the long-term and not just in the first year.

Reports:

Philip Williams and Associates.

Contacts:

James Buchholz, Wetlands Research Associates, Inc., San Rafael, CA.

Philip Williams, Philip Williams and Associates, San Francisco, CA.

Town Engineer, Town of Corte Madera, CA.

BRACUT MARSH MITIGATION BANK

Project Profile: Creation of a wetland to provide mitigation for fill in a number of smaller "pocket marshes" within an urban area.

Wetland type: Tidal salt marsh and transitional fringe.

Location: Arcata Bay in Humboldt County at Bracut on U.S. Highway 101.

Size: 2.5 hectares.

Goals of project:

To restore tidal action to diked area that had been partially filled with wood debris and river-run gravel.

Specific objectives in the plan included:

1. To establish as productive a marsh as possible and maximize the habitat value of the property;
2. To provide sufficient circulation and drainage of tidal flows to maintain a constant supply of nutrients, sediment and oxygen to the marsh system;
3. To furnish an adequate soil substrate to promote the growth and reproduction of marsh and upland vegetation; and
4. To minimize future maintenance requirements.

Project implementation involved:

1. Breaching of levee at a width sufficient to allow unrestricted tidal flow, but not wide enough to allow wave action to erode interior levees;
2. Outer levee armoring with rip-rap to provide protection from winds and storms;
3. Constructing inner islands immediately inside the dike breach to reduce erosive wave action within wetland;
4. Placing bay mud over lower portions of marsh to provide suitable substrate for marsh plant establishment;
5. Removing debris from upper portion of marsh; and
6. Planting of Spartina in areas as needed.

Judgement of success:

This project has been very controversial due to lack of establishment of marsh vegetation and possible water quality problems. The former is due to poor substrate conditions in the higher portions of the marsh where the river-run gravel forms a hard surface. The marsh was designed using an intertidal plant distribution plan taken from San Francisco Bay (Camp, Dresser and McKee (1980). It has been subsequently learned that the cordgrass within Arcata Bay is a different species than the San Francisco Bay form, has a very different elevational distribution, and is non-rhizomatous, i.e., it does not spread very rapidly from the initial plant shoot. Consequently, marsh plantings on the site have not spread significantly beyond their original location, though natural recruitment has brought in marsh plants within the lower portion of the mitigation site.

In the higher portions of the marsh, the hard surface persists and few wetland plants have become established. The exception is a rare and endangered species, Orthocarpus castillejoides var. humboldtiensis (Humboldt Bay owl's clover) which grows around temporary pools in the higher portions of the marsh. It is very prevalent throughout the site.

Water quality problems have been noted in the lower portion of the site where white filaments and strong anaerobic odors have been noted. Wood debris is prevalent immediately below the surface and gas bubbles of methane are frequent. Apparently the decay of the wood debris is resulting in poor water quality, especially during low tide when shallow ponds heat up.

Bird use on the property is variable. Most observers have noted abundant bird use in the high marsh during high tides with most species moving off-site during low tide periods. Besides roosting, no other use appears significant. A study is underway to quantify the bird use.

Significance:

This project demonstrates the difficulties involved when restoring filled wetlands. The nature of the fill can have a long-term effect on the marsh restoration. In most cases, the fill causes the native soil material to be compacted and therefore, below the level desired for the establishment of intertidal species. The problems which occur include:

1. Leaving of high organic matter fill beneath the marsh can result in decay and leaching of undesirable substances; and
2. Inappropriate substrata, especially in higher portions of the marsh where sedimentation is not significant, will reduce the ability of marsh vegetation to become established.

Several problems could have been corrected at the design phase:

1. Appropriate intertidal elevations could have been produced by over-excavating the site and filling it with dredge spoils;
2. The rip-rap placed along the levee was over-engineered and could have been reduced in scope; and
3. Better knowledge of the local intertidal distribution of plants would have produced more appropriate elevations for various species.

Finally, one must consider that marginal wetlands can provide important habitat for species not normally able to compete within a mature wetland. The presence of

Humboldt Bay owl's clover is due to the lack of other wetland vegetation and the peculiar nature of the substrate. This "accident" has provided an important habitat for a rare and endangered species.

Reports:

Camp, Dresser, and McKee (1980); Josselyn (1988a).

Contacts:

Liza Riddle, State Coastal Conservancy, Oakland, CA.

LINCOLN STREET MARSH WETLAND MITIGATION

Project Profile: Off-site mitigation for fill in a 10 acre parcel containing both wetland and upland habitats. Site selected for restoration was former fill site.

Wetland type: Tidal salt marsh with transitional upland areas.

Location: Puyallup River estuary, Pierce County, Washington.

Size: 3.9 hectares (2.2 hectares wetland and 1.7 hectares upland).

Goals of project:

Replace a wetland/upland region within an industrial complex at the Port of Tacoma with a mixture of mudflats, tidal channels, marshes, trees, grassland, and shrubland.

Specific objectives of the project were designed to:

1. Create habitat in specified ratios to support the following groups:

GROUP	AREAL PERCENTAGE OF SITE
Juvenile salmonids	50
Waterfowl	20
Shorebirds	10
Raptors	10
Small mammals	10

2. Monitor the ecological performance of the mitigation site; and

3. Maintain the site in perpetuity.

Project implementation involved:

1. Excavation of former fill and disposal off-site;

2. Grading to contour site to specific habitat types based on elevation;

3. Creation of new dike and tidal entrance to site; and

4. Planting of Lyngby's sedge, Carex lyngbyei.

Judgement of Success:

The site has been operated for too short a period (1985 to present) to make significant conclusions on its success. Thom et al. (1987) reported excellent survival of transplanted sedge shoots with a four-fold increase in the number of plants during the first growing season.

Target species were utilizing the site to a high degree and in far greater numbers than previously noted for the developed site. Juvenile salmonids were found within the main and finger channels and there appeared to be an abundance of prey resources to provide rearing habitat for juvenile fish.

The provision of a long-term monitoring program is quite important in evaluating the utilization of the mitigation site.

Significance:

The project is important as an example of establishment of detailed habitat features for specific species. Secondly, the performance of biological monitoring far exceeds that of most other initial studies.

Some problems have been reported that should be mentioned:

1. The excavated substrate had contaminants which required testing and disposal; and

2. Siltation is occurring in the entrance of the wetland and may affect tidal exchange.

Reports:

Thom et al. (1985, 1987).

Contacts:

Ron Thom, School of Fisheries, University of Washington, Seattle.

Mary Burg, Washington Department of Transportation, Seattle.

Kathy Kunz, Environmental Protection Agency, Region 10, Seattle.

CREATION AND RESTORATION OF TIDAL WETLANDS OF THE SOUTHEASTERN UNITED STATES

Stephen W. Broome
Department of Soil Science
North Carolina State University

ABSTRACT. Methods of creation and restoration of tidal wetlands in the Southeastern United States were summarized from published papers, reports, and first-hand experience. Publications by the U.S. Army Corps of Engineers which report research related to marsh habitat creation with dredged material and for shoreline erosion control were significant sources of information.

Critical aspects which should be considered in planning and implementing a tidal marsh creation or restoration project are:

- o Initial planning - Evaluate environmental impact on existing habitat, public acceptability, costs, and exposure to waves and currents that might cause erosion.
- o Elevation in relation to tide level - A surface must be created to provide the hydrologic regime to which the desired vegetation is adapted.
- o Wave climate and currents - The susceptibility of the site to erosion should be evaluated.
- o Salinity - The salinity of tidal and interstitial water determine the plant species.
- o Slope and tidal range - These factors affect the areal extent of the intertidal zone, the zonation of plant species, drainage and erosion potential.
- o Soil chemical properties - Availability of plant nutrients and the possibility of toxic contaminants should be considered.
- o Soil physical properties - These affect trafficability, i.e., bearing capacity, for planting operations and erodibility.
- o Timing of construction - Construction should be completed well in advance of optimum planting dates.
- o Cultural practices - Select the plant species adapted to environmental conditions at the site, use vigorous transplants or seedlings of local origin, plant at a spacing that will provide cover in a reasonable length of time, fertilize with N and P to enhance initial growth.
- o Maintenance - Observe the site periodically to determine the need for replanting, fertilization, wrack removal and control of undesirable plant species, excessive traffic or grazing.

Critical research needs include the following:

- o Site selection - Improved methods for predicting the probability of success on sites exposed to wave energy are needed. Methods of comparing the relative value of created tidal marsh and the habitat it displaces should be developed.
- o Revegetation - A better understanding of the environment required for optimum growth of a number of plant species is needed. Methods of creating tidal freshwater marshes need further investigation.
- o Documentation of tidal marsh development - The ecological function and structure of created or restored marshes must be more thoroughly evaluated. This information is

needed as a basis for making decisions on mitigation. Practical and economical methods are needed to evaluate success of individual marsh creation or restoration projects.

CHARACTERISTICS AND TIDAL WETLAND TYPES OF THE REGION

REGIONAL CHARACTERISTICS

The geographical area discussed in this chapter includes the coastal region of Georgia, South Carolina, North Carolina and Southeastern Virginia located between latitude 31° and 38° North (Fig. 1). The area lies in the Carolinian and Virginian provinces as defined by the U.S. Fish and Wildlife Service (Cowardin et al. 1979). The climate is humid, temperate to subtropical with mild winters, hot, humid summers and high rainfall (Table 1). The area is a part of the Atlantic Coastal Plain. This plain is composed of layered marine and non-marine sediments which were formed as the sea advanced repeatedly over the area and then withdrew (Oaks and DuBar 1974). The topography is relatively flat and gently sloping toward the ocean.

WETLAND TYPES

Tidal marshes in the region are found along low to moderate energy shorelines of estuaries. They range from narrow fringes, where tidal range is narrow, to large expanses where tidal ranges are wide and the area of intertidal land is extensive. The majority of marshes are saline (euhaline) or brackish (mixohaline) with a smaller amount of freshwater marshes at the upper reaches of tidal influence.

The regularly flooded intertidal salt marshes are dominated by nearly pure stands of smooth cordgrass (Spartina alterniflora). The high salt marsh extends from mean high water to the limit of flooding by extreme storm or spring tides. Plants that dominate the high marsh are saltmeadow cordgrass (S. patens), saltgrass (Distichlis spicata) and black needlerush (Juncus roemerianus).

The transition from salt to fresh marshes is a continuum with plant species diversity increasing as salinity decreases. Plant species characteristic of brackish marshes include black needlerush, saltmeadow cordgrass, big cordgrass (S. cynosuroides), sawgrass (Cladium jamaicense) and shrubs such as groundselbush (Baccharis halimifolia), marsh elder (Iva frutescens) and wax myrtle (Myrica cerifera). In the U.S. Fish and Wildlife Service wetland classification system (Cowardin et al. 1979), salt and brackish water tidal marshes are classified as follows: system, estuarine; subsystem, intertidal; class, emergent wetland; subclass, persistent and water chemistry, euhaline (30-40 parts per thousand (ppt) or mixohaline (0.5-30 ppt).

Tidal freshwater wetlands are located upstream from tidal salt marshes and downstream from nontidal freshwater wetlands. They are characterized by salinity less than 0.5 ppt, plant and animal communities dominated by freshwater species and daily lunar tidal fluctuation (Odum et al. 1984). While salt and brackish marshes are dominated by a few plant species, tidal freshwater marshes are characterized by a large and diverse group of plants. In the U.S. Fish and Wildlife Service classification system, persistent emergent tidal freshwater marshes are in the palustrine system and emergent wetland class. If the vegetation is non-persistent, it falls in the riverine system and emergent wetland class. Persistent emergent wetlands are dominated by plant species that remain standing through the winter until the beginning of the next growing season. Palustrine persistent emergent wetlands are characterized by such plants as cattails (Typha

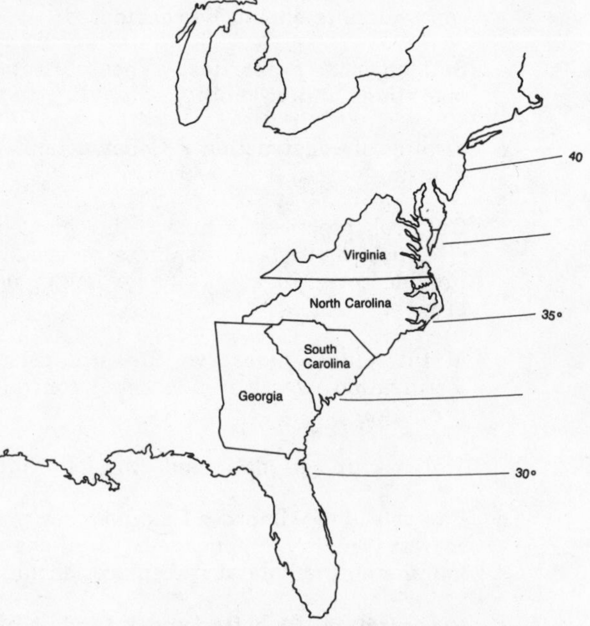

Figure 1. The geographical area discussed in this chapter is the coastal region of Virginia, North Carolina, South Carolina and Georgia in the Southeastern United States.

spp.), bulrushes (*Scirpus* spp.), sawgrass, sedges (*Carex* spp.), giant cutgrass (*Zizaniopsis miliaceae*) and common reed (*Phragmites australis*). Nonpersistent emergent wetlands are dominated by plants that fall to the surface of the substrate or are present only below the surface at the end of the growing season so that there is no obvious sign of emergent vegetation. Examples of plant species present in non-persistent wetlands are arrow arum (*Peltandra virginica*), pickerelweed (*Pontederia cordata*), arrowheads (*Sagittaria* spp.), and wild rice (*Zizania aquatica*).

Geographical Range

Regularly flooded smooth cordgrass marshes in the region are best developed and most extensive in South Carolina and Georgia where there is a wide tidal range and an abundance of silt and clay sediments supplied by rivers. According to estimates compiled by Teal (1986), 68% of the east coast's regularly flooded smooth cordgrass marshes occur in South Carolina and Georgia. There are 115,037 ha in South Carolina and 126,564 ha in Georgia, with less extensive areas in North Carolina (23,634 ha) and Virginia (33,469 ha) (Table 2). Brackish marshes make up about 33% of the total tidal marsh in the region (Table 2).

While not nearly as extensive as salt and brackish marshes, significant areas of tidal freshwater marshes are found in Virginia, South Carolina and Georgia. Tidal freshwater marshes represent only 11% of the total tidal marshes in the region. In North Carolina, inlets in the barrier islands are narrow and few in number, and the estuaries are large, resulting in a narrow lunar tidal range. Tides are irregular and controlled by wind direction and velocity; therefore, there is very little tidal freshwater marsh. The Cape Fear River in the southern part of the state is an exception. It empties directly into the Atlantic Ocean, has a one meter tidal range and 1200 ha of freshwater tidal marsh (Odum et al. 1984). Marshes bordering the Currituck and Albemarle Sounds, which are fresh to brackish and are flooded by wind-dominated tides, were classified as shallow fresh marsh by Wilson (1962). These are not tidal freshwater marshes by the definition of Odum et al. (1984) that specifies regular lunar tides as a criteria. About 70% of the tidal marshes in North Carolina are brackish and irregularly flooded.

Functions of Tidal Wetlands

Several publications have reviewed the extensive literature on wetland functions and values (Greeson et al. 1979, Tiner 1984, Daiber 1986, Mitsch and Gosselink 1986, Adamus et al. 1987). Tiner (1984) divided wetland benefits into three basic categories: (1) fish and wildlife values, (2) environmental quality values and (3) socio-economic values. Fish and wildlife values listed were habitat nursery and spawning grounds for fish and shellfish, habitat for waterfowl and other birds, and habitat for furbearers and other wildlife. Environmental quality values include water quality maintenance (such as filtering pollutants, sediment removal, oxygen production, nutrient cycling and chemical and nutrient absorption), aquatic productivity and microclimate regulation. Socio-economic values are flood control, protection from wave damage, erosion control, groundwater recharge, timber and other natural products, accumulation of peat, livestock grazing, fishing and shellfishing, hunting and trapping, recreation, aesthetics, education, and scientific research.

Adamus et al. (1987) listed the following key functions and values of wetlands: ground water recharge and discharge; floodflow alteration; sediment stabilization; sediment and toxicant retention; nutrient removal and transformation; production export; wildlife diversity and abundance; aquatic diversity and abundance; recreation; and, uniqueness and heritage.

An often overlooked function of wetlands is removal of CO_2 from the atmosphere by accumulation of organic carbon in saturated soils that inhibit decomposition (Armentano 1980). The importance of this wetland function may be more widely recognized in the future because of increasing concern about higher levels of atmospheric CO_2.

Since Odum (1961) espoused the idea that tidal marshes contribute to estuarine and coastal productivity, a great deal of research has been conducted to evaluate productivity of these marshes and their effect on the estuarine ecosystem (Daiber 1986). Because of regular or irregular tidal flooding, marshes are an integral part of the ecosystem of the adjacent water body. Tidal marshes are also among the most productive ecosystems in the world (Tiner 1984). Although primary production of the vegetation is quite high, it varies from one location to another and within a given marsh. Turner (1976) reviewed salt marsh macrophyte production along the east and Gulf Coasts of the United States and found a range from 300-2000g m^{-2} yr^{-1} of annual aboveground production with a trend of decreasing productivity from south to north. A generally positive correlation also exists between tidal amplitude and productivity that is, at least in part, due to the energy subsidy provided by the tide (Odum 1979). Other factors such as salinity and nutrient availability can modify the tidal amplitude effect. Production of roots and rhizomes generally equals or exceeds aboveground production. Belowground production of a transplanted smooth cordgrass marsh in

Table 1. Climatic data from locations along the Coast of Virginia, North Carolina, South Carolina and Georgia.

	Latitude	Mean Annual Rainfall[1] (in)	Mean Annual Temp.[1] (°F)	Mean Annual Min. Temp[2] (°F)	Ave. Frost-Free Period[3] (da)
Norfolk, VA	36° 54'	46	60	15-20	240-270
Cape Hatteras, NC	35° 16'	57	62	20-25	270-300
Morehead City, NC	34° 44'	61	63	15-20	240-270
Wilmington, NC	34° 16'	53	63	20-25	240-270
Charleston, SC	32° 47'	45	66	20-25	270
Beaufort, SC	32° 23'	47	66	20-25	270-300
Savannah, GA	32° 8'	48	67	20-25	270-300
Brunswick, GA	31° 9'	48	68	20-25	> 300

[1] U. S. Dept. of Commerce Climatological Data, Mean Annual Rainfall and Temperature 1951-1985.
[2] USDA Agric. Research Service Plant Hardiness Zones Map, 1960.
[3] Based on the period 1921-1950.

Table 2. Area of tidal marshes in Virginia, North Carolina, South Carolina, and Georgia in hectares (acres).

Marsh Type	Va[a]	NC	SC	GA	Regional Totals	% of total
Freshwater Tidal	15,814 (39,075)	1,214 (3,000)[b]	26,155 (64,531)[b]	19,040 (47,047)[b]	62,147 (153,653)	11.0
Brackish Water	36,868 (91,100)	58,418 (144,350)[c]	14,149 (34,962)[d]	75,010 (185,346)[e]	184,445 (455,758)	32.6
Salt Water	33,469 (82,700)	23,634 (58,400)[f]	135,373 (334,501)[g]	126,564 (312,736)[e]	319,040 (788,337)	56.4
Total	86,151 (212,875)	83,267 (205,750)	175,637 (433,944)	220,614 (545,129)	565,669 (1,397,748)	

[a] Silberhorn (pers. comm.), Virginia Wetlands Inventory

[b] Odum et al. 1984

[c] Wilson 1962 and Odum et al. 1984 (46,900 acres of "shallow fresh marsh" reported by Wilson minus 3,000 acres assigned to tidal fresh marsh by Odum et al. plus 100,450 acres called "irregularly flooded salt marsh" by Wilson).

[d] Tiner 1977

[e] Kundell 1986

[f] Wilson 1962

[g] Tiner 1977; total includes 50,249 acres of high salt marsh

North Carolina was estimated to be 1.1 times aboveground production (Broome et al. 1986).

A portion of the organic material produced by marsh plants is transported by tidal flushing to surrounding waters, providing a source of organic detritus that becomes a part of the estuarine food web (Teal 1962). There is considerable evidence linking primary production of wetlands to aquatic secondary production, and information suggesting that wetland destruction results in lowered production of estuarine organisms of interest to man (Odum and Skjei 1974). However, there is still controversy about the actual value of tidal marshes in estuarine ecosystems. From a review of literature, Nixon (1980) concluded that the widely accepted views of tidal marshes as large exporters of organic matter that support secondary production, and the role of marshes in nutrient cycling are not well substantiated by research. He further stated that our understanding of the interaction between coastal marshes and coastal waters is still incomplete.

Although there may be some debate and much to learn about the role of tidal marshes in coastal ecosystems, there is sufficient evidence of their value to warrant preservation and protection by government regulations. In spite of regulatory protection, disturbances or destruction of marshes may occur illegally or accidentally and permits for various types of manipulation of marshes are issued when benefits to the public outweigh adverse effects to the environment. Restoration and creation of new marshes is conducted in an attempt to mitigate damage for whatever reason it might occur.

EXTENT OF TIDAL WETLAND CREATION AND RESTORATION

TYPICAL GOALS

The goals of tidal wetland creation and/or restoration generally fall in the following categories: (1) dredged material stabilization or creation of marsh habitat using dredged material; (2) shoreline erosion control; (3) mitigation of destruction of, or adverse impact on, natural stands; and, (4) research. Many projects have a combination of two or more of these goals.

SUCCESS IN ACHIEVING GOALS

Historical Perspective

Marsh vegetation has been planted in some parts of the world for many years with quite different goals. In Europe, Spartina townsendii (the fertile form is now known as S. anglica) was planted extensively during the 1920's and 1930's, to reduce channel silting, for coastal erosion protection and to reclaim land for agriculture (Ranwell 1967). Spartina anglica was introduced to China in 1963 and since that time 30,000 ha have been planted, providing important economic, social and ecological benefits (Chung and Zhuo 1985). These benefits were reported as bird habitat, animal fodder, pasture, aquaculture, green manure, amelioration of saline soil, and land reclamation (Chung 1982). Spartina alterniflora was introduced to China in 1979 because of its potential for producing more biomass than S. anglica. After five years, successful plantings amounted to 260 ha (Zhuo and Xu 1985). The concept of marsh restoration or creation to preserve and enhance estuarine ecosystems is relatively new. Knutson et al. (1981) surveyed planted salt marshes in the United States and reported that the earliest plantings documented in the literature were in the 1950's along tidal rivers in Virginia for the purpose of stabilizing shorelines (Phillips and Eastman 1959, Sharp and Vaden 1970). Some of these sites have remained stable after more than 30 years. Knutson also discovered one shoreline planting in Virginia reported to have been planted in 1928.

Dredged Material and Shoreline Stabilization

Research on the feasibility of salt marsh development on dredged material was begun in 1969 by N.C. State University with financial support provided by the U.S. Army Corps of Engineers, Coastal Engineering Research Center (Woodhouse et al. 1972, Woodhouse et al. 1974). This work was later extended to include stabilization of eroding shorelines (Woodhouse et al. 1976). The plant species investigated were S. alterniflora and S. patens. Techniques of propagation, both vegetatively and from seed, were developed and several dredged material islands and shorelines were successfully vegetated.

Successful dredged material and shoreline stabilization studies were also carried out in the Chesapeake Bay (Garbisch et al. 1975), Galveston Bay (Dodd and Webb 1975) and San Francisco Bay (Knutson 1976). A comprehensive study of wetland habitat development with dredged material was conducted by the Dredged Material Research Program of the U.S. Army Engineer Waterways Experiment Station. Major marsh development sites were located at Windmill Point, James River, Virginia (Lunz et al. 1978); Buttermilk Sound, Georgia (Reimold et al. 1978); Apalachicola Bay, Florida (Krucynski et al. 1978); Bolivar Peninsula, Galveston Bay, Texas (Allen et al. 1978, Webb et al. 1978); Salt Pond No. 3, South San Francisco Bay, California (Morris et al. 1978); and Miller Sands, Columbia River, Oregon (Clairain et al. 1978). Results from these and other projects have demonstrated the feasibility of tidal marsh habitat development with dredged material and provided guidelines for implementation (Environmental Laboratory 1978, U.S. Army Corps of Engineers 1986). Woodhouse (1979) summarized information on coastal marsh creation in the United States discussing plant propagation, planting, fertilization, and management of the major plant species.

Knutson et al. (1982) documented the wave damping value of smooth cordgrass along a shoreline of Chesapeake Bay, Virginia. More than 50 percent of wave energy from boat wakes was dissipated within the first 2.5 m of marsh and all of the energy was dissipated within 30 m.

Conversion of Upland to Intertidal Marsh

Interest in mitigation of impacts on natural marshes led to investigations of creating marsh-creek systems on uplands graded to suitable intertidal elevations. These were areas previously used for borrow pits or undisturbed uplands (Broome et al. 1982, 1983a, 1983b; Priest and Barnard 1987). Converting upland sites to intertidal marsh requires careful attention to grading to the correct elevations and attention to soil chemical properties, particularly pH and nitrogen and phosphorus availability.

Restoration of Damaged Habitat

Mitigation of violations often involves simple removal of fill material from marshes to restore the surface to its initial elevation and to allow the vegetation to reestablish naturally. Marsh restoration efforts may also be necessary after vegetation has been destroyed by toxic chemical or oil spills (Seneca and Broome 1982). This requires removing the toxic material or delaying planting until the effects have diminished.

FACTORS AFFECTING SUCCESS OR FAILURE

The degree of success that can be achieved in developing tidal marsh habitat at a particular site may be limited by any one of a number of factors or a combination of factors. Transplanting vegetation at the correct elevation in relation to tidal regime at the site is a prerequisite to success. A second factor that often affects initial establishment and long term stability is wave stress (Knutson et al. 1981, Allen et al. 1986). This is important for shoreline erosion projects and where dredged material shorelines are exposed to long fetches. Strong currents also cause erosion and undermining of planting sites in certain locations such as along channels or inlets.

Other factors that may affect success include: tidal amplitude; slope of the area to be planted; depth of water off shore; shoreline orientation; shape of the shoreline; large boat wakes; salinity of interstitial and tidal water; sediment supply, including littoral drift; fertility status of the soil; soil physical properties, particularly texture and degree of compaction; soil erodibility; shading by trees; excessive use by domestic or feral animals and wildlife such as geese, muskrat, or nutria; and foot or vehicular traffic. Management practices also affect success. These include using viable propagules of a plant species adapted to the environment of the site, proper planting time, and fertilization. Environmental and management factors affecting success will be discussed in detail in the following section.

DESIGN OF CREATION/RESTORATION PROJECT

PRECONSTRUCTION CONSIDERATIONS

Careful and thorough planning will increase the probability of success of wetland habitat creation or restoration projects. Two important considerations are location and characteristics of the site.

Location

Location of a planting site is important with regard to logistics of equipment, supplies, and personnel. Marsh development sites are often in areas accessible only by boat, which increases the time and cost of equipment transportation. Location is also important in terms of salinity and tidal regime, which determine the plant species adapted to the site. Several factors should be taken into consideration in locating dredged material disposal sites for marsh development (Environmental Laboratory 1978, U.S. Army Corps of Engineers 1986). These include availability for disposal; capacity of the area to hold the volume of material to be dredged; proximity to the dredging project; physical and engineering features; environmental and social acceptability, such as impacts on existing habitat, disturbance of water quality and flow, and public perception of the project; and tidal fetch and current considerations that might cause erosion.

Site Characteristics

A number of site characteristics should be considered in determining the feasibility of marsh establishment and in the planning process. These are discussed in the following sections.

Elevation, Slope and Tidal Range --

The areal extent of the intertidal zone is determined by elevation, slope and tidal range. These factors determine the degree of submergence, which, in turn, affects the elevation range and zonation of plants within the marsh. The elevation required by marsh vegetation at a given site is best determined by observing and measuring the lower and upper elevation limits of a nearby natural marsh. Alternatively, trial plantings extending well below and well above the estimated limits of survival can be made to determine the elevation range (Woodhouse 1979). In irregularly flooded areas with old marshes and eroding peat soils, elevations of the entire natural marsh may be near the upper limit of transplant survival and transplants may grow well below those elevations (Broome et al. 1982).

Marsh vegetation grows on a wide range of slopes. The more gentle the slope the greater the area available for growth of intertidal vegetation. Gentle slopes also dissipate wave energy over a greater area, reducing the probability of erosion. Slopes that are too flat can result in poor surface drainage leading to waterlogging and high salinities.

Tidal range, the vertical distance between high and low water, is important in determining the area of the intertidal zone, import and export of sediments, nutrients and organic matter, drainage and zonation of vegetation. It is generally easier to establish a viable marsh in an area with a wide regular tidal range than in an area with irregular wind-driven tides. This is particularly true on exposed shorelines where wider fringes of vegetation are more resistant to erosion.

Wave Climate --

Severity of wave climate is an important factor that affects initial establishment and long term stability of marshes. Four shoreline characteristics (average fetch, longest fetch, shore configuration, and grain size of sediments) are useful indicators of wave climate severity (Knutson et al. 1981, Knutson and Inskeep 1982). Planting success is inversely related to fetch, the distance over water that wind blows to generate waves. The shoreline configuration or shape is a subjective measure of the shoreline's vulnerability to waves. For example, a cove is sheltered while a headland is more vulnerable. Grain size of beach sand is also related to wave energy. Fine-grained sands generally indicate low energy, while coarser textured sand indicates high energy. This is of course affected by the texture of sand available in a particular environment. Knutson et al. (1981) developed a numerical site evaluation form for rating potential success using these four indicators.

Boat traffic and offshore depth are two other factors that should be considered when evaluating wave climate. Boat and ship wakes are particularly significant in areas along channels. Shallow offshore water reduces severity of waves reaching the shore (Knutson and Inskeep 1982).

A shoreline erosion control study in Virginia estuaries found that using the vegetative site evaluation form effectively predicted success or failure in establishing marsh fringes along shorelines (Hardaway et al. 1984). Establishing a fringe of S. alterniflora and S. patens could be accomplished with no maintenance planting required where average fetch was less than 1.0 nautical mile (1.8 km). Along shorelines exposed to 1.0 to 3.5 nautical miles (1.8-6.5 km) plantings in coves and bays had a better chance of survival. Maintenance planting was also necessary on shorelines with this type of exposure. Where average fetches were 3.0 to 5.5 nautical miles (5.6-10.2 km), establishing marsh grasses was impractical without a permanent offshore breakwater. Marsh establishment was unsuccessful where the fetch was greater than 5.5 nautical miles (10.2 km). Effects of fetch could be modified by tidal range. Experience in North Carolina has shown that the chance of success in establishing shoreline marsh fringes increases as the regular tidal range increases, resulting in a wider intertidal zone for planting (Broome et al. 1981).

Salinity --

Salinity of the tidal water and the interstitial water determines which plant species should be planted and the type of plant community that will eventually colonize a site. Salinities of interstitial water may become too high for plant growth especially in depressions that do not drain at low tide. Clay or other restrictive layers in dredged material may cause perched water tables, also resulting in concentrated soil solutions due to evaporation. Grading old dredged material disposal sites may expose surfaces with high residual salt concentrations. Residual salt may have accumulated in a ponded area sometime in the past and then covered with spoil during a subsequent dredging operation.

Very favorable conditions for plant growth occur when seepage from adjacent uplands produces low salinity in the soil water. Such seepage may also provide a supply of plant nutrients to enhance growth.

Soil or Substrate --

Mechanical operations such as grading, shaping and planting are generally easier on sandy soils than silt or clay because of the greater bearing capacity and trafficability of sand. The disadvantage of sandy material is its low nutrient supplying capacity. This is usually not a problem where tidal waters are rich in nutrients and transport nutrient rich sediments; but, fertilizers can increase plant growth on sandy material during the establishment period. Recently deposited silts and clays are often high in nutrients but are usually soft, presenting problems for equipment operation and in anchoring plants until they are established. Silts and clays along eroding shorelines may be compact and hard, making opening and closing planting holes difficult.

Sedimentation --

A moderate amount of sedimentation from tidal and wave action, long shore drift or from uplands has a stimulating effect on growth by supplying nutrients. In the case of shorelines, accumulation of sediments prevents erosion. Excessive accumulation can damage plants and increase elevations above the normal range of intertidal vegetation. Blowing sand is often a problem on dredged material disposal sites. Sand fencing and/or vegetation should be used to protect the intertidal zone to be planted from blowing sand.

Sunlight --

Shading by trees may be a problem on some shorelines. Hardaway et al. (1984) found insufficient sunlight to be a limiting factor to marsh establishment along some creeks in Virginia.

Traffic --

Excessive foot or vehicular traffic must be excluded from the planting site.

Wildlife Predation --

Creating wildlife habitat is a principal objective of marsh creation; however, excessive use and feeding can destroy a planting, especially during the establishment period. Canada geese as well as snow geese graze on smooth cordgrass rhizomes and have been known to seriously damage new plantings. Garbisch et al. (1975) were able to minimize damage caused by Canada geese by placing a 1.3-m wide band of wire netting on the soil surface along the seaward edge of a planting. Maximizing plant density and first year production also minimized damage since geese prefer not to feed in dense tall stands of smooth cordgrass.

Dense populations of muskrats may denude large areas of brackish-water marshes (Gosselink 1984) and nutria may also be a problem. In North Carolina, muskrats selectively removed smooth cordgrass from a brackish-water planting which included big cordgrass, saltmeadow cordgrass, and black needlerush. Trapping of muskrat and nutria or exclusion by fencing may be necessary to protect plantings in some situations.

Contaminated Sediment --

Dredged material from industrial or heavily populated areas may contain contaminants such as heavy metals, pesticides, and petroleum that may be detrimental to growth of plants. An even greater concern is the potential for plant uptake and release of contaminants into the environment. Toxic material may damage organisms that feed directly on the plant material or the toxins may be passed to other organisms through the food web (U.S. Army Corps of Engineers 1986). The possibility of toxic contaminants in dredged material should be taken into consideration in areas where contamination is likely.

CRITICAL ASPECTS OF THE PROJECT PLAN

Timing of Construction

Planning should allow for obtaining permits if necessary, and construction and final grading of planting sites well in advance of optimum planting dates for the vegetation. Several weeks are required for settling of areas that receive fill material. Permitting and construction delays are not uncommon and can often cause delays beyond acceptable planting dates. This can be costly if potted seedlings have been produced, since long holding periods may result in poor quality or even death of the plants.

The optimum planting dates for intertidal vegetation in the Southeast are between April 1 and June 15. Dates earlier than April 1 increase the likelihood of storm damage before plants have taken root and field-dug plants are more difficult to obtain. Dates later than June 15 limit the length of the growing season available for plants to become established.

Construction Considerations

Perhaps the most critical aspect of creating an intertidal marsh is grading the soil surface to the elevation that provides the hydrologic regime to which the plant species of interest are adapted. This is especially critical in areas with a small tidal amplitude. The elevation range for marsh establishment at a site along the Pamlico estuary in North Carolina was only 0.37 m (0.06 to 0.43 m msl) (Broome et al. 1982); consequently, accurate and precise grading was necessary to produce a viable marsh. Slopes should be as gentle as possible while still insuring good surface runoff at low tide. Slopes in the range of 1 to 3 percent are preferable. Elevation zones occupied by a particular plant species should be determined from observation of nearby natural marshes or from trial plantings on the site. A surveyor's level may be used to relate the elevation limits of a natural marsh to a planting site or water levels on the nearest marsh may be observed. When the water level is standing at the upper limit of growth on the natural marsh, mark the waterline at the site to be planted. Repeat this procedure when the water level is standing at the lower limit of the natural marsh. In regularly flooded saline areas, the vertical range of smooth cordgrass is from about mean sea level to mean high tide and salt meadow cordgrass occupies the zone from mean high tide to the storm tide line.

Hydrology

Mitsch and Gosselink (1986) state that "Hydrology is probably the single most important determinant for the establishment and maintenance of specific types of wetlands and wetland processes". It is obvious that attempts to create or restore tidal marsh will fail without proper attention to elevation, tidal flooding, and drainage, as has been discussed previously. Factors other than tides that influence hydrology of tidal marshes are precipitation, surface inflows and outflows, groundwater, and evapotranspiration (Mitsch and Gosselink 1986).

When restoring or creating a narrow fringe of marsh along a shoreline, the important hydrologic considerations are wave climate and planting at the correct elevation in relation to tide levels. Broader and larger marsh systems require that a drainage system of channels that simulate natural creeks be installed for good tidal exchange and drainage and to provide access to fauna. Greater use by fishes, benthos, and shorebirds were reported where tidal channels were purposely created in man-made marshes (Newling and Landin 1985).

A local, natural marsh-creek system should be surveyed to determine appropriate depth, width, and spacing of drainage channels.

Other hydrologic factors affect productivity of marshes. Precipitation, runoff from the watershed, and freshwater seepage increase growth of salt marsh vegetation and affect species composition and diversity. Functions of marshes such as nutrient cycling, organic matter accumulation, import and export of organic matter and mineral nutrients, and many other chemical and physical processes are affected by hydrologic conditions.

Substrate

The physical and chemical properties of the substrate or soil are important factors in tidal marsh restoration and creation. Physical properties affect bearing capacity and trafficability which determine the equipment and methods that can be used in grading, shaping, and planting. Mechanical operations and planting are easier on sandy soils, allowing use of mechanical transplanters of the type used for tobacco or vegetable plants. Soft dredged material requires innovative planting techniques such as planting from rafts at high tide or using walkways for access (Environmental Laboratory 1978). The availability of plant nutrients in tidal marshes is related to many physical, chemical, and biological processes. Just as in upland soils, the nutrients available to plants at a given location and within a marsh are quite variable. Soil differences are mitigated to some extent by the effects of tidal inundation and the chemically reduced state of saturated soils. Tidal exchange affects soil chemical and biological processes, including deposition of sediment, influx and efflux of nutrients and flushing of toxins. Seawater is quite high in Mg, Ca, K and S and apparently provides adequate amounts of these nutrients to salt and brackish water marsh vegetation.

When soils are waterlogged, air movement is restricted, resulting in anaerobic (reduced) conditions. Oxidation- reduction processes in the soil that are important to plant nutrition are affected by anaerobic conditions (Redman and Patrick 1965). Any nitrate nitrogen present in the soil is subject to denitrification and loss to the atmosphere. Organic matter decomposition is slower and less complete and nitrogen that is released by decomposition accumulates in the ammonium form. Iron and manganese compounds are reduced to more mobile forms and sulfur is present in the sulfide form. The pH values of reduced soils tend to be buffered around neutrality. Plant available soil phosphorus is also increased by anaerobic conditions. As ferric iron (Fe^{+++}) is reduced to ferrous iron (Fe^{++}), phosphorus compounds present as ferric phosphate are released into solution (Mitsch and Gosselink 1986).

A number of studies have shown that nitrogen is a limiting factor in growth of smooth cordgrass in natural marshes (Sullivan and Daiber 1974, Valiela and Teal 1974, Broome et al. 1975). Nitrogen may also limit growth in freshwater tidal marshes (Simpson et al. 1978). Phosphorus is abundant in many fine textured sediments such as those present in marshes along the Georgia coast and provides an ample supply for the vegetation (Pomeroy et al. 1969); however, phosphorus is in short supply in some salt marsh soils. Fertilizer experiments on smooth cordgrass growing on a sandy soil at Ocracoke Island, NC indicated that P availability became a growth limiting factor as N rates were increased. Applications of N and P fertilizers produced more growth than N alone (Broome et al. 1975).

When establishing a tidal marsh, whether on dredged material, an eroding shoreline, or on a graded upland site, the response to fertilization depends on the inherent fertility of the soil and the amount of nutrients supplied by tidal inputs or other sources such as seepage, runoff, precipitation, and nitrogen fixation. Fine-textured dredged materials are often rich in nutrients and plant growth is not limited by nutrient supply (Environmental Laboratory 1978, U.S. Army Corps of Engineers 1986). Plant growth response to N and P fertilization is more likely on sandy dredged material with little silt or clay sediment being brought in by the tide and where the tidal water is low in N and P (Woodhouse et al. 1974).

Nitrogen and phosphorus are likely to be growth limiting factors along eroding shorelines. This is particularly true where shorelines have migrated to the point that the soil surface of the intertidal zone is the argillic horizon (subsoil) of an upland soil. In the Southeast, this type of soil material typically has a high P fixation capacity. It is high in hydrous oxides of iron and aluminum, exchangeable aluminum, and kaolinitic clays that are capable of sorption of large amounts of phosphorus, causing it to be unavailable or slowly available to plants (Tisdale et al. 1985). Physical properties of subsoil material make planting more difficult and may affect growth by limiting root penetration.

An experiment testing rates and sources of N and P fertilizer on transplants of smooth cordgrass along an eroding shoreline of the Neuse River in North Carolina demonstrated that adequate fertilization was necessary for successful establishment of vegetation (Broome et al. 1983c) (Fig. 2). A soil test from the site indicated no organic matter, a pH of 5.17, and the following nutrient concentrations in mg dm^{-3}: NH_4-N, 18; NO_3-N, O; P, 1.3; K, 86; Ca, 700; and Mg 291. The texture was sandy clay loam (62% sand, 13% silt, and 25% clay). Rates of N and P were tested using ammonium sulfate and concentrated superphosphate banded at the time of planting. Biomass increased with increasing rates up to 224 kg ha^{-1}N (200 lbs ac^{-1}N) and 49 kg-ha^{-1}P (100 lbs $ac^{-1}P_2O_5$) (Fig. 3). In one growing season, the unfertilized control plants produced an average biomass of 8 grams and 5 stems per plant. The highest rate of N and P produced an average biomass of 214 g per plant and 32 stems per plant. The slow-release fertilizer materials Osmocote and Mag Amp were compared to the soluble sources, ammonium sulfate and concentrated superphosphate. When the two slow-release materials were used separately, they did not produce better plant growth than the soluble materials at equal rates. The Mag Amp and Osmocote applied in combination were significantly better than the soluble materials. The slow release materials would be expected to have a greater advantage on sand which has a lower capacity to retain applied nutrients. Application of Osmocote in the planting hole has become a common practice in marsh establishment and is quite effective.

Effects of the initial fertilizer treatments were also apparent in the second growing season. Nitrogen and P were retained by the soil and/or recycled by the plants. The effects of additional fertilization were determined by topdressing one block of the experiment in June. Ammonium sulfate at the rate of 112 kg ha^{-1}N (100 lbs ac^{-1}N) and concentrated superphosphate at the rate of 49 kg ha^{-1} P (100 lbs $ac^{-1}P_2O_5$) were spread evenly on the soil surface at low tide. When harvested in October the fertilized block had an aerial biomass of 770 g m^{-2} and 527 stems m^{-2} compared to 367 g m^{-2} and 260 stems m^{-2} in the control block.

Plant nutrients are not deficient on all shorelines and fertilization may not be necessary for successfulmarsh establishment. No fertilizer was applied to a smooth cordgrass planting along the shoreline of Bogue Banks, a North Carolina barrier island (Broome et al. 1986). The substrate was sand; however, soil test results indicated a P concentration of 65 g dm^{-3}, which is much higher than the previously discussed Neuse River shoreline. The planting developed into a 15-meter wide fringe of marsh that was effective in reducing shoreline erosion

Figure 2. An unfertilized row of smooth cordgrass (left) compared with a row fertilized with N only (center) (112 kg ha^{-1} N) and N and P (right) (224 kg ha^{-1} N and 25 kg ha^{-1} P). The planting was done along the Neuse River in North Carolina on June 13 and photographed August 30.

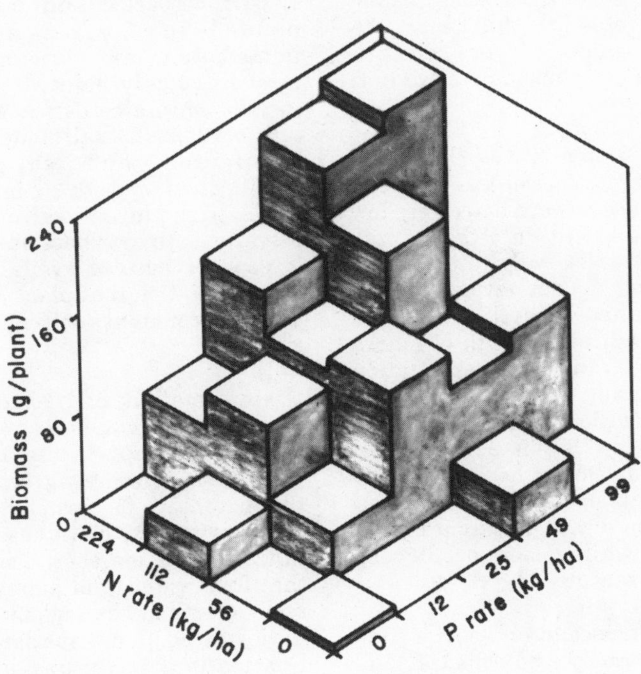

Figure 3. The effects of N and P rates on oven-dry weight of smooth cordgrass transplanted June 13 and sampled September 19.

and was comparable to a nearby natural marsh in biomass production (Fig. 4a, 4b). Possible sources of nutrients for the marsh were seepage from adjacent sandy upland soils, which included fertilized lawns and septic tank drainage fields, tidal water, and sediments deposited by tides and waves. Soil problems may be encountered when mitigation projects involve grading upland sites to elevations suitable for intertidal habitat. Cuts into the B and C horizon of soils in the southeast expose a new surface that is likely to be acid and deficient in plant nutrients. One alternative is to strip and stockpile the topsoil, grade the site and replace the topsoil to bring the surface to the correct intertidal elevation. Topsoil is higher in organic matter and has better chemical and physical properties but also may be nutrient deficient. Disadvantages are that replacing topsoil adds to the expense of site preparation, and on wooded sites roots, stumps and other debris may be returned with the topsoil, making planting more difficult. A second option is to determine the nutrient status and pH of the new surface and add fertilizers and lime if needed. Chemical deficiencies can be corrected but undesirable physical properties can make planting difficult.

A series of brackish-water marshes were established in cooperation with the phosphate mining industry in North Carolina, first with borrow pits (Fig. 5a, 5b) and then with two sites graded from pine woodland (Broome et al. 1982; 1983b; 1986). These sites provided some insight into soil fertility problems that may be encountered. Species planted were smooth cordgrass, big cordgrass, saltmeadow cordgrass and black needlerush.

Soil test results indicated low available soil P at all the sites even where topsoil was replaced (Table 3). Phosphorus levels were extremely low (1 mg dm^{-3}) at one site graded into the subsoil. Measurements of pH were generally in the range of 4.5 to 5, but these levels did not adversely affect plant growth. Fertilization with N and P was essential for establishment and growth of marsh vegetation at these sites. Broadcast application and incorporation by discing before planting was the most practical fertilization method. The best rates were 112 kg ha^{-1}N (100 lbs ac^{-1}N) from ammonium sulfate and 98 kg ha^{-1} P (200 lbs $ac^{-1}P_2O_5$). Aboveground biomass of the cordgrass was equivalent to nearby natural marshes after two growing seasons, while black needlerush required three growing seasons.

An unexpected occurrence at one of the sites was development of extremely acid soils (pH 2.5) over about 25 percent of the 2.5 ha area. These subsurface areas exposed by grading apparently contained sulfides that oxidized when exposed, thus reducing the pH. Yellow incrustations of jarosite on the surface were another indicator of acid sulfate soils, which are also known as cat clays (Bloomfield and Coulter 1973). Marsh plants did not survive in areas with pH values below 3.0. Liming at the rate of 26900 kg ha^{-1} (24,000 lbs ac^{-1}, in addition to the effects of tidal flushing and soil saturation, raised pH values above 4.0 and marsh vegetation was successfully established. The pH of acid soils tends to increase when they are flooded and become reduced (Gambrell and Patrick 1978). Problems with these kinds of soils can be avoided by keeping them saturated to prevent oxidation or by liming if oxidation occurs. A decrease in pH of soil samples upon air drying is an indicator of potential acid sulfate soils.

Soil testing should be done during the planning stages of marsh restoration or creation projects.

Revegetation

All of the site characteristics previously discussed should be taken into consideration when establishing or encouraging establishment of tidal marsh vegetation. Perhaps the two most important factors are elevation with respect to tidal regime and salinity. Elevation affects frequency and length of time of inundation, which, in turn, determines zonation of plants. Only a limited number of plant species can tolerate high salinities. Saline, regularly flooded, intertidal areas are revegetated with smooth cordgrass and saltmeadow cordgrass is planted in the occasionally flooded zone immediately above mean high water. Species most frequently planted on brackish-water sites include smooth cordgrass at the lower elevations, big cordgrass, saltmeadow cordgrass, black needlerush and saltgrass. For initial stabilization, smooth cordgrass may be planted at sites with lower salinities and at higher elevations than where it normally occurs. This is done because of availability of plants and on the assumption that plant communities adapted to the environment will eventually colonize the site.

Information and experience in freshwater wetland creation is more limited (Wolf et al. 1986) and marsh creation sites in inland waterways often revegetate naturally. There is the potential for propagation of a number of different species since species diversity is higher in freshwater marshes. Odum et al. (1984) state that freshwater tidal marshes of he Mid-Atlantic and Georgia Bight regions may contain as many as 50 to 60 plant species at a single location. Descriptions of propagation methods for a number of these species are reported by Kadlec and Wentz (1974) and the U.S. Army Engineer Waterways Experiment Station (Environmental Laboratory 1978). Garbisch

A.

B.

Figure 4. a) A shoreline at Pine Knoll Shores, North Carolina in May 1974, one month after transplanting smooth cordgrass.

b) Pine Knoll Shores after three growing seasons.

A.

B.

Figure 5. a) A borrow pit, near the Pamlico estuary in North Carolina, graded and planted with smooth cordgrass and big cordgrass in May 1980.

b) The same area, July 1, 1982.

Table 3. Results of analyses of soil samples from two brackish-water marsh creation sites in North Carolina [a]. One site was a borrow pit, the other was a pine forest upland graded to intertidal elevations.

		Mg dm^{-3}					Organic Matter
Site	pH	NH_4-N	P	K	C	Mg	(%)
Borrow Pit							
topsoil	4.4	72	5	66	349	106	3.5
subsoil	4.7	24	3	55	267	177	0.1
Graded upland	4.8	3	1	58	87	189	<0.1

[a] Analyses were done by the North Carolina Department of Agriculture Soil Testing Laboratory using the Mehlich 3 extractant to determine P, K, Ca, and Mg and Mehlich 1 for NH_4-N.

(pers. comm. 1987) has had success with establishment of softstem bulrush (Scirpus validus), arrow arum (Peltandra virginica) and pickerelweed (Pontederia cordata) at the lower elevations of tidal freshwater marshes; and common threesquare (Scirpus americanus) and rice cutgrass (Leersia oryzoides) at higher elevations. Methods were reported for successful establishment of peat-pot seedlings of pickerelweed along a shoreline in the uppermost freshwater region of Chesapeake Bay, Maryland (Garbisch and Coleman 1977). Tidal freshwater marshes can become established rapidly on suitable sites without planting. A dredged material site at Windmill Point, James River, Virginia became vegetated by natural colonization 6 months after construction (Lunz et al. 1978, Newling and Landin 1985).

Availability of plant propagules is often a problem in creation and restoration of tidal marshes. Ideally, seed or transplants should come from near the planting site because of the possibility of genetic variation. Adaptation of plants to local conditions often results in the formation of ecotypes, especially in populations with wide geographic range (Kadlec and Wentz 1974). There is considerable morphological and physiological variation among populations of smooth cordgrass (Seneca et al. 1975) and local populations are likely to be better adapted to their environmental conditions than are populations further north or south. Ecotypic variation is also documented for Typha spp. (cattails) (Kadlec and Wentz 1974). It is probably best to avoid using planting stock from other regions for most species until more is known about ecotypic variation.

Several publications are available that outline planting methods (Woodhouse et al. 1974, Garbischet al. 1975, Woodhouse 1979, Broome et al. 1981 and Lewis, R.R. 1982). Reference to these publications and personal experience were used in the summary of revegetation methods for salt and brackish-water vegetation that follows. Practical techniques of propagation and revegetation have been demonstrated for the species that are discussed and many of the planting techniques and principles described could be applied to other vegetation. Any species of plant may be transplanted to a new site at some level of effort and expense, but there are relatively few tidal wetland plant species that are routinely and economically propagated on a large scale.

Smooth Cordgrass (Spartina alterniflora) --

Revegetation with smooth cordgrass has been studied and established successfully more than any other native intertidal vegetation of the southeastern United States. It may be propagated from seeds, by digging from natural stands or produced in nurseries.

Seeding -- Seed production is most abundant in recently colonized open stands or along edges such as creek banks. Seeds mature and are ready for harvest in late September in the northern part of the region. Maturity progresses from north to south and may be as late as November in Georgia, but may vary within stands and from year to year. Harvesting should be done before seeds are lost by shattering, but not before maturity (Fig. 6). Seed heads can be clipped with knives, clippers or any mechanical aids available. Store the seedheads moist in refrigeration for 3-4 weeks before threshing. This results in easier separation of seeds from stems. The threshed seeds should be stored in plastic containers filled with seawater or artificial

Figure 6. Mature smooth cordgrass seed heads.

seawater so that the seeds are submerged. This inhibits germination during storage and loss of viability that occurs if the seeds are allowed to desiccate. Even when seeds are stored properly, storage life is only about one year; consequently, annual seed harvests are necessary.

Smooth cordgrass can be established by direct seeding on sites protected from waves, but successful germination and growth is generally limited to the upper half of the intertidal zone. Before planting, a seedbed should be prepared using a rototiller, harrow, rake or other tillage implement. Sow seeds evenly at the rate of 100 viable seeds per square meter and till again to incorporate to a depth of 2 to 3 cm. When seeding is successful, complete cover is attained by the end of the first growing season.

Direct seeding is usually the most economical method of propagation, but is impractical on sites that receive even moderate wave energy. Another negative aspect of this method is that the quantity of seeds needed for large areas are often not available. If seed supply is limited, the seeds are more effectively utilized by growing potted seedlings.

Field dug plants -- Excellent transplants can be dug from recently established stands on sandy material such as dredged material disposal sites, around inlets where sand is accumulating, or along the edges of marshes. Avoid old marshes which have a dense root mat. Plants from new stands are more vigorous, have larger stems and are easier to dig and separate. Plants are dug by loosening with a shovel and separating individual stems by hand. Good plants have large stems, with small, actively growing shoots and rhizomes attached and a well developed root system. Ideal height is about 30 cm but a wide range may be used. An individual can dig 200 to 500 plants per hour depending on conditions at the site. Removal of plants from young stands on sandy soils causes minimal damage since the areas revegetate quickly from remaining rhizomes and shoots. After plants are dug, it is important to prevent drying. Roots should be kept packed in moist sand until transplanted. A nursery site may be established if a suitable sandy, unvegetated, intertidal area in available. Such an area transplanted on one-meter centers in the spring produces good transplants for the following year's planting season.

Plugs, planting units including a soil core, root mass and associated stems, are in most cases less desirable than single stem transplants. More labor is required to harvest, transport and transplant plugs but this may be the best alternative if planting stock must be obtained from old marshes growing on peat, clay or silty soils.

Seedlings -- A good method of producing planting stock is to grow seedlings in pots or flats either in a greenhouse or outdoors during warm weather (Fig. 7). Plastic tray pack liners with 36, 5-cm square compartments are good containers for plant production (Fig. 8). Plastic is better than peat pots because roots and rhizomes are confined in individual planting units. Roots and rhizomes grow through peat pots, creating a solid mat. Seedlings may be grown in a commercial potting mix or a mixture of equal parts sand, peat and sterilized topsoil. Treating seeds with a fungicide before planting is a good practice. One effective method is soaking the seeds in a solution of 25% household bleach for 10 to 15 minutes and rinsing with tap water. Plant 5 to 10 seed per pot or compartment and keep the potting soil wet with daily watering. Flooding is not necessary. Three to four months are required to grow seedlings in a greenhouse to the appropriate size for planting in the field. Seeding in February produces good seedlings for transplanting in April or May. Fertilize as needed with Hoagland's solution or a commercial liquid or granular fertilizer according to rates recommended on the package. Diseases may be a problem when seedlings are small, particularly during long periods of cloudy weather. Fungicide sprays such as Banrot are effective in preventing disease damage to seedlings.

It has been recommended that seedlings to be transplanted in areas with salinities above 15 ppt be pre-conditioned by growing in or applying solutions of salt water (Garbisch et al. 1975). However, results of recent experiments indicated no advantage of preconditioning with salt water. Seedlings were grown in solutions of O, 10, 20, and 30 ppt sodium chloride in a greenhouse and transplanted to a dredged material site with interstitial water of 25-30 ppt and tidal inundation with water of about 30 ppt. Survival of transplants was near 100% and there was no difference in appearance and growth of plants due to the salinity pre-treatments (Broome, unpublished data).

There are several advantages to using pot-grown seedlings: (1) there is very little planting shock since an intact root system is transferred to the field and growth resumes quickly. Survival is virtually 100% on favorable sites; (2) disturbance of natural stands is avoided; (3) pot-grown seedlings provide a source of plants when suitable digging sites are not available; and (4) seedlings can be held longer than dug plants if there are delays in site preparation. Disadvantages are: (1) cost (40 to 60 cents per plant); (2) advance planning is necessary to allow time to grow plants; (3) plants growing in pots or flats are bulky and inconvenient to

Figure 7. Greenhouse production of three species of cordgrass.

Figure 8. Big cordgrass seedlings approximately three months from time of seeding.

transport. Racks or shelves in a covered trailer or truck are required for hauling large numbers of plants and transportation by boat is also inconvenient; and (4) the potting media does not contain the marsh soil flora and fauna that would be present on the roots of field-dug transplants. This may delay development of a marsh as a total ecosystem.

Transplanting techniques -- Techniques of transplanting field-dug or pot-grown plants are very similar and depend on site conditions. Mechanization is feasible on large accessible sites with soil material that will support equipment and is relatively free of stumps, roots or other debris. A farm tractor with a transplanter used for tobacco or vegetable plants is quite efficient, particularly on sandy soils (Fig. 9). Some machines may require modification of the row openers to work on certain soil materials. Models are available that accommodate potted seedlings as well as field-dug plants and fertilizer distributors may also be attached. Mechanical planters produce a more uniform planting and are much faster and more economical than hand planting.

Hand planting, however, is the method most often used for intertidal vegetation. A hole is opened with a spade or dibble to a depth of about 15 cm, the plant is inserted by a second worker, and the soil firmed around the plant (Fig. 10). As with machine planting, a better job can be done at low tide when there is no water on the surface. Portable power augers are useful for opening holes on compact soils (Garbisch et al. 1975). If slow-release fertilizer is being used, it should be placed in the hole before the plant is inserted. Adequate planting depth and firming the soil around the plant are important to prevent plants from floating out or being dislodged by wave action.

Spacing of transplants is an important consideration because it affects numbers of plants required, probability of success on exposed sites and how rapidly cover is achieved. The number of plants, cost of plants and labor requirements increase as density is increased. For example, to plant one hectare on 1 m spacing, 10,000 plants are required (4,047 plants per acre), while planting on 0.5 m spacing requires 40,000 plants per hectare (16,196 plants per acre). On exposed sites such as eroding shorelines, closer spacing increases the chance of successful establishment. Spacing of 45 to 60 cm (1.5 to 2 ft) has been found to be adequate. On favorable, protected sites, a 1-meter spacing is adequate to provide complete cover of smooth cordgrass approximately one year after transplanting. On large protected sites where rapid cover is not critical, even wider spacing may be acceptable to reduce planting costs. Seeds produced by the transplants, in addition to spread by rhizomes, are often an important factor in producing a complete cover when plants are widely spaced.

Optimum planting dates for smooth cordgrass are from April 1 to June 15, however, earlier or later planting dates may be necessary due to extenuating circumstances. Earlier planting increases the chances of storm damage and field dug plants are more difficult to obtain. Later dates shorten the growing season available for initial establishment and spread of plants, making them more susceptible to winter erosion on exposed sites.

The benefits of fertilization vary from site to site depending on the availability of nutrients in the soil. Analysis of a soil sample by the state agricultural soil testing lab is useful in determining the need for phosphorus fertilization and the pH, but soil tests are generally not effective in predicting the need for nitrogen. Transplants will benefit from fertilization during the first growing season on most sites, which is important on sites that are subject to erosion. The method of fertilization that has become most widely used (and is very effective) is placement of the slow release fertilizer Osmocote in the planting hole at the rate of about 1 oz (30 g) of material per plant. The 14-14-14 analysis with a 3-month release time is most often used, although several other analyses are available. If conventional soluble fertilizer materials are used, do not place them in direct contact with the plant root because of the salt effect. Open a hole about 5 cm from the plant for the fertilizer. Conventional fertilizer is much cheaper, but the nitrogen is available over a shorter period of time. With proper management and repeat applications conventional fertilizers can be as effective as slow release materials. Suggested amounts of fertilizer are listed in Table 4.

An alternative to fertilizing individual plants is broadcasting fertilizer before planting and incorporating it into the soil by discing. Apply 112 kg ha^{-1} (100 lbs ac^{-1}) N in the ammonium form (ammonium sulfate or urea) and 50 to 100 kg ha^{-1}P (100 to 200 lbs ac^{-1} P_2O_5) as concentrated superphosphate (45% P_2O_5). The nitrate form of N is subject to denitrification and loss under flooded conditions and it is likely that plants adapted to flooded soils are better adapted to utilizing the ammonium form of N. Additional fertilization later in the growing season or in subsequent growing seasons may be needed until the stand is established and recycling of nutrients within the marsh system is adequate. The need for additional fertilization can be determined by general appearance, color and growth of the plants. Apply the fertilizer by broadcasting evenly to the soil surface at low tide at the rate of 112 kg ha^{-1} (100 lbs ac^{-1})N and 49 kg ha^{-1}P (100 lbs ac^{-1} P_2O_5). If additional

Figure 9. Planting smooth cordgrass on a dredged material disposal area with a small tractor and mechanical transplanter.

Figure 10. Hand planting field-dug smooth cordgrass transplants.

Table 4. Suggested rates of fertilizer to apply in the planting hole (slow release fertilizers) or in a hole several inches from the plant (soluble fertilizer materials).

Analysis Fertilizer Material	N	P_2O_5	K_2O	amount per plant (g)
[a] OSMOCOTE (7-9 month release)	18	6	12	15-30
[a] OSMOCOTE (3 month release)	14	14	14	15-30
[a] Mag Amp+	7	40	6	15
[a] OSMOCOTE	14	14	14	15
[b] ammonium sulfate+	21	0	0	15
[b] concentrated superphosphate	0	46	0	10
[b] mixed fertilizers	10	10	10	25-50

[a] Slow-release materials
[b] Soluble materials

applications are needed only the N may be necessary since P is adsorbed and retained by the soil.

Salt meadow cordgrass (S. patens) --

Field-dug transplants or potted seedlings are suitable for transplanting. Vigorous plants should be dug from relatively young, open stands and held in moist sand until planted. The Soil Conservation Service has also successfully produced saltmeadow cordgrass transplants in upland field nurseries. The Cape May, New Jersey plant materials center now has trials in progress to select and release a superior cultivar.

Direct seeding is usually not a viable alternative for saltmeadow cordgrass; however, seedlings can be produced in pots or flats. Collect seeds when mature in late September, thresh and store dry in a refrigerator or cold room. Grow the seedlings in flats using the same methods as described for smooth cordgrass. Seeds tend to take longer to germinate than smooth cordgrass.

Use the same transplanting techniques as described for smooth cordgrass except for fertilization. Placing Osmocote in the planting hole is certainly effective but is not necessary because of the location of saltmeadow cordgrass above mean high water. Conventional sources of N and P can be broadcast on the soil surface at the same rates as described for smooth cordgrass. This can be done immediately before or after transplanting, or preferably, several weeks after planting when a root system has developed to utilize the nutrients more efficiently. Saltmeadow cordgrass responds vigorously to N fertilization and may benefit from several applications during a growing season.

Big Cordgrass (S. cynosuroides) --

Production of potted seedlings is the best method of propagating big cordgrass. Field-dug plants are difficult to obtain and survival of transplants is poor unless a site can be found with young seedlings 20 to 30 cm in height. These make good transplants but are rarely available. Seed should be collected when mature in mid to late September. Big Cordgrass seeds mature 2-3 weeks earlier than smooth cordgrass. Store the seed refrigerated in estuarine water or artificial seawater diluted to 10 ppt. Grow seedlings using the methods described for smooth cordgrass. Big cordgrass seedlings germinate quickly, grow rapidly and are quite vigorous.

Use the planting techniques described for smooth cordgrass. One exception is spacing if a quick cover is needed. The rhizomes of big cordgrass are shorter, resulting in slower spread and a bunchy growth habit. A spacing of 60 cm is the maximum that should be used to insure complete cover in the second growing season.

Black Needlerush (Juncus roemerianus) --

Survival of field-dug transplants is usually unsatisfactory. If this method is used, include at least one growing rhizome tip in the planting unit. Pot-grown seedlings are, however, an excellent method of propagating black needlerush. Collect clusters of seeds when mature in mid to late June and store refrigerated in paper bags. The seeds are very small and the seedlings grow slowly. Sow the seeds in November on the surface of potting soil in a flat in a greenhouse. Cover with a thin layer of potting soil. When seedlings are 3-5 cm in height, transfer individual seedlings to tray pack type flats. Fertilize and care for the seedlings as described for smooth cordgrass. They should be ready for transplanting in May or June.

Follow planting techniques as described for smooth cordgrass. Use a maximum of 60 cm spacing since black needlerush grows and spreads slowly. It responds to fertilization on soils that are very low in N and P, but its nutrient requirements for optimum growth are apparently lower than the Spartina species.

Saltgrass (Distichlis spicata) --

Saltgrass may be established by transplants, rhizomes, seeds and plugs (Environmental Laboratory 1978). Optimum transplanting dates are January through March. Seeds are harvested in the fall when mature and should be stored refrigerated because of a low-temperature-after-ripening requirement (Amen et al. 1970). Seedlings may be grown in pots or flats as described for smooth cordgrass.

Reintroduction of Fauna

It is logical to predict that, if tidal marsh habitat is created with a physical environment, hydrology and vegetation similar to natural marshes in the area, fauna that utilize such habitat will colonize the new marsh. The time required to accomplish this is unknown and undoubtedly would vary with marsh type and species of fauna. Tidal flooding is one transport mechanism which expedites and facilitates introduction of fauna to a new marsh. The roots and attached soil of field-dug transplants are a source of microbes, meio- and macrofauna. Birds, reptiles, amphibians, and mammals can migrate to the new habitat.

One can always successfully argue that there are differences between a marsh that is hundreds of years old and a recently created marsh. One obvious difference is organic matter in the soil, which may be high in nearby natural marshes and near zero in the created marsh. This difference may or may not affect the value and function of the created marsh.

One criticism of marsh creation and mitigation is that very little is known about whether created marshes function like natural marshes (Race and Christie 1982). In their review, Race and Christie found limited data comparing fauna of created and natural marshes. Cammen (1976a, 1976b) found greater macroinvertebrate biomass in natural marshes than three-year old planted marshes and there were differences in species composition. Those planted marshes were sampled after 15 years and results showed that numbers and species of infauna were similar to natural marshes. Reimold et al. (1978) recorded increases over time in numbers of crab burrows in a planted marsh in Georgia. Microbial biomass at the same site was 2 to 6 times lower than at natural areas two years after planting.

A study in Texas comparing abundance of macrofauna in transplanted marshes 2-6 years old with natural marshes, found consistently lower densities of brown shrimp, grass shrimp, pinfish, and gobies in the planted marshes. Juvenile blue crab densities were the same in both marsh types (Minello et al. 1986).

Abundance and species composition of fish and mobile invertebrates are being compared in planted and natural marshes in North Carolina as part of current research by the National Marine Fisheries Service. Sampling during the first year after planting S. alterniflora on dredged material indicated that species composition was different and the organisms were less abundant than in natural marshes. Similar results were found in a three-year-old S. alterniflora marsh created by grading an upland to intertidal elevations and transplanting. However, no difference was found in abundance or species composition of fish and mobile invertebrates utilizing a twelve-year-old planted marsh when compared to nearby natural marshes (Thayer, National Marine Fisheries Service, pers. comm.).

Newling and Landin (1985) reported results of long-term monitoring of seven Corps of Engineers marsh creation sites. The created marshes were found to be equal to nearby natural marshes in faunal species diversity and abundance, including aquatic organisms.

The structure of a man-made marsh-creek system was compared to three natural creeks in the Pamlico estuarine system in North Carolina by determining temporal changes in species composition of finfish and benthic invertebrate communities (West and Rulifson 1987). The man-made system was 2 to 4 years old over the study period. Finfish communities and benthic invertebrate communities (species composition and seasonal changes in species abundance) were also similar in the man-made and control creeks. To provide some measure of marsh function, growth and survival of spot (Leiostomus xanthurus) held in cages in each system was measured. Increase in weight of the caged spot in the man-made marsh system equalled or exceeded that of the control creeks.

More research is needed on the rate at which the faunal component colonizes created marshes and the need for introduction of certain fauna.

Buffers and Protective Structures

Wave protection structures are

beneficial where wave energy is a limiting factor in establishment and long-term stability of vegetation. Temporary protection during the establishment period, when the planting is most vulnerable to wave energy, is critical to success in many cases. Wave protection has been successfully accomplished with earthen dikes, sandbags, tire breakwaters, erosion-control mats, and plant rolls (Webb 1982; Allen et al. 1986; U.S. Army Corps of Engineers 1986).

A sandbag dike was effective in protecting a 4-ha marsh planting site at Bolivar Peninsula, Texas, but it was recommended that more cost effective alternatives be considered (Allen et al. 1978). A sand dike around the perimeter of an 8-ha freshwater marsh site at Windmill Point, Virginia provided protection for development of a freshwater tidal marsh (Environmental Laboratory 1978).

Floating or fixed tire breakwaters have been found to provide protection for vegetation. In Mobile Bay, Alabama, Smooth cordgrass sprigs behind a floating tire breakwater had a 56% survival rate, those behind a fixed breakwater a 24% survival rate, and those with no protection, a 4.0% survival rate (Allen and Webb 1982). A horizontal mat of floating tires with foam flotation was ineffective in protecting a shoreline planting of smooth cordgrass along the Neuse River in North Carolina (Broome, unpublished data). The draft, or emersed depth, of the horizontal mat was too shallow to be effective during high storm tides when most erosion occurred, while the water was too shallow for tires in the vertical position under normal conditions. The effectiveness of a floating tire breakwater is determined by its length, width and draft. The draft should be greater than half the height of significant waves (Ross 1977). A major disadvantage of using tires is that they are aesthetically undesirable and provision must be made for removal when they have served their purpose. Properly designed floating breakwaters can be towed to another site for re-use.

One effective method of protecting shoreline plantings is construction of a wooden breakwater parallel to the shore. Distance from the shore and height of the breakwater depend on water depth.

Other types of plant protection include various types of matting anchored to the soil surface and burlap plant rolls (Allen et al. 1984). Plant rolls, burlap bundles and plants sprigged through Paratex mat were found to be promising methods of erosion control by holding plants in place on dredged material in Mobile Bay, Alabama. However, a replicated experiment on three dredged material islands in North Carolina showed no benefit to using plant rolls as compared to conventionally transplanted smooth cordgrass (personal observation).

Other hazards that might require protective action include human foot and vehicular traffic, grazing animals and blowing sand. Signs and fences can be used to discourage human use and fencing or trapping and removal can be used to exclude some animals. Sand fencing and transplanting dune vegetation are effective in intercepting blowing sand.

Long-Term Management

The goal of marsh creation is to establish a wetland that is like the natural system it is designed to emulate; or, one that will become like that system through succession of the flora and fauna. Ideally such a system should be self sustaining and maintenance free. However, maintenance is often necessary to insure success, particularly in the first few years after transplanting. One of the first maintenance requirements on shorelines exposed to wave action is replacement of plants that may be washed out. Wrack or litter along drift lines should be removed if there is danger of smothering plants. Additional fertilization may be necessary on infertile sites even during the second or third growing season. Response to fertilization and rate required can be determined on a small scale on test plots before fertilizing an entire planting.

Invasion by undesirable plant species may be a problem on fresh and brackish water sites. One of the most ubiquitous weeds is common reed (*Phragmites australis*) (Daiber 1986). It is an aggressive plant found throughout the world in freshwater marshes and it is able to withstand moderate to high salinities. Common reed can quickly invade spoil or other areas where vegetation has been disrupted by mechanical or other means. Its ability to out compete and eliminate other vegetation and its low wildlife value has made it undesirable and unpopular.

One means of discouraging invasion of unwanted vegetation on marsh creation sites is to insure rapid cover of the transplanted vegetation by close spacing and other good management practices. Other control methods are cutting, draining, saltwater flushing and herbicides. Work by the Delaware Department of Natural Resources has demonstrated that the herbicides Rodeo and Roundup (glyphosphate) are effective in controlling common reed. Because of lack of specificity and the potential of environmental damage, herbicides should be used in a tidal marsh only if no other alternatives are available (Daiber 1986).

MONITORING

Monitoring created or restored marshes can range from periodic visual observations to detailed scientific measurements of structure and function of the system and comparison with similar natural systems. Monitoring is desirable to measure success of a project and to determine additional needs or inputs such as replanting, fertilization control of undesirable plants, removal of debris, control of wildlife pests, and human traffic. Choosing appropriate methods for monitoring depend on the characteristics of the site, type of vegetation, the objectives to be accomplished and time and resources available. The degree of success reported for a project is likely to affect permitting decisions on other similar projects.

A photographic record is an effective, low-cost method of monitoring a planting site. Permanent stations should be marked at the site for photographs at each visit to record growth and development of the vegetation. Aerial photography is generally more difficult and expensive to obtain, but is quite useful. Remote sensing methods have also been used to estimate biomass of wetland vegetation by correlation with spectral radiance (Daiber 1986). Photographs provide a perspective of the scope and nature of a project with which to supplement numerical data. Videotape might also provide a useful means of recording development of a planted site.

Growth measurements of vegetation are important in recording development. Zonation of vegetation in tidal marshes due to elevation with respect to tidal inundation adds to sampling problems. For example, with smooth cordgrass planted along shorelines, there are often three zones with different height and biomass. The seaward edge typically has poor growth, then a zone of good growth in the middle of the planting and a zone of poor growth at higher elevations. The pattern of growth may be reversed at other locations but there is usually some type of zonation. Because of the effects of zonation, stratified random sampling is often more appropriate than random sampling over an entire vegetation type. The stratified method involves sampling within each elevation zone that can be delineated visually. In the case of the shoreline described above one might choose to sample only the zone of maximum growth or all three zones depending on the objectives. Before deciding on any sampling scheme, consult a statistician.

The first measurement might be an estimate of survival rate four to six weeks after transplanting. This provides a measure of initial planting success, including quality of the planting material, planting methods, and the effects of wave action. It also determines the need for replanting. On large areas selected lengths of rows may be randomly chosen and counted to estimate survival.

Growth measurements at the end of the growing season are often used as a measure of success. Useful parameters include height, number of stems, cover, basal area, aboveground biomass (clipped plots), and digging and coring to measure belowground biomass. At the end of the first growing season, individual hills may still be identifiable and sampled as such. When cover is evenly distributed quadrat sampling should be used. Quadrats 0.25, 0.50 or 1.0 m^2 are often used depending on the homogeneity of the vegetation. Destructive sampling (clipping and weighing) probably provides the most accurate measure of growth. For Spartina species, clipping samples at the end of the growing season apparently has very little effect on subsequent growth; however, it should be minimized on exposed areas subject to erosion. Sampling belowground biomass can be very damaging to stands on exposed sites. Clipping does affect growth of black needlerush. Plots clipped in autumn often remain bare or are invaded by saltgrass during the next growing season.

The following measurements of plant growth were adopted by Woodhouse et al. (1974) to characterize success of plantings of Spartina:

1. Aerial dry weight - For first year growth, individual hills were randomly selected, plants were clipped at the sediment surface, dried in a forced air oven at 70°C and weighed to the nearest gram. At the end of the second growing season and thereafter, samples consisted of quadrats 0.25 m^2 in size. Only the live material (current year's growth) was retained. If other plants (invaders) are present in the samples, they should be separated by species, identified, dried and weighed. This provides a measure of plant succession over time. Peak standing crop biomass is an underestimate of net annual primary productivity (NAPP) because it does not account for mortality, decomposition, or growth occurring after the peak. Several more time consuming methods have been utilized to more accurately estimate NAPP (Shew et al. 1981).

2. Belowground dry weight - For first year transplants, roots and rhizomes attached to the clipped plants were simply dug from the soil and washed on a 2 mm screen. The plant

material remaining on the screen was oven-dried and weighed. After complete cover was achieved, belowground material was sampled by taking cores from the clipped quadrats. Cores were taken to a depth of 30 cm with a stainless steel coring tube, which had an inside diameter of 8.5 cm. Cores were separated into 0-10 and 10-30 cm lengths, washed on a 2 mm sieve, the plant material was oven-dried and weighed. Some researchers attempt to separate live and dead material.

3. Number of stems - The number of stems in a clipped sample was counted.

4. Number of flowering stems - The number of flowers or seed heads in each sample was recorded.

5. Height - The height was measured either before or after clipping. First year transplants were measured from the base of the plant to the highest point. The five tallest stem heights were measured in quadrat samples.

6. Basal area - The area occupied by stems at ground level was determined by holding the clipped stems tightly bunched and measuring their cross sectional diameter. Cross sectional area was calculated and reported per hill or per unit area.

Other useful measurements include percent cover, percent cover by species, number of colonizers, and qualitative observations such as vigor and color (U.S. Army Corps of Engineers 1986).

The number of samples required to adequately represent a given area or vegetation type varies. A general rule-of-thumb is a minimum of 15 samples. Interpreting the results requires sampling a nearby reference marsh with similar vegetation or relying on reported literature values for comparison.

The length of time to sample varies with type of vegetation and location. A three to five year time frame is reasonable to determine if growth of vegetation on a planted site is comparable to similar natural marshes. A ten-year study of a planted smooth cordgrass marsh in North Carolina showed that the aerial standing crop was greater than an adjacent natural marsh by the end of the second growing season and was equal to the natural marsh throughout the remainder of the study period. Four growing seasons were required for belowground standing crop to equal the natural marsh (Broome et al. 1986).

The broader question of whether created or restored marshes have equal value and perform the same functions as natural marshes is more difficult to evaluate. In addition to primary production, development of faunal communities and chemical and physical characteristics of the soil and nutrient cycling must be evaluated. A long-term monitoring effort has been conducted by the U.S. Army Corps of Engineers on corps-built wetlands (Landin 1984, U.S. Army Corps of Engineers 1986).

RESEARCH NEEDS

Information and research needs related to creation and restoration of tidal wetlands in the southeastern United States can be divided into the following general categories: (1) site selection, design and preparation; (2) plant propagation techniques and cultural practices; and (3) documentation of development of biological communities and functional processes in created systems. This includes determining the time required for development to occur, the successional patterns, and comparison of communities and processes with similar natural systems.

Site Selection, Design, and Preparation

A primary need in site selection is improved methods of predicting the probability of success of plantings exposed to wave energy. A better understanding is needed of how the effects of fetch on planting success are modified by depth of offshore water, tidal range, and physical properties and erodibility of the sediments. Cost effective methods for protecting plantings from wave energy need further study. These include structures such as breakwaters (floating or fixed) and design of dredged material disposal sites. Habitat displacement is an important issue to consider. Marsh creation results in the loss of upland habitat when grading is done to produce intertidal elevations. Bottom or mud-flat habitat is lost when fill is used to create marsh.

Plant Propagation Techniques and Cultural Practices

Methods for propagation and planting of smooth cordgrass, saltmeadow cordgrass, big cordgrass and black needlerush are well documented. These methods can be found in published reports and have been applied over a wide geographic area and range of conditions. There is less published material and apparently less information and experience in creating tidal freshwater marshes.

Topics that warrant further investigation are listed below:

1. Plant propagation techniques and planting methods should be improved to reduce costs.

2. A better understanding of the physical environment required for optimum growth of big cordgrass and black needlerush is needed. The elevation in relation to tidal inundation, soil conditions, and nutrient requirements are not well understood.

3. More work on source and rate of nitrogen fertilization is needed over a variety of locations and soil types. The effectiveness of urea when applied at planting and as a topdressing needs to be tested. Sulfur-coated urea might be a lower cost alternative to Osmocote as a source of slow-release N for a starter fertilizer.

4. The role of mycorrhizae in mineral nutrition and growth of marsh vegetation should be investigated. The growth of greenhouse-grown seedlings may be affected by the absence of mycorrhizae when transplanted on soils created by grading upland sites.

5. The necessity of pre-conditioning pot-grown smooth cordgrass with salt water before transplanting to field conditions where salinities are 20-30% should be tested further. This is a recommendation found in the literature; however, preliminary tests in North Carolina, both in the field and greenhouse have shown that seedlings grown without salt treatment survived well when transferred to a high-salinity environment.

6. Restrictive layers in dredged material, such as old marsh surfaces or clay layers, often cause salinities of planting sites or parts of planting sites to exceed the tolerance of marsh vegetation. This occurs when drainage is restricted and evaporation concentrates salts. Means of recognizing and correcting this problem should be investigated.

7. Methods and the necessity for planting freshwater tidal vegetation should be further investigated. Freshwater tidal marsh constitutes only 11% of the total tidal marsh in the region; consequently, less experience has been gained with this type of vegetation. Lunz et al. (1978) concluded from results at the Windmill Point, Virginia site that tidal freshwater marshes establish rapidly on suitable sites without planting. Is this generally the case, or do certain sites need to be planted to prevent invasion by undesirable species or erosion? The best plants from the standpoint of ease of propagation, planting and desirability, should be determined.

Documentation of Marsh Development

The feasibility of creating tidal marshes has been adequately demonstrated although techniques continue to be refined and improved. Despite this feasibility, the question of how these marshes compare structurally and functionally to similar natural systems has not been answered to the satisfaction of many scientists and policy makers. Many authorities take the position that new or restored wetlands ought to meet a functional test (Larson 1987) since wetlands are not only plant communities, but ecosystems that provide specific functions. Evaluating functions of created tidal wetlands is difficult in light of the controversy that still exists over the contribution of natural marshes to estuarine processes (Nixon 1980). Variability among natural marshes of the same type and between different types of marshes also makes comparisons difficult. For example, a smooth cordgrass stand established on sand in an area where natural marshes are relatively young will likely be comparable to the natural marsh for most measurements in a few years. In contrast, in areas where natural marshes are old and have accumulated peat several meters thick, a marsh created by grading an upland mineral soil or a borrow pit will lack such peat and differ from the natural marshes. Soil organic matter concentrations, soil physical and chemical properties, and hydraulic conductivity will remain measurably different for many years.

If differences between created and natural marshes are observed or measured, the question remains, are these differences important? Comparative data is important in assessing the impact of replacing natural marshes with created marshes, but decisions on permitting mitigation projects will ultimately require value judgments.

Comparative studies of created and natural marshes are needed to provide data to be used as a basis for making mitigation decisions. Perhaps of even greater importance, such studies can provide insights into the broader role of both natural and created tidal marshes in estuarine ecosystems. Created marsh systems can be used for studies beginning at time zero and following over time the development of primary production, successional changes, changes in the flora and fauna, and changes in soil chemical and physical properties. Comparisons can be made with bare unplanted areas as well as with natural marshes.

Some research topics that would provide interesting comparative data and valuable basic information are listed below:

1. Evaluating plant productivity and succession over significant time periods (5-10 years) for different marsh types and locations.

2. Determination of nutrient flux, accretion of sediments, accumulation of mineral nutrients and organic carbon, nutrient cycling, and soil development.

3. Study of the abundance and production of benthic micro- meio- and macrofauna.

4. Evaluation of the hydraulic conductivity of marsh soils and the exchange of interstitial water with estuarine water.

5. Evaluation of habitat value, particularly for fishes.

6. Develop methods for practically and economically evaluating the success of marsh creation and restoration projects by identifying key indicator species and processes to be measured.

LITERATURE CITED

Adamus, P.R., E.J. Clairain, Jr., R.D. Smith, and R.E. Young. 1987. Wetland Evaluation Technique (WET); Vol. II Methodology. Operational Draft Technical Report Y-87. U.S. Army Engineer Waterways Expt. Sta., Vicksburg, Mississippi.

Allen, H.H., E.J. Clairain, Jr., R.J. Diaz, A.W. Ford, L.J. Hunt, and B.R. Wells. 1978. Habitat Development Field Investigations, Bolivar peninsula. Marsh and Upland Habitat Development Site, Galveston Bay, Texas. Summary report. U.S. Army Engineer Waterways Expt. Sta. Tech. Rep. D-78-15.

Allen, H.H. and J.W. Webb, Jr. 1982. Influence of breakwaters on artificial salt marsh establishment on dredged material, p. 18-35. In F.J. Webb (Ed.), Proc. of the Ninth Annual Conf. on Wetland Restoration and Creation, Hillsborough Community College, Tampa, Florida.

Allen, H.H., J.W. Webb, and S.O. Shirley. 1984. Wetlands development in moderate wave-energy climates, p. 943-955. Proc. Conf. Dredging '84, Waterway, Port, Coastal and Ocean Division ASCE/Nov. 14-16, 1984. Clearwater Beach, Florida.

Allen, H.H., S.O. Shirley, and J.W. Webb. 1986. Vegetative stabilization of dredged material in moderate to high wave-energy environments for created wetlands, p. 19-35. In F.J. Webb (Ed.), Proc. of the Thirteenth Annual Conference on Wetland Restoration and Creation, Hillsborough Community College, Tampa, Florida.

Amen, R.D., G.E. Carter, and R.J. Kelly. 1970. The nature of seed dormancy and germination in the salt marsh grass Distichlis spicata. New Phytol. 69:1005-1013.

Armentano, T.V. 1980. Drainage of organic soils as a factor in the world carbon cycle. BioScience 30:825-830.

Bloomfield, C. and J.K. Coulter. 1973. Genesis and management of acid sulfate soils, p. 265-336. In N.C. Brady (Ed.), Advances in Agronomy, Vol 25. Academic Press, New York.

Broome, S.W., W.W. Woodhouse, Jr., and E.D. Seneca. 1975. The relationship of Mineral Nutrients to growth of Spartina alterniflora in North Carolina. I. Nutrient status of plants and soils in natural stands. Soil Sci. Soc. Am. Proc. 39(2):295-301.

Broome, S.W., E.D. Seneca, and W.W. Woodhouse, Jr. 1982. Establishing brackish marshes on graded upland sites in North Carolina. Wetlands 2:152-178.

Broome, S.W., E.D. Seneca, and W.W. Woodhouse, Jr. 1981. Planting Marsh Grasses for Erosion Control. UNC Sea Grant College Publication UNC-SG-81-09.

Broome, S.W., E.D. Seneca, and W.W. Woodhouse, Jr. 1983a. Creating brackish water marshes for possible mitigation of wetland disturbance, p. 350-369. In J. Hernandez (Ed.), Proceedings: First Annual Carolina Environmental Affairs Conference of the Univ. of North Carolina. Institute for Environmental Studies, Chapel Hill, North Carolina.

Broome, S.W., E.D. Seneca, and W.W. Woodhouse, Jr. 1983b. Creation of brackish water marsh habitat, p. 319-338. In D.J. Robertson (Ed.), Reclamation and the Phosphate Industry. Florida Institute of Phosphate Research, Clearwater Beach, Florida.

Broome, S.W., E.D. Seneca, and W.W. Woodhouse, Jr. 1983c. The effects of source, rate and placement of N and P fertilizers on growth of Spartina alterniflora transplants in North Carolina. Estuaries 6:212-226.

Broome, S.W., E.D. Seneca, and W.W. Woodhouse, Jr. 1986. Long-term growth and development of transplants of the salt-marsh grass Spartina alterniflora. Estuaries 9:63-74.

Broome, S.W., C.B. Craft, and E.D. Seneca. 1988.

Creation and development of brackish-water marsh habitat, p.197-205. In J. Zelazny and J.S. Feierabend (Eds.), Proceedings of a conference: Increasing ourwetland resources. Nat. Wildl. Fed., Washington, D.C.

Cammen, L.M. 1976a. Abundance and production of macroinvertebrates from natural and artificially established salt marshes in North Carolina. Am. Midl. Nat. 96:487-93.

Cammen, L.M. 1976b. Macroinvertebrate colonization of Spartina marshes artificially established on dredge spoil. Estuarine Coastal Mar. Sci. 4:357-72.

Chung, H.C. 1982. Low marshes, China, p. 131-145. In R.R. Lewis (Ed.), Creation and Restoration of Coastal Plant Communities. CRC Press, Inc., Boca Raton, Florida.

Chung, H.C. and Zhuo Rongzong. 1985. Twenty-two years of Spartina anglica Hubbard in China. Journal of Nanjing University, Research Advances in Spartina p. 31-35.

Clairain, E.J., Jr., R.A. Cole, R.J. Diaz, A.W. Ford, R.T. Huffman, L.J. Hunt, and B.R. Wells. 1978. Habitat Development Field Investigations, Miller Sands Marsh and Upland Habitat Development Site, Columbia River, Oregon. Summary report. U.S. Army Engineer Waterways Experiment Sta. Tech. Rep. D-77-38.

Cowardin, C.M., V. Carter, F.C. Golet, and E.T. LaRoe. 1979. Classification of Wetlands and Deepwater Habitats of the United States. U.S. Fish and Wildl. Serv. FWS/OBS-79/31.

Daiber, F.C. 1986. Conservation of Tidal Marshes. Van Nostrand Reinhold Co., New York.

Dodd, J.D. and J.W. Webb. 1975. Establishment of Vegetation for Shoreline Stabilization in Galveston Bay. U.S. Army Corps of Engineers, Misc. Paper 75-6.

Environmental Laboratory. 1978. Wetland Habitat Development with Dredged Material: Engineering and Plant Propagation. Technical Report DS-78-16, U.S. Army Waterways Experiment Station, Vicksburg, Mississippi.

Gambrell, R.P. and W.H. Patrick, Jr. 1978. Chemical and biological properties of anaerobic soils and sediments, p. 375-423. In D.D. Hook and R.M.M. Crawford (Eds.), Plant Life in Anaerobic Environments. Ann Arbor Sci. Pub. Inc., Ann Arbor, Michigan.

Garbisch, E.W., Jr., P.B. Woller, and R.J. McCallum. 1975. Salt Marsh Establishment and Development. U.S. Army Corps of Engineers, Technical memorandum 52.

Garbisch, E.W., Jr. and L. B. Coleman. 1977. Tidal freshwater marsh establishment in Upper Chesapeake Bay: Pontederia cordata and Peltandra Virginica, p. 285-298. In E.R. Good, D.F. Whigham, and R.L. Simpson, Freshwater Wetlands: Ecological Processes and Management Potential. Academic Press, New York.

Gosselink, J.D. 1984. The Ecology of Delta Marshes of Coastal Louisiana: A Community Profile. U.S. Fish Wildl. Serv. FWS/DBS-84/09.

Greeson, P.E., J.R. Clark, and J.E. Clark. 1979. Wetland Functions and Values: The State of Our Understanding. American Water Resources Assoc., Minneapolis, Minnesota.

Hardaway, C.S., G.R. Thomas, A.W. Zacherle, and B.K. Fowler. 1984. Vegetative Erosion Control Project: Final Report. Virginia Institute of Marine Science, Gloucester Point, Virginia.

Kadlec, J.A. and W.A. Wentz. 1974. State-of-the-art Survey and Evaluation of Marsh Plant Establishment Techniques: Induced and Natural. Vol. I: Report of Research. U.S. Army Engineer Waterways Expt. Sta. Tech. Rep. D-74-9.

Knutson, P.L. 1976. Development of Intertidal Marshlands upon Dredged Material in San Francisco Bay, p. 103-118. In Proc. of Seventh World Dredging Conference, San Francisco, California.

Knutson, P.L., J.C. Ford, and M.R. Inskeep. 1981. National survey of planted salt marshes (vegetative stabilization and wave stress). Wetlands 1:129-157.

Knutson, P.L., R.A. Brochu, W.N. Seelig, and M. Inskeep. 1982. Wave damping in Spartina alterniflora marshes. Wetlands 2:87-104.

Knutson, P.L. and M.R. Innskeep. 1982. Shore Erosion Control with Salt Marsh Vegetation. U.S. Army Coastal Engineering Research Center, Technical Aid 8203.

Krucynski, W.L., R.T. Huffman, and M.K. Vincent. 1978. Habitat Development Field Investigations, Apalachicola Bay Marsh Development Site, Apalachicola Bay, Florida. Summary Report. U.S. Army Corps of Engineers, Technical Report. D-78-32.

Kundell, J.E. and S.W. Woolf. 1986. Georgia Wetlands: Trends and Policy Options. University of GA. Carl Vinson Institute of Government, Athens, Georgia.

Landin, M.C. 1984. Habitat development using dredged material, p. 907-917. In R.W. Montgomery and J.W. Leach (Eds.), Dredging and Dredged Material Disposal Vol. 2. American Society of Civil Engineers, New York.

Landin, M.C. and C.J. Newling. 1987. Habitat Development Case Studies: Windmill Point Wetland Habitat Development Field Site, James River, Virginia, p. 76-84. In M.C. Landin (Ed.), Beneficial Uses of Dredged Material: Proceedings of the North Atlantic Regional Conference 12-14 May 1987, Baltimore, Maryland.

Larson, J.S. 1987. Wetland Creation and Restoration: An outline of the Scientific Perspective, p. 73-79. In J. Zelazny and J.S. Feierabend (Eds.), Increasing Our Wetland Resources, Proc. of a Conference held October 4-7, 1987. National Wildlife Federation, Washington, D.C.

Lewis, R.R. (Ed.). 1982. Creation and Restoration of Coastal Plant Communities. CRC Press Inc. Boca Raton, Florida.

Lunz, J.D., T.W. Zeigler, R.T. Huffman, R.J. Diaz, E.J. Clairain, and L.J. Hunt. 1978. Habitat Development Field Investigations Windmill Point Marsh Development Site James River, Virginia. Summary report. U.S. Army Engineer Waterways Expt. Sta. Tech. Rep. D-77-23.

Minello, T.J., R.J. Zimmerman, and E.J. Klima. 1986. Creation of fishery habitat in estuaries, p. 106-120. In M.C. Landin and H.K. Smith (Eds.), Beneficial Uses of Dredged Material. U.S. Army Waterways Expt. Sta. Tech. Rep. D-87-1.

Mitsch, W.J. and J.G. Gosselink. 1986. Wetlands. Van Nostrand Reinhold Co., New York.

Morris, J.H., C.L. Newcombe, R.T. Huffman, and J.S. Wilson. 1978. Habitat Development Field Investigations, Salt Pond No. 3 Marsh Development Site, South San Francisco Bay, California. Summary Report. U.S. Army Engineer Waterways Expt. Sta. Tech. Rep. D-78-57.

Newling, C.J. and M.C. Landin. 1985. Long-term Monitoring of Habitat Development at Upland and Wetland Dredged Material Disposal Sites, 1974-1982. U.S. Army Engineer Waterways Expt. Sta. Tech. Rep. D-85-5.

Nixon, S.W. 1980. Between coastal marshes and coastal waters - A review of twenty years of speculation and research on the role of salt marshes in estuarine productivity and water chemistry, p. 437-525. In P. Hamilton and K.B. MacDonald (Eds.), Estuarine and Wetland Processes with Emphasis on Modeling. Plenum, New York.

Oaks, R.Q., Jr. and J. R. Dubar. 1974. Post-Miocene Stratigraphy Central and Southern Coastal Plain. Utah State University Press, Logan, Utah.

Odum, E.P. 1961. The role of tidal marshes in estuarine production. New York State Conserv. 16:12-15.

Odum, E.P. 1979. The value of wetlands: a hierarchical approach, p. 16-25. In P.E. Greeson, J.R. Clark, and J.E. Clark (Eds.), Wetland Functions and Values: The State of Our Understanding. American Water Resources Assoc., Minneapolis, Minnesota.

Odum, W.E. and S.S. Skjei. 1974. The issues of wetlands preservation and management: A second view. Coastal Zone Manage. J. 1(2):151-163.

Odum, W.E., T.J. Smith, III, J.K. Hoover, and C.C. McIvor. 1984. The Ecology of Tidal Freshwater Marshes of the United States East Coast: A Community Profile. U.S. Fish Wildl. Serv. FWS/DBS-83/17.

Phillips, W.A. and F.D. Eastman. 1959. Riverbank stabilization in Virginia. J. of Soil and Water Cons. 14:257-259.

Pomeroy, L.R., R.E. Johannes, E.P. Odum, and B. Roffman. 1969. The Phosphorus and Zinc Cycles and Productivity in a Salt Marsh, p. 412-419. In D.J. Nelson (Ed.), Proc. of the Second National Symposium on Radioecology, U.S. Atomic Energy Commission, Washington, D.C.

Priest, W.I., III and T.A. Barnard, Jr. 1989. Plant community dynamics in a recently planted wetland bank. Wetlands in press.

Race, M.S. and D.R. Christie. 1982. Coastal zone development: mitigation marsh creation and decision making. Environ. Management 6:317-328.

Ranwell, D.S. 1967. World resources of Spartina townsendii (sensu lato) and economic use of Spartina marshland. J. Appl. Ecol. 4:239-256.

Redman, F.H. and W.H. Patrick Jr. 1965. Effect of submergence on several biological and chemical soil properties. Louisiana State Univ. Agric. Exp. Sta. Bull. No. 592.

Reimold, R.J., M.A. Hardisky, and P.C. Adams. 1978. Habitat Development Field Investigations, Buttermilk Sound Marsh Development Site, Atlantic Intracoastal Waterway, Georgia. U.S. Army, Waterways Expt. Stat. Tech. Rep. D-78-26, Vicksburg, Mississippi.

Ross, N.W. 1977. Constructing Floating Tire Breakwaters. Proc. Am. Chemical Soc. Sym. Conservation in the Rubber Industry, Chicago, Ill.

Seneca, E.D., W.W. Woodhouse, Jr., and S.W. Broome. 1975. Saltwater marsh creation, p. 427-437. In L.E. Cronin (Ed.), Estuarine Research Vol. II. Geology and Engineering, Academic Press, New York.

Seneca, E.D. and S.W. Broome. 1982. Restoration of Marsh Vegetation Impacted by the Amoco Cadiz Oil Spill and Subsequent Cleanup Operations at Ile Grande, France. Interim Rept. to Dept. of Commerce, National Oceanic and Atmospheric Administration, Washington, D.C.

Sharp, W.C. and J. Vaden. 1970. Ten-year report on sloping techniques used to stabilize eroding tidal river banks. Shore and Beach 38:31-35.

Shew, D.M., R.A. Linthurst, and E.D. Seneca. 1981. Comparison of production computation methods in a southeastern North Carolina, Spartina alterniflora salt marsh. Estuaries 4:97-109.

Simpson, R.L., D.F. Whigham, and R. Walker. 1978. Seasonal patterns of nutrient movement in a freshwater tidal marsh, p. 243-257. In R.E. Good, D.F. Whigham, and R.L. Simpson (Eds.), Freshwater Wetlands: Ecological Processes and Management Potential. Academic Press, New York.

Sullivan, M.J. and F.C. Daiber. 1974. Response in production of cordgrass, Spartina alterniflora, to inorganic nitrogen and phosphorus fertilizer. Chesapeake Sci. 15(2):121-123.

Teal, J.M. 1962. Energy flow in the salt marsh ecosystem of Georgia. Ecology 43(4): 614-624.

Teal, J.M. 1986. The Ecology of Regularly Flooded Salt Marshes of New England: A Community Profile. U.S. Fish Wildl. Serv. Biol. Rept. 85(7.4).

Tiner, R.W., Jr. 1977. An Inventory of South Carolina's Coastal Marshes. S.C. Wildl. and Mar. Res. Dept. Tech. Rep. No. 23. Charleston, South Carolina.

Tiner, R.W., Jr. 1984. Wetlands of the United States: Current Status and Recent Trends. U.S. Fish and Wildlife Service, Washington, D.C.

Tisdale, S.L., L.N. Werner, and J.D. Beaton. 1985. Soil Fertility and Fertilizers. MacMillan, New York.

Turner, R.E. 1976. Geographic variations in salt marsh macrophyte production: A review. Marine Science 20:47-68.

U.S. Army Corps of Engineers. 1986. Beneficial Uses of Dredged Material. Engineer Manual 1110-2-5026. Office, Chief of Engineers, Washington, D.C.

Valiela, I. and J.M. Teal. 1974. Nutrient limitation in salt marsh vegetation, p. 547-563. In R.J. Reimold and W.H. Queen (Eds.), Ecology of Halophytes. Academic Press, New York.

Webb, J.W., Jr. 1982. Salt marshes of the western Gulf of Mexico, p. 89-109. In R.R. Lewis III (Ed.), Creation and Restoration of Coastal Plant Communities. CRC Press Inc. Boca Raton, Florida.

Webb, J.W., J.D. Dodd, B.W. Cain, W.R. Leavens, L.R. Hossner, C. Lindau, R.R. Stickney, and H. Williamson. 1978. Habitat Development Field Investigations, Bolivar Peninsula Marsh and Upland Habitat Development Site, Galveston Bay, Texas. U.S. Army Corps of Engineers Dredged Material Research prog. Tech. Rept. D-78-15.

West, T.L. and R.R. Rulifson. 1987. Structure and function of man-made and natural wetlands in the Pamlico River estuary, North Carolina. (Abstract) Amer. Fish. Soc. Annual Meeting, Sept. 14-18, 1987. Winston Salem, North Carolina.

Wilson, K.A. 1962. North Carolina Wetlands Their Distribution and Management. North Carolina Wildlife Resources Commission.

Wolf, R.B., L.C. Lee, and R.R. Sharitz. 1986. Wetland creation and restoration in the United States from 1970 to 1985: An annotated bibliography. Wetlands 6:1-88.

Woodhouse, W.W., Jr. 1979. Building Salt Marshes along the Coasts of the Continental United States. U.S. Army Corps of Engineers, Coastal Engineering Research Center, Special Report No. 4.

Woodhouse, W.W., Jr., E.D. Seneca, and S.W. Broome. 1972. Marsh Building with Dredge Spoil in North Carolina. North Carolina State Univ. Agric. Exp. Sta. Bull. 445.

Woodhouse, W.W., Jr., E.D. Seneca, and S.W. Broome. 1974. Propagation of Spartina alterniflora for Substrate Stabilization and Salt Marsh Development. U.S. Army Coastal Engineering Research Center, Fort Belvoir, Virginia.

Woodhouse, W.W., Jr., E.D. Seneca, and S.W. Broome. 1976. Propagation and Use of Spartina alterniflora for Shoreline Erosion Abatement. U.S. Army Coastal Engineering Research Center, Ft. Belvoir, Virginia.

Zhuo Rongzong and Xu Guowan. 1985. A note on trial planting experiments of Spartina alterniflora. Journal of Nanjing University, Research Advances in Spartina. p. 352-354.

APPENDIX I: RECOMMENDED READING

Daiber, F.C. 1986. Conservation of Tidal Marshes. Van Nostrand Reinhold Company, New York.

Complete up-to-date information on management, restoration and maintenance of tidal marshes is presented with emphasis on temperate North American east coast marshes and those of western Europe. Subjects covered include natural processes of tidal marshes, water management using dikes, ditches and impoundments, vegetation management, waste treatment, dredged material for restoration, pollution and legal concerns and management concepts of the future. There is an extensive list of references that is useful for finding additional information.

Environmental Laboratory. 1978. Wetland Habitat Development with Dredged Material: Engineering and Plant Propagation. U.S. Army Waterways Expt. Sta. Vicksburg, Miss. Tech. Rep. DS-78-16.

This report is a summary of literature and research related to marsh development that was conducted by the Waterways Experiment Station. Engineering aspects presented are protective and retention structures, substrate and foundation characteristics, dredging operations, and elevation and drainage requirements. Vegetation aspects discussed are selecting plant species, collecting and storing plant materials, selecting propagules, planting, maintenance and monitoring, natural colonization, and costs. Tables of 115 plant species showing propagation methods, growth requirements, and other information are useful, although they are too superficial for application without further reading.

Greeson, P.E., J. R. Clark, and J. E. Clark (Eds.). 1979. Wetland Functions and Values: The State of our Understanding. American Water Resources Association, Minneapolis.

This book is the proceedings of the National Symposium on Wetlands held in Lake Buena Vista, Florida, Nov. 7-10, 1978. It is a comprehensive and useful volume covering the following wetland topics: conservation, management and evaluation, food chains, habitat value, hydrology, effects on water quality, aesthetic values and harvest value of wetland products.

Herner and Company. 1980. Publication Index and Retrieval System. U.S. Army Engineer Waterways Exp. St., Vicksburg Miss. Tech. Rep. DS-78-23.

Abstracts of more than 200 reports resulting from the Dredged Material Research Program are presented. The publication is useful in selecting and retrieving reports relevant to a particular project. National Technical Information numbers are provided for obtaining publications from NTIS, Springfield, VA. 22161.

Kadlec, J.A. and W.A. Wentz. 1974. State-of-the-Art Survey and Evaluation of Marsh Plant Establishment Techniques: Induced and Natural, Volume 1: Report of Research. U.S. Army Engineer Waterways Expt. Sta., Vicksburg, Mississippi.

Information on the establishment of marsh and aquatic vegetation was reviewed by searching literature and contacting knowledgeable individuals. The following topics were covered: plant species distribution, site requirements, habitat tolerances of plant species, ecotypic variation, natural establishment, propagation methods, water management, and site selection and preparation.

Knutson, P.L. and W.W. Woodhouse, Jr. 1983. Shore Stabilization with Salt Marsh Vegetation. U.S. Army, Corps of Engineers, Coastal Engineering Research Center Special Report No. 9.

Guidelines for using coastal marsh vegetation as a shore erosion control measure are presented. Criteria are provided on determining site suitability, selecting plant materials, planting procedures and specification, estimating costs, and assessing impacts.

Lewis, R.R. III. (Ed.). 1982. Creation and Restoration of Coastal Plant Communities. CRC Press Inc., Boca Raton, Florida.

Contains detailed information on planting and management of coastal vegetation. Subjects covered by the nine chapters are as follows: coastal sand dunes of the United States, Atlantic coastal marshes, salt marshes of the northeastern Gulf of Mexico, salt marshes of the western Gulf of Mexico, Pacific coastal marshes, low marshes of China, low marshes of peninsular Florida, mangrove forests, and seagrass meadows.

Mitsch, W.J. and J.G. Gosselink. 1986. Wetlands, Van Nostrand Reinhold Company, New York.

This is a comprehensive textbook covering the scientific and management aspects of freshwater and coastal wetlands, with emphasis on wetlands of the United States. The book presents a general view of principles and components of wetlands that has broad application to many wetland types and a detailed ecosystem view of dominant wetland types. Topics presented include: wetland types of the United States; hydrology, biogeochemistry, biology and ecosystem development of wetlands; tidal salt marshes; tidal fresh marshes; mangrove wetlands; freshwater marshes; northern peatlands and bogs; southern deepwater swamps; riparian wetlands; and management of wetlands. The book is a useful reference and contains an extensive list of literature cited.

Odum, W.E., T.J. Smith III, J.K Hoover, and C.C. McIvor. 1984. The Ecology of Tidal Freshwater Marshes of the United States East Coast: A Community Profile. U.S. Fish Wildl. Serv. FWS/OBS-83/17.

This report provides a thorough review of the ecology of tidal freshwater marshes from southern New England to Northern Florida. Topics discussed are plants, ecosystem processes, invertebrates, fishes, amphibians and reptiles, birds, mammals, values and management practices, and a comparison of tidal freshwater marshes and salt marshes. The report is a very useful reference for anyone interested in freshwater tidal marshes or wetlands in general.

U.S. Army Corps of Engineers. 1986. Beneficial Uses of Dredged Material. EM 1110-2-5026. Office, Chief of Engineers, Washington, D.C.

The manual contains a section on both engineering and biological aspects of wetland development on dredged material. There is detailed information on stabilization, propagation, planting costs, project design and site preparation.

Wolf, R.B., L.C. Lee, and R.R. Sharitz. 1986. Wetland creation and restoration in the United States from 1970 to 1985: an annotated bibliography. Wetlands 6:1-88.

This is an annotated bibliography that deals with creation and restoration of wetlands. Emphasis is on site engineering and preparation and plant propagation. Topics covered by the articles include site selection, planning, engineering and design, seeding, plant material selection, transplanting, harvest, fertilization, costs and maintenance. Methods for propagating about 150 plant species can be found in the articles cited.

Woodhouse, W.W., Jr. 1979. Building Salt Marshes along the Coasts of the Continental United States. U.S. Army Corps of Engineers, Coastal Engineering Research Center, Spec. Rep. No. 4.

The report is a summary of available information on salt marsh creation. Topics discussed are the value and role of marshes, the feasibility of marsh creation and the effects of salinity, slope, exposure and soils on marsh establishment. Plants suitable for marsh building are described for each region. Plant propagation, planting, fertilization, and management techniques are discussed.

Zelazny, J. and J. S. Feierabend (Eds.). 1988. Proceedings of a conference: Increasing our wetland resources. National Wildlife Federation, Washington, D.C.

The proceedings of a conference held October 4-7, 1987 in Washington, D.C. covering a wide range of wetland topics including the following: wetlands policy and management; creation and restoration; mitigation; stormwater, municipal and mine waste treatment; design and planning of restoration projects; and biological monitoring.

APPENDIX II: PROJECT PROFILES

VIRGINIA WETLAND BANK, GOOSE CREEK

Wetland Type: A regularly flooded brackish-water marsh planted with smooth cordgrass, big cordgrass, saltmeadow cordgrass, saltgrass, wax myrtle (Myrica cerifera), marsh elder (Iva frutescens) and sea myrtle (Baccharis halimnifolia).

Date planted and type of Propagule: Summer 1982; field-dug and some nursery grown plants.

Location: Goose Creek, Elizabeth River, Chesapeake Virginia off Route 191 (Jolliff Road) west of Bowers Hill interchange.

Size: 9.14 acres (3.7 ha).

Goals of Project:

The goal of the project was to establish a wetland bank for the Virginia Department of Transportation (VDOT). An accounting system was set up for the VDOT to draw on the banks resources when highway construction destroys wetlands. The accounting system was started with a credit of 7 acres (approximately 2/3 of the total area). When wetland vegetation is affected by highway construction, and mitigation is required, an equal area is subtracted from the bank.

Judgement of Success:

The project is successful. The vegetation has developed and the banking system is in use.

Significance:

The project is an apparently successful application of the mitigation banking concept. Studies of the structural and functional ecology of the marsh are being done by scientists at the Virginia Institute of Marine Science.

Reports: Priest and Barnard 1987.

Contacts: Walter I Priest, III or
Thomas A. Barnard, Jr.
Virginia Institute of Marine Science
Gloucester Point, Virginia 23062
Tel 804-642-7385

E. Duke Whedbee, Jr.
Environmental Specialist
Commonwealth of Virginia
Department of Transportation
1401 East Broad Street
Richmond, Virginia 23219
Tel 804-786-2576

WINDMILL POINT, VIRGINIA

Wetland Type: Freshwater tidal.

Date Planted and Type of Propagule: July, 1974. Sprigged and seeded with a number of freshwater marsh plants.

Location: James River, 0.4 km west of Windmill Point, Prince George County, Virginia.

Size: 9.3 ha.

Goals of Project:

Goals of the project were to investigate methods of creating freshwater tidal marsh with dredged material and to conduct studies on benthic invertebrates, fish, wildlife, plants and soil characteristics.

Judgement of Success:

The initial study was successful; however, most of the site was subsequently lost to erosion and subsidence. A combination of emergent marsh and shallow water habitat remains.

Significance:

The site is an example of creation of a large scale tidal freshwater marsh. The study is thoroughly documented in several reports by the Waterways Experiment Station. Most of the planted wetland vegetation was grazed and destroyed by wildlife (particularly Canada geese) and the upland seeded vegetation was displaced by native plant invasion. Wetland plants became well established by natural colonization leading to the conclusion that planting intertidal freshwater sites is not necessary in that area.

Reports: Landin, M.C. and C.J. Newling 1987.
Lunz et al. 1978.

Contact: Dr. M.C. Landin
U.S. Army Waterways Experiment Station
P. O. Box 631
Vicksburg, Mississippi 39180-0631

VIRGINIA VEGETATIVE EROSION CONTROL PROJECT

Wetland Type: Smooth cordgrass and saltmeadow cordgrass.

Date Planted and Type of Propagule: Late Spring of 1981 and 1982. Potted seedlings of smooth cordgrass and saltmeadow cordgrass produced by the Soil Conservation Service National Plant Materials Center in Beltsville, MD, and the Soil Conservation Service Plant Materials Center in Cape May, New Jersey.

Location: Twenty four sites along the shoreline of Virginia Chesapeake Bay and its tributaries.

Size: Various sizes.

Goals of Project:

1. To supplement previous research with detailed site analysis of the early stages of marsh development.
2. To more precisely define the physical limits of marsh planting in terms of wave stress.
3. To provide demonstration.

Judgement of Success:

The project was successful in achieving its goals.

Significance:

The study added to information needed to determine the limits of tolerance of marsh plantings to wave stress.

Report: Hardaway et al. 1984

Contact: Scott Hardaway
Virginia Institute of Marine Science
School of Marine Science
College of William and Mary
Gloucester Point, Virginia 23062

NORTH CAROLINA PHOSPHATE PROJECT AREA II

This marsh creation site was constructed by North Carolina Phosphate Corporation (Agrico) and subsequently acquired by Texasgulf Chemicals Co.

Wetland Type: Irregularly flooded (wind dominated tides) Brackish water (0-15 ppt. depending on rainfall, runoff, season, etc.). Natural marsh vegetation in the area is dominated by black Needlerush, big cordgrass, saltmeadow cordgrass and saltgrass.

Date Planted and Type of Propagule: Planted in April and May 1983 with pot-grown seedlings of big cordgrass, smooth cordgrass, saltmeadow cordgrass and black needlerush.

Location: Near Aurora, NC adjacent to South Creek, a tributary of the Pamlico River. The drainage system of the created marsh is connected to the estuary at the mouth of Drinkwater Creek.

Size: 2 ha.

Goals of the Project:

The goal of the project was to investigate the feasibility of creating brackish water marsh habitat by grading an upland site to an intertidal elevation and planting with marsh vegetation. The purpose of creating marsh was to mitigate losses of natural marsh that might occur in association with phosphate mining. The ultimate objective of North Carolina Phosphate Corporation was to obtain permits to mine through the headwaters of small tributaries, which occur as small inclusions in the mining area, in exchange for creating new marshes.

Judgement of Success:

The project was successful after correcting soil problems. The pH was below 2.5 over about 25% of the area and soil phosphorus levels were very low.

Significance:

It is an example of a large-scale mitigation project using pot-grown seedlings of brackish-water marsh vegetation. The marsh is in an irregularly flooded area with a narrow elevation range in which marsh vegetation grows. It is not threatened by erosion and offers an opportunity for long term studies of flora, fauna, structure and function of a created marsh.

Reports: Broome, Craft, and Seneca 1988.
West and Rulifson 1987

Contacts: Mr.William A. Schimming
Texasgulf Chemicals Inc.
P.O. Box 48
Aurora, NC 27806
Tel-919-322-4111

Steve Broome
Department of Soil Science
Box 7619
North Carolina State University
Raleigh, NC 27695-7619
Tel-919-737-2643

PINE KNOLL SHORES, NORTH CAROLINA

Wetland Type: A regularly flooded intertidal salt marsh that is a pure stand of smooth cordgrass below mean high water. Saltmeadow cordgrass dominates a narrow band of high marsh. The smooth cordgrass occurs as a marsh fringe about 15 m wide along the shoreline of Bogue Sound. Salinity of the estuarine water varies from 20-35 ppt. The interstitial water is often around 10 ppt. due to seepage from the upland.

Date Planted and Type of Propagule: Planted in April 1974 with field-dug transplants of smooth cordgrass.

Location: The site is along the shoreline of the barrier island of Bogue Banks in the town of Pine Knoll Shores.

Size: The marsh is about 15 m wide and extends 500 m along the shoreline. Total area is about 0.75 ha.

Goals of the Project:

Goals of the project were to test planting techniques and to determine the value of smooth cordgrass for shoreline erosion control.

Judgement of Success:

The project was very successful in preventing shoreline erosion. It has been observed over a period of 13 years.

Significance:

The planting has prevented erosion of a shoreline over a long period of time and records of biomass production and elevation of the shoreline have been maintained.

Reports: Woodhouse et al. 1976
Broome et al. 1986

Contact: Steve Broome
Department of Soil Science
Box 7619
North Carolina State University
Raleigh, NC 27695-7619
Tel 919-737-2643

EVALUATION OF MARSH VEGETATION PLANTED ON DREDGEDMATERIAL IN NORTH CAROLINA AS FISHERY HABITAT

Wetland Type: Salt water, regularly flooded (S. alterniflora) and high marsh (S. patens). The seagrasses Halodule wrightii and Zostera marina were also planted.

Date Planted and Type of Propagule: March through June 1987. Field-dug and greenhouse grown seedlings of S. alterniflora and S. patens were planted in experimental plots designed by personnel of the National Marine Fisheries Service.

Location: Three dredged material disposal sites in North Carolina were planted using the same experimental design: the Harker's Island dredge spoil island in Core Sound off marker 34 along Bardens Inlet Channel, the westernmost dredge spoil island at Swansboro, and the large dredge spoil island at New River Inlet near Sneads Ferry.

Size: The marsh planting was 185 x 30 meters at each site. The total area for the three sites is 1.67 ha.

Goals of the Project:

1. To evaluate techniques for establishing salt marsh habitat to reduce erosion and channel refilling at dredged material disposal sites.

2. To generate fishery habitat and evaluate its utilization by certain target fishery species.

Judgement of Success: Too early to determine.

Significance:

It is a statistically designed, replicated experiment that should be useful in comparing the value of planted and unplanted plots to certain target fishery species. The seagrass plantings are also an interesting feature of the project.

Contacts: Dr. Gordon Thayer or Dr. Mark Fonseca
National Marine Fisheries Service
Beaufort, North Carolina 28516
Tel 919-728-8747

Mr. Frank Yelverton
U.S. Army Engineer District, Wilmington
P. O. Box 1890
Wilmington, North Carolina 28402
Tel 919-343-4640

Dr. Stephen W. Broome
Department of Soil Science
Box 7619
North Carolina State University
Raleigh, North Carolina 27695-7619
Tel 919-737-2643

WINYAH BAY, SOUTH CAROLINA

Wetland Type: Smooth cordgrass.

Date Planted and Type of Propagule: No planting was done. Complete cover by natural colonization of smooth cordgrass is achieved approximately three years after disposal of dredged material at the proper elevation (Steve Morrison pers. comm.). The project was started in 1974 with dredging every 12-18 months.

Location: Winyah Bay, Georgetown, South Carolina.

Size: 100 acres.

Goals of Project:

The goal was to create salt marsh from open water by disposal of dredged material to replace salt marsh lost in Winyah Bay due to diking for rice fields.

Judgement of Success: Successful.

Significance:

The project is an example of salt marsh creation by disposal of dredged material at the correct elevation with no planting of vegetation. Controversy now exists over the displacement of shallow water habitat with salt marsh.

Contacts: Steve Morrison or John Carothers
Charleston District Corps of Engineers
Tel 803-724-4258

BUTTERMILK SOUND, GEORGIA

Wetland Type: Brackish water; regularly flooded.

Date Planted and Type of Propagule: Planted in June 1975. Field dug sprigs and seeds of seven plant species were planted in an experimental design. These were: Borrichia frutescens, Distichlis spicata, Iva frutescens, Juncus roemerianus, Spartina alterniflora, Spartina cynosuroides and Spartina patens.

Location: In the Atlantic Intracoastal Waterway near the mouth of the Altamaha River, Glynn County, Georgia.

Size: 2.1 ha.

Goals of Project:

The goals of the project were to determine the feasibility of establishing the seven plant species tested on dredged material. The effect of fertilizer and inundation on plant growth was determined. Water chemistry, soil chemistry, soil microbiology, invading plants, and crab burrows were monitored. Observations were made on use of the area by aquatic organisms and wildlife.

Judgement of Success: Successful.

Significance:

It is a brackish-water marsh development site with a number of experimental variables imposed. The results of the work are well documented and the U.S. Army Corps of Engineers Waterways Experiment Station continues to follow the site.

Reports: Reimold et al. 1978.
Newling and Landin 1985.

Contact: Dr. Mary C. Landin
Department of the Army
Waterways Experiment Station
Corps of Engineers
P.O. Box 631
Vicksburg, Mississippi 39180-0631

KINGS BAY, GEORGIA

Wetland Type: Regularly flooded salt marsh; smooth cordgrass.

Date Planted and Type of Propagule: To be determined.

Location: Kings Bay Naval Submarine Base, St. Marys, GA.

Size: 30 acres.

Goals of Project:

The goal of the project is to create salt marsh to mitigate losses of natural marsh related to dock construction, dredging and dredged material disposal required for construction of a submarine base.

Significance:

It is a large scale mitigation project involving creation of a smooth cordgrass marsh from an upland borrow pit. It has potential as a research site for comparing created and natural marshes.

Contact: Mr. Bob Peavy
Regulatory Branch (CESAS-OP-F)
U.S. Army Eng. District, Savannah
P.O. Box 889
Savannah, GA 31402-0889

CREATION AND RESTORATION OF COASTAL PLAIN WETLANDS IN FLORIDA

Roy R. Lewis III
Lewis Environmental Service, Inc.

ABSTRACT. Despite hundreds of mangrove and tidal marsh restoration and creation efforts in Florida over the last fifteen years, current efforts are largely more art than science. Adequate monitoring and reporting are rare, and no institutional memory exists to improve the review and monitoring process.

Based on a critical review of actual projects and the sparse literature, five factors appear most important to successful wetland establishment; these are:

1. Correct elevations for the target plant species;
2. Adequate drainage provided by gradual slopes and sufficient tidal connections; and
3. Appropriate site selection to avoid wave damage.
4. Appropriate plant materials.
5. Protection from human impacts.

OVERVIEW OF THE REGION

CHARACTERISTICS OF THE REGION

Geology

The state of Florida is the emergent portion of a large plateau called the Floridian Plateau which is a projection of the North American continent dividing the Atlantic Ocean from the Gulf of Mexico. The plateau is about equal land area and submerged plateau with depths to 100 m.

The land area of the state consists of two sedimentary provinces. The North Gulf Coast province is largely characterized by clastic sands produced by erosion of the North American continent and includes the panhandle and Big Bend areas of the state. The Florida Penisular province is characterized by a higher percentage of nonclastic sediments with increasing proportions of biologically or chemically produced carbonate materials to the south. Overlying these pleistocene sediments are holocene features derived from warm subtropical plant community growth and hardening of exposed limestone (Drew and Schomer 1984). These plant derived sediments include wetland organic peats that may reach two meters or more in depth.

Climate

Florida's climate is characterized by relatively high mean annual rainfall (122-152 cm/yr) and a humid subtropical temperature pattern (daily maximum in July of 32 °C). Mean daily maximum temperatures in the winter average 20°C. Mean daily low temperatures range from 5°C to 16°C, depending upon the portion of the state measured. Winter is the driest period with most of the rain falling in the summer months of June-September. Hard freezes are rare, normally occurring only once or twice per century. Snow occurs very rarely as flurries in the north part of the state, and even here rarely remains on the ground.

Hurricanes are common, but more frequently strike the shore in southern Florida rather than the Panhandle. Tornadoes are also common but are not as destructive as in other areas since most are of the relatively weak waterspout type (Fernald and Patton 1984).

Ecoregions

Bailey (1978) divides Florida into three coastal ecoregion provinces: Louisianian, West Indian,

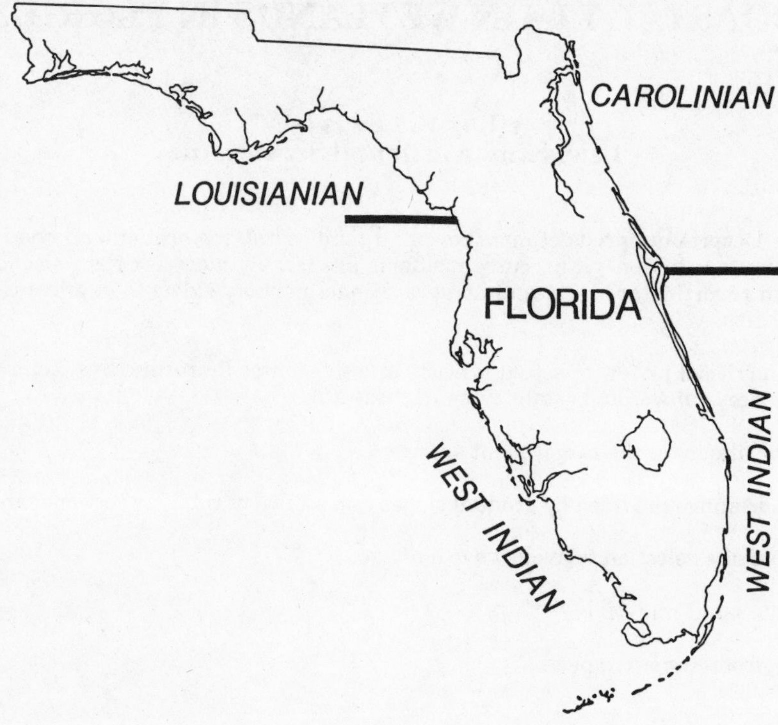

Figure 1. The marine and estuarine provinces of Florida (modified from Bailey 1978).

and Carolinian (Figure 1). The Carolinian province includes all the Atlantic coast of Florida south to Cape Canaveral and is characterized by extensive marshes, well developed barrier islands, turbid waters, small to moderate tidal ranges (0.3-2.0 m), and winter minimum temperatures of 5°C. The West Indian province extends from Cape Canaveral on the Atlantic Coast to Cedar Key on the Gulf coast and is characterized by more tropical flora and fauna, winter minimum temperatures of 16°C, and small tidal ranges (0.6-1.0 m). The Louisianian province extends from the mid-Gulf coast at Cedar Key to beyond the state line in the Panhandle, and includes Louisiana and much of coastal Texas. It is characterized by low wave energy, extensive marshes, generally temperate fauna, and small (0.6-1.0 m) tidal ranges (Cowardin et al. 1979).

WETLAND TYPES

Four estuarine or marine emergent wetland types occur in Florida. These are mangrove forest, tidal marsh, oligohaline marsh, and salt barren. In general the major mangrove forests are limited to the southern half of the state (the West Indian province area of Bailey 1978 [Figure 1]), while the major tidal marsh areas are found in the northern half of the state (the Louisianian and Carolinian provinces of Bailey 1978). This is due to the sensitivity of mangroves to freezing temperatures which limits their northern distribution (Odum et al. 1982). The cold resistant herbaceous marsh species are the dominant species is areas where mangroves are not present, but other factors such as salinity play an important role, and extensive tidal marshes are found mixed with mangroves in central and southern Florida (Durako et al. 1985).

The other two wetland types have received little study and neither their distribution around the state nor their areal cover has been determined. It is likely they are found statewide where conditions allow.

Key functions performed by these wetlands include shoreline protection, fisheries and wildlife habitat, water quality maintenance, and sources of primary production (Seaman 1985). Each of the wetland types is briefly described below to characterize major differences in functions.

Mangrove Forests

Mangrove forests in Florida are composed of four species of trees: Rhizophora mangle L. (red mangrove), Avicennia germinans (L.) L. (black mangrove), Laguncularia racemosa Gaertn. f. (white mangrove), and Conocarpus erecta L. (buttonwood). The tree species are generally distributed along a gradient in the intertidal zone with the red mangrove at the lowest elevations and the buttonwood at the highest. Forest structure in Florida is, however, not uniform and many variations on the classic zonation pattern first described by Davis (1940) occur (Snedaker 1982, Lewis et al. 1985). In addition, due to factors such as local topography, time since the last freeze event, changes in freshwater discharge, and other periodic disturbances, a given intertidal plant community can include marsh species at elevations lower or higher than the forest itself, and within windfalls or lightning strike areas of the forest. As noted, mangroves are cold sensitive tropical plants, and for this reason exhibit latitudinal zonation dependent on their cold tolerance (Lot-Hergueras et al. 1975, McMillan 1975, Lugo and Patterson-Zucca 1977, McMillan and Sherrod 1986). The black mangrove is the most cold tolerant species and extends northward along the Gulf coast to Louisiana as scattered shrubs within the predominant tidal marsh vegetation. There are approximately 273,000 ha of mangrove forest remaining in Florida, a reduction of 23% since World War II (Lewis et al. 1985).

An important characteristic of all mangroves is that their reproduction includes seedling dispersal by water and by vivipary. Vivipary means that there is no true or independent "seed", but continuous development from embryo to seedling while attached to the parent tree (Gill and Tomlinson 1969). For this reason the final reproductive unit released from the parent tree is often referred to as a "propagule" (Rabinowitz 1978).

Ecologically, mangroves are considered important as fisheries habitat (Lewis et al. 1985), as sources of detritus to support estuarine food chains (Odum et al. 1982), and as shoreline stabilizers (under limited conditions; Carlton 1974).

Tidal Marshes

Atlantic coastal marshes in Florida are dominated by _Spartina alterniflora_ Loisel. (smooth cordgrass), while Gulf coast marshes are dominated by _Juncus roemerianus_ Scheele (black needlerush). Several other plant species are common minor components of the marshes including _Spartina patens_ (Ait.) Muhl. (saltmeadow cordgrass), _Distichlis spicata_ (L.) Greene (saltgrass), and _Batis maritima_ L. (saltwort). Tidal marshes are widely distributed around the coast of Florida and often intermingle with mangrove communities (Durako et al. 1985). There are approximately 155,000 ha of tidal marsh in Florida (Lewis et al. 1985).

Reproduction takes place by waterborne seeds and asexually produced rhizomes. Due to their rhizomatous method of asexual propagation and ability to rapidly expand and anchor, some tidal marsh species such as smooth cordgrass are often pioneer colonizers of disturbed habitats and are replaced as other species such as mangroves naturally invade such habitats (Davis 1940, Lewis and Dunstan 1976, Lewis 1982a, 1982b).

Oligohaline Marshes

Though often not recognized as a distinct plant community (Durako et al. 1985), these marshes are unique in both their floral composition and their ecological role, and thus are treated here as a distinct plant community. Oligohaline is defined by Cowardin et al. (1979) as referring to "water with a salinity of 0.5 to 5.0 ppt due to ocean-derived salts" (p. 43). In Florida, oligohaline marshes are herbaceous wetlands located in tidally influenced rivers or streams where the plant community exhibits a mixture of true marine plants and typical freshwater taxa (_Typha_, _Cladium_) that tolerate low salinities. The predominant plant species of oligohaline marshes include black needlerush, _Acrostichum aureum_ L. (leather fern), _Typha domingensis_ (brackish water cattails), _Cladium jamaicense_ Crantz (sawgrass), _Scirpus robustus_ Pursh. (bulrush), and _Hymenocallis palmeri_ S. Wats. (spider lily).

Ecologically, oligohaline marshes (Rozas and Hackney 1983) and low salinity mangrove forests are becoming recognized as critical nursery habitat for such species as the _Callinectes sapidus_ (blue crab), _Centropomus undecimalis_ (snook), _Megalops atlanticus_ (tarpon), and _Elops saurus_ (ladyfish) (Odum et al. 1982, Gilmore et al. 1983, Lewis et al. 1985 [Figure 2]). Because recognition of this key role in estuarine life cycles has come only recently, much of this habitat has been lost or highly modified. The reduced amount of this habitat type may represent a limiting factor in total population sizes of some estuarine-dependent species.

Salt Barrens

The salt barren represents the upper intertidal flat which is inundated typically only by spring tides once or twice a month. This results in hypersaline conditions with seasonal expansion of typically low-growing succulent salt tolerant vegetation with lower interstitial salinities during the rainy season, and retreat with less frequent inundation and rainfall. This produces the characteristic open unvegetated patches of the salt barren substrate. These areas are also referred to as salt flats or salinas. The salt barren is typically located behind a mangrove forest or tidal marsh at a somewhat higher elevation and often occurs on exposed rock outcrops with shallow sand sediments. Although technically oligohaline marshes, salt barrens are treated here as a distinct plant community due to their unique flora and ecological value. Common plant species consist of saltwort, saltgrass, _Salicornia bigelovii_ Torr. (annual glasswort), _Salicornia virginica_ L. (perennial glasswort), _Monoanthochloe littoralis_ Engelm. (key

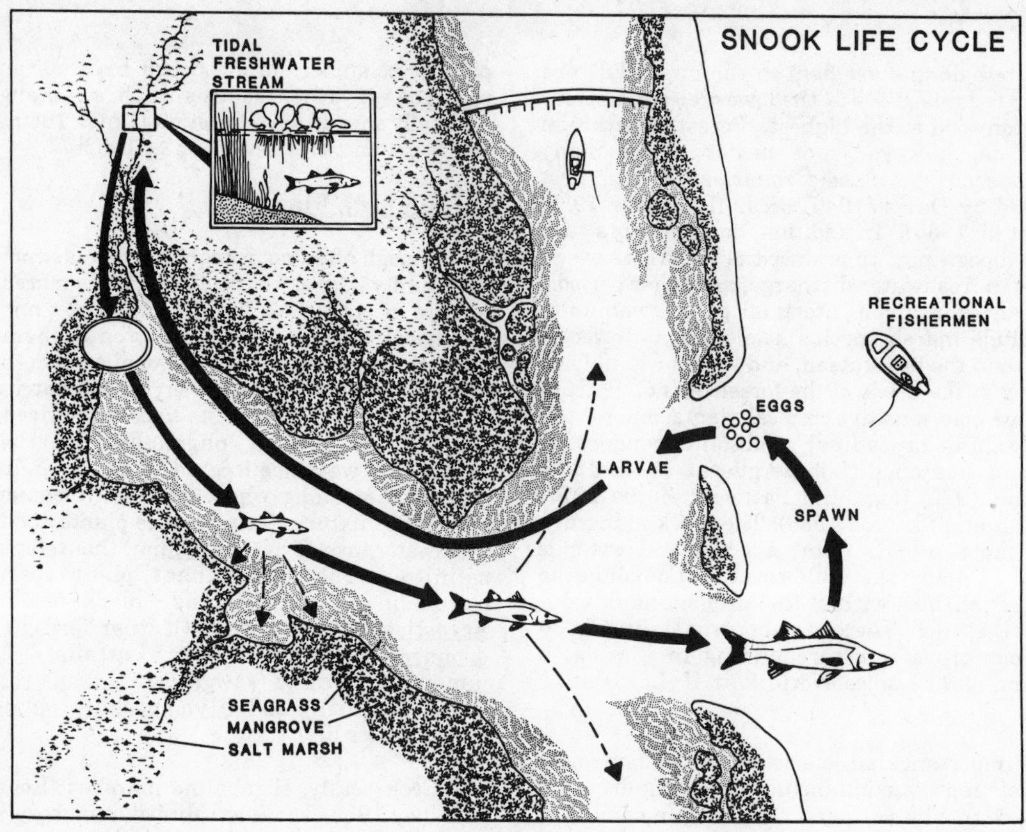

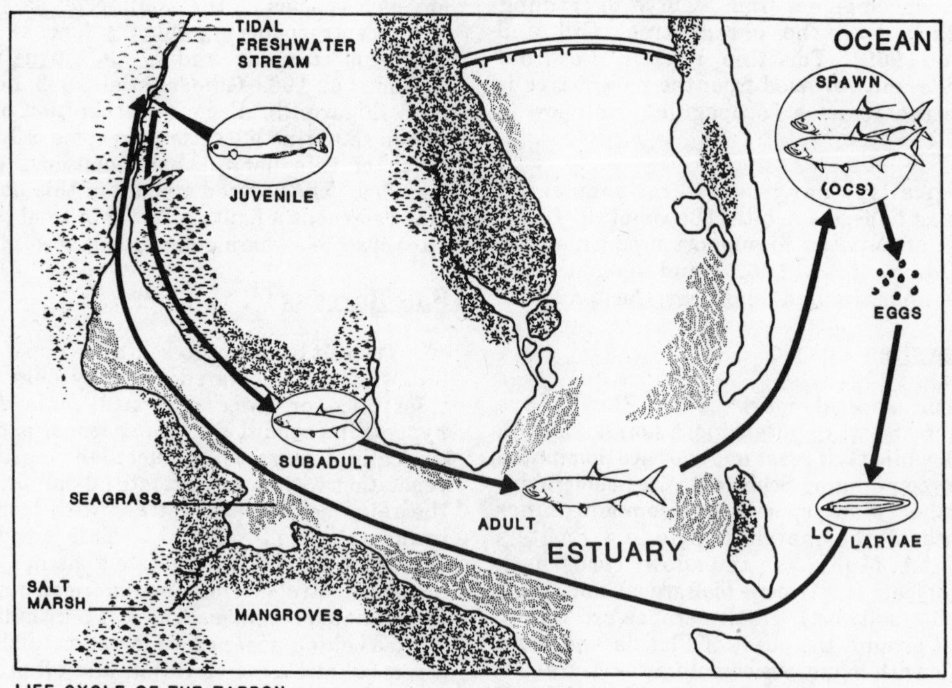

Figure 2. Life cycle of snook (top) and tarpon (bottom) in Florida, illustrating the role of oligohaline wetlands (from Lewis et al. 1985).

grass), <u>Limonium carolinianum</u> (Walt.) (sea lavender), <u>Blutaparon vermiculare</u> (L.) Mears (samphire), and <u>Sesuvium portulacastrum</u> L. (sea purslane) .

These areas have unique ecological values as seasonal feeding areas for wading birds when the lower elevation mudflats are more routinely inundated (Powell 1987), and as night feeding habitat on spring tides for snook, tarpon, and ladyfish (G. Gilmore pers. comm.). However, due to their low structural complexity and apparent lack of numerous fauna, these areas are often assumed to have low ecological value. Data now being generated strongly contradicts this assumption.

EXTENT TO WHICH CREATION/RESTORATION HAS OCCURRED

INTRODUCTION

Of the four types of wetlands discussed above, the historical emphasis of creation and restoration has been first on mangrove forests and on tidal marshes second. There are no published reports of attempts to create or restore oligohaline marshes or salt barrens, although several projects have been undertaken. In 1982, Lewis (1982b) listed 14 mangrove planting projects in Florida and Kruczynski (1982) listed two tidal marsh plantings on the Gulf coast of Florida. Since these early publications, hundreds of projects have been undertaken. Most of them have been regulatory agency or court ordered restoration projects due to illegal filling of wetlands, or mitigation projects for wetland fill or excavation allowed by permits. Lewis and Crewz (in prep.) have undertaken an analysis of the data available about these numerous projects as part of a Florida Sea Grant sponsored project (No. R/C-E-24) entitled "An analysis of the reasons for success and failure of attempts to create or restore tidal marshes and mangrove forests in Florida". A major conclusion of their two year effort has been that the database from which to draw any quantitative conclusions as to actual number of projects completed, types of plants used, and the general success of these projects is too widely scattered and unorganized to allow proper analysis. No agency contacted during the study keeps an organized file of wetland restoration and creation projects undertaken under their jurisdiction. The principal investigators were required to interview individual knowledgeable persons even to partially identify the many projects.

Following preliminary identification of projects, the available data were reviewed and 35 specific project sites were chosen for site visits and detailed analyses (Figure 3). These interviews, site analyses and the experience of the author form the bulk of the information upon which the following discussions are based. Frequent reference will be made to the available literature but it is important to understand that hundreds of "experiments" in the form of restoration and creation projects in the ground have not been subject to even routine "success" analyses (i.e., did the plants live?). This situation must be corrected and a central database created if we are to learn from our mistakes and be able to prepare adequate guidelines to assure success, and to define situations in which success is not possible.

TYPICAL GOALS OF PROJECTS

<u>Offset Adverse Environmental Impacts Through Mitigation</u>

Since 1969, Florida statutes have required a permit from the Florida Department of Environmental Regulation [FDER] in order to excavate or fill certain wetlands (Chap. 403 F.S.). The original statute did not require mitigation. Mitigation has been required on an informal basis in permit negotiations since the mid-1970s without any guidelines or performance criteria.

In 1984 the Florida Legislature passed a new wetlands protection act (Chap. 403.91) consolidating some previously scattered provisions and with new language stating that the Department of Environmental Regulation shall "consider measures ... to mitigate adverse effects". No further statutory guidance was provided. Following extensive hearings, on June 11, 1987, the FDER adopted detailed criteria outlining circumstances and conditions for mitigation. The rules have been subject to several legal challenges, which have delayed their implementation, now anticipated for September 1988. Because of this law, mitigation to offset adverse impacts of wetland loss has become the prime goal of wetland restoration and creation efforts, but the state still lacks detailed criteria as to when such effort does in fact accomplish that goal.

<u>Create Additional Habitat or Enhance Existing Habitat</u>

Aside from the permit system which has been the driving force behind the wetland restoration team and creation in Florida, some efforts have been made to reverse trends of habitat loss through restoration and creation only for the sake of creating additional fishery and wildlife habitat. These efforts have been limited by lack of funds. When a permit to fill a wetland is involved, there is often significant financial reward due to the improved development potential; the cost of mitigation is just part of the cost of doing business.

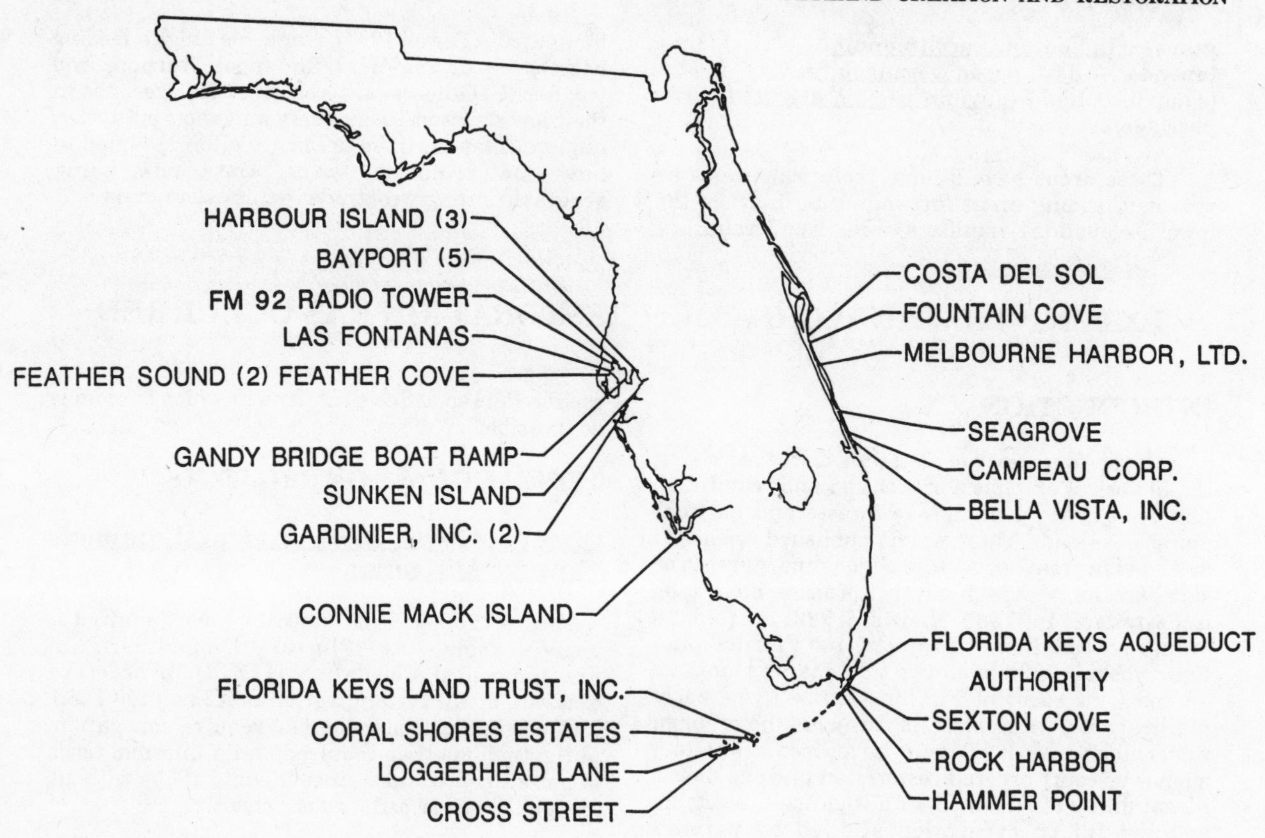

Figure 3. Field inspection sites of tidal creation/restoration locations in Florida (from Lewis and Crewz, in prep.).

In Florida, examples of marine or estuarine habitat restoration or creation not related to a permit decision are rare. Lewis et al. (1979) reported on several mitigation projects, but also included a description of the use of dredged material from a harbor deepening project in Tampa Bay to create an emergent island with an intertidal lagoon as habitat enhancement. Hoffman and Rodgers (1981a, 1981b) described in detail, the successful volunteer efforts to plant a smooth cordgrass marsh in the created lagoon of this island. Hoffman et al. (1985) listed seventeen restoration or creation projects in Tampa Bay between 1971-1981, and noted that only the one described above was for habitat restoration. Banner (1983) and Anonymous (1984) described the development of a $550,000 restoration trust fund for the Florida Keys as part of a legal settlement which, to date, has restored 4 ha of tidal wetlands at 12 sites.

More recently, the State Legislature passed a law requiring a special $300/yr license to use a gillnet for commercial fishing in certain coastal counties in Florida (Pasco, Pinellas, Manatee, Hillsborough). The money is specifically earmarked for "marine habitat restoration and research". The Florida Department of Natural Resources is managing the funds and over $200,000 is presently being spent on experimental habitat restoration of tidal marshes, mangrove forests, and seagrass meadows in Tampa Bay.

Some habitat restoration work has been attempted in Biscayne Bay, but, due to the lack of success in most attempts (Alleman 1982), current work is largely directed toward preserving existing wetlands and creating artificial reefs (Department of Environmental Resources Management 1985).

Stabilize Eroding Shorelines

Most of the shoreline erosion problems in Florida occur along the Atlantic and Gulf beaches in areas where wave energy is too high to allow for the use of intertidal vegetation for stabilization. Knutson et al. (1981) report on a national survey to identify tidal marsh planting sites for erosion control and list a total of 84 sites in 12 states. Only two of these (unspecified locations) were in Florida. Courser and Lewis (1981) report the successful stabilization of 60 m of eroding shoreline in Tampa Bay using smooth cordgrass. Teas (1977) reports that mangroves planted on an eroding shoreline in Biscayne Bay were unable to survive due to erosional forces. Smith (1982) describes attempts to control erosion at two sites in the Indian River. Some survival of herbaceous plantings of smooth cordgrass, salt hay, and seaside paspalum occurred but essentially all attempts to establish mangroves failed.

Achieve A Preset Percentage Survival For Installed Plant Materials

The goal most commonly listed as a condition for defining success of a restoration or creation project is survival of the installed plant materials at the end of a specified period. The percent survival specified is usually between 70% and 85%, although on occasion 100% survival is required. However, the method of determining such a percentage is usually not specified. Preestablished plots of varying sizes are used and are typically established in a haphazard rather than a truly random manner. An important requirement to the validity of such an approach is the preparation of an accurate description of the immediate pre- and post-planting ("time zero") conditions. It is often assumed that these conditions are identical to those outlined in a proposal or required in permit conditions. It has been the author's experience that this is rarely the case. Changes in the site due to construction problems or delays, and problems in plant material availability and timing of planting, are the norm. Because of these problems, it is impossible to say with any accuracy whether 10% or 90% of the projects achieve their preset survival goals. It is also questionable if percent survival alone is the best single criterion to measure.

Improve Water Quality

Improvement in water quality is often a verbal or written general goal of such projects, but to this author's knowledge, it has never been quantified prior to construction (e.g., increase dissolved oxygen by 10%) and then measured after construction and/or planting to determine if the goal was achieved.

Establish Similar "Habitat Values" in Wetlands Created or Restored For Mitigation

Establishment of "equivalent habitat values" are the most recent "buzz words" in goals for wetlands restoration and creation. They are meant to signify the integration of successful plant material establishment with fishery and wildlife habitat establishment and water quality improvements. In the literature, equivalency has most often been measured by qualitative measures such as species presence in conjunction with plant cover (Reimold and Cobler 1986, Dial and Deis 1986). More quantitative measures of fauna have been made by Cammen (1976) and Cammen et al. (1976) in created tidal marshes in North Carolina, and by Minello et al. (1987) in Texas. Based on the interest placed in "functional equivalency" of wetlands created or restored for mitigation and their natural counterparts (see National Wetlands Newsletter Vol. 8, No. 5, 1986), it is likely that this goal will be more frequently specified and included in permit conditions. However, problems will probably arise from the lack of uniform sampling techniques and data analysis. This will hamper comparisons between studies and determination of the "best" sampling strategy. It is hoped that the ongoing work of the National Marine Fisheries Service's Habitat Restoration Program will reduce some of these problems.

REASONS FOR SUCCESS/FAILURE

Excessive Wave Energy

Savage (1972, 1979) was one of the first to note the problems created by wave energy in preventing either volunteer or planted mangroves to survive on exposed shorelines. He felt that of all the mangrove species, the black mangrove was perhaps the best candidate for attempted shoreline stabilization projects because of its greater cold tolerance and its cable root and pneumatophore network. Carlton (1974), in summarizing the work to date, noted the repeated failures of mangrove plantings on exposed shores as reported earlier by Autry et al. (1973) and others; he questioned the value of mangroves as land builders, but left the door open for some use of mangroves as "accumulators of sediment" and possible land stabilizers. Teas (1977), following on the work of Kinch (1976) and Hannan (1976), also attempted transplanting of red, black and white mangroves to an exposed causeway site in Biscayne Bay but none survived after 24 months. He also revisited the site of a 32-year old planting of 4,100 red mangrove propagules that had initially achieved 80% survival after one year. The planter had stated that "...it seems probable that many will survive to maturity" (Davis 1940, p. 382). The inspection revealed none had survived. It is thus well documented that mangroves are not generally suitable plant materials for exposed or eroding shorelines unless some offshore protection is provided. Reark (1983) describes the planting of mangroves in constructed planters (Figure 4) where wave protection is provided. Rivers (pers. comm.) Crewz, in prep.). reports that approximately 3,000 m of exposed shoreline in the Florida Keys have been planted with mangroves behind a low riprap wall to provide protection. Both techniques appear to be effective.

A similar approach was followed by Lewis and Dunstan (1976) in studying secondary succession in disturbed mangrove areas and the natural ability of smooth cordgrass to colonize dredged material deposits and act as nurse plants (sensu MacNae 1968) to assist in the colonization of unstable or shifting sand areas. Because of the previous work of Woodhouse et al. (1974, 1976) and Garbisch et al. (1975), smooth cordgrass appeared to be better

Figure 4. Mangrove planter constructed and planted in front of existing seawalled fill, north Biscayne Bay, Dade County, Florida.

plant material than mangroves for installation on unstable shorelines in Florida (Lewis 1982a). Consequently, smooth cordgrass has since been more broadly used while mangrove plantings are limited to warmer parts of the state in more protected waters (Lewis and Crewz, unpublished data). Guidelines for evaluating potential sites for their suitability for smooth cordgrass plantings, given existing wave energy and exposure, are described by Knutson et al. (1982).

Improper Planting Elevation

Savage (1972, 1979) does not mention elevation as a critical element in the establishment of mangroves. Teas et al. (1976) state that "elevation with respect to tidal levels was a significant factor in mangrove establishment" and indicate that the red mangrove seedlings would successfully establish in low energy areas "... at elevations ranging from mean high water to mean low water", while black and white mangrove volunteers did not establish below +0.3 m mean sea level (MSL). The suggested range of red mangrove establishment is in disagreement with later work. Teas (1977) emphasizes the importance of "tidal depth of planting" as an important factor in the successful establishment of planted mangroves but offers no specifics, only repeating the advice of Pulver (1976) to plant in the usual tidal range occurrence of the species. Woodhouse (1979) incorrectly states that all three mangrove species "are found growing at elevations from slightly below mean tide level (MTL) to well above MHW [mean high water]. Where both mangroves and salt marsh occur together, the mangroves extend seaward of the salt marsh" (p. 42). Teas (1981) recommends appropriate tidal levels of "generally between mean sea level and mean high water" (p. 98), or 0.0 m to +0.4 m National Geodetic Vertical Datum (NGVD). Lewis (1982b) again notes the importance of elevation but offers no specific guidelines beyond those of Pulver (1976) and the data of Goforth and Thomas (1980) indicating higher survival of planted red mangroves above mean sea level.

This lack of specificity and errors in interpreting data has led to the incorrect conclusion by many that mangroves could be successfully planted anywhere within the intertidal zone. Good experimental evidence which establishes the best zone of planting is rare. Stephen (1984) (Figure 5) published the first data clearly showing that at the specific site location examined (Naples, Florida), the optimum tidal elevation of planted red mangroves was +0.4 m NGVD, and that volunteer black mangroves were abundant only above an elevation of +0.4 m NGVD. These data are further supported by the previously published elevation range information of Detweiler et al. (1976) from Tampa Bay, where the mean elevation of red mangroves naturally colonizing a disturbed area was +0.3 m NGVD. For white mangroves it was also +0.3 m NGVD and for black mangroves it was +0.4 m NGVD. Elevations of mangroves in an undisturbed area were very similar. Provost (1973, 1976) also emphasized the normal occurrence of tidal plants in the upper tidal range. Beever's (1986) recommended tidal ranges for planted mangroves correctly start at mean high water (about +0.3 m NGVD) but extend too high (up to +0.9 m NGVD).

Similar work is currently underway for an 80-hectare mangrove mitigation project in Broward County on Florida's east coast (Mangrove Systems, Inc. 1987). The agreed target elevation for excavation areas, in order to encourage red mangrove establishment, was +0.2 m NGVD. Ongoing experimental work indicates that a range of elevations between +0.3 and +0.5 m NGVD is more appropriate (Mangrove Systems, Inc. 1987). The inappropriate +0.2 m NGVD tidal elevation had been written into permits issued by the Florida Department of Environmental Regulation and the U.S. Army Corps of Engineers (FDER #060942909, COE #84J-2528) in spite of evidence provided by the author during permit drafting indicating that such an elevation was too low for optimum mangrove growth. Based on the experimental work cited above, the regulatory agencies agreed (in 1987) to allow modification of the permit to provide for excavation of uplands to a range of elevations between +0.3 m and +0.4 m NGVD for natural mangrove seedling recruitment. If such natural recruitment does not result in the presence of one mangrove seedling for each 4 square meters (2 m centers) over at least 50% of the area within a year, then actual planting of mangroves must take place in the bare areas. Based on the natural colonization rates observed in the test plot area (Figure 6), this density should be achieved within the allowed one year from excavation.

Salt marsh vegetation occurs throughout a tidal range similar to that of mangroves. Planted smooth cordgrass will survive and expand at a slightly

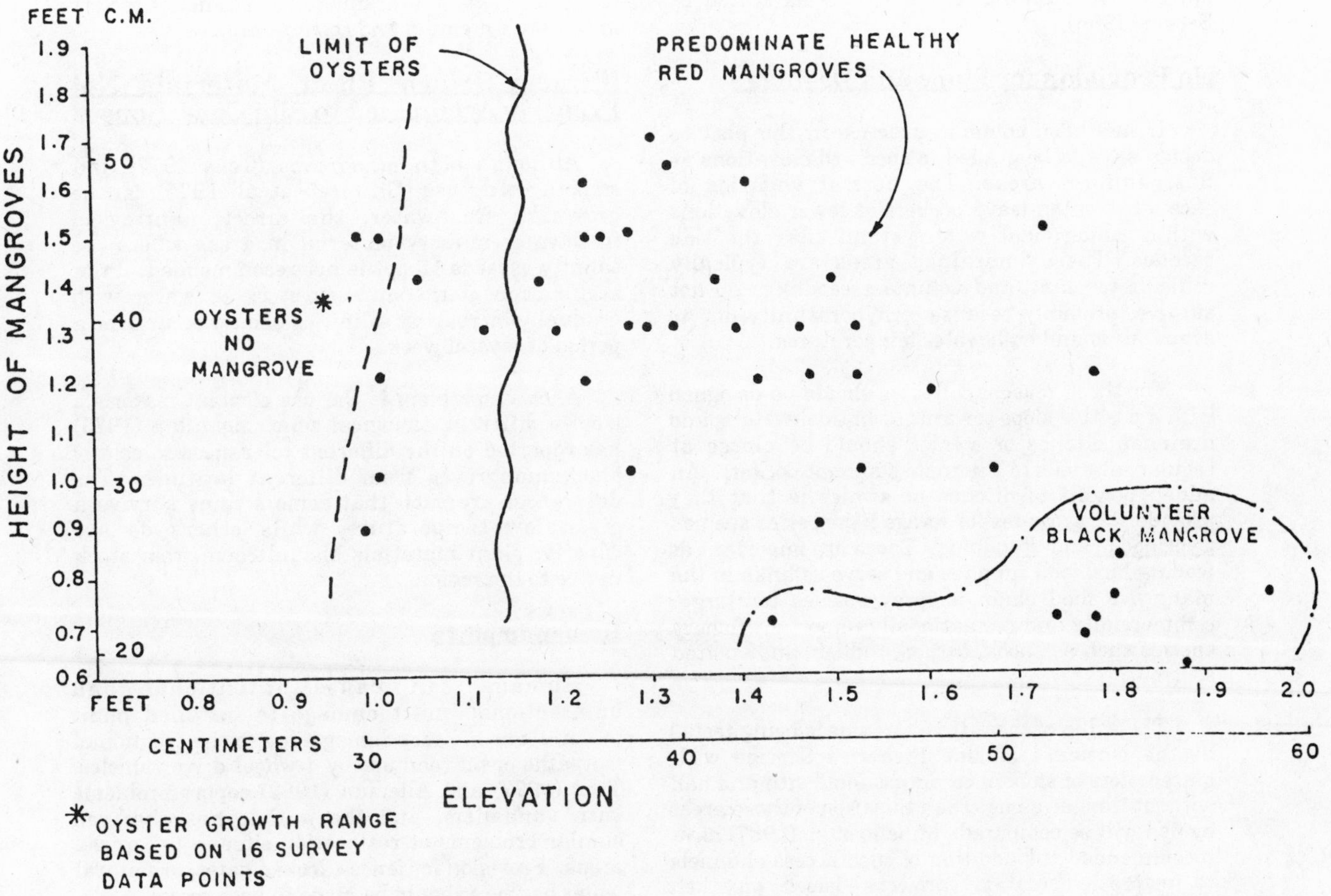

FEET C.M.
HEIGHT OF MANGROVES
LIMIT OF OYSTERS
PREDOMINATE HEALTHY RED MANGROVES
OYSTERS NO MANGROVE
VOLUNTEER BLACK MANGROVE
1.9 1.8 1.7 1.6 1.5 1.4 1.3 1.2 1.1 1.0 0.9 0.8 0.7 0.6
50 40 30 20
FEET 0.8 0.9 1.0 1.1 1.2 1.3 1.4 1.5 1.6 1.7 1.8 1.9 2.0
CENTIMETERS 30 40 50 60
ELEVATION
* OYSTER GROWTH RANGE BASED ON 16 SURVEY DATA POINTS

Figure 5. Mangrove occurrence and height related to tidal elevations (datum NGVD). From Stephen (1984).

Figure 6. Volunteer mangrove colonization 8 months after site excavation in West Lake, Broward County, Florida.

lower elevation than mangroves (+0.2 m NGVD) and is commonly found growing in deeper water in front of a mangrove fringe (Lewis 1982a). It will not survive planting at 0.0 NGVD as stated by Beever (1986).

No Provision for Slope and Drainage

It has been common practice in the past to design sites to be graded to specified elevations as flat, uniform areas. The normal vagaries of excavation often leave pockets at lower elevations within which tidal waters stand after the tide recedes. These undrained areas are typically difficult to plant, and volunteer seedlings do not survive, probably because of hypersalinity due to evaporation and high water temperatures.

For these reasons, all sites should be designed with a positive slope towards open tidal waters, and drainage ditches or swales should be placed at regular intervals to eliminate stagnant pockets. An added benefit of ditches or swales is that they provide access routes for forage fish species such as Cyprinodon and Fundulus. These are important as wading bird food supplies and serve as links in the mangrove food chain as prey species for larger commercially and recreationally important fishery species such as snook, tarpon, redfish, and spotted seatrout.

The value of these tidal streams is being tested by the National Marine Fisheries Service with paired plots of smooth cordgrass, half with and half without tidal streams. The utilization of these areas by fish will be compared. Minello et al. (1987) have recommended the addition of such access channels to restoration/creation projects based on their work in Texas. Unfortunately, the Florida Department of Environmental Regulation's draft mitigation technical manual (Beever 1986) does not recognize slope and drainage as significant factors in wetlands creation and restoration.

Nursery-Grown Plant Materials Not Properly Acclimated to Site Conditions

Although both mangroves (Teas 1977) and smooth cordgrass (Garbisch et al. 1975) can be grown in freshwater, the direct planting of freshwater nursery material in areas where the salinity exceeds 15 ppt is not recommended. Prior acclimation of the nursery stock to water with gradually increasing salinities should occur over a period of several weeks.

Another problem is the use of plant materials from a different ecological zone. McMillan (1975) has reported on the different tolerances to cold of black mangroves from different latitudes. The differences are such that some strains survive a given low temperature, while others do not. Clearly, plant materials should come from stock native to the region.

Human Impacts

Humans can cause intentional and unintentional direct damage to installed plant materials or indirect damage due to the creation of footpaths or off road use by 4-wheel drive vehicles. Teas (1977) and Alleman (1982) report problems with vandalism, and the author has observed similar problems at restoration sites in urbanized areas. Provision for fences, locked gates, or natural water barriers should be made in such areas.

DESIGN OF CREATION/RESTORATION PROJECTS

PRECONSTRUCTION CONSIDERATIONS

Location

The location of a potential site is important because of three considerations: logistics, cost, and habitat value.

Logistically, sites accessible by land are easier (and cheaper) to build and monitor than those only accessible by water. They are accessible year-round, while bad weather may increase costs associated with sites accessible only by water. The cost of a project is directly related to accessibility by the necessary heavy equipment and personnel. Land-accessible sites where excavation is necessary can cost as much as $62,000/ha to construct and plant. Costs can double or triple where only water access is available.

Sites where freshwater drainage is sufficient to produce reduced salinities in the created/restored wetland may be preferable, if the goal is to produce oligohaline estuarine nursery habitat.

Site Characteristics

Exposure to Waves--

As discussed in the section on reasons for success/failure, excessive wave energy is a common problem with survival of planted mangroves. Savage (1972) suggested that black mangroves may be more suitable than other species for controlling erosion, but no experimental evidence exists to support this hypothesis. If mangroves are the desired species, the construction of linear breakwaters or planters (Figure 4) can reduce the wave energy enough to allow plant survival. There are no published maximum fetch values for successful unprotected mangrove plantings, although Goforth and Williams (1984) describe three sites with varying degrees of exposure, and note greater success at a protected site than at an exposed one.

For tidal marsh plantings, the general guidelines of Knutson et al. (1981) and Broome (this volume) as to site selection for plantings of smooth cordgrass can be used in Florida until more specific information is available.

Tidal Range and Planned Elevation of Planting--

As described, studies have shown that the optimum planting range for both mangroves and tidal marsh plants is 1) similar to their natural range of elevations, and 2) generally falls between +0.3 m and +0.6 m NGVD in most of Florida. Local variations of tidal amplitude and type (Figure 7) are sufficiently significant to require confirmation of the range of elevation occupied by fully grown representatives of the plant species in question as close to the proposed creation/ restoration site as possible. Even then, it is best to be conservative in designing planting elevations. The lowest and highest points should be disregarded and only the middle range used. When any doubt exists, a test program similar to that of Mangrove Systems, Inc. (1987) should be instituted prior to specifying the target elevations.

Also as noted, the site should not be of uniform elevation. A design slope with drainage features (access channels) is a key to an ecologically functional project.

Salinity--

Interstitial salinities above 90 ppt are lethal to mangroves (Cintron et al. 1978). These conditions are likely to arise if proper drainage is not provided and standing pools of shallow tidal waters occur on site. Also, if the elevation of the site is such that only spring tides flood it once or twice a month, a salt barren may result due to high interstitial salinities. Based on the data of Detweiler et al. (1976), an elevation of +0.8-1.0 m NGVD would probably result in such conditions. Beever's (1986) recommendations for black mangrove plantings at MHW +0.3 m and white mangroves at MHW +0.6 m (assuming MHW at +0.5 m NGVD) are thus too high.

Another important aspect of salinity is the ecological importance of lower salinity (oligohaline) wetlands, as previously discussed. If such a habitat is planned, careful review of salinity data is essential to ensure that target salinities result in the oligohaline wetland.

Shading--

Existing terrestrial vegetation that causes shading can create problems when attempting to stabilize shorelines by planting native vegetation. In much of Florida, the introduced exotics Brazilian pepper (Schinus terebinthifolius Raddi) and Australian pine (Casuarina equisetifolia Forst) occupy disturbed shorelines. These exotics can cut off the light needed by volunteer propagules, preventing their colonization of shoreline areas. Courser and Lewis (1981) describe a successful shoreline stabilization program involving removal of Brazilian pepper.

CRITICAL ASPECTS OF THE PROJECT PLAN

Timing of Construction

For all plant materials except mangrove propagules, the optimum installation period is April through mid-June. The availability of red mangrove propagules in the numbers needed typically limits the window of planting from mid-August to mid-October. Sites being prepared to accept volunteer mangrove propagules also need to be planned for completion around this time period. Red mangrove propagules cannot be successfully stored in a dormant condition and therefore must be planted immediately after collection.

These requirements dictate that creation/ restoration sites requiring construction need to be planned for completion prior to the planting window. In addition, orders for nursery-grown plants need to be placed far enough in advance to allow for proper growth (120 days in the winter, 60 days in the summer). A "glitch factor" of at least 30 days should be factored into any construction and planting plan. A routine problem with success in creation/restoration plans is the completion of site preparation outside the optimum planting window.

Pre-Construction Quality Control

Even when accurate plans are prepared, actual site construction may not achieve the required tolerances. It is important that an as-built survey be completed before construction equipment is moved off site, so that corrections can be made quickly and inexpensively. Returning equipment to the site often produces delays, if the equipment can be brought back at all. A quick inspection of the site to check for proper drainage can be accomplished by simply watching the tide rise and fall across the site.

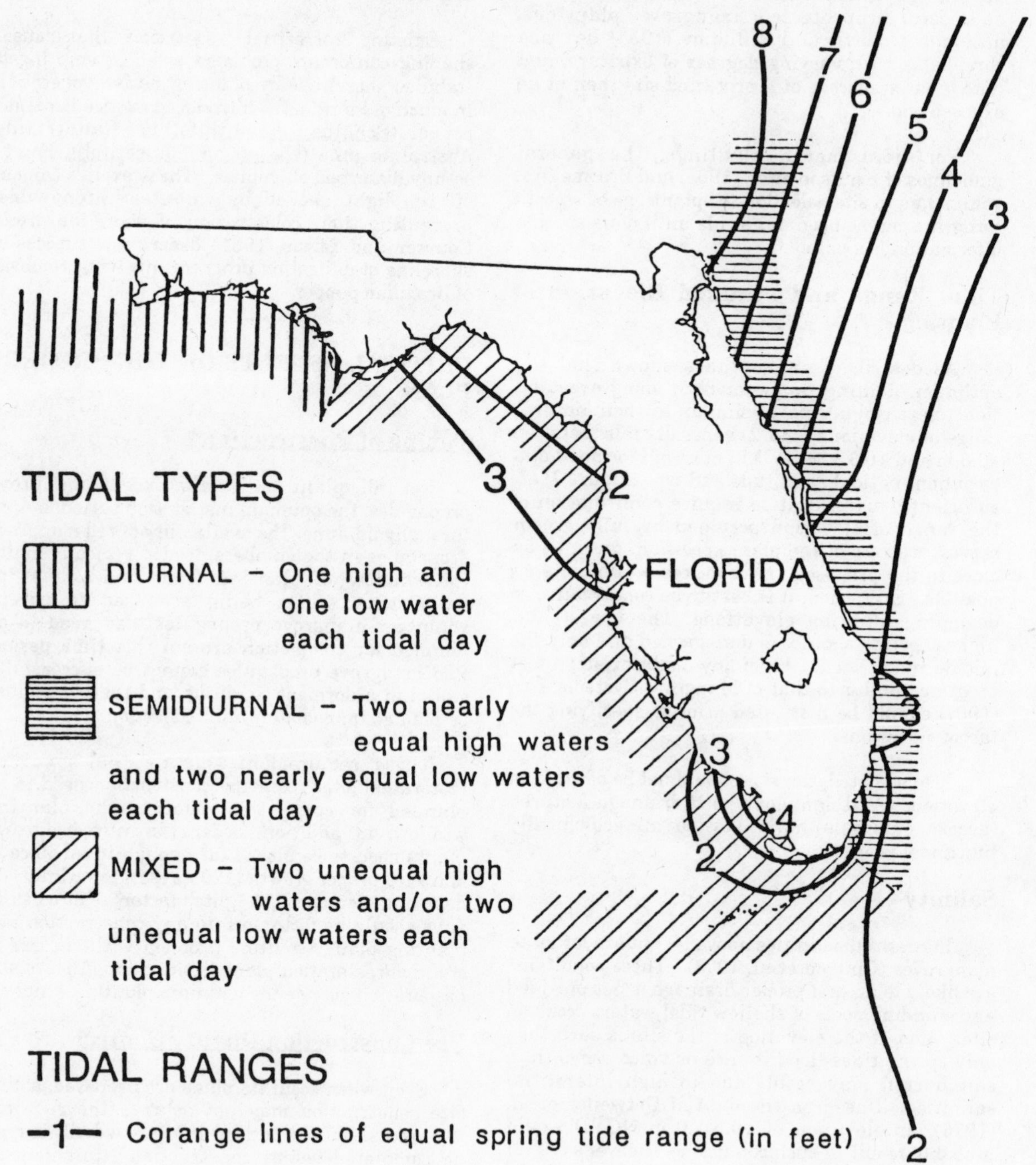

Figure 7. Tidal ranges and types in Florida. Modified from Fernald (1981).

Substrate

At a minimum, the substrate composition should be verified by simple soil auger checks. If rock or clay layers are encountered at the proposed excavation depth, the site may be unacceptable.

The need to add fertilizer to improve plant survival and growth has not been adequately tested in Florida. The general guidelines of Woodhouse and Knutson (1982) and Broome (this volume) should be used when deciding whether to use fertilizers in tidal marsh plantings. Fertilizers appear to be particularly useful in plantings on exposed shorelines where rapid growth is desirable to minimize vulnerability of new plants to wave action.

The value of adding fertilizer to mangrove plantings is not well documented. Zuberer (1977) documents the presence of nitrogen fixation activity in the roots of mangroves, while Reark (1983) argues that fertilizers were essential to the successful outcome of his project. Teas (1977) states that "Nursery mangroves of all three species were found to respond to fertilizer. Because open-water fertilization is ordinarily not practical, pre-transplanting fertilization may prove useful." (p. 56). Reark (1984) describes the addition of a soluble fertilizer to nursery-grown mangroves, but there were no controls. In fact, no controlled experimentation has been reported to demonstrate any value of added fertilizer, although Snedaker (University of Miami pers. comm.) describes experimental use of fertilizer (Agriform) on marginal sites with better survival rates.

Plant Material

Four types of mangrove plantings are available: propagules, 1-2 year old seedlings, 3-5 year old nursery-grown trees, and field-collected transplants. The 1-2 year old seedlings are most often recommended or required as plant material. They are grown from propagules harvested in the wild. In fact, Beever (1986) recommends using only one year old (one foot minimum height) nursery grown seedlings, with no reference to other plant materials.

The direct installation of red mangrove propagules (Figure 8) has been popular due to low cost and general success in protected sites. Goforth and Thomas (1980) compared red mangrove propagules and field-dug seedlings 12 to 18 months old. At the end of five years, "survival of transplanted seedlings was no more successful than that of propagules" while "the average vertical growth of seedlings ... was significantly ($p<0.001$) less than propagules" (p. 221). Stephen (1984) reports 97% survival of planted red mangrove propagules after 8 months at a large project in Naples, Florida.

Direct installation of propagules of the other three species is not practical due to their need to shed a pericarp and rest on the surface of a damp substrate for several days prior to anchoring. Broadcasting of these propagules might be successful in some projects, but Lewis and Haines (1981) report low overall successful anchoring of broadcast propagules.

The use of larger plant materials greatly increases the cost of a project (Teas 1977, Lewis 1981) and should be used only where absolutely necessary. Goforth and Thomas (1980) note that in exposed sites, transplanted 2-3 year old trees are the only successful plant material. Teas (1977) reports that all attempts to transplant large (6 m tall) mangroves failed. However, Gill (1971) reported transplanting red mangroves up to 6.5 m in height that Carlton (1974) indicates had high survival. Pulver (1976) provides guidelines for the transplanting of mangroves up to 2 m tall and reports good success.

Finally, there may be instances in which no installation of mangrove plant materials is necessary if volunteer floating propagules are numerous. The largest (80 ha) mangrove restoration project in Florida was designed to require no installed mangroves at all, and preliminary testing appears to confirm the success of this technique where the natural floating propagules are sufficiently available (Figure 6; Mangrove Systems, Inc. 1987). Given the limited availability of funds for restoration, the elimination of planting could provide extra money for more excavation and restoration of larger areas. However, each site has unique characteristics that require pilot projects to confirm the utility of this alternative.

Tidal marsh plant materials available in Florida include field-dug bare root units and plugs, and cultivated 2" pots. Unlike the situation farther north, seeds of the most frequently used species, smooth cordgrass, have not been available in large numbers in Florida due to insect and fungal damage to seed heads (Lewis, pers. obs.). Future work may reveal a viable source but one does not now exist.

Woodhouse and Knutson (1982) and Broome (this volume) provide details of cultivation of nursery units and use of bare root and plug units of smooth cordgrass that are generally applicable in Florida. Hoffman et al. (1985) describe specific projects in the Tampa Bay area in more detail.

Lewis (1983) describes one of the few projects attempting to restore a black needlerush marsh. In areas where the elevations were correct, transplanted 15 cm plugs of needlerush were generally successful. Where elevations were too low, competition from white mangrove volunteers resulted. Where elevations were too high, transplants died and more salt-tolerant (salt

Figure 8. Red mangrove (Rhizophora mangle) propagule.

barren) types of vegetation appeared. Needlerush is slower to expand than smooth cordgrass and when installed on one meter centers, appears to require 3-5 years to produce a closed stand.

Other species often transplanted in specific projects include leather fern and sawgrass. These are used in oligohaline situations. Leather fern can be transplanted well as plugs, but not as cut stems, as described by Beever (1986, p. 7). Sawgrass is available from some nurseries as cultivated units but is most often dug from the wild. Bare root units have exhibited poor survival in oligohaline environments, and plugs of 3-5 intact shoots with soil cores are recommended.

Introduction of Fauna

The establishment of a new faunal community in a created/restored system connected to tidal waters has historically been left to incidental movement of individual organisms with plant material, the natural settlement of planktonic life forms that become sessile, epibenthic or infaunal after metamorphosis (meroplankton), and the immigration of fauna, particularly fish, from adjacent wetlands. There is no evidence that active introduction of fauna would accelerate the colonization process, but controlled experimentation to answer this question has not been carried out. Based only on personal observations, the author sees no obvious need to introduce fauna because natural colonization of created/restored systems appears to be quite rapid. Nonetheless, good experimental evidence should be generated to confirm this conclusion.

Buffers, Protective Structures

Beever (1986) recommends that a buffer zone equivalent to the width of the planted area be cleared of exotic species. The author's experience has been that rapid invasion by these species will occur and that such a buffer zone, if not maintained free of exotics, will cease to be a buffer within five to ten years. In addition to clearing the leaves and stems of exotic plants, creation of a buffer zone should include destruction of the root systems with a systemic herbicide (e.g., Garlon IV) and the planting of native vegetation such as wax myrtle (Myrica cerifera L.) and marsh elder (Iva frutescens L. and Iva imbricata Walt.) to outcompete invading exotics. It is not the width of the buffer zone but the intensity of maintenance that will prevent invasion of exotics.

Foot traffic and vehicle access problems have been previously noted as problems. If a site is in an urban area where public access can be expected, the site should be fenced and legally posted.

Long Term Management

Control of exotics over the long term will require implementation of a control program as outlined above. Ownership changes are common over the long term. Future threats to restored/created sites from such changes are addressed in Florida by the routine requirement of conservation easements. The property owner retains title to the land but restricts the future use of the property by a county recorded easement. Fee simple transfer of ownership to the state is also possible for that portion of the property not needed for development.

MONITORING

What to Monitor and How

Prior to implementing a monitoring program, it is essential that measurable goals be defined. The goalsshould be reasonable and based on published literature values of parameters or on values reported in monitoring reports readily available to all parties. Because of the haphazard nature of the methods of obtaining and reporting data on wetlands in Florida, it is virtually impossible to utilize criteria for determining success much beyond percent survival, growth measurements, and areal cover measurements. This should improve as more data are obtained and centralized for review.

Problems with efforts to apply more specific criteria are illustrated by a target goal of specific numbers of invertebrate species and their density in a created/restored wetland. Baseline data from tidal wetlands in Florida is minimal and thus defining a criterion such as 30,000 organisms/m^2 as "successful" is impossible. An alternative is the comparison of a control wetland to the created/restored one. However, due to natural variations in wetlands that appear identical to an observer, variances in faunal densities as great as 50% are not uncommon. The Florida Department of Environmental Regulation's new mitigation rule establishes the use of "reference" wetlands for comparison with wetlands created/restored to meet mitigation needs. Numerical criteria of similarity have deliberately been omitted from the rule after attempts to include them received strong criticism in public hearings. Only by careful study of existing created/restored wetlands in comparison with adjacent control areas can any general criteria be established. Because this has not been done in Florida, we must work with what we have until such data is available.

What do you monitor and how? First, it is important to describe in detail, using maps and photographs, what was done at the site. How big was it? What were the slopes and elevations? What tidal benchmark was used? What types and numbers of plants were installed, and where and when? As a result of normal construction problems, the as built specifications are frequently not the same as specified in the permit. Thus, a "time zero" report detailing exactly what was done when and where is essential.

Second, some sampling regime needs to be established that will be repeated over a period of time. The typical standard for tidal marsh systems is quarterly sampling for two years. This results in reports issued at time zero (completion of construction and planting), 3, 6, 9, 12, 15, 18, 21 and 24 months for a total of nine reports. Longer times may be required for monitoring forested wetlands; a five-year minimum is recommended, and sampling intervals may lengthen as the monitoring period increases.

Third, a sampling program involving either pre-established plots or random plots determined at each monitoring inspection needs to be described and justified. Broome (this volume) supports stratified random sampling with sampling in each elevation zone. The sampling program should begin at the completion of construction and/or planting (time zero).

Fourth, each sampling should include photographs taken from the same position and angle during each monitoring episode to illustrate to the report reviewer what is happening at the site (Figure 9).

Finally, the last report should summarize all the results with appropriate graphs, and compare the results with 1) the previously established goals, and 2) literature values of parameters measured.

This monitoring program is most easily accomplished by making it a condition of the permit and requiring the permittee to pay for it. Few regulatory agencies have the staff or funds to conduct detailed compliance monitoring themselves. Simply inspecting the site once or twice during the life of the permit is an accomplishment for agency personnel, and the reports with photographs are an important compliance monitoring tool.

With regard to the parameters to be measured, percent survival of installed plant materials and/or volunteer recruitment within plots should be measured and extrapolated to describe the conditions across the surface of the created/restored area. Sample plots are important because, except in small planting areas, counting hundreds or thousands of units individually can be tedious, expensive, and error-prone. One meter square plots are typically sufficient for tidal marsh plantings. Four meter square (2 m x 2 m) plots may be better for tree species. The absolute number of plots will depend on site size; Broome (this volume) suggests fifteen as a minimum.

Each planting unit or volunteer in the plot should be measured for height and the percent cover in the plot estimated. If random units are chosen each time for measurement, a quadrat can be centered over them for measurement. For herbaceous species like smooth cordgrass, culm (stem) density can be measured using 10 cm x 10 cm quadrats. For mangroves, plant height and prop root or pneumatophore number can be noted.

Above-ground and below-ground biomass of herbaceous species can be measured by clipping at ground level, taking core samples, and separating out plant material for drying and weighing (see Broome, this volume). It is not practical to measure biomass of mangroves because this involves destructive sampling and loss of planting units.

Percent survival alone should not be the sole criterion of success. Low percent survival of rapidly expanding smooth cordgrass units may still provide 100% cover of the desired area. With mangroves, mortality of young plants is normal with competition. Pulver (1976) measured the density of mangroves in natural forests and found that as the stand height (and age) increased, the number of

A

B

Figure 9. Time sequence photographs of a planted smooth cordgrass marsh on a dredged material island (Sunken Island Extension) in Tampa Bay, Florida. A - time zero; B - 12 months.

C

D

Figure 9. C - 24 months; D - 84 months, showing mangrove invasion of marsh.

trees decreased. For example, the mean density of red mangroves decreased from 26.8 trees/m^2 for 1.2 m tall trees to 8.3 trees/m^2 for 1.9 m tall trees. This represents a survival rate of only 31%. Would that be called "successful" in a created/restored system?

Percent cover should supplement percent survival as a measurement of expansion of leaf area. Combined with growth and stem density measurements, it provides a good indication of whether a system is healthy and expanding in plant height and cover. Data are generally absent on good rates of growth and coverage, but Lewis and Crewz (in prep.) will provide some typical target values.

Faunal sampling in created/restored systems is more of an art than a science at present. A number of studies are underway in created/restored tidal wetland systems in Florida, under the auspices of the National Marine Fisheries Service (Beaufort, North Carolina), U.S. Army Corps of Engineers Waterways Experiment Station (Vicksbug, Mississippi) and Florida Department of Natural Resources (St. Petersburg). None of the results have been published at this time. Readers are encouraged to contact these agencies directly for updated publications on the subject.

Mid-Course Corrections

If, during the course of monitoring, it becomes obvious that project goals will not be met, there are two choices. One is to determine the cause of the problem and to correct it. It may be elevation, drainage, source of plant materials, etc. The other choice is to evaluate the habitat value of the system as it exists, and determine if that is sufficient to satisfy regulatory agencies. It is possible that the wetland was a failure but the project was otherwise successful. Dial and Deis (1986) describe a mitigation site on Tampa Bay where plant survival was less than 10%, yet the authors state that "the combination of mangrove, S. alterniflora, shallow subtidal and intertidal habitats ... supported the most diverse assemblages of birds observed during this study" (p. 34). In this case, the state regulatory agency decided not to require any modification of the site to improve plant survival. But when a mitigation project fails to meet specified goals, an agency may require additional creation/ restoration at another site to compensate for the lost habitat values.

If the first course of action is taken, the cost of the modifications may cause the permittee to challenge the regulatory agency's right to ask for changes in a plan it originally approved. Such questions have been raised in the case of unsuccessful creation/restoration attempts in Florida and usually, the agency has acquiesced. Careful preparation of permit conditions to provide for mid-course corrections is essential.

INFORMATION GAPS AND RESEARCH NEEDS

Centralized Data Bank

As has been stated, Lewis and Crewz (in prep.) note that the lack of a centralized database concerning historical as well as current creation/restoration projects usually hamper data analyses, and prevents comparisons of projects.

Although the Florida Department of Environmental Regulation has maintained a computerized permit tracking system for a number of years, the system has not historically recorded data on creation/restoration projects. Changes in the data entry format are now being implemented that will include a description of permit requirements. This will help provide a creation/restoration database over time, but compliance monitoring data and monitoring reports by the permit applicant will still be filed as before, at regional offices. Retrieval of this data may still be a significant problem. Equally important is the lack of data concerning previously permitted creation/restoration projects. No plans are underway to add this information to the new system, thus limiting retrieval of information about older projects, some now ten years old.

Natural Propagule Recruitment versus Planting or Transplanting

The natural recruitment rates, survival and growth of volunteer propagules needs to be tested against various sizes and densities of installed plant materials to determine the optimum densities needed for certain target coverage rates and habitat utilization. For example, it has been assumed that volunteer mangrove propagule recruitment could not match the growth or success rates of planted nursery-grown mangroves. With costs of $1,000 to $200,000 per hectare for planted mangroves (Teas 1977, Lewis 1981), significant savings could occur if natural recruitment proved effective. With limited public funds available for habitat restoration, the cost-effectiveness of particular plant materials needs to be documented.

Transplanting Larger Mangroves

As noted, the success of transplanting larger (>2 m tall) mangroves has generally not been good. The salvage of larger mangroves destined for destruction might prove valuable if larger trees could be moved successfully.

Comparable Growth Rates of Mangrove Propagules and Seedlings

Is it necessary to plant 1-2 year old nursery-grown seedlings in order to achieve more rapid cover, or will natural propagule recruitment, or planted or broadcast propagules, achieve equivalent growth? The previously described data of Goforth and Thomas (1980) needs amplification.

Rate of Faunal Recruitment

A comparison of sites of different ages in a synoptic manner is needed to determine how rapidly faunal recruitment takes place, and whether supplementing the process is necessary.

Functional Equivalency

A multi-parameter comparison of created/restored systems with several natural areas is needed to determine which functional values (habitat, water quality, primary production, etc.) can be re-established, and over what time frame.

Regional Creation/Restoration Plans

Wetland creation and/or restoration need to be examined in the context of the regional ecosystem, with the possible outcome that out-of-kind or off-site creation/restoration may be deemed acceptable due to the regional loss and scarcity of a distinct habitat type such as oligohaline marshes. A regional approach would also be useful in designing mitigation banks.

ACKNOWLEDGEMENTS

Portions of this chapter were developed under the auspices of the Florida Sea Grant Program with support from the National Oceanic and Atmospheric Administration, Office of Sea Grant, U.S. Department of Commerce, Grant No. R/C-E-24.

LITERATURE CITED

Alleman, R.W. 1982. Biscayne Bay: A Survey of Past Mitigation/Restoration Efforts. Department of Environmental Resources Management, Dade County, Florida.

Anonymous. 1984. Now you see it ... Fla. Naturalist, Fall 1984:16-17.

Autry, A.S., V. Stewart, M. Fox, and W. Hamilton. 1973. Progress report: mangrove planting for stabilization of developing shorelines. Q. J. Fla. Acad. Sci. 36 (Suppl. to No. 1):17 (abst).

Bailey, R.G. 1978. Ecoregions of the United States. U.S. Forest Service, Ogden, Utah.

Banner, A. 1983. Florida Keys environmental mitigation trust fund, p. 155-165. In F.J. Webb, Jr. (Ed.), Proc. 9th Ann. Conf. Wetlands Restoration and Creation. Hillsborough Community College, Tampa, Florida.

Beever, J.W. 1986. Mitigative Creation and Restoration of Wetland Systems--A Technical Manual for Florida. Draft Report. Florida Dept. of Environmental Regulation, Tallahassee, Florida.

Cammen, L.M. 1976. Microinvertebrate colonization of Spartina marsh artificially established on dredge spoil. Est. Coast. Mar. Sci. 4(4):357.

Cammen, L.M., E.D. Seneca, and B.J. Copeland. 1976. Animal Colonization of Salt Marshes Artificially Established on Dredge Spoil. U.S. Army Corps of Engineers TP 76-7. U.S. Army Coastal Engineering Research Center, Fort Belvoir, Virginia.

Carlton, J. 1974. Land building and stabilization by mangroves. Environ. Conserv. 1(4):285-294.

Cintron, G., A.E. Lugo, D.J. Pool, and G. Morris. 1978. Mangroves of arid environments in Puerto Rico and adjacent islands. Biotropica 10:110-121.

Courser, W.K. and R.R. Lewis. 1981. The use of marine revegetation for erosion control on the Palm River, Tampa, Florida, p. 125-136. In D.P. Cole (Ed.), Proc. 7th Ann. Conf. Restoration and Creation of Wetlands. Hillsborough Community College, Tampa, Florida.

Cowardin, L.M., V. Carter, F.G. Golet, and E.T. LaRoe. 1979. Classification of Wetlands and Deepwater Habitats of the United States. U.S. Fish Wildl. Serv., FWS/OBS-79/31.

Davis, J.H. 1940. The Ecology and Geologic Role of Mangroves in Florida. Pps. from the Tortugas Lab., Vol. 32. Carnegie Inst. Wash. Pub. NN517.

Department of Environmental Resources Management. 1985. Biscayne Bay Today--A Summary of its Physical and Biological Characteristics. Metro-Dade County, Miami, Florida.

Detweiler, T.E., F.M. Dunstan, R.R. Lewis, and W.K. Fehring. 1976. Patterns of secondary succession in a mangrove community, Tampa Bay, Florida, p. 52-81. In R.R. Lewis (Ed.), Proc. 2nd Ann. Conf. Restoration of Coastal Vegetation in Florida. Hillsborough Community College, Tampa, Florida.

Dial, R.S. and D.R. Deis. 1986. Mitigation Options for Fish and Wildlife Resources Affected by Port and Other Water Dependent Developments in Tampa Bay, Florida. U.S. Fish Wildl. Serv. Biol Rep. 86(6).

Drew, R.D. and N.S. Schomer. 1984. An Ecological Characterization of the Caloosahatchee River/Big

Cypress Watershed. U.S. Fish Wildl. Serv., FWS/OBS-82/58.2.

Durako, M.J., J.A. Browder, W. L. Kruczynski, C.B. Subrahmanyam, and R.E. Turner. 1985. Salt marsh habitat and fishery resources of Florida, p. 189-280. In W. Seaman, Jr. (Ed.), Florida Aquatic Habitat and Fishery Resources. Fla. Chapter, American Fisheries Society, Kissimmee, Florida.

Fernald, E.A. (Ed.). 1981. Atlas of Florida. Florida State University Foundation, Inc. Tallahassee, Florida.

Fernald, E.A. and D.J. Patton (Eds.). 1984. Water Resources Atlas of Florida. Florida State University Foundation, Inc. Tallahassee, Florida.

Garbisch, E.W., Jr., P.B. Woller, and R.J. McCallum. 1975. Salt Marsh Establishment and Development. U.S. Army Engineer Coastal Engineering Research Center Tech. Mem. no. 52. Fort Belvoir, Virginia.

Gill, A.M. 1971. Mangroves--is the tide of public opinion turning? Fairchild Trop. Gard. Bull. 26(2):5-9.

Gill, A.M. and P.B. Tomlinson. 1969. Studies on the growth of red mangrove (Rhizophora mangle L.): 1. Habitat and general morphology. Biotropica 1(1)1-9.

Gilmore, R.G., C.J. Donohoe, and D.W. Cooke. 1983. Observations on the distribution and biology of east-central Florida populations of the common snook, Centropomus undecimalis (Bloch). Fla. Scientist 46:313-336.

Goforth, H.W. and J.R. Thomas. 1980. Planting of red mangroves (Rhizophora mangle L.) for stabilization of marl shorelines in the Florida Keys, p. 207-230. In D.P. Cole (Ed.), Proc. 6th Ann. Conf. Restoration and Creation of Wetlands. Hillsborough Community College, Tampa, Florida.

Goforth, H.W. and M. Williams. 1984. Survival and growth of red mangroves (Rhizophora mangle L.) planted upon marl shorelines in the Florida Keys (a five year study), p. 130-148. In F.J. Webb, Jr. (Ed.), Proc. 10th Ann. Conf. Wetlands Restoration and Creation. Hillsborough Community College, Tampa, Florida.

Hannan, J. 1976. Aspects of red mangrove reforestation in Florida, p. 112-121. In R.R. Lewis (Ed.), Proc. 2nd Ann. Conf. Restoration of Coastal Vegetation in Florida. Hillsborough Community College, Tampa, Florida.

Hoffman, W.E., M.J. Durako, and R.R. Lewis. 1985. Habitat restoration in Tampa Bay, p. 636-647. In S.F. Treat, J.L. Simon, R.R. Lewis, and R.L. Whitman, Jr. (Eds.), Proc. Tampa Bay Area Scientific Information Symposium [May 1982]. Burgess Publishing Co., Minneapolis, Minnesota.

Hoffman, W.E. and J.A. Rodgers. 1981a. A cost/benefit analysis of two large coastal plantings in Tampa Bay, p. 265-278. In D.P. Cole (Ed.), Proc. 7th Ann. Conf. Wetlands Restoration and Creation. Hillsborough Community College, Tampa, Florida.

Hoffman, W.E. and J.A. Rodgers. 1981b. Cost-benefit aspects of coastal vegetation establishment in Tampa Bay, Florida. Env. Conserv. 8(1):39-43.

Kinch, J.C. 1976. Efforts in marine revegetation in artificial habitats, p. 102-111. In R.R. Lewis (Ed.), Proc. 2nd Ann. Conf. Restoration of Coastal Vegetation in Florida. Hillsborough Community College, Tampa, Florida.

Knutson, P.L., R.A. Brochu, W.N. Seelig, and M. Innskeep. 1982. Wave damping in Spartina alterniflora marshes. Wetlands 2:87-104.

Knutson, P.L., J.C. Ford, M.R. Innskeep, and J. Oyler. 1981. National survey of planted salt marshes (vegetative stabilization and wave stress). Wetlands 1:129-157.

Kruczynski, W.L. 1982. Salt marshes of northeastern Gulf of Mexico, p. 71-87. In R.R. Lewis (Ed.), Creation and Restoration of Coastal Plant Communities. CRC Press, Boca Raton, Florida.

Lewis, R.R. 1981. Economics and feasibility of mangrove restoration, p. 88-94. In P.S. Markpouts (Ed.), Proc. U.S. Fish Wildl. Serv. Workshop on Coastal Ecosystems of the Southeastern United States. Washington, D.C.

Lewis, R.R. 1982a. Low Marshes, Peninsular Florida, Ch. 7, p. 147-152. In R.R. Lewis (Ed.), Creation and Restoration of Coastal Plant Communities. CRC Press, Boca Raton, Florida.

Lewis, R.R. 1982b. Mangrove forests, Ch. 8, p. 154-171. In R. R. Lewis (Ed.), Creation and Restoration of Coastal Plant Communities. CRC Press, Boca Raton, Florida.

Lewis, R.R. 1983. Restoration of a needlerush (Juncus roemerianus Scheele) marsh following interstate highway construction. II. Results after 22 months, p. 69-83. In F.J. Webb, Jr. (Ed.), Proc. 9th Ann. Conf. Restoration and Creation of Wetlands. Hillsborough Community College, Tampa, Florida.

Lewis, R.R. and D. Crewz. In preparation. An Analysis of the Reasons for Success and Failure of Attempts to Create or Restore Tidal Marshes and Mangrove Forests in Florida. Florida Sea Grant, Univ. of Florida, Gainesville, Florida.

Lewis, R.R. and F.M. Dunstan. 1976. Possible role of Spartina alterniflora Loisel. in establishment of mangroves in Florida, p. 81-100. In R.R. Lewis (Ed.), Proc. 2nd Ann. Conf. Restoration of Coastal Vegetation in Florida. Hillsborough Community College, Tampa, Florida.

Lewis, R.R., R.G. Gilmore, D.W. Crewz, and W.E. Odum. 1985. Mangrove habitat and fishery resources of Florida, p. 281-336. In W. Seaman, Jr. (Ed.), Florida Aquatic Habitat and Fishery Resources. Fla. Chapter, American Fisheries Society. Kissimmee, Florida.

Lewis, R.R. and K.C. Haines. 1981. Large scale mangrove planting on St. Croix, U.S. Virgin Islands: second year, p. 137-148. In D.P. Cole (Ed.), Proc. 7th Ann. Conf. on Restoration and Creation of Wetlands. Hillsborough Community College, Tampa, Florida.

Lewis, R.R., C.S. Lewis, W.K. Fehring, and J.R. Rodgers. 1979. Coastal habitat mitigation in Tampa Bay, Florida, p. 136-149. In W.C. Melander and G.A. Swanson (Eds.), Proc. of the Mitigation Symposium.

General technical report RM-65. U.S. Dept. of Agriculture, Fort Collins, Colorado.

Lot-Hergueras, A., C. Vazques-Yanes, and F. Menendez L. 1975. Physiognomic and floristic changes near the northern limit of mangroves in the Gulf coast of Mexico, p. 52-61. In G. Walsh, S. Snedaker, and H. Teas (Eds.), Proc. Int. Symp. Biol. Management of Mangroves. Inst. Food. Agr. Sci., Univ. of Florida, Gainesville.

Lugo, A.E. and C. Patterson-Zucca. 1977. The impact of low temperature stress on mangrove structure and growth. Trop. Ecol. 18:149-161.

MacNae, W. 1968. A general account of the fauna and flora of mangrove swamps and forests in the Indo-Western Pacific region. Adv. Mar. Biol. 6:74-270.

Mangrove Systems, Inc. 1987. Test Area Monitoring Report no. 3. Quarter III. [West Lake Mitigation Project]. Report to Broward County Parks and Recreation Department, Ft. Lauderdale, Florida.

McMillan, C. 1975. Adaptive differences to chilling in mangrove populations, p. 62-68. In G. Walsh, S. Snedaker, and H. Teas (Eds.), Proc. Int. Symp. Biol. Management of Mangroves. Inst. Food Agr. Sci., Univ. of Florida, Gainesville.

McMillan, C. and C.L. Sherrod. 1986. The chilling tolerance of black mangrove, Avicennia germinans, from the Gulf of Mexico coast of Texas, Louisiana and Florida. Contr. Mar. Sci. 29:9-16.

Minello, T.J., R.J. Zimmerman, and E.F. Klima. 1987. Creation of fishery habitat in estuaries, p. 106-120. In M. C. Landin and H.K. Smith (Eds.), Beneficial Uses of Dredged Material. Tech. Rep. D-87-1, U.S. Army Engineer Waterways Experiment Station, Vicksburg, Mississippi.

Odum, W. E., C.C. McIvor, and T.J. Smith III. 1982. The Ecology of the Mangroves of South Florida: A Community Profile. U.S. Fish Wildl. Serv., Office of Biol. Services. FWS/OBS-81/24. Washington, D.C.

National Wetlands Newsletter, 8(5), 1986. Environmental Law Institute, Washington D.C.

Powell, G.V.N. 1987. Habitat use by wading birds in a subtropical estuary: implications of hydrography. The Auk 104:740-749.

Provost, M.W. 1973. Mean high water mark and use of tidelands in Florida. Fla. Scientist 36(1):50-66.

Provost, M.W. 1976. Tidal datum planes circumscribing salt marshes. Bull. Mar. Sci. 26(4):558-563.

Pulver, T.R. 1976. Transplant Techniques for Sapling Mangrove Trees, Rhizophora mangle, Laguncularia racemosa, and Avicennia germinans, in Florida. Fla. Dept. Nat. Resources Mar. Res. Publ. 22.

Rabinowitz, D. 1978. Dispersal properties of mangrove propagules. Biotropica 10:47-57.

Reark, J.B. 1983. An in situ fertilizer experiment using young Rhizophora, p. 166-180. In F.J. Webb, Jr. (Ed.), Proc. 9th Ann. Conf. Wetlands Restoration and Creation. Hillsborough Community College, Tampa, Florida.

Reark, J.B. 1984. Comparisons of nursery practices for growing of Rhizophora seedlings, p. 187-195. In F.J. Webb, Jr. (Ed.), Proc. 10th Ann. Conf. on Wetlands Restoration and Creation. Hillsborough Community College, Tampa, Florida.

Reimold, R.J. and S.A. Cobler. 1986. Wetland Mitigation Effectiveness. U.S. Environmental Protection Agency, Region I, Boston, Massachusetts.

Rozas, L.P. and C.T. Hackney. 1983. The importance of oligohaline wetland habitats to fisheries resources. Wetlands 3:77-89.

Savage, T. 1972. Florida Mangroves as Shoreline Stabilizers. Fla. Dept. Natural Resources Prof. Pap. Ser. No. 19.

Savage, T. 1979. The 1972 experimental mangrove planting--an update with comments on continued research needs, p. 43-71. In R.R. Lewis and D.P. Cole (Eds.), Proc. 5th Ann. Conf. Restoration of Coastal Vegetation in Florida. Hillsborough Community College, Tampa, Florida.

Seaman, W., Jr. (Ed.). 1985. Florida Aquatic Habitat and Fishery Resources. Florida Chapter, American Fisheries Society. Kissimmee, Florida.

Smith, D.C. 1982. Shore erosion control demonstrations in Florida, p. 87-98. In R.H. Stovall (Ed.), Proc. 8th Ann. Conf. Wetlands Restoration and Creation. Hillsborough Community College, Tampa, Florida.

Snedaker, S.C. 1982. Mangrove species zonation: why? In D.N. Sen and K. Rajpuorhit (Eds.), Contributions to the Ecology of Halophytes. Dr. Junk Publishers, The Hague.

Stephen, M.F. 1984. Mangrove restoration in Naples, Florida, p. 201-216. In F.J. Webb, Jr. (Ed.), Proc. 10th Ann. Conf. Wetlands Restoration and Creation. Hillsborough Community College, Tampa, Florida.

Teas, H.J. 1977. Ecology and restoration of mangrove shorelines in Florida. Environ. Conserv. 4(1):51-58.

Teas, H.J. 1981. Restoration of mangrove ecosystems, p. 95-102. In Proc. Coastal Ecosystems Workshop, U.S. Fish Wildl. Serv. FWS/OBS-80/59.

Teas, H.J., W. Jurgens, and M.C. Kimball. 1976. Plantings of red mangroves (Rhizophora mangle L.) in Charlotte and St. Lucie counties, Florida, p. 132-162. In R.R. Lewis (Ed.), Proc. 2nd Ann. Conf. Restoration of Coastal Vegetation in Florida. Hillsborough Community College, Tampa, Florida.

Woodhouse, W.W., Jr. 1979. Building Saltmarshes Along the Coasts of the Continental United States. Special Report No. 4, U.S. Army Coastal Engineering Research Center, Fort Belvoir, Virginia.

Woodhouse, W.E., Jr., and P.L. Knutson. 1982. Atlantic coastal marshes, p. 45-109. In R.R. Lewis (Ed.), Creation and Restoration of Coastal Plant Communities. CRC Press, Boca Raton, Florida.

Woodhouse, W.W., Jr., E.D. Seneca, and S.W. Broome. 1974. Propagation of Spartina alterniflora for

Substrate Stabilization and Salt Marsh Development. U.S. Army Coastal Engineering Research Center, Fort Belvoir, Virginia.

Woodhouse, W.W., Jr., E.D. Seneca, and S.W. Broome. 1976. Propagation and Use of Spartina alterniflora for Shoreline Erosion Abatement. U.S. Army Coastal Engineering Research Center, Fort Belvoir, Virginia.

Zuberer, D. 1977. Biological nitrogen fixation: a factor in the establishment of mangrove vegetation, p. 37-56. In R.R. Lewis and D.P. Cole (Eds.), Proc. 3rd Ann. Conf. Restoration of Coastal Vegetation in Florida. Hillsborough Community College, Tampa, Florida.

APPENDIX I: RECOMMENDED READING

Carlton, J. 1974. Land building and stabilization by mangroves. Environ. Conserv. 1(4):285-294.

One of the earliest comprehensive papers discussing both the new questions about mangrove land-building capabilities and planting of mangroves for shoreline stabilization.

Chapman, U.J. 1976. Mangrove Vegetation. J. Cramer, Vaduz, Germany.

Davis, J.H. 1940. The Ecology and Geologic Role of Mangroves in Florida. Carnegie Inst., Washington, Pap. Tortugas Laboratory 32(16):303-412.

The "bible" on early ecological theories about mangroves, coupled with many on-site observations and experiments, including early planting experiments and discussion of smooth cordgrass/mangrove interactions.

Detweiler, T., F.M. Dunstan, R.R. Lewis, and W.K. Fehring. 1976. Patterns of secondary succession in a mangrove community, Tampa Bay, Florida, p. 51-81. In R.R. Lewis (Ed.), Proc. 2nd Ann. Conf. Restoration of Coastal Vegetation in Florida. Hillsborough Community College, Tampa, Florida.

Durako, M.J., J.A. Browder, W.L. Kruczynski, C.B Subrahmanyam, and R.E. Turner. 1985. Salt marsh habitat and fishery resources of Florida, p. 189-280. In W. Seaman, Jr. (Ed.), Florida Aquatic Habitat and Fishery Resources. Fla. Chapter, American Fisheries Society, Kissimmee, Florida.

Getter, C.D., G. Cintron, B. Dicks, R.R. Lewis, and E.D. Seneca. 1984. The recovery and restoration of salt marshes following an oil spill, Ch. 3, p. 65-113. In J. Cairns, Jr. and A. Bulkema, Jr. (Eds.), Restoration of Habitats Impacted by Oil Spills. Butterworth Publishers, Boston, Massachusetts.

Gill, A.M. and P.B. Tomlinson. 1969. Studies on the growth of red mangrove (Rhizophora mangle L.). I. Habit and general morphology. Biotropica 1(1):1-9.

Goforth, H.W. and J.R. Thomas. 1980. Plantings of red mangroves (Rhizophora mangle L.) for stabilization of marl shorelines in the Florida Keys, p. 207-230. In D.P. Cole (Ed.), Proc. 6th Ann. Conf. Wetlands Restoration and Creation. Hillsborough Community College, Tampa, Florida.

Hamilton, L.S. and S.C. Snedaker. 1984. Handbook for Mangrove Area Management. United Nations Env. Program and East-West Center, Environment and Policy Institute. Honolulu, Hawaii.

Hoffman, W.E., M.J. Durako, and R.R. Lewis. 1985. Habitat restoration in Tampa Bay, p. 636-657. In S.F. Treat, J.L. Simon, R.R. Lewis and R.L. Whitman, Jr. (Eds.), Proc. Tampa Bay Area Scientific Information Symposium [May 1982]. Burgess Publishing Co., Minneapolis, Minnesota.

A localized listing and discussion of marine wetland restoration/creation projects on Tampa Bay with many recommendations for improving future projects and looking at restoration as a management as well as mitigation tool.

Lewis, R.R. 1981. Economics and feasibility of mangrove restoration, p. 88-94. In Proc. Coastal Ecosystems Workshop. U.S. Fish Wildl. Serv. FWS/OBS-80/59.

A summary of work through 1980 on the role of smooth cordgrass in mangrove succession and including cost estimates of planting mangroves of various sizes.

Lewis, R.R. 1982. Creation and Restoration of Coastal Plant Communities. CRC Press, Boca Raton, Florida.

A comprehensive treatment of methods of plant establishment for nine plant community types, including Gulf of Mexico marshes, peninsular Florida marshes, Atlantic coast marshes, and mangroves.

Lewis, R.R. and F.M. Dunstan. 1976. Possible role of Spartina alterniflora Loisel in establishment of mangroves in Florida, p. 82-100. In R.R. Lewis (Ed.), Proc. 2nd Ann. Conf. Restoration of Coastal Vegetation in Florida. Hillsborough Community College, Tampa, Florida.

Lewis, R.R. and F.M. Dunstan. 1975. Use of spoil islands in re-establishing mangrove communities in Tampa Bay, Florida, p. 766-775. In G. Walsh, S. Snedaker, and H.Teas (Eds.), Proc. International Symp. Biol. and Management of Mangroves, Vol. II. Gainesville, Florida.

Lewis, R.R., R.G. Gilmore, Jr., D.W. Crewz, and W.E. Odum. 1985. Mangrove habitat and fishery resources of Florida, p. 281-336. In W. Seaman, Jr. (Ed.), Florida Aquatic Habitat and Fishery Resources. Fla. Chapter, American Fisheries Society, Kissimmee, Florida.

Lewis, R.R. and C.S. Lewis. 1978. Colonial bird use and plant succession on dredged material islands in Florida. Vol. II, Patterns of Vegetation Succession. Environmental Effects Laboratory, U.S. Army Engineer Waterways Experiment Station, Vicksburg, Mississippi.

Lewis, R.R., C.S. Lewis, W.K. Fehring, and J.R. Rodgers. 1979. Coastal habitat mitigation in Tampa Bay, Florida, p. 136-149. In W.C. Melander and G.A. Swanson (Eds.), Proc. Mitigation Symp. General tech. rep. RM-65. U.D. Dept. of Agriculture, Ft. Collins, Colorado.

A classic early work describing many experimental installations of mangroves and recommending greater emphasis on planting black mangroves (*Avicennia germinans*) due to their cold tolerance and elaborate cable root network.

Lugo, A.E. and S.C. Snedaker. 1977. The ecology of mangroves. *Ann. Rev. Ecol. Syst.* 5:39-64.

Odum, W.E., C.C. McIvor, and T.J. Smith III. 1982. The Ecology of the Mangroves of South Florida: a Community Profile. U.S. Fish Wildl. Serv., Office of Biological Services, Washington, D.C. FWS/OBS-81/24.

Detailed review of the literature with emphasis on and understanding of the basic biology of mangrove ecosystems in order to manage them properly.

Pulver, T.R. 1976. Transplant Techniques for Sapling Mangrove Trees, *Rhizophora mangle*, *Laguncularia racemosa* and *Avicennia germinans*, in Florida. Fla. Dept. Natural Resources Mar. Research Publ. No. 22.

The only detailed description of transplant procedures for mangroves to 2 m in size.

Savage, T. 1972. Florida Mangroves: a Review. Fla. Dept. Natural Resources Mar. Research Leafl. Ser. Vol VII, Part 2 (Vascular Plants), No. 1.

Savage, T. 1972. Florida Mangroves as Shoreline Stabilizers. Fla. Dept. Natural Resources Prof. Pap. Ser. No. 19.

Stephen, M.F. 1984. Mangrove restoration in Naples, Florida, p. 201-216. In F.J. Webb, Jr. (Ed.), Proc. 10th Ann. Conf. Wetlands Restoration and Creation. Hillsborough Community College, Tampa, Florida.

One of the few papers establishing optimum planting elevations through observation of survival of planted red mangroves and volunteer propagules of black mangroves.

Teas, H.J. 1977. Ecology and restoration of mangrove shorelines in Florida. *Environ. Conserv.* 4:51-58.

The first comprehensive treatment of the work through 1976 on mangrove restoration; includes a description of the basic biological traits of Florida mangroves and the problem of historical losses due to dredge and fill.

Walsh, G.E. 1974. Mangroves: a review, p. 51-174. In R. J. Reimold and W.H. Queen (Eds.), Ecology of Halophytes. Academic Press, New York.

Woodhouse, W.W., E.D. Seneca, Jr., and S.W. Broome. 1974. Propagation of *Spartina alterniflora* for Substrate Stabilization and Saltmarsh Development. TM-46, U.S. Army Coastal Engineering Research Center, Fort Belvoir, Virginia.

APPENDIX II: PROJECT PROFILES

SEAGROVE

Locale: Indian River Lagoon, Indian River County.

Latitude/longitude: 27°37'10"N / 80°21'20"W.

Permit numbers: FDER 057-760-4; USCOE (none).

Age: one year.

Size: 0.2 ha (0.5 ac).

Species present:

Avicennia germinans, Baccharis halimifolia, Bacopa monnieri, Flaveria floridana, Fimbristylis spatheca, Limonium carolinianum, Laguncularia racemosa, Paspalum distichum, Rhizophora mangle, Spartina alterniflora, Sporobolus virginicus, Salicornia virginiana.

Site description:

The Seagrove site is part of a scrapedown mitigation project along the Indian River Lagoon. Part of the mitigation was a narrow fringe behind large mangroves and was not surveyed. The main site is drained by a U-shaped ditch which connects at both ends to the Indian River Lagoon; as a result, flushing is excellent. The higher portions of the marsh are completely drained at low tide. The center lower portion is not covered with smooth cordgrass as densely as the higher perimeter. Mangroves are beginning to colonize the center area.

Goals of project:

Re-establish marsh behind a mangrove fringe as mitigation.

Attainment of goals:

The project achieved its goals and is successful.

Contact: David Crewz
Florida Dept. of Natural Resources
100 8th Ave. SE
St. Petersburg, FL 33701
813/896-8626

MELBOURNE HARBOUR, LTD.

Locale: Indian River Lagoon, Brevard County.

Latitude/longitude: 28°04'36"N / 80°35'52"W.

Permit numbers: FDER 050924-4; USCOE SAJ-44.

Age: 4-5 years.

Size: approximately 0.2 ha (0.5 ac).

Species present: Sesuvium portulacastrum, Spartina alterniflora.

Site description:

The Melbourne Harbour site was designed to curtail erosion and to mitigate for development damage. Part of the site was a long fringe of smooth cordgrass fronted by coquina rock to break boat wakes and waves; construction had obliterated most of the fringe. The part surveyed was a broad area behind a berm away from ongoing construction. Apparently, the berm had accumulated after the mitigation had been completed. The elevations behind the berm were relatively constant and the vegetation evenly distributed. The site had a broad outlet to the Indian River Lagoon and flushed well and completely. Peripheral areas were planted with coastal dropseed, marsh hay, saltgrass, and red mangrove; the mangrove areas could not be located.

Goals of project: Erosion control and mitigation.

Attainment of goals: Successful.

Contact: Steve Beeman
Ecoshores, Inc.
3881 South Nova Road
Port Orange, FL 32019
904/767-6232

COSTA DEL SOL

Locale: Banana River Lagoon, Brevard County.

Latitude/longitude: 28°22'15"N / 80°36'18"W.

Permit numbers: FDER 050770284; USCOE (none).

Age: 4 years (replanted).

Size: approximately 0.4 ha (1.0 ac).

Species present:

Amaranthus sp., Ammania latifolia, Bacopa monnieri, Cyperus ligularis, Cyperus odoratus, Eleocharis albida, Eustoma exaltatum, Echinochloa walteri, Iva frutescens, Pluchea odorata, Paspalum distichum, Ruppia maritima, Spartina alterniflora, Salicornia bigelowii, Suaeda linearis, Scirpus robustus, Typha domingensis.

Site description:

The Costa del Sol site was a scrapedown mitigation to offset wetland encroachment related to nearby construction. The site was located next to a condominium development and associated stormwater runoff depressions. A large pile of soil (10 m) was located next to the site as well. The exterior of the site outside the narrow entrance had a higher elevation than the interior which had been scraped lower, ostensibly to prevent rapid filling in of the site. Apparently, the plants were installed around the margin of the site. The species found at this site indicate considerable freshwater input, and selection against saline vegetation was probably inevitable. Cattails have invaded and are aggressively replacing the saline species.

Goals of project: To provide mitigation for lost mangroves.

Attainment of goals:

Successful; salinity is lower than expected but plant diversity is high.

Contact: Lewis Environmental Services, Inc.
P.O. Box 20005
Tampa, FL 33622-0005
813/889-9684

FLORIDA KEYS WATER MAIN

Locale: Key Largo, Monroe County.

Latitude/longitude: 25°06'39"N / 80°24'52"W.

Permit numbers: FDER 13 and 44-28299; USCOE 80M-0276.

Age: 5-6 years.

Size: 3.5 ha (8.7 ac).

Species present:

Avicennia germinans, Conocarpus erecta, Fimbristylis castanea, Laguncularia racemosa, Rhizophora mangle, Spartina alterniflora.

Site description:

The Florida Keys Water Main restoration was an attempt to partially revegetate backfill following installation of a large water supply line. The disturbed mangrove area is approximately ten meters wide and parallels U.S. Highway A1A for a number of miles. An unknown number and arrangement of mangrove and smooth cordgrass units were planted along an unknown extent of the site. Therefore, we surveyed the first accessible area from the north along A1A for a predetermined distance of ten sample points at standard interplot distances (total length approximately one kilometer). The substrate ranged from muddy to an occasional rocky outcrop. The cable roots of black mangroves had trouble penetrating the substrate and remaining subterranean, indicating a hard surface just under the mud; this gave the black mangroves the appearance of having prop roots. Standing water was present along much of the length of the surveyed area and frequently was very warm. Undisturbed adjacent mangroves were approximately three meters tall, and the surface elevation under them was even, unlike the restored area where vegetation was found only on the higher elevations.

Goals of project:

To revegetate water main installation impact area with mangroves.

Attainment of goals:

Poor elevation control resulting in lack of adequate drainage from some areas (possibly due to heterogeneous fill material, leading to differential settling of substrate); did not achieve goals, although bird use is extensive.

Contact: Lewis Environmental Services, Inc.
P.O. Box 20005
Tampa, FL 33622-0005
813/889-9684

HAMMER POINT

Locale: Key Largo, Monroe County.

Latitude/longitude: 25°01'24"N / 80°30'45"W.

Permit numbers: FDER (none); USCOE 71-1176 (enforcement case).

Age: 2 years.

Size: 1.0 ha (2.5 ac).

Species present: Halodule wrightii, Rhizophora mangle.

Site description:

The Hammer Point site is a scrapedown of illegal fill along the front of a housing project situated on finger canals. The canals divide the restoration into four separate areas. The coral rock substrate was of even elevation, and at low tide the lowest elevation was under approximately 0.3 m of water; the highest elevations had some standing water but were probably exposed at the lowest of tides. The substrate was so hard that the plants had to be hosed in. Most of the red mangroves were in the sapling class (>0.3 m) because at installation they were over 1.5' tall. The low elevations designed for this restoration resulted in the herb stratum being dominated by a green alga, Batophora sp.; the presence of shoal grass also indicates that the site remains inundated permanently. Survival of red mangrove planting units at this site can be attributed, in part, to good flushing and water quality which keep the plants from being affected by increased water temperature. However, the rate of plant growth is slow because of the poor substrate.

Goals of project: Restoration of illegally filled mangrove area.

Attainment of goals:

High survival to date indicates success; long term survival is questionable.

Contact: Steve Beeman
Ecoshores, Inc.
3881 South Nova Rd.
Port Orange, FL 32018
904/767-6232

SUNKEN ISLAND

Locale: Mouth of Alafia River, Hillsborough Bay, Hillsborough County.

Latitude/longitude: 27°48'25"N / 82°26'01"W.

Permit numbers: none.

Age: 7 years.

Size: 1.7 ha (4.2 ac).

Species present:

Avicennia germinans, Blutaparon vermiculare, Laguncularia racemosa, Paspalum distichum, Rhizophora mangle, Spartina alterniflora, Suaeda linearis, Sesuvium portulacastrum.

Site description:

The Sunken Island site was an attempt to stabilize dredge spoil, and create nesting and foraging habitat for bird species utilizing this island managed by the National Audubon Society. Smooth cordgrass completely covered the planting area within three years, followed by mangrove colonization (principally black and white mangroves) which are beginning to dominate the area. The site's insular characteristics moderated freeze damage as suffered by mainland mangroves. Also, foot traffic was minimal due to protection by the Audubon Society. This project was an actual enhancement without mitigation requirements.

Goals of project:

Enhancement of open water with spoil and marsh creation.

Attainment of goals:

Marsh created for bird nesting and foraging has changed to mostly mangrove, thereby lessening habitat value for some bird species but improving it for others. Generally a success.

Contact: Lewis Environmental Services, Inc.
P.O. Box 20005
Tampa, FL 33622-0005
813/889-9684

Publications: Hoffman et al. 1985

GARDINIER MITIGATION

Locale: Archie Creek, Hillsborough County.

Latitude/longitude: 27°51'49"N / 82°23'40"W.

Permit numbers: FDER 29-42-3949; USCOE 76-074.

Age: 8 years.

Size: 1.8 ha (4.5 ac).

Species present: Spartina alterniflora, Sporobolus virginicus.

Site description:

The Gardinier site was a mitigation for offsite damage to a saltmarsh on another part of the phosphate plant's property. The center scrapedown area adjacent to the planted portion contained mostly dead smooth cordgrass, probably as a result of excessive settling or erosion of surface fines. Apparently, finer sediment from this area washed into the lower, natural area. The substrate at the perimeter was much firmer than the extremely gooey substrate in the low center area; the difference in texture may have been caused by erosion of sandy upland slopes into the planted area. The installed plants were smaller than those in the natural area "downstream" from it.

Goals of project: Mitigation for 0.6 ha of fill in marsh.

Attainment of goals: Successful.

Contact: Lewis Environmental Services, Inc.
P.O. Box 20005
Tampa, FL 33622-0005
813/889-9684

Publications: Lewis 1982a

FEATHER SOUND

Locale: Old Tampa Bay, Pinellas County.

Latitude/longitude: 27°54'15"N / 82°39'35"W.

Permit numbers: FDER 528301016; USCOE 83T-0476.

Age: 3 years.

Size: 3.1 ha (7.7 ac).

Species present:

Avicennia germinans, Laguncularia racemosa, Spartina alterniflora.

Site description:

The Feather Sound site was a restoration resulting from construction of a stormwater catchment basin. A freshwater creek runs parallel to the site. Smooth cordgrass cover was homogeneous except where mangroves were beginning to invade. The elevation range specified was 0.15-0.46 m (0.5-1.5 ft) NGVD; generally, the elevations were higher than specified. Following cordgrass establishment, the actual elevations seem more appropriate for mangrove colonization if propagules are available. Possibly, these types of areas could be hastened to mangrove status by importing propagules and distributing them over the site. Eventually, this site can be expected to be mostly mangroves, with a minor smooth cordgrass component along the creek. Another section was excavated too low and has minimal plant cover, although bird use is extensive.

Goals of project: Restoration of 3.1 ha of mangrove.

Attainment of goals:

Only 0.6 ha is now vegetated; balance has high ecological value and may revegetate on its own.

Contact: Lewis Environmental Services, Inc.
P.O. Box 2005
Tampa, FL 33622-0005
813/889-9684

CONNIE MACK ISLAND

Locale: Punta Rassa Cove, Lee County.

Latitude/longitude: 26°29'45"N / 81°59'40"W.

Permit numbers: FDER 36-24-3832; USCOE 76E-0892.

Age: 3 years.

Size: 0.3 ha (0.7 ac).

Species present:

Avicennia germinans, Laguncularia racemosa, Rhizophora mangle, Ruppia maritima.

Site description:

The Connie Mack Island site is a partial restoration of an illegal scrapedown of mature mangroves. The entire restoration was not surveyed; the unsurveyed portion was a 15' wide strip behind drastically pruned mangroves. The main part of the restoration is surrounded by a berm approximately three meters high on all sides except that facing the water. The air and water temperatures inside this pit were very intense. Many of the planted red mangroves appeared to have suffered heat stress resulting from high temperatures of standing water at low tide. The presence of widgeon grass indicates that the standing water was a permanent condition. The substrate was extremely mucky, especially in the center area closest to the water.

Goals of project: Restoration of 0.3 ha of mangroves.

Attainment of goals:

Elevations too low for good red mangrove seedling growth and survival; high temperatures of standing water may have killed many propagules; substrate density too low to give adequate support to propagules; propagule density too low to provide adequate cover in a reasonable time; berm traps heat, causing temperature stress; adjacent slopes too steep to allow habitat migration and adjustment.

Contact: David Crewz
Florida Dept. of Natural Resources
100 8th Ave. SE
St. Petersburg, FL 33701
813/896-8626

BAYPORT

Locale: Old Tampa Bay, Hillsborough County.

Latitude/longitude: 27°57'52"N / 82°33'08"W.

Permit numbers: FDER 290821843; USCOE 84W-0514.

Age: 3 years.

Size: 3.2 ha (8.0 ac).

Species present: Avicennia germinans, Spartina alterniflora.

Site description:

The five sites in the Bayport mitigation complex include a combination of successes and failures. This site is a scrapedown surrounded by a low berm with mangroves growing on and around the margins. The mitigation area is lower in places than the surrounding mangroves. A 2.4 ha (6-acre) tidal pond with channel access has been very successful. The required 1:1 mitigation ratio was far exceeded since 3.0 ha were successful.

Goals of project:

Mitigation for 0.8 ha of fill, and pond restoration.

Attainment of goals: Successful.

Contact: Lewis Environmental Service, Inc.
P.O. Box 20005
Tampa, FL 33622-0005
813/889-9684

WEST LAKE

Locale: City of Hollywood, Broward County.

Latitude/longitude: 26°2.17'N / 80°7.45'W.

Permit numbers: FDER 060942909; USCOE 84J-2528.

Age: Portions one year; balance to be done 1987-1990.

Size: Total, 80 ha (197 ac).

Species present:

Avicennia germinans, Laguncularia racemosa, Rhizophora mangle.

Site description: 600-hectare park in the City of Hollywood.

Goals of project:

To restore 80 ha of mangrove forest as mitigation for 80 ha of fill in stressed mangroves. Experimental establishment of volunteer propagule recruitment rates is being tested.

Attainment of goals: Incomplete.

Contacts: Lewis Environmental Services, Inc.
P.O. Box 20005
Tampa, FL 33622-0005
813/889-9684

Gilbert MacAdam
Broward County Parks Department
950 NW 38th St.
Oakland Park, FL 33309
305/357-8122

Publications: Mangrove Systems, Inc. 1987

APPENDIX III

404 PERMIT REVIEW CHECK LIST (FLORIDA COASTAL WETLANDS)

APPLICATION FORM

	YES	NO
Is the area of fill or excavation in wetlands clearly indicated?		
Is the type and function of the wetland to be filled or excavated described?		
Is the type and function of the created or restored wetland intended for mitagation clearly indicated?		

APPLICATION DRAWINGS

	YES	NO
Are the fill or excavation areas clearly indicated by type and acreage on a plan view?		
Do the cross-sections show elevations relative to NGVD?		
Are the elevations appropriate? Are they justified in a separate narrative?		
Is a particular tidal benchmark referenced as having been used to establish site elevations?		
Is the restoration creation area shown as being flat? (It shouldn't be).		
Is drainage of the restoration/creation area provided with a distinct tidal swale or ditch?		
Are there provisions for excluding access?		
Is the upland edge stabilized?		
Are the source(s), spacing and number of plants per unit area specified?		

APPLICATION NARRATIVE

	YES	NO
Is there a separate application narrative describing the project and the proposed wetlands mitagation? If not, inquire as to whether it was provided in the application but not forwarded to you.		
Does the narrative adequately describe the mitigation and include justification for elevations, slope, planting and monitoring reports, including reference to previous work of the consultant or published literature?		
Are you to receive copies of the monitoring reports directly from the consultant or applicant?		
Are clear success criteria stated?		
Do the methods of measuring success follow standard protocol?		
Are clear mid-course correction plans outlined with a decision date (i.e., three months post-construction)?		
Is the mitigation plan preparer's name provided? Does he have a track record? Is it satisfactory?		

CREATION AND RESTORATION OF COASTAL WETLANDS IN PUERTO RICO AND THE U.S. VIRGIN ISLANDS

Roy R. Lewis III
Lewis Environmental Service, Inc.

ABSTRACT. Major losses of coastal wetland habitat in Puerto Rico and the U.S. Virgin Islands have stimulated interest in restoring damaged areas, and in requiring mitigation in the form of wetlands creation or restoration for any future permitted losses. Unlike efforts on the mainland, documentation of efforts to date is sparse. The major emphasis to date has been on mangrove forests, but other wetland types are equally important. Our observations indicate that the opening of impounded mangrove areas works well as restoration, but creation of mangrove areas is more difficult because of the need to very carefully control the final graded elevations. The optimum target elevation for mangrove establishment appears to be +12 cm National Geodetic Vertical Datum.

Wetland restoration projects in the U.S. Virgin Islands and Puerto Rico need more centralized documentation in order to capitalize on the advances in understanding that come with each project. Some additional experimental work is also essential to determine the restoration potential of plant communities in which research has not yet been undertaken.

OVERVIEW OF THE REGION

CHARACTERISTICS OF THE REGION

Geology

Puerto Rico and the U.S. and British Virgin Islands are located east of the island of Hispaniola and north of the South American continent in the eastern Caribbean Sea (Figure 1). Geologically the islands are volcanic in origin, with Puerto Rico and the northern U.S. Virgin Islands (St. Thomas and St. John) lying on the Puerto Rican Plateau and St. Croix (the southern U.S. Virgin Island) lying to the south, separated by the Virgin Islands Basin from the Plateau. The islands are surrounded by deep water with the Puerto Rican Trench reaching 9,000 meters and the Virgin Island Basin reaching 4,400 meters.

Puerto Rico has a total land area of 886,039 hectares with three inhabited offshore islands, Viegues, Culebra, and Mona (Figure 2). The island is approximately 100 million years old. Nearly two-thirds of the land area is steep mountains. The remaining one-third contains 80% of all the level land and includes coastal lowlands produced by erosion of the mountains and deposition of alluvium at the mouths of rivers. It is in these coastal lowlands that the bulk of all fresh and saltwater wetlands in Puerto Rico are located.

The U.S. Virgin Islands consist of three islands (Figure 3), St. Croix, St. Thomas, and St. John. Their physiography is similar to that of Puerto Rico with predominant mountainous features surrounded by coastal lowlands, but they are significantly smaller. Total land area is only 34,447 hectares.

Climate

The climate of Puerto Rico and the U. S. Virgin Islands is marine subtropical with little temperature variation between summer (28°C) and winter (25°C). Rainfall varies with location, with as much as 500 cm falling in the mountain forests of Puerto Rico. Average annual rainfall for both areas is 100-125 cm with distinct dry and wet seasons. Evaporation is often greater than rainfall, and flowing streams are uncommon, except in areas downstream of the higher rainfall areas of some of the mountain ranges. Hurricanes are a prominent feature of the islands' weather patterns: 24 have passed within 80 km of the islands since 1900 (Island Resources Foundation 1977).

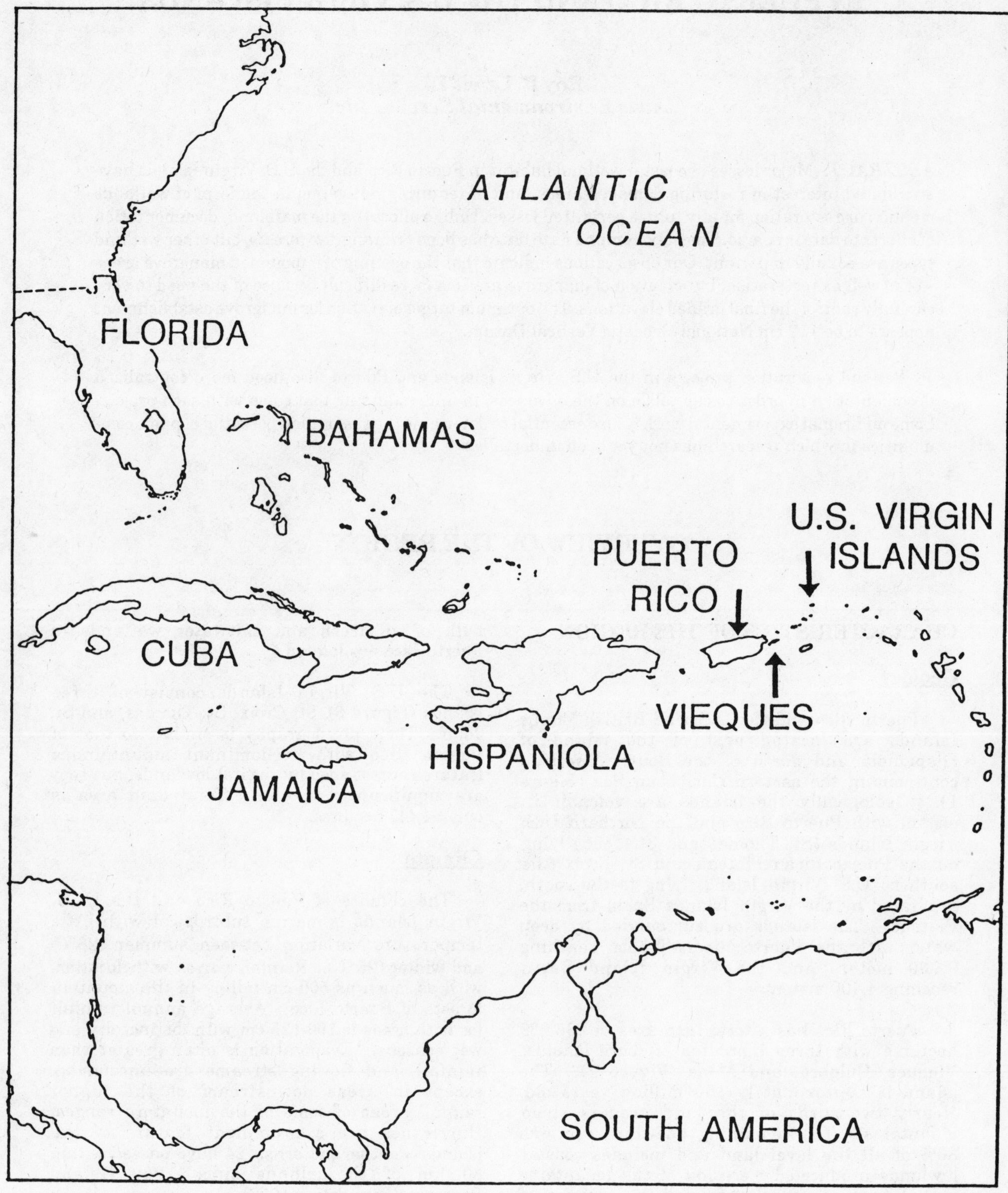

Figure 1. Puerto Rico and the Caribbean Basin.

Ecoregions

Holdridge (1947, 1967) and Ewel and Whitmore (1973) describe six life zones for the islands, two of which, the subtropical moist and subtropical dry zones, cover all the coastal areas. The subtropical moist zones have annual rainfall greater than 110 cm; the subtropical dry have less than 110 cm with minimums of approximately 50 cm, and evaporation typically exceeds rainfall.

NOAA (1977) defines seven coastal sectors of Puerto Rico (Figure 4). The south and southwest sections are subtropical dry zones, while the remaining five sectors are subtropical moist. In each of the three U.S. Virgin Islands, the eastern and southern lowlands are generally subtropical dry, while the central to western higher elevations are subtropical moist.

WETLAND TYPES

The wetland types listed here are those indicated by del Llano (1985) as being either estuarine or marine according to his revision of Cowardin et al. (1979) to fit the wetland types of Puerto Rico. They do not coincide exactly with the three marine emergent subdivisions defined by Environmental Laboratory (1978), but are felt to be more accurate by the author.

Mangrove Forest

The mangrove forests of Puerto Rico and the U.S. Virgin Islands are similar to those of Florida and are composed of the same four species: red mangrove (Rhizophora mangle L.), black mangrove (Avicennia germinans [L.] L.), white mangrove (Laguncularia racemosa Gaertn. f.) and buttonwood (Conocarpus erectus L.). Due to the lack of freezes, the canopy height is generally greater (up to 20 m), but periodic hurricanes keep maximum development from occurring (Martinez et al. 1979). In addition to serving as habitat for fish and wildlife and exporting organic matter in the form of detritus for offshore fisheries, mangroves are also important sources of wood and charcoal, particularly in Puerto Rico.

Lugo and Cintron (1975) divided the mangrove forests of Puerto Rico into two type formations, depending on whether they occurred in the subtropical moist life zone (north coast) or in the subtropical dry life zone (south coast). The basin and riverine forest types of Lugo and Snedaker (1974) are predominant on the north coast, while the fringe and overwash types are dominant on the south coast. The dwarf mangrove type has only been reported from the island of Vieques (Lewis, in press).

No good structural descriptions of the mangroves of the U.S. Virgin Islands exists, nor is an accurate acreage figure available. Mangroves in Puerto Rico presently cover 641 hectares representing only 26.4% of the original 2,431 hectares estimated to have been present (Martinez et al. 1979).

Swamp Forest

The dominant species in this community is bloodwood (Pterocarpus officinalis Jacq.) (Figure 5). This community typically occurs just landward of the mangrove forest, but is only present in the north coast type formation due to the low salt tolerance of these trees in tidal waters. The wood of these trees is prized for furniture construction and much of the original forests of this type have been eliminated due to overcutting. A subdominant species in some forests is the Puerto Rican royal palm (Roystonea boringuena Cook).

Herbaceous Swamp

A community further separated from the mangroves by the above described swamp forest community, these brackish to freshwater marshes include some typical marine plants such as leather fern (Acrostichum aureum L.), but are composed predominantly of oligohaline to freshwater plant species such as cattail (Typha domingensis Pers.), sedge (Cyperus giganteus Vahl), sawgrass (Cladium jamaicensis Crantz), and arrowhead (Sagittaria lancifolia L.) (Figure 6).

Swamp Thicket

Swamp thickets are composed of low shrubs and small trees along brackish to freshwater streams. The characteristic species are pond apple (Annona glabra L.), mahoe (Hibiscus tiliaceus L.), and leather fern.

Salt Barrens or Salinas

Salinas are areas of hyperhaline soil conditions typically found landward of a fringe of mangroves in the more arid south coast wetlands. Like similar areas in Florida, the areas have characteristic barren unvegetated areas interspersed with low ground cover species tolerant of high interstitial salinities including saltwort (Batis maritima L.), sea purselane (Sesuvium portulacastrum L. L.), and smut grass (Sporobolus virginicus L. Kunth). When flooded by rains, these are referred to as "hypersaline lagoons" (Cintron et al. 1978). Cintron et al. (1978) describe the cyclic patterns of rainfall and expansion of mangrove areas, followed by droughts, the death of mangroves, and the expansion of salt barrens.

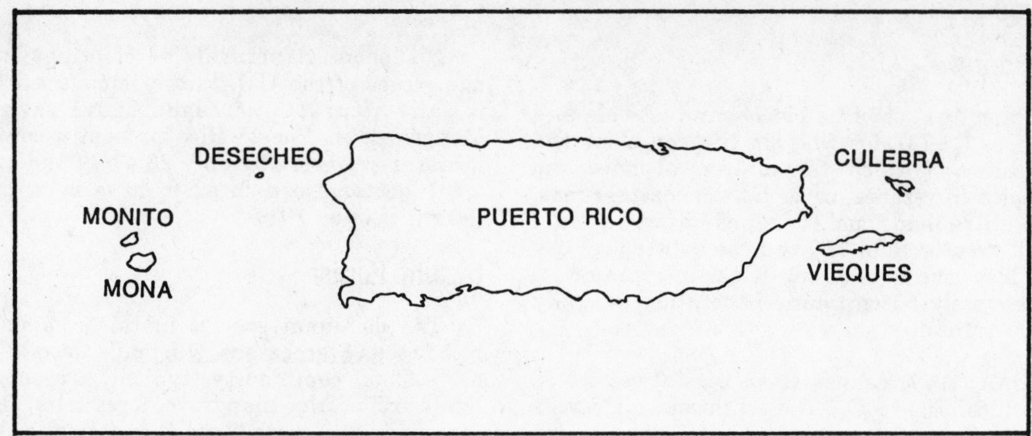

Figure 2. Puerto Rico and adjacent waters.

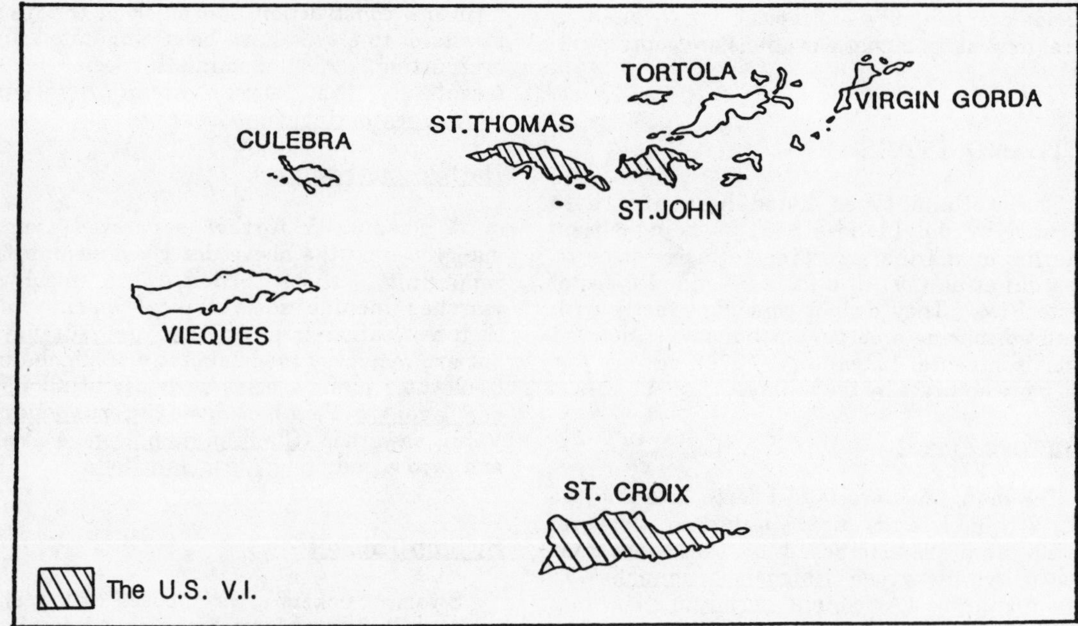

Figure 3. The U.S. Virgin Islands.

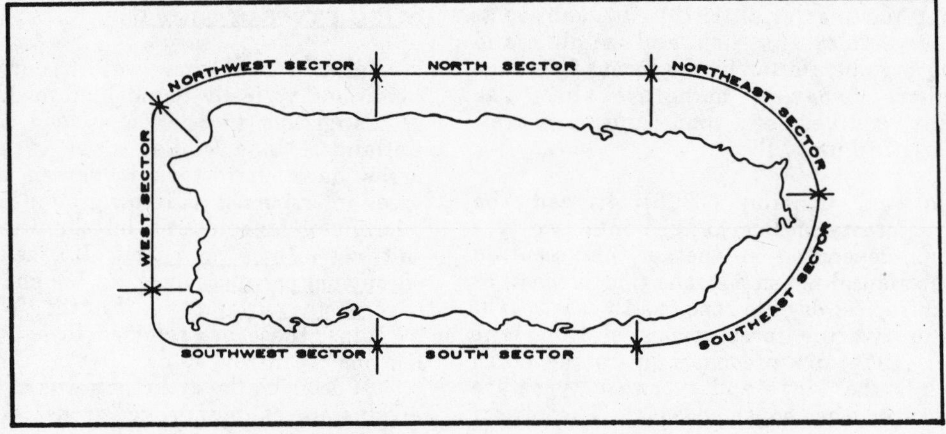

Figure 4. The seven coastal sectors of Puerto Rico (from NOAA 1977).

Figure 5 Swamp forest community dominated by bloodwood (Pterocarpus officinalis), Sabana Seca, Puerto Rico (January 1986).

Figure 6. Herbaceous marsh community dominated by cattail (Typha domingensis) and giant sedge (Cyperus giganteus), Sabana Seca, Puerto Rico (January 1986).

Figure 7. A salt pond (Great Salt Pond) on the southeast coast of St. Croix, U.S. Virgin Islands.

Salt Ponds

Salt ponds are former mangrove lagoons that have temporarily or permanently lost their free tidal connection to the ocean (Figure 7). In Puerto Rico and the U.S. Virgin Islands where the tidal range is small (24 cm, NOAA 1986, p. 233), minor blockages can dramatically change the hydrology of these lagoons. Island Resources Foundation (1977) describes the formation of salt ponds, the cycle of lowering salinities with rainfall and the increase in salinities with drought, which leads eventually to the formation of a salt barren due to evaporation. Ponds may also reopen during heavy rain storms or hurricanes, and these events may be critical to their use by estuarine species of fish and crustaceans seeking nursery habitat. They are also important wading bird and seabird feeding and nesting areas. The black mangrove is often a dominant species due to its tolerance of high interstitial salinities (Cintron et al. 1978).

Lewis (in press) describes a unique salt pond type dominated by white mangroves or buttonwood found on the island of Vieques. These ponds have apparently been isolated from tidal flow for extended periods of time; the large scale water level fluctuations due to rainfall have gradually eliminated the other mangroves which cannot adapt to prolonged flooding of their pneumatophores or prop roots.

EXTENT TO WHICH CREATION/RESTORATION HAS OCCURRED

PROBLEM OF MINIMAL DOCUMENTATION

Unlike efforts in the mainland United States, restoration and creation of wetlands in Puerto Rico and the U.S. Virgin Islands has received little interest in the past and very little published literature is available describing any such projects. The pressures of population growth and the demand for infrastructure (roads, schools, etc.) to support the population have meant that the prime efforts have been directed towards protecting what is left of the wetlands in these areas (Norton 1986, del Llano et al. 1986). For this reason, the remainder of this chapter focuses on the lessons from the few projects that are familiar to the author; these are primarily associated with mangroves. Within the last few years, a number of projects have been designed, and some permitted, that will greatly expand the information base, assuming some documentation of the projects occurs. These projects are noted at the end of this chapter; those interested should follow their progress carefully.

TYPICAL GOALS OF PROJECTS

Offset Adverse Environmental Impact Through Mitigation

Lewis (1979) and Lewis and Haines (1981) describe the planting of red and black mangrove propagules over a 3.8 hectare area on the south shore of the island of St. Croix as mitigation for the impoundment of 7.4 hectares of mangroves as part of a cooling pond. Forty percent of the red mangrove propagules are reported to have survived through the second year of planting, while survival of the black mangrove propagules is reported to be only 1-2%. Figure 8 shows the planting area and plant cover after five years.

Mangrove Systems, Inc. (unpublished) developed a mitigation plan for the Federal Aviation Administration to offset the filling of mangroves adjacent to the airport in San Juan, Puerto Rico. Two sites totalling 0.25 hectares were identified along the Suarez Canal and excavated to an elevation of +12 cm NGVD [National Geodetic Vertical Datum]. These were planted with 4,840 red mangrove propagules in October 1983. Recent inspections indicate that white mangroves have volunteered in large numbers and currently dominate the sites. Canopy height of the volunteer mangroves is now approximately 3 meters after four years. No estimate of the survival of planted propagules has been made.

No other published or unpublished reports of mitigation projects were available to the author. Several successful mangrove restoration projects have been undertaken within artificially closed salt ponds on the island of Vieques (Lewis et al. 1981, Mangrove Systems, Inc. 1985, Lewis, in press). Normal tidal circulation was restored by removing blockages and volunteer propagules have revegetated two ponds. Similar recommendations have been made for lagoons on the Roosevelt Roads Naval Station at Ceiba (Lewis 1986).

Restore Damaged Areas

There has been a great deal of interest in the possibility of restoring mangroves damaged by oil spills (Figure 9; Lewis 1983, Getter et al. 1984), by impoundment (Figure 10; Lewis et al. 1981, Birkitt 1984), or by cattle grazing (Lewis, in press). The above-described project on St. Croix was, in fact, an oil-damaged site used as mitigation for the impoundment of other

Figure 8. Mangrove restoration area on St. Croix, U.S. Virgin Islands. Top: February 1979, six months after planting, Bottom: April 1983, five and a half years after planting.

Figure 9. Mangrove forest on the south coast of St. Croix, U.S. Virgin Islands, seven years after damage by an oil spill (photographed March 1978).

Figure 10. Mangroves dead as the result of impoundment at Laguna Boca Quebrada, Vieques, Puerto Rico (December 1985).

mangroves. Lewis (in press) describes the reopening of two lagoons closed due to road construction on the island of Vieques. Both resulted in successful re-establishment of mangroves in the previously stressed areas of the lagoons. Damage due to cattle grazing on Vieques (Lewis et al. 1981) and mainland Puerto Rico (Lewis 1986) also has been reported.

SUCCESS IN ACHIEVING GOALS

The limited number of projects reviewed here restricts the general assumption that all projects to date in the islands have been successful. What projects have failed? None are known, but it is human nature to publicize successes and forget failures. A thorough search of agency files to identify additional projects that have been attempted should be undertaken and any monitoring reports located and analyzed.

REASONS FOR SUCCESS/FAILURE

Lewis (1979) and Lewis and Haines (1981) list several problems with projects: 1) physical removal due to erosion, accumulation of seagrass wrack, and floating debris; 2) eating of the planted seeds by unknown biological agents, possibly crabs; 3) death of seedlings by natural causes; and 4) planting at apparently too high an elevation.

Detailed surveys were performed prior to the planting at St. Croix in order to keep planting elevation errors to a minimum. Because the tidal range is only 24 cm (NOAA 1986) and mangroves occupy only the upper half of that range, accurate surveys and propagule placement were important. This tidal range is much narrower than in most of Florida where 60-90 cm tidal ranges are common (NOAA 1986, pp. 226-229). Based upon the success of these projects, a +12 cm NGVD elevation is recommended as the optimum for mangroves in general. As has been recommended for Florida plantings (Lewis, this volume), any excavated sites should be designed with a positive slope towards the open water to provide drainage and to eliminate pockets of standing water. Also, if possible, a tidal stream connection for fish movement into and out of the areas would enhance the fishery value of the restored system.

No projects have been reported that attempt to restore or create the other five types of coastal wetlands listed previously. Any attempts to do so should depend heavily on understanding the characteristics of the natural plant community, including ranges of elevation, interstitial salinity tolerances, and succession patterns.

DESIGN OF CREATION/RESTORATION PROJECTS

PRECONSTRUCTION CONSIDERATIONS

Location Of Project

Depending on the type of plant community to be restored/created, the location of the site should provide the appropriate hydrology for that specific plant community. If the plant community normally requires salinities of 1-5 ppt, a routine source of freshwater must be available. Therefore, attempting to establish such a community (i.e., herbaceous swamp) along the south coast of Puerto Rico might prove difficult.

Site Characteristics

In order to design a project, the following minimum information is essential:

1. What is the existing site topography related to a tidal datum?

2. What is the existing wave climate: how exposed is the site to storms?

3. What are the proposed elevations? Will they remain the same? If so, the reasons behind the absence of volunteer plants need to be analyzed. If the elevations are to be changed, what is the justification for the proposed elevations?

4. What are the sediment characteristics below the surface if excavation is to take place? Sand may be suitable, but rock is not.

5. What are the proposed plant materials? Are they routinely available? Will volunteer propagules invade the area?

6. Are any exotic plants (Australian pines) present? Will they shade the planting area? Can they be removed?

7. How is public access to be controlled?

CRITICAL ASPECTS OF THE PROJECT PLAN

Timing of Construction

The timing of construction is dependent on the type of plant materials to be used. If natural revegetation is to be allowed, excavation should be timed for completion just prior to the release of propagules; if it is a mangrove site, maximum seed availability will occur in the fall, and excavation should be completed prior to August. If planting is to occur, then spring installation is recommended. Site preparation should be completed to meet this time frame.

PRE-CONSTRUCTION QUALITY CONTROL

Even when accurate plans are prepared, actual construction may not achieve the required tolerances. It is important that an as-built survey be completed before construction equipment is moved off site, so that corrections can be made quickly and inexpensively. Requiring equipment to return to the site often produces delays, if the equipment can be brought back at all. Another quick check of the site to check for proper drainage can be accomplished by simply watching the tide rise and fall across the site.

SUBSTRATE

The design substrate composition should be verified by simple soil auger checks as a minimum. If rock or clay layers are encountered at the proposed excavation depth, the site may be unacceptable.

Detailed examination of the need to add fertilizer to improve plant survival and growth has not been tested in Puerto Rico or the U.S. Virgin Islands. The general guidelines of Woodhouse and Knutson (1982) and Broome (this volume) should be used when deciding whether to use fertilizers in herbaceous marsh plantings. Fertilizers may be useful in plantings on exposed shorelines where rapid growth is desirable to avoid extended exposure of new plants to wave action.

The response of mangrove plantings to fertilization is not well documented. Zuberer (1977) documents the presence of nitrogen fixation activity in the roots of mangroves in Florida, while Reark (1983) argues that fertilizers were essential to the successful outcome of his project. Teas (1977) states that "Nursery mangroves of all three species were found to respond to fertilizer. Because open-water fertilization ordinarily is not practical, pre-transplanting fertilization may prove useful." (p. 56). Reark (1984) describes the addition of a soluble fertilizer to nursery-grown mangroves, but there were no controls. In fact, no controlled experimentation has been conducted to demonstrate any value of added fertilizer. Such work needs to be done, but in the interim, added fertilizer does not appear essential to mangrove establishment.

PLANT MATERIAL

Mangroves are available for planting in four forms: propagules, 1-2 year old seedlings, 3-5 year old nursery-grown trees, and field-collected transplants. The 1-2 year old seedlings are the most often recommended or required plant material. They are grown from propagules harvested in the wild. In fact, Beever (1986) recommends using only "one year old (one foot minimum height) nursery grown seedlings" with no reference to the other plant materials.

The direct installation of red mangrove propagules (Figure 8) has been popular due to low cost and general success in protected sites. Goforth and Thomas (1980) compared red mangrove propagules and field-dug seedlings 12 to 18 months old. At the end of five years, "survival of transplanted seedlings was no more successful than that of propagules" while "the average vertical growth of seedlings ... was significantly ($p<0.001$) less than propagules" (p. 221). Stephen (1984) reports 97% survival of planted red mangrove propagules after 8 months at a large project in Naples, Florida while Lewis and Haines (1981) report 40% survival after one year in St. Croix.

Direct installation of propagules of the other three species is not practical due to their requirement to shed a pericarp and rest on the surface of a damp substrate for several days prior to anchoring. Broadcasting of these propagules might be successful in some projects, but Lewis and Haines (1981) report low overall successful anchoring of broadcast propagules.

The use of larger plant materials greatly increases the cost of a project (Teas 1977, Lewis 1981) and should only be used where absolutely necessary. Goforth and Thomas (1980) note that in exposed sites, transplanted 2-3 year old trees are the only successful plant material. Teas (1977) reports that all attempts to transplant large (6 m tall) mangroves in Florida failed. Gill (1971), however, reported transplanting red mangroves up to 6.5 m in height that Carlton (1974) indicates had high survival. Pulver (1976) provides guidelines to the transplanting of mangroves up to 2 m tall, and reports good success with trees to this height.

Finally, there may be instances in which no installation of mangrove plant materials is necessary. The largest (80 ha) mangrove restoration in Florida is designed to require no installed mangroves at all, and preliminary testing appears to confirm the success of this technique where natural propagule availability is sufficient (Lewis, this volume). The Suarez canal planting site previously discussed has become dominated by volunteer white mangroves. Given the limited availability of funds for restoration, the elimination of planting could provide extra money to move more soil and restore larger areas. However, each site has unique characteristics that require pilot projects to confirm the utility of this technique.

Other species often transplanted in oligohaline situations in Florida include leather fern and sawgrass. Plugs of leather fern transplant well, but not cut stems as described by Beever (1986, p. 7). Sawgrass is available from some nurseries in Florida as cultivated units but is most often dug from the wild. Bare root units have exhibited poor survival in oligohaline environments, and plugs of 3-5 intact shoots with soil core are recommended. No nurseries are known to provide these plant materials in Puerto Rico or the U.S. Virgin Islands.

REINTRODUCTION OF FAUNA

The establishment of a new faunal community in a created/restored system connected to tidal waters has historically been left to incidental moving of individual organisms with plant material, the natural settlement of planktonic life forms that become sessile, epibenthic or infaunal after metamorphosis (meroplankton), and the immigration of fauna, particularly fish, from adjacent wetlands.

There is no evidence that active reintroduction of fauna would accelerate the colonization process, but controlled experimentation to answer the question has not been done. Based only on personal observations, the author sees no obvious need to introduce fauna, because colonization of created/restored systems by fauna appears to be quite rapid. Nonetheless, good experimental evidence should be generated to answer the question.

BUFFERS, PROTECTIVE STRUCTURES

Beever (1986) recommends, for Florida, that a buffer zone equivalent to the width of the planted area be cleared of exotic species. The author's experience has been that rapid invasion by these species will occur and that such a buffer zone, if not maintained free of exotics, will cease to be a buffer within five to ten years. In addition to removing exotic plants, creation of a buffer zone should include destroying root systems with a systemic herbicide and planting native vegetation such as *Thespesia populnea* and mahoe to outcompete the exotics. It is not the width but the intensity of maintenance that will ensure that no invasion of exotics occurs in a buffer zone.

Foot traffic and vehicle access problems have been noted previously. If a site is in an urban area where public access can be expected, the site should be fenced and legally posted.

LONG TERM MANAGEMENT

Control of exotics over the long term will require implementation of a control program as outlined above. The problem of ownership change and future threats to restored/created sites is addressed in Florida by the routine requirement of a conservation easement in which the property owner retains fee simple title but restricts the future use of the property by a county recorded easement. No similar protective mechanism is known to be routinely applied in Puerto Rico or the U.S. Virgin Islands. Fee simple transfer of ownership to the state is also possible for that portion of the property not needed for development.

MONITORING

WHAT TO MONITOR AND HOW

Prior to implementing a monitoring program, it is essential that measurable goals be defined. The goals should be reasonable and based on published literature values of parameters in natural systems or on values reported in monitoring reports readily available to all parties. Due to the haphazard nature of the methods of obtaining and reporting data on wetlands in Puerto Rico and the U.S. Virgin Islands, it is virtually impossible to go much beyond percent survival, growth measurements, and areal cover measurements as reasonable criteria for determining success. This should improve as more data are obtained and centralized for review.

An example of the problem is establishing a target goal for the number of invertebrate species and their density in a created/restored wetland. Baseline data from tidal wetlands in Puerto Rico and the U.S. Virgin Islands are minimal and thus establishing a criterion such as 30,000 organisms/m^2 as "successful" is impossible. An alternative is the comparison of a control wetland to the created/restored one. Due to natural variations in wetlands that may appear identical to an observer, variances in faunal densities as great as 50% are not uncommon. Only by careful study of existing created/restored wetlands in comparison with adjacent control areas can any general criteria be established. Because this has not been done in Puerto Rico or the U.S. Virgin Islands, we must work with what we have until such data is available.

What do you monitor and how? First, it is important to describe in detail, using maps and photographs, what was done at the site. How big was it? What were the slopes and elevations? What tidal benchmark was used? What types and numbers of plants were installed, and where and when?

Second, some sampling regime needs to be established that will be repeated over a period of time. The typical standard is quarterly sampling for two years. This results in reports issued at time zero (completion of construction and planting), 3, 6, 9, 12, 15, 18, 21 and 24 months for a total of nine reports. Longer times may be required for monitoring forested wetlands and monitoring intervals may lengthen as the monitoring period increases.

Third, a sampling program involving either pre-established plots or random plots determined at each monitoring inspection needs to be described and justified. Broome (this volume) supports stratified random sampling with sampling in each elevation zone. The sampling program should begin at the completion of construction and/or planting (time zero).

Fourth, each sampling should include photographs taken from the same position and angle during each monitoring episode to illustrate to the report reviewer what is happening at the site (Figure 8).

Finally, the last report should summarize all the results with appropriate graphs, and compare the results with 1) the previously established goals, and 2) literature values for parameters measured.

This monitoring program is most easily accomplished by making it a condition of the permit and requiring the permittee to pay for it. Few federal or state regulatory agencies have the staff or funds to conduct detailed compliance monitoring themselves. Simply inspecting the site once or twice during the course of the permit is an accomplishment for agency personnel, and the reports with photographs are an important compliance monitoring tool.

With regard to the parameters to be measured, percent survival of installed plant materials and/or volunteer recruitment within plots should be measured and extrapolated to describe the conditions across the created/restored area. Plots are important because, except for small planting areas, counting hundreds or thousands of units individually can be tedious, expensive, and error-prone. One meter square plots are typically sufficient for tidal marsh plantings. Four meter square (2 m x 2 m) plots may be better for tree species. The absolute number of plots will depend on site size; Broome (this volume) suggests fifteen as a minimum.

Each planting unit or volunteer in the plot should be measured for height and the percent cover in the plot estimated. If random units are chosen each time for measurement, a quadrat can be centered over them for measurements. For herbaceous species like giant sedge, culm (stem) density can be measured using 10 cm x 10 cm quadrats. For mangroves, plant height and prop root or pneumatophore number can be noted.

Above-ground and below-ground biomass of herbaceous species can be measured by clipping at ground level, taking core samples, and separating plant material out for drying and weighing (see Broome, this volume). It is not practical to measure biomass of mangroves because this involves destructive sampling and loss of planting units.

Percent survival alone should not be the sole criterion of success. Low percent survival of rapidly expanding herbaceous marsh species may still provide 100% cover of the area desired. With mangroves, mortality of young mangroves is normal with competition. Pulver (1976) measured the density of mangroves in natural forests in Florida and found that as the stand height (and age) increased, the number of trees decreased. For example, densities of red mangroves decreased from 26.8 trees/m^2 for 1.2 m tall trees to 8.3 trees/m^2 for 1.9 m tall trees (p. 13). That represents a survival rate of only 31%; would that be called "successful" in a created/restored system?

Percent cover supplements percent survival as a measurement of expansion of leaf area. Combined with growth and stem density measurements, it provides a good indication of whether a system is healthy and expanding in plant height and cover. What are good rates of growth and coverage to look for? Data are generally absent. Lewis and Crewz (in prep.) will provide some typical target values.

Faunal sampling in created/restored systems is more of an art than a science at present. A number of studies are underway in the continental United States, under the auspices of the National Marine Fisheries Service (Beaufort, North Carolina), U.S. Army Engineer Waterways Experiment Station (Vicksburg, Mississippi) and Florida Department of Natural Resources (St. Petersburg), in created/restored tidal wetland systems in Florida. No similar studies are known to be underway in Puerto Rico or the U.S. Virgin Islands. None of the results have been published at this time. Readers are encouraged to contact these agencies directly for updated publications on the subject.

How long to monitor? Two years is normal for marsh systems. Five years should be the minimum for forested systems. If faunal monitoring is included, eight to ten years are probably necessary to document changes.

MID-COURSE CORRECTIONS

If, during the course of monitoring, it becomes obvious that the goals will not be met, there are two choices. One is to determine the cause of the problem and correct it. It may be elevation, drainage, source of plant materials, etc. The other choice is to evaluate the habitat value of the system as it exists, and determine if that is sufficient to satisfy regulatory agencies. It is possible that the wetland was a failure (no plant survival) but the project worked from other perspectives--for example, if there is intense use of shallow unvegetated areas by wading birds.

If the first course of action is taken, the cost of the modifications may cause the permittee to challenge the regulatory agency's right to ask for changes in a plan it originally approved. Such questions have been raised in the case of unsuccessful creation/restoration attempts in Florida and usually, the agency has backed down. Careful preparation of permit conditions to provide for mid-course corrections are essential.

INFORMATION GAPS AND RESEARCH NEEDS

CENTRALIZED DATA BANK

As has been stated, Lewis and Crewz (in prep.) note that the lack of a centralized database concerning historical as well as current creation/restoration projects usually hampers any data analyses, and prevents comparisons of projects. Such a condition presently exists in Puerto Rico and the U.S. Virgin Islands. If such a system is developed, it is equally important that the data concerning historical permitted creation/restoration projects be catalogued. No known plans are underway to document this information.

NATURAL PROPAGULE RECRUITMENT VERSUS PLANTING OR TRANSPLANTING

The natural recruitment, survival, and growth rates of volunteer propagules need to be tested for various sizes and densities of installed plant materials to determine the optimum densities needed for certain target coverage rates and habitat utilization. For example, it has been assumed that volunteer mangrove propagule recruitment could not match the success or growth rates of planted nursery-grown mangroves. With costs of $1,000 to $200,000 per hectare for planted mangroves (Teas 1977, Lewis 1981), significant savings could occur if natural recruitment proved effective. Particularly with limited public funds available for habitat restoration, the cost-effectiveness of particular plant materials needs to be documented.

TRANSPLANTING LARGER MANGROVES

As noted, the success of transplanting larger (>2 m tall) mangroves has generally not been good. The salvage of larger mangroves destined for destruction might prove valuable if larger trees could be moved successfully.

COMPARABLE GROWTH RATES OF MANGROVE PROPAGULES AND SEEDLINGS

Is it necessary to plant 1-2 year old nursery-grown seedlings in order to achieve more rapid cover, or will natural propagule recruitment, or planted or broadcast propagules achieve equivalent growth? The previously described data of Goforth and Thomas (1980) needs amplification.

RATE OF FAUNAL RECRUITMENT

A comparison of sites of different ages in a synoptic manner is needed to determine how rapidly faunal recruitment takes place, and whether supplementing the process is necessary.

FUNCTIONAL EQUIVALENCY

A multi-parameter comparison of created/restored systems with several natural areas is needed to determine which functional values (habitat, water quality, primary production, etc.) can be re-established, and over what time frame.

REGIONAL CREATION/RESTORATION PLANS

Wetland creation and/or restoration needs to be examined as a need of the regional ecosystem, with the possible outcome that out-of-kind creation/restoration is deemed acceptable in some instances to compensate for the loss at another site of a distinct habitat type such as swamp forest.

LITERATURE CITED

Beever, J.W. 1986. Mitigative Creation and Restoration of Wetland Systems--A Technical Manual for Florida. Draft Report. Fla. Dept. of Environmental Regulation, Tallahassee, Florida.

Birkitt, B.F. 1984. Considerations for the functional restoration of impounded wetlands, p. 44-59. In F.J. Webb, Jr. (Ed.), Proc. 10th Ann. Conf. Wetlands Restoration and Creation. Hillsborough Community College, Tampa, Florida.

Carlton, J. 1974. Land building and stabilization by mangroves. Environ. Conserv. 1:285-294.

Cintron, G., A.E. Lugo, D.J. Pool, and G. Morris. 1978. Mangroves of arid environments in Puerto Rico and adjacent islands. Biotropica 10(2):110-121.

Cowardin, L.M., V. Carter, F.C. Golet, and E.T. La Roe. 1979. Classification of Wetlands and Deepwater Habitats of the United States. U.S. Fish & Wildlife Service, Office of Biological Services 79/31.

del Llano, M. 1985. Inventario de terrenos anegadizos y habitats de aguas profundas de Puerto Rico, p. 93-111. In D.A. Scott, M. Smart, and M. Carbonell (Eds.), Report XXXI Ann. Meeting International Waterfowl Research Bureau. IWRD Slimbridge, Glos. GL27BX, England.

del Llano, M., J.A. Colon, and J.L. Chabert. 1986. Puerto Rico, p. 559-571. In D.A. Scott and M. Carbonell (compilers), A Directory of Neotropical Wetlands. IUCN Cambridge and IWRB Slimbridge, U.K.

Environmental Laboratory. 1978. Preliminary Guide to Wetlands of Puerto Rico. Technical Rep. Y-78-3. U.S. Army Engineer Waterways Experiment Station, Vicksburg, Mississippi.

Ewel, J.J. and J.L. Whitmore. 1973. The Ecological Life Zones of Puerto Rico and the U.S. Virgin Islands. Forest Service Research paper ITF-18. Rio Piedras, Puerto Rico.

Getter, C.D., G. Cintron, B. Dicks, R.R. Lewis, and E.D. Seneca. 1984. The recovery and restoration of salt marshes and mangroves following an oil spill, Ch. 3, p. 65-113. In J. Cairns, Jr. and A. Bulkema, Jr. (Eds.), Restoration of Habitats Impacted by Oil Spills. Butterworth Publishers, Boston, Massachusetts.

Gill, A.M. 1971. Mangroves--is the tide of public opinion turning? Fairchild Trop. Gard. Bull. 26:5-9.

Goforth, H.W. and J.R. Thomas. 1980. Planting of red mangroves (Rhizophora mangle L.) for stabilization of marl shorelines in the Florida Keys, p. 207-230. In D. P. Cole (Ed.), Proc. 6th Ann. Conf. Wetlands Restoration and Creation. Hillsborough Community College, Tampa, Florida.

Holdridge, L.R. 1947. Determination of world plant formations from simple climatic data. Science 105:367-368.

Holdridge, L.R. 1967. Life Zone Ecology. Provisional edition. Tropical Science Center, San Jose, Costa Rica.

Island Resources Foundation. 1977. Marine Environments of the Virgin Islands. Tech. Suppl. No. 1. Virgin Islands Planning Office, Coastal Zone Management Program.

Lewis, R.R. 1979. Large scale mangrove planting on St. Croix, U.S. Virgin Islands, p. 231-242. In D.P. Cole (Ed.), Proc. 6th Ann. Conf. Wetlands Restoration and Creation. Hillsborough Community College, Tampa, Florida.

Lewis, R.R. 1981. Economics and feasibility of mangrove restoration, p. 88-94. In P.S. Markouts (Ed.), Proc. U.S. Fish Wildl. Serv. Workshop on Coastal Ecosystems of the Southeastern United States. Washington, D.C.

Lewis, R.R. 1983. Impact of oil spills on mangrove forests, p. 171-183. In H.J. Teas (Ed.), Biology and Ecology of Mangroves. Tasks for Vegetation Science 8. Dr. W. Junk, The Hague.

Lewis, R.R. 1986. Status of Mangrove Forests, Roosevelt Roads Naval Station, Puerto Rico. Report to Ecology & Environment, Inc.

Lewis, R.R. In press. Management and restoration of mangrove forests in Puerto Rico, U.S. Virgin Islands and Florida, U.S.A. Proc. International Symp. Ecology and Conservation of the Usumacinta-Grijalva Delta, Mexico; 1987 February 2-6; Villahermosa, Tabasco, Mexico.

Lewis, R.R. and D.W. Crewz. In prep. An Analysis of the Reasons for Success or Failure of Attempts to Create or Restore Tidal marshes and Mangrove

Forests in Florida. Florida Sea Grant Project R/C-E-24, University of Florida, Gainesville, Florida.

Lewis, R.R. and K.C. Haines. 1981. Large scale mangrove planting on St. Croix, U.S. Virgin Islands: second year, p. 137-148. In D.P. Cole (Ed.), Proc. 7th Ann. Conf. Restoration and Creation of Wetlands. Hillsborough Community College, Tampa, Florida.

Lewis, R.R., R. Lombardo, B. Sorrie, G. D'Aluisio-Guerrieri, and R. Callahan. 1981. Mangrove Forests of Vieques, Puerto Rico. Vol. I, Management Report, 42 pp. Vol. II, Data Report.

Lugo, A.E. and G. Cintron. 1975. The mangrove forests of Puerto Rico and their management, p. 825-846. In G.E. Walsh, S.C. Snedaker, and H.J. Teas (Eds.), Proc. International Symp. Biology and Management of Mangroves. East-West Center, Honolulu, Hawaii.

Lugo, A.E. and S.C. Snedaker. 1974. The Ecology of Mangroves. Ann. Rev. Ecology & Systematics 5:39-64.

Mangrove Systems, Inc. 1985. Status of Mangroves on Vieques, Puerto Rico. Report to the U.S. Navy.

Martinez, R., G. Cintron, and L.A. Encarnacion. 1979. Mangroves in Puerto Rico: A Structural Inventory. Final report to the Office of Coastal Zone Management, NOAA. Dept. of Natural Resources, Area of Scientific Research, Government of Puerto Rico.

NOAA. 1977. Puerto Rico Coastal Management Program and Draft Environmental Impact Statement. U.S. Dept. of Commerce, National Oceanic & Atmospheric Admin., Office of Coastal Zone Management, Washington, D.C.

NOAA. 1986. Tide Tables 1987, East Coast of North America and South America Including Greenland. NOAA, Washington, D.C.

Norton, R.L. 1986. United States Virgin Islands, p. 585-596. In D. A. Scott and M. Carbonell (compilers), A Directory of Neotropical Wetlands. IUCN Cambridge and IWRB Slimbridge, U.K.

Pulver, T.R. 1976. Transplant Techniques for Sapling Mangrove Trees, Rhizophora mangle, Laguncularia racemosa, and Avicennia germinans, in Florida. Fla. Dept. Nat. Resources Mar. Res. Publ. 22.

Reark, J.B. 1983. An in situ fertilizer experiment using young Rhizophora, p. 166-180. In F.J. Webb, Jr. (Ed.), Proc. 9th Ann. Conf. on Wetlands Restoration and Creation. Hillsborough Community College, Tampa, Florida.

Reark, J.B. 1984. Comparisons of nursery practices for growing of Rhizophora seedlings, p. 187-195. In F.J. Webb, Jr. (Ed.), Proc. 10th Ann. Conf. on Wetlands Restoration and Creation. Hillsborough Community College, Tampa, Florida.

Stephen, M.F. 1984. Mangrove restoration in Naples, Florida, p. 201-216. In F.J. Webb, Jr. (Ed.), Proc. 10th Ann. Conf. Wetlands Restoration and Creation. Hillsborough Community College, Tampa, Florida.

Teas, H.J. 1977. Ecology and restoration of mangrove shorelines in Florida. Environ. Conserv. 4:51-58.

Woodhouse, W.E., Jr. and P.L. Knutson. 1982. Atlantic coastal marshes, Ch. 2, p. 45-109. In R.R. Lewis (Ed.), Creation and Restoration of Coastal Plant Communities. CRC Press, Boca Raton, Florida.

Zuberer, D. 1977. Biological nitrogen fixation: a factor in the establishment of mangrove vegetation, p. 37-56. In R.R. Lewis and D.P. Cole (Eds.), Proc. 3rd Ann. Conf. Restoration of Coastal Vegetation in Florida. Hillsborough Community College, Tampa, Florida.

APPENDIX I: RECOMMENDED READING

del Llano, M., J.A. Colon, and J.L. Chabert. 1986. Puerto Rico, p. 559-571. In D.A. Scott and M. Carbonell (compilers), A Directory of Neotropical Wetlands. IUCN Cambridge and IWRB Slimbridge, U.K.

Hamilton, L.S. and S.C. Snedaker. 1984. Handbook for Mangrove Area Management. United Nations International Union for the Conservation of Nature and Natural Resources, Environment and Policy Institute, East-West Center, Honolulu, Hawaii.

Island Resources Foundation. 1977. Marine Environments of the Virgin Islands. Tech. Suppl. No. 1. Virgin Islands Planning Office, Coastal Zone Management Program.

Lewis, R.R. 1981. Economics and feasibility of mangrove restoration, p. 88-94. In P.S. Markouts (Ed.), Proc. U.S. Fish Wildl. Serv. Workshop on Coastal Ecosystems of the Southeastern United States. Washington, D.C.

Lewis, R.R. (Ed.). 1982. Creation and Restoration of Coastal Plant Communities. CRC Press, Boca Raton, Florida.

Lewis, R.R. 1983. Impact of oil spills on mangroves. Proc. 2nd Ann. International Symp. on Biology and Management of Mangroves, Papua, New Guinea.

Lewis, R.R. In press. Management and restoration of mangrove forests in Puerto Rico, U.S. Virgin Islands, and Florida, U.S.A. Proc. International Symp. Ecology and Conservation of the Usumacinta-Grijalva Delta, Mexico; 1987 February 2-6; Villahermosa, Tabasco, Mexico.

Lugo, A.E. and G. Cintron. 1975. The mangrove forests of Puerto Rico and their management, p. 825-846. In G.E. Walsh, S.C. Snedaker, and H.J. Teas (Eds.), Proc. International Symposium on Biology and Management of Mangroves. East-West Center, Honolulu, Hawaii.

Martinez, R., G. Cintron, and L.A. Encarnacion. 1979. Mangroves in Puerto Rico: A Structural Inventory. Final Report to the Office of Coastal Zone Management. NOAA, Dept. of Natural Resources, Area of Scientific Research, Government of Puerto Rico.

Norton, R.L. 1986. United States Virgin Islands, p. 585-596. In D.A. Scott and M. Carbonell (compilers), A Directory of Neotropical Wetlands. IUCN Cambridge and IWRB Slimbridge, U.K.

Yanez-Aranciba, A. 1978. Taxonomy, Ecology and Structure of Fish Communities in Coastal Lagoons with Ephemeral Inlets on the Pacific Coast of Mexico. Centro. Cienc. del Mar. y Limnol., Univ. Nal. Auton. Mexico, Publ. Exp. 2:1-306.

APPENDIX II: PROJECT PROFILES

ALUCROIX CHANNEL

Locale: Port Alucroix Channel, south shore of St. Croix, U.S. Virgin Islands.

Latitude/Longitude: 17°42.00'N / 64°46.00'W.

Permit Numbers: None.

Age: Ten years.

Size: 3.8 ha.

Species Present: Rhizophora mangle, Avicennia germinans.

Site Description:

Former oil spill site used as mitigation for impoundment of 7.4 ha of mangroves by Martin Marietta Alumina. Planted with red mangroves in 1978-79.

Goals of Project: Re-establish mangrove forest as mitigation.

Judgement of Success: Successful.

Contact: Roy R. Lewis III
Lewis Environmental Services, Inc.
P.O. Box 20005
Tampa, FL 33622-0005
813/889-9684

Reports: Lewis 1979.
Lewis and Haines 1981.

SUAREZ CANAL

Locale: Suarez Canal, east of San Juan, Puerto Rico.

Latitude/Longitude: 18°26.06'N / 65°59.20'W.

Permit Numbers: None.

Age: Four years.

Size: 0.25 ha.

Species Present: Rhizophora mangle, Laguncularia racemosa.

Site Description:

Mitigation areas (two) scraped down from uplands and planted with red mangrove propagules, for the Federal Aviation Administration which had filled mangroves adjacent to San Juan Airport without permits.

Goals of Project: Create mangrove area from uplands.

Judgement of Success: Successful.

Contact: Roy R. Lewis III
Lewis Environmental Services, Inc.
P.O. Box 20005
Tampa, FL 33622-0005
813/889-9684

BAHIA TAPON

Locale: South coast of Vieques, Puerto Rico.

Latitude/Longitude: 18°06.66'N / 65°24.54'W.

Permit Numbers: None.

Age: Six years.

Size: One ha.

Species Present: Rhizophora mangle, Avicennia germinans, Laguncularia racemosa, Conocarpus erectus.

Site Description:

Former fringe forest isolated by pipeline and road. Re-opened in June 1981 by excavating to establish former tidal connection.

Goals of Project: Allow return of tidal connection and natural revegetation.

Judgement of Success: Successful.

Contact: Roy R. Lewis III
Lewis Environmental Services, Inc.
P.O. Box 20005
Tampa, FL 33622-0005
813/889-9684

Reports: Lewis et al. 1981.
Lewis in press.
Mangrove Systems, Inc. 1985.

BAHIA DE LA CHIVA

Locale: South coast of Vieques, Puerto Rico.

Latitude/Longitude: 18°06.90'N / 65°23.50'W.

Permit Numbers: None.

Age: Four years.

Size: Three ha.

Species Present: Rhizophora mangle, Avicennia germinans, Laguncularia racemosa.

Site Description:

Salt pond partially impounded due to causeway; causeway replaced by a bridge in 1983.

Goals of Project: Allow free exchange of tidal waters and drainage waters through an ephemeral opening to the ocean.

Judgement of Success: Successful.

Contact: Roy R. Lewis III
Lewis Environmental Services, Inc.
P.O. Box 20005
Tampa, FL 33622-0005
813/889-9684

Reports: Lewis et al. 1981.
Lewis in press.
Mangrove Systems, Inc. 1985.

LAGUNA BOCA QUEBRADA

Locale: Western end of Vieques, Puerto Rico.

Latitude/Longitude: 18°06.33'N / 65°34.59'W.

Permit Numbers: None.

Age: Two years.

Size: Fifteen ha.

Species Present: Rhizophora mangle, Avicennia germinans, Laguncularia racemosa.

Site Description:

Salt pond artificially isolated by road construction, causing death of 15 ha of mangroves. Road use was stopped by U.S. Navy and area was re-opened in 1985.

Goals of Project: Allow free exchange of tidal waters and drainage waters through an ephemeral opening to the ocean.

Judgement of Success: Successful.

Contact: Roy R. Lewis III
Lewis Environmental Services, Inc.
P.O. Box 20005
Tampa, FL 33622-0005
813/889-9684

Reports: Lewis et al. 1981.
Lewis in press.
Mangrove Systems, Inc. 1985.

The following project descriptions are taken from U.S. Army Corps of Engineers' Public Notices. The status of these projects is unknown but all involve some proposed wetland restoration, creation or enhancement. They deserve further investigation.

EL TUQUE LAGOON

Locale: Ponce, Puerto Rico.

Latitude/Longitude: Not provided in permit application.

Permit Numbers: 83F-5032, 85IPD-20524, 87IPM-20069.

Age: Unknown.

Size: Unknown.

Species Present: Unknown.

Site Description:

Multiple permits for fill in mangroves with some involving mangrove restoration or creation as mitigation.

Goals of Project: Unknown.

Judgement of Success: Unknown.

Contact: Juan Molina
U.S. Army Corps of Engineers
400 Fernandez Juncos Avenue
San Juan, PR 00901
809/753-4996 or 809/753-4974

SUAREZ CANAL II

Locale: Suarez Canal, near Carolina, Puerto Rico.

Latitude/Longitude: Not provided in permit application.

Permit Numbers: 87IPM-20759.

Age: Unknown.

Size: Unknown.

Species Present: Unknown.

Site Description:

After-the-fact permit for unauthorized filling of 0.12 ha of mangroves.

Goals of Project: Unknown.

Judgement of Success: Unknown.

Contact: Juan Molina
U.S. Army Corps of Engineers
400 Fernandez Juncos Avenue
San Juan, PR 00901
809/753-4996 or 809/753-4974

SOUTHGATE POND

Locale: Southgate Pond, St. Croix, U.S. Virgin Islands.

Latitude/Longitude: 17°45'30"N / 64°31'30"W.

Permit Numbers: Unknown.

Age: Unknown.

Size: Unknown.

Species Present: Unknown.

Site Description:

Proposed development with filling of 2.7 ha of wetlands in a salt pond, and pond enhancement plan as mitigation.

Goals of Project: Unknown.

Judgement of Success: Unknown.

Contact: Juan Molina
U.S. Army Corps of Engineers
400 Fernandez Juncos Avenue
San Juan, PR 00901
809/753-4996 or 809/753-4974

BACARDI CORPORATION

Locale: Catano, Puerto Rico.

Latitude/Longitude: Unknown.

Permit Numbers: 87IPM-20656.

Age: Unknown.

Size: Unknown.

Species Present: Avicennia germinans.

Site Description:

Proposed fill in 0.5 ha of wetlands; mitigation not described.

Goals of Project: Unknown.

Judgement of Success: Unknown.

Contact: Juan Molina
U.S. Army Corps of Engineers
400 Fernandez Juncos Avenue
San Juan, PR 00901
809/753-4996 or 809/753-4974

SUGAR BAY

Locale: Salt River, St. Croix, U.S. Virgin Islands.

Latitude/Longitude: 17°6.5'N /64°4.5'W.

Permit Numbers: 86IPB-20899.

Age: Unknown.

Size: Unknown.

Species Present: Unknown.

Site Description:

Planting of approximately 250 m of shoreline with red, black and white mangroves as mitigation for work related to a marine development.

Goals of Project: Unknown.

Judgement of Success: Unknown.

Contact: Juan Molina
U.S. Army Corps of Engineers
400 Fernandez Juncos Avenue
San Juan, PR 00901
809/753-4996 or 809/753-4974

TRES MONJITAS CANAL

Locale: Tres Monjitas Canal, San Juan, Puerto Rico.

Latitude/Longitude: Unknown.

Permit Numbers: 84F-1156.

Age: Unknown.

Size: Approximately seven ha.

Species Present: Unknown.

Site Description:

Mitigation associated with the Martin Pena navigation channel, Agua-Guagua project.

Goals of Project: Unknown.

Judgement of Success: Unknown.

Contacts: Juan Molina
U.S. Army Corps of Engineers
400 Fernandez Juncos Avenue
San Juan, PR 00901
809/753-4996 or 809/753-4974

Ricardo Corominas, P.E.
Redondo Construction Corp.
G.P.O. Box 4185
San Juan, PR 00936

APPENDIX III

404 PERMIT REVIEW CHECKLIST (COASTAL WETLANDS IN PUERTO RICO AND THE U.S. VIRGIN ISLANDS)

APPLICATION FORM		YES	NO
A.	Is the area of fill or excavation in wetlands clearly indicated?	—	—
B.	Is the type and function of the wetland to be filled or excavated described?	—	—
C.	Is the type and function of the created or restored wetland intended for mitigation clearly indicated?	—	—
APPLICATION DRAWINGS			
A.	Are the fill or excavation areas clearly indicated by type and acreage on a plan view?	—	—
B.	Do the cross-sections show elevations relative to NGVD?	—	—
C.	Are the elevations appropriate? Are they justified in a separate narrative?	—	—
D.	Is a particular tidal benchmark referenced as having been used to establish site elevations?	—	—
E.	Is the restoration/creation area shown as being flat? (It shouldn't be).	—	—
F.	Is drainage of the restoration/creation area provided with a distinct tidal swale or ditch?	—	—
G.	Are there provisions for excluding access by humans and grazing animals?	—	—
H.	Is the upland edge stabilized?	—	—
I.	Are the source(s), spacing and number of plants per unit area specified?	—	—
APPLICATION NARRATIVE			
A.	Is there a separate application narrative describing the project and the proposed wetlands mitigation? If not, inquire as to whether it was provided in the application but not forwarded to you.	—	—
B.	Does the narrative adequately describe the mitigation and include justification for elevations, slope, planting and monitoring reports, including reference to previous work of the consultant or published literature?	—	—
C.	Are you to receive copies of the monitoring reports directly from the consultant or applicant? If not, request direct submittal.	—	—
D.	Are clear success criteria stated?	—	—
E.	Do the methods of measuring success follow standard protocol?	—	—
F.	Are clear mid-course correction plans outlined with a decision date (i.e., three months post-construction)?	—	—
G.	Is the mitigation plan preparer's name provided? Does he have a track record? Is it satisfactory?	—	—

CREATION, RESTORATION, AND ENHANCEMENT OF MARSHES OF THE NORTHCENTRAL GULF COAST

Robert H. Chabreck
Louisiana State University Agricultural Center
School of Forestry, Wildlife, and Fisheries

ABSTRACT. Coastal marshes of the Northcentral Gulf Coast encompass over 1.2 million ha and comprise almost 50% of the coastal marshes of the United States, excluding Alaska. The region includes portions of Alabama, Mississippi, Louisiana, and Texas. Over 80% of the marshes in this region occur in Louisiana because of the influence of the Mississippi River. Salt, brackish, intermediate, and fresh marshes are well represented within the region.

Marshes have been created from dredged material deposited in shallow waters and by controlled diversion of river flow to direct sedimentation to specific sites. Plantings are seldom made on dredged material in Louisiana because of the large area to be planted and the fact that natural colonization is rapid. In fresher marshes, dredged material is left as levees after canals are dug that connect to salt water sources. Levees reduce salt water contamination and drainage of the marsh. Dredge material is usually planted in Texas, Mississippi, and Alabama to stabilize the material and hasten marsh development.

In tidal marshes, construction of weirs is the most widely used enhancement practice. Impoundments provide a mechanism for controlling water depth and salinity and regulating plant growth. But impoundments can only be constructed in marshes that will support a continuous levee system. Freshwater diversion from the Mississippi River has been used on a small scale for marsh restoration and enhancement but could be used to improve vast areas of the rapidly deteriorating marshes of southeastern Louisiana.

Precise information is needed on subsidence rates of individual localities for planning marsh creation and restoration projects. Methods for maximizing subdelta development and determining best use of dredged material are needed.

OVERVIEW OF REGION

CHARACTERISTICS OF REGION

Physiography

The Gulf Coastal Plain gently slopes toward the Gulf of Mexico and forms the coastal region of the Northcentral Gulf Coast in Alabama, Mississippi, Louisiana, and Texas. The coastal region is bordered by a broad continental shelf in the Gulf of Mexico, which contains relatively shallow water near shore. These conditions have enhanced development of marshes and barrier islands. Many rivers empty into the Gulf in this region and contain embayments that contribute to estuarine environments.

Climate

Climate of the region is temperate, with hot summers and mild winters, although several freezes occur each winter. The growing season averages about 300 days. Most of the region ishumid, and rainfall in the eastern portion averages between 140 and 150 cm (Stout 1984). Rainfall rates decrease westward along the Texas coast and range from 140 cm at the Sabine River to less than 75 cm at the Rio Grande River (Diener 1975). The Gulf coast is characterized by southerly to southeasterly prevailing winds that are an important source of atmospheric moisture. Hurricanes are common in the region and have had a detrimental effect on marshes, beaches, and barrier islands of the area.

Tides

Normal tidal range is between 30 and 60 cm, but the level of individual tides varies with the phase of the moon, direction and velocity of the wind, and other factors. Lowest tides occur during winter when strong northerly winds are present. Highest tides are associated with hurricanes, and tide levels of 1 to 2 m above normal occur in some portion of the region almost every year (Marmer 1954).

Water Salinity

Water salinity varies within the region and is largely governed for a particular site by its proximity to the Gulf of Mexico, tide water access to the site, recent rainfall rates, and river discharge. Water salinity in the adjacent Gulf of Mexico usually ranges from 20 to 25 ppt and is less than normal sea water (36 ppt). However, in the Laguna Madre of Texas or in high marsh ponds where Gulf water may be trapped, water salinity may be concentrated through evaporation and exceed that of sea water (Diener 1975).

Salinity levels are usually quite high where water from the Gulf of Mexico enters adjacent marshes. As Gulf water moves inland by tidal action, it mixes with fresher water draining from interior regions toward the Gulf. Consequently, water salinity gradually declines from the coastline to the interior reaches of the coastal marshes. Marsh vegetation varies in its tolerance to water salinity, and plants are grouped into communities having similar tolerances and are referred to as vegetation or marsh types.

Water salinity is the primary factor affecting plant distribution. Four distinct marsh types have been identified in the region (Penfound and Hathaway 1938, Chabreck 1972): salt, brackish, intermediate, and fresh. Since water salinity declines on a gradient moving inland from the Gulf of Mexico, the marsh types generally occur in bands parallel to the shoreline.

MARSH TYPES

Geographical Distribution

Coastal marshes of the Northcentral Gulf Coast encompass slightly over 1.2 million ha and comprise almost 50% of the coastal marshes in the United States, excluding Alaska (Alexander et al. 1986). Louisiana contains the largest area of marshes and comprises 81.2% of the region (Table 1). The marshes of Louisiana extend inland from the Gulf for distances ranging from 24 to 80 km and consist of the deltaic plain and the chenier plain. The deltaic plain lies in the southeastern portion of the state and makes up three-fourths of its coastal region (Chabreck 1970). Marshes of the deltaic plain were formed from deposition by the Mississippi River and are unstable and in various stages of degradation (Coleman 1966). The marsh is rapidly subsiding and eroding and being lost at a rate of over 100 km per year (Gagliano et al. 1981). An irregular shoreline with numerous large embayments characterize the area. The deltaic plain has several chains of barrier islands, which represent the outer rim of former deltas of the Mississippi River. Active land building is currently taking place in deltas of the Mississippi River and the Atchafalaya River, which is a distributory of the Mississippi River.

The chenier plain occupies the coasts of southwestern Louisiana and southeastern Texas. Marshes of the chenier plain were formed from Mississippi River sediment that was discharged into the Gulf of Mexico, carried westward by currents, and deposited against the shoreline to form marshland. Interruptions in the depositional process resulted in beach formation, and resumption of deposition caused new marshland to form seaward from the beach. This process caused several beach deposits to be stranded in the marshes. The stranded beaches are locally termed cheniers and represent a major relief feature of the area. Marshes of the chenier plain are underlain by firm clay deposits and will support the weight of levees and dredged material deposits, quite unlike the unstable subsoils of most of the deltaic plain. The chenier plain is bordered by a well developed beach, which has few openings into the Gulf of Mexico (Russell and Howe 1935).

Texas contains 15.6% of the marshes of the Northcentral Gulf Coast. The greatest area of marshland in the state lies between the Sabine River and Galveston Bay and is part of the chenier plain. Westward along the Texas coast, a well developed series of barrier islands are present, and marshes occur along the shores of bays enclosed by the offshore bars (West 1977). An extensive band of tidal flats also borders the shorelines of the bays (Diener 1975).

Mississippi contains 2.3% of the coastal marshes of the region. The greatest area of marshland is in the southwestern portion of the state and comprises a portion of the deltaic plain (Eleuterius 1973). Marshes along the Alabama coast comprise less than 1% of the Northcentral Gulf Coast marshes. In Alabama and adjacent areas of the Mississippi coast, marshes are small, disjunct, and limited to low alluvial deposits along protected bay shores and rivers (West 1977). A series of barrier islands occur offshore of both states (Crance 1971, Eleuterius 1973).

Description of Types

Since the major area of marshland in the region is in Louisiana and adjacent portions of Texas and Mississippi, most descriptions of marsh types relate to that area. The marsh types (salt, brackish, intermediate, and fresh) not only contain characteristic associations of plant species but also vary in hydrological patterns, soils, fish and wildlife values, and ecological functions performed.

Salt Marsh--

Salt marsh makes up 28.2% of the coastal marshes of the region (Table 1). This type generally occupies a narrow zone adjacent to the shoreline of the Gulf of Mexico and embayments. It is quite extensive in the deltaic plain because the broken shoreline allows tide water to move far inland.

Salt marsh has the greatest tidal fluctuation of all marsh types and contains a well developed drainage system. Water salinity averages 18.0 ppt (range: 8.1 to 29.4 ppt), and soils have a lower organic content (mean: 17.5%) than fresher types located further inland (Chabreck 1972). Vegetation within this type is salt-tolerant and is dominated by smooth cordgrass (Spartina alterniflora), salt grass (Distichlis spicata), and black rush (Juncus roemerianus).

Brackish Marsh--

Brackish marsh comprises 30% of the total marsh area of the region. This type lies inland from the salt marsh type, is further removed from the influence of saline Gulf waters, but is still subject to daily tidal action. Normal water depth exceeds that of salt marsh, and soils contain higher organic content (mean: 31.2%). Water salinity averages 8.2 ppt (range: 1.0-18.4 ppt). This marsh type characteristically contains numerous small bayous and lakes.

Brackish marsh contains greater plant diversity than salt marsh and is dominated by two perennial grasses, marshhay cordgrass (Spartina patens) and D. spicata. An important wildlife food plant of brackish marsh, Olney bulrush (Scirpus olneyi), grows best in tidal marsh free from excessive flooding, prolonged drought, and drastic salinity changes. S. olneyi is, however, crowded out by S. patens unless stands of S. olneyi are periodically burned. Wigeongrass (Ruppia maritima), the dominant submerged aquatic plant of brackish marsh, is a preferred food of ducks and coots (Fulica americana).

Intermediate Marsh--

The intermediate marsh type lies inland from the brackish type and comprises 14.7% of the marsh area of the region. Intermediate marsh receives some influence from tides, and water salinity averages 3.3 ppt (range: 0.5 to 8.3 ppt). Water levels are slightly higher than in brackish marsh, and soil organic content averages 33.9%. Plant species diversity is high, and the type contains both halophytes and freshwater species used as food by a wide variety of herbivores. S. patens dominates intermediate marsh as it does brackish marsh, but to a lesser degree. Some of the common marsh plants in intermediate zones are giant reed (Phragmites australis), narrowleaf arrowhead (Sagittaria lancifolia), and waterhyssop (Bacopa monnieri). This type also contains an abundance of submerged aquatic plants that are important foods for ducks and coots.

Fresh Marsh--

Fresh marsh makes up 27.1% of the marshes of the Northcentral Gulf Coast. It occupies the zone between the intermediate marsh and the Prairie formation or the forested wetlands in the alluvial plain of major river systems. Fresh marsh is normally free from tidal influence, and water salinity averages only 1.0 ppt (range: 0.1 to 3.4 ppt). Because of slow drainage, water depth and soil organic content (mean: 52.0%) are greatest in fresh marsh. In some fresh marshes, soil organic matter content exceeds 80%, and the substrate for plant growth is a floating organic mat referred to as "flotant" by Russell (1942). Fresh marsh supports the greatest diversity of plants and contains many species that are preferred foods of wildlife. Dominant plants include maidencane (Panicum hemitomon), spikerrush (Eleocharis spp.), S. lancifolia, and alligatorweed (Alternanthera philoxeroides). The type also contains many submerged and floating-leafed aquatic plants of value as wildlife foods. Some floating aquatics, such as water hyacinth (Eichhornia crassipes), form dense stands that block waterways and are considered pest plants.

FUNCTIONS PERFORMED

Fish and Wildlife Habitat

Coastal marshes and their associated water bodies provide valuable habitat for fish and wildlife. Salt and brackish marshes and the adjacent estuaries are important nursery grounds for many species of marine fish and crustaceans that spawn in the Gulf of Mexico (Gunter 1967). Also, some species complete their entire life cycle in estuarine waters (Chapman 1973). Water bodies associated with fresh and intermediate marshes also support abundant fisheries resources.

Coastal marshes provide important habitat for birds, mammals, reptiles, and amphibians. The number of species in each group generally decreases as the water salinity of the marsh type increases (Table 2). Groups with greatest sporting or commercial value include waterfowl, fur bearing animals, and alligators (Alligator

Table 1. Geographical distribution of marsh types in states along the Northcentral Gulf Coast.[1]

	Size of marsh types (hectares x 100)				
State	Salt	Brackish	Intermediate	Fresh	Total
Alabama	28	31	15	28	102
Mississippi	75	83	42	78	278
Louisiana	2,663	2,959	1,480	2,762	9,864
Texas	665	569	247	418	1,899
Total	3,431	3,642	1,784	3,286	12,143

[1]The total area of marsh by state is from Alexander et al. (1986). Distribution of marsh types by state is from Chabreck (1972), Diener (1975), and Crance (1971).

Table 2. Number of species of wildlife groups by marsh types along the Northcentral Gulf Coast.[1]

	Marsh types			
Wildlife groups	Salt	Brackish	Intermediate	Fresh
Birds	100	89	91	88
Mammals	8	11	11	14
Reptiles	4	16	16	24
Amphibians	0	5	6	18

[1]Source: Gosselink et al. 1979.

mississippiensis). Individuals in these groups utilize all marsh types but greatest densities are usually found in marshes with lower water salinity (Palmisano 1973, McNease and Joanen 1978, Linscombe and Kinler 1985).

Hurricane Protection

A band of marshes and barrier islands along the coast provide a buffer against hurricanes and reduces flooding of interior areas. The marshes absorb the energy of storm waves and provide a drag on the inland rush of storm waters. Coastal facilities separated from the Gulf by a wide band of marshes suffer less damage than those without protective marshes (Craig et al. 1979).

Water Quality Improvement

Coastal marshes improve water quality in estuaries by retaining pollutants or delaying their movement into estuaries. The pollutants include excess nutrients, toxic chemicals, and disease-causing micro-organisms. Some pollutants may settle out in the marsh and be converted by biochemical processes to less harmful forms. Some may remain trapped in the sediments or be taken up by plants and recycled or transported from the marsh (U.S. Congress, Office of Technology Assessment 1984).

EXTENT TO WHICH CREATION/RESTORATION/ ENHANCEMENT HAS OCCURRED

TYPES OF PROJECTS

Marsh Creation

Two types of marsh creation have traditionally been used along the Northcentral Gulf Coast. One type involves creation of marsh from dredged material deposited in shallow water during dredging operations. The other type of marsh creation involved controlled diversion of river flow to direct sedimentation to specific areas.

Much marsh has been created in the region from dredged material placement in shallow water. However, in almost all cases in Louisiana, no special treatments were applied nor attempts made to establish vegetation by planting. In Texas, Mississippi, and Alabama marsh plantings usually accompanied the material placement (Landin 1986), although many plantings were done as part of special investigations. Controlled diversion of river flow to create marsh by sedimentation has been carried out at certain locations on the lower Mississippi River. Expansion of this type of marsh creation has been recommended as a procedure for creating vast areas of new marshland. In fact, an important aspect of diverting one-third of the Mississippi River flow down the Atchafalaya River is creation of additional marsh in the recently emerged delta of the Atchafalaya River (Cunningham 1981).

Marsh Restoration

Marsh restoration primarily involves reestablishment of plants to normal conditions in a deteriorating marsh. Means by which this can be accomplished include diversion of freshwater or sediment into the marsh to offset the factors causing the deterioration. In some situations, the factors causing the deterioration can be regulated or offset by impoundment or by constructing weirs in drainage systems of the marsh. In many instances, such projects may be classified as marsh enhancement.

Diversion of freshwater into marshes is currently used, to a limited extent, at several locations along the lower Mississippi River to restore deteriorating marshes and is being considered in other areas where feasible. Weir construction has been widespread across the Louisiana coast, but marsh impoundment construction in the region has been largely restricted to southwestern Louisiana and southeastern Texas, where soil conditions permit construction of continuous levee systems.

TYPICAL GOALS FOR PROJECTS

Marsh Creation

The goals of projects to create marsh on dredged material are to optimize use of the material, expand the acreage of wetlands, and help slow the erosion process. This activity generally has been successful, but the degree of success varies throughout the region. In marshes with firm foundations, success is greater than in areas were the substrate is poorly consolidated. Without a firm base for support of deposits, the dredged material rapidly subsides and soon sinks below the surface of the water.

Plantings are rarely made on dredged

material in Louisiana, which may partially account for the poor success of marsh creation in some areas. The reason more plantings are not made is because of the large area to be planted and the fact that natural colonization is very rapid. In Texas, Mississippi, and Alabama, dredged material is usually planted to stabilize the material and hasten marsh development.

In fresh and intermediate marshes where salt water intrusion is a problem, dredged material is left as levees after canals are dug. Even where canals cross ponds or small lakes, maintenance of a levee is important. Without a protective levee, salt water may enter through the canal and kill marsh vegetation, or on low tides, the canal will excessively drain the marsh.

The goal of marsh creation by controlled diversion of river flow is to establish new marshland in an area where marsh is rapidly disappearing because of subsidence and erosion. This type of marsh creation has been used on a small scale on the Mississippi River delta. Openings made in banks of distributaries permit river water to flow during flood stages into adjacent shallow water areas. River water deposits its silt load as it spreads out into the quiet shallow waters and its velocity decreases, thus creating small subdeltas.

Success of controlled diversion depends upon proper placement of bank openings so that an adequate flow of water occurs. Also, proper water control structures are essential to prevent excessive flow and channel development. South of New Orleans where the mainline levee is not in place, marsh nourishment is occurring with visible results. A factor that may reduce the effectiveness of this type of marsh creation in the future is a reduction in sediment load carried by the Mississippi River.

Marsh Restoration

Restoring a Gulf coast marsh to its original condition is a costly and, in most cases, an impossible task. Coastal marshes are dynamic systems, and, if it were possible to completely restore a marsh, it could be restored to only one stage in its frequently changing past. Therefore, the goal of restoration should be to maximize productivity and enhance the quality and diversity of the environment within the limits of available resources and technology.

Construction of weirs in drainage systems is the most widely used restoration practice in tidal marshes. The elevation of a weir is established approximately 15 cm below that of surrounding marsh so that water will freely flow over the structure. However, the weir holds a permanent basin of water that functions as a mixing bowl with incoming tides to stabilize water salinity. Weirs greatly increase production of submerged aquatics in marsh ponds (Chabreck 1968, Larrick and Chabreck 1976) and attract ducks to the ponds (Spiller and Chabreck 1975); however, weirs do not change the plant species composition of affected marsh. Also, the weirs reduce ingress and delay egress of aquatic organisms in a marsh (Herke 1979); however, Rogers et al. (1987) found that this problem could be largely mitigated by placing a vertical slot in the weir to allow passage of aquatic organisms.

Impoundments have been constructed in coastal marshes to form wildlife management units but are restricted to marshes, such as thechenier plain, with firm soil that will support continuous levee systems. Impoundment construction requires establishment of adequate water control structures such as weirs, gated culverts, or pumps. Impoundment provides a mechanism for regulating water levels and salinities and controlling plant distribution and species composition. In many marshes that are deteriorating because of excessive salt water intrusion or drainage, impoundment offers the best and, in most cases, the only solution to marsh restoration.

Impoundments have been constructed mainly to improve habitat for ducks, but other forms of wildlife and fisheries also benefit (Chabreck 1980). However, levees that form a barrier to tidal flow also block normal movement of marine organisms and may prevent their access to the enclosed marsh and water bodies. Studies in Louisiana have disclosed that this problem can be partially corrected by opening water control structures at the proper time to allow passage of marine organisms (Davidson and Chabreck 1983).

Freshwater diversion from the Mississippi River has been used on a small scale for marsh restoration but could be used to restore vast portions of the rapidly deteriorating marshes of the deltaic plain. Culverts and siphons are used to move river water during flood stages through or over the flood protection levees. The water is allowed to flow through the marsh, thus adding sediment and nutrients and lowering water salinity in the marsh. The added sediment raises the elevation of the marsh and helps offset land loss. Reducing water salinity and adding nutrients promotes plant growth, increases plant species diversity, and improves the marsh and adjacent water bodies for fish and wildlife. A major handicap of this type of restoration is that it can only be used in marshes in drainage basins adjacent to the freshwater source. Also, areas on the landward side of flood protection levees are usually developed, and diverting river water for marsh nourishment raises not only an environmental but also socioeconomic and political issues.

DESIGN OF CREATION/RESTORATION/ENHANCEMENT PROJECTS

DREDGE SPOIL

Location and Site Characteristics

Marsh has been created with dredged material at Mobile Bay, Mississippi Sound, Southwest Pass of the Mississippi River, Atchafalaya Bay, Calcasieu Lake, Galveston Bay, Chocolate Bay, Aransas Bay, East Matagorda Bay, and various other sites along the Northcentral Gulf Coast. Project sites included areas along shorelines and in open water away from land. Dredged material was confined within dikes in some areas, but in others, only a temporary breakwater was used to protect planted material. At sites along Southwest Pass of the Mississippi River and in Atchafalaya Bay, spoil was deposited in open water without confinement levees and allowed to revegetate naturally. Most plantings on dredged material have been done on an experimental basis but have yielded valuable information for marsh creation and restoration (Landin and Webb 1986, U.S. Army Corps of Engineers 1986).

Critical Aspects of Project Plans

Construction Considerations--

Site selection for marsh creation projects along the Northcentral Gulf Coast requires field investigations and laboratory tests to evaluate location, bottom topography, wave and water energy, and substrate characteristics including consolidation and sedimentation. If dredging is involved, material may be placed in the disposal site using either hydraulic or mechanical methods. The hydraulic pipeline dredge is the type most commonly used for projects involving marsh creation or restoration. Disposal sites should be as near the excavation site as possible; however, at substantially greater cost, material can be moved through the pipeline for several kilometers with intermediate booster pumps. Other marsh development projects such as impoundments require equal care in site selection, construction design, and other critical factors.

Hydrology--

Important hydrological factors in marsh design are water salinity, tidal range, flood stages, and wave and wind action. Characteristics of wind waves such as height, fetch, period, direction, and probability of occurrence can be obtained from locally collected data. At locations where wind waves are a major consideration, early recognition of the problem may allow either relocation to alternate sites where open-water fetch in the predominant wind direction is minimized, or incorporation of protection structures into project design. Locating sites in low energy environments greatly increases the chances of project success. Tidal range and flood stages are factors that regulate elevation of sites. Water salinity is an important consideration in the selection of species for planting (U.S. Army Corps of Engineers 1986).

Substrate--

Design of substrate at sites for marsh creation requires information regarding desired elevation, slope, shape, orientation, and size of the sites. The design must allow for placement of dredge material within desired limits of required elevations, allowing for consolidation of dredged material and compaction of foundation soils. The design must include predictions of expected settlement. Substrate composed of fine clays and silts may remain in a slurry state for a significant period after placement and require a retaining structure for containment. The final elevation when such substrates dry is much more difficult to predict than when substrates are composed of sandy material, which lose water and dry quickly (U.S. Army Corps of Engineers 1986). Dredged material used as substrate for marsh creation in Galveston Bay was placed on a 0.7% slope from mean low water to over 1 m above mean high water to achieve successful elevation (Webb et al. 1986).

Producing Desired Plant Communities--

Establishment of marsh plants on dredged material can be attained by natural invasion or artificial propagation. In freshwater areas, natural invasion by marsh plants will occur within one growing season. Dredge spoil on the Mississippi and Atchafalaya River deltas that is placed several inches below mean high tide is soon colonized by delta duckpotato (Sagittaria platyphylla).

In other marsh types, several years may be required for natural invasion of marsh plants, and artificial propagation by sprigging of desired species is recommended if establishment of plant cover is desired. Along the Northcentral Gulf Coast, sprigging of S. patens is recommended for sites in intermediate and brackish marshes (Eleuterius 1974). In saltwater areas, S. alterniflora should be sprigged at elevations below mean high tide, and S. patens should be sprigged at elevations above mean high tide (Allen et al. 1978, Landin 1986). Planting material is often collected near the planting site; however, Eleuterius (1974) reported that best success was obtained when cuttings of S. patens and D. spicata were rooted in peat pellets before field planting.

Plantings in the intertidal zone on sandy dredged material along the shoreline of Bolivar Peninsula in Galveston Bay successfully established stands of S. alterniflora and S. patens. A sandbag dike was placed in the intertidal zone and along the flanks to protect a portion of the planting site from wave energy. Also, a fence was erected to exclude goats and rabbits from the site (Allen et al. 1978).

Plantings were first made in 1976, and the size and vigor of the plant community increased each year. The site withstood direct hits of hurricanes in 1983 and 1986 with no noticeable effects (Landin 1986). The density and vigor of plants inside the diked area was considerably greater than that of plants in an adjacent control area. Allen et al. (1978) believed that much of the success in establishment of the plants was related to protection provided by the sandbag dike. They also concluded that, in areas with excessively high wave energies, success is unlikely unless wave energy is dissipated.

Attempts to create marsh in Mobile Bay on a dredged material island where substrate consisted largely of clay was less successful than those in Galveston Bay. Wave energy was moderately high, and a floating tire breakwater and a fixed breakwater were installed to protect plantings of S. alterniflora. The fixed breakwater consisted of a wooden fence onto which automobile tires were mounted in a single row. Plantings were made in the intertidal zone behind both types of breakwaters and in a control area but were successful only behind the floating tire breakwater. Plant spacings of 45 cm and 90 cm were tested (the 45-cm spacing required four times more plants). The 45-cm spacing had lower plant survival but produced stands with greater density. Fertilization of plantings did not increase survival (Allen and Webb 1983).

CONTROLLED DIVERSION OF RIVER FLOW

Preconstruction Considerations

The Mississippi River has historically overflowed its banks, spread out through its many distributaries, and added vast amounts of freshwater and sediment to the Northcentral Gulf Coast. This process was essential not only for developing the spacious marshes of the region but also for their maintenance. During the early 20th century, the Mississippi River was enclosed with levees for flood protection, a process that terminated the over bank flooding. Most freshwater and sediment is now channeled down the river and deposited in the Gulf of Mexico. One-third of the flow is diverted into the Atchafalaya River where it is actively building new land in a growing delta (Cunningham 1981, Roberts and van Heerden 1982). Diversion of a portion of the flow of the Mississippi River into adjacent marshes and estuaries has been recommended as a procedure for marsh creation and restoration (Gagliano et al. 1973). Freshwater from the river would help offset encroachment of salt water into the deltaic plain marshes, enhance plant growth, and possibly facilitate a seaward advancement of marsh type boundaries. Diversion of sediment would promote subdelta development and facilitate vertical accretion of the marshes.

Critical Aspects of Project Plans

Construction Considerations--

Small-scale freshwater diversion structures have been installed at Bayou Lamoque and Violet Canal. Even though the structures are small and divert only 7 m^3/sec., improvements have been noted in wetlands in the immediate area (Chatry and Chew 1985). Opening of the Bonnet Carre Spillway for flood control on the Mississippi River in 1973 and 1975 diverted tremendous amounts of freshwater into Lake Pontchartrain, Lake Borgne, and Mississippi Sound. Changes of marsh types to less saline conditions in marshes east and south of Lake Borgne was attributed to that action (Chabreck and Linscombe 1982). Operation of the Bonnet Carre Spillway on a routine basis has been recommended for improvement of fish and wildlife habitat in the region (Fruge and Ruelle 1980). The U.S. Army Corps of Engineers, New Orleans District, is conducting studies to determine the feasibility of large-scale controlled diversion of freshwater from the Mississippi River into nearby estuarine areas. The studies have indicated that recommended fish and wildlife habitat changes could be met 9 out of 10 years with strategically placed structures. This would include a 300 m^3/sec. diversion structure in the vicinity of Davis Pond in the Barataria Basin (west of the river) and a 187 m^3/sec. structure at Caernarvon in the Breton Sound Basin (east of the river) (Chatry and Chew 1985).

Hydrology--

Freshwater diversion into an estuarine basin will cause changes in salinity regimes and enhance vegetation growth; however, some concern has been expressed regarding other impacts associated with water salinity changes. Certain harvestable resources such as oysters and brown shrimp become more accessible as salt water encroaches into formerly freshwater or low salinity habitats closer to urban areas. Unfortunately, industrial and domestic pollution are also more severe in such areas and may affect these resources (Chatry and Chew 1985).

Some concern has also been expressed about the quality of Mississippi River water introduced into estuarine systems and increased pollution that may result. However, previous introductions of Mississippi River water through the Bonnet Carre Spillway have not resulted in increased levels of contaminants (Fruge and Ruelle 1980).

Substrate–

Diversion of river flow can provide substrate for marsh creation and restoration. Marshland can be created by diverting river flow into shallow embayments favorable to the development of subdeltas. The effectiveness of sediment in building new marshland by subdelta formation is dependent upon water depth, subsidence, compaction, sea-level rise, flow regime, tidal currents, and wind. The amount of water and sediment diverted is important also, because there is a minimum diversion below which sediment will not accumulate but will be carried away by longshore currents (Gagliano et al. 1973).

Important information regarding subdelta development on the Mississippi River delta was provided by studies of four major modern subdeltas: Baptiste Collette, Cubits Gap, West Bay, and Garden Island Bay. The growth rate during active buildout ranged from 0.5 km^2/yr (Baptiste Collette) to 2.6 km^2/yr (West Bay). The average buildout rate for the four subdeltas was 1.8 km^2/yr. The efficiency of sediment retention ranged from about 50% to over 90% (average: 70%) (Gagliano et al. 1973). Marshes undergoing deterioration because of salt water intrusion, erosion, subsidence, and rising sea levels can be restored by introducing freshwater and sediment. Ponds in deteriorating coastal wetlands gradually enlarge, and the marsh occurs as islands of emergent vegetation. Sedimentation gradually fills the ponds until the substrate reaches an elevation suitable for growth of emergent plants.

Establishment of Plant Communities–

Marshes created or restored by diversion of river flow along the Northcentral Gulf Coast are naturally colonized by plants. The species that frequently become established on subdeltas are those best suited for prevailing freshwater conditions, such as S. platyphylla, common arrowhead (Sagittaria latifolia), American bulrush (Scirpus americana), common cattail (Typha latifolia), and P. australis. Marshes restored by introduced river sediment are likely to be colonized by brackish marsh plants such as S. patens, S. olneyi, and narrowleaf cattail (Typha angustifolia).

MONITORING

Monitoring of marsh creation and restoration projects should include site characteristics, dredged material placement, protective measures, plant establishment and growth, wildlife use, and other site attributes. Physical features of a site include climate, geographical location and size, topography and configuration, physical and chemical properties of the supporting subsoil, hydrology, physical and chemical properties of material to be dredged, and land use. Biological features of a site include plant communities in the area and aquatic and terrestrial animals that use the site or the surrounding area. Effective monitoring requires a multi-disciplinary team, which should include a wildlife biologist, fishery biologist, botanist, soil scientist, engineer, and land-use planner (U.S. Army Corps of Engineers 1986).

Plant establishment should be given careful consideration in the monitoring process because the success of a marsh creation and restoration project can be best measured by the plant communities that the site ultimately supports. Newling and Landin (1985) recommended two levels of monitoring to evaluate the success of plant establishment and the factors that affect plant growth. The first level of monitoring involves an annual, general reconnaissance of all sites to provide qualitative information on changes at sites that may require closer evaluation such as excessive erosion or plant mortality. Information for a general reconnaissance of sites may be determined by analysis of aerial photographs of the site.

The second level of monitoring should be intensive sampling of sites to provide quantitative data needed for research projects or for analysis of sites presenting special problems. Such monitoring requires sampling of plants and soils and should be conducted on a random basis along elevational gradients. Newling and Landin (1985) sampled 0.5-m^2 quadrats along transects in elevational strata. Information collected on individual plant species in plots included number of stems, mean height, number flowering, aboveground biomass, and total belowground biomass. Soil samples were taken in each quadrat to a depth of 25 cm and sectioned at 5-cm intervals. Each interval was tested for particle size, volatile solids, percent moisture, bulk density, pH, total Kjeldahl nitrogen, total phosphorus, and total organic carbon.

Survival data alone failed to provide complete insight into the final results of transplanting and subsequent plant growth. In ideal marsh sites, transplants often produced rapid growth and closed stands. However, the rate of growth varied among species and plots, and determining growth within and among species was difficult. Photographing plots at different time intervals was the best means of conveying growth patterns without destroying plots (Eleuterius 1974).

INFORMATION NEEDS

The Northcentral Gulf Coast has the greatest subsidence rates in the United States. Projected subsidence for the next century in Louisiana is from 60 to 90 cm and at Galveston, Texas, is 55 cm (Titus 1985). Subsidence rates for local areas in the region will vary; consequently, more precise information on individual localities is needed when planning marsh creation and restoration projects.

Diversion of vast amounts of Mississippi River water is needed to restore existing marsh that is rapidly deteriorating and being lost. Much concern is expressed regarding the quality of Mississippi River water and its impact in areas receiving the flow. During river diversion in recent years into Lake Pontchartrain and Lake Borgne via the Bonnet Carre Spillway no problems with water quality were reported. However, information is needed regarding possible impacts on water quality if vast amounts of Mississippi River water were diverted into Barataria Bay and Breton Sound. Small subdelta lobes can be created on the active delta of the Mississippi River by diverting river flow through openings in pass banks into shallow ponds and embayments. However, information is needed on the width and depth of openings and the characteristics of receiving water bodies that would maximize sediment accumulation and subdelta development. Openings that are too small may not carry enough material to be effective, and openings too large may cause erosion and channel formation in receiving water bodies.

Although dredged material has been used effectively for marsh creation, in some situations it may serve a better purpose by forming levees along canals to prevent salt water intrusion into adjacent marshes and excessive drainage of marshes. The best use of dredged material will vary with the location of the project and the amount of material available. Also, maintenance dredging will continue to produce additional material in the future for marsh creation. Information is needed on the best use of dredged material under various salinity regimes and with different project sizes.

Sprigging of marsh plants on dredged material to establish stands of vegetation is a costly and time-consuming project. Direct seeding is less costly and could be applied to some sites rather than sprigging. Additional research is needed on seeding characteristics of marsh plants and procedures for establishment of stands by direct seeding.

LITERATURE CITED

Alexander, C.E., M.A. Broutman, and D.W. Field. 1986. An Inventory of Coastal Wetlands of the USA. National Oceanic and Atmospheric Admin., Washington, D.C.

Allen, H.H. and J.W. Webb. 1983. Erosion control with salt marsh vegetation. Proc. Symp. on Coastal and Ocean Manage. 3:735-748.

Allen, H.A., E.J. Clairain, Jr., R.J. Diaz, A.W. Ford, J.L. Hunt, and B.R. Wells. 1978. Habitat Development Field Investigations Bolivar Peninsula, Marsh and Upland Habitat Development Site, Galveston Bay, Texas (Summary Report). U.S. Army Engineers, Waterways Exp. Sta. Tech. Rep. D-78-15.

Chabreck, R.H. 1968. Weirs, plugs, and artificial potholes for management of wildlife in coastal marshes, p. 178-192. In J.D. Newsom (Ed.), Proceeding of the Marsh and Estuary Management Symposium, Louisiana State University, Baton Rouge.

Chabreck, R.H. 1970. Marsh Zones and Vegetative Types in the Louisiana Coastal Marshes. Ph.D. dissertation, Louisiana State University, Baton Rouge.

Chabreck, R.H. 1972. Vegetation, Water and Soil Characteristics of the Louisiana Coastal Region. Louisiana Agric. Exp. Sta. Bull. 664. Baton Rouge.

Chabreck, R.H. 1980. Effects of marsh impoundments on coastal fish and wildlife resources, p. 1-16. In P.L. Fore and R.D. Peterson (Ed.), Proceedings of the Gulf of Mexico Coastal Ecosystems Workshop. U.S. Fish and Wildl. Serv. FWS/OBS-80/30.

Chabreck, R.H. and R.G. Linscombe. 1982. Changes in vegetative types in the Louisiana coastal marshes over a 10-year period. Proc. Louisiana Acad. Sci. 45:98-102.

Chapman, C.R. 1973. The impact of estuaries and marshes on modifying tributary runoff, p. 235-258. In R.H. Chabreck (Ed.), Proceedings of the 2nd Coastal Marsh and Estuary Management Symposium, Louisiana State University, Baton Rouge.

Chatry, M. and D. Chew. 1985. Freshwater diversion in coastal Louisiana: recommendations for development of management criteria, p. 71-84. In C.F. Bryan, P.J. Zwank, and R.H. Chabreck (Eds.), Proceedings of the 4th Coastal Marsh and Estuary Management Symposium. Louisiana State University, Baton Rouge.

Coleman, J.M. 1966. Recent coastal sedimentation: Central Louisiana Coast. Coastal Studies Series No. 17, Louisiana State University Press. Baton Rouge.

Craig, N.J., R.E. Turner, and J.W. Day, Jr. 1979. Land loss in coastal Louisiana (U.S.A.). Environ. Manage. 3:132-144.

Crance, J.H. 1971. Description of Alabama Estuarine Areas - Cooperative Gulf of Mexico Estuarine Inventory. Alabama Dept. of Cons., Dauphin Island, Alabama.

Cunningham, R. 1981. Atchafalaya delta: subaerial development environmental implications and resource potential, p. 349-365. In R.D. Cross and D.L. Williams (Eds.), Proceedings of the National Symposium on Freshwater Inflow to Estuaries, Vol. 1. U.S. Fish and Wildl. Serv. FWS/OBS-81-04.

Davidson, R.B. and R.H. Chabreck. 1983. Fish, wildlife, and recreational values of brackish marsh impoundments, p. 89-114. In R. J. Varnell (Ed.), Proceedings of the Water Quality and Wetlands Management Conference. Louisiana Environmental Prof. Assoc., New Orleans.

Diener, R.A. 1975. Cooperative Gulf of Mexico Estuarine Inventory and Study--Texas: Area Description. National Oceanic and Atmospheric Admin. Tech. Rep. NMFS, Circ. 393.

Eleuterius, L.N. 1973. The marshes of Mississippi, p. 147-190. In J.Y. Christmas (Ed.), Cooperative Gulf of Mexico Estuarine Inventory and Study, Mississippi. Gulf Coast Res. Lab., Ocean Springs, Mississippi.

Eleuterius, L.N. 1974. A Study of Plant Establishment on Dredge Spoil in Mississippi Sound and Adjacent Waters. Gulf Coast Research Lab., Ocean Springs, Mississippi.

Fruge, D.W. and R. Ruelle. 1980. A Planning-Aid Report on the Mississippi and Louisiana Estuarine Study. U.S. Fish and Wildl. Service, Lafayette, Louisiana.

Gagliano, S.W., P. Light, and R.E. Becker. 1973. Controlled Diversion in the Mississippi River Delta System: An Approach to Environmental Management. Louisiana State University Center for Wetland Resources Rep. No. 8.

Gagliano, S.W., K.J. Meyer-Arendt, and K.M. Wicker. 1981. Land loss in the Mississippi deltaic plain. Trans. Annu. Meeting Gulf Coast Geol. Soc. 31:295-300.

Gosselink, J.G., C.L. Cordes, and J.W. Parsons. 1979. An Ecological Characterization Study of the Chenier Plain Ecosystem of Louisiana and Texas. U.S. Fish and Wildl. Serv. FWS/OBS-78-9. Washington, D.C.

Gunter, G. 1967. Some relationships of estuaries to the fisheries of the Gulf of Mexico, p. 621-638. In G.H. Lauff (Ed.), Estuaries. Am. Assoc. for the Advancement of Sci. Publ. No. 83, Washington, D.C.

Herke, W.H. 1979. Some effects of semi-impoundment on coastal Louisiana fish and crustacean nursery usage, p. 325-346. In J.W. Day, Jr., D.D. Culley, Jr., R.E. Turner, and A.J. Mumphrey, Jr. (Eds.), Proceedings of the 3rd Coastal Marsh and Estuary Management Symposium. Louisiana State University. Baton Rouge.

Landin, M.C. 1986. Wetland beneficial use applications of dredged material disposal sites. Proceedings of the Annual Conference on Wetlands Restoration and Creation, Hillsborough Community College, Tampa, Florida 13:118-129.

Landin, M.C. and J.W. Webb. 1988. Wetland development and restoration as part of Corps of Engineers programs: case studies, p. 388. In J.A. Kusler, M.L. Quammen and G. Brooks (Eds.), Proceedings of the Wetlands Mitigation Symposium, Mitigation of Impacts and Losses. Assoc. of State Wetland Mgrs., Berne, New York.

Larrick, W.J., Jr. and R.H. Chabreck. 1976. The effects of weirs on aquatic vegetation along the Louisiana coast. Proc. Annu. Conf. Southeast. Assoc. Game and Fish Comm. 30:581-589.

Linscombe, G. and N. Kinler. 1985. Fur harvest distribution in coastal Louisiana, p. 187-199. In C.F. Bryan, P.J. Zwank, and R.H. Chabreck (Eds.), Proceedings of the 4th Coastal Marsh and Estuary Management Symposium, Louisiana State University, Baton Rouge.

McNease, L. and T. Joanen. 1978. Distribution and relative abundance of the alligator in Louisiana coastal marshes. Proc. Annu. Conf. Southeast. Assoc. Game and Fish Comm. 32:182-186.

Marmer, H.A. 1954. Tides and sea level in the Gulf of Mexico, p. 101-118. In P.S. Galtsoff (Ed.), Gulf of Mexico: Its Origin, Waters, and Marine Life. U.S. Fish and Wildl. Serv. Fishery Bull. 89. Washington, D.C.

Newling, C.J. and M.C. Landin. 1985. Long-term Monitoring of Habitat Development at Upland and Wetland Dredged Material Disposal Sites, 1974-1982. U.S. Army Engineer Waterways Experiment Station Tech. Rep. D-85-5.

Palmisano, A.W. 1973. Habitat preferences of waterfowl and fur animals in the northern Gulf coast marshes, p. 163-190. In R.H. Chabreck (Ed.), Proceedings of the 2nd Coastal Marsh and Estuary Management Symposium. Louisiana State University, Baton Rouge.

Penfound, W.T. and E.S. Hathaway. 1938. Plant communities in the marshland of southeastern Louisiana. Ecol. Monogr. 8:1-56.

Roberts, H.H. and I.L. van Heerden. 1982. Reversal of coastal erosion by rapid sedimentation: the Atchafalaya delta (south-central Louisiana), p. 214-231. In D.F. Boesch (Ed.), Proceedings of the Conference on Coastal Erosion and Modification in Louisiana: Causes, Consequences, and Options. U.S. Fish and Wildl. Serv. FWS/OBS-82-59.

Rogers, B.D., W.H. Herke, and E.E. Knudsen. 1987. Investigation of a Weir-Design Alternative for Coastal Fisheries Benefit. Louisiana State University Agric. Center, Baton Rouge.

Russell, R.J. 1942. Flotant. Geogr. Rev. 32:74-98.

Russell, R.J. and H.V. Howe. 1935. Cheniers of southwestern Louisiana. Geogr. Rev. 25:449-461.

Spiller, S.F. and R.H. Chabreck. 1975. Wildlife populations in coastal marshes influenced by weirs. Proc. Annu. Conf. Southeast. Assoc. Game and Fish Comm. 29:518-525.

Stout, J.P. 1984. The Ecology of Irregularly Flooded Salt Marshes of the Northeastern Gulf of Mexico: A Community Profile. U.S. Fish and Wildl. Serv. Biol. Rep. 85(7.1).

Titus, J.G. 1985. How to Estimate Future Sea Level Rise in Particular Communities. Environmental Protection Agency, Washington, D.C.

U.S. Army Corps of Engineers. 1986. Beneficial Uses of Dredged Material. EM1110-2-5026, Office, Chief of Engineers. Washington, D.C.

U.S. Congress, Office of Technology Assessment. 1984. Wetlands: Their Use and Regulation. U.S. Gov. Printing Off., Washington, D.C.

Webb, J.W., M.C. Landin, and H.H. Allen. 1988. Approaches and techniques for wetland development and restoration of dredged material disposal sites, p. 132. In J.A. Kusler, M.L. Quammen, and G. Brooks (Eds.), Proceedings of a National Wetland Symposium, Mitigation of Impacts and Losses, Assoc. of State Wetland Mgrs., Berne, New York.

West, R.C. 1977. Tidal salt marsh and mangal formations of Middle and South America, p. 193-213. In V.J. Chapman (Ed.), Ecosystems of the World, I: Wet Coastal Ecosystems. Elsevier Scientific Publ. Co., New York.

APPENDIX I: RECOMMENDED READINGS

Allen, H.H., E.J. Clairain, Jr., R.J. Diaz, A.W. Ford, J.L. Hunt, and B.R. Wells. 1978. Habitat Development Field Investigations Bolivar Peninsula, Marsh and Upland Habitat Development Site, Galveston Bay, Texas (Summary Report). U.S. Army Engineers, Waterways Exp. Sta. Tech. Rep. D-78-15.

Field investigations were conducted at Bolivar Peninsula, Galveston Bay, Texas, to test the feasibility and impact of developing marsh on dredged material. The investigation provided baseline information before habitat development and evaluated post-development operations. S. alterniflora and S. patens were used to evaluate fertilizer treatments and planting methods. Wildlife use of habitat developed was compared with adjacent control areas.

Allen, H.H., S.O. Shirley, and J.W. Webb. 1986. Vegetative stabilization of dredged material in moderate to high wave-energy environments for created wetlands, p. 19-35. In Proc. Annu. Conf. on Wetland Restoration and Creation. Hillsborough Community College, Tampa, Florida.

Tests were conducted for protecting plantings of S. alterniflora by using four protection techniques including large sandbags, tire breakwaters with plants wrapped in long burlap rolls, and plants sprigged into a woven mat and then laid and anchored to the substrate. Tests were conducted at dredge disposal sites in Mobile Bay and Mississippi Sound in Alabama, Southwest Pass on the Mississippi River Delta in Louisiana, and Galveston Bay in Texas.

Allen, H.H., J.W. Webb, and S.O. Shirley. 1984. Wetlands development in moderate wave-energy climates, p. 943-955. In Proc. of the Conf. Dredging '84. Waterway, Port, Coastal and Ocean Division ASCE. Clearwater Beach, Florida.

S. alterniflora was planted on Theodore Island in Mobile Bay, Alabama, to provide shoreline stabilization. Establishment of plantings was more successful along shorelines where protection techniques involving breakwaters were used.

Chatry, M. and D. Chew. 1985. Freshwater diversion in coastal Louisiana: recommendations for development of management criteria, p. 71-84. In C.F. Bryan, P.J. Zwank, and R.H. Chabreck (Eds.), Proceedings of the 4th Coastal Marsh and Estuary Management Symposium. Louisiana State University, Baton Rouge.

Controlled diversion of Mississippi River water into estuarine areas of southeastern Louisiana is planned to enhance vegetative growth, reduce land loss, and increase production of fish and wildlife by establishing favorable salinity conditions. Priorities must be established for estuarine basins to be affected and resources to be managed by the program. Pre- and post-operational monitoring of environmental conditions will be necessary to identify optimum salinities and develop operational plans for achieving optimum salinities.

Cunningham, R. 1981. Atchafalaya delta: subaerial development environment implications and resource potential, p. 349-365. In R.D. Cross and D.L. Williams (Eds.), Proceedings of the National Symposium on Freshwater Inflow to Estuaries, Vol. 1. U.S. Fish and Wildl. Serv. FWS/OBS-81-04.

Approximately 30% of the Mississippi River flow has been diverted down the Atchafalaya River since 1963. Sedimentation has filled many lakes in the Atchafalaya Basin, and record flooding from 1973 to 1975 created 32 km^2 of new land by sedimentation in Atchafalaya Bay.

Eleuterius, L.N. 1974. A Study of Plant Establishment on Dredge Spoil in Mississippi Sound and Adjacent Waters. Gulf Coast Research Lab., Ocean Springs, Mississippi.

Species characteristics and transplanting techniques were evaluated for establishment of submerged aquatics on dredge spoil and barren sea bottoms and emergent vascular plants on dredge spoil and dunes. Successful establishment of plant cover was largely affected by conditions at the planting site.

Fruge, D.W. and R. Ruelle. 1980. A Planning Aid Report on the Mississippi and Louisiana Estuarine Area Study. U.S. Fish and Wildl. Serv., Lafayette, Louisiana.

The value of fish and wildlife resources and factors affecting their abundance in coastal areas of southeastern Louisiana and Mississippi are discussed. The impacts of freshwater introduction from the Mississippi River into estuarine systems and procedures for minimizing adverse impacts are described.

Gagliano, S.M., P. Light, and R.E. Becker. 1973. Controlled diversion in the Mississippi River Delta System: An Approach to Environmental Management. Louisiana State University Center for Wetland Resources Rep. No. 8.

Diversion of Mississippi River water along lower reaches for subdelta deposition would be capable of creating 32 km^2 of new marshland per year. Freshwater introduced at the upper end of interdistributary estuarine systems could be used to offset salt water intrusion and introduce nutrients.

Kruczynski, W.L. 1982. Salt marshes of the northeastern Gulf of Mexico, p. 71-87. In R.R. Lewis, III (Ed.), Creation and Restoration of Coastal Plant Communities. CRC Press, Inc., Boca Raton, Florida.

The characteristics of marsh along the coastal regions of Alabama and Mississippi are discussed, and studies on establishment of marsh communities on dredge spoil in the area are summarized.

Landin, M.C. 1986. The success story of Gaillard Island, a Corps confined disposal facility. Proc. of the Dredging Seminar and Western Dredging Assoc. Conf., Baltimore, Maryland. 19:41-54.

Gaillard Island, a 526-ha confined disposal facility in Mobile Bay, was monitored to determine plant colonization and wildlife use. The island is an important nesting site for seabirds. Aquatic habitat in the island's interior is used by waterfowl and shorebirds. Shorelines were stabilized by planting S. alterniflora and using erosion control fabrics.

Landin, M.C. and J.W. Webb. 1988. Wetland development and restoration as part of Corps of Engineers programs: case studies, p. 388. In J.A. Kusler, M.L. Quammen and G. Brooks (Eds.), Proceedings of the Wetlands Mitigation Symposium, Mitigation of Impacts and Losses. Assoc. of State Wetland Mgrs., Berne, New York.

Marsh creation programs at selected dredged material disposal sites were described including Bolivar Peninsula, Chocolate Bay, Stedman Island, and East Matagorda Bay in Texas.

Morton, R.A. 1982. Effects of coastal structures on shoreline stabilization and land loss--the Texas experience, p. 177-186. In D.F. Boesch (Ed.), Proceedings of the Conference of Coastal Erosion and Wetland Modification in Louisiana: Causes, consequences, and Options. U.S. Fish and Wildl. Serv. FWS/OBS-82-59.

Seawalls and bulkheads constructed for shoreline protection may not always be successful and in some instances may increase erosion of adjacent property. Costs for structures may exceed the value of land being protected.

Newling, C.J. and M.C. Landin. 1985. Long-term Monitoring of Habitat Development at Upland and Wetland Dredged Material Disposal Sites 1974-1982. U.S. Army Engineers Waterway Exp. Sta. Tech. Rep. D-85-5.

Wetland habitat development projects were monitored at six dredged material disposal sites. After 8 years, all sites have developed and stabilized and are considered highly successful. No maintenance has been performed, and wildlife use exceeds that occurring in nearby control areas.

U.S. Army Corps of Engineers. 1986. Beneficial Uses of Dredged Material. EM1110-2-5026. Washington, D.C.

Techniques and engineering procedures for creation and restoration of marsh and aquatic habitat on dredged material disposal sites are described, plus other beneficial uses of dredged material. Includes information on species propagation and planting, site selection, site design, soil problems, monitoring, contaminants, and legal and other considerations.

Webb, J.W., Jr. 1982. Salt marshes of the western Gulf of Mexico, p. 89-109. In R.R. Lewis (Ed.), Creation and Restoration of Coastal Plant Communities. CRC Press, Inc., Boca Raton, Florida.

The characteristics of marsh along the Louisiana and Texas coasts are described. Specific projects involving marsh creation and restoration are reviewed, and recommended techniques for planting dredged material disposal sites are discussed.

APPENDIX II: PROJECT PROFILES

GAILLARD ISLAND, ALABAMA

Wetland Type: Diked, confined dredged material disposal facility.

Location: Lower Mobile Bay.

Size: 526 ha.

Goals of Project:

The site was built in 1981 and is used for disposal of material from maintenance dredging. Salt marsh habitat has been developed in a portion of the area. A 324-ha containment pond is used by local crabbers and fishermen. The entire site provides diverse habitat for fish and wildlife, and over 25,000 sea and wading birds nest on the island.

Significance:

Dike erosion and subsidence problems have necessitated stabilization efforts. Because of severe wave erosion, various techniques such as floating-tire breakwaters, plant rolls, and erosion control mats were employed to facilitate plant establishment for erosion control. A long-term management plan incorporating environmental and engineering strategies is being developed for the island.

Contact: Mary C. Landin
U.S. Army Engineer Waterways
Experiment Station
Vicksburg, MS 39180

BOLIVAR PENINSULA, TEXAS

Wetland Type: Salt marsh.

Location: Bolivar Peninsula in lower Galveston Bay, Texas.

Size: 8 ha.

Goals of Project:

Sandy dredged material was hydraulically placed along the shoreline to create salt marsh habitat in 1974 and 1981. The project site in 1974 was graded and fenced, and a sandbag dike was constructed in the intertidal zone to provide protection from wave energy. The site was planted with S. alterniflora and S. patens. The 1981 site was stabilized with erosion control biostabilization techniques.

Significance:

Plantings became well established in the intertidal zone and at higher elevations and demonstrated that vegetation could be used to stabilize sandy dredged material. The site withstood a direct hit from Hurricane Alicia in 1983. Research will continue through 1990.

Contact: Hollis H. Allen, Mary C. Landin,
or James W. Webb
U.S. Army Engineer Waterways
Experiment Station
Vicksburg, MS 39180

SOUTHWEST PASS, LOUISIANA

Wetland Type: Delta marsh.

Location: Active delta of the Mississippi River.

Size: 445 ha when completed (depends upon channel sediments to be removed; area created may actually be larger).

Goals of Project:

Dredged material from the Southwest Pass shipping channel is being pumped over the west levee and allowed to colonize naturally with marsh and aquatic plants. Along the east bank, a floating-tire breakwater and a fixed breakwater were constructed and dredged material was planted with S. alterniflora. The east bank has direct exposure to the Gulf of Mexico.

Significance:

The east bank site was monitored after planting in May 1985. Sediment accumulated rapidly behind each breakwater and unprotected shores eroded. Sediment accumulation was so rapid that plantings were covered and had to be replanted in July. Three separate hurricanes struck the site in late summer and fall of 1985 and completely destroyed all breakwaters. Vegetation colonization is rapidly occurring on the west bank and is still being monitored. Large quantities of material is placed there on a nearly continuous basis, and the head of the discharge pipe is moved as needed to keep the material at an intertidal elevation.

Contact: Mary C. Landin or Hollis H. Allen
U.S. Army Engineer Waterways
Experiment Station
Vicksburg, MS 39180

COFFEE ISLAND, ALABAMA

Wetland Type: Salt marsh.

Location: Mississippi Sound about 16 km south of Bayou La Batre, Alabama.

Size: Unknown but relatively small.

Goals of Project:

Dredged material was placed along the east face of Coffee Island to protect an existing salt marsh from wave energy. The dredged material was planted with S. alterniflora for stabilization.

Significance:

Plant rolls were placed end to end and seaward of

transplants and stabilized the exposed dredged material face. After 2 years, the site was accreting sediment and protecting the island and salt marsh from further erosion.

Contact: Hollis H. Allen
U.S. Army Engineer Waterways
Experiment Station
Vicksburg, MS 39180

STEDMAN ISLAND, TEXAS

Wetland Type: Salt marsh.

Location: Stedman Island in Aransas Bay, Texas.

Size: 5 ha.

Goals of Project:

A dredged material spillover adjacent to a disposal site was planted with S. alterniflora to create salt marsh habitat.

Significance:

After 2 years, 95% of the site was covered with vegetation. However, after 39 months, plants began to die at lower elevations, and no growth was noted the following year. At higher sites, all plants appeared healthy. Possible causes for death of plants were high salinity, sulfide concentrations, inadequate aeration, nitrogen limitations, chemical pollution, or combinations of these factors.

Contact: James W. Webb or Mary C. Landin
U.S. Army Engineer Waterways
Experiment Station
Vicksburg, MS 39180

TENNECO MANAGEMENT UNIT, LOUISIANA

Wetland Type: Fresh to brackish marsh.

Location: Terrebonne Parish, Louisiana.

Size: 2,800 ha.

Goal of Project:

The project will reintroduce freshwater and sediment flow, improve water circulation, and reduce saltwater intrusion through a structural water management plan.

Significance:

This project is part of a mitigation bank designed to enhance fish and wildlife habitat in an area of coastal marsh that is rapidly deteriorating because of salt water intrusion.

Contact: David M. Soileau
U.S. Fish and Wildlife Service
Lafayette, LA 70502

LAKE BORGNE CANAL FRESHWATER DIVERSION, LOUISIANA

Wetland Type: Brackish marsh.

Location: Lake Borgne Canal and Mississippi River at Violet, Louisiana.

Size: 8,000 ha.

Goals of Project:

To introduce freshwater from the Mississippi River through two 125-cm pipes that will siphon water over the protection levee of the river and discharge it into the Lake Borgne Canal.

Significance:

Flood protection levees along the Mississippi River prevent traditional flow of river water into adjacent marshes. As a result, salt water intrusion is causing a gradual deterioration of the marshes and increasing erosion rates. Introduction of freshwater from the river will help offset this process.

Contact: Sherwood M. Gagliano
Coastal Environments, Inc.
Baton Rouge, LA

CALCASIEU LAKE SHORELINE EROSION CONTROL, LOUISIANA

Wetland Type: Brackish marsh.

Location: Southeastern shore of Calcasieu Lake, Louisiana.

Size: Test plantings were made along the shoreline.

Goals of Project:

Experimental plantings of S. alterniflora were made along the shoreline of Calcasieu Lake in an attempt to find a vegetative means of reducing shoreline erosion. The shoreline of the lake is currently eroding at a rate of 3 m/yr.

Significance:

S. alterniflora will protect shorelines where the species can be successfully transplanted. Test plantings should be made to determine site suitability. Where the site is suitable, single-stemmed plants should be placed 60 cm apart in rows and fertilized with a time-release fertilizer tablet.

Contact: Jack R. Cutshall
Soil Conservation Service
Alexandria, LA

BACKFILLING AND PLUGGING OF CANALS, LOUISIANA

Wetland Type: Brackish marsh.

Location: Vermilion Parish, Louisiana about 13 km south of Delcambre.

Size: Canal about 180 m long and 12 m wide.

Goals of Project:

To partially restore marsh sites where access canals have been dredged to reach oil and gas drilling locations. Abandoned sites are partially restored by returning the spoil to the canal and placing a plug or dam across the canal entrance.

Significance:

Success of restoration depends upon maintenance of a well constructed dam across the entrance to the canal. Also, extreme care must be used in replacing the spoil so that excessive amounts of material are not removed from the shoreline, thus forming a depression or basin. Spoil deposits are insufficient to refill the canal, but aquatic plants often invade the site because of reduced depth. Emergent plants will revegetate the former spoil disposal site.

Contact: Donald Moore
National Marine Fisheries Service
Galveston, TX

REVEGETATION OF PIPELINE SITE, LOUISIANA AND TEXAS

Wetland Type: Brackish and salt marsh.

Location: East (Louisiana) and west (Texas) of Sabine Pass near the mouth of the Sabine River.

Size: Pipeline about 13 km long.

Goals of Project:

To evaluate revegetation rate of pipeline construction site where single ditching and double ditching were employed and to determine if transplanting facilitated recovery of the site.

Significance:

Double ditching involved refilling the pipeline ditch with spoil in the order removed so that the topsoil was replaced last. Double ditched sites became revegetated faster than sites single ditched where spoil was mixed. Transplanting S. alterniflora, S. patens, and D. spicata did not increase revegetation rates.

Contact: Robert H. Chabreck
Louisiana State University
Baton Rouge, LA 70803

MILLER LAKE WEIR, LOUISIANA

Wetland Type: Brackish marsh.

Location: Rockefeller Wildlife Refuge, Cameron Parish, Louisiana.

Size: Weirs provide water management in an area of about 1,200 ha of marsh.

Goals of Project:

Two weirs were constructed in drainage outlets of the marsh to stabilize water levels, prevent drastic water salinity changes, and improve the area as habitat for waterfowl.

Significance:

Weirs prevent excessive drainage of marshes during low tides and maintain a basin of water which functions to dilute highly saline tidewater that enters the area. During certain low tides in winter, as much as 80% of the marsh ponds are drained.

Contact: Ted Joanen
Louisiana Department of Wildlife and Fisheries
Grand Chenier, LA 70643

BIRD ISLAND WEIR, LOUISIANA

Wetland Type: Brackish marsh.

Location: Marsh Island Wildlife Refuge, Iberia Parish, Louisiana.

Size: This weir provides water management on an area of about 1,600 ha of marsh.

Goals of Project:

The weir stabilizes water levels and salinities in the marsh and aquatic habitats affected and improves the area as winter habitat for migratory waterfowl.

Significance:

The weir reduces the rate of tidal exchange in the marsh and stabilizes water levels. Production of aquatic vegetation in marsh ponds and lakes controlled by the weir was three times greater than in nearby free-flowing systems. Aquatic plants are an important food source for wintering ducks, and duck use was significantly greater in areas affected by weirs.

Contact: Greg Linscombe
Louisiana Department of Wildlife and Fisheries
New Iberia, LA

FRESHWATER DIVERSION FROM MISSISSIPPI RIVER, LOUISIANA

Wetland Types: Coastal marshes and estuaries of the Deltaic Plain.

Location: Freshwater diversion sites are proposed for the Barataria Bay and Breton Sound basins.

Size: The area to be affected by the proposed project includes the vast basins on both sides of the Mississippi River south of New Orleans.

Goals of Project:

The project is described by the U.S. Army Corps of Engineers, New Orleans District, as the Louisiana Coastal Area Study and is designed primarily as a salinity management program. The project would restore and enhance vast areas of coastal marshland that is rapidly deteriorating because of salt water intrusion.

Significance:

Land loss in the Louisiana coastal marshes is estimated to be 100 km^2/yr, and the proposed project area is undergoing greatest losses. Freshwater introduction into the basins would be designed to reestablish historical zonations of salinity patterns to maximize plant growth and enhance fish and wildlife productivity.

Contact: Dennis Chew
U.S. Army Corps of Engineers,
New Orleans District
New Orleans, LA 70160

ROCKEFELLER REFUGE IMPOUNDMENTS, LOUISIANA

Wetland Types: Intermediate and brackish marsh.

Location: Rockefeller Wildlife Refuge, Cameron Parish, Louisiana.

Size: Includes 10 separate impoundments ranging in size from 160 to 1,600 ha.

Goals of Project:

The impoundments provide water management systems for regulating water depths and salinity to produce desired plant communities and improve habitat for waterfowl. Impoundments are managed as fresh- and brackish-water units. Some units are permanently flooded, and in others water levels are manipulated to produce annual grasses and sedges.

Significance:

Impoundments provide the most effective means of enhancing wildlife habitat in coastal marshes. Approximately one-half of the refuge is under impoundment management, and approximately 80% of the waterfowl on the refuge occur within the impoundments.

Contact: Ted Joanen
Louisiana Department of Wildlife and Fisheries
Grand Chenier, LA 70643

LACASSINE NATIONAL WILDLIFE REFUGE IMPOUNDMENT, LOUISIANA

Wetland Type: Freshwater marsh.

Location: Lacassine National Wildlife Refuge, Cameron Parish, Louisiana.

Size: Approximately 8,000 ha.

Goals of Project:

To provide resting and feeding habitat and a sanctuary for migratory waterfowl in a vast agricultural region heavily utilized by the birds.

Significance:

The impoundment is managed as a permanently flooded freshwater basin with water depth ranging from 30 to 90 cm. Waterfowl habitat within the impoundment is greatly superior to that of adjacent natural marsh. The impoundment also produces an abundance of freshwater fishes and alligators and attracts various types of birds in addition to waterfowl.

Contact: Bobby Brown
U.S. Fish and Wildlife Service
Lacassine National Wildlife Refuge
Lake Arthur, LA

SHORELINE AND BARRIER ISLAND RESTORATION, LOUISIANA

Wetland Type: Beaches and barrier islands.

Location: Southeastern Louisiana.

Size: Barrier islands that have been or will be included are Grand Isle, East Timbalier Island, Timbalier Island, and Isles Dernieres.

Goals of Project:

Restoration planned is to raise the average barrier island elevation by sand dredging, rebuild back-barrier marshes, and revegetate the sites. These projects will reduce erosion of barrier islands and reduce destruction by hurricanes.

Significance:

Beach restoration at Grand Isle was successful in reducing the erosion experienced by other beaches and barrier islands in the vicinity when struck by 3 hurricanes during 1985. Barrier islands offer the first line of defense in protecting coastal marshes and communities from the destructive forces of hurricanes.

Contact: Shea Penland
Louisiana Geological Survey
Baton Rouge, LA

CREATION AND RESTORATION OF COASTAL WETLANDS OF THE NORTHEASTERN UNITED STATES

Joseph K. Shisler
Environmental Connection, Inc.

ABSTRACT. The wetlands of the coastal zone of the northeast have been managed since the colonization of the United States. Restoration work associated with mitigation of impacts has been going on in the region for over twenty years. Despite this history, there has not been an extensive evaluation of these projects to determine their success and how they function.

The mitigation process should be directed towards a management approach that is concerned with the total system instead of just the "vegetated" wetland. Goals should be based upon a wetland system's requirements within a watershed or region. The use of adjacent wetlands as models is critical in this process. Monitoring the created or restored wetlands can provide an important database which can be used in planning future projects. Goals, clearly defined in the design process, will promote meaningful evaluations.

REGIONAL CHARACTERISTICS

The geographical area discussed in this chapter covers Maine to northeastern Virginia and includes the Acadian and Virginian provinces defined by Cowardin et al. (1979). The Acadian Province extends from Avalon Peninsula to Cape Cod and is dominated by boreal biota. This province contains a heavily indented and frequently rocky shoreline influenced by a large tide range. The Virginian Province is the transition zone between the Acadian and Carolinian provinces and is dominated by temporal species and moderate tide ranges. The coastal region exhibits pronounced seasonal temperature fluctuations. Extreme variations in seawater temperature, which is warmest in August through September and coolest in December to March, are among the greatest in the world (Sanders 1968).

Cooper (1974) defined two of the major wetland groups associated with the region, the New England and Atlantic coastal wetlands. The Atlantic coastal wetlands are built upon sands of the outer coastal plain, while the substrate of New England wetlands is glacial till with rocks. Coastal wetland systems become smaller and isolated north of Boston compared to the more expansive wetland systems of the Chesapeake and Delaware estuaries. Annual climatic changes, a major factor of the north, have impacted the wetlands in the form of ice (Teal 1986). Alexander et al. (1986) have indicated that there are over 17 million acres of coastal wetlands in this region (Table 1).

TYPES OF WETLANDS

A number of factors determine the types of wetlands and their locations. Penfound (1952) identifies the most important physical factors as: (1) water depth; (2) fluctuation of water levels; (3) soil moisture; and (4) salinity. Frey and Basan (1978) present a detailed list of reasons for the differences in coastal marshes, including: (1) character and diversity of the indigenous flora; (2) effects of climatic, hydrographic, and edaphic factors upon this flora; (3) availability, composition, mode of deposition, and compaction of sediments, both organic and inorganic; (4) organism-substrate interrelationships, including burrowing animals and their prowess plant in affecting marsh growth; (5) topography and aerial extent of the depositional surface; (6) range of tides; (7) wave and current energy; and (8) tectonic and eustatic stability of the coastal area. Tiner (1985a, 1985b) also summarizes the role of all these factors affecting coastal wetlands. These and other factors (e.g., location, size, vegetation sources, problem species, maintenance, and economics) are major considerations in the development of functioning wetland systems.

Recent publications have addressed the various types of northeastern coastal wetlands and their functions (e.g., Hill and Shearin 1979, Nixon and Oviatt 1973, Niering and Warren 1980, McCormick and Somes 1982, Simpson et al. 1983a, Nixon 1982, Daiber 1982, 1986, Odum et al. 1984, Mitsch and Gosselink 1986, Teal 1986, Tiner 1985a, 1985b). These wetlands have been some of the most intensively studied systems in the United States, but many of their complex functions and the impacts of human encroachment still have not been fully explored. Less research associated with wetland creation and restoration has been conducted in the northeast as compared to the mid-Atlantic, Florida, Gulf, and Pacific coasts. How-

Table 1. Coastal Wetlands of the Northeast.

	Wetland Acres (x 1000 acres)				
State	Salt Marsh	Fresh Water	Tidal Flats	Swamp	Total Acreage
Maine	166	257	583	250	1,256
New Hampshire	75	N/A	N/A	N/A	75
Massachusetts	481	151	415	249	1,296
Rhode Island	79	0	0	571	650
Connecticut	166	N/A	N/A	N/A	166
New York	267	34	N/A	N/A	301
New Jersey	2,174	217	486	4,723	7,600
Delaware	781	71	113	1,234	2,199
Maryland	1,636	256	18	194	2,104
Virginia	1,523	200	N/A	N/A	1,723
Totals	7,348	1.194	1,615	7,221	17,378

Source: Alexander et al. 1986

ever, the northeast possibly has the greatest concentration of restoration and creation sites in the United States.

ESTUARINE WETLANDS

Coastal wetlands are primarily limited by the influences of tide and salinity. In relation to salinity, they can be subdivided into: (1) polyhaline--strongly saline areas [18-30 parts per thousand (ppt)], (2) mesohaline--moderate salinity areas (5-18 ppt), and (3) oligohaline--slightly brackish areas (0.5-5 ppt) (Figure 1) (Cowardin et al. 1979). The diversity of vegetation species increases with a decrease in salinity. Major estuarine wetland types are: (1) intertidal flats, (2) emergent wetlands, and (3) scrub-shrub wetlands (Tiner 1985a, 1985b).

SALT MARSHES

Salt marshes are emergent wetlands located in the polyhaline zone of the estuary. Tidal inundation results in the formation of two major vegetation zonations, high and low marsh. The low marsh area is located between mean low water (MLW) and mean high water (MHW) and is dominated by tall form Spartina alterniflora throughout the region (Figure 2). McCormick and Somes (1982) describe a short growth form of S. alterniflora as a saline low marsh type in Maryland. Recently, Kennard et al. (1983) has shown that the tall form S. alterniflora is an accurate indicator for the landward extent of mean high tide. There is an increase in the number of species associated with the salt marsh systems towards the southern limits of the region. Species such as Spartina cynosuroides, Juncus roemerianus and Scirpus robustus often become dominant species associated with the wetland systems south of Delaware.

In the northern section of the region, the high marsh is dominated by short form S. alterniflora and S. patens, but other species associated with the high marsh area include Distichlis spicata and Juncus gerardii. There is an increase in the number of vegetation species towards the ecotonal edge (Nixon 1982). Miller and Egler (1950) identified some 150 species in the Wequetequock-Pawcatuck marshes in Connecticut. In many marshes, these vegetation species form a complex mosaic, rather than distinct zones, as a result of minor elevational changes. In the southern section of the region, McCormick and Somes (1982) identify three major dominants in saline high marshes in Maryland as, Spartina patens/Distichlis spicata, Iva frutescens/Baccharis halimifolia, and Juncus roemerianus.

The alteration of marshes for mosquito control and other reasons increases the diversity of the marsh. Iva frutescens, Panicum virgatum, Baccharis halimifolia, and Phragmites australis, have colonized dredged material associated with alterations of the marsh (Miller and Egler 1950). The development of the open marsh water management technique has minimized changes in vegetation associations and increased standing crops of certain species (Shisler and Jobbins 1977, Meredith and Saveikis 1987).

Pools and pannes are associated with the high marsh areas (Figure 2). Pools vary in depth and may be devoid of vegetation or vegetated by Ruppia maritima. Pannes are slight depressions in the high marsh that may or may not be vegetated as a result of extreme temperatures and/or salinity. Plant species, such as Salicornia bigelovii or S. virginica, may be associated with these depressions.

Tidal range affects the percentage of high and low marsh (Provost 1973) and the growth of S. alterniflora (Odum 1974, Shisler and Charette 1984a). Tides in the region range from shallow wind blown tides in Barnegat Bay, New Jersey to tidal ranges of over 8 meters along the coast of Maine.

BRACKISH WATER WETLANDS

Brackish wetlands are located in the mesohaline zone (5 to 18 ppt) associated with estuarine systems that are seasonally exposed to a wide range of salinities. The increased run-off in the spring decreases salinity, while low flow in the late summer increases the salinity. Salinity oscillation creates the transitional zone between the fresh water and estuarine systems. Larger wetland systems are found in the southern sections of the region due to the more gentle topography. Tiner (1985a) identifies four major plant communities in New Jersey brackish marshes: (1) Typha angustifolia; (2) Spartina cynosuroides; (3) Phragmites australis; and (4) Scirpus americanus. In Maryland, McCormick and Somes (1982) define ten types of emergent wetland vegetation associations: (1) Spartina alterniflora; (2) S. patens/Distichlis spicata; (3) Iva frutescens/Baccharis halimifolia; (4) Juncus roemerianus; (5) Typha spp.; (6) Hibiscus spp.; (7) Panicum virgatum; (8) Scirpus spp.; (9) Spartina cynosuroides; and (10) Phragmites australis. Brackish water marshes comprise 58% of the tidal wetlands in Maryland (McCormick and Somes 1982).

OLIGOHALINE WETLANDS

The zone between mesohaline and oligohaline is exposed to a minimum amount of salt (5.0 to 0.5 ppt). These systems contain the highest diversity of all the estuarine wetlands as a result of their location on the upper periphery of the salinity continuum (Tiner 1985a, 1985b).

ESTUARINE SCRUB-SHRUB WETLANDS

Scrub-shrub wetlands are located along the ecotonal edge of the uplands. They are not

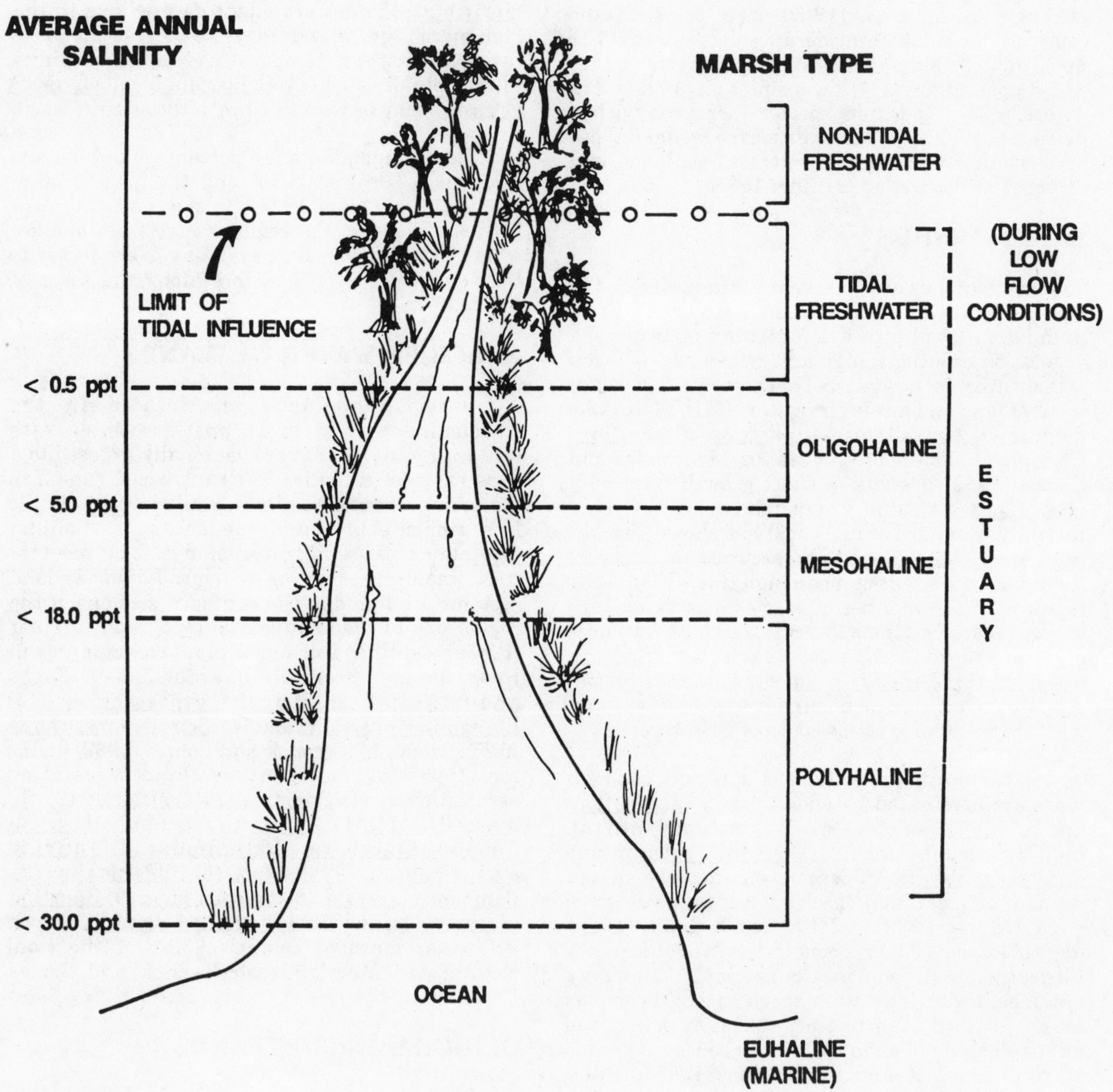

Figure 1. The average annual salinity within the major tidal wetland types (from Odum et al. 1984, based on terminology from Cowardin et al. 1979).

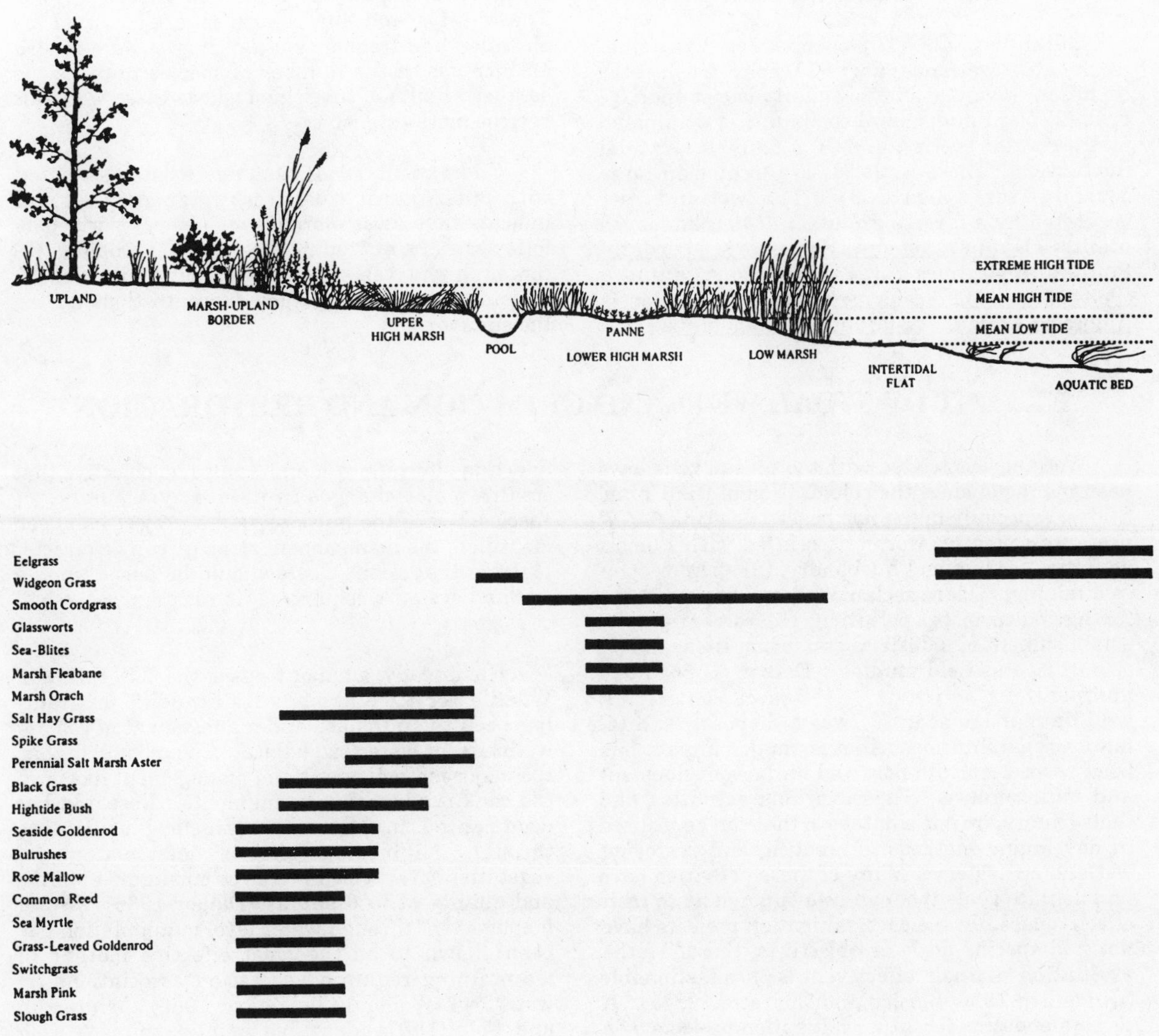

Figure 2. Generalized distribution of vegetation in a salt marsh (from Tiner 1985a).

extensive systems throughout the region, but become more extensive towards the southern areas of the region where they are dominated by I. frutescens and B. halimifolia. In oligohaline areas of Delaware and Maryland, Myrica cerifera may form a shrub thicket at the transition zone between oligohaline tidal marshes and the freshwater tidal swamps (Tiner pers. comm.).

TIDAL FRESH WATER WETLANDS

Odum et al. (1984) characterize tidal freshwater wetlands as: (1) near freshwater conditions (average annual salinity is less than 0.5 ppt); (2) plant and animal communities dominated by freshwater species; and (3) a daily, lunar tidal fluctuation. These wetlands are located in large coastal rivers (Figure 3). The wetlands are vegetated by a diverse group of: (1) broad-leaved plants (Nuphar advena, Pontederia cordata, Peltandra virginica); (2) herbaceous annuals (Polygonum spp., Bidens laevis, Impatiens capensis, Ambrosia trifida, Amaranthus cannabinus); (3) annual and perennial sedges, rushes and grasses (Zizania aquatica, Spartina cynosuroides, Scirpus spp., Eleocharis spp., Cyperus spp., Carex spp.); and (4) grasslike plants or shrub-form herbs (Acorus calamus, Typha spp., Hibiscus moscheutos, Sium suave); and (5) hydrophytic shrubs (Cephalanthus occidentalis, Myrica cerifera, Rosa palustris, Salix spp., Cornus amomum) (Figure 4) (Odum et al. 1984, Tiner 1987). As with estuarine systems, two vegetation zones may be recognizable. These high and low marshes are defined by elevation and frequency of flooding. There is also an increase in the number of species towards the southern limits of the region where these wetlands become more extensive.

Odum et al. (1984) summarize data associated with mid-Atlantic tidal freshwater marshes and indicate that these marshes may be categorized as follows: (1) eutrophic or hyper-eutrophic; (2) contain high levels of suspended sediments; and (3) may have depressed oxygen concentrations during the summer.

EXTENT OF TIDAL WETLAND CREATION AND RESTORATION

Wetlands associated with the coastal zone have been managed since the colonization of the United States, especially in the northeast. Beeftink (1977) presents seven pressures associated with human activities: (1) animal husbandry; (2) strip or open cast mining; (3) land reclamation and improvement for agriculture; (4) pollution; (5) recreation; (6) establishment of industrial and urban sites; and (7) scientific and field studies. Daiber (1986) adds another four activities: (1) insect control; (2) wildlife management; (3) waste disposal; and (4) marsh rehabilitation. Another major impact has been associated with port and harbor development and maintenance. These various activities and their history provide a database that can be utilized in developing methods of creating and restoring wetland ecosystems. Many of these activities have had definite goals that can be evaluated as to their effectiveness. In the past, mitigation projects have not had specific goals or objectives, therefore the evaluation of their effectiveness is questionable (Quammen 1986, Shisler and Charette 1984a). A recommendation for future mitigation projects by a number of researchers has been that they contain definite goals so that they can be properly evaluated (Charette et al. 1985).

The management of wetland systems is a complex issue that has to be based on a number of parameters to determine the goals and objectives of the plan. For example, natural salt marshes are usually low in avian diversity and biomass, while impoundments (both low and high level) can often produce high biomass and diversity values (Burger et al. 1982, Daiber 1982, 1986). The alteration of a Phragmites australis or salt hay (a mixture of high marsh plants species) impounded wetland into a Spartina alterniflora marsh is also an objective that creates major changes in its use by a number of species. Once the purpose(s) of a project has been identified, the management strategy can be directed towards that goal. Goals should be based upon a wetland system's requirements within a watershed or region.

Historically, a major focus of the U.S. Fish and Wildlife Service and many state wildlife programs has been the purchase and management of coastal wetlands for waterfowl habitat. To accomplish this, the major method of wetland management has been the construction of impoundments. Research has documented increases in waterfowl utilization through the interspersion of open water and vegetative cover which produces maximum quantity and quality of food supply (Daiber 1986). Water management through water level manipulation has been shown to be the most effective method of maintaining required vegetation associations for waterfowl (Weller 1978, Daiber 1986). Whiteman and Cole (1987) identified habitat changes with impoundments over time due to non-management in Delaware. Impoundments created in fresh water and brackish water areas undergo a series of changes in soil and water chemistry, vegetation, and invertebrates, which cause waterfowl populations to stabilize within a few years and eventually decline (Whitman and Cole 1987). It also is important to note that management of impoundments for waterfowl occurred at the expense of other avian species (Andrews 1987) as well as certain fish, shellfish, and other animals.

Impoundments were also constructed for

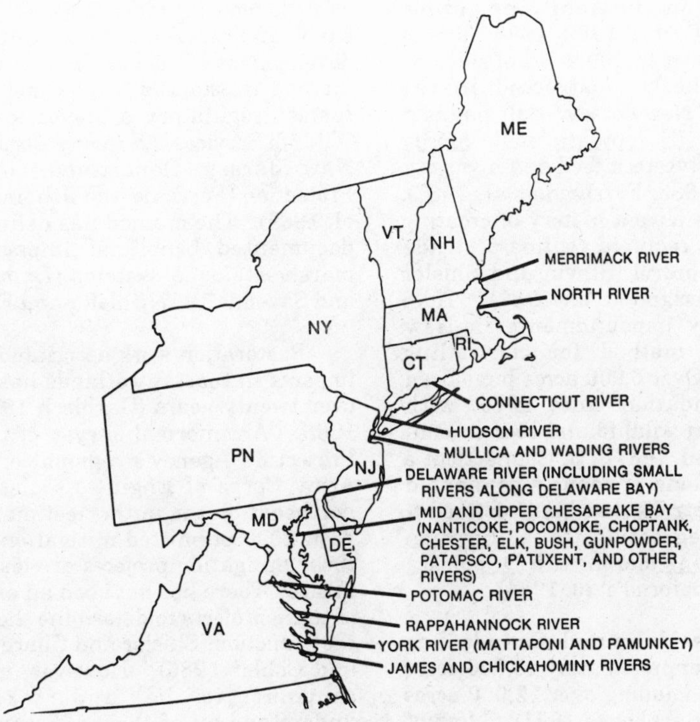

Figure 3. Representative freshwater tidal wetlands over 500 acres in size (from Odum et al. 1984).

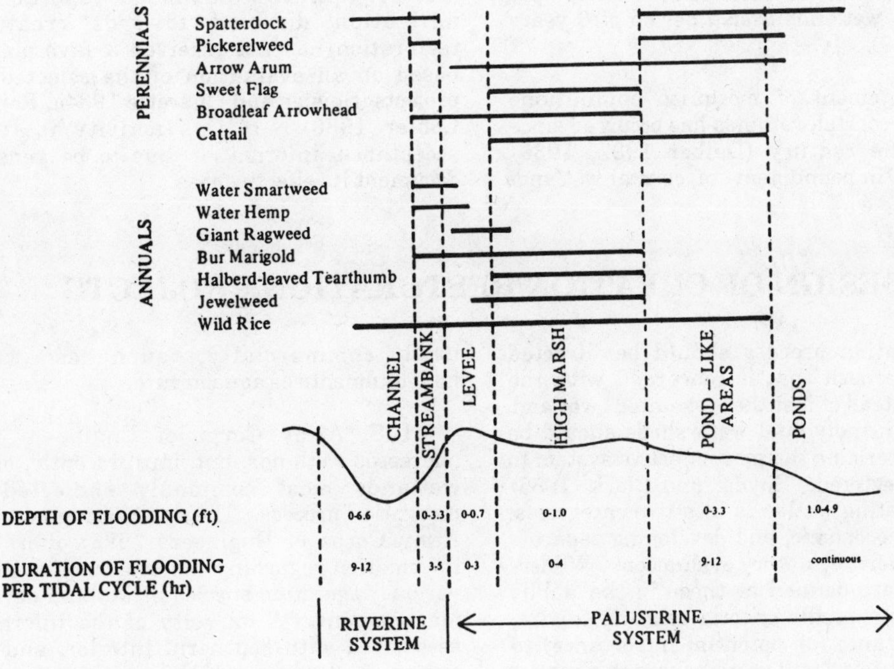

Figure 4. Generalized distribution of vegetation in a freshwater tidal wetland (from Tiner 1985a, adapted from Simpson et al. 1983).

muskrats, salt hay production, and other agricultural products (Daiber 1982, 1986, Mitsch and Gosselink 1986). Over 11,690 acres of salt hay impoundments have been constructed in the Delaware Bay region of New Jersey. Salt hay is a mixture of S. patens, D. spicata, and Juncus gerardii that serves as livestock feed and a variety of other uses (Daiber 1986, Ferrigno et al. 1987). Salt hay impoundments have a history of creating mosquito problems that required routine pesticide applications for their control (Slavin and Shisler 1983, Daiber 1986, Ferrigno et al. 1987). Tidal restoration of salt hay impoundments in New Jersey is the major method for controlling mosquitoes (Figure 5). Over 6,900 acres have been restored to tidal inundation since 1970, with documented increases in wildlife and recreational utilization (Ferrigno et al. 1987). In Connecticut a 25-acre impounded wetland previously dominated by dense stands of P. australis has been restored to tide changing the vegetation to a Spartina alterniflora and S. patens marsh with D. spicata and Salicornia spp. (Bongiorno et al. 1984).

Since 1965, Ducks Unlimited Canada has constructed or restored approximately 45,000 acres of Canadian wetlands, including over 12,000 acres in the coastal regions (Barkhouse 1987). Most of the restoration involved diked abandoned agricultural areas which had no tidal inundation. Water control structure installation was the method utilized to control water levels and vegetation growth. These restored wetlands provide valuable wildlife habitat, but ecological processes proceed rapidly, and management is required to maintain this productivity. These conclusions are based upon the study of 34 wetlands over a period of 6 years (Barkhouse 1987).

The management of mosquito populations associated with coastal wetlands has occurred since the turn of the century (Daiber 1982, 1986). Drainage and impoundment of coastal wetlands was the major engineering method employed in the early 1900's (Bourn and Cottam 1950). The development of the open marsh water management method in the 1960's has met with approval by certain regulatory agencies, e.g., U.S. Fish and Wildlife Service, U.S. Army Corps of Engineers, and New Jersey Department of Environmental Protection (Ferrigno and Jobbins 1968, Ferrigno et al. 1969). The method has definite objectives with documented beneficial impacts upon the salt marsh/estuarine systems (Daiber 1986, Meredith and Saveikis 1987, Shisler and Ferrigno 1987).

Restoration work associated with mitigation of impacts in coastal wetlands has been going on for over twenty years (Garbisch 1977, Charette et al. 1985). An informal survey of the Environmental Protection Agency's regional offices and the U.S. Army Corps of Engineer's district offices in the northeast by the author estimated that there had been 2000 permitted mitigation projects. Most of these mitigation projects are less than a few acres in size. There has not been an extensive evaluation of these projects to determine their success and how they function (Shisler and Charette 1984a, Reimold and Cobler 1986). However, mitigation projects continue to be undertaken without an understanding of their effectiveness. A detailed evaluation of northeastern mitigated wetland projects in various environmental conditions could provide the needed data base to design an effective management strategy.

The information associated with the evaluation of mitigated wetland systems in the northeast is limited. In two published reports, wetland mitigation directed towards creation and restoration has not received a favorable review based on an evaluation of the selected existing projects (Shisler and Charette 1984a, Reimold and Cobler 1986). If this activity is to receive acceptance, information has to be generated to document its effectiveness.

DESIGN OF CREATION/RESTORATION PROJECTS

The mitigation process should be directed towards an approach that is concerned with the total system instead of just the "vegetated" wetland. Wetlands within individual watersheds should be evaluated to determine the most effective system to be created or restored. Snyder and Clark (1985) propose segregating wetlands into two categories, wilderness and economic, and developing separate strategies for their regulatory evaluation. Wilderness wetlands are defined as those in the public sector that meet specific criteria relating to size, location, importance (or potential importance) to consequential species, and scarcity on a regional or national basis. Economic wetlands are impounded, impacted, and/or privately owned wetlands that are in the final throes of corruption and that will be used commercially, such as waterfowl impoundments or aquaculture.

U.S. Army Corps of Engineers research addresses wetlands, but, until recently, only those wetlands most commonly subjected to the mitigation process. The engineering manual (U.S. Army Corps of Engineers 1986) offers detailed information regarding the construction and use of various vegetation species in wetland development management. A majority of the information is associated with southern, interior, and western systems and larger wetland creation projects (over 5 acres or more in size).

Knowledge of the usual wetland development

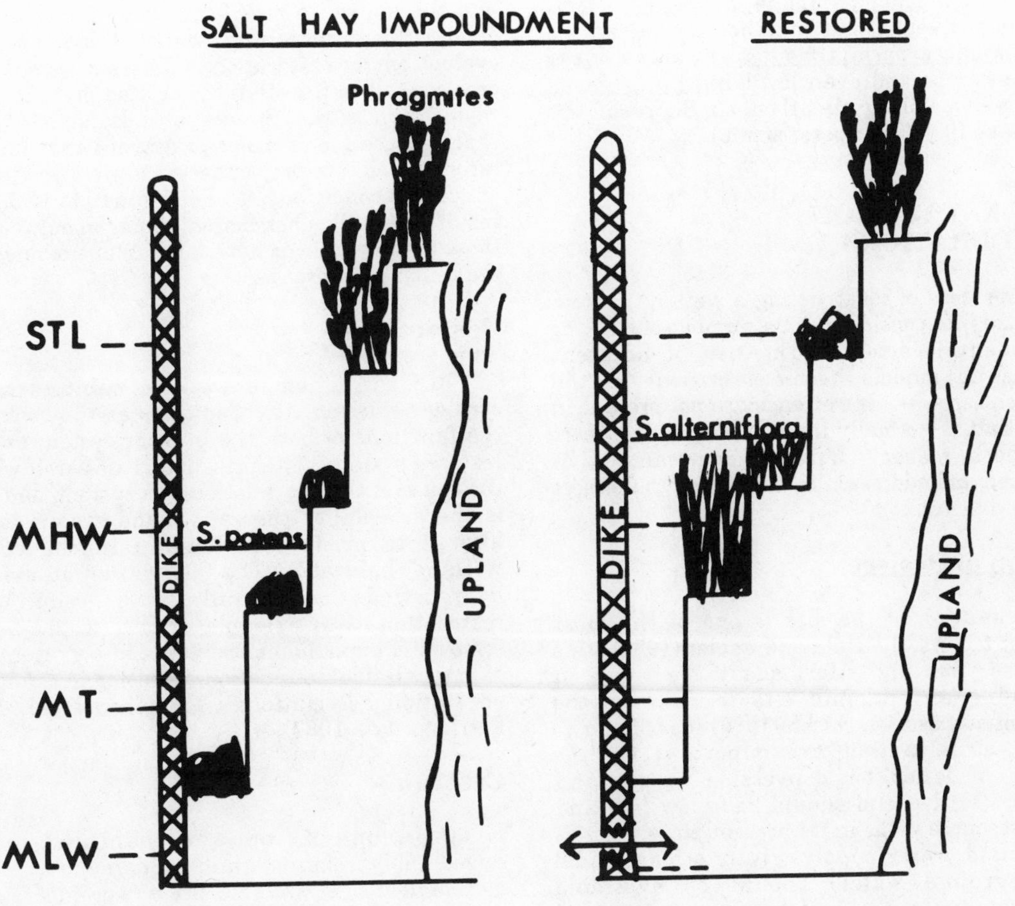

Figure 5. Salt hay impoundment before and after tidal restoration (STL = storm tide level, MHW = mean high water, MT = mean tide level, MLW = mean low water).

process is important for an understanding of wetland restoration and creation (Odum 1988). Research associated with the development of coastal wetlands in the northeast is important for establishing a data base to be used in restoration/creation projects for the region. Northeast coastal marshes developed as the result of coastal submergence and sedimentation (Bloom and Stuiver 1963, Bloom and Ellis 1965, Redfield 1972, Orson et al. 1987). Natural coastal wetland development is in shallow areas usually associated with low energy environments, often behind barrier islands and along river systems. Sediments begin to drop out of the water column in the estuarine systems where they form deltas and intertidal flats. Frey and Basan (1978) present three physiographic stages in marsh maturation: (1) youthful marsh, where low-marsh environments constitute most of the total area; (2) mature marsh, where areas of high and low marsh are approximately equal; and (3) old marsh, where high marsh comprises most of the area. Odum (1988) characterizes the early development stages as being dominated by opportunistic species with soils composed of mineral material and the subsurface hydrology and chemistry similar to early stages of soil development. As the wetland ages the plant community becomes diversified on soils with higher organic content and the subsurface hydrology and chemistry changes in response to the wetlands soils.

Coastal wetlands are advancing into low-lying ecotonal areas as a result of sea-level rise (Psuty 1986a, 1986b, Kana et al. 1988). During the invasion of the ecotonal edge, areas of palustrine forest are replaced by scrub-shrub and eventually high marsh habitats. In some areas of eastern Maine, salt marshes are palustrine scrub-shrub wetlands (e.g., Chamaedaphne calyculata bogs), while further south, salt marshes are advancing upon Chamacecyparis thyoides swamps, Acer rubrum swamps and even low-lying Pinus taeda forest such as along the Eastern Shore of the Chesapeake Bay (Tiner pers. comm.).

PRECONSTRUCTION CONSIDERATIONS

The process of constructing a wetland system has to take into consideration a number of existing environmental factors. The use of adjacent wetlands as models becomes critical in the assessment process. If wetlands are not present in an area that is normally inundated by tides, there has to be a reason. If the reason can not be determined and addressed in the design, the project is destined to fail.

Location of Project

The location of the site is one of the most important factors. Reimold and Cobler (1986) state that more time and effort should be spent in determining an optimum site to increase the chances of success. Garbisch (1986) identifies two principal criteria that are important for the selection of lands for conversion to wetland habitats: (1) the land should have low fish and wildlife resource value in its present state; and (2) an adequate water supply (river, stream, tidal source, ground water) should be available. Unfortunately, the most suitable areas for wetland creation are usually those that already have high biological productivity (e.g., shallow water habitats). Shisler and Charette (1984a) determined that location explained the greatest amount of variability in their study of mitigated marshes in New Jersey.

Restoration/creation adjacent to functioning wetland systems offers the greatest chance of success and is recommended. The created wetland system should, ideally, be in the same watershed as the system that is being altered so that there is not a net loss of wetlands within the watershed. The creation of wetlands from upland sites will increase or maintain the total acreage of wetlands within a given watershed. Increased wetland acreage will create changes in the hydrological regimes of the watershed and, therefore, the watershed should be evaluated for impacts.

The use of equipment and the transportation of materials for the construction and possible maintenance of the restored/created wetland should be addressed in some detail in each project.

Site Characteristics

The selection of a suitable wetland creation site will depend upon the existing site characteristics and the ability to modify these characteristics to produce a functioning wetland system. An evaluation for restoration is different from that for creation, since a wetland is or was present at the restoration site. Snyder and Landrum (1987) reported the major areas of concern that must be incorporated into project design are: (1) evaluation of current conditions; (2) determination of desired results; and (3) construction management. Within these areas of concern are a multitude of questions that must be addressed.

Restoration--

Reconstruction of wetland habitats has to consider existing site conditions as they relate to the functions of both the present system and the restored system. The principal issues are: what is the cause of the degradation of a system, and what is the probability that a wetland system can be altered to produce and maintain the desired wetland habitat? The alteration of existing hydroperiods may be all that is required for restoration. An example of this would be the removal of impoundment dikes to allow resumption of normal tidal inundation, resulting in changes in vegetation associations (Bongiorno et al. 1984, Ferrigno et al. 1987).

Creation--

Creation of wetland habitats differs considerably from restoration and, in most cases, is more difficult. If wetlands are or were not present at a given site, then major limitations may exist which have to be overcome to create suitable conditions. These alterations may or may not be economically or environmentally acceptable. The advisability of destroying mature upland forest habitats or other critical habitats in an area to create small and questionably functionable wetlands has to be seriously considered.

CRITICAL ASPECTS OF THE PROJECT PLAN

The recent U.S. Army Corps of Engineers engineer manual (U.S. Army Corps of Engineers 1986) separates the design of a wetland habitat utilizing dredge material into four parts: (1) location; (2) elevation; (3) orientation and shape; and (4) size. These criteria can also be applied to the construction of any wetland habitat.

Timing of Construction

Certain times of the year are more conducive for the successful construction and/or restoration of wetland systems, especially in the northeastern region. Climatic conditions limit access, use of equipment, revegetation procedures, and the success of these procedures. The effects of the construction activities on the habitats of amphibians, fish, and mammals have to be considered.

Construction Considerations

Construction considerations vary with individual site conditions. A given set of designs will not be applicable to every site and wetland type. The use of experienced personnel and the knowledge of how adjacent wetland systems are functioning will provide an understanding of what should be included in the new wetland system. Techniques, methods, and recommendations may be totally different between marsh restoration and creation projects within a given area because of site conditions. Garbisch (1986) stated that the single important factor in wetland creation is elevation. For this reason, a topographic and/or bathymetric survey of the site at 0.5 foot contours should be undertaken and the location of proposed plantings should be recorded with relation to prospective water levels. After a detailed survey of the site is prepared, the plan can be designed. There has to be enough flexibility in the design so that as the existing system reacts, modifications can be included in the project without total redesign. An "as built" plan submitted after the system is complete and functioning can be more important than the construction plans.

Site conditions should be continually monitored by knowledgeable personnel during the construction phase. This monitoring ensures that construction personnel are building the project as designed or conceptually planned. Small deviations from the design plan may result in drastic change in vegetation associations and possible project failure. Also, the adherence to a plan that will not function within given site conditions is destined to fail. Elevations should be periodically checked with adjacent reference wetland systems to ensure stability of the wetland.

Slopes

Slopes are another major consideration in the development of wetland habitats. Wetland plant species must be able to stabilize the slopes and maintain coverage over a period of time to control erosion. Of the four projects evaluated by Reimold and Cobler (1986) in New England, three had problems because the slopes were too steep. They recommended slopes ranging from 1:5 to 1:15 for increasing wetland vegetation diversity and decreasing erosion potential.

Elevations

The elevations of the restored/created wetland are one of the most critical considerations in project design and construction. Final elevations of the site will be affected by settlement and consolidation of the substrate, therefore these factors must be considered in project design. Shisler and Charette (1984a) reported that relative elevations of constructed marshes in New Jersey were too low to support adequate growth of planted vegetation when compared to adjacent wetlands. Determination of the final elevation is critical and should be based upon the elevational requirements of the desired habitat. Shisler and Charette (1984b) identified consolidation of sand around sewage pipes buried in wetland habitats as causing standing water problems and loss of emergent vegetation cover (Figure 6).

The zonation of the marsh plant species is related to elevations with respect to tidal ranges and water levels. A number of studies have addressed the impacts of elevation and hydrology associated with wetland systems (U.S. Army Corps of Engineers 1986). An annotated bibliography and review of publications dealing with vegetation and elevation data in the northeast region would be useful for preparing and reviewing of permit applications.

Size

Size of the created/restored wetland will affect the use of the wetland by certain species. Wetlands created adjacent to functioning systems will more easily develop wetland functions and assimilate associated fauna. Snyder (1987) presented an overview in the holistic approach to wetland creation and restoration that should be considered in assessing the size of the system. A productive wetland system in most cases requires a combination of open water and ecotonal and upland habitats. The construction of only the emergent wetland vegetation component will not result in a viable system and will not be comparable to the natural system in many cases. Garbisch (1986) recommends construction of tidal channels in a created wetland system for the control of litter and for their value in the exchange of nutrients, increased habitat diversity, and elimination of mosquito breeding. U.S. Army Corps of Engineers (1986) recommends tidal channels to increase tidal circulation and to increase marsh productivity. Tidal channels in high salt marsh habitats have been used for over 20 years in New Jersey for mosquito control with increases in diversity of wetland species and productivity (Shisler and Jobbins 1977, Daiber 1986, Meredith and Saveikis 1987). The extension of a functional system by the use of tidal creeks into the created system will create higher rates of success. These systems can be relatively small, since they are extensions of functioning systems, e.g., fringe wetlands.

Small isolated wetlands may have limited use because of the size requirements of various wetland species. Therefore, it is important to determine the

SEDIMENT COMPACTION

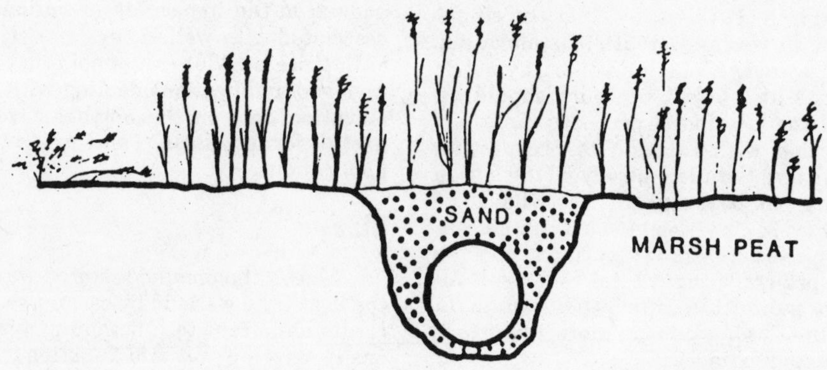

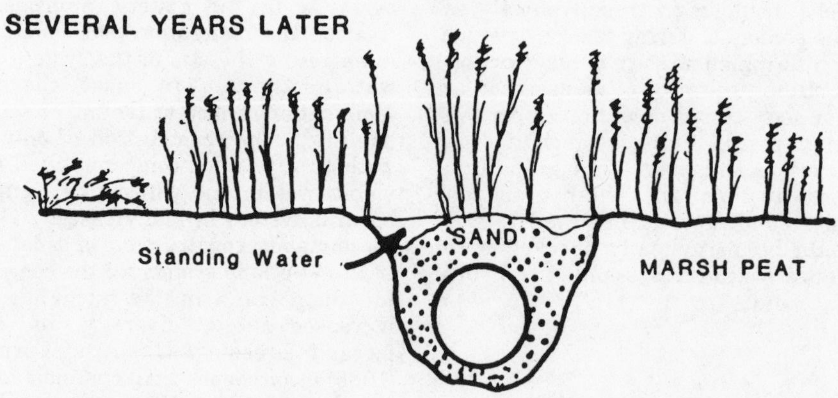

Figure 6. Sediment compaction over time from the use of sand as a replacement for marsh peat (from Shisler and Charette 1984b).

size limitation of certain wetland faunal species which potentially use the created wetland. Other factors associated with size are maintenance, economics, and availability of the site.

The concept that vegetation is the most important component in the development of wetland systems is questionable. Natural wetland systems are far more complex. For example, the water in the system has its origin from areas outside the hydric vegetation zone. The use of the wetlands by fish, avian, and mammal populations has to be evaluated in the created and restored systems. The evaluation of the success of a created wetland is usually determined by the success of the planted species, for example an 85 percent survival. The creation of the habitat is, however, more critical. For example, a created, emergent wetland may consist of many indigenous species that were associated with the sediment or brought into the system during construction or with the tides than were planted.

Waterfowl biologists consider 50% open water a guideline in determining suitable waterfowl habitat (Weller 1978, Daiber 1986). The open water component contributes to spatial heterogeneity and is important to wetland functions. The aquatic habitat provides the medium for exchange of nutrients and populations between the wetland system and adjacent areas. The creation of extensive wetland systems without allowing for water circulation may allow standing water on the surface. Standing water on the surface may change vegetation composition and create mosquito breeding habitat. These areas will retain water long enough, in some cases 7 days, for the mosquito to complete the aquatic stage of its life cycle.

Ice

Ice has had an impact upon northeastern coastal wetland systems. Ice can impact large sections of the marsh by rafting and erosion. It can also cause changes in the elevation of the marsh which affects vegetation associations (Teal 1986). The movement of ice by tidal action results in geomorphological effects upon sediment through erosion, transport, and accretion. Boulders weighing several tons have been transported considerable distances by ice in Barnstable Harbor (Redfield 1972). The destruction of sections of gobi mats used in a New Jersey marsh creation project was attributed to ice (Shisler and Charette 1984a, b).

Wetland creation and restoration projects should not be constructed where ice floes are possible due to wind and currents. The ice accumulation and movement during winter periods would likely eliminate newly constructed wetlands (Reimold and Cobler 1986).

Animal Populations

Animal populations can affect the success of a wetland project by their presence and feeding behavior. Both Branta canadensis and Chen hyperborea (Canada and Snow Geese) affect the growth of a created/restored wetland (Smith and Odum 1981). Branta canadensis has become a major nuisance species in New Jersey suburban areas where it heavily utilizes stormwater facilities and feeds on adjacent lawns.

Geese will usually not alight in tall vegetation. They prefer to land in open water and then proceed towards the emergent wetland and upland areas. Garbisch (1986) suggests the construction of a fence along the open water edge of the wetland consisting of posts connected by nylon line (ca. 1/8 inch diameter) rails spaced every 6 inches, from 6 inches above low water and high water levels. The design of wetland systems with a shrub border will limit geese utilization of the upland edge.

Toxic Materials

Recent publications have addressed the recycling of heavy metals by wetland vegetation (Breteler et al. 1981, Sanders and Osman 1985, Kraus et al. 1986). The impacts of recycling toxic materials has to be addressed through additional research, especially if wetlands are to be considered as stormwater management facilities for water quality control. In any event, the presence of toxins at potential restoration and creation sites must be examined prior to site selection and construction activities.

Hydrology

The driving force in the development of any wetland system is the hydrology. If the correct hydrology is not present or is insufficient, the desired project will fail. The determination of the hydrological regime for the site then becomes the most critical consideration. The use of adjacent tidal wetlands as models becomes important in the assessment of the project design.

Two hydrological differences are associated with coastal wetlands, namely the regularly flooded zone and irregularly flooded zone. The regularly flooded zone is flooded at least once a day by tides and is known as the low marsh. The higher elevations, or high marsh, are flooded for brief periods, during storms and spring tides. The irregularly flooded zone may be termed seasonally flooded-tidal or temporarily flooded-tidal in freshwater tidal areas (Tiner 1985a, b). Odum et al. (1984) concludes that the hydrology of tidal freshwater marshes and associated streams and rivers has been poorly studied. The factor of tide range is critical, especially when there is a small tide range (less than a foot), which makes construction and grading of the substrate to control unwanted species more difficult.

The impacts of hydro-period on various wetland species associated with tidal freshwater systems have not been well documented in the northeast.

Substrate

Garbisch (1986) provides the following specifications for acceptable wetland substrate: the substrate shall consist of a minimum of one foot in depth of clean inorganic/organic materials of which 80-90% by weight will pass through a No. 10 sieve. From their field inspection of a number of created wetland projects in New Jersey and a literature review, Shisler and Charette (1984a) proposed the ranking of substrate as: (1) natural marsh peat; (2) clay and silty clay; (3) estuarine sediments (dredge material); and (4) sand. They also reported some problems with stability and the colonization of sand by certain wetland vegetation species and recommended that sand not be placed on existing marsh peat. Wetland substrates may present problems with handling and application in areas subjected to tides, but usually the benefit of the seed bank and its supplement to the planted vegetation outweighs these problems. Research should address procedures of handling natural wetland substrates to ensure viability of the seed bank.

Revegetation

The success of site revegetation is determined by a number of conditions, but primarily by salinity, elevation, and hydrological factors. Most of the research in the northeast has been directed towards salt marsh species, with limited research concerning the brackish and freshwater wetland systems. The documentation of revegetation in other wetland systems is an important research topic.

The time of year to revegetate an area is critical in the northeast. If seed bank material is to be used it should be in place before the beginning of the growing season. The use of transplants can occur throughout the growing season, ideally in the spring, but should be in place at least one month before the first frost to allow root establishment. Planting in the summer months may expose vegetation to extreme temperatures, salinity, and dryness due to lack of tidal inundation. Most of the research has been directed towards the mid-Atlantic states and south, where transplanting can occur at any time of the year that the ground is not frozen (Garbisch 1986). In the northeastern region, seasonal conditions make transplanting ineffective except during the growing season.

The most successful and expensive method of wetland revegetation is the use of peat-potted plants, plugs, sprigs, and dormant underground plant parts (tubers, bulbs, rhizomes) (Garbisch 1986). The selection of species for revegetation is important, but a source of the material must be availableeither commercially or naturally. Major considerations in vegetation selection are: (1) availability and cost of material; (2) collection and handling ease; (3) storage ease; (4) planting ease; (5) disease; (6) urgency of need for vegetative cover; and (7) site elevation (U.S. Army Corps of Engineers 1986).

New Jersey Department of Environmental Protection recommends the replacement of P. australis dominated wetlands with S. alterniflora, and new freshwater emergent wetlands planted with P. virginica and Sagittaria spp. (Kantor and Charette 1986). The restoration/creation of high marsh habitats is not recommended due to the number of failures in the State (Shisler and Charette 1984a, Charette et al. 1985).

The use of adjacent wetland areas as donor sites for certain plant species may limit possible impacts of site selection on vegetation populations. The use of these wetland resources in existing wetland creation projects has to be undertaken with an understanding of the donor wetlands. The removal of material should be done in a manner that does not effect the existing wetland and its function. Garbisch (1986) recommends a checkerboard technique to avoid the disruption of single large areas of wetlands. Care has to be taken to not create locations ideal for attack by faunal populations, such as Branta canadensis. Openings in the surface may create sites for root predation by avian and mammal populations that would alter the existing wetland habitat.

Natural colonization of vegetation has been shown in certain cases to be more productive than a planted site (Shisler and Charette 1984b). The importance of natural revegetation is that existing gene stock is utilized. Natural revegetation can occur in freshwater tidal wetlands within several months, while in more saline conditions, several years may be required for total revegetation (U.S. Army Corps of Engineers 1986).

Garbisch (1986) lists commercial sources for plant materials for the United States. Of the sixteen sources listed, only four are in the northeast region. The use of plant material grown outside the region will create problems of adaptation of the material. Additional commercial plant material sources are needed since there will be increased pressure upon existing sources to supply plant material. The use of adjacent wetlands as sources for plant material may become a more viable option.

Fertilization

Fertilization is important in establishing certain planted species. A number of experiments have been carried out on the impacts of fertilizers on various species in the mid-Atlantic region and south. Garbisch (1986) recommends a side-dressing (placed below the surface) with a controlled release fertilizer at the time of planting. Osmocote is a controlled release fertilizer that performs well in saline waters and under saturated soil conditions. Very small burlap sacks containing the fertilizer are placed beneath the transplant for underwater planting (Garbisch 1986).

Faunal Populations

No data on the placement of faunal populations into restored/created areas has been documented in coastal areas. Natural colonization in the northeastern wetlands has been documented in altered systems in a number of publications (Ferrigno 1970, Shisler and Jobbins 1977, Shisler and Charette 1984a, Daiber 1982, 1986). Bontje (1988) documented a 100 percent increase in avian species and 700 percent increase in numbers associated with a partially completed 63-acre restored marsh in comparison with an adjacent unrestored marsh.

Buffers

Limited research in the northeast has been directed to the establishment of buffers zones for the protection of wetland habitats, especially in created/restored wetlands (ASWM 1988). Limited data are available as to the type and size of buffer zones and their value. Individual states and various commissions have instituted width regulations with regard to the type of wetlands and the presence of endangered species. The New Jersey Pinelands model (Roman et al. 1985, Roman and Good 1983) is the only real model in the northeast, but it is based on limited data.

Long-Term Management

The objective of restoration/creation projects is to build wetland systems without causing negative impacts upon other ecosystems. Constructed systems should meet the objectives of the project without the need for extensive maintenance programs.

Coastal wetland systems are destined to change over time because of their dynamic nature. These pulsed systems are characterized by oscillating water levels and, under such conditions, a single self-perpetuating ecosystem is often unrealistic (Niering 1987). Design plans for the restoration and creation of wetland systems have to consider this dynamic environment. Areas subjected to daily tidal inundation are the easiest to maintain in vegetation homeostasis, while those constructed above regular tidal inundation create habitats that will change with minor alterations ofenvironmental conditions. Niering et al. (1977) document five vegetation changes within a period of 500-1000 years.

Phragmites australis is considered the dominant vegetation species associated with disturbed wetlands. The species dominates about one-third of Delaware's coastal marshes (Jones and Lehman 1987) and the wetlands along the Delaware and Hudson Rivers. The importance of the species in the overall function of the wetland systems of the United States has not been identified. Phragmites australis has one of the highest standing crops associated with wetland systems. In dense homogeneous stands, the plant will militate against waterfowl, waterbird, and furbearer populations by replacing desirable food plants and reducing habitat heterogeneity, and open water space (Buttery and Lambert 1965, Ward 1968, Vogl 1973, Jones and Lehman 1986). This species also causes fire hazards, impedes water flow, penetrates and clogs underground pipelines, restricts access, and provides roosts for destructive blackbirds (Beck 1971, Riemer 1976, Ricciuti 1982). Phragmites australis, however, is effective in erosion and water quality control and is commercially harvested outside the United States.

Phragmites australis should not be recommended as a revegetation species in coastal areas and should be strictly controlled in recently constructed wetland projects. The use of physical control methods (mowing, plowing, disking) and burning are ineffective, and actually facilitate the plant's spread and propagation (Garbisch 1986). Both Dowpon and Rodeo have been shown to be an effective means of control (Garbisch 1986, Jones and Lehman 1987).

Both Typha latifolia, T. angustifolia, and Lythrum salicaria, also produce monotypic stands that limit utilization of wetland systems by certain other wetland species. These species are usually associated with disturbed wetlands and many restored and created wetland projects with such species can behave like disturbed wetlands (Odum 1988). Wetland vegetation species that naturally colonize wetlands (i.e., pioneer species) are not recommended, since they will naturally occur over time and usually dominate the wetland in a relatively short period.

MONITORING

Monitoring of restored/created wetlands can provide an important database upon which recommendations can be made for future projects. These evaluations should include a combination of detailed analysis of the habitat parameters and general field observations for several years. The use of photographs from fixed locations to document changes in wetland habitats is an inexpensive method that provides an important database for comparison. Detailed analysis of various habitat components becomes cumbersome for the uninitiated in wetland science and can lead to ineffective results.

Parameters to Monitor

The monitoring of habitat characteristics distinguishing low intertidal marsh, high marsh, emergent wetlands, scrub-shrub, etc. may be better for general evaluation than detailed biological survey information. Standard accepted habitat sampling procedures, species composition, and growth measurements should be used.

Evaluations should be performed by experienced personnel with an understanding of the various types of wetlands and their functions. The use of detailed evaluation instruments for before and after assessments are not recommended by Golet (1986). Conclusions from these assessments will present a variety of results which may be difficult to interpret and present erroneous conclusions. If designed habitats are present and functioning, the system may be considered as having reached its objective. If these systems are not present, then a project may be considered as not meeting its objective.

How to do it?

Periodic sampling by either the regional representatives or consultants could provide indications of mitigation effectiveness. The importance of follow-up evaluations are that they supply detailed information as to the success rate of certain types of management methods within individual regions. The evaluation should be directed towards assessing the presence of a wetland habitats that were to be created. Presently, in the northeast, exists the greatest concentration of wetland mitigation projects, yet evaluations are limited (Shisler and Charette 1984a, Reimold and Cobler 1986).

How to Interpret the Results?

Goals and objectives must be clearly defined during design of the mitigation projects, if the created or restored system is going to be evaluated in a meaningful way. Historically, the major problem with the evaluation of mitigation projects has been the absence of goals associated with the evaluated projects (Quammen 1986).

RESEARCH NEEDS

The coastal wetlands of the northeast have been heavily impacted by man and the restoration of these systems should become a major goal of the mitigation process. The continual use of mitigation as a component of federal and state permitting programs must be based on an effective process to be successful. To implement this process, there should be a detailed evaluation of the individual systems to determine the critical wetland habitats within the regions and watersheds. The need for data associated with wetland systems and their functions is imperative considering the increasing number of neophyte marsh builders and wetland management programs. Shisler and Charette (1984a) recommend that an ecological management approach be used in future New Jersey mitigation projects. This approach would include the use of experimental projects to test methods and techniques and record study results. The evaluation of past projects offers ideal starting points, but future projects will have to be innovative in design and application. The use of these future projects as experiments will allow the private sector and permit applicants to become involved. The costs of these experiments can be minimized and to a large extent, borne or shared by the private sector. If it does not work, the contractor could be held totally responsible.

INFORMATION GAPS

Major information gaps are present in the current knowledge of the restoration and creation of northeastern coastal wetlands. Some of the gaps can be filled through detailed review of existing wetland projects and the supplementation of these data with additional research. Others will require research projects specifically designed to address specific questions.

Most of the restoration and creation research has been conducted outside of the northeast region and has been associated with large projects. The northeastern region is unique. Its history, development pressures, and climatic conditions make the application of research results from other regions of limited value.

Research needs include, but are not limited to the following:

1. **Tidal range impacts on the vegetation associations.** Detailed information concerning the distribution of major wetland species associated with tidal wetland habitats of the northeast is not available in a single publication. Tiner (1987) has addressed the geographic range and preliminary habitat data of 150 plus coastal wetland plant species, but additional data are needed as to the hydrological requirements and success of these species. The work of several researchers (McCormick and Somes 1982, Nixon 1982, Odum et al. 1984, Tiner 1985a, b, Teal 1986), has documented state and regional wetland systems. It should be expanded. A manual addressing the species and their location in reference to the tidal range and major environmental conditions may help correct some of the problems of misidentification of species by habitat by consultants and approval by regulators.

2. **Transplants vs. natural succession.** A major consideration with wetland creation and restoration is natural succession of vegetation associations within the systems. Natural revegetation and succession may be a feasible alternative to transplanting. Transplanting may be an effective method of controlling certain vegetation pest species, if planting takes place at the appropriate time. The timing of transplanting versus natural succession requires additional research.

3. **Potted transplants vs. natural (donor) transplants.** A major requirement of many mitigation projects has been the use of potted transplants. However, the use of adjacent wetlands as donor sources allows for the transfer of indigenous bacterial communities associated with the root zones of the transplants. These local transplants are acclimated to the local environmental conditions. Other seed sources may also be transplanted in the process. Research comparing these methods could be conducted.

4. **Transplanting time of year.** Some publications (U.S. Army Corps of Engineers 1986, Garbisch 1986) state that transplanting can occur any time during the year when the ground is not frozen, but in the northeast there are seasonal limitations. Research associated with the seasonal impacts of transplanting and the success of individual species would be beneficial. Most of the research to date is associated with salt marsh systems (predominantly S. alterniflora), with limited information available concerning the other wetland species found in coastal wetland types.

5. **Individual species requirements.** There is a need for basic ecological research associated with common coastal wetland species. Salt marsh species (S. alterniflora and S. patens) have received the bulk of emphasis in the literature, particularly as they relate to wetland creation. The restoration of wetland systems will have to address other species, since many of the degraded wetland systems are located along major estuarine systems with a variety of environmental conditions. The northeast region contains extensive degraded wetlands in the brackish to freshwater zones that offer numerous restoration opportunities.

6. **Habitat requirements of endangered and threatened species.** A large percentage of the endangered and threatened species of an area are associated with wetland habitats. Sixty-four percent of the 249 vegetative species identified by New Jersey as endangered and threatened grow in wetland or aquatic habitats (Tiner 1985a). At the federal level, 17 of 20 Delaware and 23 of 25 New Jersey plant species considered endangered or threatened are associated with wetlands (Tiner 1985a, b). How can wetland habitats be restored/created and maintained for these species? Basic research associated with these species and their ecological requirements may assist with the creation and restoration of habitats needed for their survival.

7. **Toxic materials and their movements.** Recent research has documented biological transfer and amplification of toxic material by certain wetland plant species. Research is needed to determine which wetland plant species and/or wetland systems amplify toxic materials. The use of wetlands as stormwater management facilities may increase toxic loading associated with run-off and ultimately the biological amplification of toxic materials. Are there design criteria that could be implemented in the construction and restoration of wetlands that would trap toxic materials before entering the biological system?

8. **Wetlands as stormwater management facilities.** Wetlands are being created as a major component in stormwater management programs. These wetlands are being used for water quality maintenance. Research determining the impacts of stormwater run-off on the various components of wetlands is needed. Do these stormwater run-off wetlands function as biological amplifiers of heavy metals?

9. **Forested wetlands.** The major wetland type in the northeast is palustrine forested wetland, for which there is a very limited scientific data

base. A small percentage of these are tidally influenced, with even less available data. Basic research concerning the functioning of the palustrine systems is needed.

10. **Maintenance of wetlands systems.** Restoration and creation of wetland habitats without detailed maintenance plans may create problems for local governments in keeping these systems functioning. Research addressing wetland habitats and maintenance requirements would be useful.

11. **Pest and nuisance species.** Certain wetland habitats become ideal locations for the production of mosquito populations and other nuisance species. Usually, restored/created wetlands juxtapose human populations and can provide breeding areas and refuges for nuisance species populations, such as mosquitoes, rats and raccoons. Research associated with the design of restored and created wetlands which minimize the production of nuisance species should be addressed.

12. **Stockpiling of wetland material.** Many restoration/creation projects make use of the organic soils of the wetland filled by the development activity. The stockpiling of the wetland materials on site and their eventual use in mitigation would provide an important seed bank, gene stock, nutrients, etc. required by the created system. A major problem with stockpiling is the creation of cat clays, acid conditions, and oxygenation of the previously saturated soil. Research associated with the methods and consequences of stockpiling various wetland sediments is needed.

13. **Size of a wetland system.** The size of a wetland is important and usually not considered in mitigation. The construction of isolated small systems may not be effective. What are the requirements of the wetland species and system as a unit? Does an individual species require an acre or several acres as its habitat? What are the functioning units (vegetation, open water, buffer) of various species or groups?

14. **Wetland inventories.** The development of individual wetlands within a watershed without an understanding of the total system may not be effective. The total system should be inventoried to determine the most suitable wetland rehabilitation and creation projects.

LIMITATIONS OF KNOWLEDGE

The limitations of our knowledge of wetland management creates complex issues. There is a need to combine the various data sources from both the "gray" and "published" scientific literature to provide a better understanding of wetland management. Wetland management is nothing new (Daiber 1986, Maltby 1988), although some organizations think so. Wetland management has been carried out for thousands of years. A major problem is the application of old methods by the new group of wetland management personnel. Wetland management does not have to be reinvented, just fine-tuned and applied with definite objectives and goals. Detailed evaluation of sites that have been restored or created would provide a database that could be applied to future mitigation projects. The data would also suggest the direction of future research project endeavors.

LITERATURE CITED

Alexander, C.E., M.A. Broutman, and D.W. Field. 1986. An Inventory of Coastal Wetlands of the USA. National Oceanic and Atmospheric Administration. U.S. Dept. of Commerce, Washington, D.C.

Andrews, R. 1987. Other waterbirds and their use of coastal wetlands in the northeastern United States, p. 71-80. In W.R. Whitman and W.H. Meredith (Eds.), Waterfowl and Wetlands Symposium: Proceedings of a Symposium on Waterfowl and Wetlands Management in the Coastal Zone of the Atlantic Flyway. Delaware Coastal Management Program, Delaware Dept. of Natural Resources and Environmental Control, Dover, Delaware.

ASWM Proceedings: National Wetlands Symposium. Mitigation of Impacts and Losses, p. 433-435.

Barkhouse, H.P. 1987. Management related study of man-made wetlands located in coastal regions of the maritime provinces, Canada, p. 82-97. In W.R. Whitman and W.H. Meredith (Eds.), Waterfowl and Wetlands Symposium: Proceedings of a Symposium on Waterfowl and Wetlands Management in the Coastal Zone of the Atlantic Flyway. Delaware Coastal Management Program, Delaware Dept. of Natural Resources and Environmental Control, Dover, Delaware.

Beck, R.A. 1971. Phragmites control for urban, industrial, and wildlife needs. Proc. Northeast Weed Control Conf. 25:89-90.

Beeftink, W.G. 1977. Salt marshes, p. 93-121. In R.S.K. Barnes (Ed.), The Coastline. John Wiley & Sons, London.

Bloom, A.L. and M. Stuiver. 1963. Submergence of the Connecticut coast. Science 139:332-334.

Bloom, A.L. and C.W. Ellis, Jr. 1965. Postglacial Stratigraphy and Morphology of Coastal Connecticut. Conn. Geol. Nat. Hist. Survey Guidebook No. 1. Hartford, Connecticut.

Bongiorno, S.F., J.R. Krautman, T.J. Steinke, S. Kawa-Raymond, and D. Warner. 1984. A study of the restoration in Pine Creek salt marsh, Fairfield, Connecticut, p. 10-23. In F.J. Webb (Ed.), Proceedings of the Eleventh Annual Conference on Wetlands Restoration and Creation. Hillsborough Community College, Tampa, Florida.

Bontje, M.P. 1988. The application of science and engineering to restore a salt marsh, 1987, p. 267-273. In J. Zelazny and J.S. Feierabend (Eds.), Proceedings of a Conference: Increasing Our Resources. National Wildlife Federation, Washington, D.C.

Bourn, W.S. and C. Cottam. 1950. Some Biological Effects of Ditching Tidewater Marshes. U.S. Fish and Wildlife Service. Res. Rept. 19.

Breteler, R.J., I. Valiela, and J.M. Teal. 1981. Bioavailability of mercury in several north-eastern U.S. Spartina ecosystems. Estuar. Coast. Shelf Sci. 12:155-166.

Burger, J., J. Shisler, and F.H. Lesser. 1982. Avian utilization on six salt marshes in New Jersey. Biol. Conserv. 223:187-212.

Buttery, B.R. and J.M. Lambert. 1965. Competition between Glyceria maxima and Phragmites communis in the region of Surlingham Broad. I. The competition mechanism. J. Ecol. 53:163-181.

Candeletti, T.M. and F.H. Lesser. 1977. Mosquito control techniques on Island Beach State Park. Proc. NJ Mosq. Control Assoc. 64:39-42.

Charette, D.J., J.K. Shisler, and R. Kantor. 1985. Guidelines for the mitigation of tidal salt marshes in New Jersey, p. 941-960. In O.T. Magoon, H. Converse, D. Miner, D. Clark, and L.T. Thomas (Eds.), Coastal Zone 85. Am. Soc. Of Civil Eng. New York, New York.

Copper, A.W. 1974. Salt marshes, p. 55-99. In H.T. Odum, B.J. Copeland, E.A. McMahan (Eds.), Coastal Ecological Systems on the United States. Vol. II. Conservation Foundation, Washington, D.C.

Cowardin, L.M., V. Carter, F.C. Golet, and E.T. LaRoe. 1979. Classification of Wetlands and Deepwater Habitats of the United States. U.S. Fish and Wildlife Service. FWS/OBS-79/31.

Daiber, F.C. 1982. Animals of the Tidal Marsh. Van Nostrand Reinhold Company. New York, New York.

Daiber, F. 1986. Conservation of Tidal Marshes. Van Nostrand Reinhold Company. New York, New York.

Ferrigno, F. and D.M. Jobbins. 1968. Open Marsh Water Management. Proc. of NJ Mosq. Exterm. Assoc. 55:104-115.

Ferrigno, F., L.G. McNamara, and D.M. Jobbins. 1969. Ecological approach for improved management of coastal meadowlands. Proc. NJ Mosq. Exterm. Assoc. 56:188-203.

Ferrigno, F. 1970. Preliminary effects of open marsh water management on the vegetation and organisms of the salt marsh. Proc. of NJ Mosq. Exterm. Assoc. 57:79-94.

Ferrigno, F., J.K. Shisler, J. Hansen, and P. Slavin. 1987. Tidal restoration of salt hay impoundments, p. 284-297. In W.R. Whitman and W.H. Meredith (Eds.), Waterfowl and Wetlands Symposium: Proceedings of a Symposium on Waterfowl and Wetlands Management in the Coastal Zone of the Atlantic Flyway. Delaware Coastal Management Program, Delaware Dept. of Natural Resources and Environmental Control, Dover, Delaware.

Frey, R.W. and P.B. Basan. 1978. Coastal Salt Marshes, p. 101-169. In R.A. Davis, Jr. (Ed.), Coastal Sedimentary Environments. Springer-Verlag, New York, New York.

Garbisch, E.W. 1977. Recent and Planned Marsh Establishment Work Throughout The Contiguous United States A Survey and Basic Guidelines. U.S. Army Engineers Waterways Exp. Sta. Rept. D-77-3.

Garbisch, E.W. 1986. Highway and Wetlands: Compensating Wetland Losses. U.S. Dept. of Transportation, Federal Highway Administration Report No. FHWA-IP-86-22.

Golet, F.C. 1986. Critical issues in wetland mitigation: a scientific perspective. National Wetlands Newsletter 8(5):3-6.

Hill, D.E. and A.E. Shearin. 1979. Tidal Marshes of Connecticut and Rhode Island. Connecticut Agricultural Experiment Station, New Haven, Connecticut. Bulletin 709.

Jones, W.L. and W.C. Lehman. 1986. Phragmites control with aerial applications of glyphosate in Delaware. Trans. Northeast Fish Wildl. Conf. 43:15-24.

Jones, W.L. and W.C. Lehman. 1987. Phragmites control and revegetation following aerial applications of glyphosate in Delaware, p. 185-199. In W.R. Whitman and W.H. Meredith (Eds.), Waterfowl and Wetlands Symposium: Proceedings of a Symposium on Waterfowl and Wetlands Management in the Coastal Zone of the Atlantic Flyway. Delaware Coastal Management Program, Delaware Dept. of Natural Resources and Environmental Control, Dover, Delaware.

Kana, T.W., W.C. Eiser, B.J. Baca, and M.L. Williams. 1988. New Jersey Case Study, p. 61-86. In J.G. Titus (Ed.), Greenhouse effect, sea level rise and coastal wetlands. U.S.Environmental Protection Agency. Office of Policy Analysis. Washington, D.C. EPA-230-05-86-013.

Kantor, R.A. and D.J. Charette. 1986. Wetland mitigation in New Jersey's coastal management program. National Wetlands Newsletter 8(5):14-15.

Kennard, W.C., M.W. Lefor, and D.L. Civco. 1983. Analysis of Coastal Marsh Ecosystems: Effects of Tides on Vegetational Changes. Univ. of Connecticut, Institute of Water Resources, Storrs, Connecticut. Res. Proj. Tech. Completion Rept. B-014 CONN.

Kraus, M.L., P. Weiss, and J.H. Crow. 1986. The excretion of heavy metals by the salt marsh cord grass, Spartina alterniflora and Spartina's role in mercury cycling. Mar. Envirn. Res. 20:307-316.

Maltby, E. 1988. Wetland resources and future prospects - an international perspective, p. 3-14. In J. Zelazny and J.S. Feierabend (Eds.), Proceedings of a Conference: Increasing Our Resources. National Wildlife Federation, Washington, D.C.

McCormick, J. and H.A. Somes. 1982. The Coastal Wetlands of Maryland. Maryland Department of Natural Resources.

McNeil, P. 1979. Island Beach State Park mosquito control project: a progress report. Proc. NJ Mosq. Control Assoc. 66:123-127.

Meredith, W.H. and D.E. Saveikis. 1987. Effects of open marsh water management (OMWM) on bird populations of a Delaware tidal marsh, and OMWM's use in waterbird habitat restoration and enhancement, p. 299-321. In W.R. Whitman and W.H. Meredith (Eds.), Waterfowl and Wetlands Symposium: Proceedings of a Symposium on Waterfowl and Wetlands Management in the Coastal Zone of the Atlantic Flyway. Delaware Coastal Management Program, Delaware Dept. of Natural Resources and Environmental Control, Dover, Delaware.

Mitsch, W.J. and J.G. Gosselink. 1986. Wetlands. Van Nostrand Reinhold Company. New York, New York.

Miller, W.R. and R.E. Egler. 1950. Vegetation of the Wequetequork-Pawcatuck tidal marshes, Connecticut. Ecol. Monogr. 20:143-172.

Niering, W.A., R.S. Warren, and C.G. Weymouth. 1977. Our dynamic tidal marshes; vegetation changes as revealed by peat analysis. Conn. Arboretum Bull. 22.

Niering, W.A. 1987. Wetlands hydrology and vegetation dynamics. National Wetlands Newsletter 9(2):9-11.

Niering, W. and R.S. Warren. 1980. Vegetation patterns and process in New England salt marshes. BioScience 30:301-307.

Nixon, S.W. and C.A. Oviatt. 1973. Ecology of a New England salt marsh. Ecol. Monogr. 43:464-498.

Nixon, S.W. 1982. The Ecology of New England High Salt Marshes: A Community Profile. U.S. Fish and Wildlife Service. FWS/OBS-81/55.

Odum, E.P. 1974. Halophytes, energetics and ecosystems, p. 599-602. In Reimold, R.J. and W.H. Queen (Eds.), Halophytes. Academic Press, Inc. New York, New York.

Odum, W.E. 1988. Predicting ecosystem development following creation and restoration of wetlands, p. 67-70. In J. Zelazny and J.S. Feierabend (Eds.), Proceedings of a Conference: Increasing Our Resources. National Wildlife Federation, Washington, D.C.

Odum, W.E., T.J. Smith III, J.K. Hoover, and C.C. McIvor. 1984. The Ecology of Tidal Freshwater Marshes of the United States East Coast: A Community Profile. U.S. Fish and Wildlife Service FWS/OBS-83/17.

Orson, R.A., R.S. Warren, and W.A. Niering. 1987. Development of tidal marsh in a New England river valley. Estuaries 10:20-27.

Quammen, M.L. 1986. Measuring the success of wetlands mitigation. National Wetlands Newsletter 8(5):6-8.

Penfound, W.T. 1952. Southern swamps and marshes. Bot. Rev. 18:413-446.

Provost, M.W. 1973. Mean high water mark and use of tidelands in Florida. Florida Scientist 36:50-66.

Psuty, N.P. 1986a. Impacts of impending sea-level rise scenarios: the New Jersey barrier island responses. Bull. NJ Acad. Sci. 31:29-36.

Psuty, N.P. 1986b. Holocene sea level in New Jersey. Physical Geography 7:156-167.

Redfield, A.C. 1972. Development of a New England salt marsh. Ecol. Monogr. 42:201-237.

Reimold, R.J., and S.A. Cobler. 1986. Wetlands Mitigation Effectiveness. U.S. Environmental Protection Agency Contract No. 68-04--0015. Boston, Massachusetts.

Ricciuti, E.R. 1982. The all too common, common reed. Audubon 85:65-66.

Riemer, D.N. 1976. Long-term effects of glyphosate applications to phragmites. J. Aquat. Plant Manage. 14:39-43.

Roman, C.T., R.A. Zampella, and A.Z. Jaworski. 1985. Wetland boundaries in the New Jersey Pinelands: Ecological relationships and delineations. Water Resources Bulletin 21:1005-1012.

Roman, C.T. and R.E. Good. 1983. Wetlands of the New Jersey Pinelands: Values, Functions, Impacts and a Proposed Buffer Delineation Model. Division of Pinelands Research, Center for Coastal and Environmental Studies, Rutgers University, New Brunswick, New Jersey.

Sanders, H.L. 1968. Marine benthic diversity: a comparative study. Am. Nat. 102:243-282.

Sanders, J.G. and R.W. Osman. 1985. Arsenic incorporation in a salt marsh ecosystem. Estuar. Coastal Shelf Sci. 20:387-392.

Shisler, J.K. and D.M. Jobbins. 1977. Salt marsh productivity as effected by the selected ditching technique, open marsh water management. Mosquito News 37:631-636.

Shisler, J.K. and D. Charette. 1984a. Evaluation of Artificial Salt Marshes in New Jersey. New Jersey Agricultural Experiment Station Publication Number P-40502-01-84.

Shisler, J.K. and D. Charette. 1984b. Mitigated marshes and the creation of potential mosquito habitats. Proc. NJ Mosq. Control Assoc. 71:78-87.

Shisler, J.K. and F. Ferrigno. 1987. The impacts of water management for mosquito control on waterfowl populations in New Jersey, p. 269-282. In W.R. Whitman and W.H. Meredith (Eds.), Waterfowl and Wetlands Symposium: Proceedings of a Symposium on Waterfowl and Wetlands Management in the Coastal Zone of the Atlantic Flyway. Delaware Coastal Management Program, Delaware Dept. of Natural Resources and Environmental Control, Dover, Delaware.

Simpson, R.L., R.E. Good, M.A. Leck, and D.F. Whigham. 1983a. The ecology of freshwater tidal wetlands. BioScience 33:255-259.

Simpson, R.L., R.E. Good, R. Walker, and B.R. Frasco. 1983b. The role of Delaware River freshwater tidal wetlands in the retention of nutrients and heavy metals. J. Environ. Quality 12:42-48.

Slavin, P. and J.K. Shisler. 1983. Avian utilization of a tidally restored salt hay farm. Biol. Conserv. 26:271-285.

Smith, T.J. and W.E. Odum. 1981. The effects of grazing by Snow Geese on coastal salt marshes. Ecology 62:98-106.

Snyder, R.M. and J.R. Clark. 1985. Coastal wetland restoration developers challenge. Proceedings of the Coastal and Ocean Management Coastal Zone 85 4:339-350.

Snyder, R.M. and F.R. Landrum. 1987. Estuarine rehabitiation - a management perspective. Coastal Zone Management (In press).

Snyder, R.M. 1987. Wetland rehabilitation the holistic approach. p. 10-18. In J.A. Kusler, M.L. Quammen, and G. Brooks (Eds.), Proceeding of the National Wetland Symposium: Mitigation of Impacts and Losses. Assoc. State Wetland Managers, Chester, Vermont.

Teal, J.M. 1986. The Ecology of Regularly Flooded Salt Marshes of New England: A Community Profile. U.S. Fish Wildl. Serv. Biol. Rep. 85(7.4).

Tiner, R.W., Jr. 1985a. Wetlands of New Jersey. U.S. Fish and Wildlife Service, National Wetlands Inventory, Newton Corner, Massachusetts.

Tiner, R.W., Jr. 1985b. Wetlands of Delaware. U.S. Fish and Wildlife Service, National Wetlands Inventory, Newton Corner, Massachusetts and Delaware Department of Natural Resources and Environmental Control, Wetlands Section, Dover, Delaware Cooperative Publication.

Tiner, R.W. 1987. A Field Guide to Coastal Wetland Plants of the Northeastern United States. University of Massachusetts Press, Amherst, Massachusetts.

U.S. Army Corps of Engineers. 1986. Beneficial Uses of Dredged Material. Engineers Manual 1110-2-5026. Office, Chief of Engineers, Washington, D.C.

Vogl, R.J. 1973. Effects of fire on the plants and animals of a Florida wetland. Am. Midl. Nat. 89:334-347.

Ward, P. 1968. Fire in relation to waterfowl habitat of the Delta marshes. Proc. Tall Timbers Fire Ecol. Conf. 8:254-267.

Weller, M.W. 1978. Management of freshwater marshes for wildlife, p. 267-284. In R.E. Good, D.E. Whigham, and P.L. Simpson (Eds.), Freshwater Wetlands: Ecological Processes and Management Potential. Academic Press, New York.

Whitman, W.R. and R.V. Cole. 1987. Ecological conditions and implications for waterfowl management in selected coastal impoundments of Delaware, p. 99-119. In W.R. Whitman and W.H. Meredith (Eds.), Waterfowl and Wetlands Symposium: Proceedings of a Symposium on Waterfowl and Wetlands Management in the Coastal Zone of the Atlantic Flyway. Delaware Coastal Management Program, Delaware Dept. of Natural Resources and Environmental Control, Dover, Delaware.

APPENDIX I: RECOMMENDED READING

A number of individual articles could be recommended as reading material, but only actual texts were included. These texts and reports provide an ideal sources for additional reading and references concerning wetland management in the northeast.

Daiber, F. 1982. Animals of the Tidal Marsh. Van Nostrand Reinhold Company. New York.

A comprehensive textbook summarizing the literature as it pertains to the biology and natural history of those animals characteristic of tidal marshes. The text covers from protozoa through mammals as they relate the major factors of the tidal marsh. An extensive reference section is provided that is very useful.

Daiber, F. 1986. Conservation of Tidal Marshes. Van Nostrand Reinhold Company. New York.

The text is the best summary of literature from various sources associated with management of tidal wetland ecosystems. The text includes a section "Dredged material for wetland restoration" which summarizes wetland creation literature from the U.S. Army Corps of Engineers dredging and wetlands research programs and several other sources. Other sections deal with vegetation management, water management, and the future wetland management concept. An extensive reference section, over 40 pages, is provided.

Duncan, W.H. and M.B. Duncan. 1987. The Smithsonian Guide to Seaside Plants of the Gulf and Atlantic Coasts. Smithsonian Institution Press. Washington, DC.

An important field guide to the vegetation of the Atlantic and Gulf coasts of the United States. Color photographs of 588 species along with another 361 species are described in detailed.

Garbisch, E.W. 1986. Highway and Wetlands: Compensating Wetland Losses. U.S. Dept. of Transportation, Federal Highway Administration Report No. FHWA-IP-86-22.

An important practical guide for wetland creation and restoration. Concepts, methods, and general specifications for restoration and/or creation of wetland habitats are presented. The publication includes a number of photographs and detailed drawings showing the fundamental methods. A number of wetland plants are shown in line drawings along with pertinent information concerning habitats, geographical range, commercial sources, recommended propagules, and site seeding potential. A small bibliography is provided that addresses major literature sources related to the various subjects. A list of commercial plant sources is also provided.

Lewis, R.R. 1982. Creation and Restoration of Coastal Plant Communities. CRC Press Inc. Boca Raton, Florida.

One of the first text that addresses the restoration/creation of coastal systems. A number of papers are presented that address management programs from the mid-Atlantic southward and in China. A list of world wide commercial plant sources is provided along with list of professional societies and journals.

McCormick, J. and H.A. Somes. 1982. The Coastal Wetlands of Maryland. Maryland Department of Natural Resources.

A detailed evaluation of the coastal tidal wetlands of Maryland that was performed as a major task for the Department of Natural Resources in implementing Maryland's Wetlands Act. It provides detailed information on the types of wetland vegetation and their functions. Detailed data are presented on the distribution, composition (by county and watershed), values and research needs of the wetland types in Maryland. A method evaluation of wetland sites is present. A very good publication to understand the various coastal wetland systems and method of statewide evaluation.

Mitsch, W.J. and J.G. Gosselink. 1986. Wetlands. Van Nostrand Reinhold Company. New York.

A comprehensive text concerned with both freshwater and coastal wetlands of the United States. The text formulates wetland information into a number of chapters that addresses the various components associated with the individual systems. An extensive reference section with over 40 pages of citations. A very useful wetland reference text.

Niering, W.A. 1985. National Audubon Society Nature Guides. Wetlands. Alfred A. Knopf, Inc. New York, New York.

A general field guide of the wetlands of the United States that contains a lot of information that can be use by everyone. It includes color photographs of trees, wildflowers, fishes, insects, birds and mammals associated with the wetlands and their distribution. An introduction addresses the major wetlands and there functions.

Nixon, S.W. 1982. The Ecology of New England High Salt Marshes: A Community Profile. U.S. Fish and Wildlife Service. FWS/OBS-81/55.

A community profile report which summarizes the literature of the New England high marsh habitat. The report is broken down into several chapters that provide basic information concerning the development, zonation, functions and human impacts. A reference section lists major citations associated with high marshes in northeast.

Odum. W.E., T.J. Smith III, J.K. Hoover, and C.C. McIvor. 1984. The Ecology of Tidal Freshwater Marshes of the United States East Coast: A Community Profile. U.S. Fish and Wildlife Service. FWS/OBS-83/17.

A community profile report addressing freshwater tidal marshes on the east coast, from Maine to northern Florida. Chapters discuss geographical location, development, community components (plants, invertebrates, fishes, amphibians and reptiles, birds, mammals), ecosystem process, management practices. An important chapter compares the tidal freshwater marshes with salt marshes and non-tidal wetlands. An extensive reference section, 17 pages, with additional appendices listing the major components of the freshwater tidal wetlands with references. A very useful text.

Reimold, R.J. and S.A. Cobler. 1985. Wetlands Mitigation Effectiveness. U.S. Environmental Protection Agency Contract No. 68-04-0015. Boston, Massachusetts.

The report provides a detailed evaluation of the five mitigation sites in the U.S. Environmental Protection Agency Region I, of which only two are associated with tidal coastal wetlands. Provides information concerning problems with the mitigation process with possible solutions.

Shisler, J.K. and D.J. Charette. 1984. Evaluation of Artificial Salt Marshes in New Jersey. New Jersey Agricultural Experiment Station Publication Number P-40502-01-84.

The report evaluates wetland mitigation sites in New Jersey associated with the Division of Coastal Resource program. A total 30 projects were located and field survey. Eight were selected for quantitative sampling and comparison. A number parameters are measured and analyzed between mitigated sites and natural wetlands. Recommendations and possible guidelines are presented.

Teal, J.M. 1986. The Ecology of Regularly Flooded Salt Marshes of New England: A Community Profile. U.S. Fish Wildl. Serv. Biol. Rep. 85(7.4).

Another community profile report that summarizes and synthesizes information on the ecology of intertidal, regularly flooded Spartina alterniflora marshes of New England. The research at the Great Sippewissett Salt Marsh in Falmouth, Massachusetts is focus of the report.

Tiner, R.W., Jr. 1985a. Wetlands of New Jersey. U.S. Fish and Wildlife Service, National Wetlands Inventory, Newton Corner, Massachusetts.

A detail summary of the wetlands of New Jersey. The reasons and methods of wetland inventory and classification system are discussed detailed. Other chapters discuss in detailed the formation and hydrology, soils, plant communities, values, trends and wetland protection in New Jersey. Extensive reference sections are at the end of each chapter. An important source of information concerning the wetlands of New Jersey.

Tiner, R.W., Jr. 1985b. Wetlands of Delaware. U.S. Fish and Wildlife Service, National Wetlands Inventory, Newton Corner, Massachusetts and Delaware Department of Natural Resources and Environmental Control, Wetlands Section, Dover, Delaware. Cooperative Publication.

A detail summary of the literature of wetlands of Delaware as compared to available information. The text is similar to the New Jersey publication in format and information. An important source of information concerning Delaware wetlands.

Tiner, R.W. 1987. A Field Guide to Coastal Wetland plants of the Northeastern United States. University of Massachusetts Press, Amherst, Massachusetts.

The publication is a field guide to northeastern coastal wetlands vegetation, designed for nonspecialists. More than 150 plants are fully described and illustrated with line drawing, and over 130 additional plants are referenced as similar species with distinguishing characteristics. An overview of wetland ecology in the northeast is provided along with a series of maps that identifying major wetland systems in the states. An appendix lists state and federal agencies and private environmental groups that deal with wetland protection.

U.S. Army Corps of Engineers. 1986. Beneficial Uses of Dredged Material. Engineer Manual No. 1110-2-5026. Office, Chief of Engineers, Washington, DC.

A comprehensive report summarizing the beneficial uses of dredged material in the United States. The report contains chapters on habitat development, wetland habitats, island habitats, aquatic habitats, beaches and beach nourishment, and monitoring studies that are useful in coastal wetland restoration and creation aspects. A detailed list of recommended propagules and techniques for selected marsh species is presented in the appendix. The appendix also includes the notes on the general collection, handling, and planting techniques with additional remarks concerning habitat, water level and food value.

Wolf, R.B., L.C. Lee, and R.R. Sharitz. 1986. Wetland creation and restoration in the United States from 1970 to 1985: an annotated bibliography. Wetlands 6:1-88.

The bibliography provides information concerning the engineering, preparation and plant propagation.

APPENDIX II: PROJECT PROFILES

Wetland Type: Salt marsh, S. alterniflora.

Location: Atlantis Cove Rd., Lagoon Blvd., and Harbor Beach Blvd. (Half Moon Cove), Brigantine, Atlantic County, NJ.

Size: 0.8 acres.

Goals of Project:

Creation of salt marsh for destruction of 0.95 acres of State regulated wetlands.

Judgement of Success:

Shisler and Charette (1984a) in 1983 field inspection identified that vegetation growth has been successful on most of the site. Peat potted S. alterniflora plants were planted in June-July 1981 on 18-inch centers. Dredged material was placed on fabric cloth at an elevation (plateau) determined from adjacent wetland. Gobi bricks were used to anchor the cloth. Problems have developed with the use of the Gobi brick from wave energy, ice, and vandalism. There has been an accumulation of litter, killing the vegetation in sections. Increased diversity associated with the gobi bricks act as rocky substrate.

Significance:

The use of Gobi bricks and the creation of a fringe marsh on a plateau.

Reports: NJ Division of Coastal Resources Permit No.77-0034-2.

Shisler and Charette (1984a).

Wetland Type: High marsh, S. patens.

Location: Section 1: between Route 72 and Bay Ave., Manahawkin, NJ. Section 2: between Railroad Ave. and Bay Ave., Manahawkin, NJ.

Size: combined 1 acre.

Goals of Project:

The creation of a high marsh habitat. Sand was used as a backfill material and leveled to the elevation of the congruent marsh. Peat-potted S. patens transplants were used on a 2-ft. grid in 1977.

Judgement of Success:

Spartina alterniflora was identified as the major vegetation species in a 1983 survey of the site (Shisler and Charette 1984a). Problems associated with the site were relative elevations and poor sediment characteristics.

Significance:

Pipeline alterations are major disturbances associated with wetland habitats.

Reports: NJ Division of Coastal Resources Permit No. w74-10-073.

Shisler and Charette (1984a).

Wetland Type: Brackish marsh.

Location: Stratford Land Improvement Corporation (SLIC) Site, Stratford, CT.

Size: approximately 20 acres.

Goals of Project:

An after-the-fact application for Section 404 permit. Mitigation included the construction of a 1.5 acre fresh to brackish pond and dike alterations to increase tidal circulation for wetland enhancement.

Judgement of Success:

Metcalf & Eddy biologists visited the site in 1984, and concluded that the area was dominated by salt-stressed P. australis, and that the fresh to brackish pond was inappropriate mitigation for the habitat destroyed. Recommendations were to plant S. patens, remove P. australis debris, and increase the tidal circulation.

Significance:

A New England wetland project of enhancement and restoration that was completed as an after-the-fact permit.

Reports: U.S. Army Corps of Engineers Permit No. CT-ANSO-83-013.

Reimold and Cobler (1986).

Wetland Type: Salt to brackish marsh (high low).

Location: Taylors Point in Bourne, Barnstable County, Massachusetts.

Size: 1.8 acres.

Goals of Project:

The mitigation project would create 1/7 acre of high marsh and 1/5 acre of low marsh.

Judgement of Success:

The site was field inspected by Metcalf & Eddy biologists in 1984 and 1985. They documented that the site was basically devoid of vegetation except is some areas that supported isolated stands of S. alterniflora, S. patens, and D. spicata. Their conclusion was that the lack of success was the result of: (1) the planting of the marsh too early in the season when environmental conditions were still harsh; (2) the intertidal elevation at which the plants were planted; (3) the large year-round Canada goose population feeding on the root and rhizomes; (4) the occurrence of icing during the winter months; (5) the small size of the transplants and the possibility that they were not planted with sufficient root/rhizome material to assure complete anchoring of the plants at the time of transplant; and (6) the absence of a temporary protective offshore bar to decrease wave action.

Significance:

The use of stocked piled material and transplanting of wetland species.

Reports: U.S. Army Corps of Engineers Permit No. MA -POCA-81-384.

Reimold and Cobler (1986).

Wetland Type: Restoration of coastal wetland.

Location: Island Beach State Park, New Jersey.

Size: Two systems, 1 and 5 acres.

Goals of Project:

To create open water systems with a fringe marsh in a Phragmites australis dominated wetlands. The open water would serve as a reservoir for indigenous fish populations for the control of mosquito populations. Work was done in 1977 using an amphibious dragline.

Judgement of Success:

The area is routinely inspected by the Ocean County Mosquito Control Commission for mosquito breeding. No pesticides have been utilized for the control of mosquito populations since the project was completed in the 1977. Increases in waterfowl, vegetation diversity, and fish populations have been observed (Figures 7 & 8).

Significance:

The restoration of P. australis dominated wetland system in a sandy substrate.

Reports: Ocean County Mosquito Control Commission, Barnegat, New Jersey.

Candeletti, T.H. and F.H. Lesser. 1977.

McNeil, P. 1979.

Wetland Type: Impoundment of P. australis to tidal marsh.

Location: Fairfield, Connecticut.

Size: 10 hectares.

Goals of Project:

Restoration of a Phragmites australis impoundment into a S. alterniflora, S. patens, D. spicata marsh by the opening of the dike.

Judgement of Success:

In 1983 the impoundment vegetation composition of the impoundment was evaluated by researchers (Bongiorno et al 1984). They recorded that after 3 years of tidal inundation the marsh was quickly colonized by S. alterniflora, S. patens, D. spicata, and Salicornia spp.

Significance:

The documentation of a tidally restored wetland system in New England. Data demonstrates natural succession of vegetation species in an impoundment area.

Reports: Bongiorno et al. 1984.

Wetland Type: Salt hay impoundments to tidal wetlands.

Location: Delaware Bay, New Jersey.

Size: 6900 acres.

Goals of Project:

Tidal restoration of salt hay impoundments has three objectives: (1) control mosquito populations; (2) elimination of insecticides for mosquito control on the impoundments; and (3) enhancement of estuarine food chain organisms. It involves the state purchase of salt hay impoundments from willing sellers. Once purchased, the county mosquito commission removes the ditch plugs, and/or dikes subjecting the previously enclosed area to tidal inundation.

Judgement of Success:

A number of publications have addressed the success of the method in restoring previously impounded wetlands into salt marsh systems.

Significance:

A large scale program that has met its objectives with beneficial results. The program has been in operation since the late 1960's.

Reports: Slavin and Shisler. 1983.

Ferrigno et al. 1987.

Wetland Type: Salt marsh management.

Location: Coastal marshes.

Size: Various.

Goals of Project:

Open Marsh Water Management was a method developed on coastal marshes for the control of mosquito populations without harmful effects on the salt marshes-estuarine system. The method had three objectives: (1) control mosquito populations; (2) elimination of insecticides for mosquito control; and (3) enhancement of estuarine food chain organisms. The method included the duplication of those habitats of the salt marsh that did not produce mosquitoes.

Judgement of Success:

A number of publications have addressed the success of the method in various states along the east and west coast.

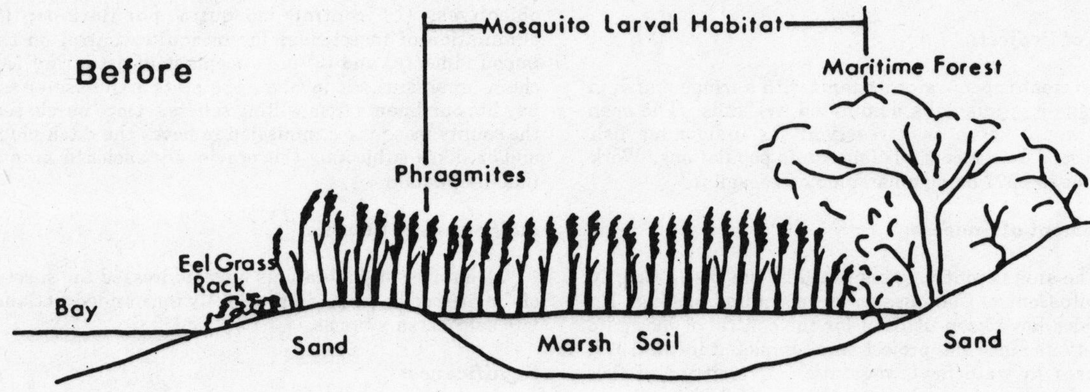

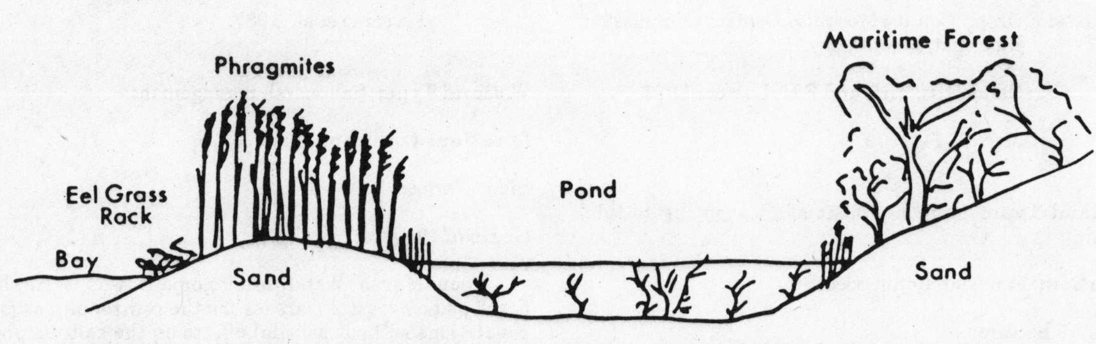

Figure 7. Cross section of a barrier island before and after habitat restoration of the Phragmites australis area on Island Beach State Park, New Jersey.

Figure 8. The changes in vegetation over five growing seasons on the two experimental areas on Island Beach State Park, New Jersey.

Significance:

A large-scale program that has met its objectives with beneficial results associated with coastal wetlands. The program has been in operation late 1960's.

Reports: Daiber. 1982, 1986.

Ferrigno and Jobbins. 1968.

Ferrigno et al. 1969.

Ferrigno. 1970.

Meredith and Saveikis. 1987.

Shisler and Jobbins. 1977.

Shisler and Ferrigno. 1987.

Wetland Type: Salt marsh, S. alterniflora.

Location: Adjacent to the New Jersey Turnpike, Hackensack, New Jersey.

Size: 63 acres.

Goals of Project:

The restoration a 63-acres monoculture Phragmites australis marsh into a Spartina alterniflora marsh. The site is located along the New Jersey Turnpike where the Cromakill Creek and Hackensack River meet. The plan was to restore the site with tidal creeks and flats.

Judgement of Success:

The site consists of three habitats: 10-15 percent is open water/mud flats; 10 percent dry berms; and 75 percent cordgrass meadows. There has been a documented increase in bird species (32 vs. 16) and numbers (1592 vs. 204). Benthic invertebrates tripled in numbers and doubled in species. No noticeable changes on mammalian, herptilian, and fish populations were observed.

Significance:

A large private restoration project with positive results.

Reports: Bontje 1988.

REGIONAL ANALYSIS OF THE CREATION AND RESTORATION OF SEAGRASS SYSTEMS

Mark S. Fonseca[1]
National Marine Fisheries Service
National Oceanic and Atmospheric Administration
Beaufort Laboratory

ABSTRACT. Seagrasses occur in most coastal, marine regions and are highly productive habitats. They are not a traditional wetland type but do meet the criteria for protection of aquatic habitat under Section 404 of the Clean Water Act. Including seagrass acreage would increase national wetland acreage approximately 17 percent. Seagrasses are described under six Ecoregions for management. Adequate water clarity for light transmission is required for restoration and survival of seagrass meadows.

Goals and performance guidelines for seagrass restoration and creation projects have historically been inappropriate. Consequently, seagrass restoration has never prevented a net loss in habitat. Suggested goals to prevent such losses include: development of persistent cover, generation of equivalent acreage or increased acreage, replacement with the same seagrass species, and restoration of secondary (faunal) production. These goals are to be differentiated from measures of density and percent survival. Monitoring for cover and persistence should continue for 3 years.

Site selection is a complex problem. The primary choice for restoration sites should be areas previously impacted or lost. The secondary choices should be perturbed aquatic areas irrespective of their previous plant community or uplands which can be excavated and converted to seagrass habitat. Population growth rate determines the species chosen for the restoration. Inclusion of specific conditions in the permit will enhance the probability of project success.

Research needs include: defining functional restoration, compiling population growth and coverage rates by Ecoregion, examining the resource role of mixed species plantings, determining the impact of substituting pioneer for climax species on faunal composition and abundance, evaluating the substitution of other species (e.g., mangroves, salt marshes) on cumulative damage to habitat resources when suitable sites cannot be found for seagrass planting, developing culture techniques for propagule development, exploring transplant optimization techniques such as the use of fertilizers, and delineating seagrass habitat boundaries. Most important would be the implementation of a consistent policy on seagrass restoration and management among resource agencies wherein restoration technique, monitoring, and performance and compliance guidelines would be standardized.

OVERVIEW OF REGION OR WETLAND TYPE DISCUSSED

This chapter will discuss a single habitat type--seagrass. Seagrass is a submerged meadow composed of one or more seagrass species. It occurs in most coastal, marine regions of the United States. Seagrasses have been shown to be highly productive and serve as important spawning, nursery, feeding, and refuge habitat for numerous estuarine and marine fauna (Thayer et al. 1984, Phillips 1984, Zieman 1982, Kenworthy et al. 1988). Many seagrass bed fauna are economically valuable.

Seagrasses have not traditionally been considered a wetland type, but rather have been defined as vegetated shallows. However, they meet most of the criteria applied to other wetland species such as saturated soil, and inundation, and are a vegetation type adapted for life in

[1]The views expressed in this chapter are the author's own and do not necessarily reflect the views or policies of the National Marine Fisheries Service or the National Oceanic and Atmospheric Administration.

saturated soil conditions.

Some wetland managers feel that seagrasses must be separated from traditional wetland species. The Environmental Protection Agency's "Guidelines for specification of disposal sites for dredged or fill material" (Federal Register 1980) describes the discrepancy between wetland definitions and recognizes that other Federal and State laws include vegetated shallows as wetlands. At the very least, seagrasses fall under the definition of shellfish beds and fishery areas established in Section 404(c) of the Clean Water Act. The technical definition of seagrass as a traditional wetland species is beyond the scope of this chapter. However, given their proximity and known biological linkage to other wetland systems (e.g., saltmarshes, mangroves), it is critical that seagrasses be considered in wetland management.

There are approximately 60 species of seagrass in the world. Of these at least 12 occur in the United States and adjacent waters (Table 1). Seagrasses are angiosperms; flowering plants which have evolved to a marine environment from a terrestrial existence. Similar to other flowering plants, seagrasses are characterized by vascularized leaf, root, and rhizome (tiller or runner) systems which set them apart from nonvascular algae or seaweeds (especially rhizophytic algae) for which they are often mistaken.

Some of the different growth forms of various seagrass species are shown in Figure 1. These plants differ from most other wetland plants in that they lead an almost exclusively subtidal existence (even down to 100 meters in clear, tropical waters), carry on both sexual and asexual reproduction in the water, reside in either brackish or, for the most part, marine salinities, and utilize the water column for support.

The subtidal existence of these plants sets them apart from other coastal vegetation in that they are completely dependent on water clarity for adequate light for photosynthesis and, thus, survival. Water clarity (low turbidity) is critical for the restoration and management of this ecosystem as has been seen in Chesapeake Bay (Orth and Moore 1981), San Francisco Bay (pers. obs.), and Tampa Bay (Lewis et al. 1985). Without adequate water clarity at a restoration site, or the preservation of water clarity over existing seagrass meadows, the entire concept of seagrass restoration and management becomes academic.

Seagrass ecosystems, intricately linked to the maintenance of living marine resources, have generally been disregarded in coastal management programs including coastal wetland inventories and protection efforts. However, recent but rather incomplete surveys have shown that there are at least 6.2 million acres of seagrass habitat in the southeast U.S. alone (Continental Shelf Associates, Inc. and Martel Laboratories, Inc. 1985, J.C. Zieman, pers. comm., R.L. Ferguson, pers. comm.). This acreage is not included in current estimates of wetland abundance by any agency despite the classification of seagrass as a wetland under the Cowardin system (Cowardin et al. 1979). This exclusion ignores the role of seagrass as a critical habitat type (Thayer et al. 1984, Phillips 1984, and Zieman 1982). If this acreage were included, it would constitute an approximate 19 percent increase in the wetland acreage of the southeast U.S. Since a majority of the nation's coastal wetlands occur in the southeast, seagrass acreage constitutes an increase of nearly 17 percent in our total national wetland area (for the lower 48 states). A lack of aerial photography with sufficient water depth penetration to map seagrass beds is generally acknowledged as the historic reason for not including seagrasses in wetland inventories.

CHARACTERISTICS OF REGIONS

As stated earlier, this overview of seagrasses is of national scope, encompassing several ecologically defined regions, or Ecoregions. Common to all these Ecoregions is the subtidal or nominally intertidal nature of seagrasses. Different tidal amplitudes, predictable by latitude and varied by local basin geomorphologies, have helped generate characteristic seagrass distributions. Tidal amplitudes, taken together with differing temperature, salinity, light tolerance, and life histories of the various seagrass species may be described as forming six characteristic seagrass Ecoregions. The definition of these Ecoregions is based on seagrass population growth and abundance data presented in Phillips (1984), Thayer et al. (1984), Fonseca (1987), and Fonseca et al. (1987a, b, c, 1988). The Ecoregions are 1) the northeast coast, north of Chesapeake Bay; 2) southeastern Temperate coast, Chesapeake Bay and south through Georgia (although there is no seagrass reported in South Carolina and Georgia; the only two coastal marine States without seagrass); 3) portions of Florida and the Gulf coast north of 28° latitude; 4) portions of Florida and U.S. Caribbean territories south of 28° N latitude; 5) the west coast, including Alaska (until more is known about restoration of seagrass beds in Alaska, they are grouped with the west coast in general, it is likely they will warrant assignation of new ecoregional definitions as data are acquired); and 6) the Hawaiian islands and Pacific territories.

Table 1. List of seagrass species by family, genus and species, and common names (if given) that are found in the United States and adjacent waters.

Family and Species	Common Name
Hydrocharitaceae	
Enhalus acoroides Royle*	
Halophila decipiens Ostenfeld	paddle grass
Halophila engelmanni Ascherson	star grass
Halophila johnsonii Eiseman	Johnson's seagrass
Thalassia testudinum Konig	turtlegrass
Potamogetonaceae	
Halodule wrightii Ascherson	shoalgrass
Phyllospadix scouleri Hook	surf grass
Phyllospadix torreyi S. Watson	surf grass
Ruppia maritima L.	widgeongrass
Syringodium filiforme Kutz	manatee grass
Zostera japonica Aschers. et Graebner	
Zostera marina L.	eelgrass

* One verbal report from Hawaii.

These Ecoregions do not necessarily match the distribution of an individual seagrass species. Rather, the seasonal growth characteristics of a given species may be generalized within these regional definitions. This is a critical factor in predicting seagrass restoration performance, planting season, and thus, conservation of resource values.

A further subdivision of an Ecoregion may be made when we consider the settings in which seagrasses are found. These settings are important because they can define the vulnerability of a given seagrass bed to different anthropogenic impacts. To illustrate this point, consider that there are five basic settings in which seagrass occurs: intertidal, rocky intertidal, subtidal estuarine, subtidal coastal or near shore (<10 meters), and deepwater (>10 meters) (Table 2). Intertidal and rocky intertidal beds are particularly sensitive to oil fouling. Intertidal, subtidal estuarine, and subtidal coastal beds are all particularly sensitive to dredging impacts where rocky intertidal beds (Phyllospadix sp. in particular) are not. All of these beds are extremely sensitive to degradation in water quality, particularly increased turbidity. Therefore, many non-permit associated pollution sources such as storm water runoff may have substantial and lasting effects on seagrass productivity and coverage (Fonseca et al. 1987c).

WETLAND TYPES TO BE DISCUSSED

Geographical Range

It is important to distinguish between the geographical range of different seagrass species

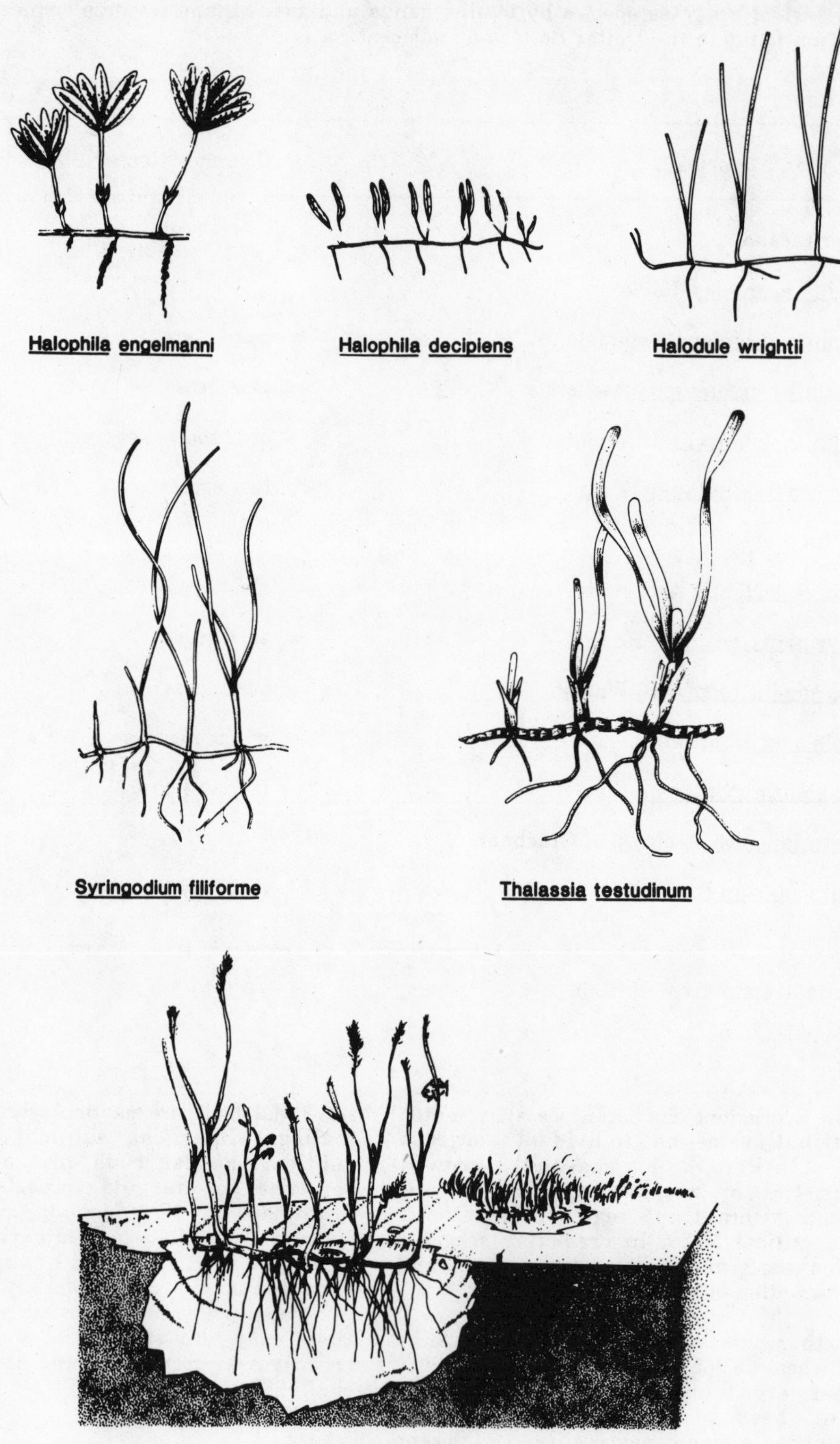

Figure 1. An illustration of the many growth forms that seagrasses of the United States exhibit. Redrawn from originals by R. Zieman and M. Fonseca. Courtesy of NMFS/NOAA, Beaufort, NC.

Table 2. Seagrass Ecoregions of the United States with known habitat settings within those regions.

ECOREGION	SETTINGS
1) Northeast coast, north of Chesapeake Bay	Intertidal subtidal estuarine coastal/nearshore (< 10 m)
2) Southeast temperate coast, Chesapeake Bay through Georgia	intertidal subtidal estuarine
3) Florida and Gulf Coast, north of 28° N latitude	subtidal estuarine coastal/nearshore deepwater (> 10 m)
4) Florida and Caribbean south of 28° N latitude	subtidal estuarine coastal/nearshore deepwater
5) West coast, including Alaska	intertidal rocky intertidal subtidal estuarine coastal/nearshore
6) Hawaiian Islands and Pacific jurisdiction	subtidal estuarine coastal/nearshore deepwater

and the Ecoregional definitions set forth above. Such a distinction is necessary where, for example, with eelgrass (Zostera marina) the geographical range extends from near the Arctic circle on both coasts of the U.S. south to North Carolina on the east coast and to the Gulf of California on the west coast. Across this extraordinary latitudinal range, this species exhibits both annual and perennial growth, with growth peaks either in the summer or in the fall and spring (Thayer et al. 1984, Phillips 1984). This variation in growth season and life history points out the need for different planting times, projected coverage rates, and thus very different performance and compliance criteria (Fonseca et al. 1982, 1984, 1985, 1987c, 1988).

This type of geographical versus Ecoregional distinction in planning and implementation of seagrass restorations has been utilized for the seagrasses of Florida (Fonseca et al. 1987c). In that paper the population growth and coverage rate of transplanted turtlegrass (Thalassia testudinum), manatee grass (Syringodium filiforme), and shoalgrass (Halodule wrightii) were compared between north and south Florida. In the case of shoalgrass, a comparison was made with North Carolina sites as well. Although the comparisons were made within the geographical range of each species, the conclusions were that south Florida seagrasses fell not only under a different model of growth, but could be planted year-round. Northern

Florida and North Carolina plantings of shoalgrass (the one species common to both States) were similar in their performance. Therefore, common generalizations as to the timing of their restoration were appropriate.

Another equally important factor regarding the geographic distribution of seagrass may be found where seagrasses exhibit different tolerance ranges for such environmental conditions as temperature and salinity across their geographic range. Eelgrass is suspected to have developed races (genetically defined adaptation) in response to those conditions (Backman 1984). The existence of these races in eelgrass has been used as the basis for selecting transplanting stock to match the conditions at a planting site (Backman 1984).

On a smaller scale, the geographical distribution of seagrasses among areas of different currents and wave exposures pose problems in site selection, performance and compliance evaluation (Fonseca and Fisher 1986, Fonseca et al. 1988). The synecology of seagrass beds in high current and/or wave areas is such that along channel banks and shoals, beds typically form in discrete, mounded patches, or "leopard skin" distribution (den Hartog 1971). In quiescent areas, seagrasses form a more continuous cover, resembling what one generally conceives of as a meadow. The exception to this is when there is insufficient unconsolidated sediment on top of underlying bedrock for the plants to root. In these instances, even though the area may be a quiet backwater, seagrasses will only be able to grow in depressions in the bedrock where sufficient sediments exist (a minimum of about 5 cm for some genus such as Halophila or +20 cm for Thalassia).

These unvegetated areas are typically but erroneously selected as areas to receive seagrass transplants. Areas that either have patchy beds or do not have any existing seagrass cover at all are also selected for transplanting. The problem of using such sites for transplanting was addressed by Fredette et al. (1985),

> "One of the first needs for a successful seagrass transplant is selection of a suitable site. To the nonspecialist, it may not be intuitively obvious that not all shallow-water sites barren of seagrass will suffice. On the other hand, people familiar with a need to have some type of proper environmental conditions might ask, 'If seagrass does not currently exist at the site, what makes you believe it can be successfully established?' This valid question demands an answer based on thorough investigations of the physical attributes of a site.
>
> If, after examination the site appears suitable, the question that still remains is 'Why does it not support seagrass?' One answer might be that the site historically had supported seagrass, but an environmental (e.g., hurricane, extreme temperatures) or human-induced (e.g., pollution, hydraulic changes) disturbance resulted in loss."

Therefore, if there are no adequate site history records, and environmental surveys of appropriate parameters and duration are lacking, there is no scientific justification that would recommend a site for planting.

On an autecological scale, there exists a further consideration for a planting operation regarding the life history of the different seagrass species within an Ecoregion. Not all seagrasses have the same growth habit. den Hartog (1970) described most of the known seagrasses in the world and their different forms, while Tomlinson (1974) described the relation between the asexual (branching) reproduction of seagrasses and their high per shoot productivity (see next section). The usefulness of this seemingly esoteric information to practical management problems is, in fact. profound. Although all seagrasses produce new leaf material from existing shoots at astonishing rates (Zieman and Wetzel 1980), not all seagrasses add new shoots (via vegetative reproduction) at an equally astonishing rate. In fact, many seagrasses do not add new shoots for weeks at a time, whereas other species of seagrass add new shoots daily. Those species that add new shoots rapidly have a concomitantly high bottom coverage rate, a factor of paramount importance to seagrass restoration. Those species that are slow in their addition of shoots are too slow in coverage of the bottom for effective habitat restoration on their own.

In much of the temperate and boreal United States, there is often only one or two seagrass species from which to choose for restoration operations (Ecoregions 1 and 5, Table 2). In other Ecoregions, there are often several species from which to choose. In order to quickly stabilize the sediment at a site, the fastest-covering species suitable to that Ecoregion should be chosen (Fonseca et al. 1987c). If the species that was lost was one of the slow-covering species, plantings of this seagrass may be interspersed among the faster-growing species to compress (in time) the successional process (e.g., planting Halodule interspersed with Thalassia) (Derrenbacker and Lewis 1982).

Slower-covering genera such as Thalassia,

Enhalus, Phyllospadix, and all species at the edge of their geographic range present a particularly difficult problem in restoration attempts. As development occurs in the coastal zone, the older, climax meadows are more often the ones damaged. Because they are slow-growing, those meadows represent decades of development. Even artificially-assisted restoration of these systems, such as transplanting, will take many years.

In summary, the concept of Ecoregions describing seagrass growth crosses the geographic ranges of the species themselves. Generalizations regarding planting unit performance and compliance with permit standards may be more accurate if made within the Ecoregion as opposed to the geographic ranges of the species themselves. Another basis for the Ecoregional assessment is the suspected existence of races of seagrass that are specifically adapted to certain environmental conditions. If these races could be identified, then selection of planting stock could be matched with the conditions at a planting site, potentially improving restoration performance. The potential for making such identifications should be enhanced by working within Ecoregions. Even applying the Ecoregion and race concepts requires operational consideration of synecological and autecological aspects of seagrass responses under transplantation. These factors control the degree of cover and the rate at which that cover may be achieved. All these factors together contribute to our ability to set realistic performance and compliance standards.

Key Functions Performed

Wood et al. (1969) described seven basic functional roles for tropical seagrasses. Thayer et al. (1984) added three additional functions for temperate systems based on research findings between 1969 and 1984. The functions described in these two papers are universally applicable to all seagrasses. In an abbreviated format, the functions are:

1) a high rate of leaf growth,

2) the support of large numbers of epiphytic organisms (which are grazed extensively and may be of comparable biomass to the leaves themselves),

3) the leaf production of large quantities of organic material which decomposes in the meadow or is transported to adjacent systems. Since few organisms graze directly on the living seagrasses, and the detritus formed from leaves supports a complex food web,

4) shoots, by retarding or slowing currents, enhance sediment stability and increase the accumulation of organic and inorganic material,

5) roots, by binding sediments, reduce erosion and preserve sediment microflora,

6) plants and detritus production influence nutrient cycling between sediments and overlying waters,

7) decomposition of roots and rhizomes provides a significant and long-term source of nutrients for sediment microheterotrophs,

8) roots and leaves provide horizontal and vertical complexity which, coupled with abundant and varied food resources, leads to densities of fauna generally exceeding those in unvegetated habitats, and

9) movement of water and fauna transports living and dead organic matter (particulate and dissolved) out of seagrass systems to adjacent habitats.

Much of the knowledge of seagrass systems centers around their function as a primary producer. Less is known about how the different meadow formations (or the mosaic of different seagrass species) act to support the secondary (faunal) production of the system (Heck 1979). The role of restored seagrass meadows in providing similar resource value is, unfortunately, unknown.

A typical question that is posed by those charged with managing seagrass systems is "how much can we afford to lose?". Such a question often cannot be answered scientifically on a regional scale until it is too late to preserve the resource value of the system. In order to avoid such a result, efforts may be made to require replanting wherever seagrasses are destroyed. These seagrass beds are created under the assumption that the created meadow provides the same resource value as the natural meadow for which it was intended to compensate. This may be true. However, the questions of relative resource value and the rate of development of created vs. natural seagrass meadows are only now being addressed scientifically and the answers are not yet clear.

EXTENT TO WHICH CREATION/RESTORATION HAS OCCURRED

TYPICAL GOALS FOR PROJECTS

Goals for seagrass restoration vary widely, depending on the perspective of the project proponents. Permit applicants under Section 404 of the Clean Water Act often find themselves dealing with a subject about which they know little (seagrass restoration), and for which there is no clear guidance. This is largely due to the lack of consistent policy on the subject between, or indeed, even within local, State and Federal resource agencies.

There have been few quantitative goals established for individual seagrass restoration projects. Goals have been of a highly qualitative nature, such as "successful restoration of the seagrass community". Although one usually has a visual image of a "successful" seagrass restoration as a thick, lush meadow, success criteria typically fall wide of defining a seagrass meadow with functional habitat recovery. With this lack of functional guidance, goals provided by the private sector have, for the most part, been biologically unsound.

A measurement of percent survival, blade density, biomass, or percent survival are too frequently adopted as a single project goal (refer to Appendix II). Consequently, post-project data have usually been collected on percent survival, blade density, and percent success. However, this group of parameters do not provide useful, functional definitions of success. For example, there is typically a loss of plants and whole planting units immediately after planting. Sixty to 90 days after planting, the loss of plants and planting units should cease, indicating that the remaining plantings have become rooted. At this point, the plantings are also less prone to erosion. For slow-growing species, however, measurements of planting unit survival may do little to convey how the meadow is developing. For example, if turtlegrass plantings on one-meter centers have 100 percent survival, but take years to begin to coalesce. The high survival count will not reflect the extremely slow rate of habitat development.

Blade density too may increase after restoration simply by individual shoots doubling or tripling the number of standing leaves without any additional coverage of the bottom. Blade and shoot density, along with biomass estimates vary widely within and between natural beds, both spatially and temporally. Blade density and biomass are products of local gradients of environmental conditions, a complex relationship that is not a readily predictable or controllable process, and therefore, a poor basis on which to define success. Also, the term "percent success" is a dimensionless parameter with "success" still undefined, making an extremely nebulous definition for habitat restoration. Use of such vague terminology invites non-compliance with permit conditions.

There are other less obvious problems with using these measures of planting success. One of the prime goals of restoring seagrass beds is to enhance and restore faunal diversity and abundance. Unfortunately, faunal development of created seagrass beds is only now being evaluated.

Goals for seagrass restorations or seagrass mitigation plans need rigorous definition. Goals may be generically defined and are applicable across Ecoregional boundaries. Suggested goals are:

1) development of <u>persistent</u> vegetative cover (cover being defined as the area where rhizomes overlap),

2) equivalent acreage of vegetative cover gained for cover acreage lost (where with naturally patchy distributions, such as in wave-exposed areas, more planting area may be required to accommodate the sum total of vegetative cover),

3) increase in acreage where possible,

4) eventual replacement of same seagrass species as were lost, and

5) development of faunal population structure and abundance in the new bed equivalent to natural, reference beds.

These goals may only be met through a careful consideration of several project design criteria. These are discussed below. As it turns out, creation and restoration of seagrass systems is not at all a trivial exercise.

SUCCESS IN ACHIEVING GOALS

With reference to the goals defined above (1-5), goal 1 has frequently been achieved. But major failures such as the attempted 200 acre plantings in Biscayne Bay have overshadowed the relatively small numbers of successes. Goals 2 and 3 have, to the author's knowledge, never been met. There has never been a seagrass restoration that has prevented a net loss of habitat (Fonseca et al. 1988). Goal 4 is generally only applicable to subtropical and tropical areas where there occur several sympatric species of seagrass. When climax species such as turtlegrass have been lost, there have been only rare instances of recovery. In most instances it was not possible to prove that transplanted seagrasses effected the recovery as opposed to natural colonization of the site. It should be noted that success in recovering climax species is so rare that out-of-kind replacement with another seagrass species is encouraged (Derrenbacker and Lewis 1982, Fonseca et al. 1987c). For this reason it has been recommended that impacts to these climax species should be avoided at all costs (Fonseca et al 1987c). There is no evidence that Goal 5 has ever occurred. Recovery of seagrass-associated fauna on a per unit area of bed may be occurring (Homziak et al. 1982, McLaughlin et al 1983, Fonseca 1987). However, since there has been virtually no net recovery of seagrass acreage on local or regional bases, it is highly unlikely that system-level recovery of faunal structure, abundance, and function has occurred to any degree.

REASONS FOR SUCCESS/FAILURES

Where successes have occurred (defined by meeting the above-stated goals), they have been the result of appropriate design criteria. Conversely, failures have been masked by improper success criteria, site selection, technique, and monitoring. As a result, mitigation of seagrass losses is sometimes viewed in the same manner as salt marsh creation, which is technically less difficult and has a good track record of success.

On a larger scale, the major reason for failure may stem from the lack of consistent terminology (and in some instances, policy) by resource agencies on the subject of seagrass restoration. Without enforcement of ecologically sound and biologically relevant success criteria, seagrass restoration will not become a predictable management tool. Enforcement of these success criteria have been achieved through strictly written permit conditions (Thayer et al. 1985).

DESIGN OF CREATION/RESTORATION PROJECTS

PRECONSTRUCTION CONSIDERATIONS

Location of Project

If the area has returned to conditions that will support seagrasses, planting sites should be located on the impacted area. Examples of such sites include backfilled access canals, pipelines, and power line crossings. If a site has been altered to the point where it will no longer support seagrass, then the restoration site should be located in the same water body, and as near to the impacted site as possible. If such a site cannot be found, then anthropogenically impacted areas in communicating water bodies should be selected. This selection should be made with the realization that compensation has not been provided for local system losses.

Site Characteristics

A basic requirement in selecting a restoration site is knowing its environmental history. This addresses the point raised earlier in a quote from Fredette et al. (1985): "If seagrass does not currently exist at the (chosen) site, what makes you believe it can be successfully established?". If this information is not immediately available, it should be obtained or determined that it does, indeed, not exist. The only alternative to this approach that may be reliably employed is a commitment to a scientifically valid environmental monitoring project to evaluate temperature, salinity, currents, sediments, and especially light penetration to the bottom. Since this is not a technical guide to seagrass meadow creation, description of environmental monitoring is beyond the scope of discussion. However, when environmental monitoring is required, the accurate measure of photosynthetically active radiation reaching the bottom in both time and space over a planting site is pivotal in evaluating site suitability.Very small restorations (<1000 m^2) do not require pre-planting environmental monitoring since it costs less to plant them than to monitor environmental conditions.

Sites should not be selected among existing, naturally patchy seagrass meadows (Fonseca et al. 1988). Some seagrass beds exist naturally in this configuration due to the existence of an annual population, their colonization of sediment--filled pockets in surface bedrock, a hydrodynamically active setting, or some combination thereof. Although a seagrass may be transplanted in some of these areas and may temporarily proliferate, the environmental factors creating the patchiness will shortly revert the area to pre-planting levels of patchiness. Thus, only a temporary pulse of productivity would be achieved and no persistent increase in seagrass acreage would be generated.

An important point is that to maintain a

given level of seagrass coverage in say, wave-swept areas, there must be a given level of unvegetated area as well. The seagrass cover and its associated inter-patch spaces must be regarded in toto as seagrass habitat; one cannot be maintained without the other. This eventually leads to the question "at what level of patchiness does one perceive a seagrass habitat?". Empirical observations by the author of aerial photographs at a scale of 1:24,000 suggests that when the ratio of average patch diameter to interspace distance exceeds 50:1, seagrass habitat continuity is difficult to discern. At this point, patches could be dealt with as individual units, rather than as parts of a larger seagrass habitat. The 50:1 ratio will change with the scale of observation and the factors controlling habitat boundary recognition deserve further study.

Apart from between-patch spaces, sites should also not be selected on broader, naturally-occurring unvegetated areas. Such a selection would displace another habitat, and it is likely that a natural balance of vegetation and non-vegetated areas is required for maintenance of specific marine resource values. Recent research findings suggest that unvegetated areas are zones of significantly higher predation, which mobilizes energy up the food chain (McIvor 1987, Rozas 1987). Under this scenario, vegetated areas act as a reservoir of prey items that are less energetically costly for predators to obtain in unvegetated areas. Thus, transplanting into these areas in an attempt to reduce patchiness, or altering seagrass species composition is hypothesized to negatively affect trophic transfer of energy for some fauna. Without clear evidence as guidance, modifications of bed patchiness is essentially a form of unisectoral management, likely benefiting some species to the detriment of others. Such an approach should be conducted only as experimental research, not a management alternative.

The following choices for restoration sites are given in order of preference:

1) restore areas previously impacted by poor water quality that once had seagrass, and the water quality has improved,

2) convert filled or dredged areas that were once seagrass meadows back to original elevation and transplant onto them (some subtropical areas, such as the Florida Keys, may colonize on their own with pioneer seagrass species, e.g., shoalgrass),

3) convert filled or dredged areas, irrespective of their previous plant community, to a suitable elevation for seagrass,

4) convert uplands to seagrass habitat.

The first two options are ecologically sound since they will act to restore the lost balance of plant communities in an area. The third option may not always perform that function and along with the fourth option creates conflicts with terrestrial habitat values. However, in some instances those uplands may be already zoned for development and will be lost as a natural resource, or, in many cases, what may appear to be original upland habitat is actually an older filled area. This is particularly true in urbanized estuaries.

There are cases where a "public interest", or "water dependent" project is approved and none of the above choices are available. As matters stand, the site selection conflict is emerging as the single most difficult aspect of seagrass mitigation projects. The only portion of the permit process that has changed, however, is that as permits for seagrass-conflicting projects are issued, resource agencies are realizing they are permitting another net loss of habitat. This issue of site selection is the most significant "red flag" that 404 administrators will encounter in seagrass mitigation projects.

CRITICAL ASPECTS OF THE PROJECT PLAN

Timing of Construction

Timing of restoration is key to determining the timing of any site construction or modification. This is where the Ecoregions concept discussed above is particularly useful. Timing is determined by the life history of the seagrasses in the area, which is, in turn, the basis for the Ecoregional definitions in Table 2. There are, however, several factors related to the life history of seagrasses that place constraints on a straightforward determination of planting dates. One factor is that after some seagrasses flower, the reproductive shoot dies. If a restoration operation uses many flowering shoots, then many shoots will die before giving rise to new, vegetative shoots. Seeds that may be set by transplanting the flowering shoots are an extremely tenuous means of revegetating an area because of low germination rates and the variability of seed set and retention. Unless otherwise demonstrated through field surveys, natural seeding or vegetative fragment recruitment should not be counted on to provide significant coverage. Planting should be done after the flowering shoots can be identified, or one should over-plant by increasing the number of shoots per planting unit by roughly 30 percent.

Knowledge of the flowering factor may be a valuable asset in planning a seagrass restoration, though such data are not always available. The existence of annual forms (plants growing from seed and flowering the same year, as opposed to the typically perennial form) within an Ecoregion may preclude seagrass restoration altogether. If a seagrass shoot has a one-year life history, then there will be little vegetative cover within a season, but much seed setting. Since no proven methods exist at this time to restore most seagrasses from seed, restoration or creation of recurring, annual populations remains technically unfeasible. Research on Z. marina in Chesapeake Bay is continuing to resolve this problem for this species (R.J. Orth and K.A. Moore, pers. comm.)

Timing of construction, which is based on the actual timing of planting, must consider utilization of the planting site by fauna. For example, setting of herring eggs on eelgrass in the Pacific northwest or shorebird nesting on dredge material islands are natural phenomena that may preclude the disruptive activity of site engineering, donor site harvest, or even human presence.

Timing of construction must consider the lead time needed to evaluate, create, and stabilize a site prior to recommended transplanting times. Construction timing may also be affected by the availability of planting stock, especially in the case of salvage operations (a case where destruction of a seagrass bed is permitted, such as bridge construction, and the seagrass may be salvaged and used as planting stock at a new site). These times vary on a case by case basis, but, construction timing should always consider the planting times given below. In Ecoregion 1 (northeast U.S.), planting should be performed in the spring. In Ecoregion 2 (southeast temperate coast), planting of shoalgrass should be done in the spring, while eelgrass should be planted in the fall. In Ecoregion 3 (north Florida and U.S. Gulf of Mexico), planting should be done in the spring. In Ecoregion 4 (south Florida and Caribbean) planting may be done any time of the year, although spring plantings may provide slightly faster coverage. In Ecoregion 5 (west coast) planting should be done in the spring, while in Ecoregion 6 (Hawaii and Pacific) there are no data on transplants to the author's knowledge, but a spring planting is the educated guess.

Construction Considerations

Very few sites have been constructed specifically for seagrass restoration. Most sites where construction (or some kind of terrain alteration) has taken place are the result of illegal fill removal or refilling dredged areas. In these cases, ownership of the planting site has not been an issue. In selecting off-site planting areas, bottom ownership has sometimes been a critical issue.

From the relatively few sites that have been created for seagrass planting, adequate depth of unconsolidated sediment, and availability of chemically uncontaminated sediment has been of prime importance. Most seagrasses require 5-20 cm of sediment depth. The exceptions are the surf grasses which grow in crevices among rocks as well as in unconsolidated sediment.

If a site is specifically constructed for seagrasses, then considerations of wave energy, tidal currents, and flushing effects on temperature and salinity have to be considered. Since light is of utmost importance in seagrass growth and survival, a site should be constructed to minimize sediment resuspension (i.e., reduce wave energy) while maintaining appropriate temperature regimes. Environmental tolerance ranges for the various species are described in Phillips (1984), Thayer et al. (1984), and Zieman (1982). Contouring of a site should provide the natural, ambient tidal range elevations for the given seagrass species. Slopes should be gentle and the site devoid of angular unconformities that would refract waves and focus wave energy on the planting site.

One of the most commonly asked questions in planning a seagrass site, whether for construction of a new site or selection of an existing one, is "how deep should it be?". This relates to the above mentioned environmental tolerance ranges. Even if a site were constructed or selected that had perfect temperature, salinity, wave energy, sediment depth, etc., without sufficient light, the seagrass plantings would die.

Seagrass survival does not depend solely on maximum light intensity. Rather, it has been demonstrated that seagrass survival depends on the amount of time in a day that light levels are above a critical intensity, or the photosaturation period (Dennison 1987, Dennison and Alberte 1985, 1986). In other words, seagrasses can only use so much light per unit time, making it necessary for them to have sufficient light over a longer period of time (e.g., Z. marina: 12 hours at 200 microEinsteins/m^2/sec) to balance their metabolism. It is not easy to obtain a reliable measure of this condition since it requires special equipment and relatively long-term monitoring to discern the light characteristics of a given site. Such a determination is difficult if the site has not yet been constructed and nearby areas must be monitored while plans are being made.

Fortunately, because of the low cost of the instrument, Secchi depth measurements may be correlated with the lower limit of seagrass growth (Kenworthy, pers. comm.). Secchi measurements require extensive field monitoring to average out conditions at a given site. In lieu of these data, plans should provide depths equivalent to adjacent seagrass beds. If no beds are adjacent to the site, then a statistically valid monitoring of the light conditions is prudent.

Hydrology

As mentioned above, the wave and current regime of a site influences the quantity of light reaching the plantings through sediment resuspension. A more immediate effect of these forces is to move the substrate, buffeting and sometimes eroding the planting units from the bottom (Fonseca et al. 1985). Erosion of 1-2 mm/day can cause a 50 percent loss in planting units during the first 60 days after planting (Fonseca et al. 1985) whereas burial of more than two-thirds of the photosynthetic tissue appears to be fatal (author, pers. obs.). Site hydrology also affects the form in which a meadow develops (Fonseca et al. 1983). In high current areas, patchy beds develop. While the patches and between-patch area may be correctly termed a seagrass habitat, performance of planting is measured by the actual area of seagrass cover (the area where the long shoots overlap). Therefore in high current or wave areas, a greater area will need to be planted to obtain the desired total area of bottom coverage. In quiescent areas, a relatively continuous meadow will often be attained, and planning of planting acreage is straightforward.

Prolonged freshwater inflow and lowering of salinities in seagrass beds is often fatal to the plants. Widgeongrass and shoalgrass are two euryhaline exceptions to this rule.

Substrate

Sediments within the sand size range and smaller have not been shown to limit seagrass growth per se. Other co-varying factors such as nutrient supply, currents, and in the case of fine-grained sediments, resuspension and light reduction, have been related to seagrass growth. Transplanting bare-root sprigs (Derrenbacker and Lewis 1982, Fonseca et al. 1982, 1984) has been criticized as being of limited use in highly organic sediments (Alberte, pers. comm.) although successful restorations have been performed in high organic sediments in North Carolina (Kenworthy et al. 1980). Concomitantly, low light conditions which do not allow the plant to generate and transport oxygen to the roots to fight sulfide intrusion may explain the poor performance of some restorations placed in sediments with high organic content (M. Josselyn, pers. comm.).

Fertilization of plantings may be useful in areas with little or no organic matter in the sediment. Depending on the sediment type, phosphorous (in carbonate sediments) or nitrogen (in siliceous sediments) may be limiting. This is an area of intense debate in seagrass research (Bulthuis and Woelkering 1981, Short 1983 a,b, Short and McRoy 1984, Pulich 1985, Williams 1987). Some studies have been completed specifically on fertilizer effects on seagrass transplants (Orth and Moore 1982 a,b, Fonseca et al., 1987b) but the results are equivocal, based on inconsistent performance and assessment of the fertilizers used. At present, fertilization is considered to be acceptable in low (<2 percent) organic sediments, but no reduction in planting intensity is recommended (Fonseca et al. 1987b).

Revegetation

There is a large body of information on revegetation and transplanting projects in general. The reader is referred to Appendix I, particularly the review by Fonseca et al. (1988) for specifics regarding the subject. By way of a general summary, seagrass revegetation projects are conducted largely with wild-harvested, vegetative shoots. A timely paper by Lewis (1987) describes the hazards of damaging existing natural beds and the options available in the State of Florida for alternative planting stock. Previous work by Riner (1976) and more recently by Roberts et al. (1984) have pioneered research into the use of seeds for temperate species. Turtlegrass has been easily transplanted by seed for many years (Thorhaug 1974). Some work is being conducted on cultivation of turtlegrass seeds, but this process still is only a grow-out procedure, dependent on the harvest of wild seed.

Seagrass transplants are typically fragile and must be handled with care. The plants are extremely susceptible to desiccation and must be kept soaked, or preferably, in ambient temperature water during the whole planting process. Large plants also are prone to breakage. The technique of transplanting is well-documented and involves using bare-root shoots stapled into the bottom, or plugs (see review, Appendix I). While the technology is well-developed, its application is not.

Of the 12 species that occur in the United States, only 7 have been reported as being transplanted (Halophila decipiens, T. testudinum, H. wrightii, Ruppia maritima, S. filiforme, Zostera japonica, and Z. marina). Of these, only T. testudinum, H. wrightii, S. filiforme, and Z. marina have enough quantitative planting data to consider their use in revegetation projects. These four species probably constitute the majority of seagrass cover in the United States, but the others have not enjoyed similar scrutiny and managers are not in a position to declare their relative importance. For example, it is now known that there are at least one million acres of Halophila off the west coast of Florida (Continental Shelf Associates, Inc. and Martel Laboratories 1985). This species has been transplanted on one occasion at depths over 10 meters (author, unpubl. data).

Reintroduction of Fauna

Reintroduction of fauna is usually a passive process in seagrass revegetation projects. There have been some attempts to introduce scallops to restored Z. marina areas in Long Island, New York, but these have not been successful (Chris Smith, NY Sea Grant Extension Service, pers. comm.). In a wild, uncontained system, the relatively sedentary scallop is one of the few candidates for introduction to a created seagrass bed. There is no information known at this writing concerning the use of seagrass in contained systems for commercial animal production purposes.

There is little known about the rate at which created seagrass beds take on the faunal composition or abundance of their natural counterparts (Homziak et al. 1982, McLaughlin et al. 1983, Fonseca 1987). Studies are now underway that indicate that in monospecific, temperate seagrass communities, faunal composition and abundance comparable to natural beds may be reached in two years if the plantings persist and achieve cover similar to natural beds (Fonseca 1987).

Buffers, Protective Structures

Limiting impacts to revegetated areas is critical during the initial establishment period of seagrass restoration (first 60-90 days). Any upland sources of sediment should be contained as these may create persistent turbidity or actually bury the restoration. Offshore berms or sandbars have been employed to provide protection from waves, but these must be constructed so as to not impound water and raise in-meadow temperatures. Floating-tire wave breaks have promise in this regard. Artificial grass has been considered as a means of retarding water and sediment movement until plantings are established, but this method is untested. Previous models of plastic grasses have suffered from epiphytic fouling, causing them to fall to the sediment surface. Recent advances in epiphyte-sloughing forms of artificial grass for use in freshwater systems makes this concept worthy of further investigation.

Long Term Management

With long term management the ownership question comes into play. Unless the planting site is owned by the permitting agency, vis-a-vis the Federal or State agency, it is conceivable that long-term use conflicts could arise that would compromise the intent of the revegetation project and permit issuance. Because permits have a relatively short life span in comparison to the life span of a seagrass meadow, these conflicts have not yet emerged, although restored beds are susceptible to destruction.

A particularly vexing destructive process which can easily negate long-term management plans is propeller scarring. For the most part, this is the result of misjudgment of water depth and accidental grounding of small craft. However, with the advent of hydraulic trim controls on outboards, small craft frequently chance short cuts over shallow areas, many of which contain seagrass. The aboveground foliage can be removed by the propeller and in severe cases, cutting of the rhizome occurs. Zieman (1976) has demonstrated the persistence of these impacts for years in some seagrass beds, and large areas of the seagrass beds in the Florida Keys have been eliminated by prop scarring (C. Kruer, pers. comm.). Kruer has suggested that improved channel marking would eliminate much of this destruction. Without such aids to navigation around beds or restoration sites, long-term management for many of the shallower seagrass beds will be difficult.

Maintenance of seagrasses adjacent to dredged material islands is another problem. Maintenance dredging operations routinely require fresh deposition of material on existing disposal sites. If seagrass restorations have been established adjacent to these sites, careful engineering is required to prevent their subsequent destruction. Innovative placement of material can assure the persistence of the planted areas although recent thin-layer application are not promising. The concept of preventing subsequent impacts to created seagrass beds is new, and while of merit, has not been explored at a management level.

MONITORING

What to Monitor

A major shortcoming in seagrass planting as a management tool has been poor monitoring or the lack of monitoring. Without monitoring, there can be no objective assessment of restoration performance and permit compliance (assuming that the permit had appropriate conditions in the first place).

Monitoring specifications have been proposed in at least two publications (Fonseca et al. 1987c, 1988). These publications point out that no one data type can stand alone in a monitoring program.

Several factors must be considered to lead to

an ecologically valid characterization of seagrass restoration success. The number of planting units that survive should be recorded. This may be expressed as a percentage of the original number, but the actual whole number is critical as well. A random (as opposed to arbitrary) sample of the average number of shoots and area covered (m^2) per planting unit should be recorded until coalescence (the point where individual planting units grow together and the planting unit origin of individual shoots cannot be readily observed). The number of surviving planting units may then be multiplied by the average area per planting unit to determine the area covered on the planting site. The data from pre-coalescence surveys may be compared with existing data to assess performance relative to other, local plantings by plotting the average number of shoots per planting unit (not area) over time. The comparison may be statistical or visual (which often suffices to detect grossly different population growth rates).

Shoot addition is recommended over area addition as a measure of transplant performance (contrasted with compliance). For example, in high current areas shoots grow more densely, and measuring shoot addition is a more accurate means of assessing the asexual reproductive vigor of the plantings. After coalescence, the area of bottom covered should be surveyed using randomized grid samples (Fonseca et al. 1985). These data may be collected over time to assess persistence of the planting as well as total seagrass coverage, both of which are the measurement of compliance.

Population growth and coverage data do not exist for all seagrass species in all Ecoregions at this time. Ecoregions 1, 5, and 6 are particularly lacking in these data. If collection of these data could be instituted, then regional offices would quickly develop the capability to objectively and efficiently deal with seagrass mitigation projects.

How To Do It

It is often desirable to secure funding for the planting from those responsible ahead of permit issuance. This is because experience has shown that after a permit is issued and a project completed, it is difficult, short of legal action, to get the planting done. Performance bonds or letters of credit from applicants to contractors have been used with success in this regard. In any event, points of finance, as well as the technical language of restoration technique, site selection and monitoring are critical elements of a 404 permit.

Random samples should be collected on survival, number of shoots, and area covered per planting unit. If a planting site is sufficiently small, all planting units should be surveyed for presence or absence (survival survey). The existence of a single short shoot on a planting unit indicates survival. If a site is large, then randomly selected rows or subsections (area in m^2) should be sampled. Because each row or subsection is actually the level of replication, at least 10 replicates should be performed at the level at which one wishes to generalize one's findings (e.g., over the whole planting site). At the very least, stabilization of the running mean of survival or shoots and area per planting unit should be obtained as a measure of statistical adequacy.

Presence or absence, and number of short shoots per planting unit are straightforward measures, although they usually require snorkeling or SCUBA diving to assess (a factor that is surprisingly not considered, or equipped for, by many attempting these data collections). The area covered by a planting unit may be measured by recording the average of two perpendicular width measurements (in meters) of the planting unit over the bottom. These numbers are averaged, divided by 2, squared, and multiplied by pi to compute the area of a circle (pi r^2), and in this case, the planting unit. This procedure tends to give a higher value than use of a quadrat, criss-crossed with string on 5 cm centers, that is laid over the planting unit. In this case, the number of 5 X 5 cm grids (or half grids if there are only 1 or 2 shoots in the 5 X 5 cm grid) that have seagrass shoots are totaled and converted to square meters of cover for the planting unit. The quadrat method is more appropriate for seagrasses that propagate in long runners (e.g., shoalgrass), and do not form a clear radial growth pattern (e.g., eelgrass). An individual can be trained to perform these counts in a few hours, and can count individual planting units in 5-10 minutes or less at early stages of a restoration's development.

How Long To Do It

Monitoring of shoot numbers and area covered per planting unit should proceed quarterly for the first year after planting and biannually thereafter for two more years (a total of three years). After planting units begin to coalesce and the planting unit from which shoots originated can no longer be discerned, areal coverage data should be recorded and counts on a planting unit basis suspended.

How To Interpret The Results

The population growth and coverage data may be compared periodically with published values (dependent on species and Ecoregion) as a relative indicator of performance. More important, the computations described above allow a direct comparison on a unit area basis of planted versus lost acreage (average area/planting unit times number of surviving planting units). Success may then be based on whether the appropriate ratio of coverage (e.g., 1:1, or 2:1) has been generated; a quantitative measure commensurate with ecological function. If the restoration project is for mitigation, then compliance may thereby be interpreted as both acreage generated and the unassisted persistence of that acreage over time (the three year period). The persistence issue is also critical. If the planting does not persist, then resource values have experienced a net loss and the project has not been effective.

Mid-course Corrections - What Can Go Awry?

In a seagrass restoration, just about anything can go wrong. Typical problems are, in order of frequency: natural physical disruption (i.e., storms and associated waves and sediment movement), biogenic impacts (smothering by macroalgae, decapod excavations, and grazing, e.g., pinfish), and anthropogenic impacts (overzealous clammers and errant boat operators, i.e., motorboat prop scarring). These impacts are part of the risk in restoring seagrass systems.

A common problem is impact of a project upon adjacent beds. For example, a permit may be issued for laying of an underground pipeline with replanting of seagrass in the backfilled area over the pipe. But, during the operation, maneuvering of barges and other activities, cause erosion and burial of adjacent seagrass habitat not specifically identified in the permit. The permit should contain language identifying such potential impacts and develop contingency plans for the mitigation of adjacent habitat loss.

If losses are detected early enough in the planting season, additional planting units may be added as a mid-course correction. If there are fewer than 90 days left before the first major seasonal decline of local, natural grasses, replanting should be postponed until the next year. If replanting is performed, then the monitoring clock should be reset to zero. Otherwise, a site could experience chronic planting failures without any impetus to change procedures.

Another mid-course correction may be needed upon examination of the population growth and coverage data. If the population growth is within expected limits for the species and Ecoregion, then one may be assured that the observed coverage rate is the best that can be expected for that site. If the coverage rate is lower than expected while the shoot rate is as predicted, then the projected timetable for grow-out should be lengthened appropriately. Although this would not necessarily change the permit conditions or length of commitment of the applicant, it allows an objective evaluation that is fair to all parties. In other words, the restoration is performing as well as can be expected and the anticipated coverage should be reached, albeit at a later date. Low shoot generation and coverage rates, as well as large losses of whole planting units indicate that a restoration is in trouble. Timetables may again be altered, but replanting may be warranted or a new site sought if the rate of shoot addition to planting units was not significantly different from zero after the first year of growth.

Chronic failure of plantings need to be carefully examined. It is important to distinguish between acts of nature and acts of incompetence or non-compliance. For example, if a permit applicant does everything asked of him by resource agencies and still cannot, after three years, come up with the acreage required, is the applicant required to finance planting in perpetuity? There needs to be some measure of agency responsibility in the site selection and approval process that prevents this situation. At what point has the applicant fulfilled the permit requirements? In reality, seagrass restoration has been and will continue to be a risky management option. This point becomes more profound if one considers that a terrestrial crop cannot be guaranteed, despite millennia of collective practice. The agencies involved should proceed with a clear realization that they are taking a calculated risk in their ability to prevent a net loss in habitat. If chronic failure is due to chronic non-compliance with established (in the permit) procedures, then existing, formal methods of ensuring compliance should be instituted.

INFORMATION GAPS AND RESEARCH NEEDS: LIMITATIONS OF KNOWLEDGE

Throughout the text of this chapter, limitations in our knowledge of seagrass systems and their restoration has been implicitly and explicitly stated. The information for management purposes fits closely with the basic research needs identified by investigators relating to basic autecological and synecological functions of seagrass systems. There at least are nine information gaps that require immediate attention for effective management of these systems:

1) A definition and evaluation of "functional restoration" of a seagrass bed must be made. Is it just a floristic survey? Will faunal abundance follow? Will ecologically and economically valuable functions be realized, and if so, how long will it take?

2) Population growth and coverage rates should be compiled for seagrasses in all regions so that Ecoregion boundaries may be better defined. Parent data sets on transplant performance should be centrally compiled on an Ecoregional basis as a measure of performance. Areas lacking in these data should have experimental plots initiated and monitored. Species lacking these data are Enhalus acoroides, all Halophila species, eelgrass on the west and northeast coasts, all seagrasses in the northern Gulf of Mexico (S. filiforme, T. testudinum, H. wrightii, Ruppia maritima, and all Halophila species), Z. japonica, and all Phyllospadix species.

3) The resource role of mixed species plantings should be evaluated.

4) The impact of substituting pioneer for climax species in "compressed successional" transplanting (Derrenbacker and Lewis 1982) on faunal composition and abundance should be investigated for resource maintenance.

5) The substitution of other species (e.g., mangroves, salt marshes) when suitable sites cannot be found for seagrass planting should be evaluated for their potential cumulative damage to habitat resources.

6) Culture techniques for propagule development (seed and tissue culture) should be refined, bypassing the need to damage donor beds when salvage is not available.

7) Transplant optimization techniques should be explored, especially the use of fertilizers.

8) A consistent definition of seagrass habitat boundaries should be developed.

9) The development of a consistent national policy on seagrass management should be explored. Special consideration should be given to artificially-propagated seagrass meadows. Site evaluation methodologies, especially for light availability, should be standardized and adopted by all resource agencies. The cumulative impact of small-scale, piecemeal loss of seagrass by deliberate (e.g., dredging) and accidental (prop-scarring) impacts should be considered in local and Ecoregional assessments for the protection and maintenance of this valuable habitat.

ACKNOWLEDGMENTS

I would like to thank the many anonymous reviewers for their insightful and helpful comments. Thanks are given also to Jon Kusler and Mary Kentula for both their guidance and extraordinary effort in seeing this document to completion. Thanks are extended to Jud Kenworthy and Gordon Thayer whose stimulating discussion, debate, and research over the years on the subjects of seagrass ecology and restoration provided much of the impetus and citable works on the subject. Sandy Wyllie-Echeverria researched and wrote the original draft of Project Profile 3. Special thanks are given to Carolyn Currin for providing conclusive, critical evaluation at many points during the writing and final editing of the manuscript.

LITERATURE CITED

Ashe, D.M. 1982. Fish and wildlife mitigation: description and analysis of estuarine applications. Coastal Zone Manage. J. 10:1-52.

Backman, T.W.H. 1984. Phenotypic expressions of Zostera marina L. ecotypes in Puget Sound, Washington. Ph.D. Dissertation, Univ. Washington, Seattle.

Bulthuis, D.A., and W.J. Woelkerling. 1981. Effects of in situ nitrogen and phosphorus enrichment of the sediments on the seagrass Heterozostera tasmanica in Western Port, Victoria, Australia: a decade of observations. Aq. Bot. 19:343-367.

Churchill, A.C., A.E. Cok, and M.I. Riner. 1978. Stabilization of subtidal sediments by the transplantation of the seagrass Zostera marina L. N.Y. Sea Grant Rep. NYSSGP-RS-78-15.

Clark, W. 1984. North Carolina's mitigation policy: a new challenge. Natl. Wetlands Newsletter Nov.-Dec.:13-16.

Continental Shelf Associates, Inc. and Martel Laboratories, Inc. 1985. Florida Big Bend Seagrass Habitat Study Narrative Report. A Final Report by Continental Shelf Associates, Inc. submitted to the Mineral Management Service, Metairie, LA. Contract No. 14-12-0001-30188.

Cowardin, L.M., V. Carter, F.C. Golet, and E.T. LaRoe. 1979. Classification of Wetland and Deepwater Habitats of the United States. U.S. Fish and Wildlife Service, Washington, D.C.

den Hartog, C. 1970. The Seagrasses of the World. North-Holland Publishing Co., Amsterdam.

den Hartog, C. 1971. The dynamic aspect in the ecology of sea-grass communities. Thalassia jugosl. 7:101-112.

Dennison, W.C. 1987. Effects of light on seagrasses: photosynthesis, growth and depth distribution. Aq. Bot. 27:15-26.

Dennison, W.C. and R.S. Alberte. 1985. Role of daily light period in the depth distribution of Zostera marina (eelgrass). Mar. Ecol. Prog. Ser. 25:51-61.

Dennison, W.C. and R.S. Alberte. 1986. Growth and photosynthesis of Zostera marina (eelgrass) transplants along a depth gradient. J. Exp. Mar. Biol. Ecol. 101:257-282.

Derrenbacker, J.A. and R.R. Lewis. 1982. Seagrass habitat restoration in Lake Surprise, Florida Keys, p. 132-154. In R.H. Stoval (Ed.), Proc. 9th Ann. Conf. on Wetlands Restoration and Creation. Hillsborough Comm. Coll., Tampa, Florida.

Federal Register. 1980. Guidelines for specification of disposal sites for dredged or fill material. 45(249):85336-85357.

Fonseca, M.S. 1987. Habitat development applications: use of seagrass transplanting for habitat development on dredged material, p. 145-150. In M.C. Landin and H.K. Smith (Eds.), Beneficial Uses of Dredged Material; Proc. of the 1st Inter-agency Workshop, 7-9 Oct., 1986, Pensacola, FL. USACE Tech. Rep. D-87-1.

Fonseca, M.S., W.J. Kenworthy, and G.W. Thayer. 1982. A Low Cost Planting Technique for Eelgrass (Zostera marina L.). Coastal Engineering Technical Aid No. 82-6. U.S. Army Engineer Coastal Engineering Research Center, Ft. Belvoir, Virginia.

Fonseca, M.S., J.C. Zieman, G.W. Thayer, and J.S. Fisher. 1983. The role of current velocity in structuring seagrass meadows. Estuarine Coastal Shelf Sci. 17:367-380.

Fonseca, M.S., W.J. Kenworthy, K.M. Cheap, C.A. Currin, and G.W. Thayer. 1984. A Low Cost Transplanting Technique for Shoalgrass (Halodule wrightii) and Manatee Grass (Syringodium filiforme). Instruction Report EL-84-1. U.S. Army Engineer Waterways Experiment Station, Vicksburg, Mississippi.

Fonseca, M.S., W.J. Kenworthy, G.W. Thayer, D.Y. Heller, and K.M. Cheap. 1985. Transplanting of the Seagrasses Zostera marina and Halodule wrightii for Sediment Stabilization and Habitat Development on the East Coast of the United States. Technical Report EL-85-9. U.S. Army Engineer Waterways Experiment Station, Vicksburg, Mississippi.

Fonseca, M.S., and J.S. Fisher. 1986. A comparison of canopy friction and sediment movement between four species of seagrass with reference to their ecology and distribution. Mar. Ecol. Prog. Ser. 29:15-22.

Fonseca, M.S., W.J. Kenworthy, and G.W. Thayer. 1987a. Transplanting of the Seagrasses Halodule wrightii, Syringodium filiforme, and Thalassia testudinum for Habitat Development in the Southeast Region of the United States. Technical Report EL-87-8. U.S. Army Engineer Waterways Experiment Station, Vicksburg, Mississippi.

Fonseca, M.S., W.J. Kenworthy, K.A. Rittmaster, and G.W. Thayer. 1987b. The Use of Fertilizer to Enhance Transplants of the Seagrasses Zostera marina and Halodule wrightii. Technical Report EL-87-12. U.S. Army Engineer Waterways Experiment Station, Vicksburg, Mississippi.

Fonseca, M.S., G.W. Thayer, and W.J. Kenworthy. 1987c. The use of ecological data in the implementation and management of seagrass restorations, p. 175-187. In M.D. Durako, R.C. Phillips, and R.R. Lewis (Eds.), Proc. Symp. on Subtropical-Tropical Seagrasses of the Southeastern United States. Fl. Mar. Publ. Ser. No. 42.

Fonseca, M.S., W.J. Kenworthy, and G.W. Thayer. 1988. Restoration and management of seagrass systems: a review, p. 353-368. In D.D. Hook, W.H. McKee, Jr., H.K. Smith, J. Gregory, V.G. Burrell, Jr., M.R. DeVoe, R.E. Sojka, S. Gilbert, R. Banks, L.H. Stolzy, C. Brooks, T.D. Matthews, and T.H. Shear (Eds.), The Ecology and Management of Wetlands. Vol. 2: Management, Use and Value of Wetlands. Timber Press, Portland, Oregon.

Fredette, T.J., M.S. Fonseca, W.J. Kenworthy, and G.W. Thayer. 1985. Seagrass Transplanting: 10 Years of Army Corps of Engineers Research, p. 121-134. In F.J. Webb (Ed.), Proc. of the 12th Ann. Conf. on Wetlands Restoration and Creation. Hillsborough Comm. Coll., Tampa, Florida.

Heck, K.L. 1979. Some determinants of the composition and abundance of motile macroinvertebrate species in tropical and temperate turtlegrass (Thalassia testudinum) meadows. J. Biogeography 6:183-200.

Homziak, J., M.S. Fonseca, and W.J. Kenworthy. 1982. Macrobenthic community structure in a transplanted eelgrass (Zostera marina) meadow. Mar. Ecol. Prog. Ser. 9:211-221.

Kenworthy, W.J., M.S. Fonseca, J. Homziak, and G.W. Thayer. 1980. Development of a transplanted seagrass (Zostera marina) meadow in Back Sound, Carteret County, North Carolina, p. 175-193. In D.P. Cole (Ed.), Proc. of the 7th Ann. Conf. on the Restoration and Creation of Wetlands. Hillsborough Comm. Coll., Tampa, Florida.

Kenworthy, W.J., G.W. Thayer, and M.S. Fonseca. 1988. The utilization of seagrass meadows by fishery organisms, p. 548-560. In D. D. Hook, W.H. McKee, Jr., H.K. Smith, J. Gregory, V.G. Burrell, Jr., M.R. DeVoe, R.E. Sojka, S. Gilbert, R. Banks, L.H. Stolzy, C. Brooks, T.D. Matthews, and T.H. Shear (Eds.), The Ecology and Management of Wetlands. Vol. 1: Ecology of Wetlands. Timber Press, Portland, Oregon.

Lewis, R.R. 1987. The restoration and creation of seagrass meadows in the southeast United States, p. 153-173. In M.J. Durako, R.C. Phillips, and R.R. Lewis (Eds.), Proc. Symp. on Subtropical-Tropical Seagrasses of the Southeastern United States. Fl. Mar. Res. Publ. Ser. No. 42.

Lewis, R.R., M.D. Durako, M.J. Moffler, and R.C. Phillips. 1985. Seagrass meadows of Tampa Bay: a review, p. 210-246. In S.F. Treat, J.L. Simon, R.R. Lewis, and R.L. Whitman (Eds.), Proc. Tampa Bay Area Scientific Information Symp. (May 1982). Burgess Publ. Co., Minneapolis, Minnesota.

Mangrove Systems, Inc. 1985. Combined Final Report, Florida Keys Restoration Project. Florida Department of Environmental Regulation, Tallahassee, Florida.

McIvor, C.C. 1987. Marsh fish community structure: roles of geomorphology and salinity. Ph.D. Dissertation. Univ. Virginia, Charlottesville, Virginia.

McLaughlin, P.A., S.A. Treat, A. Thorhaug, and R. Lemaitre. 1983. A restored seagrass (Thalassia) bed and its animal community. Environ. Cons. 10:247-254.

Orth, R.J. and K.A. Moore. 1981. Submerged aquatic vegetation of the Chesapeake Bay: past, present, and future. Trans. N. Am. Wildl. Nat. Resour. Conf. 46:271-283.

Orth, R.J. and K.A. Moore. 1982a. The biology and propagation of Zostera marina, eelgrass, in the Chesapeake Bay, Virginia. Va. Inst. Mar. Sci. Spec. Rep. Appl. Mar. Sci. Ocean Eng. 265.

Orth, R.J. and K.A. Moore. 1982b. The effect of fertilizers on transplanted eelgrass Zostera marina in the Chesapeake Bay, p. 104-131. In F.J. Webb (Ed.), Proc. of the 9th Ann. Conf. on Wetlands Restoration and Creation. Hillsborough Community College, Tampa, Florida.

Phillips, R.C. 1984. The Ecology of Eelgrass Meadows in the Pacific Northwest: A Community Profile. U.S. Fish Wildl. Serv. FWS/OBS-84/24.

Pulich, W. 1985. Seasonal growth dynamics of Ruppia maritima Aschers in southern Texas and evaluation of sediment fertility types. Aq. Bot. 23:53-66.

Riner, M.I. 1976. A study on methods, techniques and growth characteristics for transplanted portions of eelgrass (Zostera marina). M.S. Thesis, Adelphi Univ., Garden City, New York.

Rozas, L.P. 1987. Submerged plant beds and tidal freshwater marshes: nekton community structure and interactions. Ph.D. Dissertation, Univ. Virginia, Charlottesville, Virginia.

Roberts, M.H., R.J. Orth, and K.A. Moore. 1984. Growth of Zostera marina seedlings under laboratory conditions of nutrient enrichment. Aq. Bot. 20:321-328.

Short, F.T. 1983a. The response of interstitial ammonium in eelgrass (Zostera marina L.) beds to environmental perturbations. J. Exp. Mar. Biol. Ecol. 68:195-208.

Short, F.T. 1983b. The seagrass, Zostera marina : plant morphology and bed structure in relation to sediment ammonium in Izembek Lagoon, Alaska. Aq. Bot. 16:149-161.

Short, F.T. and C.P. McRoy. 1984. Nitrogen uptake by leaves and roots of the seagrass Zostera marina L. Bot. Mar. 27:547-555.

Thayer, G.W., W.J. Kenworthy, and M.S. Fonseca. 1984. The Ecology of Seagrass Meadows of the Atlantic Coast: A Community Profile. U.S. Fish. Wildl. Serv. FWS/OBS-84/02.

Thayer, G.W., M.S. Fonseca, and W.J. Kenworthy. 1985. Restoration of seagrass meadows for enhancement of near shore productivity, p. 259-278. In N.L. Choa and W. Kirby-Smith (Eds.), Int'l Symp. on the Util. of the Coast. Zone. Planning, Pollution, and Productivity, Rio Grande, Brazil, 1982.

Thayer, G.W., M.S. Fonseca, and W.J. Kenworthy. 1986. Wetland mitigation and restoration in the southeast United States and two lessons from seagrass mitigation, p. 95-118. In Estuarine Management Practices. Proc. 2nd Nat'l Est. Res. Symp., Baton Rouge, Louisiana, 1985.

Thorhaug, A. 1974. Transplantation of the seagrass Thalassia testudinum Konig. Aquaculture 4:177-183.

Tomlinson, P.B. 1974. Vegetative morphology and meristem dependence - the foundation of productivity in seagrasses. Aquaculture 4:107-130.

Williams. S.L. 1987. Competition between the seagrasses Thalassia testudinum and Syringodium filiforme in a Caribbean lagoon. Mar. Ecol. Prog. Ser. 35:91-98.

Wood, E.J.F., W.E. Odum, and J.C. Zieman. 1969. Influence of sea grasses on the productivity of coastal lagoons, p. 495-502. In A. Ayala Castanares and F.B. Phleger, (Eds.), Coastal Lagoons. Universidad Nacional Autonoma de Mexico, Ciudad Universitaria, Mexico, D.F.

Zieman, J.C. 1976. The ecological effects of physical damage from motorboats on turtle grass beds in southern Florida. Aq. Bot. 2:127-139.

Zieman, J.C. 1982. The Ecology of the Seagrasses of South Florida: A Community Profile. U.S. Fish. Wildl. Serv. FWS/OBS-82/25.

Zieman, J.C. and R.G. Wetzel. 1980. Productivity in seagrasses: methods and rates, p. 87-116. In R.C. Phillips and C.P. McRoy (Eds.), Handbook of Seagrass Biology, An Ecosystem Perspective. Garland STPM Press, New York.

APPENDIX I: RECOMMENDED READING

The following community profiles on seagrass published by U.S. Fish and Wildlife Service:

Phillips, R.C. 1984. The Ecology of Eelgrass Meadows in the Pacific Northwest: A Community Profile. U.S. Fish Wildl. Serv. FWS/OBS-84/24.

Thayer, G.W., W.J. Kenworthy, and M.S. Fonseca. 1984. The Ecology of Eelgrass Meadows of the Atlantic Coast: A Community Profile. U.S. Fish Wildl. Serv. FWS\OBS-84\02. Reprinted 1985.

Zieman, J.C. 1982. The Ecology of the Seagrasses of South Florida: A Community Profile. U.S. Fish Wildl. Serv. FWS/OBS-82/25.

Specific readings on seagrass restoration and management:

Churchill, Cok, and Riner (1978).

Fonseca et al. (1982; 1984; 1985; 1987c; 1988).

Lewis, R.R. (1987).

APPENDIX II: PROJECT PROFILES

Project profiles 1 and 2 are quoted in their entirety with minor editing from Thayer et al. (1986) with permission of the authors.

EAST COAST

In Thayer et al. (1986) p. 108; example 1.

In December 1983, the North Carolina Coastal Resources Commission (CRC) adopted a mitigation policy, which applies, in part, to seagrasses. This policy requires that adverse impacts to coastal lands and waters be mitigated or reduced through proper planning, careful site selection, compliance with local standards for development, and creation or restoration of coastal resources. Shortly after promulgation of this policy, a project was submitted to the North Carolina Office of Coastal Management (OCM) that requested the removal of salt marsh and seagrass for construction of a marina. This project eventually was granted mitigation status by CRC, meaning that there was sufficient public benefit and water dependency to consider mitigation alternatives to compensate for the wetland loss.

By April 1984, the authors had been asked to participate as representatives of the National Marine Fisheries Service in a review of the seagrass mitigation plan and to make recommendations not only on the plan but also on subsequent seagrass mitigation efforts. As part of this process, numerous meetings were held with state and federal agencies to apprise them of available data on seagrass restoration technology. These data were derived from a cooperative research program on the restoration of seagrasses between the National Marine Fisheries Service (Beaufort Laboratory) and the U.S. Army Corps of Engineers (Waterways Experiment Station, Environmental Laboratory).

At this point it became clear that though a policy had been adopted by CRC, no technical guidelines had been developed to implement it. In essence, CRC had stated that the concept of mitigation was acceptable, but no direction on specific and acceptable actions had been provided. The policy lacked specific directions concerning site selection criteria, acceptable resource trade-offs, performance and compliance standards, accepted methodology for monitoring, and reporting on the above. The lack of any such guidance on mitigation severely compromised the ability of state and federal agencies to enforce the Coastal Area Management Act, the Fisheries Conservation and Management Act, and the National Environmental Policy Act. This first mitigation proposal received by OCM had no guidelines by which to control the project. The agencies and the applicant were then forced to develop mitigation guidelines and a mitigation plan for the marina project at the same time.

The first problem encountered centered on the inadequacy of a resource inventory of the impact site. A cursory inspection by the applicant misidentified the seagrass species present (*Halodule wrightii* actually was present, but *Zostera marina* was reported). The spatial and temporal separation of these species in North Carolina strongly supports the argument that the meadows are not ecological equivalents. The restoration process for *H. wrightii* is also different than that for *Z. marina*. This point of ecological equivalency was contested by the applicant and, in one sense, rightfully so. Data simply do not exist on ecological equivalency among species of seagrasses.

The resource agencies, however, had to make a decision based on the best available information and ecological principles. The fact that two seagrasses are separate species--with each one having distinct environmental requirements for growth (different seasons), different life histories, and different depth ranges and morphologies--supported the contention that unique ecological functions may be supported by each species in the estuarine system. The decision on the part of the agencies' to promulgate this more conservative view was a statement that we must make ecological decisions based on ecological data, and, lacking those data, any action that may compromise the integrity of habitat function must be denied. Such an approach is totally consistent with the North Carolina mitigation policy that emphasizes ecosystem protection and enhancement (Clark 1984) and is emphasized by other work (Ashe 1982).

Another aspect of the ecosystem function concept arose when off-site mitigation was proposed for this project. The initial proposal called for on-site mitigation using an adjacent area at that time devoid of seagrass. This site was rejected by the resource agencies after a time series of aerial photographs demonstrated a perpetual lack of seagrass cover. The applicant had claimed that the site was barren as a result of previous dredging of a channel, which was consistent with agency requirements for selecting a disturbed site for restoration. Because aerial photographs revealed that cover was absent prior to the channel dredging, it was concluded that the site was naturally and chronically without seagrass cover and any planting would run a high risk of failure. At best, the plantings at the proposed site would be a temporary pulse in system productivity since they would likely fail, providing inadequate compensation for the impact site meadows that had persisted through many years.

Once species, acceptable sites, and transplanting procedures were verified and approved, it was quickly realized that there were no provisions for monitoring the site to ascertain performance and compliance with mitigation standards. In fact, there were no standards. Fortunately, there was research on seagrass restoration in the area so that guidelines could be developed based on testable data.

GULF OF MEXICO

In Thayer et al. 1986. p. 110. example 2.

Examples of the use of research data on seagrass restoration to mitigate construction-related damage exist. One is the restoration of seagrass meadows (65.8 acres) that were damaged or destroyed during the construction of replacement bridges through the

Florida Keys (Mangrove Systems, Inc., 1985). Regulatory agencies were provided with a thorough discussion of the value of the affected seagrass meadows. As a consequence, steps were taken to accurately determine the extent of damage and the technology available to restore these areas.

The project followed the four interrelated aspects that have been shown to be critical to the success of a mitigation effort: (1) site resource inventory; (2) transplanting technology, (3) site selection; and (4) monitoring and performance evaluation.

The site and resource inventory, which employed ground-truth methods to verify aerial photography, allowed the categorization of impact areas and non-restorable areas altered so as to no longer support seagrass. This categorization of restorable and non-restorable habitat was made based on environmental criteria important to the growth and development of the seagrasses used, particularly the criteria of sufficient sediment depth. Mitigation plans based on environmental requirements of the target species such as this one are rare and should be encouraged. Two areas of 30.5 and 35.3 acres were determined to be unrestorable and restorable, respectively. The inventory also identified additional disturbed areas as planting sites that were unrelated to bridge construction but were available for seagrass mitigation. The availability of these sites (17.0 acres) may have gone unnoticed had the inventory effort not been made. As a consequence, the restoration ratio reached 0.8:1 as opposed to the 0.54:1 that would have occurred had these areas not been identified.

Transplanting technology and site selection were related to the site inventory. Observations made during the inventory suggested the need for suitable anchoring devices for appropriate species. Available technology was employed to meet these criteria (Fonseca et al. 1982, Derrenbacker and Lewis 1982). Selection of sites started with all available on-site (affected by construction) plantable areas. After these areas were eliminated as choices, other disturbed areas in the immediate vicinity were considered.

In this mitigation, site selection was made much easier since even the unrelated impact sites had either previously supported seagrass, or were contiguous with existing meadows. More important, each site had a definable source of impact that had since been alleviated.

An important aspect was the establishment of a comprehensive monitoring of the seagrass growth in both planted and control (unplanted) areas. Data were collected not only on survival of transplant units, but also on the rate of coverage. The use of a coverage criteria rather than other non-repeatable methods (e.g., leaf length) allowed verification of performance over time that was mutually beneficial to the contractor as well as to the agency determining compliance. The contractor was able to accurately estimate performance and, thus, efficiently plan for replanting or selecting alternative sites where needed. The monitoring agency was able to have a quantifiable (and more importantly, verifiable) means of determining compliance. By the end of August 1984, 47.54 acres of seagrass had been planted with almost 73 percent at acceptable coverage levels. This overall success and coverage is, in large measure, the result of proper site evaluation and application of techniques appropriate both for the sites and the plants used.

Finally the cohesive nature of these four actions (site survey, site selection, appropriate technology, and monitoring program) has provided an information set that has proved repeatable in other areas. The ability to apply this information elsewhere in other unrelated projects has enhanced the original value of the project significantly by adding to guidelines for planting on a wider geographical basis.

WEST COAST

In 1986, Wright-Schuchart Harbor Company submitted a proposal to the City of Eureka to construct an oil platform module assembly yard on the western shoreline of Humboldt Bay. Humboldt Bay and the City of Eureka are located on the Pacific coast of northwest California. Accordingly, the City of Eureka issued a coastal development permit for the upland portion of the project which fell within their jurisdiction. Final approval of the project, however, rested with the California Coastal Commission whose jurisdiction applied to the portion of the project located in the waters of Humboldt Bay. The California Coastal Initiative Act of 1972 led to the creation of the California Coastal Commission in 1976. The responsibilities of the Commission include managing the resources of the coastal zone and ensuring coastal access.

In the review of Wright-Schuchart's application and the City of Eureka's coastal development permit, the Commission staff noted that the construction and dredging associated with the project would result in a net loss of 0.43 acres of eelgrass (Zostera marina) habitat. The staff identified this habitat as a valuable coastal resource, and therefore, project approval was granted with the condition that the applicant submit a plan to mitigate the loss prior to project construction. Final review and approval of this plan would be conducted by the California Department of Fish and Game.

In accordance with the condition of the permit, the applicants sought the services of an environmental consulting firm and requested that the firm develop a transplant plan. What follows is the Coastal Commission's response to the Eelgrass Transplanting Plan that was prepared.

"In consultation with the California Department of Fish and Game, the applicant has submitted an Eelgrass Transplanting Plan prepared by a local firm, April 1986. This preliminary plan proposed to transplant the eelgrass from the project site to three recipient sites nearby. The total acreage for these three sites is 0.93 acres. This represents a mitigation replacement of 2.17 to 1. All three sites are located on the edges of sub-tidal channels on either side of Woodley Island in currently unvegetated areas. All sites also have adjacent eelgrass beds. While it is not known why the recipient sites do not have eelgrass, past studies have shown that eelgrass can be successfully transplanted. The plan contains a transplanting and monitoring program, however specifics are still being developed. Department of Fish and Game staff have reviewed this plan and believe that it has a reasonable chance of success. Given a final plan which details the transplanting program, including timing, monitoring and performance standards, that is reviewed and approved by the California Department of Fish and Game, the project is consistent with Section 30233(a) of the Coastal Act and with the City's LCP (Local Costal Program) policies. The project is the least environmentally damaging alternative which has been mitigated, as conditioned, to the maximum extent feasible."

Before moving to a critical analysis of the Commission's report, some background statements are in order. First, Section 30233(a) of the Coastal Act states in part that:

> "(a) The diking, filling, or dredging of open coastal waters, wetlands, estuaries, and lakes shall be permitted in accordance with other applicable provisions of this division, where there is no feasible less environmentally damaging alternative, and where feasible mitigation measures have been provided to minimize adverse environmental effects, and shall be limited to the following:
>
> (1) New or expanded port, energy, and coastal-dependent industrial facilities, including commercial fishing facilities."

Second, a Local Coastal Program (LCP) is developed by the local decision making body (City, County, etc.) and is approved by the Coastal Commission. The area of jurisdiction covered by an LCP is restricted to the upland area surrounding the coastal waters, and estuaries in question. The Commission, however, has jurisdiction over the area below the mean high tide line. Third, the Commission is required to use the expertise of The California Department of Fish and Game to evaluate projects regarding resources where mitigation is required. This is mandated by the Coastal Act.

At first glance, the mitigation required by the Coastal Commission appeared adequate. For example, the replacement ratio of 2.17 to 1 indicates that there could possibly be a net habitat gain. Also, the fact that "currently unvegetated areas" would be planted, suggests that valuable habitat might be created in areaswhere none had existed. This also suggests that soft bottom habitat is not valuable. Finally, the requirement that California Fish and Game both review and approve the project also seemed adequate. On close examination, however, there were some significant problems. These problems include site selection, monitoring and performance standards, and "up front mitigation". The mitigation is described as "up front" because the transplant took place before the construction began.

The site selection process bears further review. The Commission states that the sites suggested by the applicant in "consultation with The California Department of Fish and Game" are "in currently unvegetated areas". The Commission did not consider this site selection as a conflict with other habitat types. They further state that "While it is not known why the recipient sites do not have eelgrass, past studies have shown that eelgrass can be successfully transplanted". No other criteria regarding site selection is mentioned. Therefore, the only criteria for site selection was the absence of vegetation.

Although it was not entirely clear how the criteria for site selection was developed, the six-month report, following the transplant, states:

> "While looking for suitable planting sites in Humboldt Bay, it was noted that there were very few areas available. Areas which would support eelgrass growth already had growth on them. After discussions withthe Department of Fish and Game it was decided to plant in areas between or just above existing eelgrass beds or in areas where vegetation was sparse."

Ironically, the consulting firm explains the contradiction not stated by the Coastal Commission. Namely that "Areas which would support eelgrass already had growth on them". Even after noting this fact, however, the transplant was conducted under the Coastal Commission's criteria of site selection.

It is also important to note that The Coastal Commission does not give a reference for the statement that "past studies have shown that eelgrass can be successfully transplanted". Although several seagrass meadows in the southeastern region of the United States have been transplanted with relative success (Fonseca et al. 1985), the success rate of similar transplants in Northern California has been relatively low (author's personal observation).

The discussion involving monitoring and performance standards for the project indicated that no monitoring or performance standard guidelines are presented by the Commission. It would appear that the environmental consulting firm in consultation with the Department of Fish and Game had been given the responsibility of designing as well as implementing the monitoring program. Although it made sense for the consultants to design the monitoring program, it was inappropriate for them to establish the performance standards. This becomes apparent upon examination of the company's recommendations.

> "Typically, (the consulting firm) uses a survival rate of 80% for terrestrial revegetation and land restoration project (sic). Due to the seasonal variation in density and the extreme variation in survival rates for past projects, any set percentage of survival would be suspect. Based on past projects, the best estimate of a realistic survival rate over the two year monitoring period would be 50%.
>
> Density is also typically used as a measure of success in terrestrial vegetation restoration projects. Densities vary markedly in the eelgrass beds of Humboldt Bay, however, making comparison difficult. To be consistent with the above standard of a 50% survival rate over two years, we recommend a density standard of 50% that of the control (donor) sites."

Since there are no references given, we cannot report on the past projects. More important, however, there are no methods given describing how the data regarding relative density of Humboldt Bay eelgrass beds was obtained. The Coastal Commission had not done adequate research to determine the usefulness of the performance standards promulgated by the consulting company. It should also be mentioned that a 50% survival rate drops the 2.17 to 1 mitigation ratio to 1.09 to 1.

Regarding the monitoring of the transplant the consulting company presented the following:

> "Post-project monitoring will begin immediately upon completion of the transplant and continue for at least two years. For the first six months, monitoring of physical parameters will be done twice a month. For the remainder of the program, monitoring will be done on a monthly basis. Monitoring of shoot growth will be done seasonally (except during winter when growth is at a minimum). Additionally, monitoring will be conducted after the first major storm event following the transplant.
>
> Three sites will be used as controls. One small site will be near to the harvest site. This site will be a control for the transplanting method. The parameters (author's note: it is presumed they are referring to environmental parameters) at this site should be essentially the same as the harvest site due to the proximity of the two. The variable at this control site is the fact that the material has been transplanted. Consequently, plant losses or die-backs at this site will indicate that the transplanting methods were flawed.
>
> The other control sites are within the existing eelgrass beds on the south side of Indian Island. These sites are near to or adjacent to the transplant sites. The progress of the transplanted material will be measured against the established beds."

It was unclear how this monitoring program would yield data of sufficient quality to evaluate the transplant. No systematic plan or sampling technique was presented. Some general notions about "control sites" are mentioned but it is not clear how these "control sites" in the statistically defined nature of the term were to be selected or evaluated. Although it is claimed that "monitoring of shoot growth" will be undertaken, the methodology for this data collection is not explained and "growth" is not defined (this may be taken to mean productivity as to shoot addition rate). There is no reference to parameters such as cover, survival, or addition of new shoots per planting unit. Although the consulting firm intended to conduct an extensive monitoring program, this was not clear from their description of that program. Apparently this issue was never raised by the Coastal Commission. Also, loss of plants was to be attributed to planting methods without consideration of unsuitability of the environment.

Perhaps the most ironic consequence of the Humboldt Bay Eelgrass Transplant Plan was that the project proposed by the City of Eureka and Wright-Schuchart Harbor Co. was never constructed. The eelgrass transplant did, however, take place and to the consulting firm's credit, monitoring is taking place.

In summary, habitat that the Coastal Commission identified as being a valuable coastal resource may be lost. The entire project represents inadequate planning and lack of agency guidance from the beginning. In retrospect, agencies mandated to manage and preserve the natural resources should have taken a more assertive role in this emerging area of wetland mitigation. Most alarming was the complete failure of "up-front mitigation", a practice widely believed to be a panacea (the author included) for seagrass mitigation practices.

CREATION AND RESTORATION OF FORESTED WETLAND VEGETATION IN THE SOUTHEASTERN UNITED STATES

Andre F. Clewell
A. F. Clewell, Inc.

Russ Lea, Director
North Carolina State Hardwood Research Cooperative
School of Natural Resources
North Carolina State University

ABSTRACT. This chapter describes forested wetland creation and restoration project experience and establishment methods in the region from Virginia to Arkansas south to Florida and Louisiana. In contrast to marshes, forest replacement is more complex and requires a much longer development period. A wide variety of forest establishment techniques have been employed, some with initial success but none of them proven. Most projects began during the 1980's and are too new for critical evaluation. Most of these projects pertain to bottomland hardwood and cypress replacement. The two most significant trends in project activity have been the direct seeding of oaks on abandoned croplands and the replacement of all trees and sometimes the undergrowth at reclaimed surface mines. Although some young projects appear promising in terms of species composition and structure, it is still too early to assess functional equivalency.

Project success depends largely on judicious planning and careful execution. The most critical factor for all projects is to achieve adequate hydrological conditions. Other important factors may include substrate stability, availability of adequate soil rooting volume and fertility, and the control of herbivores and competitive weeds. A checklist of these and other important issues is appended for the benefit of personnel who prepare project plans and review permit applications.

Success criteria for evaluating extant projects throughout the southeast are either inadequately conceived or usually lacking. Emphasis needs to be placed upon the presence of preferred species (i.e., indigenous trees and undergrowth characteristic of mature stands of the community being replaced) and on the attainment of a threshold density of trees that are at least 2 meters tall. Once such a stand of trees is attained, survival is virtually assured and little else could be done that would further expedite project success. At that point, release from regulatory liability should be seriously considered.

Several critical information gaps were identified:

1. The sylvicultural literature does not cover all aspects of wetland tree establishment. Further investigation is warranted.

2. The conditions conducive to effective natural regeneration need to be elucidated.

3. Techniques for undergrowth establishment should be developed. Although the undergrowth accounts for more than 90 percent of forest species composition, its intentional introduction has scarcely been attempted.

4. Baseline ecological and floristic studies need expansion for certain plant communities and regions, otherwise project planning will be inadequate.

5. Research is needed to determine if successful forest replacement in terms of structure and species composition will provide the functional services of the original ecosystem that is being replaced.

6. Most extant projects are not being monitored. The time is ripe for a coordinated southeastern regional monitoring effort.

INTRODUCTION - MAGNITUDE OF BOTTOMLAND FOREST LOSS

Eighty-one percent of the terrain that originally supported bottomland forests in the United States has been converted to other land uses. The southeastern United States sustained about 92 percent of all forested wetland losses during a period from the mid-1950's to the mid-1970's, mostly from clearing and drainage for agriculture (Haynes and Moore 1988). The economic and ecological profundity of these losses has been tardily recognized. Many functional services ceased with bottomland forest removal, including timber production, flood abatement, food chain support (particularly detrital export to estuaries), improvement of water quality through nutrient and pollutant filtering and organic matter transformations, sediment retention, wildlife and endangered species habitat, and others. Appreciable attempts at recovering bottomland forests has begun only within the present decade.

MARSH VS. FOREST REPLACEMENT

Most national and regional efforts to recover wetlands have been directed at marsh ecosystems and involve the creation or restoration of replacement marshes. Bottomland forest replacement strongly contrasts with marsh replacement in scope and approach. Marsh replacement is accomplished within a few years at most, sometimes resulting in functional equivalency and a close approximation of original marsh vegetation. Techniques for marsh replacement are relatively well known and widely accepted.

In contrast, bottomland forest replacement requires decades, and techniques being used are not yet developed with repeatable precision. Functional equivalency has yet to be addressed for bottomland forest projects. Efforts thus far have been directed at the complex task of vegetational establishment. Our presentation, therefore, focuses on vegetational composition and structure, rather than ecosystem functions or services rendered by the replacement plantings. Until such time as functional aspects have been carefully assessed, we make the working assumption that ecosystem function is intimately related to the vegetation, that is, composition and structure. We feel that this approach is reasonable ecologically and pragmatic in terms of fostering the mitigation projects needed to offset the loss of bottomland forest regionally.

Forest replacement is not nearly as dramatic as marsh replacement. For at least several years, young trees co-occupy the terrain with brush and weedy herbs, causing new project sites to appear disheveled. Plants that will persist and eventually contribute to a replacement forest are initially small in stature and present only a fraction of the phytomass of a mature bottomland forest. It is nearly impossible to document success in these immature stands in a way that appeases skeptics. Prolonged establishment periods and largely unproven methodologies for bottomland forest replacement have generated caution from both the regulatory community and agencies supporting forest projects. Environmental permitting is generally more complex than for marsh restoration, because project goals are less easily defined and require more time for attainment.

FOREST REPLACEMENT OVERVIEW

In the southeastern United States, there have been two concentrations of effort in bottomland forest replacement. The first has been the reforestation of bottomlands that were cleared for agriculture and later abandoned, especially in the Mississippi Delta. The focus there has been to establish a forest canopy of selected tree species, particularly of oaks and other heavy-seeded trees with limited dispersal. Trees of other species and all undergrowth plants are ignored or are expected to become established by natural regeneration. The overriding concern is to produce a tree canopy over large tracts of land. The second kind of bottomland forest replacement has been associated with surface mining, primarily for phosphate in central Florida. There, the projects are intensive and restricted to small tracts on reclaimed lands. Plantings attempt not only to replace the full spectrum of tree species but also undergrowth components, with considerable attention given to establishing the appropriate hydrology and hastening soil development.

APPROACHES TO FOREST RECOVERY

The two main approaches to bottomland forest recovery are restoration and creation. Most tree planting projects on abandoned farmlands of the Mississippi Delta are examples of "restoration". In these projects, soils and hydrology are largely intact, and the principal task is to reconstitute the former vegetation. Most surface mining projects and upland conversion to wetlands represent bottomland forest "creation", whereby the entire habitat must be engineered, including abiotic components and vegetation. Creation projects are often more complex than restoration projects. Nonetheless, there are no fundamental differences between such projects in terms of planning, revegetation techniques, and monitoring. Differences between restoration and creation projects generally reflect the physical site attributes more than the basic approaches to establishing the vegetation.

Enhancement (or rehabilitation) is a third approach to bottomland forest recovery. During enhancement, young or degraded forests undergo stand improvement (thinning, interplanting, drainage, competition removal, etc.) to improve growth, species composition, or a specific function or service. Timber stand improvement and certain wildlife management activities are frequently practiced in southeastern bottomlands. Such enhancement projects are peripheral to the thrust of our presentation, which focuses on restoration and creation.

OVERVIEW OF REGION - CHARACTERISTICS OF THE REGION

The region covered in this chapter ranges from Virginia to Arkansas south to Louisiana and Florida and includes the states of Alabama, Arkansas, Florida, Georgia, Kentucky, Louisiana, Mississippi, North Carolina, South Carolina, Tennessee, and Virginia. The climate is generally warm-temperate with prolonged growing seasons, mild winters, high humidity, and seasonally distributed precipitation that overall exceeds evaporation. Physiographically, most of the region lies within the Atlantic Coastal Plain, Gulf Coastal Plain, Mississippi Embayment, and Piedmont. These four provinces contain a wealth of forested wetlands, both in terms of variety and acreage. Many of these wetlands extend up the numerous river valleys into the southern Appalachian, Ouachita, and Ozark mountain regions, where cooler climates prevail. Parts of Kentucky and Tennessee occupy the Interior Plateau province.

Soils of the Coastal Plain and Piedmont are mostly sands and weathered clay loams, respectively, and are generally deficient in plant nutrients. In contrast, the alluvium along larger streams is quite fertile and supports luxuriant and floristically rich forests. Much of this alluvium originated in the highlands, where erosional processes carried nutrient-rich soil particles into streams. Not all southeastern bottomlands are fertile. Some are peaty and acid, including many isolated swamps, tributary headwaters, and the bottomlands associated with backwaters and smaller streams. Forests in such habitats are relatively depauperate floristically and are usually less productive than those along alluvial streams.

FORESTED WETLAND TYPES

Several broadly defined forested wetland vegetation types are recognized in this chapter and are briefly described below. All are palustrine forested wetlands, according to Cowardin et al. (1979). Many occupy hydric habitats, in which substrates are at least seasonally flooded by river overflow or saturated by groundwater seepage. Some occupy mesic habitats, which are only temporarily flooded or saturated. Plants of mesic habitats must tolerate flooding or saturation a few days or weeks at a time but ordinarily enjoy moist, aerated soils. Appendix I lists scientific equivalents for vernacular names of plants used in the text.

1. **Muck-swamps.** Baldcypress and/or water tupelo (or sometimes Ogeechee tupelo) swamps on fertile floodplains of larger alluvial streams of the Coastal Plain, where flooding is prolonged and often deep.

2. **Cypress Heads or Strands.** Pondcypress, often in combination with swamp tupelo, growing on the Coastal Plain in peaty, acid isolated ponds ("cypress heads"), where inundation is prolonged and often deep, and within shallow, slowly moving streams draining bogs or peaty, acid swamps ("cypress strands").

3. **Bottomland Hardwoods.** Mainly deciduous, dicotyledonous trees (e.g., red maple, river birch, water hickory, green ash, swamp cottonwood, sycamore, overcup oak, willow oak, sweetgum, and elms, to name just a few), often containing several dominant species in a given stand and growing on fertile alluvial floodplains subject to seasonal flooding. Also occurring in valleys above usual elevations of overbank flooding, where groundwater maintains constantly high soil moisture. Common throughout the Southeast.

4. **Mesic Riverine Forest.** The extension of bottomland hardwood forest on higher terraces and levees of flood plains and protected valley walls. Consisting of those typical bottomland hardwood species that are less tolerant of frequent flooding or more tolerant of well drained soils (e.g., water oak, laurel oak, cherrybark oak) and often some evergreen hardwoods (e.g., southern magnolia, live oak), and/or some conifers (e.g., loblolly pine, spruce pine). Common throughout the Southeast. Portions of this forest may not be classified as jurisdictional wetlands under the Clean Water Act.

5. **Bay Swamps.** Broadleafed, coriaceous, evergreen trees such as sweetbay, swamp bay, or loblolly bay, and sometimes conifers (especially slash pine) occupying peaty, acid headwater swamps of streams ("bayheads"), colluvial swamps along tannic blackwater streams, seepages in some ravines, those back swamps of larger floodplains that are ordinarily unaffected by river overflow, and isolated depressions within uplands. Bay swamps often encircle cypress heads, where soils are less flooded. In peninsular Florida, red maples, swamp tupelos, and other deciduous hardwoods are often intermixed with the evergreen bays. Bay swamps are abundant in, and essentially limited to the Coastal Plain.

6. **Peat Swamps.** Titi, gallberry and other hollies, fetterbush and other ericads, and often conifers (especially pond pine), regionally called "pocosins", "Carolina bays", or "titi swamps", which occupy the often expansive areas of deep peat accumulation and the banks of small, blackwater streams. Generally consisting of shrubs and small trees on slightly elevated terrain surrounding bay swamps, these shrub bogs suffer fires that spread into them during particularly dry years from surrounding pine flatwoods or herb bogs. Recovery is primarily by coppice-sprouting without an intervening seral stage. Abundant in, and essentially limited to the Coastal Plain.

7. **White Cedar Swamps.** White cedar, growing in monotypic stands in deep, peaty, acid, more-or-less isolated, headwater swamps near the coast in Virginia and North Carolina and growing as conspicuous and often dominant trees within bay swamps and cypress strands in panhandle Florida. A few, isolated white cedar forests also occur in northern peninsular Florida and along the toe of slopes in the fall line sandhill region of North Carolina, South Carolina, and Georgia.

8. **Wet Flats.** Swamp tupelo, pondcypress, slash pine, and red maple, singly or in combination, growing on sandy to clayey surface soils, underlain by a plastic horizon that severely restricts filtration. This type occupies isolated wet depressions between streams in the coastal plain from South Carolina to Florida and westward to Louisiana and Arkansas and shares many plant species with bay swamps and cypress heads.

KEY FUNCTIONS PERFORMED

Southeastern forested wetlands provide a variety of functions, depending upon the type of wetland. Some of the more important functions are listed alphabetically below:

1. Aesthetics, in terms of sensory experience.

2. Air quality improvement by forest trees filtering particulates from adjacent urban, agricultural, and industrial areas.

3. Crawfish, finfish, and shellfish production.

4. Detrital transformations and export of tree leaves and other particulate organic matter, which forms the basis of fresh water and estuarine food chains that support shell and fin fisheries.

5. Flood abatement and concomitant flood control, by resistance of stream flow offered by trees and by the retention of stormwater runoff in isolated systems.

6. Honey production.

7. Maintenance of ecosystem functions in terms of biotic diversity, food chain support, stream flow mediation, and water quality transformation, and filtering.

8. Noise abatement by forest trees in urban areas.

9. Recreational opportunities.

10. Sediment retention.

11. Sinks for pollutants and excess nutrients.

12. Timber production, particularly bottomland hardwoods and cypress.

13. Water storage on floodplains, which contributes to stream flow in dry seasons.

14. Water storage in isolated systems, which contributes to the groundwater and thus to soil moisture of surrounding uplands in dry seasons.

15. Wildlife habitat, including for some endangered species and rookeries for wading birds.

CREATION/RESTORATION PROJECT EXPERIENCE - GOALS OF FORESTED WETLAND CREATION/RESTORATION

The goals of project work have been to:

1. Create a forest that resembles in species composition and physiognomy a locally indigenous forested wetland community on sites that did not previously support that community or that had been drastically altered, e.g., surface-mined and reclaimed land.

2. Restore the same type of forested wetland vegetation, which was previously removed, without much disturbance to the soil or hydrologic regime. (The distinction between "creation" and "restoration", although sometimes subtle, should not be confused.)

3. Enhance an existing forest to accelerate seral processes or to improve a particular function or service, e.g., to provide suitable habitat for an endangered species.

The ideal of creation/restoration projects is to duplicate an original forest stand in terms of species composition, structure, and function. This goal can only be approximated, because natural forests are themselves in constant flux. The ideal can be satisfactorily approached, though, with prudent project planning and execution. Duplication of original forests may not always satisfy current functional needs or local land use plans. In such instances, an altered forest community should be designed. Altered forest restoration/creation may represent the only option at those sites where land use activities have modified soils or water balances to the point that duplication is impossible. With the exception of temporary cover and nurse crops, only plants of preferred species should be intentionally planted, unless there is project-specific justification to the contrary. Preferred species are defined as those indigenous species that are typical of mature, undisturbed, local stands of the community being restored. Excluded are naturalized exotics and those species that normally occur in association with canopy disturbances or systems under stress (Clewell and Shuey 1985).

Species introductions can be active (intentional seeding or planting) and/or passive (regeneration from "volunteer" colonization by means of natural dissemination of seeds and spores). The forest products industry in the Southeast depends almost exclusively on natural regeneration which follows timber harvesting of bottomland hardwoods and cypress (Figure 1). Many trees regenerate from coppice sprouts from stumps, and other trees are replaced by advanced regeneration (i.e., saplings that were not harvested), seeds in place, or from seeds from mature timber nearby. For full-scale projects, provision should be made to introduce undergrowth (herbs, shrubs, understory trees), as well as potential overstory trees. Undergrowth replacement can be accomplished concomitantly with tree establishment at some project sites. At other sites, undergrowth replacement may have to be postponed until potential overstory trees are released from competition by means of herbicides or other treatments that would benefit undergrowth.

EXTENT OF PROJECT WORK

There have been relatively few intentional efforts to create or restore forested wetlands in the Southeast. Most projects are still in the planning stage or have been "in the ground" for only a year or two and are too young for assessment. However, large acreages of tree farms, intended to create a canopy of one or few tree species, have been planted across the entire Southeast and can be useful in evaluating species-site relationships and estimating performance for creation projects.

There is a dearth of published information on forested wetland projects, as revealed by a recent bibliography (Wolf et al. 1986) and in the issues of "Restoration and Management Notes." Monitoring reports are scarce, even in the "gray". We made a thorough, but not necessarily exhaustive, effort to identify extant projects. We contacted environmental personnel at all southeastern regional offices of the U.S. Environmental Protection Agency and the U.S. Army Corps of Engineers (including Waterways Experiment Station), key personnel in the U.S. Fish and Wildlife Service, Soil Conservation Service, Forest Service, various state agencies (departments of transportation, natural resources, wildlife), universities, some consulting firms actively engaged in restoration work, and organizations such as Ducks Unlimited. This search yielded some information, but most was unsuitable for inclusion in this document.

Just before this manuscript went to press, an annotated bibliography was published on the reestablishment of bottomland hardwood forests, which cited several reports not referenced herein (Haynes et al. 1988). Abstracts of those reports elaborated on topics that we have treated and apparently did not introduce entirely novel themes or project descriptions.

Figure 1. Natural regeneration of bottomland hardwoods (oak, hackberry, hickory, ash) at least 40 years old on a levee in Tensas Parish, Louisiana. Photo by M. Landin.

KINDS OF PROJECT WORK

There are two major kinds of on-going forested wetland creation/restoration projects in the Southeast, (1) phosphate mine reclamation and (2) reforestation of bottomland hardwoods on floodplains that had been cleared for row crops. Other types of projects include:

1. Borrow pit reclamation in South Carolina (Kormanik and Schultz 1985).

2. Shellrock mine reclamation in Florida (Posey et al. 1984).

3. Coal mine reclamation in the southern Appalachians (Starnes 1985).

4. Wildlife habitat enhancement by oak hammock plantings in Florida pine flatwoods (Moore 1980).

5. Cypress swamp creation in the Mississippi embayment (Bull 1949, Peters and Holcombe 1951, Gunderson 1984, page 439 in Mitsch and Gosselink 1986).

6. Artificial retention pond (for stormwater runoff at industrial and residential developments) tree plantings in Florida (S. Godley pers. comm. 1987).

7. Highway corridor reforestation across river bottoms by state road departments in Alabama (J. Shill pers. comm.) and Arkansas (W. Richardson pers. comm.).

8. Shoreline erosion control with planted bottomland hardwoods and cypress at reservoirs (Silker 1948, Anding 1988).

9. Cooling reservoir (for thermal discharge from a nuclear reactor) shoreline restoration with cypress-hardwood plantings in South Carolina (Wein et al. 1987).

10. White cedar swamp restoration in the Big Dismal Swamp of Virginia-North Carolina (Carter 1987).

11. Enhancement of second-growth hardwood hammocks in the Florida Everglades by selective herbicidal removal of Brazilian peppertrees (Ewel et al. 1982).

12. Urban lake shoreline enhancement by planting cypress and bay (Cox 1987).

13. Sustained-yield timber management in commercial forests. We do not consider these operations as "restoration", although they provide valuable information on reforestation techniques (Malac and Heeren 1979).

MINE RECLAMATION PROJECTS

Many forest creation projects occur at phosphate mines in central Florida. Some attempts have been made to recreate headwater streams and their attendant riverine forests in surface-mined and physically reclaimed land. Other projects are designed to provide forests on reclaimed land bordering mine-pit lakes and reclaimed marshes. Although tree planting on phosphate-mined lands began earlier, the first project to create a forest along a stream flowing on reclaimed land started in 1980 at Sink Branch, shown in Figure 2 (Robertson 1984). After 6 years, planted bottomland hardwood trees (containerized nursery stock) averaged 3.4 meters tall (Gurr & Associates, Inc. 1986). Dogleg Branch restoration began in 1983 and was the first project that was specifically designed to restore undergrowth as well as trees (Clewell and Shuey 1985, Clewell 1986a). The largest project is Agrico Swamp, which covers 20 hectares of bottomland hardwoods and bayhead (Erwin 1985, Erwin et al. 1985). This project site borders natural wetlands along a nearby stream.

Several other forested projects have been initiated, e.g., Miller et al. (1985) and Clewell (1986b), and most were listed by Ruesch (1983). Haynes (1984), Robertson (1985), Marion (1986), and Robertson (in Harrell 1987) provided overviews of wetland restoration associated with phosphate mining. Some phosphate mine projects restored riverine forests within unmined corridors such as riverine forests that were cleared for power lines, pipelines, and dragline crossings (Clewell 1986c,d). Most project reports are available through inter-library loan from the Florida Institute of Phosphate Research, 1855 W. Main St., Bartow, Florida 33830; (813) 533-0983.

The extensive literature on the revegetation of coal surface-mined lands contains relatively little information on reforestation, and most of that pertains to non-wetland, upland reclamation sites. Often, the steep valleys typical of many coal-mined sites in the Appalachians offer little opportunity for forested wetland creation (Starnes 1985).

BOTTOMLAND HARDWOOD CREATION ON FALLOW FIELDS

Restoration of former bottomland hardwood forests is being accomplished on floodplains which had been cleared since about 1950 and row-cropped, primarily with soybeans. Most project sites occur in the Mississippi River delta or along tributaries of that river. The objective of reforestation is the reestablishment of forest canopy, particularly by oaks and other heavy-seeded trees that could not volunteer readily by natural dissemination from adjacent forests.

Undergrowth and some tree species are expected to volunteer passively. Undergrowth plants usually are not introduced intentionally, although some regulatory authorities encourage such plantings.

The largest project is currently underway near Monroe, Louisiana, where 1,821 hectares have been purchased by the state and are being reforested to create a corridor between existing wildlife management areas (Harris 1985). At Panther Swamp National Wildlife Refuge in Mississippi, 445 hectares are being reforested (Anonymous 1986). At the Malmaison Wildlife Management Area in Mississippi, 405 hectares are being reforested (Anonymous 1984). In 1985-86, 218 hectares of open land were reforested at the Tensas River National Wildlife Refuge in Louisiana, and another 809 hectares of marginal crop and pasture lands are scheduled (Larry Moore pers. comm. 1988).

Haynes and Moore (1988) summarized bottomland hardwood restoration projects in 1987 for all southeastern National Wildlife Refuges. They identified projects at 12 refuges. The prevalent species planted were Nuttall oak, cherrybark oak, willow oak, water oak, and pecan. Direct seeding of acorns began six years previous (ca. 1981), and seedlings were planted at older projects. They reported that fertilizers and cultural practices such as mowing, disking, and herbicidal applications were seldom used. Three projects exceeded 10 years of age; the oldest was 19 years old and contained oaks 12 meters tall with diameters of 23 to 27 centimeters. Research plots to test reforestation techniques have been established in several locations, notably at the Delta Experimental Forest in Mississippi by the U.S. Forest Service (Johnson and Krinard 1985a, b, Krinard and Kennedy 1987). Other experimental projects have been installed in southwestern Tennessee (Waldrop et al. 1982), at the Thistlethwaite Game Management Area in Louisiana (Toliver 1986-87), and are being planted primarily with oaks both as acorns and seedlings.

NEW PROGRAMS

Two new programs will likely result in bottomland reforestation; however, no specifics are yet available. One is the Conservation Reserve Program of the U.S. Department of Agriculture, Soil Conservation Service. The other stems from a provision in the 1987 Farm Bill, whereby defaulted farmland may be reforested by the U.S. Fish and Wildlife Service. Mark Brown of the University of Florida (pers. comm. 1988) is calling for forest plantings along drainage canals which are presently maintained in herbaceous vegetation at great expense. He also suggests that construction of ponds intended as effluent sinks would make excellent habitats for planted forests because water tables are carefully maintained.

VEGETATION TYPES BEING RESTORED

Existing projects in the Southeast focus primarily on bottomland hardwoods and secondarily on cypress restoration. Will Conner of Louisiana State University (pers. comm. 1988) reports that the state of Louisiana began restoring bald cypress forests by planting seedlings beginning in 1983. The first planting has attained a mean height of 3.2 meters in 5 years with about 95 percent survival. Phosphate mine projects are often mixtures of deciduous bottomland hardwoods, evergreen hardwoods typical of bayheads, and cypresses. Such mixtures reflect strand and low hammock vegetation in peninsular Florida.

We are unaware of any projects attempting to restore shrub bogs. We are also unaware of any restoration of cane, a bamboo-like grass. Dense stands called "cane breaks" formerly interrupted bottomland hardwood forests throughout much of the Southeast and were heavily frequented by wildlife. Land management practices have essentially eliminated cane breaks, and we encourage the restoration of this vanishing community.

PROJECT SUCCESS

Most creation projects are still too recent to predict their ultimate success. Performance targets for most projects are necessarily imprecise, owing to an absence of stated project goals. Success criteria have not been identified for most projects and were inappropriately conceived for others. There has been an implicit attitude that success criteria will become self-evident, once the planted trees mature. Monitoring data are inadequate or absent for all but a few creation projects.

In spite of these drawbacks, existing data and general observations shed light on the suitability of various restoration techniques and allow optimism for the success of some projects. The photograph of Sink Branch (Figure 2) reveals a closed canopy of relatively tall, vigorously growing mixed hardwoods. Trees at several younger phosphate project sites are emerging above the brush cover. Undergrowth establishment has been recorded at some projects (Clewell 1986a).

Bottomland hardwood test plots in the Mississippi Embayment have demonstrated impressive tree growth, mostly on fallow cropland on floodplains. One such study

Figure 2. Riverine forest creation along Sink Branch on reclaimed, phosphate mined land in central Florida. Sweetgum, oaks, and other hardwoods are in their eighth year, after being planted as containerized nursery stock.

(Broadfoot and Krinard 1961) provided photographs of plots of planted cypress and bottomland hardwoods representing 11 species, which were from 17 to 25 years old and which averaged from 13 to 19 meters tall, depending on the species. Six-year-old eastern cottonwoods were also pictured, which averaged 16 meters tall. These photos graphically demonstrated that rapid vegetational restoration is possible. Krinard and Johnson (1976) reported that cypress trees planted 21 years previously averaged 15 centimeters in diameter, maximum 35 centimeters. Krinard and Kennedy (1987) reported average heights for trees planted 15 years previously: cottonwood 18 meters, sycamore 12 meters, green ash 11 meters, sweetgum 9 meters, Nuttall oak 8 meters, and sweet-pecan 7 meters. Figure 3 shows 45 year old hardwoods established by the Tennessee Valley Authority to stabilize road crossings at a reservoir.

Haynes and Moore (1988) suggested that planted bottomland hardwood forests on abandoned farmland could become self-regenerating communities in 40 to 60 years. Toliver (1986-87) predicted saw-sized timber could be harvested from similar restoration projects in 35 to 60 years.

We are sufficiently impressed by the existing evidence to state that forested wetland creation/restoration projects that are carefully planned and executed will be successful in terms of plant species establishment and physiognomonic traits. Success in terms of functional equivalency to natural forested wetlands has not, however, been documented. We feel confident that a close correlation exists between forest form (i.e., composition and physiognomy) and most functional attributes. Cairns (1985, 1986) previously suggested this

Figure 3. Water tupelo (left) and willow oak (right) planted 45 years ago by the Tennessee Valley Authority (TVA) along Kentucky Lake.

relationship but with reservation. Our confidence is supported by published results and by our personal observations and experience at project sites. This optimism is further supported by the prevalence of cleared lands throughout the Southeast that have regained their forest cover solely by natural regeneration. In other words, intentional restoration activities were not always required for success, as long as the integrity of the physical environment was maintained and propagule delivery was adequate.

We believe that the eight forested wetland types listed earlier can be restored, but not necessarily under all existing habitat conditions. Problems with submergence and stability would presumably preclude restoration of deep-water cypress-tupelo and white cedar forests; however, these community types may be established as shallow water systems. Peat swamp restoration, other than tree farming, has not been attempted, but we know of no unsurmountable obstacles to such restoration. Bay swamp creation is more difficult than deciduous bottomland hardwood creation on phosphate-mined lands, because soil moisture levels and organic matter requirements are more critical. In general, project success is more a matter of proper design and implementation than of forest type.

AGENCY INVOLVEMENT

Many forested wetland restoration projects are required as mitigation for permitted activities. Permit negotiations are adversarial by nature. Once a permit is issued, it behooves agency personnel to become partners in the project, offering expert advice in project design and providing frequent surveillance. By doing so, project engineers gain the benefit of prior agency experience, and agency personnel improve their expertise from surveillance activities.

CRITICAL FACTORS

Six critical factors interact to determine whether or not a project will be successful. They are: hydrology, substrate stabilization, rooting volume, soil fertility, control of noxious plants, and herbivore control. Hydrology is universally the critical factor. The other factors vary in importance from project to project. In the next section, we will suggest methods for reducing the adverse impacts which may be posed by these factors.

Hydrology

Forested wetlands are closely controlled by hydrology, in terms of annual inundations and/or soil moisture regimes. Engineers, hydrologists, and soil scientists must cooperate to determine whether water delivery timing, depth, and quality are synchronous with the natural systems being emulated. If any of these aspects become asynchronous with the life cycles or growth requirements of the species being established, then planted vegetation will become stressed and subject to limited performance or, in the extreme case, mortality. Whenever possible, control should be asserted over adjacent waters that dictate the hydrological regimen. Flashboard risers, flap gates, retention ponds, and spillways are examples of engineering alternatives that should be considered for implementation before revegetation begins. Such alternatives will mitigate losses from storms, river floods, and other stochastic events that cannot be foreseen. Newly planted vegetation is particularly susceptible to water stress, especially when seedlings of species that are adapted to shaded swamps are planted in exposed project sites. Supplemental water may be released as needed through control structures until young plantings are adequately rooted.

Substrate Stabilization

Project sites are often open and subject to erosion, which hinders establishment of trees and desirable undergrowth. Eroded sediments may accumulate, blocking drainage, smothering vegetation, and reducing water quality. Topographic relief must be planned with substrate stabilization in mind, and final grading must be done with considerable care. Without such care, the reclaimed stream may become gullied, causing the project to be forfeited because of inconsistent water delivery across the project site. Project engineers should expect to make repairs to control erosion and deposition during the first 6 to 18 months.

Rooting Volume

Roots of planted trees must have an adequate volume of soil in which to gain anchorage and to exploit moisture and nutrients. Factors limiting rooting volume are the depth to the wet season water table and mechanical resistance in terms of bulk density and compaction. Roots require oxygen for metabolism. Oxygen is abundant in aerated soils but is soon depleted in saturated soils. As a result, root growth is generally restricted to soil horizons above the water table. In seasons of active growth, roots of most trees die within a few days following oxygen depletion, and tree mortality quickly follows root death. Trees of only a few species are capable of translocating oxygen to their roots and into the rhizosphere and therefore surviving for extended periods in waterlogged substrates. Most wetland trees are necessarily shallow-rooted for that reason. If project sites have high water tables, trees may be unable to attain sufficient anchorage and will eventually topple. If soils are infertile, the volume of aerated soil may be too small to supply adequate nutrients for growth. With regard to mechanical resistance, roots of some trees are physically unable to penetrate clays or other soils with bulk density values exceeding 1.6. Density is increased by compaction caused by heavy equipment at project sites (Holland and Phelps 1986). At forest creation sites, organic matter is also generally lacking, and other soil properties and macrofauna that contribute favorable structure and fertility for plant growth are diminished. The existing mineral substrate may require conditioning prior to tree planting. A common problem at mine project sites during dry seasons is the hardening of otherwise sandy substrate by relatively small increments of clay. Plant roots are unable to penetrate these "crusted" soils, causing moisture stress.

Soil Fertility

Native substrate fertility varies considerably with the project site. Fertilizer supplements are usually necessary. Otherwise, trees may languish too long as saplings and become suppressed by weeds. Strategies for application may be required to prevent weeds from being the major beneficiaries of fertilizer amendments. Adjustments of pH are sometimes necessary by means of amendments of lime.

Noxious Plant Control

Noxious or "nuisance" species include (1) aggressive colonizers or "weeds" of open environments, such as Johnson grass, giant ragweed, saltbush, and primrose willow, (2) perennial turf grasses, such as burmuda grass and particularly bahia grass, (3) certain perennial cover crops, especially tall fescue, (4) naturalized exotics that compete successfully with indigenous vegetation, including kudzu, honeysuckle, Chinaberry, Brazilian peppertree, and (5) preferred species that may proliferate to the point that they suppress young overstory trees, e.g., box elder, wild grapes, blackberries, greenbriers, hempvine, morning-glory.

The Surface Mining Control and Reclamation Act (PL95-87) mandates herbaceous plantings for sedimentation control. However, tall fescue and other widely planted cover plants substantially reduce the survival and growth of tree seedlings (Klemp et al. 1986). Vogel (1980) determined that cover, particularly grasses, should be removed (scalping, herbicide, cultivation) prior to tree planting. Leguminous cover crops are beneficial, as long as they do not suppress initial seedling growth, because they contribute nitrogen and stimulate the development of soil fauna.

Bahia grass and burmuda grass are widely planted throughout most of the Southeast. They have been used as cover crops at project sites or have seeded onto project sites from nearby fields. Both species, particularly bahiagrass, are strongly competitive for nutrients and moisture, and both are known to be allelopathic (Fisher and Adrian 1981, Whitcomb 1981). Both grasses are low-growing and preempt space in which taller weeds could grow. Smaller saplings suffer both from competition and from the lack of shade ordinarily provided by taller weeds. Perennial turf grasses represent a major threat at any project site.

On the other hand, tall weeds, such as dog fennel and broomsedge, are beneficial, because they shelter young trees and desirable undergrowth from sun and wind without crowding them. We should be tolerant of such species, because they directly parallel old field succession, whereby young trees are protected by plants of these same species (Kurz 1944, Kay et al. 1978). For this reason, many weeds are beneficial, although their profusion may hide young trees from cursory view and may unjustifiably influence performance reviews by regulatory personnel.

Herbivore Control

Herbivores often inflict heavy damage to planted trees. Squirrels and chipmunks exhume direct-seeded acorns, and rodents girdle seedlings (Johnson and Krinard 1985b). Raccoons, rabbits, and deer enjoy free meals of tree seedlings (Toliver 1986-87, Haynes and Moore 1988). In winter, newly planted seedlings provide an attractive supplement to scarce food supplies for nuisance animals. Beavers can essentially "clearcut" tree seedlings the night after they are planted. Beavers may also drown planted seedlings by blocking drainage in culverts and at other points of constricted flow. Among the worst transgressors are nutria, a semi-aquatic fur-bearing rodent introduced from South America in the 1930's which relishes tap roots of cypress seedlings (Conner et al. 1986, Anonymous 1988). Conner (pers. comm. 1988) reported that nutria in Louisiana consumed approximately 1,500 of 2,000 cypress seedlings within a few days after they were planted in 1987.

DESIGN OF CREATION/RESTORATION PROJECTS

PRE-CONSTRUCTION CONSIDERATIONS

Project Location

Although a replacement wetland is ideally placed in the same location as the original wetland, there are instances when other placements are more desirable within the immediate watershed for several reasons:

1. To take advantage of optimal topography or hydrology created by land reclamation.
2. To allow the direct transfer of topsoil from a forest being removed to a replacement forest creation site without stockpiling.
3. To reduce the distance to a forest that will serve as a seed source of preferred species.
4. To facilitate proposed land uses.
5. To avoid concentrations of contaminants that are known to exist on-site.
6. To provide a wildlife corridor between natural areas.
7. To coordinate with approved regional watershed planning concepts.

The last two items may serve to mitigate cumulative impacts caused by prior land uses.

Project Size

Wetlands can only comprise that portion of a watershed where the surface water is available or the groundwater table lies at or near the soil surface. In other words, the watershed controls the size and water balance of wetlands. If a restoration project occupies a greater acreage than the original forest being restored, then the restored forest will be proportionately less hydric and more mesic. For that reason, project size should not exceed the acreage of the original forest, unless the water balance is concomitantly improved. Designs to improve the water balance for an enlarged project should assess potential secondary and cumulative impacts to other environmental systems in the watershed.

Site Characteristics Planning

Site characteristics planning must begin by answering the question of whether or not the project site will support the kind of restoration proposed, in terms of relief, exposure, hydrology, soils and fertility, erosion potential, and seed banks. Recent or proposed drainage projects within the immediate watershed require particular study, because an altered water table may cause project failure.

Water Management

The regulation of stream flow is often desirable or essential during the initial years. The stream in question may be a river that seasonally overflows onto a project site or a created stream within a project site. Weirs or other control structures may be available to regulate flows. If so, flow can be augmented in dry seasons to encourage plant growth. Likewise, peak flows in wet seasons can be diverted to prevent erosion. Without controls, successful reforestation has proven to be quite difficult in Mississippi River bottomlands in Louisiana. Attempts to replace hardwoods with cypress have sometimes been unsuccessful, due to the absence of control structures to prevent inundation of seedlings.

Channel erosion along newly created streams is always a consideration in designing reclamation projects. To reduce potential channel erosion, headwaters could be placed adjacent to impoundments. An impoundment would receive much runoff that would otherwise pass directly into the new stream and cause erosion and sedimentation problems. Soils in the restored wetland next to the new stream would stay wet from groundwater seepage under hydrostatic pressure from the impoundment. This concept is under consideration for mining projects and holds considerable promise (King et al. 1985).

Sometimes hydrologic conditions have been altered off site and must be accommodated in the restoration plan. For example, operators of reservoirs upstream may release asynchronous discharges, causing floods in normal dry seasons, or they may reduce seasonally high flows in wet seasons. Although such scheduling of discharges may assist in tree establishment, a continuation would likely prove deleterious to forest ecosystem function and development. In another example, some power generating facilities produce thermal discharges. Species selected for thermally enhanced project sites must be favored by high temperatures and must not require prolonged winter dormancy (Sharitz and Lee 1985).

Proper watershed management is a necessary component of any wetlands creation project. If watershed activities are not coordinated with the revegetation efforts, serious problems may develop that require expensive engineering solutions. For example, approximately 45 hectares of phosphate-mined land was physically reclaimed to pre-mining elevations and planted with a temporary cover crop. Concurrently, a new stream and 2 hectares of attendant bottomland forest were created within this new watershed (Clewell 1986b). Once the upland cover crop died, it was not immediately replaced. Runoff from thunderstorms was no longer retarded by the cover crop and passed unchecked into the new stream. As a result, flash floods occurred and caused channel erosion. Repeated repairs were required to stabilize the stream channel. Eroded sediments had to be removed with difficulty from sites in which trees had already been planted. In retrospect, revegetation activities of the surrounding watershedshould have been coordinated with wetland creation activities.

SITE PREPARATION AND PLANNING

Site Preparation Principles

The goal of site preparation is to configure and stabilize a physical habitat in which a restored/created forest can be established that will function to provide multiple ecological services. For many restoration projects, little or no site preparation is needed. On reclaimed land or at highly disturbed sites, contouring and stream creation may be required. Site preparation should create land forms that resemble natural features. Tall dikes, for example, will appear unnatural indefinitely and should be installed only if no other alternative is available. Structural alternatives should be designed, if possible, so that they may be removed when no longer needed (e.g., weirs), or so that they will subside in a few years and blend into the landscape (e.g., berms and stream deflectors, described below).

Substrate stabilization is essential for newly prepared sites, in order to prevent sheet erosion or gullying and to allow newly planted trees to become rooted. Temporary cover crops may provide sufficient control at some sites. Where erosion potentials are greater, a series of planting techniques are available, and detailed instructions for their use have been provided by Allen and Klimas (1986). Plantings are generally less expensive and less intrusive than engineering alternatives. The plants used, though, should either represent species typical of the forest being restored or should not be expected to persist indefinitely at the site.

Regardless of the cover crop or erosion plantings to be used, they should be sown or planted as soon as the site is available. A delay not only invites erosion but also allows undesirable weeds to become established and will compete with planted trees and preferred undergrowth plants. Weed prevention must be taken seriously. Once established, weeds may require unexpectedly costly control. Weeds will reduce tree survival, impair tree growth, and will delay project release from regulatory liability.

Contouring Strategies

Contouring must be accomplished carefully with regard to elevation and seasonal fluctuations of the water table. If water table movements cannot be predicted, final grading may have to be delayed until groundwater measurements are available from piezometers. In general, original contours should be reestablished, unless other factors take precedence, such as availability of fill materials, safety requirements, proposed land use criteria, and design for flood abatement.

Contours should be as gentle as possible. Sharp breaks in topographic continuity invite gullying. Permanent repair may require burying a culvert along the length of a gully across a topographic discontinuity.

Sheet erosion must be controlled by establishing a fast-growing cover crop, such as winter ryegrass or (in summer) millet. Spot-seeding may be necessary later in the season for bare areas. Perennial cover crops may also have to be planted unless native herbaceous plants colonize the site during the first year. If vegetative cover is insufficient, low berms (0.5 meters or less) can be plowed on slopes parallel to topographic contours. Runoff collects behind berms and infiltrates into the soil. Berms were effective at one project, where they were needed for the first year (Clewell and Shuey 1985). Subsequent rains have since eroded them nearly to the natural grade.

For terraces subject to seasonal flooding, we recommend an uneven or corrugated surface, like that prepared with a bedding plow. Later, trees will be planted on the elevated microsites where rooting volume is favorable. Construction engineers and bulldozer operators may require convincing to do what appears to be an untidy job.

Stream Channel Construction

If stream channels are to be created within the wetland, they should be relatively broad and shallowly parabolic in profile. This configuration forces the water column at peak flow to spread out and experience maximum friction with the bottom (Fig. 4). Flows are deterred by friction and their erosive forces spent. Deflectors can be positioned every 30 meters or so to initiate meandering. A deflector can be constructed by piling about 7 logs perpendicular to the flow of the stream. Sand or hay bales should be placed behind each deflector to prevent its being undercut by the stream. Meandering increases the length of the stream between any two points. Friction of the water column with the channel is thereby increased, and the erosive force of the current is reduced accordingly.

Topographic breaks in the stream channel should be avoided or will require riprap, boulders, or other devices to prevent gullying. If strong flows are expected, the channel may have to be lined with cobbles, packed clay, or other materials to prevent gullying (Starnes 1985). For lesser flows, effective erosion control is realized by sprigging stoloniferous marsh plants and allowing them to proliferate across an entire channel in seasons of low flow. Suitable plantings may include pennywort, spikerush, pickerelweed, and bulrush.

Where vegetation alone is inadequate, erosion matting can be effective. Mats of excelsior, sandwiched between nylon netting, are easily pegged to the soil. Plants can readily grow through the mesh, replacing it as it biodegrades.

Soil Conditioning

Substrates at newly created project sites may be inadequate in their structure, fertility, and organic matter content. Tree planting might best be delayed until soil conditioning activities are completed. These activities may include green manuring, by alternately growing and disking fast-growing covercrops. Other possibilities are to spread and disk organic matter, such as straw, bark, wood fiber, or sludge. Straw was successful

Figure 4. Recently created stream segment of Dogleg Branch on reclaimed, phosphate mined land in central Florida.

in preventing soil crusting in microplot studies (Best et al. 1983). At one project site, willows and wax myrtles had previously colonized an abandoned phosphatic clay settling pond. These shrubs were herbicided, chopped, and disked into the soft substrate in preparation for tree planting (Ericson and Mills, 1986). At a new site designed by A.F. Clewell, water hyacinths were disked into a mineral substrate in 1988 to provide organic matter. Trees are scheduled to be planted later that year. Starch gel polymers have also been used as soil amendments (Clewell and Shuey 1985) and are gaining in popularity. They improve moisture retention and diminish soil crusting. Subsoil rippers have been used effectively in conjunction with tree planting on clayey soils (Powell et al 1986). Rippers augment rooting volume in compacted soils and "fold in" surface organic matter into the planting slit.

Sludge amendments should be considered as a means for introducing or attracting an abundance of soil micro-fauna. Composted sludge may be the preferable form because it would also introduce wood fiber. We encourage sludge application both as a soil amendment and as positive method of waste disposal. State and local standards may limit or preclude sludge application near water bodies. Sludge applications should be limited to aerated substrates which favor soil microfauna.

Soil fertility may be improved with amendments of sludge or fertilizers and by the planting of leguminous cover crops, whose symbionts contribute nitrogen. Highly acid soils may require extensive conditioning. Lee et al. (1983) conditioned pyritic soil in Mississippi with amendments of crushed limestone, rock phosphate, and chicken manure. The pH was improved from 2.9 to 5.5, and the survival of planted trees and shrubs was correspondingly improved.

Introductions of mycorrhizal fungi have been attempted, primarily to enhance phosphorous availability. Tests have shown that planted trees in the Southeast give little or no response to mycorrhizal inoculation, either because of the abundance of native mycorrhizal fungi already in the soil (Schoenholtz et al. 1986) or because of the adequate availability of phosphorus in the soil (Wallace and Best 1983). Some workers have demonstrated a positive response (Wallace et al. 1984), but others have reported that initial responses were temporary, and untreated trees eventually attained the same size (Rice et al. 1982).

If vines or perennial turf grasses are abundant, they should be largely or entirely removed concurrently with soil conditioning. A combination of herbicidal treatment and disking may be required. Honeysuckle, grapes, kudzu, greenbriers, and ground nut all rampantly proliferate and smother young trees. Hempvines massively enshroud saplings, causing 2.5-meter-tall cypresses to bend until their tops touch the ground (Clewell 1987). Some vines are preferred species but must be temporarily considered as noxious until planted trees attain sufficient height to withstand their competition.

Timing of Project Activities

Final grading of project sites should be scheduled, if possible, in the dry season, in order to engage heavy equipment in otherwise boggy terrain and to minimize the erosion of loose substrates (Lea 1988). Orders of plant materials must be made early. Tree seedlings and other nursery stock are often unavailable unless special-ordered and contract-grown as much as a year in advance. Plantings should be scheduled for the most appropriate season for each species, particularly bare root seedlings. Tree planting should be delayed as needed to accommodate soil conditioning procedures or the removal of vines and rhizomatous turf grasses. Introductions of the more vulnerable species, particularly undergrowth, may best be delayed a year or more and interplanted when the initially harsh conditions of the open site are ameliorated by an established vegetative cover.

Species Selection

The problem of species selection and planting stock is not trivial. Plants representing hundreds of woody species comprise forested wetlands in the Southeast. Inappropriate choices of species may result in non-attainment of project goals. Mortality of trees and undergrowth plantings may be unacceptably high, growth may be retarded, and capture of the site may be inadequate by the intended vegetation.

Reconnaissance of surrounding plant communities provides a valuable first step in deciding basic components of the forest type to be established. Baseline ecological studies or published accounts of local vegetation provide a detailed basis for planning the vegetational composition at a project site. Lists of indigenous species should be prepared from reconnaissance, baseline studies, and appropriate literature. From these lists should be culled all exotic introductions and those weedy, short-lived colonizers typical only of forest gaps (from tree fall) and disturbances (e.g., along trails). These colonizing species will appear all too readily without assistance.

Remaining species on the list are the preferred species, mentioned earlier. They represent the trees and undergrowth typical of mature, undisturbed forest vegetation. Project plans should call for the introduction of as many of the preferred species as possible. The only non-preferred species to be planted are temporary cover and nurse crops.

Nursery or cutting stock should be derived from regional sources to improve the chances of introducing ecotypes that are adapted to local climate and soils. If local sources are not available, planting stock is most likely to be adaptive if secured from a region north of the project site, according to results of provenance tests on sweetgum (Stubblefield 1984). We suggest that planting stock from the south would be less cold-hardy and that stock from the east or west would be adapted to different precipitation regimes.

The on-site environmental conditions necessary for species survival will vary along gradients, particularly for soil water balance. Only a minority of species are adapted to a plethora of site conditions that can be successfully planted over a broad range of environmental gradients. When species are indiscriminately melded across a project, the resulting vegetation will not resemble baseline forests, and plantings will suffer undue mortality and low vigor. Several references (e.g., Teskey and Hinkley 1977) supply information on species tolerances, which the project manager may use in committing species to those environmental gradients at the project site. The resulting created forest may have less species overlap along gradients than in natural systems. The dominant species in natural forests have the luxury of establishment on microsites for which they are specially adapted and that may be unavailable at projects sites, e.g., on decaying logs. Undergrowth plants are strongly influenced by flooding and moisture availability, and hydric sites generally have less species diversity than mesic sites (Bell 1974, Clewell et al. 1982).

Not all species intended for introduction can be planted successfully at new project sites. Seral development may be prerequisite for some species. A targeted mixed hardwood community may have to develop beneath an initial canopy of pioneer species, such as willows or cottonwoods. Some of the desired hardwoods may penetrate this canopy within a few years, while the emergence of others may be delayed for several decades until the gap replacement process begins.

Nurse Crops

One method of hastening seral development is to introduce nurse crops prior to, or concurrently with the establishment of preferred species. Nurse crops may provide shade, preempt space occupied by highly competitive species such as turf grasses, contribute humus from abundantly produced leaf litter, or produce nitrogen through their symbionts. Species suitable as nurse crops are those that will grow rapidly and are either relatively short-lived or can be harvested economically upon serving their purpose. The planting of a nurse crop requires a knowledge of stand successional patterns as well as clear institutional memory to implement the later phases of this option.

Black locust and European alder are commonly planted on coal surface-mine spoils. Although they are not planted specifically as nurse species, they fulfill that purpose. They begin to die after 17 to 20 years as the result of competition from potential overstory trees that volunteer beneath their cover (Thompson et al. 1986).

Cottonwoods are potential nurse species. They grow rapidly, provide shade, contribute much humus, and eventually succumb to the competition of hardwoods that are interplanted with them. Volunteer willows are being intentionally used as nurse species (J. G. Sampson pers. comm.). Slash pines could serve as a nurse species. Their seedlings survive and grow faster than hardwoods when planted in competition with bahia grass. Hardwood seedlings could be interplanted later, as the pines shade-out the grass. If the pines were not desirable as canopy trees, they could be harvested for sale as fence posts. Wax myrtle is another potential nurse species, because of its nitrogen-fixation and the shelter provided by its dense, evergreen foliage (Clewell 1986e).

NATURAL REFORESTATION

Natural Regeneration

In many instances, natural regeneration of preferred, bottomland forest species may be passively employed to reclaim degraded or perturbed wetlands. Natural regeneration might also be judiciously incorporated into certain forest creation plans, as long as back-up plantings were required in case of seed failures. A seed source must be contiguous with the project or the site must be available to flood waters that transport seeds. There are examples of mitigation sites that have been overrun with volunteer vegetation, which out-performed planted stock. Clewell (1986f) reported 8-year-old natural regeneration, primarily of red maples and sweetgums with a density of 15,324 trees per hectare, adjoining a "seed wall", i.e., the edge of the nearest reproductively mature forest facing the site.

Other documentation of natural regeneration was provided by Farmer et al. (1982), Rushton (1983), and Wade (1986). In many cases, the naturally regenerated forest is more desirable, because the plants of the component species are precisely distributed according to the environmental gradients and microsites to which they are best adapted. On the other hand, species richness from natural regeneration may be inadequate, especially if the seeds reaching the project site represent only a few species. Supplementary plantings of additional species may be appropriate in such instances.

Sites which should be considered for natural regeneration are:

1. Those that are narrow (no greater than two tree heights from the surrounding seed wall).

2. Those exposed to flood waters bearing seeds.

3. Those in which the original soil and hydrological regimen have suffered little or no alteration. Undisturbed soils are often reforested from coppice sprouts of cut trees or from seed banks.

Natural regeneration usually attracts an abundance of noxious species, particularly vines, which may interfere with the establishment of preferred species. Noxious species, from the standpoint of timber management, include boxelder and river birch, which may capture the site, particularly if the adjacent seed wall is depauperate in seed supply of preferred species. Conner et al. (1986) reported that natural regeneration is largely unsuccessful in restoring cleared stands of cypress in the Southeast; maples and other hardwoods tend to replace the cypress. Too many factors must exist simultaneously for successful germination and establishment. Erratic flooding interrupts that constellation of factors, in part due to anthropomorphic activities associated with flood control, road construction, and petroleum exploration.

When natural regeneration is prescribed for restoration projects, sites should be disked or the existing vegetation otherwise largely removed prior to the dormant season. If overbank flooding does not occur or there is a seed failure in the seed wall, the procedure may have to be repeated the next year. Natural regeneration may require several years before full stocking is realized, especially in those areas remote from the seed wall or isolated from flood waters that carry seeds.

Direct Seeding

On the fertile floodplains of the Mississippi River, natural regeneration may be negligible for oaks and other large-seeded trees such as sweet-pecan, relative to the suite of more competitive species that rapidly colonize exposed substrates.

Direct seeding by planting acorns (manually or by machine) has proven quite effective in reforesting bottomlands, where oaks are absent (Johnson and Krinard 1985a,b). In contrast, direct broadcast seeding of hardwoods has met with mixed success (Tackett and Graves 1979, 1983; Cross et al. 1981; Wittwer et al. 1981) and is not recommended. At best, broadcast seeding generally results in patchy, overstocked, monotypic stands of low diversity.

Pelletized seeds are being developed for direct seeding, in which the coatings contain fertilizers, fungicides, and other substances designed to enhance successful germination (Almagro et. al. 1987).

ARTIFICIAL REFORESTATION

Many options exist for artificial regeneration, including planting of:

1. Bare root seedlings.

2. Containerized seedlings.

3. Stem cuttings.

4. Transplanted saplings or larger trees, usually by tree spade.

Bare Root Seedlings

The standard stock for forest plantations has been bare root seedlings. The technology for producing them is well developed, and they survive and grow well in moist substrates. Seedlings should be grown from local sources, preferably those trees whose seeds have been progeny-tested to determine acceptable growth performance.

Good quality seedlings should possess a multitude of attributes in order to survive and grow well. For details, the inexperienced planter should consult Williams and Hanks (1976). From our personal experience, we offer the following recommendations:

1. Seedlings should be hardened or fully dormant when planted.

2. Preferably, seedlings should be planted directly; otherwise they should be properly chilled to 1 to 4 degrees C and stored after lifting.

3. Seedlings that have begun to flush with new growth are a poor risk and should not be planted.

4. Seedlings should have well developed terminal buds, and the roots should be highly

fibrous to facilitate initial absorption of water and nutrients when outplanted.

5. When planted, hardwood seedlings are typically much larger than pine seedlings, and hence they cannot be planted with the same speed or ease. Many planting failures have resulted from a field boss encouraging higher production by root and top pruning. If a seedling has been properly handled in the nursery, no root pruning is necessary, and planting crews should be discouraged from doing so.

6. In terms of physiological characteristics, seedlings should be grown in ways that promote optimal carbohydrate and mineral reserves. Seedlings should be free of symptoms of mineral deficiencies and should be turgid. Weak seedlings cannot cope with physiological drought that occurs before their roots re-establish contact with soil at the planting site.

7. Seedling size is the most important characteristic for hardwoods. In our experience, stem caliper at the root collar should be between 10 to 13 mm and never less than 6 mm. Most nurseries do not cull seedlings during packaging. Inferior seedlings should be discarded by planting personnel and not counted as planted. The cost of bare root seedlings is low, relative to other project costs. Project managers have no economic rationale for cutting corners and skimping on quality control by planting inferior seedlings.

8. Seedlings suffer the cumulative insults wrought upon them from the lifting and packing operation to their shipping, handling, and field planting. Seedlings should be in transit as short a time as possible, to reduce the time they are subject to molding and desiccation. The planter should be encouraged to accept a consignment of seedlings at the nursery, in order to eliminate delays in shipment. Bags of seedlings should be transported in the shade and protected from desiccating wind and kept at 1 to 4 degrees C, making certain to avoid freezing temperatures. Initial growth reflects the care provided to seedlings at every stage of handling.

Containerized Seedlings

Container-grown seedlings have been successfully established in sites that are too harsh for adequate survival of bare root seedlings. They may be planted further into the growing season than can bare root seedlings. Several types of containers have evolved in nursery practice, and all have potential in wetland mitigation:

1. "Tubelings", also called "plugs" or "tray trees", each consisting of a tree seedling and a minimal soil mass enclosed by the root system in plastic or styrofoam containers, which may be removed before shipping.

2. Paper sacks or other biodegradable containers that are planted along with the seedlings.

3. Systems in which the container and the growth medium are one and the same, such as molded peat or wood fiber.

4. Gallon-sized (usually) plastic pots or bags, each with a tree from about 7 to 24 months old, which is delivered in its container and planted with its potting medium intact.

The technology for containerized stock is expanding because such stock best survives mishandling and inhospitable site conditions. However, container-grown seedlings cost several times that of bare root seedlings, are more difficult to plant, and should be limited to unsuitable sites for bare rootstock because of economic constraints.

In spite of their endurance relative to other nursery stock, containerized seedlings must be transported with extreme care in summer. Temperatures may rise above 35°C in closed trucks, and transpirational stress may be lethal to trees carried in open trucks. If containerized trees sit in the sun for a few hours before planting, transpiration may deplete all soil moisture, and their black plastic pots may absorb enough heat to kill roots. We have seen well laid plans for plantings go awry when a brief thunder shower prevented hot trucks from being unloaded. The trucks became stuck in the rain-softened substrate so that trees could not be unloaded where they could be stored temporarily in the shade for a few hours before planting.

Cuttings

Certain hardwood trees lend themselves to being propagated vegetatively by rooted cuttings. Poplars, willows, sycamore, green ash, sweetgum, and several others will grow from such cuttings. Hardwood cuttings or "whips" are sections about 30-55 cm long, harvested from one-year-old twigs in the dormant season and stored in plastic bags just above freezing until planting in the spring. Optimum whip diameters range from 8 to 13 mm.

Willow and poplar whips are safely out-planted. For other species, greater survival and early growth is achieved if cuttings are pre-rooted in a nursery and then handled as bare root seedlings. Normally, cuttings are planted vertically, the stem apex flush with the soil surface or with no more than 3 cm exposed, the buds pointing up. Allen and Klimas (1986) provided additional details on planting whips along shorelines.

Saplings

Small saplings that are less than three years old have occasionally been transplanted from natural forests to project sites (e.g., Cox 1987). Transplanting large saplings is often prohibitively expensive and is likely to produce mixed results on mitigation sites. Transplanted trees sometimes suffer high mortality, even if properly balled, bagged, and pruned. Saplings often fail to grow for several years because of the severe reduction in root biomass upon lifting and the interruption of optimum growing conditions that were provided in the nursery or in the undisturbed forest interior.

Seedlings planted correctly will probably grow to the same size as tree-spaded saplings in 5 years. A notable exception is cabbage palm, mature specimens of which can be successfully transplanted in moist but well aerated soils.

The most extensive use of transplanting is at two projects in Palm Beach County, Florida, where entire forests of mature cabbage palms, dahoon hollies, red maples, laurel oaks, and slash pines were tree-spaded to new locations (Posey et al. 1984).

Planting Options

From our observations throughout the region, the successful plantings are those that have had proper site preparation prior to planting, followed by surveillance after planting to insure that the post-planting maintenance activities are properly and timely implemented. For example, a season of severe weed competition without control can negate hundreds of man hours of effort.

Intensive site preparation is also essential for several reasons:

1. Planting and maintenance will be safer and less expensive.

2. Debris will not scour the planting site at flood stage.

3. Access to trafficking the site by foot or machinery will be easier.

4. Post-establishment monitoring will be facilitated.

A desirable treatment is a late growing season harrowing to control weeds and brush prior to planting and to loosen crusted or impacted soils. If fertilizers are to be applied, they should be timed with the planting effort.

The options available for tree planting are many and varied, depending on the type and size of the planting stock, the size and land form of the project site, the presence of stumps and debris, and the availability of equipment and labor. For large acreages, unrooted cuttings or seedlings can be planted expediently using continuous-furrow planting machines. Where more control of species is required, cuttings and seedlings may be planted by hand using any of a variety of dibbles, hoedads (a commercially available tree planting hoe), shovels, or other hand tools.

Planting crews can be contracted either on a negotiated fixed fee with first-year survival rates guaranteed or on a variable scale, where the price per hectare can be adjusted within a range to reflect the quality of the planting job. In either case, responsibility for quality control ultimately lies with the project manager.

When planting, the seedlings or cuttings should be put in a hole large enough to accommodate the entire root system without recurving or "J-rooting", and at a depth equal to or slightly above the root collar. Exposed roots and J-rooting reduce vigor and invite pests and disease. Seedlings should be planted erect. In our experience, hardwoods exhibit less apical dominance than conifers and may not straighten up if planted with a lean exceeding 10 percent of perpendicular. Leaning seedlings will either sprout a new leader from near ground level or die. Finally, the seedling should be heeled into the ground, firmly packing soil around the roots. A firm tug on the shoot will indicate whether or not a seedling has been properly planted.

In the Southeast, the preferred time to plant is from January to March. Seedlings should not be planted in frozen soil. A good day for planting will be cool but above freezing, with little or no wind. Seedlings should never be carried with their roots exposed. On cold or windy days, fine roots desiccate in the time required to remove the seedling from the bag and place it in the ground. On marginally acceptable days, roots should be protected in canvas planting bags or in buckets of water.

Pitfalls

The degree of success, relative to pre-determined project goals, is generally proportional to the amount of on-site supervision by qualified professionals during site preparation and planting and to the frequency of monitoring during the first few months thereafter. We strongly recommend that permit conditions require competent supervision and monitoring. We have listed some grievous errors, made at forested wetland creation sites in the absence of supervisory personnel, that emphasize the importance of our recommendation:

1. Vigorous saplings were loaded at a nursery into open trucks and delivered to the project site dead from wind-burn and desiccation. They were planted, because nobody told the planting crew otherwise.

2. Potted trees were delivered on a Friday afternoon and allowed to roast in the direct summer sun before being planted dead on Monday.

3. Gallon-sized trees were removed from flat-bottomed pots and were planted in holes dug with pointed spades. Air pockets remained beneath their root-balls and stressed or killed many saplings.

4. Nurseries shipped trees of the wrong species. They were planted.

5. Mesic trees were planted in hydric sites.

6. Cuttings of willows and cottonwoods were planted upside down.

7. Project sites were not fenced or staked, and work crews planted up to 40 percent of their tree seedlings on adjacent land.

8. Gallon-sized stock was ordered without specifications for hardening and was heavily fertilized at the nursery shortly before being shipped. The tree roots continued to exploit the potting mix and did not extend into project soils. Upon the arrival of the dry season, the root systems were insufficient to obtain adequate soil moisture, and the trees suffered stress and retarded growth.

9. Rhizomatous turf grasses were sown as cover crops at several sites. Many or most gallon-sized trees succumbed to competition within a year of being planted.

NON-ARBOREAL VEGETATION

Importance of Undergrowth

Visually, taller trees characterize forests. Most plant species in southeastern forests, though, are the herbs, shrubs, woody vines, and small trees, which collectively comprise the undergrowth. For example, of the 409 species of vascular plants recorded in one riverine forest, only 36 (8.8%) were trees and the other 373 species comprised the undergrowth (Clewell et al. 1982). Floristically, the undergrowth characterizes the forest and should be considered in forest creation projects. Otherwise, the project may become little more than a tree farm. Besides representing an important floristic element, the undergrowth contributes functional values, including wildlife food and cover (Harlow and Jones 1965), critical habitat for predacious arthropods that control insect populations (Altieri and Whitcomb 1980 and references), nutrient cycling (Bormann and Likens 1979), and soil stabilization and sediment trapping at flood stage.

Most projects tacitly assume that the undergrowth will voluntarily appear. This assumption was not supported by an extensive survey of abandoned phosphate mines, which were allowed to revegetate in a manner similar to old field succession (Florida Bureau of Geology 1980). In this survey, there were 27 sites that had been abandoned from 45 to 70 years, which:

1. Covered at least 20 hectares.

2. Contained forests with at least 20 meters-square of basal area per hectare--all representing natural regeneration. The mean number of undergrowth plant species per stand was 63 and the maximum was only 92, many or most of them weeds and exotics that persisted on forest edges. These numbers of species pale in comparison to the above-mentioned 373 undergrowth species recorded in only 4.6 hectares of an undisturbed riverine forest within the same mining district.

The purposeful introduction of undergrowth plants can be accomplished in four ways:

1. Direct planting (sprigging) from natural forests.

2. Out-planting of nursery-grown stock.

3. "Topsoiling" or "mulching" with topsoil from a donor forest.

4. Transferring blocks of forest topsoil intact with a tree spade from a donor forest. Donor forests may already be permitted for mining or other development, and the use of their topsoil represents the conservation of a resource at a replacement wetland, which would otherwise be lost.

The richness of seed banks from forest soils has been demonstrated in microplot studies (Farmer et al. 1982; Wade 1986) and in pilot plots (Clewell 1983). At one project site (Clewell 1986a), topsoil transfer was accomplished with scrapper pans and bulldozers, which thoroughly mixed the soil. Plants of many undergrowth species sprouted from this topsoil. Plants of some undergrowth species of the donor forest appeared at the project site only at the bases of trees that were transplanted with a tree spade. Apparently, not all species can be transferred successfully, unless the soil is kept intact. Undergrowth transfers that survived the harsh, exposed conditions of the project site were those planted in constantly moist or wet soil, in which tall weeds quickly grew and provided partial shade.

Once weed growth is suppressed by the initial tree canopy closure, undergrowth plants are expected to proliferate. For that reason, it is not essential to establish large numbers of undergrowth plants. Instead, we recommend that a few plants each of selected undergrowth species be evenly distributed throughout a project site. Some undergrowth species may be introduced immediately (Figure 5). Other species introductions may have to be delayed a few years, until shaded openings appear beneath thickets of larger saplings.

Topsoiling is advantageous, because pedogenic benefits accrue apart from undergrowth transfer. These benefits include the widespread introduction of mycorrhizal fungi and soil micro-fauna, as well as organic matter that enhances moisture retention, crumb structure, and cation exchange capacity. Availability of a donor site or cost considerations may necessitate spot introductions of topsoil, sprigging, or dependence on natural regeneration of undergrowth species.

A passive method of augmenting preferred species is to attract seed vectors, i.e., birds or other animals that consume seeds and pass some of them unharmed in their feces. King et al. (1985) recommended the installation of brush piles and poles (for bird perches) to attract animals to project sites. Such installations have been recently attempted, and tall dead trees were "transplanted" as bird perches at one site, much to the amusement of the work crew (J. Sampson pers. comm.). We applaud these novel approaches and hope that they prove effective. Nonetheless, we caution that brush piles could attract rabbits and cotton rats which may cause serious damage by eating tree seedlings and saplings.

POST-PLANTING MANAGEMENT

Protection

Project sites should be fenced whenever possible prior to revegetation. Unfenced sites are subject to cattle grazing, dirt bikes, RV vehicles, etc. Control of herbivores should be considered. Indirect methods of control are usually the most cost-effective. For example, weed control around young trees can discourage browsing by rodents which are reluctant to venture beyond cover. A frequently disked buffer zone around the project area will deter rodent trespass and will also provide a fire break. Large debris piles harbor a variety of pests and should be eliminated. Direct methods of herbivore control are sometimes unavoidable. For example, Louisiana foresters planted 10,000 cypress seedlings in 1988, each enclosed in a chicken wire sheath to discourage nutrias (Will Conner pers. comm.), even though such sheaths are not always effective (Conner and Toliver 1987, Conner 1988).

The insects standing by to ravage any project are legion--defoliators, borers, sapsucking insects, gall producing insects, etc. A complete discussion of them is beyond the scope of this chapter. Rather, our discussion of this potential problem is to encourage frequent vigilance, so that remedies can be applied to save plantings. Vigorous planting stock usually overcomes insect depredations.

Maintenance

A major goal of planning should be to minimize the need for maintenance activities. The value of careful planning and competent supervision of site preparation and planting in eliminating unanticipated maintenance costs and accelerating project release cannot be over-emphasized. Unlike production forestry, total removal of competition is not necessary as long as acceptable survival and resistance to climatic extremes, diseases, and pests are obtained. Intensive competition control can double or triple initial growth of planted trees. But such efforts may not be economically justified in mitigation projects. Instead, acceptable competition control can often be attained by site preparation.

Post-planting weed control, if necessary, can be achieved by mowing or disking between trees or with pre-emergent herbicides applied in circles or strips centered on trees. The most benefit, in terms of tree survival and growth, is derived from first-year weed suppression; however, substantial benefits may be accrued by a several year program of control. Herbicidal applications should be prescribed with the goals of minimizing losses of preferred undergrowth and protecting water quality.

Hardwoods are generally quite sensitive to herbicides, and extreme care should be exercised in their use. Direct contact with systemic herbicides must be avoided. Foliage and green bark must be protected. The effectiveness of pre-emergent chemicals is based largely on their retention in the soil above the root zone of planted stock. Heavy rains or disruption of the soil

Figure 5. Ferns and shrubs planted the previous year in a moist, semi-open site to augment preferred species distribution along reclaimed Dogleg Branch on phosphate mined land in central Florida.

profile can cause excessive leaching of herbicides into the soil, resulting in high mortality, especially in newly planted stock. Equipment must be calibrated to assure herbicidal application at recommended rates. Even if all precautions are heeded, risk of injury to young plants cannot be completely eliminated. Ultimately, it is the experience of the operator under familiar conditions that provides the best knowledge of herbicidal behavior.

The USDA, Forest Service (1988) issued detailed evaluations of several herbicides, including glyphosate, which is relatively immobilized in soil and which biodegrades rapidly. In all instances, herbicides should be applied by a licensed operator who is familiar with best management practices for using herbicides in wetlands.

Other common maintenance activities include repairs to prevent erosion and sedimentation that occur within the first year or two of the project, to replant trees as needed to achieve prescribed densities of stocking, and to repair fences. Replacement trees require the same cultural care as the initial trees; otherwise, the replanting effort may fail.

Fire protection should be considered. Fire breaks may need annual harrowing. Herbicidal removal of highly flammable, noxious species may be judicious, e.g., colonies of congon grass. Manual removal of weeds by machete or weed eaters is futile and results in the inadvertent loss of at least some planted stock or other preferred vegetation.

MONITORING

Monitoring Functions

The reasons for monitoring are to alert the project engineer to additional project activities that may be needed and to allow agency personnel to know when a project can be released from regulatory purview. Two modes of monitoring are needed:

1. Inspection tours.

2. Quantitative sampling events.

Inspection tours allow the identification of problems as they arise, so that maintenance can be prescribed and implemented promptly. Such tours may be needed weekly for the first few weeks after final earth-moving activities in order to check for erosion and sedimentation and to make sure any water control structures and irrigation equipment are properly functioning. Thereafter, inspection tours should be made monthly until the project is released.

Quantitative sampling allows the periodic assessment of the degree to which each of the specific project goals has been achieved. These goals should be carefully developed during initial planning, before any project site activity begins. The goals should be stated in terms of success criteria that can be objectively measured by monitoring protocols.

Quantitative monitoring events should occur annually for at least the first 5 years and longer if major project activities are continuing. Quantitative monitoring may not be needed thereafter, except for the final event, which must document that all success criteria have been attained. Then the project should be released from regulatory liability.

Monitoring data should be incorporated in reports that include photos. The first report should describe the project site, list permit requirements, list success criteria, describe site preparation and initial planting activities, describe quantitative monitoring methods, and present initial monitoring data. Subsequent reports should present monitoring data and describe maintenance activities undertaken since the issuance of the previous report. Each report should evaluate monitoring data in terms of the degree of attainment of success criteria.

Reports should be made accessible to restoration professionals by deposition in the library of a research facility or institution actively involved in restoration work.

Success Criteria

The development of success criteria (performance standards) is necessarily site-specific and depends on project goals, logistical and legal constraints, proposed land use, and the resources of the local environment. Success criteria should allow project release from regulatory purview and liability as soon as the project can withstand the stresses of the local environment (floods, droughts, etc.). That time will usually correspond with the time of initial canopy closure and should arrive within five or six years for a carefully conceived and well executed project. Thereafter, any further activities would not likely produce substantive benefit to the project or alter the inertia of seral development. If the project can no longer be significantly manipulated, then it should be released.

Intermediate operations, such as pre-commercial thinning, may accelerate seral development or improve timber values. These operations, though, are ancillary to successful forest establishment. They may be opted by the owner or land management agency after regulatory release, in accordance with long term land use plans. We recommend no intermediate forest practices until a bottomland hardwood forest is 20 years old because natural succession processes better dictate which tree species are better suited for a mitigation site than do project personnel.

Because of changes in agency personnel and priorities, institutional memory may not extend beyond five or six years. Thereafter, the project could become an institutional liability. Of equal importance, the owner cannot be expected to carry the costs of a project indefinitely, and a later change of ownership may complicate the interpretation of project responsibilities. For these reasons, we recommend the earliest possible release. Early release, though, demands careful project planning and critical surveillance which should be the focus of agency and owner interest during project development.

Agency personnel, who are responsible for ensuring that projects are indeed successful, are understandably reluctant to release projects without overwhelming cause. We urge careful scrutiny of existing projects with the object of testing the validity of our rationale for proposing early release.

Reference Wetlands

For marsh restoration/creation projects, reference wetlands are sometimes used as a standard for measuring particular ecosystem functions, such as providing acceptable water quality or wildlife habitat. Water quality parameters are measured. Macro-invertebrate and wildlife populations are sampled. Data are compared with reference data.

Forest projects differ from marsh projects. Trees are not planted with the object of providing immediate water quality benefits. If water quality is a potential problem at phosphate mine projects, the site is reclaimed initially as a marsh, into which trees are planted. Trees and preferred forest undergrowth will eventually and gradually assume the function of providing good water quality from marsh plantings. Figure 6 shows that process at a site where maidencane, softrush, etc., were initially introduced. Three months later, after marsh vegetation captured the site, the trees were planted. The trees will not fully assume the water quality function from marsh vegetation for at least several more years.

Most of the other functions of forested wetlands require full forest development before they can be compared with reference wetlands. For that reason, reference wetlands for forest projects are necessarily limited for functional comparisons, but they may be employed for certain vegetational comparisons, particularly preferred species composition. Forests are much more complex than marshes in terms of composition and physiognomy. With complexity comes ecosystem stability. Added stability brings with it a margin of safety in equating vegetational composition and structure with functional attributes.

We advise against the adoption of success criteria that require direct comparisons with one specific, natural, "reference" wetland, which is a specific requirement in some, recently issued permits. Climax theorists have imbued three generations of ecologists with the idea that a given environment allows little variation between pristine stands of the same community. Even though classical climax theory is outdated, its archetypal nuances stubbornly linger. A pertinent example is the presumption that relatively high similarity indices exist between any two stands of the same community. Experience dictates to the contrary. Similarity values for canopy trees were calculated for 71 pairs of stands within a mature, undisturbed, contiguous riverine forest (Clewell 1986a). Similarity values ranged from 0.04 (virtually no species in common between two sites) to 0.98 (virtually all species in common and with the same relative densities). Differences in undergrowth paralleled disparities in arboreal composition. Inter-stand discrepancies reflected a multiplicity of stochastic events, such as accidents of dispersal (initial floristic composition phenomena) or localized effects of fire or weather-related phenomena.

Similarity indices are invalid measures of project success for several reasons, not the least of which is the disparity in similarity between natural stands and thus the subjectivity in selecting a particular reference wetland. Another reason is that natural volunteer seeding could drastically alter tree species composition and density at a project site specifically planted to match a particular reference wetland. A third reason is the disparity in seral stage between recently installed projects and indefinitely old reference stands.

In lieu of specific reference wetlands and similarity indices, we recommend a comparison of species composition at the project site with its generalized forest ecosystem, as it naturally occurs in that locality. The developing forest at the project site should fit into the gamut of variation known for that forest type. To that end, lists of preferred species are critical. The new canopy should consist only of preferred tree species. Their relative densities should approximate those in regional forests of the same type, with allowances for natural regeneration and disproportionate natural thinning.

Undergrowth plants representing a wide array of preferred species should be well distributed throughout the project site. Their frequency is more important than their density or cover initially, because they will proliferate in response to forest canopy development at the expense of the initial weed flora. For that reason, the presence of weeds and most other non-preferred species can be ignored, unless their competition is interfering with tree establishment.

Monitoring Protocols

The only focus of a monitoring program should be to answer the specific questions posed by the success criteria. Efficient protocols should be adopted that avoid the elaboration of impressive but unessential or duplicative data. Data collection and analyses should be accomplished in ways that document trends. For these reasons, field methods should be consistent, and permanent sampling stations are preferred over the random selection of sampling points at each monitoring event.

We offer the following five recommendations for success criteria, with the understanding that these will often require modification or additions concomitant with the goals and conditions of individual projects.

Figure 6. Bayhead swamp at Hall Branch, created on reclaimed, phosphate mined land in central Florida. Pondcypress, sweetbay, popash, etc. were planted as 45 cm tall (approximately) containerized nursery stock 31 months prior to the photo. Maidencane, the profusely growing grass in the foreground, was among the marsh plants introduced at this site three months before trees were planted.

Public Law 95-87 specified additional success criteria, as discussed by Chambers and Brown (1983), for certain mine reclamation projects.

1. The watershed area within the same ownership shall be functioning in a manner that is consistent with project goals. For example, the watershed should be reclaimed to its approximate original grade and suitably vegetated with a cover crop prior to project initiation.

2. The substrate shall be stabilized and any erosion shall not greatly exceed that expected under normal circumstances in natural forests similar to that being restored or created.

3. There shall be a density of at least 980 potential overstory trees per hectare (400 per acre) that are at least 2 meters tall. No hectare-sized area shall contain less than 860 trees (350 trees per acre), regardless of height. All trees shall be preferred species (as defined above) and shall have been rooted at the project site for at least 12 months. They shall include a prescribed number of species (depending on baseline information) that have a minimal density of at least 25 trees per hectare per species overall at the project site. They shall occur in proper zonation, e.g., hydric trees in wet sites.

4. There shall be an adequate representation of undergrowth species. (The specifics depend on the forest type and baseline data and must be tailored to individual projects.) At a minimum, there shall be at least 10 preferred undergrowth species in each 0.4 hectare-sized

area of the project site, all represented by plants that have been rooted at the project site for at least 12 months.

5. Streams and standing water bodies shall be of sufficient water quality so as not to inhibit reforestation or interfere with the attainment of other success criteria.

For the watershed requirement, we recommend monthly readings of a few, strategically placed shallow piezometers wells and staff gauges to monitor ground and surface water in the project site and its immediate watershed. Readings can be made during monthly project tours. Without these readings, it may be difficult to determine whether or not the water balance is adequate to attain project goals.

For the substrate stability requirement, narrative descriptions and photographs will often suffice. If sedimentation threatens to be a problem, a sediment measuring protocol should be implemented.

For tree density monitoring, we recommend a suitable number of randomly selected and permanently staked transect lines that traverse the entire project site perpendicular to stream flow or radiating from the center of an isolated wetland. Trees that fall within a prescribed distance on either side of the line are tallied by species and measured for height.

Large project sites should be clearly staked into sectors, so that intra-stand variations in tree density and other parameters can be determined easily. Should tree densities fall slightly below predetermined thresholds, the potential for precocious seed production should be evaluated before additional plantings are prescribed. Oaks sometimes produce acorns five years after being planted, and maples can produce seeds profusely at that age. Three-year-old cypress trees frequently produce cones, and dahoon hollies regularly fruit in the same year in which they were planted. These young trees may augment existing densities sufficiently without additional tree plantings.

For undergrowth, a floristic list of preferred species can be made from reconnaissance within each sector.

For water quality, EPA-approved methods should be followed. Stream fauna sampling data, if required, should be interpreted in context of the stage of forest development. Streams at new project sites are not expected to exhibit faunal characteristics of shaded, detritus-driven streams within mature forests.

Water quality monitoring may need to be done only in the first and last years, unless unusual circumstances prevail. For example, if mine process water is used at project sites, it should first be tested to see if it contains petroleum derivatives or flocculants that could kill vegetation.

INFORMATION GAPS AND RESEARCH NEEDS

The silvicultural data base is extensive, and most important facts are already known with regard to planting and growing trees of key species. The Florida Institute of Phosphate Research embarked on an ambitious tree establishment program in 1988 to test the effectiveness of various nurse crops, soil amendments, cultural techniques, and competition reduction measures, and to document the extent of precocious seed production within project sites. This program should supplement the arsenal of techniques already available to forest restoration specialists. David Robertson (in Harrell 1987) called for more information on direct seeding and tubeling technology.

Besides filling gaps in sylvicultural technology, five topics of importance await consideration:

1. The extent and circumstances need to be documented by which natural dissemination can contribute to forest regeneration. Very little has been published that addresses that topic. Unless more is known about the potentials for volunteer seeding of trees, little can be done to exploit this natural process.

2. Techniques for undergrowth establishment need to be developed with species-specific precision. We have emphasized the floristic importance of the undergrowth. We have recommended limited interplanting of undergrowth species upon the untested assumption that plants of these species will eventually proliferate into a diverse, naturally occurring undergrowth. This method needs testing in a variety of forest types.

3. The existing data baseline needs expansion for southeastern forest types in terms of composition, form, and function. Too much of our knowledge is based on narrative evidence or on more detailed studies of sites that do not represent the geographic range of a forest type. Many ecological studies present only tree data, or they concentrate the riches of the undergrowth into a few, sterile ecological index values.

4. The assumption needs to be tested that functional values are directly related to species composition and forest structure. Through the extensive work of Ewell and

Odum (1984) and their many associates, we can begin to make form/function assessments for cypress domes. Other bottomland forest types await study.

5. Existing projects should be monitored rigorously, with the intention of documenting species composition and forest physiognomy with respect to age and history of project activities. In addition, the degree to which various functional services are being provided should be assessed. A monitoring team should be assembled to identify key projects throughout the Southeast and thoroughly examine them.

LITERATURE CITED

Allen, H.H. and C.V. Klimas. 1986. Reservoir Shoreline Revegetation Guidelines. Technical Report E-86-13, U.S. Army Engineer Waterways Experiment Station, Vicksburg, Mississippi.

Almagro, G.A., D.F. Martin and M.J. Perez-Cruet. 1987. Pelletization of pine and oak seeds: Implications for land reclamation. Florida Scientist 50:13-20.

Altieri, M.A., and W.H. Whitcomb. 1980. Weed manipulation for insect pest management in corn. Environmental Management 4:483-489.

Anding, G. 1988. Riparian zone management associated with reservoir projects in the Vicksburg District, p. 22-25. In C.O. Martin and H.H. Allen (Eds.), Proceedings of the U.S. Army Corps of Engineers Riparian Zone Restoration and Management Workshop 24-27 February 1986. Misc. Paper EL-88-3, U.S. Army Engineer Waterway Experiment Station, Vicksburg, Mississippi.

Anonymous. June, 1984. Turning farmland into forests, p. 10-11. In Woodlands for Wildlife, Mississippi Dept. Wildlife Conservation.

Anonymous. 1986. Results of oaks direct seeding are promising. Tree Talk 7(2):9, 11.

Anonymous. 1988. Biocontrol mistake? Aquaphyte (Center for Aquatic Plants, University of Florida, Gainesville, Florida) 8:1-2.

Bell, D.T. 1974. Studies on the ecology of a streamside forest: composition and distribution of vegetation beneath the tree canopy. Bull. Torrey Bot. Club 101:14-20.

Best, G.R., P.M. Wallace, W.J. Dunn, and J.A. Feiertag. 1983. Enhancing ecological succession: 4. Growth, density, and species richness of forest communities established from seed on amended overburden soils, p. 377-383. In Proceedings, 1983 Symposium on Surface Mining, Hydrology, Sedimentology and Reclamation, University of Kentucky, Lexington, Kentucky.

Bormann, F.H., and G.E. Likens. 1979. Pattern and Process in a Forested Ecosystem. Springer-Verlag, New York.

Broadfoot, W.M., and R.M. Krinard. 1961. Growth of hardwood plantations on bottoms in loess areas. Tree Planters' Notes 48:3-8.

Bull, H. 1949. Cypress planting in southern Louisiana. Southern Lumberman 179(2248):227-230.

Cairns, J., Jr. 1985. Facing some awkward questions concerning rehabilitation management practices on mined lands, p. 9-17. In R.P. Brooks, D.E. Samuel, and J.B. Hill (Eds.), Wetlands and Water Management on Mine Lands. School of Forest Resources, Pennsylvania State University, State College, Pennsylvania.

Cairns, J., Jr. 1986. Restoration, reclamation, and regeneration of degraded or destroyed ecosystems, p.465-484. In M.E. Soule (Ed.), Conservation Biology: Science of Diversity. Sinaur Publ., Ann Arbor, Michigan.

Carter, A.R. 1987. Cedar restoration in the Dismal Swamp of Virginia and North Carolina, p. 323-325. In A.D. Laderman (Ed.), Atlantic White Cedar Wetlands. Westview Press, Boulder/London.

Chambers, J.C., and R.W. Brown. 1983. Methods for Vegetation Sampling and Analysis on Revegetated Mined Lands. U.S. Department of Agriculture, Forest Service, General Technical Report INT-151.

Clewell, A.F. 1983. Riverine forest restoration efforts on reclaimed mines at Brewster Phosphates, central Florida, p. 122-133. In D.J. Robertson (Ed.), Reclamation and the Phosphate Industry, Florida Institute of Phosphate Research, Bartow, Florida. Publ. No. 03-036-010.

Clewell, A.F. 1985. Guide to the Vascular Plants of the Florida Panhandle. University Presses of Florida/Florida State University Press, Tallahassee, Florida.

Clewell, A.F. 1986a. Vegetational Restoration at Dogleg and Lizard Branch Reclamation Areas, Fourth Semi-Annual Report, Autumn, 1986. Brewster Phosphates, Lakeland, Florida.

Clewell, A.F. 1986b. Vegetational Restoration at Hall Branch Reclamation Area, Autumn, 1986. Brewster Phosphates, Lakeland, Florida.

Clewell, A.F. 1986c. Alafia River Crossing "A" at Lonesome Mine, Statistical Report, Summer, 1986. Brewster Phosphates, Bradley, Florida.

Clewell, A.F. 1986d. Dragline crossing "B" restoration at Lonesome Mine, Statistical Report, Summer, 1986. Brewster Phosphates, Bradley, Florida.

Clewell, A.F. 1986e. Reforestation Trials, 6th Year Results of Bare Root Tree Seedling Establishment and the Mulching Technique for Sand Pine Scrub Restoration. Brewster Phosphates, Lakeland, Florida.

Clewell, A.F. 1986f. Assessment of 5.75 Acres North of Old Dike at McMullen Branch Restoration Area. Brewster Phosphates, Bradley, Florida.

Clewell, A.F. 1987. Vegetational Restoration at Hall Branch Reclamation Area, 1987 Monitoring Report. Brewster Phosphates, Lakeland, Florida.

Clewell, A.F. 1988. Bottomland hardwood forest creation along new headwater streams, p. 404-407. In J.A. Kusler, M.L. Quammen, and G. Brooks (Eds.), National Wetland Symposium: Mitigation of Impacts and Losses. Association of State Wetland Managers, Berne, New York.

Clewell, A.F., J.A. Goolsby and A.G. Shuey. 1982. Riverine forests of the South Prong Alafia River system, Florida. Wetlands 2:21-72.

Clewell, A.F. and A.G. Shuey. 1985. Vegetational Restoration at Dogleg and Lizard Branch Reclamation Areas, Second Semi-Annual Report, Autumn, 1985. Brewster Phosphates, Bradley, Florida.

Conner, W.H. 1988. Natural and Artificial Regeneration of Baldcypress in the Barataria and Lake Verret Basins of Louisiana. Ph.D. dissertation, Louisiana State University, Baton Rouge, Louisiana.

Conner, W.H. and J.R. Toliver. 1987. Vexar seedling protectors did not reduce nutria damage to baldcypress seedlings. U.S. Forest Service, Tree Planters Notes 38(3): 26-29.

Conner, W.H., J.R. Toliver, and F.H. Sklar. 1986. Natural regeneration of baldcypress (Taxodium distichum [L.] Rich.) in a Louisiana Swamp. *Forest* Ecology and Management 14:305-317.

Cowardin, L.M., V. Carter, F.C. Golet, and E.T. LaRoe. 1979. Classification of Wetland and Deepwater Habitats of the United States. U.S. Fish and Wildlife Service, FWS/OBS-79/31.

Cox, R.M. 1987. Revegetation projects in Orlando lakes. Aquaphyte (Center for Aquatic Plants, University of Florida, Gainesville, Florida) 7:15.

Cross, E.A., F.C. Gabrielson, and D.K. Bradshaw. 1981. Some effects of vegetative competition and fertilizer on growth, survival, and tip moth damage in loblolly pine planted on alkaline shale surface mine spoil, p. 59-63. In Proceedings, 1981 Symposium on Surface Mining, Hydrology, Sedimentology and Reclamation, University of Kentucky, Lexington, Kentucky.

Ericson, W.A. and J. Mills. 1986. Florida's non-mandatory reclamation of phosphate lands, current regulations and their implementation, p. 191-196. In Proceedings 1986 National Symposium on Mining, Hydrology, Sedimentology, and Reclamation, University of Kentucky, Lexington, Kentucky.

Erwin, K.L. 1985. Fort Green Reclamation Project, Third Annual Report, 1985. Agrico Chemical Company, Mulberry, Florida.

Erwin, K.L., G.R. Best, W.J. Dunn, and P.M. Wallace. 1985. Marsh and forested wetland reclamation of a central Florida phosphate mine. Wetlands 4:87-104.

Ewel, J.J., D.S. Ojima, D.A. Karl, and W.F. DeBusk. 1982. Schinus in Successional Ecosystems of Everglades National Park. Report T-676, South Florida Research Center, Everglades National Park, Homestead, Florida.

Ewel, K.C. and H.T. Odum (Eds.), 1984. Cypress Swamps. University Presses of Florida/University of Florida Press, Gainesville, Florida.

Farmer, R.E., Jr., M. Cunningham, and M.A. Barnhill. 1982. First-year development of plant communities originating from forest topsoils placed on southern Appalachian minesoils. J. Applied Ecology 19:283-294.

Fisher, R.F. and F. Adrian. 1981. Bahia grass impairs slash pine seedling growth. Tree Planters' Notes (1):19-21.

Florida Bureau of Geology. 1980. Evaluation of Pre-July 1, 1975 Disturbed Phosphate Lands, Appendix F. Florida Department of Natural Resources, Tallahassee, Florida.

Gunderson, L.H. 1984. Regeneration of cypress in logged and burned strands of Corkscrew Swamp Sanctuary, Florida, p. 349-357. In K.C. Ewel and H.T. Odum (Eds.), Cypress Swamps. University of Florida Press, Gainesville, Florida.

Gurr & Associates, Inc. 1986. 1986 Annual Report, Wetlands Vegetation Monitoring, Mobil's Stream Reclamation Projects. Mobil Mining and Minerals Company, Nichols, Florida.

Harlow, R.F. and F.K. Jones. 1965. The White-Tailed Deer in Florida. Technical Bulletin No. 9, Florida Game and Fresh Water Fish Commission, Tallahassee, Florida.

Harrell, J.B. 1987. The Development of Techniques for the Use of Trees in the Reclamation of Phosphate Lands, Final Report. Florida Institute of Phosphate Research, Bartow, Florida, Publication No. 03-001-049.

Harris, S.A. 1985. Sportsmen's paradise regained. Louisiana Conservationist 37(5):24-25.

Haynes, R.J. 1984. Summary of wetlands reestablishment on surface-mined lands in Florida, p. 357-362. In Proceedings, 1984 Symposium on Surface Mining, Hydrology, Sedimentology, and Reclamation, University of Kentucky, Lexington, Kentucky.

Haynes, R.J., J.A. Allen, and E.C. Pendleton. 1988. Reestablishment of Bottomland Hardwood Forests on Disturbed Sites: An Annotated Bibliography. U.S. Fish and Wildlife Service, Biological Report 88(42).

Haynes, R.J. and L. Moore. 1988. Reestablishment of bottomland hardwoods within national wildlife refuges in the Southeast, p. 95-103. In J. Zelazny and J.S. Feierabend (Eds.), Increasing Our Wetland Resources. National Wildlife Federation, Washington, D.C.

Holland, L.J., and L.B. Phelps. 1986. Topsoil compaction during reclamation: field studies, p. 55-62. In Proceedings 1986 National Symposium on Mining, Hydrology, Sedimentology, and Reclamation, University of Kentucky, Lexington, Kentucky.

Johnson, R.L. and R.M. Krinard. 1985a. Oak Seeding on an Adverse Field Site. U.S. Department of Agriculture, Forest Service, Research Note SO-319.

Johnson, R.L. and R.M. Krinard. 1985b. Regeneration of oaks by direct seeding, p. 56-65. In Proceedings, Third Symposium of Southeastern Hardwoods. U.S. Department of Agriculture, Forest Service, Southern Forest Experiment Station, New Orleans, Louisiana.

Kay, C.A.R., A.F. Clewell, and E.W. Ashler. 1978. Vegetative cover in a fallow field: response to season of soil disturbance. Bulletin Torrey Botanical Club 105:143-147.

King, T., R. Stout, and T. Gilbert. 1985. Habitat Reclamation Guidelines, a Series of Recommendations for Fish and Wildlife Habitat Enhancement on Phosphate Mined Land and Other Disturbed Sites. Florida Game and Fresh Water Fish Commission, Bartow, Florida.

Klemp, M.T., J.L. Torbert, J.A. Burger, and S.H. Schoenholtz. 1986. The effect of fertilization and chemical weed control on the growth of three pine species on an abandoned bench and return-to-contour site in southwestern Virginia, p. 77-81. In Proceedings, 1986 Symposium on Surface Mining, Hydrology, Sedimentology, and Reclamation, University of Kentucky, Lexington, Kentucky.

Kormanik, P.P. and R.C. Schultz. 1985. Significance of Sewage Sludge Amendments to Borrow Pit Reclamation with Sweetgum and Fescue. U.S. Department of Agriculture, Forest Service, Southeastern Forest Experiment Station, Research Note SE-329.

Krinard, R.M. and R.L. Johnson. 1976. 21-year Growth and Development of Baldcypress Planted on a Flood-Prone Site. U.S. Department of Agriculture, Forest Service, Research Note SO-217.

Krinard, R.M., and H.E. Kennedy, Jr. 1987. Fifteen-Year Growth of Six Planted Hardwood Species on Sharkey Clay Soil. U.S. Department of Agriculture, Forest Service, Research Note SO-336.

Kurz, H. 1944. Secondary forest succession in the Tallahassee Red Hills. Proceedings Florida Academy of Science 7:1-100.

Lea, R. 1988. Forest management impacts on bottomland hardwoods, p. 156-158. In J.A. Kusler, M.L. Quammen, and G. Brooks (Eds.), National Wetland Symposium: Mitigation of Impacts and Losses. Association of State Wetland Managers, Berne, New York.

Lee, C.R., J.G. Skogerboe, D.L. Brannon, J.W. Linkinhoker, and S.P. Faulkner. 1983. Vegetative restoration of pyritic soils, p. 271-274. In Proceedings, 1983 Symposium on Surface Mining, Hydrology, Sedimentology and Reclamation, University of Kentucky, Lexington, Kentucky.

Little, E.L., Jr. 1979. Checklist of United States Trees (Native and Naturalized). U.S. Department of Agriculture, Forest Service, Agricultural Handbook No. 541.

Lyle, E.S., Jr. 1987. Surface Mine Reclamation Manual. Elsevier Science Publishing Co., Inc., New York.

Malac, B.F. and R.D. Heeren. 1979. Hardwood plantation management. So. J. Applied For. 3(1): 3-6.

Marion, W.R. 1986. Phosphate Mining: Regulations, Reclamation and Revegetation. Florida Institute of Phosphate Research, Bartow, Florida, Publ. No. 03-043-040.

Miller, H.A., J.G. Sampson, and C.S. Lotspeich. 1985. Wetlands reclamation using sand-clay mix from phosphate mines, p. 193-200. In F.J. Webb (Ed.) 1985, Proceedings of the 12th Annual Conference on Wetland Restoration and Creation, Hillsborough Community College, Tampa, Florida.

Mitsch, W.J. and J.G. Gosselink. 1986. Wetlands. Van Nostrand Reinhold Co., New York.

Moore, W.H. 1980. Survival and Growth of Oaks Planted for Wildlife in the Flatwoods. U.S. Department of Agriculture, Forest Service, Research Note SE-286.

Peters, M.A. and E. Holcombe. 1951. Bottomland cypress planting recommended for flooded areas by soil conservationists. Forests and People 1(2): 18, 32-33.

Posey, D.M., Jr., D.C. Goforth, and P. Painter. 1984. Ravenwood shellrock mine: wetland and upland restoration and creation, p. 127-134. In F.J. Webb (Ed.), 1984 Proceedings, 11th Annual Conference on Wetland Restoration and Creation, Hillsborough Community College, Tampa, Florida.

Powell, J.L., R.B. Gray, J.B. Ellis, D. Williamson, and R.I. Barnhisel. 1986. Successful reforestation by use of large-sized and high-quality native hardwood planting stocks, p. 169-172. In Proceedings, 1986 National Meeting, American Society for Surface Mining and Reclamation, Jackson, Mississippi.

Rice, C.W., R.I. Barnhisel, and J.L. Powell. 1982. The establishment of loblolly pine (Pinus taeda L.) seedlings on mined land with Pistolithus tinctorius ectomycorrhizae, p. 527-234. In Proceedings, 1982 Symposium on Surface Mining Hydrology, Sedimentation and Reclamation, University of Kentucky, Lexington, Kentucky.

Robertson, D.J. 1984. Sink Branch: Stream relocation and forested wetland reclamation by the Florida phosphate industry, p. 135-151. In F.J. Webb (Ed.), Proceedings, Eleventh Annual Conference on Wetland Restoration and Creation, Hillsborough Community College, Tampa, Florida.

Robertson, D.J. 1985. Freshwater Wetland Reclamation in Florida, an Overview. Florida Institute of Phosphate Research, Bartow, Florida.

Ruesch, K.J. 1983. A Survey of Wetland Reclamation Projects in the Florida Phosphate Industry. Florida Institute of Phosphate Research, Bartow, Florida, Publ. No. 03-019-011.

Rushton, B.T. 1983. Examples of natural wetland succession as a reclamation alternative. In D.J. Robertson (Ed.), Reclamation and the Phosphate Industry, Florida Institute of Phosphate Research, Bartow, Florida, Publ. No. 03-036-010.

Schoenholtz, S.H., J.L. Torbert, and J.A. Burger. 1986. Factors affecting seedling ectomycorrhizae on reclaimed surface mines in Virginia, p. 89-93. In Proceedings, 1986 Symposium on Mining, Hydrology, Sedimentology and Reclamation, University of Kentucky, Lexington, Kentucky.

Sharitz, R.R. and L.C. Lee. 1985. Recovery processes in southeastern riverine wetlands, p. 499-501. In Riparian Ecosystems and Their Management: Reconciling Conflicting Uses. U.S. Department of Agriculture, Forest Service Report RM-120.

Silker, T.H. 1948. Planting of water-tolerant trees along margins of fluctuating-level reservoirs. Iowa State Journal of Science 22: 431-447.

Starnes, L.B. 1985. Aquatic community response to techniques utilized to reclaim eastern U.S. coal surface mine-impacted streams, p. 193-222. In J.A. Gore (Ed.), The Restoration of Rivers and Streams, Theories and Experience. Butterworth Publishers, Boston.

Stubblefield, G.W., III. 1984. Patterns of Geographic Variation in Sweetgum in the Southern United States. Doctoral Thesis, North Carolina State University, Raleigh, North Carolina.

Tackett, E.M. and D.H. Graves. 1979. Direct-seeding of commercial trees on surface-mine spoil, p. 209-212. In Proceedings, 1979 Symposium on Surface Mining, Hydrology, Sedimentology and Reclamation, University of Kentucky, Lexington, Kentucky.

Tackett, E.M. and D.H. Graves. 1983. Evaluation of direct-seeding of tree species on surface mine spoils after five years, p. 437-441. In Proceedings, 1983 Symposium on Surface Mining, Hydrology, Sedimentology and Reclamation, University of Kentucky, Lexington, Kentucky.

Teskey, R.O., and T.M. Hinkley. 1977. Impact of Water Level Changes on Woody Riparian Wetland Communities. Volume II: Southern Forest Region. U.S. Fish and Wildlife Service, FWS/OBS-77/59.

Thompson, R.L., W.G. Vogel, G.L. Wade, and B.L. Rafaill. 1986. Development of natural and planted vegetation on surface mines in southeastern Kentucky, p. 145-153. In Proceedings, National Meeting, American Society for Surface Mining and Reclamation, Jackson, Mississippi.

Toliver, J.R. 1986-1987. Survival and growth of hardwoods planted on abandoned fields. Louisiana Agriculture 29(2):10-1.

U.S. Department of Agriculture, Forest Service. 1988. Vegetation Management in the Coastal Plain/ Piedmont, Appendix A. Management Bulletin R8-MB 15, Atlanta, Georgia.

Vogel, W.G. 1980. Revegetating surface-mined lands with herbaceous and woody species together, p. 117-126. In Trees for Reclamation. U.S. Department of Agriculture, Forest Service, General Technical Report NE-61.

Wade, G.L. 1986. Forest topsoil seed banks for introducing native species in eastern surface-mine reclamation, p. 155-164. In J. Harper and B. Plass (Eds.), Proceedings, 1986 National Meeting, American Society for Surface Mining and Reclamation, Jackson, Mississippi.

Waldrop, T.A., E.R. Buckner, and A.E. Houston. 1982. Suitable trees for the bottomlands, p. 157-160. In E.P. Jones, Jr. (Ed.), Proceedings, Second Biennial Southern Silvicultural Research Conference, Atlanta, Georgia.

Wallace, P.M. and G.R. Best. 1983. Enhancing ecological succession: 6. Succession of vegetation, soils and mycorrhizal fungi following strip mining for phosphate, p. 385-394. In Proceedings, 1983 Symposium on Surface Mining, Hydrology, Sedimentology and Reclamation, University of Kentucky, Lexington, Kentucky.

Wallace, P.M., G.R. Best, J.A. Feiertag and K.M. Kervin. 1984. Mycorrhizae enhance growth of sweetgum (Liquidambar styraciflua) in phosphate mined overburden soils. In Proceedings, 1984 Symposium on Surface Mining, Hydrology, Sedimentology and Reclamation, University of Kentucky, Lexington, Kentucky.

Wein, G.R., S. Kroeger, and G.J. Pierce. 1987. Lacustrine vegetation establishment within a cooling reservoir, p. 206-216. In F.J. Webb (Ed.), 1987 Proceedings of the 14th Annual Conference on Wetland Restoration and Creation. Hillsborough Community College, Tampa, Florida.

Wharton, C.H., W.M. Kitchens, E.C. Pendleton, and T.W. Sipe. 1982. The Ecology of Bottomland Hardwood Swamps of the Southeast: A Community Profile. U.S. Fish & Wildlife Service, FWS/OBS-81/37.

Williams, R.D., and S.H. Hanks. 1976. Hardwood Nurseryman's Guide. U.S. Department of Agriculture, Forest Service. Agricultural Handbook No. 473.

Wittwer, R.F., S.B. Carpenter, and D.H. Graves. 1981. Survival and growth of oaks and Virginia pine three years after direct seeding on mine spoils, p. 1-4. In Proceedings, Symposium on Surface Mining, Hydrology, Sedimentology and Reclamation, University of Kentucky, Lexington, Kentucky.

Wolf, R.B., L.C. Lee, and R.R. Sharitz. 1986. Wetland creation and restoration in the United States from 1970 to 1985: an annotated bibliography. Wetlands 6(1):1-88.

APPENDIX I: COMMON AND SCIENTIFIC PLANT NAMES

Scientific equivalents of common names used in this chapter are listed below. Authorities for scientific names and sources of common names are from Clewell (1985) and Little (1979).

Common Name	Scientific Name
Bahia grass	Paspalum notatum
Baldcypress	Taxodium distichum
Blackberries	Rubus spp.
Black Locust	Robinia pseudoacacia
Box Elder	Acer negundo
Brazilian Peppertree	Schinus terebinthifolius
Broomsedge	Andropogon glomeratus, A. virginicus
Bulrush	Scirpus spp.
Bermuda grass	Cynodon dactylon
Cabbage Palm	Sabal palmetto
Cane	Arundinaria gigantea
Cattails	Typha spp.
Chinaberry	Melia azedarach
Cogongrass	Imperata cylindrica
Cottonwood, Eastern	Populus deltoides
Cottonwood, Swamp	Populus heterophylla
Dahoon Holly	Ilex cassine
Dog Fennel	Eupatorium capillifolium
Elm	Ulmus spp.
European Alder	Alnus glutinosa
Fetterbush	Lyonia lucida
Gallberry	Ilex glabra
Giant Ragweed	Ambrosia trifida
Grapes	Vitis spp.
Green Ash	Fraxinus pennsylvanica
Greenbriers	Smilax spp.
Ground Nut	Apios americana
Hempvine	Mikania spp.
Honeysuckle	Lonicera spp.
Johnsongrass	Sorghum halepense
Kudzu	Pueraria lobata

Loblolly Bay	Gordonia lasianthus
Maidencane	Panicum hemitomon
Millet	Brachiaria ramosa
Morning-glory	Convolvulaceae spp.
Oak, Cherrybark	Quercus falcata var. pagodifolia
Laurel	Quercus laurifolia
Live	Quercus virginiana
Nuttall	Quercus nuttallii
Overcup	Quercus lyrata
Water	Quercus nigra
Willow	Quercus phellos
Ogeechee Tupelo	Nyssa ogeche
Pennywort	Hydrocotyle spp.
Pickerelweed	Pontederia cordata
Pond Pine	Pinus serotina
Pondcypress	Taxodium ascendens
Primrose Willow	Ludwigia peruviana
Red Maple	Acer rubrum
Ryegrass, Winter	Lolium perenne
River Birch	Betula nigra
Saltbush	Baccharis halimifolia
Slash Pine	Pinus elliottii
Softrush	Juncus effusus
Southern Magnolia	Magnolia grandiflora
Spikerush	Eleocharis spp.
Spruce Pine	Pinus glabra
Swamp Tupelo	Nyssa biflora
Sweetbay	Magnolia virginiana
Sweetgum	Liquidambar styraciflua
Sweet-Pecan	Carya illinoensis
Sycamore	Platanus occidentalis
Water Hickory	Carya aquatica
Water Tupelo	Nyssa aquatica
Wax Myrtle	Myrica cerifera
White Cedar	Chamaecyparis thyoides
Willow	Salix caroliniana

Tall Fescue	Festuca arundinacea
Titi	Cyrilla racemiflora or Cliftonia monophylla

APPENDIX II: CHECKLIST FOR PERMITTING PERSONNEL

This checklist is designed to flag those issues that should be addressed in permit applications. For elaboration, see Clewell (1988).

Personnel	**Yes**	**No**
1. Have appropriate skilled professionals prepared the project plan?	—	—
2. Will they provide on-site supervision of site-preparation and planting? Of monitoring?	—	—
Forest Type To Be Created/Restored		
1. Does the forest type occur locally?	—	—
2. Is there adequate baseline data to plan project goals (species composition, tree relative densities, data on water tables, stream flows, soil conditions, topography)?	—	—
3. Is there a source of the necessary plant materials to install this project?	—	—
4. What is the proposed land use after project release? Is it consistent with project goals? Will the released project be protected indefinitely?	—	—
Project Siting		
1. Does the soil or substrate on the site have the properties to support the vegetation targeted for restoration or creation?	—	—
2. Does the proposed project conform to existing topographical and hydrological conditions?	—	—
3. Will the project be connected hydrologically with natural waters if it is advantageous to do so?	—	—
4. Are there any off-site constraints, such as watershed obstructions, recent or proposed drainage plans, asynchronous or thermal discharges up stream?	—	—
5. Is there any regional development plan which may affect siting?	—	—
6. Is an adequate buffer included to isolate and protect the project, if needed?	—	—
7. Will the site be fenced or staked?	—	—
8. Will the water quality be adequate to support the project biota?	—	—
9. Has natural regeneration been considered? Are seed walls sufficiently close to the project, or will flood waters bearing seeds reach the project?	—	—
Site Preparation		
1. Is there a need for, and provisions to, manipulate hydrology early in the project (weirs, other control structures)?	—	—
2. Will slopes be contoured to minimize erosion? If not, are there provisions to prevent erosion (berms, cover crops, etc.)?	—	—
3. Will stream channels be designed to minimize erosion and encourage meandering? (Deflectors, sprigging stoloniferous plants, engineering options, etc.).	—	—

Yes No

4. Will irrigation be needed initially? If so, how will it be provided?

5. Are there any noxious plants on site that should be removed, such as rhizomatous grasses, vines, or exotics?

6. Are the soils adequate in terms of fertility, rooting volume, bulk density, crusting, or will conditioning be needed (subsoil ripping, pH adjustment, amendments of fertilizers or leguminous cover crops, incorporation of organic matter, etc.)?

7. If sludge is to be used as a soil amendment, have appropriate permits been obtained?

8. Will topsoil be conserved? If so, can it be spread on the project site without prior stockpiling, to conserve its seed bank and other propagules?

Project Installation

1. Will a competent supervisor be on-site at all times during final grading and vegetation planting?

2. Will there be provisions to harden nursery stock before delivery in terms of reduced fertilization and watering?

3. Is there sufficient lead time to contract-grow trees?

4. Will trees be delivered for planting at the appropriate time?

5. Is the growing stock (seeds, bare root seedlings, tubelings, etc.) appropriate for the site conditions in terms of adequate moisture and competition from cover crops or weeds?

6. Are there provisions for introducing preferred undergrowth species?

Maintenance

1. Does the site plan allow for unanticipated maintenance activities during the first 2 years, in terms of erosion control, sediment removal, and replanting of trees or other plant materials that did not survive?

2. Are there strategies for minimizing competition from noxious species, such as specific site-preparation activities, nurse crops, or rapid tree-growing regimes? If not, are there provisions for weed removal?

Monitoring

1. Have adequate success criteria been drafted that pertain to project goals?

2. Has a monitoring protocol been devised that answers questions posed by the success criteria?

3. Will groundwater piezometers be installed and staff gauges be placed in water bodies, so that monthly water levels can be monitored?

4. Will there be sufficient inspection tours to allow for necessary maintenance or remedial activities?

5. Will reports be submitted and made publicly available?

APPENDIX III: RECOMMENDED READING

Personnel responsible for evaluating permit applications should be familiar with the wetland communities proposed for creation/restoration. There are amazingly few, adequate descriptions of southeastern fresh water wetlands. Perhaps the initial choices would be the works of Mitsch and Gosselink (1986) and Wharton et al. (1982).

Forested wetland creation/mitigation is such a new field that no appropriate texts are available that comprehensively introduce this topic. Several references for coal mine reclamation provide a feel for the kinds of activities required for forested projects. The recent text by Lyle (1987) would be suitable. The only other suggestion we have would be to review recent symposium volumes, such as the National Symposia on Surface Mining, Hydrology, Sedimentology and Reclamation, sponsored by the University of Kentucky.

FRESHWATER MARSH CREATION AND RESTORATION IN THE SOUTHEAST

Kevin L. Erwin
Kevin. L. Erwin Consulting Ecologist, Inc.

ABSTRACT. Freshwater marsh habitat has been created or restored in the southeast to mitigate the environmental impacts associated with development activity and to provide enhancement of water quality. These projects vary in size, design, and function, and are inadequately discussed in the literature. There is a question of whether agency-mandated mitigation projects have actually been implemented and to what extent these created freshwater marshes are providing the desired wetland functions.

Key elements to successfully constructing a functional freshwater marsh system include: (1) realistic goals and measurable success criteria; (2) proper pre-construction design evaluation including a hydrological analysis; (3) contour design; (4) construction technique; (5) proper water quality; (6) compatibility of adjacent existing and future land uses; (7) appropriate substrate characteristics; (8) re-vegetation techniques; (9) re-introduction of fauna; (10) upland buffers and protective structures; (11) supervision by an experienced professional; (12) post-construction long term management plan; and (13) monitoring and reporting criteria.

The monitoring required must be adequate in scope to determine the success or failure to meet project goals. A typical monitoring plan for a created freshwater marsh should include: (1) a post-construction, pre-planting survey of project contours and elevations; (2) ground and surface water elevation data collection; (3) water quality data collection; (4) biological monitoring including, but not limited to, fish and macroinvertebrate data collection; (5) evaluation of vegetation species diversity, percent cover, and frequency; and (6) wildlife utilization.

Critical information gaps and research needs can be divided into the following categories: (1) site selection and design; (2) project construction techniques; (3) comparative studies of the biological communities and processes in created and natural systems; and (4) the role of uplands and transitional habitats.

INTRODUCTION

In preparing this chapter, a search was conducted for documentation of nonforested, interior, freshwater wetland projects across the Southeast United States. From the distribution of reports in the literature and a lack of response to a questionnaire widely distributed to regulatory agencies, consultants and scientists, it may be concluded that most of the marsh creation and restoration projects have occurred within the State of Florida. This is not to say that freshwater marsh creation and restoration projects do not exist elsewhere within the region, but they were not brought to the attention of this reviewer.

The search revealed that a majority of projects were started in the mid 1970's. They consist primarily of enforcement-related restoration projects and development-related mitigation. Enforcement-related projects usually occur as a result of unauthorized activity in a wetland. Restoration is required to rehabilitate the habitat and, presumably, related wetland functions.

Development projects involving mitigation include: residential developments; surface mining; and highway, marina, and dock construction. Wetland creation or enhancement is often required where these activities are expected to result in adverse alteration of a wetland. The intended result is no net loss of habitat and functions.

Another type of project resulting in wetland creation deals with utilization of the created habitat for stormwater runoff or treated domestic waste effluent. Wetland creation projects for stormwater treatment are usually associated with construction of impervious surfaces such as parking lots and highways. The ability of some wetlands to provide treatment of polluted surface water and secondary or tertiary treated domestic waste effluent is well documented in the literature citations (Richardson et al. 1978, Kadlec 1979, Kadlec and Tilton 1979, Tilton and Kadlec 1979, Ewel 1976, Ewel and Odum 1978, 1979, 1984, Spangler et al. 1977, Fetter et al. 1978).

Marshes are also being created as detention areas within impacted rural landscapes. Farmland in Florida is rapidly being converted to development. Natural water detention and flow/drainage areas are often non-existent on these lands in the amounts and locations required for development, and developers are planning to create wetlands for these purposes. The same methods used to create wetlands on reclaimed phosphate mines (discussed later) are readily adaptable. Several of these wetland drainage systems should be operational in the next few years and, if properly designed and monitored, should reveal new information on wetland creation techniques and values.

Many acres of wetlands are also being built within the water storage areas of new citrus groves in south and central Florida to offset the drainage impacts on wetlands within the groves. Wetlands have several attributes that cause them to have major influences on chemicals that flow through them (Sather and Smith, 1984). The need for water quality enhancement combined with the loss of wetland habitats and the regulatory agencies' reluctance to utilize natural wetland systems, has led to the construction of marshes for these purposes.

Finally, wetlands are often created simply as a result of construction activity.

The vast majority of these marsh creation and restoration projects have been undertaken as a result of regulatory permit conditions. Very few projects utilizing scientific experimental designs have been undertaken to date. Projects described in the special conditions of a permit are often poorly designed or lack sufficient detail so that from the beginning they are probably doomed to failure. These projects are often designed through negotiation between an applicant and the regulatory agency, resulting in a compromise which may not establish reasonable goals or require any monitoring that would provide a reviewer with the documentation needed to determine the success or failure of the project. Race and Christie (1982) reviewed mitigation projects involving wetland creation and questioned the effectiveness of artificially created marshes to provide biological and hydrological functions and societal values of natural marshes. Quammen (1986) notes that several subsequent studies following that of Race and Christie have evaluated whether agency-mandated mitigation projects have actually been implemented. Few have evaluated how well the artificial marshes are functioning. A matter of great concern is the apparent lack of follow-up on these permitted projects. To remedy this problem, this reviewer suggests that in the future, a list of all projects be compiled by regulatory agencies and certain projects selected for a periodic evaluation. Such a process would, at the very least, provide some useful information regarding the science of restoration and creation, and reveal any inadequacies of project design and monitoring. Figure 1 is a suggested format for the tracking of wetland mitigation projects.

WETLAND TYPES

Freshwater marshes in the southeast are a very diverse group of habitats. They are all dominated by sedges and grasses which are adapted to saturated soil conditions. Otherwise, marshes differ in their hydrology, geologic origin, and size. This discussion addresses all marshes that are located inland and upstream of tidal influence. The term marsh in this chapter will also be used to characterize the following habitat types (Mitsch & Gosselink 1986) which for the purposes of this discussion contain more similarities than significant differences:

Bog	A peat accumulating wetland that has no significant inflows or outflows and supports acidophilic mosses, particularly sphagnum.
Fen	A peat accumulating wetland that receives some drainage from surrounding mineral soil and usually supports marsh-like vegetation.
Peatland	A generic term for any wetland that accumulates partially decayed plant matter.
Wet prairie	Similar to marsh.
Reed swamp	Marsh dominated by <u>Phragmites</u> (common reed).
Wet meadow	Grassland with waterlogged soil near the surface, but without standing water for most of the year.
Slough	An elongated marsh often bisected by a creek with slowly flowing surface water.
Pot hole	Shallow marsh-like pond.
Playa	A term used in the Southwest United States for a marsh-like pond similar to a pothole, but with a different geologic origin.

A thorough description and history of wetland definition, classification, and inventory, including

Projects | Wetland Mitigation

Permit #: ______________________ ______________________

County/State: ______________________ Reviewer: ______________________

Description of Wetland Type: ______________________

____

Acres Destroyed: ______________________ Acres Created: ______________________

Acres Restored: ______________________ Acres Enhanced: ______________________

Other: ______________ Acres of Upland Compensation: ______________________

Upland Type: ______________________

Mitigation Method: ______________________

Mitigation Site: ______________________

Species Used: ______________________

Control Depth: ______________________ Estimated Cost: ______________________

Estimated Commencement: ______________________ Deadline: ______________________

Estimated Completion: ______________________

Wetland Construction Contractor and Supervisor: ______________________

______________________ Phone: ______________________

Comments: ______________________

____

Monitoring and Management

Plan Rec'd: ______________________ Approved: ______________________

Duration: ______________________ Monitoring Contractor: ______________________

Phone: ______________________ Sampling Mehods: ______________________

____

Date Reports Due: ______________________ Baseline Req'd: Y/N ______________________

Date Received: ______________________

Report:

#1 Initial/Baseline Rec'd: ______________________

#2 Due: ____________ Rec'd: ____________ #6 Due: ____________ Rec'd: ____________

#3 Due: ____________ Rec'd: ____________ #7 Due: ____________ Rec'd: ____________

#4 Due: ____________ Rec'd: ____________ #8 Due: ____________ Rec'd: ____________

#5 Due: ____________ Rec'd: ____________ #9 Due: ____________ Rec'd: ____________

#10 Due: ____________ Rec'd: ____________

Date Inspected: ______________________ Reviewer: ______________________

Action Required: ______________________

Comments: ______________________

____

____

Figure 1. A form for tracking wetland mitigation projects.

the use of these terms, can be found in Mitsch and Gosselink (1986).

Five basic types of nonforested, interior, freshwater wetlands or marshes are found in the southeast:

Riverine — Those associated with shoreline fringes of non-tidal streams and rivers.

Isolated — Marshes completely surrounded by upland habitat.

Headwater systems or seepage zones — Marshes formed as a result of ground water seepage exposed at the base of a upland recharge zone where soils are often saturated but inundation may be infrequent.

Slough or flow-way systems — Broad, shallow, slow flowing systems such as the Florida Everglades.

Lake fringe — Marshes common to the shorelines and shallow interior zones of lakes.

The types of marshes described above reflect tremendous diversity in both geomorphology and physiognomy. Species richness and diversity often range widely depending on the geomorphology or zonation (vegetation) of a particular system, its hydrology, and the extent of human induced or natural perturbation.

Most of our wetlands, including marshes, have been altered by human influence, mostly in the form of direct impacts of habitat development such as dredge and fill related activity, or hydraulic manipulation resulting in altered hydroperiods. As the high, well-drained, upland sites, historically preferred for development activities, become scarce, greater pressure will be put on wetlands. The transitional areas will be affected most with perhaps greater damage resulting via destruction of adjacent upland habitat (Figure 2). Many of those wetlands preserved within the developed areas of the region should be considered urbanized wetland systems. Because of the urban related impacts on the system's hydrology, the wetlands are currently undergoing significant changes in character and function. In the future this will lead to significant alteration of those systems and, perhaps, a complete loss of their preferred values such as wildlife habitat.

KEY FUNCTIONS PERFORMED

As different as inland marshes are with regard to physiognomy and geomorphology, the basic functions and values attributable to these habitat types are generally similar. The Federal Highway Administration's wetland functional assessment methodology recognizes eleven wetland functions (Adamus 1983): groundwater recharge, groundwater discharge, flood storage, shoreline anchoring, sediment trapping, nutrient retention, food chain support, fishery habitat, wildlife habitat, active recreation, and passive recreation and heritage. Inland marshes can perform all of these functions (depending upon the circumstances).

These functions are described in the proceedings of a national symposium on wetlands held in 1978 (Greeson, Clark, and Clark 1979) and in Reppert et al. 1979, Larson 1982, Adamus 1983, Sather and Smith 1984, Gosselink 1984, Mitsch and Gosselink 1986, and Kusler and Riexinger 1985.

THE EXTENT TO WHICH CREATION/RESTORATION HAS OCCURRED

GENERAL REASONS FOR MARSH CREATION OR RESTORATION

As noted above, marsh creation and restoration in the Southeast has occurred primarily as a result of a permit condition in response to a need for enhancement or mitigation such as: mitigation under Section 404, reclamation under the Surface Mining Act, or as mitigation for dredge and fill activity permitted by local or state government, e.g., under Chapter 17-4 of the Florida Administrative Code and Henderson Wetlands Act of the State of Florida. The largest and more thoroughly documented projects have been undertaken as a result of surface mining in Florida. The Florida Department of Natural Resources requires reclamation of similar habitat as a permit condition for all surface mining activity.

Two of the largest wetland public works projects currently underway are the Kissimmee River and Everglades restoration projects. Each of

A

B

Figure 2. The regulation of isolated wetlands results in the "preservation" of the wetland without regard to its adjacent landscape. A: Original setting of the wetland; B: Setting of the wetland after development of the adjacent area.

these projects is a major effort. Both are being undertaken by the State of Florida. At this point the projects are taking two distinct courses.

The Kissimmee River project is an effort to restore the hydroperiod and inundation to the channelized Kissimmee River and marsh flood plain by placement of dams in the manmade channel. This will cause reflooding of the historic floodplain. The objective of the Kissimmee River Run Revitalization is to return full flow to a total of 43.443 kilometers, or about one third of the original meandering course of the 158 kilometer river. This includes restoring flow to 35.4 kilometers of oxbows and re-establishing 520 hectares of wetlands. At present, most of the river's flow bypasses oxbows. The plan will divert the full flow of the river into selected old river runs located alongside existing water control structures. Under pre-channelization "flood conditions", excess flow will be routed through the main canal channel while maintaining full flowing oxbows.

The Everglades project seeks to restore pre-1900 conditions to the Everglades systems. Although the plan involves a wide range of areas for action, most activity to date has been expended for land acquisition to create buffers for the Everglades National Park/Big Cypress Preserve system. This buffer is for watershed purposes and for the protection of critical habitat of the Florida panther.

Neither of these projects involves the actual construction or restoration of marsh habitat. Their goals are related to the enhancement of existing wetlands by various structural and nonstructural means.

TYPICAL GOALS FOR PROJECTS

The major initial shortcoming of wetland creation or restoration projects is a failure to identify realistic goals. Since the majority of projects in the region were not undertaken by wetland managers but by wetland regulators and contractors, the implied goal is reclamation of the habitat impacted by permitted activities. Typical stated goals for projects often include the creation of wildlife habitats, where a nonspecific design is aimed at providing any type of marsh or aquatic system that would be suitable for waterfowl and wading birds. This broad goal usually means that the developer of the project and the regulatory agency will accept whatever marsh vegetation is created and the biota it attracts.

Recently the State of Florida has adopted a much more specific "type for type" mitigation policy which includes replacement ratio guidelines. This policy leans toward restoration of a specific habitat including soils, hydrophytes, macrobenthos, and wildlife. However, complete "restoration" of any type of system, including marsh, down to minute details is impossible and, therefore, unrealistic.

Most of the documents reviewed in the literature were of a nontechnical nature and did not state project goals. Technical information on marsh creation was implied in the titles, but not provided in the texts. Many of these papers were reports of permitted projects lacking detailed design and/or monitoring information.

The largest number of papers dealing with wetland creation and restoration in the Southeast come from the last decade of the Proceedings of Annual Conferences on Wetland Restoration and Creation at Hillsborough Community College, Florida (Webb 1982 through 1987, Cole 1979, 1981).

Broader inventories and studies of projects in the Southeast also generally left the goals unclear. Wolf et al. (1986) provided the best statement of goals in their summary of projects. Unfortunately, the majority of these were either tidal wetlands or not within the Southeast United States. The Wolfe and Sharitz effort underscores the scarcity of well-planned, goal-oriented freshwater marsh creation projects within the Southeast region.

SUCCESS IN ACHIEVING GOALS

The majority of reports or papers describing projects do not state goals or criteria for judging success. Most projects are driven by permit conditions where the success is determined by the extent of the cover by herbaceous species and diversity. Most endemic species are acceptable within some arbitrary guidelines for restricting the percent cover of certain species the regulatory agencies consider problematic such as cattail (_Typha_ spp.), willow (_Salix_ spp.), primrose willow (_Ludwigia_ spp.), Brazilian pepper (_Schinus terebinthifolius_), and Melaleuca (_Melaleuca quinquenervia_).

Recently, however, many local, state, and federal agencies have placed great emphasis on the restoration of specific wetland features and determine success or failure on the ability of the project developer to create a "mirror image" of the original wetland. Undocumented accounts of failure to meet these specific guidelines are numerous. These failures are not unexpected, given the impossibility of creating "mirror images", and considering the inappropriate design criteria and lack of monitoring often associated with projects.

The desired overall goal should be to create a functional wetland containing the desired hydrology and a satisfactory coverage of a community of preferred plant species. If this goal is met, faunal requirements are usually attainable.

Recently some regulatory agencies such as the Florida Department of Environmental Regulation, have begun to use "reference wetlands" as an approach for determining success with a marsh creation or restoration project. The inherent variability of seemingly similar marshes has been shown to be characterized by dissimilar macrobenthic (aquatic insect) communities even within wetlands of similar hydrophyte structure (Erwin, unpublished data). The scientific community generally rejects the notion that a reference marsh concept is satisfactory. The author believes that the concept should be used only for evaluating the structural and functional attributes of a particular habitat or system.

In general, the Southeast region offers abundant plant material, adequate rainfall, and a long growing season. This provides the developer of a marsh creation project with the basic tools necessary to construct a functional marsh in a relatively short period of time (three to five growing seasons). This is illustrated by one of the largest and probably the best documented marsh creation projects, the Agrico Fort Green Phosphate Mine Reclamation Project. This project was undertaken in 1982 and has been the subject of numerous research articles by the author (Erwin 1983-1988, Erwin et al. 1984, Erwin and Bartleson 1985, Erwin and Best 1985). Simply stated, the goal was to create a marsh, over 30 hectares in size, utilizing a variety of construction techniques (Appendix I). This area, along with nearby natural areas, has been monitored closely to determine whether it is possible to create a naturally functioning marsh system. Sixty hectares of marsh have actually been created with 24 hectares of marsh planted with trees.

Currently, goal-oriented review of projects developed as a condition to permits are discontinued at the end of the permit life. In many cases where monitoring is required for less than three years, not enough data will be collected to evaluate the success or failure of the project in meeting the selected goals.

One goal which is significant and often overlooked is the ability of a created system to function on its own as part of a larger ecosystem for many years. Our current regulatory process calls for the protection of wetlands with little or no ability to provide for watershed and adjacent upland habitat integrity. The consideration of adjacent landscapes is important and should be included in each project. Wetland mitigation should not be undertaken at the expense of valuable and dwindling upland habitats which may have high wildlife or other values but lack protection. It is also possible to create a functioning freshwater marsh in the short term and, through poor planning within the upland portions of the watershed, cause the long term demise of the created wetland. Unfortunately, this is probably the fate of most urban wetlands preserved or mitigated in our present regulatory system.

In conclusion, when setting goals flexibility and unlimited creativity should be employed. "Type for type" habitat replacement should be considered but will not be appropriate in cases where such replacement is not technically possible or where another type of wetland has greater value or more regional significance.

In order to determine the success or failure of a project it must be possible to determine, through some type of accepted evaluation process, whether the project has been able to attain certain goals. The reasons for marsh creation project failures include the following (which will be discussed later in greater detail:

1. Goals were not properly identified or they are unrealistic;

2. No practical methodology was available or applied to determine the degree of goal attainment;

3. Geohydrology was improper;

4. Evaluation and understanding of the wetland area to be mitigated was inadequate;

5. Watershed evaluation was inadequate;

6. Monitoring was unsatisfactory or not enforced by the permitting agency;

7. Project was not constructed as designed due to contractor misunderstandings and/or lack of supervision by a knowledgeable expert;

8. Handling of mulch or plant materials was improper;

9. The created wetland area was not maintained free of problematic exotics such as Melaleuca;

10. The area was not kept free of nuisance animals such as feral hogs or from overgrazing by cattle;

11. An entity was not identified to be responsible for undertaking a long term management plan with available funding required to insure future success;

12. A long term management plan for the watershed in which the wetland lies was not adopted (inappropriate future land uses surrounding the created wetland may alter the habitat's character completely, if not eliminate it all together);

13. The design was not attainable with the budget available;

14. Soil types were unsatisfactory; and

15. Water sources were unsatisfactory or contaminated.

The project developer as well as the regulator should: identify quantifiable, realistic goals; maintain direct supervision through the construction and monitoring process; utilize flexibility in post construction modifications where necessary; and if problems or failures result, be able to enforce permit conditions when necessary.

DESIGN OF RESTORATION PROJECTS

PRE-CONSTRUCTION CONSIDERATIONS

The primary premise of the regulator and project designer should be that the wetland to be mitigated and the wetland to be created are integral parts of an ecosystem.

Hydrology is the single most important element necessary for the maintenance or creation of specific wetland systems and their functions. Hydraulic conditions can directly modify or change chemical and physical properties such as nutrient availability, degree of substrate anoxia, sediment properties, and pH. Water inputs are invariably the dominant source of nutrients to wetlands and water outflows often remove biotic and abiotic material from wetlands (Mitsch and Gosselink 1986). Wetland fauna and flora will almost always respond to slight changes in hydrologic conditions with substantial changes in species richness, diversity, and productivity. Thus an abrupt and usually significant change in the functional integrity of a marsh will result from an alteration in its hydrology. Therefore, due to the importance of hydrology to both the mitigated and created wetlands, the watershed and surrounding geomorphology of the subject area must be thoroughly evaluated.

To select appropriate goals for the marsh creation or restoration project, a thorough analysis of the wetland under consideration for mitigation must first be undertaken. The initial step of this evaluation process should be a determination of the limits and nature of the watershed in which the marsh is located. When looking at this "big picture" the watershed should be evaluated at least generally with respect to vegetation communities, wildlife, geomorphology, surface and groundwater conditions, water quality, and surrounding land use. The evaluation of the surrounding land use should not only include existing facilities such as residential and industrial development but also planned land use changes which can often be estimated by identifying existing zoning within the area. This evaluation from a watershed perspective will also assist during the process of quantification of the wetland's functions. For example, the subject wetland may be one of a kind within the region or it may be very common. The wetland may be of a relatively pristine condition or may be found to be impacted through past modifications or current activities such as the discharge of polluted surface water runoff. Any mitigation plan submitted to the regulatory agency without benefit of some information on the watershed and the proposed mitigated wetland system's role is incomplete and should be rejected.

With the evaluation of the proposed mitigated wetland's watershed system underway, the particular characteristics and functions of the original marsh should then be examined. A thorough survey of soils, hydrology, vegetation, and wildlife is also desirable. This evaluation can be done on either a qualitative or quantitative basis, depending upon the specific needs. For example, the wetland evaluation should lean strongly toward a quantitative evaluation and an extended consideration of the possibility of successful mitigation in the first place (Erwin this volume) if (1) previous watershed analysis has placed a high value on the wetland to be mitigated because of the relative scarcity of similar systems within the region, (2) it is critical habitat for an endangered or threatened wildlife species, or (3) it is a habitat where successful creation or restoration has not been documented. If success of other similar projects has not yet been demonstrated, this would be an opportunity for the agency to consider preconstruction mitigation with a requirement that success be attained prior to the alteration of the wetland to be mitigated. Failure to successfully evaluate the wetland to be mitigated and the watershed system in which it lies will complicate the ability of the agency to judge any future success of the project.

The location and configuration of the created marsh project is extremely important because, without an adequate water supply, the project will fail. The values of a particular marsh system are often dependent upon their relationship to the adjacent landscape. For example, small isolated marshes are extremely common throughout central and south Florida, often occurring in densities greater than six per square kilometer. Wildlife such as waterfowl and wading birds may require groups of several isolated marshes in order to meet their needs for feeding and reproduction. The diversity of these isolated systems with regard to their geomorphology, hydroperiod, macrophyte and macrofaunal communities provide long term support to wildlife where changing conditions over many years make some of these marshes, for one reason or another, more suitable to certain species than others (Milton Weller, Texas A&M University pers. comm.). In some isolated wetland systems (i.e., prairie potholes), groundwater recharge is related to the edge:volume ratio of the wetland. Therefore, the regional significance of groups of these marshes is an important factor to be considered.

The regional site-specific evaluations discussed above for the wetland proposed to be mitigated and its respective watershed should be used to screen future possible sites for the marsh creation project. If a five hectare marsh with a 50 hectare watershed is to be mitigated by constructing a marsh of similar size in the same region with a 30 hectare watershed, there may be a problem. These types of comparisons should be made throughout the evaluation process for future project sites.

As mentioned earlier, another important factor for selection of a site for a marsh creation project is the existing and future projected land uses within the watershed in which the wetland will be constructed. The existing and future land uses within the subject area should be evaluated in the site selection process and determined to be compatible with the goals to be attained by the created wetland. It will also be necessary to provide some assurances that the watershed of the wetland to be created will remain intact, and not be significantly altered with regard to size, flow, and water quality characteristics by existing facilities or the construction of new projects.

The marsh creation project should not be located in an area where future development pressures may call for the alteration or elimination of the created marsh for some future land use. It may be argued that if a marsh was created and satisfactorily attained the goals the first time, it could be created a second time. But this could be a problem, particularly in high growth areas such as south Florida where land for development is becoming increasingly scarce. The regulatory agency should consider protecting the marsh creation project from future alteration by some legal means such as a conservation easement.

CRITICAL ASPECTS OF THE PROJECT PLAN

Hydrology

The project should be constructed in an area of suitable land use with an adequate watershed to provide the proper hydroperiod, and time and degree of inundation required to meet the established goals. In cases where a marsh is not being created as a result of similar marsh destruction, applicable data may be limited or nonexistent. In this case, it may be necessary to require the development of a water budget or model to assure that the watershed is adequate to create the proper hydroperiods and depths and degree of inundation to attain the goals of the selected habitat type. Hydrological modeling and water budgeting, even with quantitative analysis, is often inaccurate to some degree. Due to this fact and also because it is often difficult to maintain exacting control during construction, some flexibility should be allowed to make adjustments such as changing the elevation of the water control structure inlet or outlet mechanisms. This fine tuning of the system is often desirable and necessary in order to achieve the preferred results. There is nothing wrong with this practice which unfortunately is often discouraged by regulatory agencies because they are unable to monitor the operation of the control structures.

We must learn to regard wetland creation as an inexact science where flexibility is usually required to achieve the desired goals. The applicant is often required to submit a detailed surface water management plan or hydrological analysis for flood protection in an area where residential development is proposed. In this instance, a post-construction manipulation of a wetland which increases the water level by as little as one foot could require additional placement of fill in the residential development (which can kill upland trees) and/or the redesign of its drainage system. In such a case, lowering the elevation of the marsh may be more appropriate although revegetation may be required.

Another critical aspect of the project plan is a requirement that the marsh to be created be maintainable as a "stand alone" system. The water to be supplied to the created marsh should be supplied via low energy means such as ground or surface water or flow from an adjacent, natural, or properly designed manmade system. In no case should high energy "demand" systems such as pumps be utilized as the primary source of water for the project. These systems are expensive, energy intensive, difficult to maintain, and difficult to regulate.

Contour Design

The contour of a marsh creation project should be determined by the hydrological and watershed analyses and the goals of the project. If the goal is to create habitat for a specific type of wildlife such as waterfowl, or even a particular species, special attention must be given to the contours of the project. Water levels should normally fluctuate within the project, and the depths and hydroperiod must be conducive to creating the habitat required by the preferred species. Wading birds, dabbling ducks, and many species of fish could be attracted within one properly designed project where a variety of contours will create satisfactory water levels, macrophyte communities, and bottom conditions. Where the goal of a particular project is creation of habitat for a specific species, the design contours of the project become increasingly important. For instance, the endangered woodstork (Mycticorax americana) has suffered greatly from wetland habitat loss and alteration of hydrology of many of its remaining wetland feeding areas. This species depends on decreasing water levels during the winter/spring breeding season in south and central Florida to concentrate its food supply of small fish and crustaceans into smaller areas where its tactile feeding methods are successful. A proposed freshwater marsh created to provide woodstork habitat would then be required not only to mimic a natural hydroperiod, but also provide contours where wading is possible and pools of concentrated fish and crustaceans will form during its breeding season.

Water Quality

The quality of the water to be discharged into the marsh project must be compatible with the intended use or function of the wetland. The compatibility of the land use within the basin in which the wetland is to be created is critical, as well as the future land uses that can be expected to ultimately surround the project area.

Construction Considerations

Adequate planning and design of any wetland creation or enhancement project can easily be defeated by problems during the construction process. Two ways to reduce construction problems are: (1) Educate the contractor on the design specifications and other requirements such as logistics. This usually can be facilitated by one or more pre-construction meetings both in the office and on-site, and (2) Provide adequate supervision by a qualified expert during the critical phases of construction, preferably the wetland scientist originally involved in the project's design. When the construction plans are reviewed, the agency should require the applicant to: detail contours of ± 3cm; provide a construction schedule and phasing plan (for large projects); and specify the degree of supervision that will be provided by an expert in wetland creation or reclamation (particularly on large, more complex projects).

The timing of construction is important, but often overlooked. In the case of a freshwater marsh where seedbank material (suitable wetland mulch from a donor wetland) or plantings are to be utilized, the completion of earth work should be targeted for the early wet season where in situ desiccation can be substantially reduced. The construction of large projects greater than four hectares in size should be coordinated to prevent extended gaps of time between the completion of contouring and the planting or mulching and final reflooding. Large, exposed, unstabilized areas are subject to erosion and the colonization of problematic exotics such as Melaleuca or monocultures of less desirable aggressive species such as Typha.

Substrate

Substrate is usually overstated as a limiting factor. Substrate type is not usually critical, but can affect the hydrology because it is porous or poorly drained. In general, marsh construction can be completed with satisfactory results if other critical aspects of the project plan, such as hydrology and construction, are correctly planned. Most seedbank, planting techniques, and plant colonization can be successful on a variety of substrates. However, the project designer and reviewer should give attention to the substrate when the wetland is being specifically created to fulfill a particular water quality enhancement goal. Here the nature of the substrate may have as great, if not more important, value than vegetation in removing and binding certain pollutants.

Revegetation

The type of revegetation needed is dependent upon the site specific goals of each project. If the goal is to create a marsh closely resembling, in an overall sense, the marsh proposed for alteration, mulching should be investigated. In this process, the wetland which is to be destroyed or altered would be stripped of its surficial layer of substrate containing above ground plant parts, root structures, and viable seed material. Depending upon the timing of the project construction, this mulch material would either be moved directly to the finished contours of the creation project or temporarily stockpiled for later use. The use of wetland mulch in the creation of freshwater marsh is well documented (Erwin 1983 through 1987, Erwin and Best 1985, Erwin 1985, Dunn and Best 1983). These references should be consulted with respect to the intricacies of using this technique. When properly executed, the mulching technique can yield satisfactory results. However, during the stripping process from the donor wetland site, materials will be thoroughly mixed, resulting in revegetation of the project site by most if not all of the desired species, but with a different cover frequency and pattern of distribution than that in the donor site.

In some cases, mulching is not the appropriate revegetation technique, particularly where a donor site is unsuitable or unavailable. In addition, one of the drawbacks to mulching is the difficulty, in some instances, to place desired species in exact locations or zones of the created wetland.

Planting is accepted and widely used, and a proven alternative to mulching. The species selected for planting depend directly upon the created wetland's hydrology, and the intended goals. When planting is the preferred method of vegetating a freshwater marsh creation project, the following points must be considered:

1. Will the species of plants in numbers required be available when needed during construction periods? Materials should be planted during the wet season or plans for irrigation should be made.

2. Can all of the materials be planted within a reasonable time to improve chances for survival?

3. Have the proper locations for the various species to be planted been identified on the project's site to insure proper hydroperiod and degree of inundation?

4. Finally, will the process be supervised by a knowledgeable individual responsible for maintaining quality control during this process?

In most cases where planting is the preferred method for vegetating a project, it is necessary to provide some degree of substrate stabilization to reduce erosion. A variety of grass seed mixtures will satisfactorily serve this purpose with reseeding encouraged in areas as long as necessary. Animals such as grazing cattle and feral hogs can cause significant damage to a recently planted marsh project. These and other nuisance animals should be excluded from a project site as long as necessary.

In some instances, mulching or the mechanical revegetation of a project area is not necessary and volunteer colonization may be used. This is a limited case, usually where a marsh is being created adjacent to an existing marsh and the area being created is not overly large so that the volunteer colonization of the site by desired species would take too long to meet the desired goals of the project. Temporary stabilization of the substrate may still be necessary to prevent erosion and encourage colonization by the preferred plants. This option of volunteer colonization of a project site should not be considered when problematic exotics such as Melaleuca are found within or adjacent to the site and would be considered a threat to colonization by the preferred marsh species. Many natural and some created forested and nonforested wetlands have been overrun by problematic species such as Melaleuca and/or monocultures of cattail where substrates were exposed for an extended period of time prior to planting, allowing settlement of wind dispersed seeds and subsequent colonization.

Reintroduction of Fauna

If the goals of the marsh creation project are met, namely satisfactory site location, an adequate hydroperiod, and a successful revegetation, reintroduction of desired fauna should not be necessary. In some cases, the stocking of fish, herps and amphibians may be required in isolated marsh systems where some aquatic habitat has been constructed. In some cases where the impacted wetland is a rare type, geographically isolated, or harboring protected and/or uncommon species, trapping and relocation of the organisms to the constructed wetland site may be required.

Buffers and Protective Structures

In many instances, an upland buffer surrounding the created wetland and/or upland corridor connecting it to adjacent habitats including other wetlands, is necessary to optimize wetland values within the created wetland and adjacent natural systems. However, creating a narrow upland buffer around a marsh may be of limited value in an urbanized setting where the buffer may be subjected to human degradation. Wetlands and uplands function as components of highly variable ecosystems. We should recognize suitable upland habitat as an important factor when assessing the values and functions of both the natural wetland and the marsh project. The consideration of landscape ecology is second only to hydrology when creating a freshwater marsh. Whereas hydrology is the ultimate limiting factor on the ability to create the wetland, it is the landscape ecology of the system within which the project is located that will determine the created wetland's success in providing the desired wildlife habitat and other functions.

When an upland buffer or upland corridor is created, it is usually necessary to prevent undesirable human intrusion. Fences and/or native landscaping should be considered. In urbanized settings where suitable upland vegetation is not present for use as a buffer or corridor, openwater systems such as lakes could be utilized as long as hydroperiod and water levels are adequate to prevent negative impacts on the marsh. In some cases, marsh projects could be created adjacent to a properly designed lake not only as a buffer but also as an extremely valuable littoral zone design feature of the lake. Manmade lakes often lack a suitable source of carbon and shallow areas of emergent vegetation which are important fish and wildlife habitat. Properly designed freshwater marsh projects increase the value of manmade lakes by becoming a positive design feature and improving the habitat value and water quality. The location of freshwater marsh projects together with existing or proposed manmade lakes is preferable to isolating lakes and wetlands.

Long Term Management

Long term management should be considered when planning a project and reflected in the design so that a low maintenance approach to management is possible. The wetland project should be owned by an entity which will have the finances available to implement future management practices such as control of exotic plant species. It will also be necessary to protect the project area from undesirable uses of adjacent lands. Although not usually considered as a design feature of a wetland creation project, the identification of an appropriate entity responsible for implementing long term management is very important.

In most cases, public ownership is preferable to private ownership since a public agency will be more likely to maintain the project in a manner consistent with project goals. The question of the wetland project's future ownership should be dealt with during the review process. In some instances, such as large reclamation projects, private ownership could be retained and protection afforded through use of conservation easements and deed restrictions. At the present time, most marsh creation projects are without long term management plans and contain no opportunity for funding for future maintenance that will probably be required.

The regulatory agency must be prepared to provide enforcement of the management plan when necessary. As previously discussed, too often wetland creation projects are developed as part of a permit and forgotten before the project is completed since many of these marsh creation projects will undoubtedly outlive the permits which led to their creation. The agency should create a mechanism requiring the proper management of the project with regular reporting made to the permitting agency. In most cases, the burden should be placed upon the permittee for the cost of management, including inspections and generation of reports or studies. Management reports should be tied to the identifiable goals of the project.

MONITORING

Monitoring of marsh creation or enhancement projects should be undertaken so that, eventually, a final determination may be rendered on the project's success or failure in attaining desired goals. A monitoring program should include data requirements, evaluation criteria, and methods for reporting with goal evaluation in mind. The key items to be monitored are usually the site's hydrology, flora, and fauna. Achieving satisfactory hydrology and vegetation will usually lead to the accomplishment of additional goals, such as wildlife utilization. Often the marsh creation project may have a more specific goal, such as utilization of the site by a particular species of wildlife or improvement of water quality. In this case, the monitoring of the project must be more specific and intense.

A number of agencies in addition to the Environmental Protection Agency and U.S. Army Corps of Engineers require monitoring as a normal part of the permitting process. These include city and county governments, water management districts, and other state environmental regulatory agencies. With the large number of permits being issued simultaneously by regulatory agencies, there is a need for consolidation of monitoring and evaluation requirements. Unfortunately there is little or no follow-up evaluation of data collected in monitoring efforts to see if they are in fact properly collected. Also, data collection may not be designed to provide the requisite data for evaluation of goal attainment. A suggested format for tracking wetland mitigation projects is provided in Figure 1. There is also a critical shortage of baseline data on natural systems (reference wetlands) to provide comparative information.

A discussion of evaluation techniques to be incorporated in monitoring requirements is provided in Erwin (see Volume II). A standardized monitoring plan is suggested below. It should be tailored to each project's goals.

1. **Elevation and Contours.** The contours and elevations should be surveyed to determine degree of compliance with the design. These measurements can be taken qualitatively via simple observation by experienced personnel. However, large complex projects require quantitative evaluation utilizing topography provided by ground survey (spot elevations) or aerial photography.

2. **Ground and Surface Measurements.** The use of staff gauges, piezometers, and/or constant water level recorders is often imperative to determine whether the desired hydrological conditions have been established within the project. The collection of this data

and evaluation of the vegetation will determine whether any "fine tuning" of contours, watershed area, or control structure elevation (such as adding or removing a board in the notch of a weir) is required. It is often difficult to estimate the desired pre-construction hydrological parameters of the project site. Therefore, the proper collection of post-construction hydrological data and the ability to fine tune the project is extremely important. Ground and surface water levels may be required to be taken at one or more locations upstream of the project within the subject watershed, within the project and/or at its discharge point. Baseline data collected in a natural marsh of similar size and physiognomy is most helpful where a lack of useful baseline data exists or when attainment of a more specific goal pertaining to a particular habitat type or function is desired.

3. **Water Quality.** Water quality monitoring requirements vary depending upon site specific conditions, such as the existence of contamination sources within the watershed or where treatment of a particular pollutant is desired within the project. Water quality data should be used in conjunction with vegetation and macroinvertebrate analysis. Many species of wildlife, such as waterfowl, will utilize a contaminated wetland site. For this reason the collection and evaluation of water quality data within a project is particularly important when the project is located within a watershed containing possible sources of contamination. Monitoring data should be collected, at a minimum, on a seasonal basis from wet season to dry season. The most common deficiency in water quality monitoring is the lack of quality control during collection and analysis of the samples. Measures should be implemented to identify and control such problems.

Maintaining some degree of flexibility in the water quality monitoring plan may be appropriate. Often it is necessary to begin monitoring for a wide range of parameters where it is believed contaminants are present and/or a lack of useful baseline data exists. As these data are collected, regular evaluations should be made with regard to the adequacy of the monitoring plan. After one or two years of monitoring, it is often appropriate to change the parameters monitored and/or decrease the amount of sampling because problems that were expected failed to materialize. Concerns regarding other parameters could become significant and cause an intensification of the effort. The permit conditions should always reflect this flexibility in the proposed monitoring plan.

4. **Biological Monitoring.** Dragnet sweeps are generally the most useful qualitative method of sampling small fish and macroinvertebrates, particularly when baseline data is established for similar systems. Quality control during sample collection, evaluation, and data analysis is extremely important. Sample station locations and methodologies should not vary from sampling event to sampling event. Site conditions such as water depth, temperature, water quality, and flood conditions should be recorded during each sampling event as these parameters will often influence the macrobenthic community. Without this information the reviewer is forced to use his or her imagination when evaluating the data. Recent reports such as Erwin (1985) and Erwin (1987) stress the importance of collecting macrobenthic samples in each macrophyte community present in the marsh system to obtain a representative characterization of the site. Similarity of species is often low between the various macrophyte communities. In addition, net sweeps should be taken within each macrophyte community for approximately 20 minutes in order to ascertain the effectiveness of the selected quantitative methodology in collecting the species found within the site (see Erwin Volume II).

5. **Vegetation.** The degree of monitoring required will depend on the amount of baseline information needed for evaluation of the type of system being created. Information regarding species diversity, distribution, or frequency for representative natural habitats (reference wetlands) and the marsh creation project aid in the determination of project success. The data should be collected with the project goals, criteria for success, and degree of desired compliance in mind. In a review of wetland evaluation techniques Erwin (see Volume II) suggests several methods of qualitative and quantitative data analysis which are appropriate for monitoring reference, created, or restored wetland sites. Some combination of species richness, frequency, percent cover and bare ground should be documented. Generally, percent cover and species richness is the most useful information for determining success. Bare ground is often widespread and common within freshwater marshes and will change from season to season and year to year as will the floristic composition or zonation of a particular marsh. Direct comparisons are possible from one marsh to the next, whether natural or manmade.

However, seasonal changes are normal and as long as the desired species are present within acceptable ranges of coverage, the project can usually be considered successful with regard to the vegetation.

Since species composition and cover

usually change dramatically from season to season, the monitoring should take place, at a minimum, early and late in the growing season (twice a year). Changing climatic factors such as excessive rainfall or droughts can alter the previously established patterns and should not be regarded as a problem. For example, the Agrico Swamp West project in central Florida (Erwin 1985) yielded dramatic changes in floristic composition during a severe drought. However, the ability of the created marsh to withstand the normal natural perturbations was confirmed the following year when hydrological climatic conditions returned to normal and the seedbank present in the system responded appropriately, displacing the temporary upland invaders with the typical marsh species (Erwin 1986).

Methods such as ground and aerial photography and mapping of sites should be used whenever possible when more quantitative data collection is not appropriate. Aerial photographs with appropriate groundtruthing can determine the existence and aerial extent of dominant species and are easily used to determine achievement of goals because comparisons can be made on species present and coverage. The primary productivity of the wetland places an upper limit on the size of animal populations within the system. Any reduction in the size of the plant community generally will have adverse repercussions on wildlife population sizes, so that plants are an obvious major focus for study even when the animal species are of primary concern.

WILDLIFE UTILIZATION

The most widely used and generally successful method of evaluating wildlife utilization of a natural or created site is observation. Observations should be made during the correct season, time of day, and over a satisfactory number of events by qualified personnel. Once again, where more specific goals have been established with regard to a particular species, more intense monitoring may be required which may involve quantitative surveys of the project to determine number of nests per hectare or breeding pairs per season, etc. Wildlife utilization of a creation or restoration project is almost always one of the specified or implied goals, but actual monitoring or observation of wildlife utilization is often lacking in the permit conditions (where it should be required). Special consideration should be given to endangered species, threatened species, or wildlife species of special concern (listed species). The natural wetland should be evaluated with respect to: (1) utilization by listed species at the present time, or (2) its future suitability for utilization by these species. Factors leading to the present or expected usage of the habitat by listed species (such as open water areas for waterfowl) should be thoroughly documented. In addition, the wetland's proximity to other wetlands or certain types of upland habitat might dictate its degree of utilization by certain species of wildlife.

INFORMATION GAPS AND RESEARCH NEEDS

Information gaps and research needs related to the creation and restoration of freshwater marshes can be divided into the following categories: (1) site selection and design; (2) project construction techniques; (3) comparative studies of the biological communities and processes in created and natural systems; and (4) the role of upland/transitional habitats.

SITE SELECTION AND DESIGN

There is a need for information related to the suitability of wetland creation in urbanized landscapes. The subject of landscape ecology needs to be evaluated with regard to the impact of surrounding land uses on natural and created freshwater marsh systems. Given the fact that we have lost over 116 million acres of an original estimated total of 215 million acres of wetlands (Tiner 1984), the feasibility and success of the restoration of wetlands in urbanized landscapes should be a high priority and not just considered as mitigation. The development of cost effective designs and construction methods is needed to encourage more wetland restoration.

PROJECT CONSTRUCTION TECHNIQUES

The cost of wetland construction is usually in proportion to the size of the created marsh and construction techniques. Data on the success of mulching vs. various planting techniques is needed along with per hectare costs for each method. When planting is required, the most suitable species should be identified to meet various objectives that may be important, including but not limited to; type of wildlife habitat, maximum rate of cover, preferred species composition, and design constraints such as substrate, hydroperiod, and degree of inundation.

COMPARATIVE STUDIES OF CREATED AND NATURAL SYSTEMS

The author believes that the feasibility of creating freshwater marsh habitat has been demonstrated although refinement of techniques is needed. Regulatory agencies and scientists still question how these wetlands compare in structure, function, and value to comparable natural systems (Race and Christie 1982). The high degree of variability among the different types of freshwater marsh systems, both natural and created, makes most comparisons difficult. However, comparative studies are needed. Comparative studies should evaluate successional changes in flora, fauna, soil, and water chemistry over time including, but not limited to, macrobenthic community development, role of mycorrizae, nutrient flux and cycling, soil formation, plant productivity, and fish and wildlife habitat value. Also, the functional difference in organic soil accumulation between an old natural marsh and a recently created marsh should be evaluated.

ROLE OF UPLAND AND TRANSITIONAL HABITATS

The relationship of adjacent transitional and upland habitats to the wetland system's functions and values has been demonstrated but requires further study. Adjacent upland habitats should be included in the above mentioned comparative studies. We may never fully quantify all aspects of the wetland-upland relationship, but we need to determine how and where the created wetland should be located within the upland, how much (in general) buffer is needed, and whether some upland habitat creation or enhancement is also required to enable a particular created wetland to function as desired (Figure 2).

LITERATURE CITED

Adamus, P.R. 1983. A Method for Wetland Functional Assessment. Volume II. The Method. U.S. Department of Transportation, Federal Highway Administration. Office of Research, Environmental Division, Washington, D.C. 20590. (No. FHWA-IP-82-24).

Brown, M.T. and H.T. Odum. 1985. Studies of a Method of Wetland Reconstruction Following Phosphate Mining. Final Report. Florida Institute of Phosphate Research, Publication #03-022-032.

Cole, D. (Ed.). 1979. Proceedings of the 6th Annual Conference on Wetlands Restoration and Creation. Hillsborough Community College, Tampa, Florida.

Cole, D. (Ed.). 1980. Proceedings of the 7th Annual Conference on Wetlands Restoration and Creation. Hillsborough Community College, Tampa, FLorida.

Cole, D. (Ed.). 1981. Proceedings of the 8th Annual Conference on Wetlands Restoration and Creation. Hillsborough Community College, Tampa, Florida.

Dunn, W.J. and G.R. Best. 1983. Enhancing ecological succession: 5. seed bank survey of some Florida marshes and the role of seed banks in marsh reclamation. In Proceedings, National Symposium on Surface Mining, Hydrology, Sedimentology, and Reclamation, Office of Continuing Education, University of Kentucky, Lexington, Kentucky.

Erwin, K.L. 1983. Agrico Fort Green Reclamation Project, First Annual Report. Agrico Mining Company, Mulberry, Florida.

Erwin, K.L. 1984. Agrico Fort Green Reclamation Project, Second Annual Report. Agrico Mining Company, Mulberry, Florida.

Erwin, K.L. 1985. Agrico Fort Green Reclamation Project, Third Annual Report. Agrico Mining Company, Mulberry, Florida.

Erwin, K.L. 1986. Agrico Fort Green Reclamation Project, Fourth Annual Report. Agrico Mining Company, Mulberry, Florida.

Erwin, K.L. 1987. Agrico Fort Green Reclamation Project, Fifth Annual Report. Agrico Mining Company, Mulberry, Florida.

Erwin, K.L. 1988. Agrico Fort Green Reclamation Project, Sixth Annual Report. Agrico Mining Company, Mulberry, Florida.

Erwin, K.L. and F.D. Bartleson. 1985. Water quality within a central Florida phosphate surface mined reclaimed wetland, p. 84-95. In F.J. Webb, Jr. (Ed.), Proceedings of The 12th Annual Conference on Wetland Restoration and Creation. Hillsborough Community College Environmental Studies Center, Tampa, Florida. May 16-17.

Erwin, K.L. and G.R. Best. 1985. Marsh community development in a central Florida phosphate surface-mined reclaimed wetland. Wetlands 5:155-166.

Erwin, K.L., G.R. Best, W.J. Dunn, and P.M. Wallace. 1984. Marsh and forested wetland reclamation of a central Florida phosphate mine, p. 87-103. Wetlands 4:87-103.

Ewel, K.C. 1976. Effects of sewage effluent on ecosystem dynamics in cypress domes, p. 169-195. In D.L. Tilton, R.H. Kadlec, and C.J. Richardson (Eds.), Freshwater Wetlands and Sewage Effluent Disposal. University of Michigan, Ann Arbor.

Ewel, K.C. and H.T. Odum. 1978. Cypress swamps for nutrient removal and wastewater recycling, p. 181-198. In M.P. Wanielista and W.W. Eckenfelder, Jr. (Eds.), Advances in Water and Wastewater Treatment

Biological Nutrient Removal. Ann Arbor Sci. Publ., Inc. Ann Arbor, Michigan.

Ewel, K.C. and H.T. Odum. 1979. Cypress domes: nature's tertiary treatment filter, p. 103-114. In W.E. Sopper and S.N. Kerr (Eds.), Utilization of Municipal Sewage Effluent and Sludge on Forest and Disturbed Land. The Pennsylvania State University Press, University Park, Pennsylvania.

Ewel K.C. and H.T. Odum (Eds.). 1984. Cypress Swamps. University Presses of Florida, Gainesville.

Fetter, Jr., C.W., W.E. Sloey, and F.L. Spangler. 1978. Use of a natural marsh for wastewater polishing. J. Water Pollution Control Fed. 50:290-307.

Gosselink, J.G. 1984. The Ecology of Delta Marshes of Coastal Louisiana: A Community Profile. U.S. Fish and Wildlife Service, Biological Services FWS/OBS-84/09. Washington, D.C.

Greeson, P.E., J.R. Clark, and J.E. Clark (Eds.). 1979. Wetland Functions and Values: The State of Our Understanding, Proceedings of the National Symposium on Wetlands, Lake Buena Vista, Florida, American Water Resources Association Tech. Publ. TPS 79-2. Minneapolis, Minnesota.

Kadlec, R.H. 1979. Wetlands for tertiary treatment, p. 490-540. In P.E. Greeson, J.R. Clark, and J.E. Clark (Eds.), Wetland Functions and Values: The State of Our Understanding. American Water Resources Association TPS 79-2. Minneapolis, Minnesota.

Kadlec, R.H. and D.L. Tilton. 1979. The use of freshwater wetlands as a wastewater treatment alternative. CRC Crit. Rev. Environ. Control 9:185-212.

Kevin L. Erwin Consulting Ecologist, Inc. 1989. First Annual Wetland Mitigation Monitoring Report for the Charlotte County Correctional Institution.

Kusler J.A. and P. Riexinger (Eds.). 1985. Proceedings of the National Wetland Assessment Symposium. Association of State Wetland Managers, Berne, New York.

Larson, J.S. 1982. Understanding the ecological values of wetlands, p. 108-118. In Research on Fish and Wildlife Habitat. EPA-600/8-82-002. U.S. Environmental Protection Agency, Washington, D.C.

Mitsch, W.J. and J.G. Gosselink. 1986. Wetlands. Van Nostrand Reinhold Company Inc., New York.

Quammen, M.L. 1986. Measuring the success of wetlands mitigation, p. 242-245. In J.A. Kusler, M.L. Quammen, and G. Brooks (Eds.), Proceedings of the National Wetlands Symposium, Mitigation of Impacts and Losses. Association of State Wetland Managers, Berne, New York.

Race, M.S. and D.R. Christie. 1982. Coastal zone development: mitigation, marsh creation, and decision making. Environmental Management 6:317-328.

Reppert, R.T., G. Sigleo, E. Stakniv, L. Messman, and C. Myer. 1979. Wetlands Values: Concepts and Methods for Wetlands Evaluation. IWR Research Report 79-R-1, U.S. Army Engineer Institute for Water Resources, Fort Belvoir, Virginia.

Richardson, C.J., D.L. Tilton, J.A. Kadlec, J.P.M. Chamie, andW.A. Wentz. 1978. Nutrient dynamics of northern wetland ecosystems, p. 217-241. In R.E. Good, D.F. Whigham and R.L. Simpson, (Eds.), Freshwater Wetlands--Ecological Processes and Management Potential. Academic Press, New York.

Sather, J.H. and R.D. Smith. 1984. An Overview of Major WetlandFunctions and Values. NWS/OBS-84/18. U.S. Department of the Interior, Fish and Wildlife Service, Washington, D.C.

Spangler, F.L., C.W. Fetter, Jr., and W.E. Sloey. 1977. Phosphorus accumulation-discharge cycles in marshes. Water Resour. Bull. 13:1191-1201.

Tilton, D.L. and R.H. Kadlec. 1979. The utilization of a freshwater wetland for nutrient removal from secondary treated wastewater effluent. J. Environ. Qual. 8:328-334.

Tiner, R.W. 1984. Wetlands of the United States: Current Status and Recent Trends. National Wetland Inventory, U.S. Fish and Wildl. Serv., Washington, D.C.

Webb, Jr., F.J. (Ed.). 1982. Proceedings of the 9th Annual Conference on Wetlands Restoration and Creation. Hillsborough Community College, Tampa, Florida.

Webb, Jr., F.J. (Ed.). 1983. Proceedings of the 10th Annual Conference on Wetlands Restoration and Creation. Hillsborough Community College, Tampa, Florida.

Webb, Jr., F.J. (Ed.). 1984. Proceedings of the 11th Annual Conference on Wetlands Restoration and Creation. Hillsborough Community College, Tampa, Florida.

Webb, Jr., F.J. (Ed.). 1985. Proceedings of the 12th Annual Conference on Wetlands Restoration and Creation. Hillsborough Community College, Tampa, Florida.

Webb, Jr., F.J. (Ed.). 1986. Proceedings of the 13th Annual Conference on Wetlands Restoration and Creation. Hillsborough Community College, Tampa, Florida.

Webb, Jr., F.J. (Ed.). 1987. Proceedings of the 14th Annual Conference on Wetlands Restoration and Creation. Hillsborough Community College, Tampa, Florida.

Wolf, R.B., L.C. Lee, and R.R. Sharitz. 1986. Wetland creation in the United States from 1970-1985: an annotated bibliography. Wetlands 6:1-78.

APPENDIX I: PROJECT PROFILES

PHOSPHATE MINE WETLAND RECLAMATION, AGRICO SWAMP WEST

Location:

Agrico Swamp West is located adjacent to the western boundary of the flood plain of Payne Creek at Agrico's Fort Green Mine in southwest Polk County, in central Florida. The reclamation plan includes a 60.75 hectare experimental wetland and 87.48 hectares of contiguous uplands in the watershed (Figure 3). The reclamation site was originally pine flatwoods and rangeland with some mixed forest (Figure 4) before it was mined in 1978 and 1979 (Figure 5).

Goals of Project:

The goal of the Agrico Swamp West reclamation project was to reclaim a "high quality" wetland ecosystem. This goal required the development of a design that, based upon ecological principles, is self maintaining and in harmony with natural systems.

The reclamation project was designed and constructed to create freshwater marsh, hardwood swamp, open water, and upland habitats. The design specifically lends itself to an intensive monitoring program for the evaluation of the various tree planting and marsh establishment methods, biological integrity, and the quality and quantity of ground and surface waters within the project. The ecosystem engineering applied to the design of the project introduced a variety of relatively new concepts to ecosystem reclamation and monitoring of surface mined lands and will continue to develop data over the next several years. This should aid in the evaluation of these techniques and the types of improvements required.

Construction Technique:

Due to the removal of the ore body during mining, sufficient material did not exist in the area to achieve design elevations. Sand tailings were therefore pumped between the overburden spoil piles over a period of eight months to provide the backfill material necessary to construct the planned elevations (Figure 6). The backfilling operation was completed in February, 1981. Earth moving began in March, 1981 with bulldozers and scrapers used to redistribute the sand tailings as planned. Overburden was spread over the sand to provide a 0.3 meter thick cap. As the earth moving progressed, the excess water displaced by the sand and overburden was pumped into the mine's water recirculation system to maintain satisfactory operating conditions.

The project area was contoured so that all drainage flows from the west toward the east. The levee constructed along the eastern boundary of the project impounds the drainage from the 148.23 hectare watershed to form wetlands at the design elevation. Two swale outlets were constructed in the levee to allow the overflow discharge of water from the wetland into the flood plain of Payne Creek. The elevation of the wetland along the base of the levee is +118 feet mean sea level (MSL). The elevation rises gradually to an elevation of +121 feet MSL along the western boundary of the wetland and less gradually westward across the upland portion of the project to an elevation of +134 feet MSL. A water budget for the project was developed to evaluate the disposition of storage, inflow, and outflow of water within the project area during a typical year.

Ponds were constructed within the wetlands with bottom elevations of approximately 108 feet MSL to maintain open water areas all year round. Small, shallow depressions were constructed randomly throughout the fluctuating water zone to retain water and harbor fish populations during periods of low water. Two lakes were also constructed in the uplands which overflow via swales eastward into the wetlands (Figure 7).

Methods:

The freshwater marsh created at Agrico Swamp West utilized two restoration techniques, resulting in the establishment of two initially contrasting wetland habitats. One marsh habitat was created by using a mulch (Figure 8) from a nearby freshwater marsh. The mulch contained seed and root material from the native wetlands. This procedure of mulching has now been incorporated into a number of wetland reclamation plans (Erwin, K.L. 1988, Brown and Odum 1985). The other marsh habitat was created with overburden soils (Figure 9). This area was recontoured and allowed to vegetate naturally. The overall goal of this study is to determine the optimal method for establishing high diversity, late successional marsh ecosystems immediately after mining and recontouring. The ultimate purpose is to demonstrate the successful reclamation of a freshwater marsh.

Monitoring Vegetation:

Monitoring of the marsh is performed to characterize the vegetation found in the two wetland areas both immediately and for several years after reclamation. Percent cover values and species richness have been monitored since the fall of 1982 (Figures 10 and 11).

The study continues to provide information regarding the rate and direction of marsh succession under various reclamation schemes and natural perturbations such as droughts and winter freezes. The monitoring program is designed to determine to what extent reclamation with or without mulching can meet the reclamation criteria of establishing late successional perennials as well as controlling aggressive weedy species. Information derived from this study has served as a useful tool for developing marsh reclamation guidelines (Erwin 1988).

The goal of this monitoring program is to determine to what extent reclamation with or without mulching can meet the reclamation criteria of establishing late successional perennials as well as controlling aggressive weedy species. The herbaceous vegetation was monitored using the following modified line-intercept technique. Table 1 provides the number and percent cover of marsh species in mulched and overburden areas. Sixty one

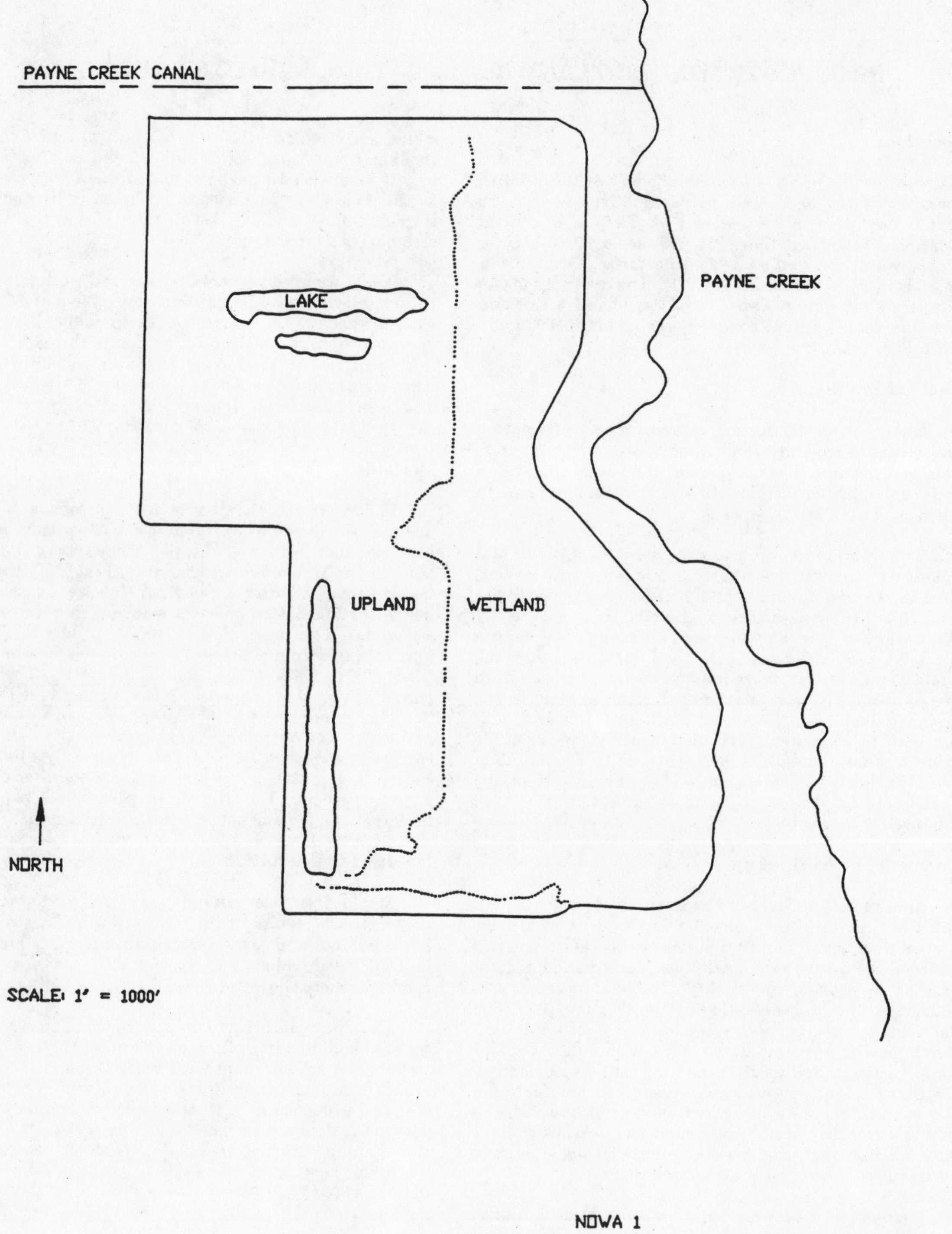

Figure 3. Agrico Swamp West Wetland Reclamation Project.

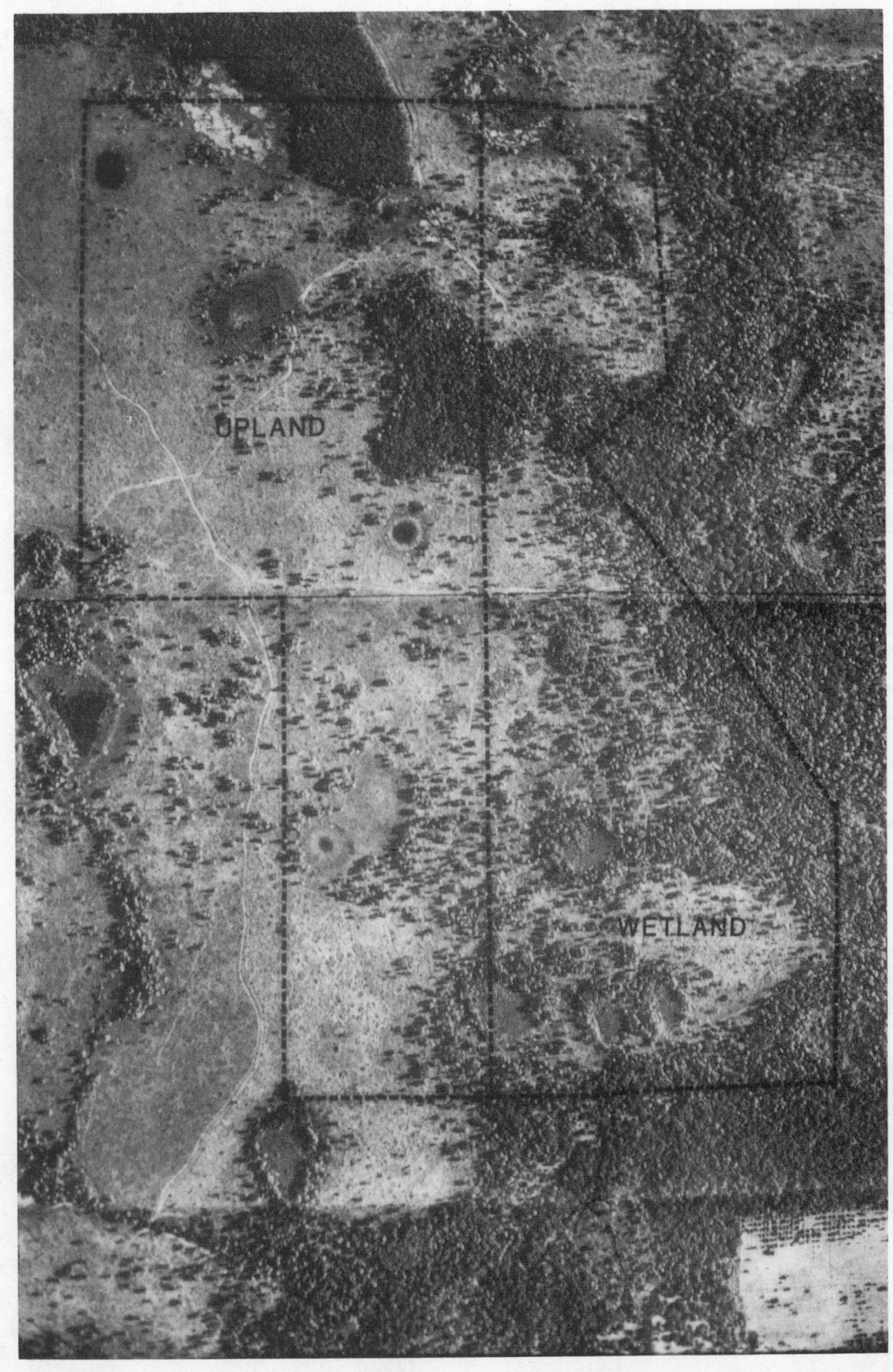

Figure 4. Agrico Swamp West area prior to surface mining.

Figure 5. Two views of the Agrico Swamp West project area after surface mining (1980).

Figure 6. Backfilling mine cuts during reclamation (1981).

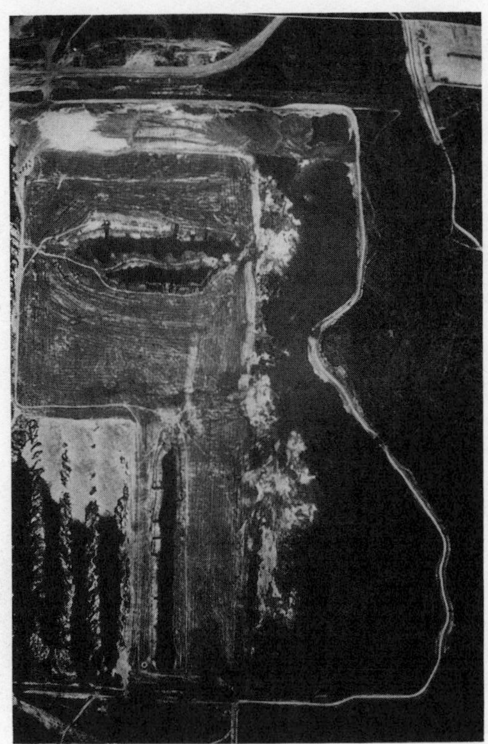

Figure 7. Aerial photo of Agrico Swamp West following reclamation in 1984.

Figure 8. Freshly spread wetland mulch in Agrico Swamp West (1982).

Figure 9. Agrico Swamp West wetland area not inoculated with wetland mulch (1982).

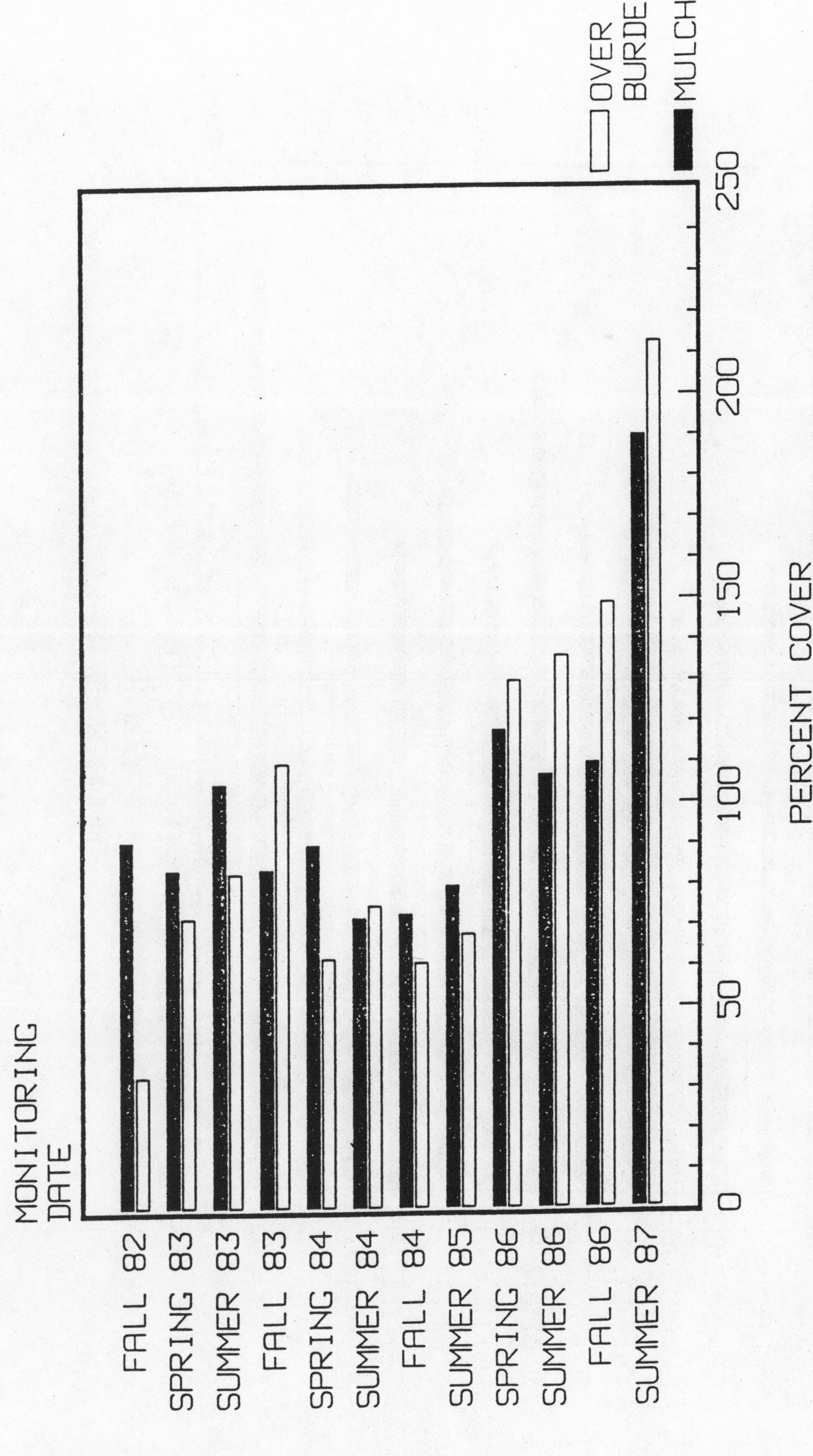

Figure 10. Percent cover of vegetation in mulched and overburden areas of Agrico Swamp west from Fall 1982 through summer 1987.

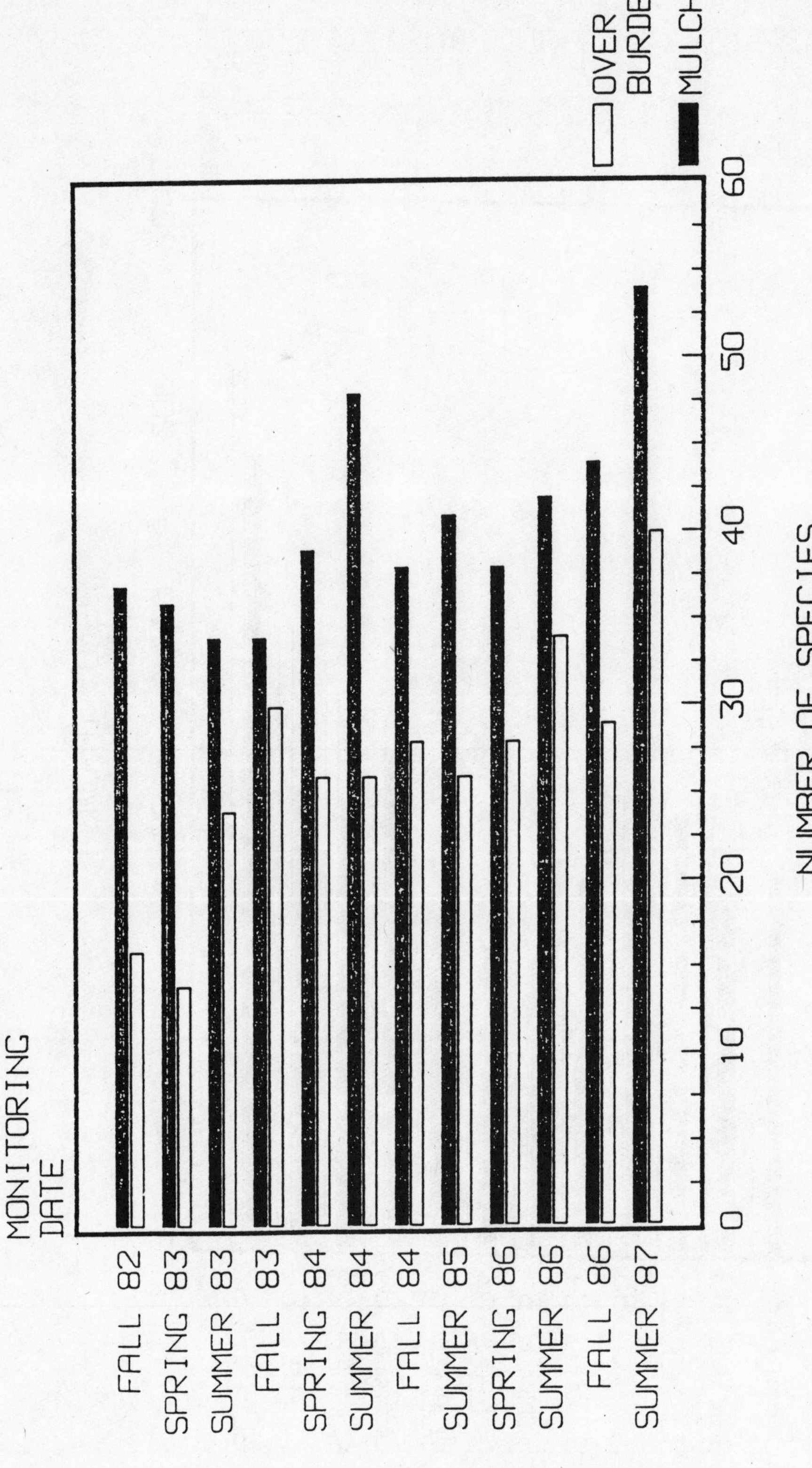

Figure 11. Species richness of vegetation in mulched and overburden areas of Agrico Swamp west from Fall 1982 through Summer 1987.

Table 1. Number and percent cover of marsh species in mulched (M) and overburden (O.B.) areas in Agrico Swamp West. Percent cover in this table does not take into consideration the area of non-covered ground (see text for discussion).

	# OF SPECIES		PERCENT COVER	
	M	O.B.	M	O.B.
Fall 1982	37	16	91	33
Spring 1983	36	14	84	72
Summer 1983	34	24	105	83
Fall 1983	34	30	84	110
Spring 1984	39	26	72	75
Fall 1984	38	28	73	61
Summer 1985	41	26	80	68
Spring 1986	38	28	118	130
Summer 1986	42	34	107	136
Fall 1986	44	29	110	149
Summer 1987	54	40	190	213

species were encountered within the project during 1987 (Figures 12 and 13). Species richness in the mulched and overburden areas during the summer of 1987 was 54 and 40, respectively. The most distinctive difference in plant distribution between overburden and mulched areas is shown by Pontederia cordata (30% more dominant in mulched), Wolffiella gladiata (24% more dominant in mulched), Sagittaria lancifolia (18% more dominant in mulched), and Typha domingensis (17% more dominant in overburden). The project has developed a seed bank of its own including those areas that did not receive the mulched treatment. It appears that mulching is valuable in establishing a rapid cover of preferred species.

Similar studies of natural marsh areas undertaken in 1988 and 1989 will allow for a short term comparison of species richness and cover values between natural marsh sites and a relatively stable reclaimed marsh.

Macrobenthic Monitoring:

The biological monitoring program is designed to complement the forest and marsh community reclamation and water quality monitoring of the project. The objective is to develop a model of the long term trends in biological community development in a reclaimed wetland ecosystem.

Macroinvertebrate composition and abundance are recorded from the substrata and macrophytes (leaves and stems) of both natural and created wetlands to determine the influence of macrophytes and water quality on the benthic community. Many of the species found on the macrophytes were also found in the substrata, but the compartmentalization of species between various macrophyte types was considerable. All four sample seasons (November, February, May, and August) show a considerable variability between macrophyte types and locations. In 1987, a total of 46 taxa were collected in core samples from natural marshes, and the number present in individual samples ranged from 7 to 22 (mean 11.5). Densities ranged from 926 to 7,501/m^2 (mean 3,925), and Shannon-Weaver diversity ranged from 1.72 to 2.95 (mean 2.39). Sixty-seven taxa were recorded from the created wetland, samples ranged from 4 to 33 (mean 16.4). Densities and diversity ranged from 926 to 18,196/m^2 (mean 5,736) and 0.98 to 4.20 (mean 2.69), respectively. In general, aquatic macrophyte habitats harbored a more species rich assemblage than did the openwater substrata, but densities in the open water zone fell within the range of densities recorded for the various macrophytes.

Based on the sampling methods utilized in this study, it is apparent that even though marshes, both natural and created, may share various species to a lesser or greater degree, and be similar in densities and diversity, each is unique with regard to the structure of its macroinvertebrate community.

The third year of biological monitoring (1987) of the six-year-old created marsh shows a well developed macroinvertebrate utilization of the wetland's substrata and macrophytic components. The collection of large numbers of organisms and taxa demonstrates that a rich, diverse benthic community has developed within the site.

Water Quality Monitoring:

The water quality monitoring of Agrico Swamp West is designed to assess the surface and groundwater quality on-site as well as in the receiving waters of Payne Creek.

Surface water and groundwater within the marsh are of good quality and the area has apparently stabilized from the effects of previous mining and reclamation. Surface waters in the wetland meet the ultimate test by supporting large and diverse populations of macroinvertebrates, fish, and wildlife. The only water quality test parameter generally nonconforming to state water quality standards continues to be pH. The high pH values obtained in openwater samples do not appear to be causing any adverse effects and in fact may be responsible for the binding of phosphorus, and perhaps fluoride and other elements thereby enhancing water quality. The high pH of the openwater areas is not affecting the groundwater or Payne Creek.

Fish and Wildlife Monitoring:

Agrico Swamp West has developed into exceptional fish and wildlife habitat. Eighty-three species of birds have been observed within the wetland during normal sampling activities. The site is dominated by waterfowl in the winter and wading birds during the spring. Ten species of fish were collected by electric shocking in the spring of 1986. Many large-mouth bass were too large to be sufficiently stunned for capture. All fish collected were identified and measured.

Judgement of Success:

The data collected during the last four years within and adjacent to Agrico Swamp West confirms the project's apparent success. It is evident that the marsh and macroinvertebrate community is well developed and positively reacting to natural environmental stresses. In addition, the project's water quality is excellent and represents no problem to the receiving waters of Payne Creek or the area's groundwater supplies. The documentation of the fish and wildlife utilization of Agrico Swamp within the last two years confirms the project's value as a wetland wildlife resource. Comparative studies in 1988 and 1989 of Agrico Swamp West and selected natural wetlands will enable some comparisons to be made on the richness, abundance, and diversity of these systems.

Contact: Kevin L. Erwin
Kevin L. Erwin Consulting Ecologist, Inc.
2077 Bayside Parkway
Fort Myers, FL 33901
(813) 337-1505

AGRICO 8.4 ACRE WETLAND

Location:

The 8.4 acre (3.402 hectare) wetland is located at Agrico's Fort Green Mine in the vicinity of Agrico Swamp West. The site was mined in late 1983. Reclamation efforts began within 90 days of mining by pumping tailings into the area in January 1984. The tailings were graded and capped with overburden to complete

Figure 12. Marsh vegetation within a mulched section of Agrico Swamp West (1984).

Figure 13. Close up of marsh vegetation within a mulched section of Agrico Swamp West (1984).

recontouring within one year of the date of ore extraction.

Construction Technique:

Revegetation was initiated by spreading mulch from donor wetlands onto the recontoured surface of the permit area. As with Agrico Swamp West, trees were planted in certain areas. This facet of the project will not be reported here (see Erwin 1987). Revegetation activities were completed in March 1986.

Goals of Project:

The goal of this project is the successful reclamation of a palustrine ecosystem after mining activity as determined by meeting the following specific conditions:

1. Mulching has resulted in a multi-specific herbaceous assemblage with a similarity to undisturbed areas.

2. Transplanted and seeded hardwood species are viable, surviving and have attained the abundance and species content of at least 400 trees per acre. Cover measurement shall be restricted to those trees exceeding the herbaceous stratum in height and those indigenous species that contribute to the overstory of the mature riverine forest of Payne Creek.

3. Vegetation is naturally reproducing (tree species excluded).

4. Floral diversity and similarity indices are comparable to those of similar undisturbed off-site communities.

Monitoring Vegetation:

Monitoring of the 8.4 acre marsh community was initiated in May 1986. Three established transects (Figure 14) were monitored during June (spring), August (summer), and December (fall) 1987. The rationale for this sampling strategy was to develop seasonal baseline data corresponding: (1) to the early growing season tri-period; (2) to develop a maximum biomass data base for the summer wet season, and (3) to establish a record of all seasonal species and normal decline of cover/biomasses associated with late fall dormancy. The herbaceous vegetation is monitored using the modified line-intercept technique. Occurrence of non-vegetated areas (Bare ground) throughout the transects has been given the same consideration as vegetative cover. Coverages based totally upon species occurrences (which often total much greater than 100%) may no longer be an acceptable method for determining reclamation success. Results indicate that of the ten most frequent (Table 2) species, two are typically considered to be upland/transitional species, while the remaining eight occur in transitional to inundated areas. Groundsel (Baccharis halimifolia) and broomsedge (Andropogon glomeratus) are the dominant upland/transitional species in the area. Duckweed (Lemna valdiviana) is generally restricted to continuously inundated areas while Polygonum hydropiperoides is typically common within transitional wetland areas. Three of the dominant top ten species, maidencane (Panicum hemitomon), pickerel weed (Pontedaria cordata) and cattail (Typha domingensis), are considered to be tolerant of relatively deep water situations, however, these species (especially maidencane) can tolerate a wide variety of saturated or inundated conditions. The most dominant species on the site is bogrush (Juncus effusus) which is generally associated with areas which range from transitional zones of fluctuation to areas of continual shallow inundation.

The present array of dominant species and the zones of inundation noted during sampling indicate that the wetland is being managed as a shallow, inundated freshwater marsh with a minimal zone of fluctuation and extended hydroperiod. Macrobenthic invertebrate and water quality monitoring commenced in 1988. This data is currently being evaluated.

Judgement of Success:

Current trends indicate the wetland is achieving the completion of successful restoration goals. Continued monitoring will insure the stated specific conditions for the creation of the 8.4 acre wetland are met. The data collected to date indicates that the goal of establishing a well developed, diverse, and reproducing marsh at this site has probably been attained.

Contact: Kevin L. Erwin
Kevin L. Erwin Consulting Ecologist, Inc.
2077 Bayside Parkway
Fort Myers, FL 33901
(813) 337-1505

CHARLOTTE COUNTY CORRECTIONAL FACILITY WETLAND MITIGATION PROJECT

Location:

The Charlotte County Correctional Facility is located in south central Charlotte County approximately 13 kilometers north of Fort Myers, Florida. The facility was developed on a 112 hectare parcel containing pasture, pine flatwoods and isolated freshwater marshes that had been impacted by Interstate 75 construction, drainage, and agricultural use. The development of the State prison impacted 19.6 hectares of isolated freshwater marsh and wet prairie habitats.

Construction Technique:

Prior to development a baseline evaluation of topography, vegetation, hydrology, wildlife, and macroinvertebrates was conducted within the wetland areas to be impacted. Wetland substrates (mulch) were evaluated and the depth at which viable seeds, roots, and tubers were found was recorded for all areas. This layer of mulch was then stripped and stockpiled in October, 1987. In early 1988 a 19.6 hectare area (Figure 15) of improved pasture was excavated to the design elevations and contours. The stockpiled wetland mulch was spread and the created marsh area was allowed to fill with water. The created marsh is a part of the facility's surface water management system. Some permanently inundated "pond" areas and uplands were incorporated into the marsh design to create greater habitat diversity. All wetland reclamation activities were completed in June, 1988.

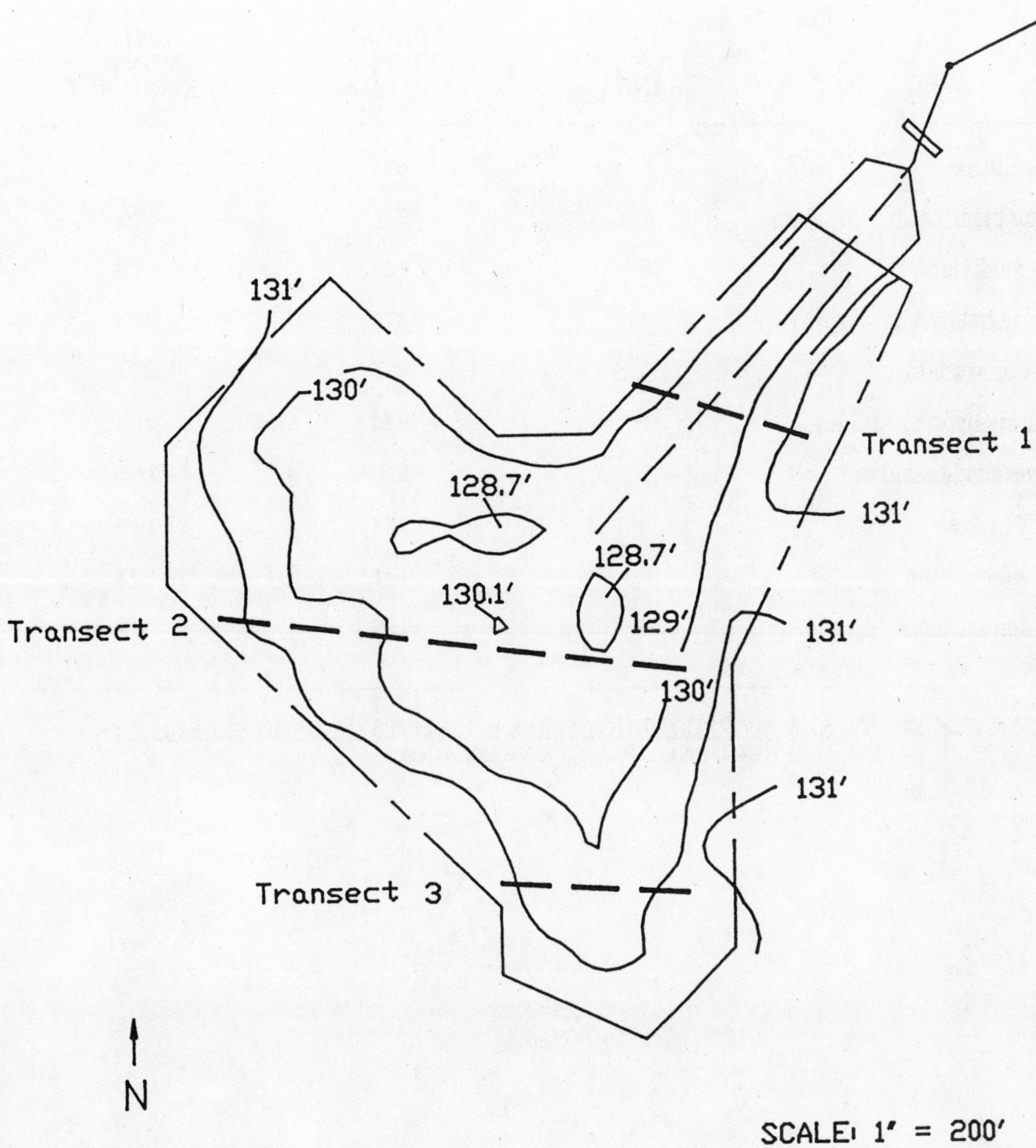

Figure 14. Location of three vegetation monitoring transects at the Agrico 8.4 Acre Wetland Reclamation Project

Table 2. Mean frequency of the ten most frequently occurring species at the Agrico 8.4 acre wetland for the 1987 growing season.

TAXON	RANK	ANNUAL FREQUENCY TOTAL	1987 MEAN FREQUENCY*
Juncus effusus	1	823	61%
Panicum hemitomon	2	766	57%
Lemna valdiviana	3	424	31%
Baccharis halimifolia	4	379	28%
Pontederia cordata	5	296	22%
Polygonum hydropiperoides	6	284	21%
Andropogon glomeratus	7	210	16%
Cyperus haspan	8	164	12%
Paspalum notatum	9	141	10%
Typha domingensis	10	139	10%

$$\text{* MEAN FREQUENCY} = \frac{\text{SUM OF FREQUENCY FOR EACH PLOT FOR EACH SEASON}}{\text{450 FREQUENCY PLOTS X 3 SEASONS}}$$

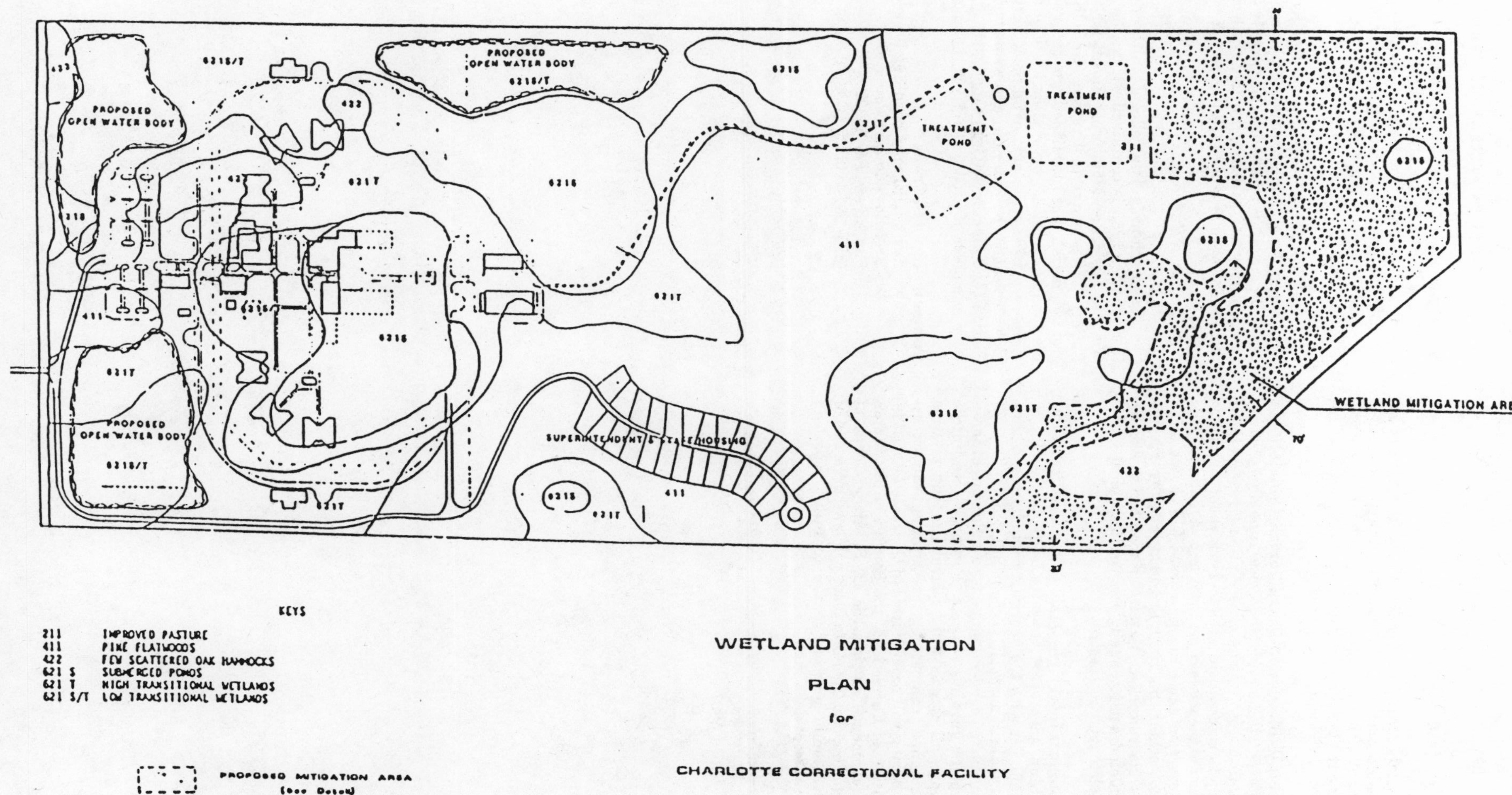

Figure 15. Site plan for the Charlotte County Correctional Facility.

Goals of Project:

The goal of this project is the successful reclamation of an isolated marsh system as mitigation for permitted development of marsh and wet prairie habitats. Post-development wetland monitoring commencing in November, 1988 includes:

1. **Wetland Floristic Characterization**--Multiple quadrats will be evaluated twice annually (dry/wet season) in the wetland conservation (preserve) and mitigation (creation) areas. Permanent quadrats located in the areas of proposed impact (baseline), the preservation and mitigation areas will characterize each wetland macrophyte community by species cover classification. Vegetation monitoring will be conducted semi-annually for a minimum of three complete growing seasons.

2. **Biological Integrity**--Qualitative evaluation of the wetland macroinvertebrates will be made in conjunction with the vegetation monitoring within each wetland macrophyte community. Sampling will be conducted semi-annually, for a minimum of three complete growing seasons. For each qualitative sample, all species will be identified and relative abundance computed. An annual report on monitoring will be submitted to the State of Florida Division of Corrections, the South Florida Water Management District, the U. S. Army Corps of Engineers, and Charlotte County following the end of each annual growing season to fulfill reporting requirements.

3. **Hydrological Data**--Baseline hydrological data will be collected in wetlands where impacts are to be made and preserved wetlands by the establishment of staff gauges. Regular monitoring of staff gauges will be conducted post-development in the preserved and created wetland areas.

In addition to the vegetation and macroinvertebrate monitoring, fixed point panoramic photographs will be taken at regular intervals and weekly water level readings provided. Observed wildlife utilization in all of the wetlands will be recorded and reported.

Judgement of Success:

The monitoring data collected in the created marsh in November 1988 is contained within the First annual Wetland Mitigation Monitoring Report for the Charlotte County Correctional Institution (KLECE 1989). The created marsh has developed an extensive cover of macrophytes from the wetland mulch (Figure 16). Water levels have been acceptable. Wildlife utilization has been high particularly by waterfowl and wading birds including Florida sandhill cranes (Grus canadensis pratensis) and large numnbers of woodstorks (Mycteria americana).

Contact: Kevin L. Erwin
Kevin L. Erwin Consulting Ecologist, Inc.
2077 Bayside Parkway
Fort Myers, FL 33901
(813) 337-1505

Figure 16. Photographs from the same photo location station showing development of vegetation cover from mulched areas of the Charlotte County Correctional Facility wetland mitigation project (A: 6/16/88, B: 7/5/88, C: 11/11/88).

RESTORATION AND CREATION OF PALUSTRINE WETLANDS ASSOCIATED WITH RIVERINE SYSTEMS OF THE GLACIATED NORTHEAST

Dennis J. Lowry
IEP, Inc.

ABSTRACT. Published information on freshwater wetland creation in the glaciated northeastern United States is largely limited to five sources: 1) the results of a workshop held at the University of Massachusetts in 1986 assessing the science base for mitigating freshwater wetland alterations (Larson and Neill 1987); 2) reports on wetland creation efforts associated with a highway project in New York (Pierce 1983, Pierce and Amerson 1982); 3) a report to the Environmental Protection Agency Region I in 1986 examining three mitigation sites (Reimold and Cobler 1986); 4) several earlier papers examining man-made wildlife marshes in New York (Benson and Foley 1956, Cook and Powers 1958, and Lathwell et al. 1969); and 5) a number of scattered papers primarily from symposium proceedings (Butts 1988, Golet 1986, Lowry et al. 1988, and Peters 1988). There is more experience in constructing wetlands than this literature base would indicate, however most of the creation projects have not been documented in published form or in readily available reports. There has been no compilation of the experience obtained from most of the wetland creation projects in the region and there appears to be a general lack of detailed monitoring which would provide data necessary for assessment of results. Most projects seemed to have occurred in New York, Massachusetts, and Connecticut, with relatively few in the other northeastern states.

Long-term, comprehensive studies evaluating the functions of created freshwater wetlands in the region are not presently being conducted. There is, therefore, a need to document the ability of such areas to provide a range of ecological and hydrological functions, rather than just serving as sites where wetland plants grow and that waterfowl visit.

In evaluating future projects involving wetland creation as mitigation for wetland loss in the region, the following critical points should be emphasized:

1. The project proposal should provide an assessment of the wetland functions which may be destroyed, the reliability with which they may be replaced, and the risks if they cannot be adequately replaced.

2. Goals should be developed based upon the most significant functions. Relatively simple, tangible goals (e.g., % plant cover) may be appropriate in permits, but the goals need to be the result of a thought process focused on replacing wetland functions.

3. Proper consideration of hydrology is the most critical factor affecting the success of projects. It is necessary to understand the hydrogeologic setting and water budget of the created area, to have contractors accurately carry out the plans, and to have the means to adjust for errors in design (e.g., water level control measures).

4. The majority of projects to date have attempted to create marsh/open water habitats, for reasons explained in the text. Therefore our present capability to create other wetland types, particularly swamps, fens, and bogs, is more in question.

5. Every attempt should be made to replace lost wetland in the same hydrogeologic unit and reach of the riverine system associated with the original wetland. The level of detail required in data collection and assessment should increase when this cannot be achieved.

6. An understanding of the area where the wetland is proposed to be created is also needed, both to know what is being lost as well as its capability to provide intended functions.

7. Detailed consideration of a number of logistical constraints is always necessary. These may include: hydrologic controls, machinery needs, availability of plant stock and soils, sediment and erosion control, wildlife predation, and barriers to human intrusion.

8. Monitoring requirements should depend upon the functions determined to be of most significance at the assessment stage, the extent or proportion of the existing wetland which will be lost, and whether the proposed restoration or creation will be "on-site" with vegetation and soils similar to the existing wetland, or "off-site" under different conditions.

INTRODUCTION - REGIONAL OVERVIEW

The glaciated northeast, for the purpose of this chapter, refers to the portion of the United States east of Ohio which was subjected to the Wisconsin glaciation. This includes New England, most of New York, northeastern Pennsylvania and northern New Jersey. Due to similarities in hydrogeologic settings and ecological communities, much of the chapter may also be applicable to those portions of the north-central states covered by the Wisconsin glaciation. This is largely due to the overwhelming influence which the last glaciation, and the resulting surficial geologic conditions, had on wetland occurrence, form and function (Motts and O'Brien 1981; Novitzki 1981).

The types of wetland covered by this chapter are principally palustrine forested, scrub-shrub, and emergent wetlands which are hydrologically (i.e., by surface water) connected to riverine (and possibly lacustrine) systems (Cowardin et al. 1979). Portions of the chapter may also be applicable to emergent wetlands of riverine and lacustrine systems, as well as aquatic beds of these three freshwater systems. The critical distinguishing features of these wetlands are: (1) they are dominated by vascular hydrophytes, and, (2) they are not hydrologically isolated from watercourses. They typically have both inflowing and outflowing surface water. The extent of such wetlands in the region has not been specifically documented, however estimates range between five and 25% of the total land and water area (Tiner 1984).

Geologic and hydrogeologic settings of the region are varied. This is due largely to the pre-glacial bedrock-controlled topography and subsequent range of depositional environments and resulting surficial deposits. Since the hydrogeologic setting of a wetland in association with its physiographic and topographic location, largely determines wetland hydrology, knowledge of wetland geology is essential to understanding wetland functions and attempting to mitigate for loss of those functions (O'Brien 1987, Hollands et al. 1987, Peters 1988).

Wetland hydrogeologic classification systems specific to the glaciated northeast have been developed by Motts and O'Brien (1981) and Hollands (1987). The categories developed by Novitzki for Wisconsin wetlands are also applicable to this region. Golet and Larson (1974) partly incorporate this into their classification of freshwater wetlands of the area with a "site type" rating.

The majority of the palustrine wetland considered in this chapter is probably an expression of the high water table in stratified sands and gravels. However, wetlands perched on dense glacial till are more abundant than regional inventories typically indicate: in Massachusetts, 48% and 32% of the wetland area is underlain by stratified drift and till, respectively (Heeley 1973). Lake-bottom deposits and alluvium also frequently support palustrine wetlands associated with watercourses.

Wetland soil types also vary, ranging from poorly drained mineral soils to organic soils (histosols and histic epipedons) of varying thicknesses and degrees of decomposition. Although no comprehensive inventory of the relative extent of the various soil types exists, most surface horizons are probably sapric organics or high organic content silt loams (mucky silts). Soil types and properties reflect the hydrologic and biological environment present during their development. Soil types and properties, in turn, influence those components in many ways (e.g., regulating the rate of ground and surface water movement, influencing water chemistry, etc.). The soil component, therefore, requires consideration in mitigation efforts (Maltby 1987, Veneman 1987) because it affects wetland functions.

At present, forested wetland is the most abundant vegetative type among the palustrine wetlands of the northeast (Tiner, U.S. Fish & Wildl. Serv., pers. comm. 1988, Golet and Parkhurst 1981), perhaps comprising as much as 60% of the palustrine wetland area. Forested wetlands are most often dominated by deciduous trees, with red maple (*Acer rubrum*) the most widespread and abundant species. A comprehensive review of these wetlands is provided by Lezberg (in prep.). The scrub-shrub class is typically dominated by broad-leaved deciduous species, while the emergent wetlands are frequently dominated by robust persistent herbaceous hydrophytes. Types of scrub-shrub wetlands are distinguished locally as sapling or shrub swamps, shrub fen or carr, or shrub bog; emergent wetlands may be distinguished as shallow or deep marsh, wet meadow, fen, or bog.

The climate of the northeast favors wetland development. High annual precipitation is distributed evenly throughout the year. Mean annual precipitation typically exceeds 40 inches (101 cm) and each month has an average rainfall of 3-4 inches; extended droughts are uncommon. In contrast, evapotranspiration is in the r ange of 20-26 inches per year. This provides abundant water for runoff and/or ground water recharge (Motts and O'Brien 1981; NOAA 1979). This relationship does not hold true for the north-central states where annual precipitation exceeds evapotranspiration by a smaller magnitude (Novitzki 1981; NOAA 1979).

The palustrine wetlands bordering riverine systems provide numerous functions depending upon the setting and other characteristics (Larson 1976, Larson and Neill 1987). Key functions are flood storage, water quality maintenance, ground water protection, fish and wildlife habitat, primary productivity, recreation, and aesthetics.

EXTENT TO WHICH CREATION/RESTORATION HAS OCCURRED

Since there are few published documents which describe wetland creation/restoration efforts in the region, much of the following summary is based upon the personal experience of the author.

Palustrine wetlands associated with watercourses have been created by man in the glaciated northeast principally by three main activities: 1) creation of wildlife habitat by federal or state fish and wildlife agencies or private groups such as Ducks Unlimited (many of these are enhancement projects of pre-existing wetlands); 2) inadvertent creation by construction projects such as highways; and 3) intentional creation to mitigate for the loss of wetlands caused by development, most often commercial development or highway construction. A number of stormwater detention facilities have also inadvertently developed wetland plant communities.

Wetlands created by the third process are probably much less common than those created by the first two, yet they are the primary subject of this chapter. The degree to which information or success rates can be transferred from the first two to the third is debatable, since the goals, settings, and resources vary considerably. Golet (1986) raises a clear distinction between wetlands created and managed on public lands by public agencies for specific purposes versus "a myriad of unrelated mitigation projects scattered across the landscape". Nevertheless, wetlands created at such sites as the Great Meadows National Wildlife Refuge in Concord, Massachusetts, or enhanced such as the Great Swamp impoundment in South Kingstown, Rhode Island, are evidence that some projects at some sites can create or enhance wetlands (at least for some values). It is important to note, however, that the setting of such sites can be carefully chosen, and not restricted to the proximity of an associated development.

Projects of the third type are rapidly expanding in number. Well over 200 small-scale (<2 acres, most <0.5 acre) wetlands have been created in Massachusetts since 1983 (Nickerson, Tufts Univ., pers. comm. 1988). Most of these are probably connected in some manner to riverine systems because of the regulatory requirements for replacing "bordering vegetated wetlands" in that state (310 CMR 10.00). Lowry et al. (1988) provide two case studies of such projects. Tufts University is presently reviewing the status of such replacement wetlands in Massachusetts.

A number of wetland replacement projects have also been carried out in Connecticut, with the most notable associated with the Central Connecticut Expressway. An early review of that site by Reimold and Cobler (1986) rated the mitigation as "ineffective"; however more recent assessments are more favorable (Lefor, Univ. of Connecticut, pers. comm. 1988). Butts (1988) also provides a review of this project as well as several other wetland replacement efforts associated with highway construction in Connecticut.

A number of projects have been constructed in New York state as well, although no compilation is available (Reixinger, N.Y. Dept. of Env. Cons., pers. comm. 1987). The largest wetland mitigation effort in the northeast appears to be that associated with the Southern Tier Expressway in the Allegheny River Valley where 78 acres are being created. Some of the early work for this project is reported by Pierce and Amerson (1982) and Pierce (1983). Prior to final design and construction of the wetlands, a two-year demonstration project was conducted to examine the potential success of several environmental (water depth, soil type) and vegetation planting treatments. A comprehensive description of this work is beyond the scope of this chapter, however, one is provided by Southern Tier Consulting (1987).

Only a few projects appear to have occurred in the glaciated portion of New Jersey to date, with the Hackensack Meadowlands the notable exception. The New Jersey Department of Transportation is in the planning stages of several large replacement wetlands and have a pilot project underway to explore design options.

At least a few inland wetland creation projects have been constructed in New Hampshire. Two of them were reviewed in one of the few published reports evaluating the success of wetland mitigation projects in the region (Reimold and Cobler 1986). What appears to be the first inland wetland creation effort in Rhode Island is presently (1988) under construction in the northern part of the State where roughly 7 acres is being established to mitigate for wetland filled by the Woonsocket Industrial Highway (Ellis, Rhode Island Dept. of Env. Manage., pers. comm. 1988). No citations for freshwater wetland creation projects in Vermont or Maine were encountered.

GOALS

As the result of regulatory review, the primary goal of most projects is to provide a wetland with an area approximately equal to the wetland which will be lost. Beyond this, the goals (if stated) are usually directed at creation of habitat through the establishment of certain vegetative types. Creation of marsh/open water habitat, possibly with some interspersed shrub growth, is most common. Such habitat is popular for several reasons: 1) projects are usually proposed in areas where forested wetlands are more abundant, thus wildlife benefits are cited or claimed by diversifying habitat conditions, 2) construction logistics are less complex and less expensive, 3) the availability of commercial plant stock is greatest for marsh emergents, and 4) there is little experience in attempting to establish forested, bog, or fen wetlands.

A large percentage of the small projects in Massachusetts amount to adjusting the configuration of the wetland, essentially filling in one location and excavating out a similar size area in an upland adjacent to the same wetland. Thus, many engineering firms have developed "typical wetland replacement area details", such as that shown in Figure 1.

Actual planting of wetland vegetation apparently has been infrequent, and appears to be decreasing. Instead, most projects now rely on natural colonization or growth of existing propagules in soils dredged from the area to be altered and transported to the new area. An attempt is made to establish the proper water regime in order to set the stage for such colonization. Often the only planting is the sowing of an erosion-control seed mixture. Planting of nursery stock of tubers of emergents or small shrubs is infrequent. When it does occur, nursery stock of emergent species is usually obtained from sources in either the north-central states or from the mid-Atlantic region since local sources are scarce. In the New York studies, control plots which were not planted established similar vegetative cover to planted plots, although dispersal of propagules from planted plots was possible. The researchers concluded that, "Although wetland vegetation may become more abundant in a shorter time if plantings are made, the additional cost must be weighed against the fact that wetland vegetation will become established naturally provided a proper environment is available" (Southern Tier Consulting 1987). These studies also found that an effective planting technique was to plant cores of wetland soil from nearby wetlands.

Next to the goal of establishing vegetative cover, probably the most frequent goal is that of providing compensatory flood storage. This may involve straight-forward cut-and-fill procedures within specific elevation intervals, or may require more complex flood routing calculations with hydraulic controls.

Occasionally, the goal of improving water quality maintenance functions is expressed. This may be accomplished by dispersing surface water flow through a non-channelized emergent wetland or by extending surface water detention time within the wetland. The habitat function is, however, most often the focus; goals of maintaining or improving other functions, particularly those related to ground water, are infrequently addressed.

Restoration, as literally defined, has been rare. At best, this has involved removing fill placed illegally in wetlands. Several larger-scale restoration projects are underway or proposed, but the proposed characteristics for the restored wetland are considerably different than pre-existing conditions, at least those encountered within recent times. "Enhancement" designs have been more common, but as noted by Golet (1986), they frequently involve creating marsh and open

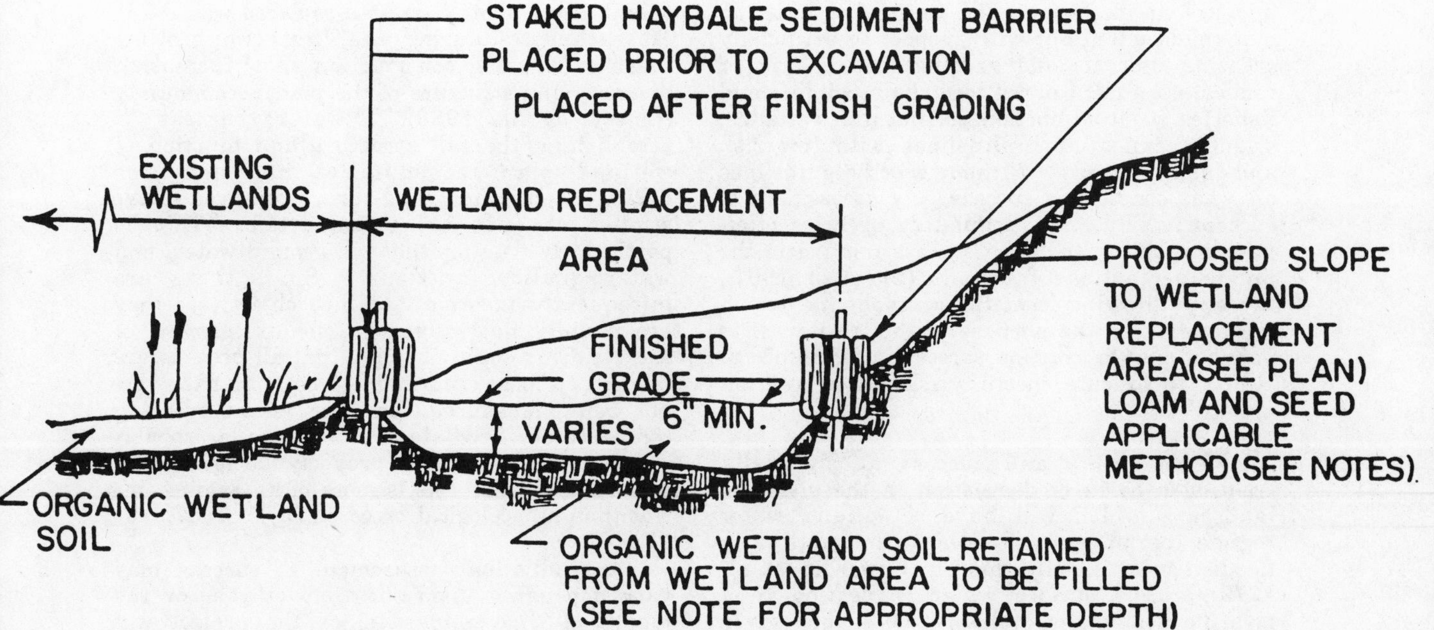

Figure 1. An example of a typical wetland replacement area detail used by engineering firms.

water habitat out of forested wetlands, often citing wildlife benefits due to the diversification of habitat. Actual replacement, creation, or restoration of forested wetland has rarely been attempted, nor has it been attempted for fens or bogs. Experience, and probably our present capability, is restricted to establishing marsh/open water, wet meadow, and some shrub wetland plant communities.

SUCCESS

As noted by Larson (1987), "the main criterion of wetland replication success today appears to be growth of wetland plants on man-made sites. But with respect to artificially creating detritus and grazing food chains to replicate the full suite of food chain, wildlife and fisheries habitat functions of natural wetlands, science cannot offer guidelines with low risk and high certainty". Without specific guidelines on what constitutes success, any assessment of success is subjective. Regulatory agencies often have the ability to define success relative to the public perception of benefit (Sheehan 1987), thereby allowing specific components to be measured. For the purpose of this review, it is possible only to examine success in terms of the ability to execute specific tasks in a creation project.

Man's ability and success at physically creating a basin or depression in the glaciated landscape which will have a wetland water regime (permanently flooded to intermittently flooded or artificially flooded; Cowardin et al. 1979), and a surface water connection to a riverine system, is probably quite good given some understanding of the hydrogeologic setting, local water table, and/or the surface watershed of the basin. The success at creating the specific water regime needed to support a certain wetland plant community is obviously less, and typically requires the provision for artificial water level control measures (e.g., V-notched weir with flashboard control).

The success rate of establishing a predominance of wetland vegetation within a basin appears to vary considerably from site to site. Given the development of a wetland water regime, it seems intuitive that with time, some wetland vegetation will colonize a site. But, establishing the desired plant community depends primarily on creating the proper hydrology, and examples of failures by this criteria exist. Examples of failures of commercially available root stock of emergents and shrubs also exist.

Permanence of the plant community has not been measured in many situations for more than a few years. Work on artificial marshes in New York noted a decline in wetland vegetation over several years, attributed in part to chemical changes in flooded soils (Benson and Foley 1956, Cook and Powers 1958, Lathwell et al. 1969). Quantitative plant data on a replaced wetland in Massachusetts during the first two growing seasons following construction show increasing diversity and structure of the plant community (Lowry et al. 1988). However, success at establishing the full structure and function of wetland plant communities has not been documented and long-term comprehensive studies are not being conducted. This is particularly true for the soil, ground water, and water quality functions. Since these are infrequently incorporated into objectives, they are usually not judged in the assessment of success. This does not imply that all projects are failures by these criteria, but simply that they are not being measured. The success at creating compensatory flood storage functions is probably good if the system is properly designed and constructed, since this is more of an engineering feat than an ecological process (Daylor 1987).

An individual assessment of success may vary depending upon the time of year or the length of time elapsed since the project was constructed. Projects may be judged too soon after construction or not evaluated for a sufficient time period to determine results. Plant communities typically require several growing seasons to be sufficiently established to judge their eventual status. The only published assessment of wetland creation effectiveness in the northeast examined three sites from 2-24 months after construction, and rated two sites "ineffective" and one "marginally successful" (Reimold and Cobler 1986); the latter site was examined after two years, the other two within one year of construction, indicating a relationship between success rating and time elapsed since construction.

It is generally agreed that the hydrologic component is the most important in creating wetland. In the glaciated northeast, success or failure at creating suitable hydrology depends on the ability to understand the hydrogeologic conditions and to model and control the water budget of the wetland. Secondly, it depends on the ability of site contractors to understand and carry out the intended plans. And thirdly, it depends upon whether means were incorporated into the plan to allow for modifications in the inflow-outflow conditions to adjust for errors in the design. Reasons for success or failure at growing wetland vegetation appear to be unknown. Studies measuring the range of parameters which may influence vegetative growth in this region are not being conducted.

DESIGN OF CREATION/RESTORATION PROJECTS

PRECONSTRUCTION CONSIDERATIONS

A fundamental component of any project proposing to replace or restore a wetland is a demonstration that the area proposed to be lost or restored has been closely examined and assessed. A detailed inventory of the existing wetland's hydrogeologic, hydrologic, soil, and biological characteristics, aimed at assessing the wetland's functions, is necessary. It is necessary to know what will be lost before deciding the potential for (and means to) replace or mitigate for the loss. Functions and values of the existing wetland which are indicated to be most important to the watershed, region, or public perception often become those which the replacement design is most strongly directed at replacing. The level of detail in the inventory should be site-specific, possibly starting with a published assessment methodology which requires basic data gathering (e.g., Adamus 1983) and then proceeding to detailed site data as deemed appropriate (e.g., Larson and Neill 1987). These "preconstruction" considerations are really project review considerations. They address the wetland functions which will be destroyed by a proposed activity, the reliability with which they can be replaced, and the risks (public detriments) if they cannot be adequately replaced.

For example, the significance of wetland hydrogeologic setting should be considered. For a wetland (or portion of one) proposed to be filled, a knowledge of the setting should be demonstrated, geared toward understanding the ground and surface water interactions (recharge-discharge relationships) on a seasonal basis. Further, the significance of those functions to the associated watershed and ground water system (e.g., downstream water supply) should be evaluated. Preliminary data collection and analysis should indicate whether more detailed data is required (Hollands et al. 1987). In most cases, the level of detail required in the data gathering and assessment should probably be related to the extent or proportion of existing wetland proposed to be lost, and whether it will be replaced in the same hydrogeologic setting and reach of the associated waterway with similar vegetation and soils.

Similarly, soil profiles of the existing wetland are needed with an assessment because they influence ground and surface water interactions, water quality maintenance, flood storage and shoreline erosion functions (Maltby 1987). Again, additional data should be required as appropriate, such as that in Table 1 from Veneman (1987). As stated by Maltby (1987), "more than any other part of a wetland system, the organic horizon component must be the most difficult to re-create, and for practical purposes this may have to be regarded as impossible". Therefore, the significance of altered stratigraphy of deep organics, where present, needs to be considered.

Biological characteristics need to be inventoried to assess habitat and water quality functions, and to identify the presence of species of concern or floristic assemblages unique to the area or region. Finally, all this information needs to be integrated to develop some understanding of the interactions between the hydrogeologic, hydrologic, edaphic, chemical, and biologic components. Even with these data, however, there is an even more difficult task of relating the significance of specific functions within the context of the wetland's watershed or on a more regional scale.

A similar understanding of the characteristics of the area where wetland creation is proposed is necessary, both to understand what is being lost or altered in that area as well as the capability of the area to give rise to the intended conditions. Test borings should be conducted to understand subsurface stratigraphy as it relates to hydrogeologic properties and to gather water table information. An estimated water budget under proposed conditions should be considered to aid in determining final water regimes.

The feasibility of constructing the project in the desired location should be considered from a logistical viewpoint as well. Hydrologic controls (e.g., diversions, temporary drawdowns) necessary to facilitate construction, machinery necessary to accomplish the work, source and suitability of vegetation, and soils to be used as substrate all need to be evaluated.

CRITICAL ASPECTS OF THE PROJECT PLAN

Project plans should be reviewed with the objective of determining the probability of replacing the functions and values of the

Table 1: Data on soils required for evaluation of the mitigation of specific wetland functions as defined by Adamus and Stockwell (1983) (from Veneman 1987). (x) = parameter is required to assess specific wetland function.

Importance Level	Parameter	Soils		Wetland Functions										Monitoring Required
		Organic	Mineral	I	II	III	IV	V	VI	VII	VIII	IX	X	
I. Minimum Data Required	General description	x	x	x	x	x	x	x	x	x	x	x	x	x
	Soil profile	x	x	x	x	x	x	x	x	x	x	x	x	x
	Soil survey information	x	x	x	x	x	x	x	x	x	x	x	x	x
	Physical parameters	x	x	x	x	x	x	x						x
	Chemical parameters	x	x					x	x					x
II. Detailed Data Required	Fiber content	x	x	x		x	x			x				x
	P-retention	x	x			x		x	x					x
	Pore water analysis	x	x					x	x					x
	Alkalinity exch. acidity	x	x					x	x	x				x
	Seedbank capacity	x	x			x	x	x	x	x	x	x	x	x
	Soil organisms	x	x						x	x				x
III. Highly Detailed Data Required	Clay Character-ization		x			x						x		
	Microbes:													
	Decomposition	x	x					x	x		x			x
	Identification	x	x					x	x					x
	Heavy metals	x	x					x	x					x
	Pesticides	x	x					x	x					x
	Gas analysis	x	x					x	x	x	x		x	
	Peat features	x											x	
	Temperature regime	x	x						x					

Wetland Functions:
I = Ground water recharge and discharge,
II = flood storage and desynchronization,
III = Shoreline anchoring and dissipation of erosive forces,
IV = Sediment trapping,
V = Nutrient retention and removal,
VI = Food chain support,
VII = Habitat for fisheries,
VIII = Habitat for wildlife,

wetland, not just the probability of growing wetland vegetation. While it may be appropriate to establish relatively simple goals (e.g., establishing 75% vegetative cover within two growing seasons, etc.), an understanding of the existing wetland and its functions, and the need to recreate that wetland in a similar setting in an attempt to replace those functions, should not be abandoned. If this is done, more tangible goals for establishing wetland hydrology, soils, and vegetation may be sufficient for project design.

As noted, probably the most critical aspect of any wetland creation or restoration plan is that of hydrology. It should be well documented that the proposed hydrologic regime can be established. Initially, this requires demonstrating that the hydrogeologic setting is conducive to the proper ground water regime, or that the surface watershed will be sufficient to drive the system. In situations where the creation is an extension of an existing wetland and proposed conditions are similar to those in the existing wetland, establishing similar grades on suitable soils should be sufficient to create proper hydrology. This is probably an optimum situation, since there will be an unrestricted hydraulic connection to an existing wetland. Every attempt should be made to replace the wetland within the same hydrogeologic unit and reach of the riverine system associated with the original area.

In situations where more restricted hydraulic connections are proposed, modeling of the water budget and provision for surface water level controls to correct for inaccurate predictions are possible requirements. V-notched weir outlets having the capability to insert flashboards to adjust water levels within intervals of several inches are often advisable, although eventually a more permanent invert may be desired.

At a basic level, the suitability of the soils or substrate to provide a proper growing medium for the proposed vegetation under the intended water regimes should be examined. This should include nutritional status and other chemical parameters under the potential redox states likely to develop. The thickness of the soils to be deposited should be specified. Soils should be sufficient to support the intended vegetation or provide other functions such as ground water discharge control or pollution attenuation. The significance of the changes in soil stratigraphy as they influence ground and surface water interactions should be examined. Excavations should proceed to subgrades based upon a knowledge of how thick wetland soils will be when they are deposited. Additional issues requiring attention include: 1) the ability to physically handle and grade the soils to the elevations called for (e.g., is there a need for low ground-pressure machinery?), 2) the need for water level control during deposition of the soils, 3) water quality concerns during drawdown and movement of organic or fine-grained mineral sediments, and 4) sedimentation and erosion control measures needed pending establishment of vegetative cover. In general, proposed contouring should reflect the most gentle slopes possible; ideally they should be less than 3% and rarely exceed 8%.

Revegetation proposals should be limited to the use of species indigenous to the region which are compatible with the planned hydrologic and soil conditions. Particular chemistry requirements of proposed species should be considered (e.g., pH requirements or limitations). Commercial stock should, preferentially, be from northeastern nurseries since the success of many species appears to decline with shipping. Transplanting from nearby wetlands is advisable if the source is sufficient to allow the needed quantities by transplanting from random, dispersed locations within the wetlands. This option has the advantage of including the entire biological system associated with the root systems.

Wherever feasible, an attempt should be made to either transplant as much vegetation from the original wetland as possible (particularly shrubs) or to transport the upper 6-12 inches of soil from this area (separately from the remaining soil profile) and re-deposit it as the surface horizon for the created wetland. This will enable the existing propagules to regenerate quickly. Even if this is not feasible, the planting of tubers for emergents is often questionable. Given the proper soils and hydrology, such species should colonize the area within two to three growing seasons. However, if more rapid revegetation is desired, planting tubers is appropriate, typically at spacings of 18-36 inches. If rapid establishment of temporary cover is needed, a fast-growing annual grass (e.g., millet) or a perennial grass which is acceptable to include in the plant community can be planted. Because of the length of time shrubs require to colonize a site, planting them is worthwhile. The planting of large saplings of wetland tree species may also be desirable, to encourage development of shrub and forested plant communities. However, it is probably impractical to attempt establishing mature trees. If ground cover is not established by the end of the first growing season, exposed soil surfaces should be straw-mulched or comparably covered (netted if inundated and potentially subject to flowing water) to minimize erosion during the non-growing season.

Wildlife predation, particularly by Canada geese and muskrat, has proven to be a major concern at some sites during the revegetation process (pers. exp.). The probability of these species impairing the success of the developing wetland community should be a legitimate concern in permitting processes. Proposals should address control strategies for such predation in certain high-risk locations.

Provision should be included in permits for replanting in the event of predation, poor initial growth, or growth for only a few years. Generally, replanting measures should be limited to major failures (>30%?) of planted stock to avoid disturbing the surviving population. Similarly, to avoid disturbance of the developing vegetation, physical barriers to human intrusion, such as snow-fencing, may be warranted in high-use areas. Providing a vegetated buffer (e.g., shrub thickets) between the replacement area and surrounding developed land is often needed in the long run, and should be considered as part of the mitigation plan.

The preferred time of year for construction is site-specific depending on hydrologic factors, breeding of fish and wildlife, logistical constraints (e.g., working in frozen organic soils), optimum times for planting, downstream concerns, etc. Sites subject to major flood events should obviously be left alone during times of high flood potential. Construction should be timed to minimize impacts to breeding activities of sensitive fish and wildlife species. The optimum time for planting most of the emergent species appears to be early spring; shrubs and saplings are best planted in either spring or fall, but summer plantings may occur if sufficient watering can be provided.

Active reintroduction of fauna for most projects is probably not necessary, as natural colonization will usually occur as (if) conditions become suitable, however, this should be documented by monitoring. There may be specific situations where particular species, perhaps of a concerned status, should be reintroduced.

Ideally, wetland creation/restoration projects will not be dependent upon long-term, continued management; it may be unrealistic to believe that many private projects will implement such plans effectively. The optimum plan sets the conditions (hydrology, soils, elevations) for wetland development, perhaps with some initial adjustments of water regime via outlet control modifications. To the extent possible, control of "undesirable" exotics (Lythrumsalicaria, Phragmites australis) is an appropriate goal. Maintenance of a "successional sere" (or particular plant community) by, for example, continual cropping of naturally colonizing wetland tree species, is probably not appropriate; projects which infer an enhancement of wetland functions by replacing forested conditions with emergent habitat requiring such management are of questionable value.

MONITORING

Monitoring of the replacement/restoration wetland should be a requirement of every project. What to monitor and the level of detail are, again, site-specific. They depend largely on the functions determined to be most significant at the assessment stage. The sliding-scale approach presented in Larson and Neill (1987) seems appropriate: three levels of data requirements are proposed for each of the hydrology, soils and vegetation components, ranging from minimum to highly detailed data (see Table 1 for soils example). The objective of the monitoring should be to develop an understanding of the functions being provided over time by the replacement area in comparison to those which were provided by the lost area.

At a minimum, plant species composition, density and cover data should be obtained yearly for an extended period, perhaps for a five-year period following construction. Caution should be taken, however, not to relate vegetative growth success to success at replacing wetland functions (Larson 1987). Data on water regimes, water chemistry, soil conditions, ground and surface water interactions (e.g., nested peizometers), and wildlife use should be considered. Monitoring requirements should be greater in situations where there is little prior experience to draw upon, such as in creating/restoring forested wetlands. The duration of such monitoring also needs to be extended.

Deficiencies in the created or restored wetland indicated by the monitoring may require implementation of corrective measures. These again are site-specific; however, flexibility should be built into any permit to enable such measures.

INFORMATION GAPS AND RESEARCH NEEDS

Long-term research needs to be conducted on virtually every aspect of wetland creation/restoration to begin to answer the many gaps in our present knowledge, primarily concerning the significance of existing wetland functions and our ability to create those functions at different locations. From the hydrogeologic/hydrologic perspective, further research is needed to examine the role of wetlands in ground and surface water interactions in different hydrogeologic settings of the glaciated northeast, and then to determine when this role becomes critical to watershed functions and whether it can be replaced by moving the wetland in the landscape. Studies such as those conducted by O'Brien (1987) on two wetlands in eastern Massachusetts need to be repeated on other sites. Because of the uniqueness of each wetland setting in the glaciated landscape, however, the transferability of data from one site to another is always in question. As summarized by O'Brien (1987), "the difficult task of ground water investigation is to define the role of the wetland in the larger ground water regime, to show how alteration will affect that ground water regime, and to predict the effect of the mitigation on the ground water".

On a more applied level, information is needed concerning the success rate of creating wetlands in different hydrogeologic settings. Is the potential for success greater in stratified sand and gravel through excavation into the saturated zone and creation of a water-table wetland, or by creating perched conditions on low-permeability tills and relying on surface water inputs to drive the system?

In terms of soils, we need to know in what situations it is critical to use true organics to maintain recharge-discharge relationships and water quality functions. The significance of disturbed organic soil profiles in relation to these processes is another question. Further, we need to know more about the chemical changes and resulting water quality concerns and/or impediments to plant growth which occur when wetland soils are dewatered, physically disturbed, and subjected to possibly different water regimes and therefore different redox patterns.

Much research is needed on why planting measures are successful on one site and fail on another. In what situations is fertilization or liming necessary? Horticultural expertise is needed in propagation and planting of wetland species to improve the success rate of this most basic of goals. Questions also remain concerning when natural colonization of vegetation is preferable to planting of commercial stock or transplanting from nearby wetlands. Finally, to what extent should success of the created/restored wetland be assessed by the success of the vegetative component: is there a direct correlation between vegetative composition and structure and the presence or degree of other wetland functions?

LITERATURE CITED

Adamus, P. 1983. A Method For Wetland Functional Assessment. USDOT FHWA Report No. FHWA-IP-82-24.

Butts, M.P. 1988. Status of wetland creation/mitigation projects on state highway projects in Connecticut, p. 13-18. In M.W. Lefor and W.C. Kennard (Eds.), Proceedings of the Fourth Connecticut Institute of Water Resources Wetlands Conference, "Wetlands Creation and Restoration", November 15, 1986, Univ. of Connecticut, Storrs, Connecticut.

Benson, D. and D. Foley. 1956. Waterfowl use of small, man-made wildlife marshes in New York State. N.Y. Fish and Game Journal 3(2):218-224.

Cook, A.H. and C.F. Powers. 1958. Early biochemical changes in the soils and waters of artificially created marshes in New York. N.Y. Fish and Game Journal 5(1):9-65.

Cowardin, L.M., V. Carter, F.C. Golet, and E.T. LaRoe. 1979. Classification of Wetlands and Deepwater Habitats of the United States. U.S. Fish and Wildl. Serv. FWS/OBS-79/31.

Daylor, F.F. 1987. Engineering considerations in wetlands mitigation, p. 101-114. In J.S. Larson and C. Neill (Eds.), Mitigating Freshwater Wetland Alterations in the Glaciated Northeastern United States: An Assessment of the Science Base. Proceedings of a workshop at the Univ. of Massachusetts, Amherst, Sept. 29-30, 1986. Univ. of Massachusetts Environmental Institute Publ. No. 87-1.

Golet, F.C. 1986. Critical issues in wetland mitigation: a scientific perspective. National Wetlands Newsletter 8(5):3-6.

Golet, F.C. and J.S. Larson. 1974. Classification of Freshwater Wetlands in the Glaciated Northeast. U.S. Fish and Wildl. Serv. Resource Publ. 116. Bureau of Sport Fisheries and Wildlife, Washington, D.C.

Golet, F.C. and J.A. Parkhurst. 1981. Freshwater wetland dynamics in South Kingston, Rhode Island.

Environ. Manage. 5:245-251.

Heeley, R.W. 1973. Hydrogeology of wetlands in Massachusetts. M.S. Thesis, Univ.of Massachusetts, Amherst.

Hollands, G. 1987. Hydrogeologic classification of wetlands in glaciated regions. National Wetlands Newsletter 9(2):6-9.

Hollands, G., G.E. Hollis, and J.S Larson. 1987. Science base for freshwater wetland mitigation in the glaciated northeastern United States: hydrology, p. 131-143. In J.S. Larson and C. Neill (Eds.), Mitigating Freshwater Wetland Alterations in the Glaciated Northeastern United States: An Assessment of the Science Base. Proceedings of a workshop at the Univ. of Massachusetts, Amherst, Sept. 29-30, 1986. Univ. of Massachusetts Environmental Institute Publ. No. 87-1.

Larson, J.S. 1976. Models for Assessment of Freshwater Wetlands. Water Resources Research Ctr., Univ. of Massachusetts, Amherst. Publ. No. 32.

Larson, J.S. 1987. Wetland mitigation in the glaciated northeast: risks and uncertainties, p. 4-16. In J.S. Larson and C.S. Neill (Eds.), Mitigating Freshwater Wetland Alterations in the Glaciated Northeastern United States: An Assessment of the Science Base. Proceedings of a workshop at the Univ. of Massachusetts, Amherst, Sept. 29-30, 1986. Univ. of Massachusetts Environmental Institute Publ. No. 87-1.

Larson, J.S. and C. Neill (Eds.). 1987. Mitigating Freshwater Wetland Alterations in the Glaciated Northeastern United States: An Assessment of the Science Base. Proceedings of a workshop held at Univ. of Mass., Amherst. September 29-30, 1986. The Environmental Institute Publ. No. 87-1.

Lathwell, D.J., H.F. Mulligan, and D.R. Boudin. 1969. Chemical properties, physical properties and plant growth in twenty artificial marshes. N.Y. Fish and Game Journal 16(2):158-183.

Lezberg, A. (in prep.). The Ecology and Conservation of Northeastern Deciduous Forested Wetlands. Massachusetts Audubon Society Environmental Science Dept.

Lowry, D.J., E.R. Sorenson, and D.M. Titus. 1988. Wetland replacement in Massachusetts: regulatory approach and case studies, p. 35-56. In M.W. LeFor and W.C. Kennard (Eds.), Proceedings of the Fourth Connecticut Institute of Water Resources Wetlands Conference, "Wetlands Creation and Restoration", November 15, 1986, Storrs, Connecticut.

Maltby, E. 1987. Soils science base for freshwater wetland mitigation in the northeastern United States, p. 17-52. In J.S. Larson and C. Neill (Eds.), Mitigating Freshwater Wetland Alterations in the Glaciated Northeastern United States: An Assessment of the Science Base. Proceedings of a workshop at the Univ. of Massachusetts, Amherst, Sept. 29-30, 1986. Univ. of Massachusetts Environmental Institute Publ. No. 87-1.

Motts, W.S. and A.L. O'Brien. 1981. Geology and Hydrology of Wetlands in Massachusetts. Pub. No. 123. Water Resources Research Center, Univ. of Mass., Amherst.

NOAA. 1979. Climatic Atlas of the United States. Washington, D.C.

Novitzki, R.P. 1981. Hydrology of Wisconsin Wetlands. U.S. Geol. Surv. and Univ. Wisconsin-Extension Geol. and Natural Hist. Surv. Info. Circ. 40.

O'Brien, A.L. 1977. Hydrology of two small wetland basins in eastern Massachusetts. Water Resources Bulletin 13(2):325-340.

O'Brien, A.L. 1987. Hydrology and the construction of mitigating wetland, p. 82-100. In J.S. Larson and C.S. Neill (Eds.), Mitigating Freshwater Wetland in the Glaciated Northeastern United States: An Assessment of the Science Base. Proceedings of a workshop at the Univ. of Massachusetts, Amherst, Sept. 29-30, 1986. The Environmental Institute Publ. No. 87-1.

Peters, C.R. 1988. The significance of hydrogeology to the mitigation of functions in freshwater wetlands of the glaciated northeast. In J.A. Kusler, M.L. Quammen, and G.Brooks (Eds.), National Wetland Symposium: Mitigation of Impacts and Losses. Association of State Wetland Managers, Berne, New York.

Pierce, G.J. 1983. New York State Department of Transportation Wetland Construction. National Wetlands Newsletter 5(6): 12-13.

Pierce, G.J. and A.B. Amerson. 1982. A pilot project for wetlands construction on the floodplain of the Allegheny River in Cattaraugus County, New York, p. 140-153. In R.H. Stoval (ed.), Proceedings of the Eighth Annual Conference on Wetlands Restoration and Creation. Hillsborough Community College, Tampa, Florida.

Reimold, R.J. and S.A. Cobler. 1986. Wetlands Mitigation Effectiveness. A Report to the EPA Region I, Contract No. 68-04-0015.

Sheehan, M.J. 1987. Regulating Success. New England Division, Army Corps of Engineers. Unpubl. draft manuscript.

Southern Tier Consulting. 1987. Wetland Demonstration Project, Allegheny River Floodplain. State of New York Dept. of Transportation, Federal Highway Administration contract #D250336-CPIN 5119.01.321.

Tiner, R.W. 1984. Wetlands of the United States: Current Status and Recent Trends, U.S. Fish and Wildlife Service, National Wetland Inventory Project, Washington, D.C.

Veneman, P.L.M. 1987. Science base for freshwater wetland mitigation in the northeastern United States: soils, p. 115-121. In J.S. Larson and C. Neill (Eds.), Mitigating Freshwater Wetland Alterations in the Glaciated Northeastern United States: An Assessment of the Science Base. Proceedings of a workshop at the Univ. of Massachusetts, Amherst, Sept. 29-30, 1986. Univ. of Massachusetts Environmental Institute Publ. No. 87-1.

APPENDIX I: PROJECT PROFILES

The following brief project profiles are representative of wetland creation projects in the northeast. They are obviously not intended to be a comprehensive listing, nor was any attempt made to critically review the status/success of the sites.

ANDOVER BUSINESS PARK, ANDOVER, MASSACHUSETTS

Project Purpose: Construction of office park; wetland fill-in occurred for access road, additional wetland alteration occurred for flood-storage compensation.

Wetland Type/Area Lost: 1.5 acres palustrine scrub-shrub & palustrine emergent marsh with channelized stream.

Wetland Type/Area Created: 1.5 acres palustrine emergent marsh with non-channelized surface flow.

Procedure: Excavation, re-grading of original soils, no plantings, water level control via culverts.

Status: Completed spring of 1984. Monitoring of plant species composition and cover for 3 years. See Lowry et al. (1988).

COULTER DRIVE ACCESS ROAD, CONCORD, MASSACHUSETTS

Project Purpose: Construction of access road.

Wetland Type/Area Lost: 6000 square feet palustrine scrub-shrub, 9000 square feet palustrine emergent marsh, 5000 square feet palustrine open water.

Wetland Type/Area Created: 12000 square feet palustrine scrub-shrub, 14000 square feet palustrine emergent marsh, 3000 square feet palustrine open water.

Procedure: Excavation, re-grading of original soils for emergent and shrub portions (no soils placed in open water area), planting of shrubs but not emergents.

Status: Completed spring of 1984; plant species composition data obtained for 3 years. See Lowry et al. (1988).

SKY MEADOW, DUNSTABLE, MASSACHUSETTS & NASHUA, NEW HAMPSHIRE

Project Purpose: Construction of golf course and condominiums. Wetland filling/excavation occurred for golf course.

Wetland Type/Area Lost: Approximately 8-10 acres of palustrine forested on relatively deep (up to 30 feet) peat.

Wetland Type/Area Created: Restoration plan proposes to establish 3 acres of palustrine emergent marsh, 2 acres of palustrine scrub-shrub, and 1 acre palustrine aquatic bed around 3 acres palustrine open water.

Procedure: Re-grading with drag-line equipment. Planting of over 30,000 tubers of emergents and 1000 shrubs.

Status: Restoration completed during summer of 1988. Monitoring of plant and wildlife communities required by 404 permit.

SYFELD SITE, KEENE, NEW HAMPSHIRE

Project Purpose: Construction of shopping center.

Wetland Type/Area Lost: Estimated to be 10-13 acres of palustrine forested and palustrine open water.

Wetland Type/Area Created: 5 acres of palustrine open water.

Procedure: Excavation.

Status: Unknown.

DIGITAL EQUIPMENT CORPORATION, LITTLETON, MASSACHUSETTS

Project Purpose: Office park construction; wetland created out of construction-phase sedimentation pond to accept and treat treated sewage effluent and parking lot runoff.

Wetland Type/Area Lost: None.

Wetland Type/Area Created: Approximately 1 acre palustrine emergent marsh.

Procedure: Placement of progressively finer mineral substrate (rock to gravel to sand) in deep pond, covered with geotextile fabric and then organic soils; planted with emergents.

Status: Project completed circa 1985. Monitoring unknown, although wetland appears healthy.

FAFARD COMPANIES, MILFORD, MASSACHUSETTS

Project Purpose: Office park construction; wetland

filling for access road.

Wetland Type/Area Lost: Acre palustrine forested.

Wetland Type/Area Created: Palustrine open water and palustrine emergent marsh, approximately 1 acre.

Procedure: Excavation followed by replacement of original wetland soil.

Status: Project partially completed in 1985, additional wetland area under construction in 1987 and 1988.

TAMPOSI AND NASH, NASHUA, NEW HAMPSHIRE

Project Purpose: Construction of industrial park.

Wetland Type/Area Lost: 1.9 acres palustrine emergent marsh.

Wetland Type/Area Created: 1.7 acres palustrine emergent marsh.

Procedure: Unknown.

Status: Constructed fall of 1984; present condition unknown.

NEW LONDON AND NEWINGTON, CONNECTICUT, CONNECTICUT DEPARTMENT OF TRANPORTATION

Project Purpose: Construction of highway.

Wetland Type/Area Lost: 20 acres palustrine forested/palustrine scrub-shrub/palustrine emergent marsh.

Wetland Type/Area Created: 22 acres palustrine open water/palustrine emergent marsh.

Procedure: Excavations.

Status: Constructed in summer of 1985; see Butts (1988).

PORTSMOUTH HOSPITAL, PORTSMOUTH, NEW HAMPSHIRE

Project Purpose: Construction of new hospital.

Wetland type/Area Lost: 4 acres palustrine emergent marsh.

Wetland Type/Area Created: Palustrine emergent marsh/palustrine open water.

Procedure: Excavation followed by re-grading of soils from original wetland.

Status: Constructed in 1986. Plant species composition data being obtained.

SOUTHERN TIER EXPRESSWAY/ALLEGHENY RIVER VALLEY, NEW YORK

Project Purpose: Highway construction.

Wetland Type/Area Lost: 43 acres.

Wetland Type/Area Created: Palustrine emergent marsh/palustrine open water: 78 acres.

Procedure: See Pierce 1983.

Status: See Southern Tier Consulting 1987.

REGIONAL ANALYSIS OF THE CREATION AND RESTORATION OF KETTLE AND POTHOLE WETLANDS

Garrett G. Hollands
IEP, Inc.

ABSTRACT. Kettles are topographic basins created by a variety of glacial processes and occur randomly throughout glaciated regions. They are associated with both permeable and impermeable deposits. Kettle wetlands can have complex hydrology but are divided into two general hydrologic types: those having no inlet or outlet streams, and those associated with surface water streams. Kettle ground water hydrology is generally described as that associated with permeable deposits where ground water is an important part of their water balance, and that associated with low permeability deposits where ground water is not the dominant element of their water balance. Complex relationships of surface water, ground water, water chemistry and other hydrologic elements combine to create water balances. This has been documented in the Prairie Potholes region where site specific hydrologic research has been conducted. Specialized soils and vegetation occur in kettles with unique hydrology. Kettle wetlands have wetland functions similar to other freshwater wetland types.

Kettle-like wetlands have been created by man for a variety of purposes. Creation of kettles for mitigation has occurred at only a few locations. Renovation of Prairie Potholes has occurred with success.

Creating kettle wetlands is similar to other types of freshwater wetland creation, except where unique vegetation and hydrology are involved and replication may be a complex, technical effort. Identification of limiting factors is critical to wetland creation. Typical factors important to kettle wetlands are: surface water hydrology, ground water hydrology, stratigraphy, soils, and water chemistry. Depending upon the goals of the project, other limiting factors may include: nuisance animals, long term maintenance/monitoring, lack of funds, and disposals of excavated soil.

The primary concern in creating kettle wetlands is the establishment of the proper hydrology. This normally requires mid-course corrections in design during construction to establish proper post-construction hydrology.

Critical research needs include studies on microstratigraphy, geochemical processes, the properties of organic soil, and the details of hydrology.

INTRODUCTION

There is little literature available concerning the subject of wetland creation and restoration. Literature available specific to kettle and pothole wetland restoration and creation is even more scarce. This chapter defines what kettles and pothole wetlands are and cites the small amount of literature available to describe these wetlands. The portions of this chapter which discuss wetland creation and restoration are based upon interviews with those people who have had actual experience, primarily in restoration, and my own experience in creating kettle and through-flowing wetlands in Massachusetts.

DEFINITION OF WETLAND TYPES

The word "kettle" is a glacial geologic term which applies to basins created by ablation of buried glacier ice in glacial drift (Flint 1971, American Geological Institute 1972). The term "kettlehole" is an improper term. The term "pothole" is synonymous with the term "kettle" and is generally used in the prairie states. Other terms such as "ice-block cast" (Kaye 1960) and "circular disintegration ridge" (Clayton 1967) are also used throughout the geologic literature to further subdivide and classify kettles. It is sufficient to say that there are a large number of types of kettles which were formed in highly variable sedimentary environments associated with wasting glaciers. Kettles are a subtype ice-disintegration feature (Flint 1971) and are specifically related to buried ice.

Three basic types of kettles (Flint 1971) occur: (1) those associated with coarse-grained stratified drift; (2) those associated with glacial till; and (3) those associated with a combination of till and stratified drift, both fine and course grained. Kettles occur less commonly in till than in stratified drift. Kettles in stratified drift tend to have regular shapes, whereas kettles in till are generally irregular.

Flint (1971) recognized three sources of kettle origin: (1) buried ice projecting through stratified drift; (2) buried ice below and covered with stratified drift, and (3) ice buried within or on top of stratified drift. Each type creates a different depth of depression from deepest to shallowest, respectively. The first may have till exposed in the bottom whereas the bottoms of the latter two generally contain stratified drift.

Flint (1971) also stated that....

> "Kettles are peculiar to the terminal zone of a glacier where thinning is actively in progress. Many occur at the proximal bases of end moraines where thin glacier termini stagnated and became detached and covered with till, possibly by overriding ice from upstream. In some areas a peripheral belt of glacier ice many kilometers in width becomes stagnant and separates into isolated masses chiefly through meltwater ablation. Stratified drift deposited upon and between such masses creates extensive complexes of kettles."

Flint (1971) chose to differentiate depressions in which till predominates from those predominated by stratified drift. He subdivided depressions where till predominates into two broad classes: hummocky ablation drift and disintegration ridges. The term "ground moraine" has been used by some geologists to map areas where these features are common.

Hummocky ablation drift is a random assemblage of hummocks, ridges, basins, and small plateaus, without pronounced parallelism and without significant form or orientation (Flint 1971). Two processes of creation of hummocky ablation drift were described. Ablation drift may form on top of wasting glacier ice and be slowly dropped onto the land surface as the ice melts. The second way occurs as the glacier melts upwards from its base. A combination of the two processes was probably common. Many prairie potholes occur in this type of terrain.

Disintegration ridges are orderly rather than chaotic features, generally long, straight, or curved ridges of till or stratified drift and in some cases associated with valleys. Some are circular or ring shaped (Clayton 1967). They are the result of the filling of crevices and other openings in disintegrating ice, by both down-wasting and upward-wasting processes.

More complex kettles occur such as ice-walled lakes (Clayton and Cherry 1967). They were formed when lakes occurred on top of wasting glacier ice. Subsequent melting of the ice created depressions rimmed with gravel and underlain by fine-grained lacustrine sediments. They are common features of central North America.

The American Geological Institute (1972) defines kettles in the following manner:

> "kettle [glac geol]: A steep-sided, usually basin or bowl-shaped hole or depression without surface drainage in glacial-drift deposits (esp. outwash and kame), often containing a lake or swamp, and believed to have formed by the melting of a large, detached block of stagnant ice (left behind by a retreating glacier) that had been wholly or partly buried in the glacial drift. A kettle is usually 10-15 m deep, and 30-150 m in diameter. Cf: pothole [glac geol.] Syn: kettle hole; kettle basin."

Whereas most kettles are associated with continental glaciation, kettles are also associated with end moraines and outwash of mountain glaciation. Some kettle-like features are associated with rock glaciers and other periglacial features, while others have been created by eolian processes (deflation basins) or animal activities (buffalo wallows). Many kettles have been created by glacial activities but modified by the eolian activities of deflation and deposition, and animal activities including those of man.

In summary, ice disintegration features, of which kettles are the topographically negative or depression feature, result from the separation and disintegration of a marginal belt of ice. This condition occurs where and when the ice is thin and stagnant or very slowly flowing. In the northern Great Plains, till is the dominant material of kettles. In areas of more regional relief, such as the New England, Mid Atlantic and Mid West states, stratified drift predominates.

For the purposes of this chapter the terms "kettle" and "pothole" are synonymous. They refer to any depression which was formed by glacial, periglacial, or eolian processes, and occur on glacial deposits of the Pleistocene Epoch.

GEOGRAPHIC REGION

Kettle wetlands occur only in those portions of the United States which were glaciated during the Pleistocene Epoch (2 million to 10,000 year B.P.). This area includes all or portions of Maine, New Hampshire, Vermont, Massachusetts, Rhode Island, Connecticut, New York, New Jersey, Pennsylvania, Ohio, Indiana, Illinois, Wisconsin, Minnesota, Iowa, Kansas, Nebraska, North Dakota, South Dakota, Montana, Idaho, Washington, Colorado, Utah, Oregon, and California. Alaska is not included in this discussion but contains numerous kettles of glacial and periglacial origin.

HYDROLOGY

Hydrology is believed to be the most dominant limiting factor controlling the occurrence of a wetland (Hollands, Hollis, and Larson 1987); therefore emphasis on hydrology is given in this chapter. Classification of kettle hydrology has been most specific in the prairie pothole region of the northern Great Plains (Meyborm 1966, Sloan 1970, Steward and Kantrud 1971, Eisenlohr et al. 1972, Malo 1975, Millar 1976, Winter and Carr 1980, Winter 1983). Kettles have been included in other hydrologic and vegetative classifications of wetlands (Shaw and Fredine 1956, Heely 1973, Golet and Larson 1973, Hollands and Mulica 1978, Cowardin et al. 1979, Novitzki 1982, Hollands 1987).

A number of attempts have been made to simplify the hydrogeologic classification of wetlands, but they also point out the complexity of wetland hydrology. Kettle wetlands are included in some hydrogeologic wetland classifications, although none are specific to kettles. (Steward and Kantrud 1971, Heeley 1973, Hollands and Mulica 1978, Motts and O'Brien 1981, Winter 1981, Novitzki 1982, Hollands 1987).

All of the nontidal water regime modifiers of the National Wetland Classification (Cowardin et al. 1979) apply to kettles. Both saline and mixosaline water chemistry modifiers also apply. The modifiers for pH (acid, circumneutral, and alkaline) may apply to appropriate kettles.

Two general types of kettle hydrology occur; those associated with surface streams, and those which are hydrologically isolated kettles.

KETTLES WITH NO INFLOWING OR OUTFLOWING SURFACE WATER STREAMS

Kettles with no inflowing or outflowing surface water are not through-flowing systems. They generally have small water budgets not dominated by flowing surface water. They can be divided into three subtypes, as described below.

Water-Table Kettles

Water-table kettles are commonly associated with stratified drift but also occur in till. Their depressions penetrate into the water-table and ground water dominates their water budget. Depending upon their water-table level fluctuations and surrounding elevation of the ground water-table, they can fluctuate from recharge to discharge conditions. Their cover type varies from clear open water to densely wooded swamps, including all the possible vegetation types in between.

Wetlands on Low Permeability Deposits

These kettles commonly are associated with till but also occur on stratified drift where micro-stratigraphic layers serve as a "perching" or low-permeability layer. The term "perched" is a controversial term to ground water geologists. Truly perched conditions seldom occur, wherein unsaturated sediments are below the saturated base of the wetland. Low permeability deposits below the wetland cause slow downward water movement, so that flooding or saturated soils persist in the wetland long enough to give rise to a community of hydrophytes. The water budget of these kettles is dominated by direct precipitation, interflow, and surface water runoff.

Kettles Fluctuating Between Water-Table Dominated and Precipitation Dominated Water Balances

Some kettles seasonally fluctuate from a water-table dominated water balance to one dominated by precipitation. These occur both in till and stratified drift, and respond to the relative elevation of the water-table to the bottom of the kettle and the yearly range of water-table fluctuations.

KETTLES ASSOCIATED WITH SURFACE WATER STREAMS

Kettles With Both an Inlet and an Outlet

These kettles are connected to the water-table and perennial or ephemeral inlet and outlet streams. Surface water or ground water may be the dominant water budget element. Multiple inlets and, less commonly, multiple outlets may occur. These are generally complex hydrologic systems where a variety of combinations of water budget components is possible. Some of these systems are identical to riverine wetlands and others (ones with ephemeral inlets and outlets) are more similar to "perched" kettles with no inlets or outlets.

Kettles With Only an Outlet

These kettles generally occur in coarse-grained stratified drift as water-table discharge wetlands. Their outlet may be perennial or ephemeral. They occur less commonly in till where the outlet stream is generally ephemeral.

Kettle With Only an Inlet

These kettles are predominantly water-table recharge features which occur in coarse-grained stratified drift. The inlet stream generally is ephemeral but in rare cases the inlet stream is perennial.

THE WATER BUDGET

A wide variety of water budgets may occur for each of the six general kettle-hydrologic categories described above. A water budget is defined as:

Input = Output

$$PPt + SWi + GWi + IF + RO + Et = SWo + GWo + S$$

where:

PPt = Precipitation as rainfall or snowfall directly on the wetland surface

SWi = Streamflow in

SWo = Streamflow out

GWi = Ground water discharge into wetland

GWo = Ground water recharge out of wetland

IF = Interflow or horizontal shallow ground water flowabove the water-table

RO = Surface water runoff

Et = Evapotranspiration

S = Water storage within wetland as surface water or soil water.

Hydrologic situations vary with only slight modifications of one or more of the elements of the water budget. In addition, as vegetation and organic soils develop in the wetland, the water budget will be modified (Gosselink and Turner 1978).

The hydrology of individual prairie potholes is variable (Steward and Kantrud 1971). However the typical pothole has cyclic hydrology similar to that of kettle wetlands on till found in New England. Many such kettles have a yearly cycle fluctuating from a surface water pond in the spring, dry land in the summer and fall, and returning to a pond in the following late winter and spring (Malo 1975). A pond located on dense deposits such as till may result in a ground water mound which recharges the water-table (Winter 1983).

In summary, the hydrology of a kettle wetland is complex and site specific. A large number of combinations of water budget components may exist. No simple, all-encompassing statements can be made. To determine or predict the water budget of either a naturally-occurring kettle or man-made depression requires detailed site specific data collection and surface water and ground water modeling. The Hydrology Panel of the National Wetlands Values Assessment Workshop (U.S. Fish and Wildlife Service 1983) made the following statement:

"It is the opinion of the Hydrology Panel that water is the primary and critical driving force underlying the creation and maintenance of wetlands and that a knowledge of wetland hydrology is basic to an understanding of all wetland functions. There has been little substantive work done on the hydrology of wetlands, and we lack the knowledge needed to evaluate hydrologic processes in wetlands without careful measurements. Continuing research has resulted in a questioning of many of the basic assumptions previously held by hydrologists, especially in the areas of ground water, generation of runoff, and storm peaks. Therefore, an evaluation based on an examination of recent literature may be misleading. Furthermore, hydrology, unlike some functions of wetlands, cannot be directly observed or easily sampled. Hydrologic processes must be carefully measured for a long enough period to ensure that the measurements are meaningful and that uncertainty or error limits can be included. Water budgets are very important, but underlying assumptions and inherent errors must be identified. The state-of-the-art in wetland hydrology is not such that we can make definitive statements about recharge, discharge, or evapotranspiration from maps or site visits. We cannot extrapolate from the results of a few comprehensive wetland hydrology studies to all wetlands because of the complexity and variety of the hydrologic systems involved. More research to provide an improved capability to quantify and describe basic processes, such as evapotranspiration, recharge, and discharge, would improve our capability to measure and assess wetland functions."

SOILS

Many types of wetland (hydric) soils can be found in kettles, ranging from sapric, fibric, and hemic organic soils to hydric mineral soils. Thick organic soils are generally associated with water-table wetlands, while hydric mineral soils are predominantly associated with ephemeral kettles. The type of soil is determined by the kettle's hydrology, vegetation, and the import and export (retention) of organic matter. Hydric soils of the glaciated Northeast have been described by Tiner and Veneman (1987).

VEGETATION

Any type of wetland vegetation adaptable to the climatic region where the kettle is located may occur. As with other wetlands, the vegetation is primarily determined by hydrology (hydroperiod and chemistry). Kettles in the Northeast are predominantly red maple wooded swamps or bogs. In the Midwest they contain mostly shrub/scrub, bog, or coniferous forest communities. In the northern Great Plains states they contain a variety of marsh communities and unique vegetation such as "willow rings".

Kettle wetlands include riverine, lacustrine, and palustrine types as characterized by the National Wetland Classification System (Cowardin et al. 1979) although palustrine wetlands predominate. Most of the various subsystems and classes may occur. Along the New England coastline (Long Island to New York) estuarine kettles also occur, but this chapter does not discuss these.

The more specialized plant communities found in prairie potholes are best described in Steward and Kantrud (1972). In some cases, such as the raised bogs in Washington County, Maine, kettles may have served as the "seeds" within which bogs originated and paludification occurred to the point where the bogs grew well beyond the original kettle basins.

KEY FUNCTIONS

Kettle wetlands, because of their variability, have the potential to perform all of the ten functions described by Adamus (1983). On the other hand, the many kettles which are isolated surface water systems (not riverine) have less value for those functions dependent upon a through-flowing surface water system such as "flood storage" and "food chain support". These isolated wetlands also do not have the potential for additive value (i.e., contributing to the greater value of downstream ecosystems), but they may be site specifically important (Davis et al. 1981). Isolated kettles may retain all the water that enters them, whereas water in riverine systems drains back to the river. In some regions these isolated wetlands have considerable additive value as part of a ground water recharge/discharge regional aquifer system (Winter and Carr 1980, Winter 1983) or their spacial interspersion may be valuable to wildlife (Brown and Dinsmore 1986). Site specific investigations should be made to determine what functions a specific kettle wetland may perform.

EXTENT TO WHICH CREATION/RESTORATION HAS OCCURRED

Creation of kettle-like wetlands has taken place for reasons other than compensating for wetlands which have been filled. Many kettle-like wetlands have been created by accident. The following discussion illustrates the kinds of kettle-like wetlands which have been created.

FARM PONDS

Numerous farm ponds have been and continue to be created. Some are quite old, but most have been constructed since the 1950's when government sponsored programs began to aid farmers in creating ponds for cattle and other purposes. They have evolved into various vegetation types (Dane 1959). Extensive literature is available from the Soil Conservation Service (SCS) and the U.S. Fish and Wildlife Service (USFWS) on creating multipurpose farm ponds. The literature describes how to locate and construct a pond so that it has a viable water balance. Many of these ponds have very low-water budgets and are hydrologically isolated. Many have developed extensive wetland vegetation communities. USFWS has investigated the wetlands which have been developing in some of these farm ponds (Tiner pers. comm. 1987).

GRAVEL PIT AND QUARRY PONDS

In many places gravel and rock excavations have created ponds in both low permeability and water-table hydrogeologic situations. Vegetative communities of a wide variety of types have developed within these ponds following abandonment.

IRRIGATION PONDS

Ponds dug into the water-table to provide water for irrigation purposes are widespread in agricultural areas. Vegetative communities have colonized both active and abandoned irrigation ponds.

GRADING ACTIVITIES

Land grading associated with urban development and agriculture is used primarily to remove areas where ponded water occurs. In some cases the land grading fails and actually creates ponded water, which, in turn, becomes a vegetated wetland.

ROAD CONSTRUCTION

Highway, driveway, and railroad construction in some places have inadvertently trapped water behind embankments. In some locations the necessary culverts were not installed, or were incorrectly installed, undersized, or have become plugged. The resulting ponded water has allowed vegetated wetlands to become established in previously upland areas.

EXCAVATION BY USE OF EXPLOSIVES

Military bombing and artillery ranges occur within the United States, where craters are blasted into the earth during training exercises. Many of these craters intersect the water-table. Some have created depressions in which the bottoms are compacted by the force of the blast.

Both "perched" and water-table wetland vegetative communities have formed in these craters. Elsewhere, for example in Vietnam, hundreds of thousands of such "pothole" wetlands have been created by bombing. Many are used today for agricultural purposes, others contain wetlands used by fish and wildlife. Explosives have also been employed to create pothole wetlands for wildlife habitat creation and management (Mathiak 1965, Strohmeger and Frederickson 1967).

CRANBERRY BOGS

Prior to the early 1970's, most cranberry bogs were created by converting a natural wetland into a managed cranberry bog. Investigations conducted by the USFWS (Tiner pers. comm. 1988) show that since 1977 approximately 60% of new cranberry bogs have been created from upland areas. Complex dike and water-control structures and activities are associated with commercially-operated bogs. Many created bogs have been abandoned and have become shrub or wooded swamps. Numerous examples of these occur in southeastern Massachusetts.

WETLANDS WHICH ARE SIMILAR TO KETTLE WETLANDS CREATED FOR WETLAND MITIGATION PURPOSES

Wetlands have also been created to compensate for permitted wetland losses. Examples include:

Route 25, Bourne, Massachusetts

The construction of Route 25 in 1986 and 1987 in Bourne, Massachusetts, by the Massachusetts Department of Public Works, filled approximately nine acres of viable cranberry bog. After an extensive adjudicatory hearing, the Massachusetts Department of Environmental Quality Engineering issued a Final Order of Conditions under the Massachusetts Wetlands Protection Act (MGL Chapter 131, Section 40) which required the creation of nine acres of replacement wetlands. This new wetland contains no inlet or outlet and is a water-table wetland. While it was intended only to create a shallow marsh, the resulting wetland should be similar to wetlands found in kettles in the surrounding outwash plain.

Coultor Road, Concord, Massachusetts

A roadway constructed by the Town of Concord filled one half acre of wetland consisting of mostly shrub swamp and a small irrigation pond. As required by the Massachusetts Wetland Protection Act (MGL Chapter 131, Section 40), a replacement wetland was created by excavation into the water-table. It was designed to contain vegetation communities of shrub swamp and shallow marsh as well as open water. The wetland has survived three growing seasons and the vegetation communities have been successfully established with the exception of the shrub swamp, which was killed by unusually high water during the summer of 1987. This illustrates the difficulty in controlling or predicting water levels in wetlands having no outlet control, and the sensitivity of woody plants to water levels.

Prairie Potholes in North Dakota, South Dakota, and Montana

Approximately 200 acres of prairie pothole wetland have been created by the U.S. Bureau of Reclamation at the Indian Hill Tract in North Dakota. The project was an area of drained potholes where the hydrology of the wetlands was restored. Revegetation has successfully been completed. Ducks Unlimited, Inc. has been restoring prairie potholes in numerous locations in the Dakotas, and has successfully created new wetlands from upland in the National Grassland of South Dakota and Montana. The Northern Prairie Research Station, Jamestown, North Dakota has restored prairie potholes for the last ten years in the pothole region of the northern Great Plains. Many types of potholes have been restored to accomplish a number of goals. Other organizations which have attempted to create or restore prairie potholes have been state highway departments, the Federal Highway Administration, and state fish and game departments.

TYPICAL GOALS FOR PROJECT AND "SUCCESS"

Project goals should be as simple and as specific as possible, with clearly stated criteria to measure success. For example, a goal could be no more than to create ten acres of shallow marsh. Although, the goal is, perhaps, overly simplistic, success would, in this case, be judged by measuring the mitigation area to determine if ten acres of wetland was created and determining if that wetland was a shallow marsh. Conversely, multiple goals requiring complex monitoring and analysis of data may result in an inability to determine success.

If the goal is to recreate a wetland at a different location, then it is necessary to assess in detail the wetland which will be lost because a comparative analysis of the new versus old wetland will be required. A detailed inventory of all elements of the existing wetland should be undertaken. These data should be used to assess the functions of the wetland (Larson 1987). Current wetland assessment methods (i.e., Adamus 1983, Hollands and Magee 1986) are not capable of detailed assessment of the function of individual wetlands, but are useful for regional wetland assessments. Larson (1987) states. . .

> "the role of wetlands in flood-water detention can be estimated and replicated with reasonable certainty and sufficiently low risk so as to provide useful guidance for replicating this function when wetland losses are truly unavoidable. But the science base of knowledge is too incomplete to support assertions that artificial wetlands will provide the other functions of natural wetlands, especially those associated with water supply, water quality and nutrient transformations."

In any case, the function or value of the wetland planned for destruction should be quantified. For example, if 900,000 acre-feet of natural valley wetland flood storage is to be taken by filling and the goal is to recreate the same amount of natural valley flood storage in the mitigation wetland, then the creation of 900,000 acre-feet of new flood storage at an equal elevation to the original wetland is measurable and evidence of success. It is not necessary to monitor the next 100-year flood to see if the flood-storage mitigation actually maintained flood-stage elevations. Such a measurement not only would be very difficult to perform and analyze, but could require decades before the data were collected!

Many of the functions which the literature ascribes to wetlands are poorly defined, and may be extremely difficult to quantify or qualify by expensive and detailed field examinations. Predicting that a new wetland will function similar to a naturally occurring wetland is very difficult and determining the success of a replacement wetland even more difficult (Larson and Neill 1987). Each step introduces new problems, infuses new assumptions, and increases the probability of error. Another method is to duplicate, as closely as possible, the water surface elevations, soils types, and inundation frequencies without planting of vegetation. If this is successfully done then an assumption can be made that the area will become a wetland and it will recreate the lost wetland.

In-kind replication is not possible for all elements of a natural wetland. In its fullest sense, in-kind replication implies that the replacement wetland will be identical to the original wetland. This is impossible (Golet 1986). One can not recreate identically a complex wetland that is the product of glacial geologic processes, and 14,000 years of plant, soil, and hydrologic evolution. The only in-kind replication possible must have a very simple definition, such as creating the same type of dominant vegetation class (i.e., Red Maple Wooded Swamp = Red Maple Wooded Swamp). To carry the term in-kind further than such broad and simple criteria is impossible.

It is important to recognize that a clearly defined goal is a very important part of the regulatory process. Failure to achieve a goal could be the basis for court action, and defense of the goal criteria/definition is a critical court issue. Failure to pick realistic goals which could be defensible in court may result in enforcement failure.

A key goal in any replication/restoration project should be to create a site hydrology which will persist in perpetuity with little or no maintenance, and which will support the desired vegetation community.

REASONS FOR FAILURE/SUCCESS

Determining that a project is a failure or success requires well defined and measurable goals. Without them, measuring success or failure is an arbitrary and personal judgment. In the world of high-cost land development and courtroom proceedings, goals must be specific, quantifiable, and reproducible to be legally defensible.

The following is a partial list of reasons for project failure; project success results from proper attention to each:

1. Goals not properly identified,
2. Lack of information on the lost wetland,
3. Geohydrology not correctly created (e.g., no low permeability layer created or one which leaks, or a water-table which is not understood properly and is too deep or too shallow),
4. Water budget not understood (e.g., too little or too much water),
5. Improper soils,
6. Improper water chemistry (e.g., saline),
7. Improper planting (e.g., wrong plant species or density),
8. Improper maintenance,
9. Nuisance animals (e.g., geese),
10. Not constructed as designed--lack of inspection,

11. No effective monitoring,

12. No enforcement by government agencies,

13. Lack of funds--economics,

14. No method to evaluate degree to which goals have been achieved,

15. No long-term management plan and funds to maintain the system once success has been achieved (e.g., no unconditional guarantee), and

16. Failure to identify and address "limiting factors."

Success is greatly increased for restoration projects where the wetland basin and soil remain, but where the hydrology was altered and can be restored. For example, with a drained prairie pothole, simply removing drainage devices, (e.g., ditches) and restoring the "plug" results in re-establishment of the wetland's hydrology (Meeks pers. comm. 1987). Revegetation may occur naturally or limited planting may be required, especially if a certain plant community is desired.

REASONS FOR RESTORATION/REPLICATION

Most restoration of kettle wetlands has occurred in North and South Dakota and has been performed by Ducks Unlimited, the U.S. Bureau of Reclamation, and the USFWS. These projects have been primarily for restoration of waterfowl habitat, in particular, brood and nesting habitat for ducks. While other functions have also been desired, waterfowl habitat is dominant (Meeks pers. comm. 1987). Another value resulting from the pothole restoration is creation of habitat for other upland and wetland dependent game and non-game species. Opportunities for recreational activities such as hunting, trapping, fishing, and bird watching are values which have also been enhanced (McCabe, pers. comm. 1987).

In Massachusetts the goals for kettle-type wetland replication have not been wildlife-oriented, as wildlife habitat only became a state protectable wetland function on November 1, 1987. The Massachusetts Wetlands Protection Act does not require in-kind replication but only the replacement of wetland function in a "similar manner" (Dept. Env. Qual. Eng. 1983). Kettle-type wetlands have been created to provide for the functions of public and private water supply (fire ponds and farm ponds), flood control (natural-valley flood storage), prevention of pollution (water quality improvement), and fisheries (fish ponds).

DESIGN OF CREATION/RESTORATION PROJECTS

PRE-CONSTRUCTION CONSIDERATIONS

It is very important at the pre-construction stage to determine the natural resource elements of the original wetland, and of the wetland to be restored or renovated (Larson 1987). This determination should include a detailed inventory of geology, hydrology, soils, vegetation and wildlife. Following this inventory, the functions of the wetland should be determined quantitatively or qualitatively, as appropriate. Failure to do so will prevent or, at a minimum, greatly hinder design and later judgment of success.

This basic inventory will establish criteria to use in screening possible sites for replication. It is also useful in determining limiting factors (Meeks pers. comm. 1987). A limiting factor is an element of the site's natural resource elements that influences success. Limiting factors may vary regionally, e.g., the availability of water in the National Grasslands where precipitation averages 12 inches per year. Therefore, obtaining a site where the proper water budget is available (large enough watershed) is critical. Water may not be the limiting factor in southeastern Massachusetts where abundant rainfall (42 inches per year) and shallow ground water are available. In Massachusetts, the cost of land, averaging $80-100,000 per buildable acre, is in many cases the limiting factor.

Limiting factors may also be goal-related. If the goal is to create wetlands to increase nesting habitat and brood rearing areas (e.g., Ducks Unlimited projects in the northern Great Plains), then the limiting factor to success may be designing sites so that predators such as fox, raccoon, skunk, and coyote can be prevented from destroying nests and broods. Restored potholes for duck production without predator control results in failures to meet the goal of increased ducks (Meeks pers. comm. 1987).

One of the major, often overlooked,

preconstruction considerations is the cost of construction, monitoring, and maintenance over a long period of time. A recent project in North Attleborough, Massachusetts cost approximately $1.6 million in construction cost alone for approximately two acres of new wetland. This does not include the cost of the land and consultant fees. The site was particularly expensive because of the need to blast bedrock. Land ownership may also change and the new owner may not be aware of his responsibility for monitoring and maintenance, or the cost of these activities.

SITE CHARACTERISTICS

The new wetland site must have enough general elements to offer a high probability of creating the hydrogeology desired. Hydrology is, in general, the most common limiting factor. Emphasis in the site selection process should be given to hydrologic considerations.

CRITICAL ASPECTS OF THE PROJECT PLAN

Surficial geology/stratigraphy

Surficial geology must be determined because it is the basis for any hydrogeologic analysis (O'Brien 1987). For example, it is important to determine whether the site is predominantly till or stratified drift. In many cases, surficial geologic maps prepared by the U.S. Geological Survey or state geological surveys exist. For any replication site, test borings or pits must be dug to determine site stratigraphy. Both vertical and horizontal stratigraphic variations in glacial deposits are the norm. Permeability changes can be critical to wetland hydrology. Many wetlands contain micro-stratigraphy consisting of thin-low permeability units which retard vertical water movement. Failure to identify these units can lead to complete failure to achieve the desired wetland hydrology.

Hydrology

As previously noted, hydrology is the most common limiting factor. For the purpose of this chapter, a detailed discussion of hydrologic elements is not possible. A number of site specific hydrologic elements exist for wetlands or potential wetland sites. Site-specific water-budget analysis is strongly suggested. Quantitative analyses are preferred, but occasionally qualitative estimates of water budget components may be sufficient.

Wetland Basin Design

The shape of a proposed or restored wetland should be based upon hydrogeologic and water-budget analysis, site characteristics, and the intended goals. For a water-table dominated wetland, the wetland basin must be excavated into the water-table deep enough to attain the desired hydroperiod for the intended vegetation community. If the primary goal is to provide fish habitat, the water depth needed for the target species and their required life cycles must be achieved. In some locations sufficient water depth below thick winter ice is required to allow over-wintering survival. The size and depth of the wetland basin, in many cases, must be related to the size of the contributing watershed to insure a sufficient water budget.

Dewatering

Creation of wetlands by excavation into the water-table commonly requires dewatering or control of ground water inflow during construction. This is especially true for wetlands with no outlet. Restoration of such wetlands may require outlet control structures such as dikes, dams, and weirs. Management of excess water in the intended wetland area and in any excavations for outlet-control structures may require dewatering. Dewatering methods include the use of ditches, surface water pumps, and drawdown wells. However, disposal of the removed water may create downstream impacts on water quantity, chemistry, and sedimentation. Erosion control is also an integral part of dewatering.

Watering

For many situations, especially for low permeability wetlands, it is necessary to provide additional water to create the required degree of saturation to insure plant-growth success. Commonly, construction occurs during the dry months of July through November. Wetland vegetation may be planted during these dry months when sufficient natural water is unavailable, but this is not advisable. A source of artificial water such as trucked-in water, the use of hydrants, piping, or installation of wells may be needed, and this is usually very expensive.

Timing of Construction

Timing of construction activities should reflect intended wetland goals. Limiting factors should be identified and considered in the project schedule. The construction of the replicate wetland should usually occur during the time when likelihood of success is optimal, not when construction schedules are best served. However, cost related to timing variations should be examined. Construction during different periods of the year may require different construction methods, some of which could cause possible wetland failure. Normally projects are timed for construction during the dry season when earth moving is the least problematic and excess water and erosion control is minimal. Otherwise, special and expensive equipment will be necessary to work in wet conditions (e.g., barges, draglines, all-terrain excavators).

Construction Considerations

Cutting, Stumping and Grubbing--

Cutting, stumping and grubbing are the first steps in many land excavation processes, especially when upland is to be converted to wetland. They may also be necessary in some restoration projects where wooded swamps are to be restored as open water or shallow marsh. Disposal of trees, stumps, and boulders can be difficult, particularly in cases where removal from the site is required. On-site stump dumps may be required and may need to be properly located and designed. In some cases, such as open water wetlands where warm water fish habitat is desired, stumps may be left as hiding places for fish or for topographic variations. In some cases, stumps can be used in the wetland design to create nesting islands.

Erosion and Sediment Control--

An erosion and sediment control plan should be prepared and implemented during construction, and as long thereafter as necessary. All project phases typically require erosion and sediment control. In some rare cases the process must be continued for the life of the project. Sediment damage to downstream areas can be considerable and could severely impact downstream ecosystems.

Excavation--

Excavations to create a kettle wetland can be wide and deep, removing large quantities of material. Disposal of this excess material may be a problem. The excavated material should become part of the cut and fill budget for the total project. Commonly the earth removed for the kettle basin is suitable as construction fill elsewhere on site, or can be sold to offset costs of the replication. In some cases, bedrock may be encountered and ripping or blasting may be required. Disposal of excess rock presents additional problems. Cost of rock excavation is much higher than unconsolidated material excavation. The importance of depth to bedrock data, obtained by borings, test pits, or seismic profiling is critical and commonly a limiting factor. Excess rock in some projects can be designed into the wetland, used as shoreline protection, islands for wildlife nesting, reefs for fish habitat, or to create microtopography within the wetland.

Side Slope Stabilization--

Excavation side slopes may fail and slump if the stability of earth slope is not determined. Excavation into the water-table, followed by dewatering, commonly results in slumping. Slopes where water levels fluctuate up and down seasonally or during maintenance are prone to slumping and should be designed to withstand the fluctuations.

All of the above discussion applies to wetland restoration projects, particularly to the construction of outlet controls, dams, and dikes. Restoration projects predominantly require much less construction activity than do replication projects. Restoration generally consists of "plugging the drain" and allowing water to raise to previous levels. More detailed design specifications are obtainable in various civil engineering sources.

Soils

The need for placing wetland-organic soils in the bottom of the replication wetland is debatable, depending upon the goals desired. Obtaining organic soils is not always possible or can be very expensive. Reuse of organic soils from the lost wetland area is preferred, but also presents problems of excavation, stockpiling, soil chemistry changes, organic matter decomposition, sediment control, and water quality protection. Placing saturated organic soils into a new wetland area is extremely difficult, commonly requiring low load-bearing tracked equipment, which is not always obtainable. Stabilization of these soils is also a problem.

The advantages of reusing wetland soils include the fact that organic soils generally maintain saturated conditions which helps wetland plants survive droughts. They also contain indigenous seed banks and root stock which insures rapid revegetation with appropriate plants. However, undesirable plants (such as purple loosestrife) may also be contained in the soils, creating management problems which may become limiting factors in achieving goals.

Revegetation

Vegetation is site-specific. If the goal is to replicate the lost wetland in-kind it may be impossible to achieve or it may take a number of years. An eastern red maple wooded swamp, for example, cannot be said to have achieved in-kind status until the trees have obtained an age equal to those lost and all other characteristics of the swamp are similar (Golet 1986). Some types of vegetation such as floating bog communities may be impossible or very difficult to create. The hydroperiod of the desired vegetation community must first be established before any hope of creating the desired community is achieved. If the proper hydrology and chemistry is created then the desired vegetative community should respond. Commonly, revegetation occurs very rapidly following the creation of wetland hydrology on fresh water sites, even with no planting or seeding. In isolated kettles without inflowing surface water, natural revegetation may not occur due to lack of an inflowing seed source. In these situations, seeding and planting is necessary.

In most cases, use of a temporary erosion-controlling grass seed mixture is desired to stabilize the organic or hydric-mineral soils. Control of nuisance animals and insects may be required and, in some cases, may be a limiting factor requiring considerable thought and expense.

Water Chemistry

Some kettles such as saline potholes and acid bogs have very unique water chemistries (Steward and Kantrud 1971) which are the result of age and complex histories. Reproducing these chemistries will be extremely difficult, if not impossible. Projects with goals to replicate wetlands with unique water chemistries should be reviewed carefully.

It should be much easier to achieve unique water chemistry through renovation projects, especially those where the original wetland basin, soils, and hydrology are largely intact. Simple replacement of the plug and passage of time should give rise to a unique water chemistry similar to the original wetland.

Reintroduction of Fauna

The key to faunal reintroduction is creation of similar habitat. Actual transportation of fauna to the new wetland is normally not necessary. If creation of a fishery is a goal, stocking may be needed, especially for game fish.

Buffers

Lack of a correct buffer from conflicting land uses can be a limiting factor preventing achievement of a particular goal. Buffers are most important to wildlife, especially for species dependent upon surrounding upland for much of their life cycle (i.e., salamanders). Physical buffers such as woods, fences, hedges, and/or open water may be necessary to prevent predation or human intrusion during and after construction. Buffers may also be important if aesthetic goals are to be achieved. Buffers may also be desirable to protect the wetland from erosion or chemical wash-off from adjacent uplands.

Long-term management

Long-term management is an important element in achieving many wetland goals. Since wetlands are dynamic landscape elements, changes within them are expected but may be contrary to desired goals. If the goal is to create a wet meadow within a low permeability kettle in Massachusetts, that plant community must be managed as a wet meadow, for it will rapidly change to a shrub swamp or wood swamp if woody vegetation is not controlled. Management is also important since the best plans and construction may not meet the site limitations. Changes in hydrology may occur, and soil may settle and consolidate when saturated. The project should be designed as "maintenance free" as possible. A project heavily dependent upon significant long-term maintenance activities should be thoroughly reviewed before a permit is issued.

A number of problems occur with wetland management. The first is land ownership. The wetland replication/restoration site must be dedicated legally as wetland forever. Restrictions or easements must be placed and recorded upon the site's deed so that subsequent land owners are aware of their responsibilities and limitations. Private ownership of replicated wetland is less likely to maintain wetland goals than public ownership (Golet 1986). When possible, wetland replication/restoration projects should be deeded or leased to a government agency, especially to one with the interest and funding to maintain the wetland.

Funds must be available for long-term maintenance of wetlands. Without the proper funds, maintenance cannot occur. Maintenance must be an integral part of a monitoring program in order to assure success.

Enforcement is a very important part of a mitigation program. Too often the replication/ restoration project is approved as mitigation in a permit application. Upon issuance of the permit, the project is forgotten by the regulatory agency and no inspection is conducted to insure even the most simple compliance with the replication plan. This is primarily because of the lack of agency funds and manpower. Enforcement can be facilitated by requiring submission of as-built-plans to the regulatory agency, by requiring monitoring with periodic reports, and by requiring that funding be made available to an independent source to conduct inspections for the regulatory agency. The burden of cost of inspection should be placed on the permittee, not the agency. Enforcement must be tied into definable goals established in the permit.

MONITORING

Two factors must be monitored: (1) the success in achieving desired goals, and (2) the limiting factors that impact the site. Monitoring plans should be designed to collect the specific data needed to measure success.

The key items to be monitored are, at a minimum, hydrology and revegetation. Other items which may be monitored are physical, chemical, and biological changes. Generally, if success is achieved for these factors, other values such as wildlife habitat, water-quality improvement, and flood control functions will also be achieved. If the wetland project has been designed for very specific goals, such as an increase in nesting pairs of ducks or a decrease in heavy metals from urban runoff, then very specific, detailed monitoring may be required.

If the wetland is a result of a permit process, the monitoring plan should be developed and, in some cases, implemented prior to issuance of the permit. A method to provide adequate funding for the life of the monitoring plan is also needed. Provision should also be made for possible changes in the plan to respond to alterations to the wetland needed to achieve success. This burden should be borne by the permit applicant.

It is suggested that monitoring of vegetation be intensively conducted for the first two growing seasons, and corrective modifications made as necessary. Monitoring of vegetation should then occur at 3 to 5 year intervals. For some types of goals, such as water-quality maintenance, monitoring for the life of the project may be needed. Wetlands such as prairie potholes, may require monitoring through not only a typical water year but also during a 10-year hydrologic cycle, as these wetlands are the product of droughts and floods as well as average conditions.

The results from a monitoring program should be analyzed to determine success by both the permittee and the regulatory agency. The raw data plus data analysis should be provided. In some cases duplicate samples (i.e., for water quality) may be desirable so that the agency can provide its own analysis. Monitoring has a very important role in enforcement, in so much that monitoring data may be key to determining "success" or "failure" in litigation. Without monitoring, enforcement may be impossible in many situations.

MID-COURSE CORRECTIONS

Commonly during the construction or monitoring phases, it is discovered that the project as designed will not accomplish specific goals. This is often because of site-specific surficial geology or hydrology conditions and inadequate pre-project data collection. For example, the post-excavation elevation of the ground water-table is very difficult to predict. Actual water-table elevations are commonly not determined until excavation is completed. The science of hydrogeology is not precise enough to predict water levels to less than one foot, and in most cases less than two feet. Small changes in water depth in kettles can create too dry or too wet conditions for the desired vegetation. Too often, this lack of understanding is attributed to a drought, flood, or unusually high water-tables and the failure to observe long-term cyclic water data for the site. The "rare event" is very commonly used as an excuse to explain our failure to understand and predict hydrology.

In new wetlands ground water wells, properly screened, are probably necessary and must be installed and monitored during construction to establish final excavation grades. Surface water level gauge readings and stream-flow measurements must be obtained to define water budget components. Controlled outlets with the ability to raise and lower water levels may be critical in either a new or restored wetland. This may be necessary not only to achieve short-term goals, but to maintain desired goals over a long period where management of water levels may be needed.

Mid-course changes may create limiting factors which make achieving desired goals impossible. The old goals may have to be modified or abandoned and new goals established.

CONCLUSION

INFORMATION GAPS AND RESEARCH NEEDS

The lack of information on wetland science and general research needs are described by Larson and Neill (1987). The following are data specific to kettle wetlands.

MICROSTRATIGRAPHY AND HISTORY OF WETLAND DEVELOPMENT

Little has been done to examine why wetlands have developed in various kettles. Some kettles contain wetlands and others do not, a situation which occurs side by side in many locations. Kettle wetlands are the result of 10,000-14,000 years of evolution during a variety of climates, vegetation communities, and geologic processes. They are not simply the product of one event in time. To replicate or restore them to "original" conditions requires an understanding of this history.

Commonly, the limiting factor which allows a wetland to form in one kettle and not in another when the dominant geologic deposit is permeable stratified drift is the presence of small impermeable stratigraphic units which retard downward movement of water out of the kettle. Unpublished research conducted by the author in Iceland indicates eolian-fine sand and silt deposition in kettles in highly permeable outwash acts as a retarding layer. Eolian activity during deglaciation and periglacial periods were important events in New England and the northern Great Plains, which greatly impacted the permeability of the upper-most geologic deposits of the kettles. This eolian layer is normally of less significance in kettles formed in low permeability glacial till.

GEOCHEMICAL PROCESSES

Geochemical processes, particularly the accumulation of iron and manganese cement within units underlying organic soils, decrease permeabilities and increase detention of water in wetlands. More research is needed in this area.

ORGANIC SOIL

The water storage and drainage through organic soils, and the geochemical and biological activities within those soils are poorly understood. Organic soils are believed to be an integral part of many of the functions of kettle wetlands and are in need of much research.

HYDROLOGY

One of the most poorly understood aspects of wetlands is their hydrology (Novitzki 1987). For kettle wetlands there is a specific need to understand their relationship to regional ground water systems including recharge and discharge.

LITERATURE CITED

Adamus, P.R. 1983. A Method for Wetland Functional Assessment, Volumes I and II. Offices of Research, Development and Technology, Federal Highway Administration, U.S. Department of Transportation, Washington, D.C. 20590, FHWA-IP-82-24.

American Geological Institute. 1972. Glossary of Geology. Washington, D.C.

Brown, M. and J.D. Dinsmore. 1986. Implications of marsh size and isolation for marsh management. Jour. of Wild. Manag. 50:392-397.

Clayton, L. 1967. Stagnant-glacier features of the Missouri Coteau in North Dakota, p. 25-46. In L. Clayton and T.F. Freers (Eds.), Glacial Geology of the Missouri Coteau and Adjacent Areas. North Dakota Geological Survey, Miscellaneous Series 30, 18th Annual Midwest Friends of the Pleistocene Guidebook.

Clayton, L. and J.A. Cherry. 1967. Pleistocene superglacial and ice-walled lakes of west-central North America. In L. Clayton and T.F. Freers (Eds.), Glacial Geology of the Missouri Coteau and Adjacent Areas. North Dakota Geological Survey, Miscellaneous Series 30, 18th Annual Midwest Friends of the Pleistocene Guidebook.

Cowardin, L.M., V. Carter, F.C. Golet, and E.T. LaRoe. 1979. Classification of Wetlands and Deepwater Habitats of the United States: Office of Biological Services, U.S. Fish and Wildl. Serv., FWS/OBS-79/31.

Dane, C.W. 1959. Succession of aquatic plants in small artificial marshes in New York state: N.Y. Fish and Game Jour. 6:57-76.

Davis, C.B., J.L. Baker, A.G. Van der Valk, and C.E. Beer. 1981. Prairie pothole marshes as traps for nitrogen and phosphorus in agricultural runoff, p. 153-164. In F.B. Richardson (Ed.), Selected Proceedings of the Midwest Conference on Wetland

Values and Management's Minnesota Water Planning Board, Minneapolis, Minnesota.

Department of Environmental Quality Engineering. 1983. Regulations of the Massachusetts Wetlands Protection Act (MGL 131040). 310 CMR 10.00 of the Massachusetts Environmental Codes.

Eisenlohr, W.S., Jr. and others. 1972. Hydrologic Investigations of Prairie Potholes in North Dakota, 1959-68. U.S. Geological Survey, Professional paper 585-A.

Flint, R.F. 1971. Glacial and Quaternary Geology. John Wiley and Sons, New York.

Golet, F.C. 1986. Critical issues in wetland mitigation--a scientific perspective. National Wetlands Newsletter 8(5):3-6.

Golet, F.C. and J.S. Larson. 1973. Classification of Freshwater Wetlands in the Glaciated Northeast. Resource Publication 16, U.S. Bureau of Sport Fisheries and Wildlife, Washington, D.C.

Gosselink, J.G. and R.E. Turner. 1978. The role of hydrology in freshwater ecosystems, p. 66-79. In R.E. Good, D.F. Wiigham, and R.L. Simpson (Eds.), Freshwater Wetlands: Ecological Processes and Management Potentials. Academic Press, New York.

Heeley, R.W. 1973. Hydrogeology of wetlands in Massachusetts. M.S. Thesis, University of Massachusetts, Amherst.

Hollands, G.G. 1987. Hydrogeologic classification of wetlands in glaciated regions. National Wetlands Newsletter 9(2):6-9.

Hollands, G.G., G.E. Hollis, and J.S. Larson. 1987. Science base for freshwater wetland mitigation in the glaciated northeastern United States: Hydrology, p. 131-143. In J.S. Larson and C. Neill (Eds.), Mitigating Freshwater Wetland Alterations in the Glaciated Northeastern United States: An Assessment of the Science Base. The Environmental Institute, University of Massachusetts at Amherst, Publication 87-1.

Hollands, G.G. and W.S. Mulica. 1978. Application of morphological sequence mapping of surficial geological deposits to water resource and wetland investigations in eastern Massachusetts. Geological Society of America, Abstracts with Program, Northeastern Section, 10(23):470.

Hollands, G.G. and D.W. Magee. 1986. A method for assessing the functions of wetlands, p. 108-118. In J.A. Kusler and P. Riexinger (Eds.), Proceedings of the National Wetland Assessment Symposium, Association of State Wetlands Managers, Chester, Vermont.

Kaye, C.A. 1960. Surficial geology of the Kingston quadrangle, Rhode Island. U.S. Geological Survey Bull. 10:71-I.

Larson, J.S. 1987. Wetland mitigation in the glaciated northeast: risks and uncertainties, p. 4-16. In J.S. Larson and C.Neill (Eds.), Mitigating Freshwater Wetland Alterations in the Glaciated Northeastern United States: An Assessment of the Science Base. The Environmental Institute, University of Massachusetts at Amherst, publication 87-1.

Larson, J.S. and C. Neill (Eds.). 1987. Mitigating Freshwater Wetland Alterations in the Glaciated Northeastern United States: An Assessment of the Science Base. The Environmental Institute, University of Massachusetts at Amherst, publication 87-1.

Malo, D.D. 1975. Geomorphic, Pedologic, and Hydrologic Interactions in a Closed Drainage System. Ph.D. Dissertation, North Dakota State University, Fargo, North Dakota.

Mathiak, H. 1965. Pothole Blasting for Wildlife: Wisconsin Conservation Department, Madison, Wisconsin, Publication 352.

Meyborm, P. 1966. Unsteady ground water flow near a willow ring in hummocky moraine. Journal of Hydrology 4:38-62.

Millar, J.B. 1976. Wetland Classification in Western Canada: A Guide to Marshes and Shallow Open Water Wetlands in the Grasslands and Parklands of the Prairie Provinces. Canadian Wildlife Service Report, Ser. 37.

Motts, W.S. and A.L. O'Brien. 1981. Geology and Hydrology of Wetlands in Massachusetts. Water Resources Research Center, University of Massachusetts at Amherst, Publication 123.

Novitzki, R.P. 1987. Some observations on our understanding of hydrologic function. National Wetlands Newsletter 9(2):3-6.

Novitzki, R.P. 1982. Hydrology of Wisconsin's Wetlands. U.S. Geological Survey, Information Circular 40.

O'Brien, A.L. 1987. Hydrology and the construction of a mitigating wetland, p. 82-100. In J.S. Larson and C. Neill (Eds.), Mitigating Freshwater Wetland Alterations in the Glaciated Northeastern United States. An Assessment of the Science Base. The Environmental Institute, University of Massachusetts at Amherst, publication 87-1.

Shaw, S.P. and C.G. Fredine. 1956. Wetlands of the United States. U.S. Fish and Wildl. Serv., Circular 39.

Sloan, C.E. 1970. Prairie Potholes and the Watertable. U.S. Geological Survey Professional Paper 700-B.

Steward, R.E. and H.A. Kantrud. 1971. Classification of Natural Ponds and Lakes in the Glaciated Prairie Region. U.S. Fish and Wildl. Serv., Resource Publication 92.

Steward, R.E. and H.A. Kantrud. 1972. Vegetation of Prairie Potholes, North Dakota, in Relation to Quality of Water and Other Environmental Factors. U.S. Geological Survey Professional Paper 585-D.

Strohmeyer, D.S. and L.H. Fredrickson. 1967. An evaluation of dynamited potholes in northwest Iowa. Journal of Wildlife Management.

Tiner, R.W. and P.L. Veneman. 1987. Hydric Soils of New England. Coop. Extension Bulletin C-183, University of Massachussetts, Amherst, Massachusetts.

U.S. Fish and Wildlife Service. 1983. Proceedings of the National Wetlands Values Assessment Workshop. U.S. Department of the Interior, Washington, D.C.

Winter, T.C. 1981. Uncertainties in estimating the water balance of lakes. American Water Resources Bulletin 17(1):82-115.

Winter, T.C. 1983. The interaction of lakes with variably saturated porous media. Water Resources Research 19(5):1203-1218.

Winter, T.C. and M.R. Carr. 1980. Hydrologic Setting of Wetlands in the Cottonwood Lake Area, Stutsman County, North Dakota. U.S. Geological Survey, Water-Resources Investigations 80-99.H.

APPENDIX I: PROJECT PROFILES

I-495/ROUTE 25, BOURNE, MASSACHUSETTS

Project Purpose:

The construction of Route 25 in Bourne resulted in the taking of nine acres of active cranberry bog. The Massachusetts Department of Public Works did not propose to replicate the lost wetlands. The Massachusetts Department of Environmental Quality Engineering issued a Superseding Order of Conditions allowing the taking without mitigation, but this was appealed. Following a lengthy and detailed adjudicatory hearing, a Final Order of Conditions was issued finding that the cranberry bogs were significant for prevention of pollution and flood control. Mitigation was required, in part, the replication of nine acres of shallow marsh in an existing upland area. The primary goal of the replication was to create wetlands capable of renovating highway runoff. This is an example of a large excavation to produce a kettle-like water-table wetland, but not as a replacement for a lost kettle.

Wetland Type/Area Lost: Nine acres of highly productive cranberry bog, part of a riverine system.

Wetland Type/Area Created: Nine plus acres of cattail dominated shallow marsh in a kettle-like depression with no outlet.

Procedure:

Numerous test borings and pits were dug to determine site surficial geology and ground water level. Observation wells were installed and monitored. These data were placed into a complex ground water computer model and ground water flow directions, rates, and levels were predicted. Correlation with U.S. Geological Survey observation wells were made to calibrate the model. Excavation of permeable sand and gravel was conducted using standard earth moving equipment in 1987. Final grades were created and ground water flooded the bottom of the depression. Ground water elevations higher than predicted were encountered and attributed to "unusual conditions". Mid-course actions were needed to modify water levels. No outlet for the depression was designed, so no artificial water level manipulations were possible. The area was planted with marsh and wet meadow vegetation in the summer of 1987 and has completed one growing season. However, significant problems were encountered with establishment of the desired vegetation.

COULTER DRIVE, CONCORD, MASSACHUSETTS

Project Purpose:

An access road to an industrial park was proposed to solve traffic safety problems. This would result in the loss of shrub swamp, shallow marsh, open water, and a small irrigation pond dug into the water-table. All of the area was within the upper portions of the 100-year floodplain of the Assabet River. The Town of Concord Conservation Commission, under the Massachusetts Wetland Protection Act, found the wetland to be significant for flood control, storm damage prevention, public and private water supply, ground water supply, and prevention of pollution. In-kind replication defined as equal area of wetland type was required. The goal was to create a wetland which functioned in "a similar manner" to the one lost. Maintenance of natural-valley flood storage was a primary goal. Under the Concord Wetland Conservancy District local bylaw, wildlife habitat was also an interest and a goal.

Wetland Type/Area Lost: 6000+ square feet of shrub swamp (mostly buckthorn-dominated); 9000+ square feet of shallow marsh; 5000 square feet of open water; and an abandoned irrigation pond for past agricultural activities. The pond was dug into the water-table. The vegetated wetlands were associated with spring high-water tables and surface water flooding.

Wetland Type/Area Created: 12,000 square feet of shrub swamp consisting of a variety of domesticated nursery shrubs, specifically highbush blueberry and pepperbush; 14,000+ square feet of shallow marsh, mostly cattail; and 3000+ square feet of open water.

Procedure:

Water-table elevations were determined with test pits. An excavation to include compensatory flood storage was designed into the excavation. Original soils were stockpiled and used in final grading of all areas but the open water. Shrubs were planted above the average high-water level, but no emergent marsh species were planted. All bare-soil areas were hydroseeded with a standard erosion control seed mixture.

Status:

The area was completed in the spring of 1984. Excellent growth occurred in the first growing season. Wetland plants colonized the shallow marsh areas and deep marsh plants colonized a portion of the open water. After four growing seasons all wetland areas, except open water, are densely vegetated. In the growing season of 1987 unusually high water levels resulted in flooding of the shrub swamp portion of the wetland. This killed approximately 80% of the planted shrubs. If the wetland had a controlled outlet and water level monitoring had occurred, the water level could have been dropped to prevent the damage to the shrubs. This points out the sensitivity of woody plants to flooding. Projects where woody wetland plants are critical to success require detailed hydrologic analysis and the ability to prevent extended flooding. Mallard ducks have used the area for brood rearing.

INDIAN HILLS, NORTH DAKOTA

Project Purpose:

The area consisted of approximately 200 acres of drained prairie potholes on a slope of approximately 200 feet in relief. Type 1, 3 and 4 (Shaw and Fredine 1956) potholes occurred on the site. The upper potholes drained into the mid-slope potholes which, in turn, drained into the lower most and the largest pothole which had no outlet. The purpose was to create pothole wildlife habitat in a multipurpose recreational area where hunting of both upland and wetland game species was to occur. This area was formerly known as the Fahl-Wackerly Tract and was drained for agricultural purposes. Both shallow (less than a foot) and deep (approximately 20 feet) drainage ditches had been used to drain the natural potholes. The potholes had been plowed and used for agricultural purposes.

Wetland Type/Area Altered: Approximately 200 acres of Type 1, 3 and 4 prairie potholes.

Wetland Type/Area Restored: Approximately 200 acres of Type 1, 3 and 4 prairie potholes.

Procedure:

A variety of dikes with controlled outlets were used to "plug" the drainage ditches. This work was conducted and designed by the Bureau of Reclamation, Bismarck, North Dakota.

Status:

The work has been completed and pothole hydrology recreated. The potholes have become revegetated naturally, and appear to be viable pothole wetlands.

DUCKS UNLIMITED PROJECTS IN NORTH DAKOTA, SOUTH DAKOTA AND MONTANA

Project Purpose:

Ducks Unlimited has been restoring and, to a much lesser degree, creating prairie pothole wetlands to increase duck production through the creation of nesting and brood rearing habitat. Some sites occur where potholes have all but been destroyed, such as in "black desert" areas, where only their basin topography exists.

Wetland Type/Area Lost: All types of potholes have been lost.

Wetland Type/Area Restored: All types of potholes have been restored.

Procedure:

All types of restoration procedures have been used, ranging from simple plugging of small ditches to extensive and large dikes, dams, and outlet structures. The methods used were site specific and determined by the limiting factor of the site, usually water availability. The cheapest method was normally used. Water-level management has been used to control nuisance vegetation and increase duck productivity. In the National Grasslands, wetlands have been created where none existed by the damming of shallow swales. In this area, where precipitation averages 12 inches per year, water budget and topography are limiting factors. All designs for duck production required attention to be given to predator control, which can be a limiting factor. Identification of predator type and methods to keep predators from the nesting areas was a key to success. Various predator controls such as islands, peninsular cut-offs, and electric fencing have been successfully used. A high degree of success (i.e., two broods per wetland acre) have been measured by various investigators using a variety of techniques.

REGIONAL ANALYSIS OF FRINGE WETLANDS IN THE MIDWEST: CREATION AND RESTORATION

Daniel A. Levine and Daniel E. Willard
School of Public and Environmental Affairs
Indiana University

ABSTRACT. "Fringe" wetlands (as we define the term for the purposes of this chapter) are those found along lakes and reservoirs. They are scattered along the shorelines of the Great Lakes and the Tennessee Valley Authority reservoir system, and abundant along numerous small lakes and reservoirs in the Midwest. Their prevalence here suggests that Midwestern administrators of the regulatory program under Section 404 of the Clean Water Act will be confronted with permits which impact fringe wetlands.

Fringe wetlands affect water quality through their influence on nutrient cycling, sedimentation, and heavy metal movement. They also stabilize shorelines by minimizing the erosive forces of waves and seiches. These wetlands provide important habitat for fish and wildlife. All of these functions are influenced by the water level fluctuations characteristic of fringe wetlands.

We found very few documented cases of fringe wetland mitigation under Section 404. Section 404 projects rarely stated specific goals, other than to create a wetland to mitigate for the loss of an existing one. We recommend that project goals be clearly stated in the permit and be specific enough to provide a set of criteria with which to evaluate success. Specific goals also dictate the project's design considerations. We recommend that these considerations also be clearly outlined in the permit and include the following:

- A justification for the site location;
- A description of the site characteristics prior to mitigation, including substrate, elevations (1 ft. intervals), and water levels and fluctuations;
- Construction plans detailing how the site will be modified to suit project goals; for example, diking to control water levels suitable for the target species;
- A list of target species (scientific and common names) consistent with project goals, a revegetation plan outlining the type and source of propagules used, the planting methods, densities, and timing;
- A long-term management plan which covers water level control (if applicable), nuisance species control, a financial plan, and identification of the responsible party(s); and
- A complete monitoring plan, including what will be measured, when and how often it will be measured, how the measurements will be taken, and how monitoring results will be interpreted.

A clear statement of project goals and design considerations will allow the 404 administrator to critically evaluate the permit and the project's likelihood of successful mitigation.

To further ensure the success of mitigation, we recommend the following. First, the establishment of a fringe wetland should not be attempted where the fetch is greater than 13 km unless a dike is constructed to reduce wave action. Second, revegetation should utilize a combination of both natural (i.e., seed bank) and artificial (i.e., transplants) methods. This could provide both immediate cover and a backup source of propagules. These propagules could ensure some type of vegetative cover if the transplants die as a result of changing environmental conditions. We also recommend that the transplants come from stock ecotypically-adapted to site conditions.

Further research is needed 1) to determine how fringe wetland plant species and plant communities as a whole influence water quality, 2) to determine how water level fluctuations affect species composition and nutrient cycling in these wetlands, 3) to quantify shoreline stabilization needs and functions, and 4) to develop ecotypically-adapted planting stocks.

INTRODUCTION

Approximately 50% of the wetlands in the Midwest have been drained over the past 100 years, primarily for agricultural development (Tiner 1984). A significant proportion of the remaining 50% are "fringe" wetlands. A fringe wetland (as we use the term) is a wetland found along the edge of a lake or reservoir. The purpose of this chapter is to help Section 404 administrators review permits for activities which will impact fringe wetlands.

To provide background, we first describe the physiographic and climatic characteristics of the Midwest region. We then define fringe wetlands and discuss the functions of these wetlands, including water quality, shorelines, and fish and wildlife habitat. We feel that understanding these functions and how they are affected by water level fluctuations is imperative for sound permit decisions. In the next section of this chapter, we discuss creation and restoration projects for fringe wetlands in the Midwest. This discussion addresses project goals (or the lack thereof), success in achieving these goals, and the reasons for project successes or failures. After this, we list critical elements of the permits, and discuss these elements including preconstruction considerations, critical aspects of the creation plan, and monitoring programs. Finally, we conclude with an evaluation of information gaps and research needs for fringe wetlands in this region.

DESCRIPTION OF THE MIDWEST

The Midwest region, for the purposes of this chapter, includes the states of Illinois, Indiana, Iowa, Kentucky, Michigan, Missouri, Ohio, Tennessee, and Wisconsin. Fenneman (1970) described this region as the Central Lowland. It is bordered on the east by the Appalachian Plateau, on the south by the Interior Highlands (Ozark Plateau), and on the west by the Great Plains. We will consider the United States-Canadian border across the Great Lakes as the northern edge of this region.

Topographically, the Midwest region slopes down from the east at 1000 feet above mean sea level (MSL) to 500 feet above MSL at the Mississippi River and also down from the west at 2000 feet above MSL to the Mississippi River. Much of the region's terrain was created by glacial processes. Each of the recent glacial advancements pushed far southward into the Midwest. However, the southern half of Missouri, the southern tip of Indiana, southeast Ohio, Kentucky, and Tennessee were not glaciated. The major hydrologic features of the Midwest include the Great Lakes (excluding Lake Ontario) and the Mississippi, Ohio, and Missouri Rivers. Much of the region is dotted with natural lakes in the glaciated areas and man-made reservoirs of various sizes in the unglaciated south.

Annual rainfall for this region ranges from 20-30 inches in the north to 50-60 inches in the south. Most of the region receives around 40 inches of rain annually. Freeze free days range from 90 days in northern Michigan to 180-210 days in Kentucky and Tennessee. The central states average 150 freeze free days annually.

The soils of the Midwest region are predominantly in the order Alfisol (Brady, 1984). However, other soil orders are also represented. Spodosols cover the northern portions of both Wisconsin and Michigan. Mollisols cover most of Iowa, the northern part of Illinois, the northwest corner of Missouri, and parts of central Kentucky. Inceptisols are found in the unglaciated regions of Ohio, Kentucky, and Tennessee.

Eastern deciduous forest is the predominant type of natural vegetation in the Midwest (Braun, 1950). The northern part of the region is characterized by northern hardwoods, dominated by hemlock (Tsuga canadensis) and white pine (Pinus strobus). The southern unglaciated portion is dominated by mixed mesophytic forests, characterized by sugar maple (Acer saccharum), beech (Fagus grandifolia), tulip poplar (Liriodendron tulipifera), and white and northern red oak (Quercus alba and Q. borealis). Beech and sugar maple dominate the central part. The western part is dominated by oak-hickory forest, characterized by several species of Quercus and Carya. The prairie grass big bluestem (Andropogon gerardii) is also prevalent there.

Since the Midwest region is a mosaic of lakes, reservoirs, and rivers, all wetland types are found here except estuarine and marine. However, this chapter will discuss fringe wetlands only. See Willard et al. (this publication) for a description of riparian wetlands in the Midwest.

FRINGE WETLAND CHARACTERISTICS

For the purposes of this chapter, however, we define fringe wetlands as all those found along the edge of lakes and reservoirs. According to Brinson (M. Brinson, 1988, East Carolina State University, pers. comm.), a fringe wetland is a wetland in which water flow is perpendicular to the zonation of vegetation. However, as we use the term, fringe includes wetlands on lakes and reservoirs where water flow may not be perpendicular to the zonation pattern of the vegetation. We also consider wetlands along lake and reservoir margins which have been diked as fringe wetlands. Applying the Fish and Wildlife Service's Wetland Classification scheme (Cowardin et al. 1979), fringe wetlands are generally Lacustrine-Littoral (L2) wetlands and those Palustrine (P) wetlands that adhere to the definition given above.

Fringe wetlands are widely distributed throughout the Midwest. For example, 42,840 hectares of coastal wetlands were identified along Michigan's Great Lakes shoreline alone in 1975 (Jaworski and Raphael 1978). Furthermore, the Tennessee Valley Authority (TVA) reservoir system has 71,227 hectares of shoreline which could support fringe wetlands (Fowler and Maddox 1974). These figures are a small percentage of the total fringe wetland acreage in the Midwest, since inland lakes and reservoirs outside the TVA system have not been quantified.

The hydrology of fringe wetlands is dominated by fluctuating water levels. The magnitude of the natural fluctuation depends upon the size of the lake or reservoir, whether the lake has a natural outlet or outflow, if groundwater or evaporation are controlled, the fetch, and climatic cycles. Water levels are also influenced by water level management devices such as dikes and dams.

Vegetation types within a fringe wetland range from emergent to submerged. Emergent plants include cattail (Typha), common reed (Phragmites), bulrushes (Scirpus), sedges (Carex), bur-reed (Sparganium), wild rice (Zizania aquatica), and pickerel weed (Pontederia cordata). Floating leaf plants include American lotus (Nelumbo lutea), water lilies (Nuphar and Nymphaea), and floating pondweed (Potamogeton natans). Submerged plants include waterweed (Elodea canadensis), water-milfoil (Myriophyllum), wild celery (Vallisneria americana), and pondweed (Potamogeton).

Wetland plants continually respond to water level fluctuations. Accordingly, delineating fringe wetlands is difficult, a boundary drawn one year may not be accurate the following year. Furthermore, fringe wetlands often gradate at the landward extension into another type of wetland (e.g., a wet meadow). We recommend delineating an area in which a fringe wetland **may** occur given known water level fluctuations. This delineation can be drawn at the mean or historic high water mark and the lakeward edge of the littoral zone.

FRINGE WETLAND FUNCTIONS

Fringe wetlands provide many important functions. They improve water quality by acting as sinks for nutrients, by filtering suspended solids, and by absorbing heavy metals. They stabilize shorelines, minimizing the erosive forces of waves, seiches, and boat wakes. They provide important food sources, nesting sites, nurseries, and refuges for fish and wildlife. These functions highlight the ecological, recreational, and economic value of fringe wetlands.

WATER QUALITY

Fringe wetlands influence the water quality of lakes and reservoirs by affecting nutrient cycling, sedimentation, and heavy metal movement. Their influence is unquantified, but may be substantial. For example, a map of Great Lakes water quality indicates that areas of extensive fringe wetland loss correlate spatially with regions of high littoral eutrophication (Jaworski and Raphael 1979). This may be more than a coincidence since wetlands have been effective at improving water quality where they have been used for wastewater treatment (Spangler et al. 1976, Nichols 1983, Willenbring 1985).

Eutrophication is primarily caused by Phosphorus (P) and Nitrogen (N) loading. Fringe wetlands can act as both P or N sinks (Toth 1972, Nute 1977, van der Valk et al. 1979, Kelley 1985) and sources (Bender and Correll 1974, Lee et al. 1975). Uptake and release rates and their timing depend upon sediment characteristics, water chemistry, and the species composition of the wetland.

Sediments, particularly inorganic sediments, play significant roles in the ability of wetlands to retain P (King 1985). Edzwald (1977) noted that the amount of P adsorbed was dependent upon both clay type (illite, montmorillonite, and kaolinite) and pH. He hypothesized that the type and amount of clay in sediment would determine P adsorption potential. The adsorption potential is also a function of the water nutrient concentration and the redox potential at the sediment-water interface (King 1985).

The uptake and release of P and N vary seasonally (Kelley et al. 1985). For example, emergent and submerged plants are P sinks during the growing season (Schults and Maleug 1971, Klopatek 1975, Klopatek 1978). During this time, they pump P from the sediments and the water. Klopatek (1975) found a P uptake rate of 5.3 g P/m^2/year for river bulrush (Scirpus fluviatillis). Phosphorus uptake rates for broad-leaved cattail (Typha latifolia) can range from 0.019 g/m^2/year (Boyd 1970a) to 3.5 g/m^2/year (Prentki et al. 1978). In contrast to these species, Elodea canadensis recycles nutrients continually and can contribute up to 50% of the internal P load of a lake (Moore et al. 1984).

As plants senesce, P is released from their decomposing tissues, and they become P sources to the water. A substantial percentage of the tissue P is released quickly. For example, Boyd (1970a) found a 50% loss of P in dead broad-leaved cattail tissue in 20 days.

With time and increased accumulation, however, this decomposing tissue can become a P sink. This phenomenon occurs where decomposition is slow (Davis and van der Valk 1978) and where the water P concentration is high (Davis and Harris 1978). Davis and van der Valk (1978) found that one-year old fallen and standing litter from river bulrush and narrow-leaved cattail (Typha augustifolia) accumulated 192 and 230 percent, respectively, of the total P of fresh standing litter.

Submerged and emergent plants can also act as a N sink during the growing season. Klopatek (1978) reported a N uptake rate of 20.8 g N/m^2/year for river bulrush. For this species, the N uptake rate is greater than the P uptake rate. Nitrogen release from dead plant litter appears to be much slower than for P, taking months (Boyd 1970b) or years (Chamie 1976).

Nutrient cycling in fringe wetlands is also affected by water level fluctuations (Klopatek 1978, Kelley et al. 1985). Water levels affect the plant species composition of wetlands which, as indicated above, influences nutrient cycling. Kelley et al. (1985) found that more N and P were released into the water during high water periods. During this period, the vegetation was dominated by emergent macrophytes. These included bur-reed (Sparganium eurycarpum), giant or soft-stemmed bulrush (Scirpus validus), and broad-leaved cattail. During low water periods, vegetation reverted to a wet meadow dominated by bluejoint (Calamagrostis canadensis), tussock sedge (Carex stricta), and Carex aquatilis. During this period, N and P were released via decomposition but were stored in the soil. Hence, these nutrients were not released into the water, but they remained in the wetland system.

The mere presence or absence of water will also affect nutrient cycling (Bentley 1969, Amundson 1970, Lee et al. 1975, Klopatek 1978). For example, Klopatek (1978) reported that drainage of unvegetated marsh soils resulted in a release of large quantities of organic-N and NO_3. He attributed this release to an increased decomposition rate in the exposed substrate. Reflooding this wetland resulted in a net input of N into the marsh, with soil N increasing significantly within 1 year.

The release of nutrients from vegetation into a lake can control plankton community composition and the timing of planktonic production (Landers and Lottes 1983, Moore et al. 1984). For example, Landers and Lottes (1983) observed that eurasian water-milfoil (Myriophyllum spicatum) went through several die-off periods during the growing season. Each die-off sequence resulted in a large release of P. This, in turn, resulted in an increase in phytoplankton production and a change in species composition from green to bluegreen algae.

Regulators must understand the ecological implications of the nutrient uptake and release processes of fringe wetlands. Loucks (1981) and Weiler et al. (1979) described a model (WINGRA III) which simulates wetland/lake nutrient interactions. The model simulates the response of several trophic levels to changes in wetland nutrient cycling processes as affected by macrophyte management alternatives. This model is useful in identifying important relationships within a complex aquatic ecosystem as well as providing a tool with which to compare the outcomes of various management alternatives.

In addition to their nutrient cycling functions, fringe wetlands also filter suspended solids from the water column. A wetland receives particulate matter via runoff from surrounding uplands, litterfall from vegetation, and transport from channels, tides, and wave action (Kadlec and Kadlec 1978). For instance, a Great Lake fringe wetland may contain significant quantities of detritus, clay particles, and sands which were all deposited by wave action. Fringe wetlands may also serve as a source of suspended solids (Semkin et al. 1976). Whether a wetland is a source or sink for suspended solids depends upon the density of the vegetation, the type of suspended solids, hydrology, and morphology of the wetland. Dean (1978) described how these processes interact. This will be discussed further under "Shoreline Stabilization".

Fringe wetlands also play a role in the processing of heavy metals and organic contaminants. Wetland plants absorb and soils adsorb heavy metals (Banus et al. 1975, Seidel 1976, Wolverton et al. 1976). Seidel (1976) determined the uptake rates for 15 metals by ten different wetland species, including tussock sedge, narrow-leaved cattail, and common reed (Phragmites australis). Kadlec and Kadlec (1978) suggested that heavy metals may be cycled into the wetland system and subsequently contaminate higher trophic levels. Lunz et al. (1978), however, found that Cadmium (Cd), Chromium (Cr), Lead (Pb), and Zinc (Zn) present in the sediment of a manmade marsh were not absorbed by barnyard grass or Japanese millet (Echinochloa crusgalli), narrow-leaved cattail, or arrow arum (Peltandra virginica). They noted that soil conditions **typical** of marshes, such as moisture saturation, high organic matter content, near neutral pH, and low oxygen concentration, cause metals to be present in insoluble forms. This restricts the transfer of these metals into plant tissue. Understanding how these parameters affect metal mobility is essential and should be a priority research goal in the future.

SHORELINE PROTECTION AND STABILIZATION

Another function of fringe wetlands is to protect and stabilize shorelines (Garbisch 1977, Allen 1978, Dean 1978). Dean (1978) described how wetland vegetation provides such protection and stabilization. For example, the sediment-root matrix increased the durability of the shoreline. Furthermore, emergent and submergent plants dampened both wave action and the frictional forces of longshore currents, thereby reducing the energy hitting the shore. Finally, the vegetation caused sand and other material to be stored in nearshore dunes. Through this mechanism, they reduced the loss of these materials to wind erosion.

Extreme water level fluctuations severely hamper the ability of a fringe wetland to stabilize shorelines primarily by making it difficult for vegetation to become established and maintain itself. However, some species succeed temporarily under these conditions. Hoffman (1977) described how introduced reed canary grass (Phalaris arundinacea), Garrison creeping foxtail (Alopecurus arundinaceus), common reed, giant bulrush, and broad-leaved cattail became established and survived for 1-3 years in two reservoirs whose water level fluctuations averaged 3.5 m over a 6-year period. Furthermore, high water periods may also create new shorelines above a zone of existing protective vegetation.

FISH AND WILDLIFE HABITAT

Fringe wetlands are an important habitat for fish and wildlife throughout the Midwest. Over three million migratory waterfowl which travel through the Great Lakes area each year depend on suitable wetland habitat (Great Lakes Basin Commission 1975a). Many waterfowl species use these fringe wetlands for nesting (Table 1), resting, and feeding (Jaworski et al. 1980). Furthermore, fish depend on these wetland ecosystems for spawning, feeding, and shelter (Tilton and Schwegler 1978, Jaworski et al. 1980, Mitsch and Gosselink 1986). A partial list of Midwestern fish species and their use of wetlands is provided in Table 2. Fringe wetlands also provide important habitat to furbearers such as muskrat (Ondatra zibethica), racoon (Procyon lotor), and beaver (Castor canadensis) (Tilton and Schwegler 1978).

Water level fluctuations greatly influence the ability of a fringe wetland to provide fish and wildlife habitat (Liston and Chubb 1985, McNicholl 1985, and Prince 1985). The presence, absence, and depth of water can determine if and where particular species will feed and nest (McNicholl 1985).

Water level fluctuations have both positive and negative effects on fish habitat. Keith (1975) outlined several positive effects of high water levels:

o Shoreline terrestrial vegetation is flooded,

Table 1. Bird species that nest in fringe wetlands. Based on Prince (1985).

Common Names	Scientific Names
Least Bittern	Ixobrychus exilis
Pied-billed Grebe	Podilymbus podiceps
Canada Goose	Branta canadensis
Mallard	Anas platyrhynchos
Black Duck	Anas ribripes
Green-winged Teal	Anas crecca
Redhead	Aythya americana
Virginia Rail	Rallus limicola
Sora	Porzana carolina
Common Moorhen or Gallinule	Gallinual chloropus
American Coot	Fulica americana
Wilson's phalarope	Phalaropus tricolor
Black Tern	Chlidonias niger
Marsh Wren	Cistothorus sp.
Common Yellowthroat	Geothlypis trichas
Red-winged Blackbird	Agelaius phoeniceus
Yellow-headed Blackbird	Xanthocephalus xanthocephalus
Common Grackle	Quiscalus quiscula
Song Sparrow	Melospiza melodia
Swamp Sparrow	Melospiza georgiana

Table 2. Midwestern fish species which use fringe wetlands. Based on Barnes (1980).

Common Name	Scientific Name	Wetlands Habitat	Used for Reproduction	Commercial Value
spotted gar	Lepisosteus oculutus	*	*	n
longnose gar	Lepisosteus osseus	*	*	n
bowfin	Amia calva	*	*	n
gizzard shad	Dorosoma cepedianum	*	-	f
grass pickerel	Esox americanus	*	*	n
northern pike	Esox lucius	*	*	s
muskellunge	Esox masquinongy	*	*	s
central mudminnow	Umbra limi	*	*	f
golfish	Carassius auratus	*	*	n
carp	Cyprinus carpio	*	*	s
brassy minnow	Hybognathus hankinsoni	*	*	f
golden shiner	Notemigonus crysoleucas	*	*	f
pugnose shiner	Notropis anogenus	*	?	n
pugnose minnow	Notropis emiliae	*	?	f
blackfin shiner	Notropis heterodon	*	?	f
blacknose shiner	Notropis heterolepis	*	-	f
fathead minnow	Pimephales promelas	*	*	f
white sucker	Catostomus commersoni	*	-	f
lake chubsucker	Erimyzon sucetta	*	*	f
black bullhead	Ictalurus melas	*	*	s
yellow bullhead	Ictalurus natalis	*	*	s
brown bullhead	Ictalurus nebulosus	*	*	s
tadpole madtom	Noturus gyrinus	*	*	f
pirate perch	Aphredoderus sayanus	*	-	f
banded killifish	Fundulus diaphanus	*	*	n
starhead topminnow	Fundulus notti	*	?	f
brook stickleback	Culea inconstans	*	*	n
brook silverside	Labidesthes sicculus	*	*	f
rock bass	Ambloplites rupestris	-	-	s
green sunfish	Lepomis cyanellus	*	*	s
pumpkinseed	Lepomis gibbosus	*	*	s
bluegill	Lepomis macrochirus	*	-	s
smallmouth bass	Micropterus dolomieui	-	-	s
largemouth bass	Micropterus salmoides	*	*	s
black crappie	Pomoxis nigomaculatus	*	*	s
white crappie	Pomoxis annularis	*	*	s
Iowa darter	Etheostoma exile	*	*	f
yellow perch	Perca flavescens	*	*	s

* = species is often found in or requires aquatic vegetation for reproduction
? = relationship unknown
\- = species is rarely found in or does not require aquatic vegetation for reproduction.
n = species has no commercial value.
f = species may be an important forage fish.
s = species is an important sport fish.

which initiates the vegetation's death, decomposition, and subsequent release of nutrients. This increases the overall productivity of the system;

- Fish food organisms (e.g., worms and terrestrial insects) are quickly added to the water;
- New cover and habitat for shoreline species are created; and
- An area of water is created that is sparsely populated with fish, which could stimulate reproduction and growth as fish attempt to fill this "void".

Higher water levels did, in fact, increase populations of largemouth bass (Micropterus salmoides) (Keith 1975, Miranda et al. 1984) and white crappie (Pomoxis annularis) (Beam 1983). The timing of water level fluctuations is important. Beam (1983) emphasized that if water levels fluctuated during the spawning season, young-of-year white crappie numbers were reduced. Carbine (1943) and Hassler (1970) found similar results for northern pike (Esox lucius). Water level fluctuations during spawning periods could also affect the reproductive success of the other species which depend on aquatic vegetation for spawning (Table 2).

Water level fluctuations can also impact waterfowl. For example, birds species which nest on dry ground or floating leaves are highly susceptible to rising water levels (McNicholl 1979, Weller 1981, McNicholl 1985). Furthermore, wetland inundation can make some food sources unavailable to ducks and geese. Water level fluctuations are sometimes necessary, however, for the habitat to function effectively. Without periodic flooding, marshes can become choked with vegetation (Weller 1981). Waterfowl generally prefer a mixture of open water and emergent vegetation. Weller (1981) found that the highest number of waterfowl are found in a habitat that has between 50-75% open water and 25-50% scattered patches of emergent vegetation. Periodic flooding of wetlands prevents vegetation from becoming overgrown.

The economic value of the functions described above can be substantial. Jaworski et al. (1980) estimated that Michigan's coastal wetlands were worth $1210/hectare/year for fishing, hunting, trapping, and recreation. This converts to almost $52 million annually (1977 dollars) for the 42,839 hectares of coastal wetlands in Michigan. Sullivan (1976) estimated that wetlands could provide about $8650/ha/year of water quality treatment in the form of P removal as well as secondary and tertiary treatment. In addition, the Great Lakes Basin Commission (1975b) predicted that in the year 2000 there would be almost $22 million worth of wave damage along the 2,807 kilometers of U.S. shorelines of the Great Lakes with recreational, agricultural, and undeveloped land uses. This suggests a fringe wetland is worth over $7,500/km in shoreline protection, assuming complete protection. This value increases to approximately $40,000/km for the protection of shorelines with commercial, industrial, and residential land uses. We don't suggest that wetland creation for shoreline protection is feasible for the entire shoreline of the Great Lakes, but it should be considered for some circumstances.

EXTENT OF CREATION AND RESTORATION

Section 404 of the Clean Water Act (33 U.S.C. 1344) has been interpreted to require mitigation for wetlands lost due to filling. Filling typically occurs during road, housing, marina, and resort construction or during dredge spoil disposal. Mitigation projects permitted to date has ranged from reducing the acreage destroyed to complete creation of a new wetland with an area ratio of ten acres of created wetland for every one that was lost.

We found very few documented cases where fringe wetlands were created for mitigation under Section 404. Where they were created, the goal of the projects was generally stated as the creation of a wetland to mitigate for the loss of an existing one. More specific goals were rarely provided, although fish and wildlife habitat was often an implicit goal. For example, the Harbour Project on Sandusky Bay, Ohio, created over 30 hectares of fish and wildlife habitat.

Since specific goals were rarely provided for mitigation projects, evaluating their success is difficult. Without goals, it is uncertain how to choose a set of criteria with which to evaluate success. We discuss the importance of goals further in the "Design Consideration" section below.

Mitigation under Section 404 has not been the only reason for creating or restoring fringe wetlands. They have been created by state Departments of Natural Resources or Conservation to improve fish and wildlife habitat. Wildlife habitat improvement projects have been undertaken for at least 50 years, given the strong incentive from the commercial and recreational interest in fishing, hunting, trapping, and bird watching throughout the Midwest.

Projects such as these have apparently been "successful", as measured by an increased density of commercial or game fish, fish size, the number of breeding pairs of waterfowl, or the total hunting hours provided by an area. Water level control capability is perhaps the most important factor contributing to the success of fish or waterfowl habitat improvement projects. Without water level control, a project may fail to establish or maintain desired vegetation, adequate water quality, desired shoreline lengths, or optimum open water/vegetation ratio.

Shoreline stabilization has also been a goal of fringe wetland creation or restoration. Although shoreline stabilization projects are well documented for marine fringe wetlands, there is little information concerning freshwater fringe wetlands, particularly in the Midwest. Wave energy and water levels determine the success of these projects. Garbisch (1977) discourages planting along a shoreline that is exposed to more than a 13 km fetch in marine systems and recommends this as a "rule of thumb" for freshwater systems. Above this value, the wave energy becomes too great. A project on two South Dakota reservoirs displayed the temporary nature of success on reservoirs with large water level fluctuations (Hoffman 1977). Vegetation established to protect the shorelines was destroyed during an extended high water period. This project indicated that maintaining long-term shoreline stabilization may require periodic revegetation and the use of species adapted to water level fluctuations.

Slurry pond reclamation has also become a goal of fringe wetland creation and restoration projects in the Midwest. This is due mainly to the recent reclamation laws. Coal mining practices have generally been considered environmentally adverse. Klimstra and Nawrot (1985), however, claimed that mining activities have added to the total wetland area in Illinois and the Midwest in general. Since the 1800's, more than 5,666 hectares of wetlands have resulted directly from coal mining. These wetlands have become established through natural colonization. Because of this, mining companies are now reclaiming their slurry ponds by creating wetlands. The primary cause of failure to artificially revegetating these ponds is the use of commercial stocks which are not ecotypically adapted to the harsh environment of slurry ponds. The Cooperative Wildlife Research Laboratory in Carbondale, Illinois, has developed a nursery for wetland plants adapted to slurry pond environments. This has greatly increased the success of these projects.

For complete description of each project, see Appendix II.

DESIGN CONSIDERATIONS

Wetland creation and restoration projects initiated as mitigation for wetland losses should incorporate a number of design considerations into the actual permit. These can be categorized as: 1) pre-construction considerations; 2) critical aspects of the plan itself; and 3) monitoring of the project. The actual specifics will depend on the project goals. Hence, the project goals must be clearly stated in the permit and specific enough to provide a set of criteria with which to evaluate success. Specific goals will determine wetland configuration, how and what vegetation to plant, timing of construction, and ultimately every phase of a project. For example, a project with a goal of creating five one hectare plots of food habitat for canvasback ducks will dictate the use of different plants and planting times and different morphometric requirements than a project whose goal is to create a fringe wetland to protect one kilometer of shoreline. Precisely stated goals also help to define a monitoring plan. If the lake or reservoir has a lake association, the association should be consulted when the permittee is identifying goals of the project.

Having stated the goals of a project, the permit should include the following:

- o A delineation of the location of the mitigation site;
- o A description of the characteristics of the site before and after mitigation including: substrate, elevations (1 ft. intervals), water levels, and range of water level fluctuations;
- o Schedule of excavation work;
- o A description of the vegetation plans including species (scientific and common names), propagules used, source of propagules, planting densities, time of planting, water depth of planting, and type of substrate planted. A map of the plant zones should also be included;

- o Type, rate, and timing of fertilizer application;
- o Description of protective structures, including water level control structures, how and when they are installed, and if or when they are removed;
- o If applicable, a long-term management plan, including a water level management plan, how it will be financed, and who will be the responsible party;
- o A complete monitoring plan, including what will be measured, when and how often it will be measured, how the measurements will be taken, and how monitoring results will be interpreted; and
- o A list of criteria with which to judge success, including at least the % of successful plant establishments.

The permit for the Harbour Project on Sandusky Bay (U.S. Army Corps of Engineers (ACOE) Permit No. 82-475-3 revision A, 1982, Buffalo District) is a good example of what a permit should include. Design characteristics of the dike and wetland were complete, timing of each phase of the development was outlined, and monitoring aimed at determining the fish habitat value of the created wetland was required. The permit did lack a clear statement of the goals, however. Therefore, a monitoring program had no specific criteria with which to gauge success.

PRE-CONSTRUCTION CONSIDERATIONS

Location

Wetland creation projects can be located on- or off-site. With an off-site approach, the first step in the location process is to identify all available sites on the reservoir or lakeshore which are of sufficient size to meet project goals. With an on-site approach, the focus is upon the site itself. In selecting the actual site, the first choice should be an area which historically supported a wetland but was filled or lost due to water quality degradation. Removing the fill from an old wetland or improving water quality to restore suitable water conditions can sometimes make wetland restoration in this location easier than at a site where a wetland never existed. The second choice could be an area along the edge of a lake or within the lake itself which can be impounded. The Harbour Project and the Iowa Department of Conservation subimpoundments are good examples of the success of this technique. For a large lake or reservoir, the last location choice should be an upland site along the lakeshore which can be converted into a wetland via excavation. This method would likely require the construction of a permanent breakwater to provide a low wave-energy environment unless a natural cove, beach or barrier exists.

Site Characteristics

Having identified an appropriate site for wetland creation/restoration, the permittee must evaluate the site characteristics. These include substrate, hydrology, and water quality parameters. Substrates should be analyzed for texture, nutrient levels, and contamination by heavy metals or organics. When substrate is to be taken from the destroyed wetland, it also should be analyzed for the presence of a viable seed bank (as described below under "Revegetation"). Hydrologic characteristics which should be evaluated include wave actions and currents, water depth, potential degree of sheltering from wind and waves, and extent and periodicity of water level fluctuations. Anticipated water level fluctuations are of particular importance. Finally, water quality parameters such as turbidity, alkalinity, pH, and nutrient, heavy metal and organic contaminant concentrations should be considered.

Construction

Water level control capability is an important factor in many Midwest restoration or creation projects. Diking has been commonly used along the Great Lakes. Forty percent of the wetlands along the U.S. Lake Erie shoreline have been diked to allow water level control (Jaworski and Raphael 1979). Diking has also been performed by various state Departments of Natural Resources to create wildlife refuges and by hunting clubs to provide habitat for waterfowl. These efforts may have protected many of the Great Lakes wetlands from recent high lake levels.

Slope angles are important in dike construction. Slopes of 3:1 to 5:1 have been successful in maintaining the integrity of the dike (Farmes 1985). In addition, the top and lakeward toe of the dike should be supported with riprap. The amount of riprap needed depends upon the magnitude and direction of the waves. Dikes for the Sandusky Bay project were constructed using these design characteristics.

For dredge spoil disposal operations, the size of the containment area must be determined. Size will depend upon the amount of spoil to be disposed, the nature of the spoil, and the type of environment to be created within the containment site.

Once dikes have been created, the area

within the dike must be contoured. A varied topography within the impounded area is preferred. In the Sandusky Bay project, the 37.2 ha impoundment was divided into four cells of various sizes. The bottom of the cells were contoured to create a mosaic of islands and channels using dredge material. The Iowa Department of Conservation (IDOC) employed a similar scheme for their subimpoundments on four large reservoirs (Moore and Pfeiffer 1985). The diked areas were developed to provide a habitat combination of 50% palustrine emergent wetland (PEM) and 50% open water (POW), creating a "hemi-marsh". This condition was generally sought as the ideal habitat for maximizing waterfowl use. IDOC also maximized the shoreline lengths within these areas to provide the greatest density of waterfowl breeding territories.

Protection from grazing animals may be necessary during establishment of vegetation. Hoffman (1977) mentions the use of a barbed-wire fence to prevent cattle from grazing on recently transplanted grass and emergent species. Furthermore, snow-fencing has commonly been used to exclude geese and muskrats from establishment areas.

Hydrology

The establishment and survival of wetland species is determined by water levels and fluctuations (Warburton et al. 1985, Keough 1987). Water levels must be considered when implementing both planting and wildlife management strategies.

Where water level fluctuations cannot be controlled, plant species must be chosen which are tolerant of anticipated fluctuations. Table 3 lists some Midwestern species and their water level fluctuation tolerances. The TVA uses Japanese millet, common buckwheat (Fagopyrum esculentum), and Italian ryegrass (Lolium multiflorum) in their reservoirs where water level fluctuations are severe (Fowler and Maddox 1974, Fowler and Hammer 1976).

Water levels can be manipulated in some instances to provide a desirable habitat for target species (Knighton 1985). The IDOC manipulated water levels in four subimpoundments to provide water depths appropriate for establishing waterfowl food stock species (e.g., smartweed, Polygonum; sedges, Carex sp.; and millet, Echinochloa sp.) and for various species of waterfowl (e.g., giant Canada goose, Branta canadensis maxima).

Problems may be encountered if the water level control structures fail. For example, water levels became too high in a subimpoundment built at the Hawkeye Wildlife Area in Iowa (M.W. Weller, 1987, Texas A&M, pers. comm.). This resulted in an open water pond instead of a mixture of PEM and POW. This habitat failed to meet the goal of the project: the attraction of a large number of waterfowl.

The successful establishment of wetland vegetation is also influenced by waves and ice. The erosive forces of waves and ice can prevent plant establishment or destroy existing vegetation. Breakwater structures can lessen these forces. To provide a low wave-energy environment for soon-to-be-planted areas, the Wisconsin Department of Natural Resources (WDNR) has constructed two types of breakwaters (Berge 1987). A tire breakwater two meters wide (2 m = 7 tires) and 61 m long was placed in approximately one meter of water. The tires were not flush with the bottom and were slightly exposed at the surface. The structure allowed establishment of wild celery. A 30.5 m geoweb breakwater was also installed. Geoweb is a thick plastic honeycombed wall 6.1 m long, 1.2 m wide, and 20.3 cm thick. Each length is comprised of numerous 10.2 X 20.3 cm rectangular cells. The geoweb barrier was not successful in this application, because it was not heavy enough to remain stationary against two foot waves (Berge, 1988, WDNR, pers. comm.).

Substrate

Substrate also influences the success of plant establishment. Plants will fail if the substrate is too hard or too soft (Kadlec and Wentz 1974). The type of substrate also dictates which species will survive at a site. Tables 3 and 4 list some Midwestern wetland species and their preferences for both substrate moisture content and type.

Substrates may have to be prepared prior to planting. For example, Hoffman (1977) plowed and disked the substrate of 13 South Dakota reservoir sites to improve the substrate's physical characteristics. Fertilizer applications may also be needed to overcome substrate nutrient deficiencies. The TVA applied a N-P-K (6-12-12) fertilizer to the exposed mud banks and slopes of their reservoirs during seeding (Fowler and Maddox 1974, Fowler and Hammer 1976). Local U.S. Department of Agriculture Extension Agencies may be helpful in determining appropriate nutrient ratios for fertilizers for each site.

Two techniques have been used to create suitable substrate for vegetation. An Army Corps of Engineers (St. Louis District) project on Carlyle Lake, Illinois, used tire breakwaters to accumulate sediment in an area where bulrush was to be planted. However, the success of these structures has been varied and is not well documented. In Sandusky Bay, dredge materials

Table 3. Aquatic macrophytes tolerant of water level fluctuation. Based on Martin and Uhler (1939).

Common Names	Scientific Names
Soils Always Quite Moist	
Pondweed	*Potamogeton americanus*
Pondweed	*Potamogeton gramineus*
Bur-reeds	*Sparganium* sp.
Seaside arrow grass	*Triglochin maritimum*
Arrowheads	*Sagittaria* sp.
Fowl-meadow grass	*Glyceria striata*
Spike grasses	*Distichlis* sp.
Rice cutgrass	*Leersia oryzoides*
Square-stemmed spike-rush	*Eleocharis quadragulata*
Dwarf spike-rush	*Eleocharis parvula*
NA	*Eleocharis acicularis*
Olney three-square	*Scirpus americanus*
Hard-stemmed bullrush	*Scirpus acutus*
Beak rushes	*Rhynchospora* sp.
Pickeral weed	*Potendaria cordata*
Soils Sometimes Dry	
Switchgrass	*Panicum dichotomiflorum*
Water millet	*Echinochloa* sp.
Yellow nutsedge	*Cyperus esculentus*
Tearthumb/Smartweeds	*Polygonum* sp.

Table 4. Midwestern wetland species substrate and depth preferences and methods of propagation. Synthesized from Kadlec and Wentz (1974).

Species	Preferred Substrate[1]	Depth Range (cm)[2]	Methods of Propagation[3]
Annuals			
Bidens sp. O,Sa,Si	O, Sa, Si	-	S
Echinochloa sp.	Sa	30	S
Najas flexilis	Bl,Si,Sa,OO	30 - 800	T,W,S
N. marina	OO	-	T,W,S
Zannichellia palustris	Si	30 - 150	S
Zizania aquatica	O,C	5 - 180	S
Perennials			
Brasenia schreberi	OO,O,L	<180	T,R,S
Ceratophyllum demersum	O	30 - 150	W
Distichlis spicata	Si-C,Sa,C	-	T,R,S
Elodea canadensis	OO,BO,C	30 - 300	T,W,C
Heteranthera dubia	O	-	T,S
Leersia oryzoides	--	wet soil	T,R
Myriophyllum alterniflorum	all	-	T,W,S,C
M. exalbescens	O,Sa	-	T,W,S,C
M. heterophyllum	O	-	T,W,S,C
Nuphar advena	OO	<300	T,R,S
N. variegatum	O,Sa,Si	100 - 300	T,R,S
N. rubrodiscum	O	-	T,R,S
N. microphyllum	O	<300	T,R,S
Nelumbo lutea	Si,C-O	30 - 150	T,Tu,S
Nymphaea odorata	OO,O,P,Sa	<300	T,R,S
Phragmites australis	BO,Sa,Si C,P-Sa	-100 - +200	T,R,S
Potamegeton amplifolius	O,Sa,Si	100 - 800	T,R,C,S
P. foliosus	--	60 - 180	T,S
P. gramineus	Sa,Si,O,G	0 - 800	T,R,C,S
P. natans	Bl,P,Sa,Si OO,Cp	90 - 120	T,R,S
P. pectinatus	Si,Sa,O	5 - 300	T,R,C,Tu,S
P. perfoliatus	C,Si,Bl,BO,Sa	60 - 240	T,R,C,S
P. pusillus	Sa,Si,O	60 - 800	T,S
P. spirillus	Sa,Si,O	0 - 300	T,W,S
P. zosteriformis	O	800	T,S
Ruppia maritima	Si,Sa,O	30-300 below MLW	T,R,C,S
Sagittaria latifolia	O	< 30	T,U,S
S. platyphylla	--	< 30	T,Tu,S
Scirpus acutus	all	<150	T,R,S
S. americanus	Sa,C	< 60	T,R,Tu,S
S. californicus	Sa,Si	<180	T,R,S
S. fluviatilis	--	< 50	T,R,Tu,S
S. olneyi	O,C	-7 - +120	T,R,S
S. robustus	O,C	-15 - +120	T,R,S
S. validus	Sa,C,Ma	<120	T,R,S
Sparganium americanum	--	< 30	T,R,S
S. eurycarpum	--	<120	T,R,S
S. fluctuans	OO	<180	T,R,S
S. minimum	BO,P,Bl	-	T,R,S
Typha augustifolia	O,Bl	100	T,R,S
T. glauca	O	< 60	T,R,S
T. latifolia	P,O	< 30	T,R,S
Vallisneria americana	Sa-O,Si	30 - 300	T,Tu,S
Zostera marina	SM,Sa,G	30 - 180	T,R,S
Annuals and Perennials			
Carex sp. P,O,BO	< 15	T,R,S	
Cyperus sp.	Sa,C,Si-L	< 30	T,W,R,Tu,S
Unknown			
Eleocharis acicularis	P,O,Sa,Si	<120	T,R,Tu,S
E. equisetoides	O	-	T,R,Tu,S
Ele. palustris	Sa,Si	< 50	T,R,Tu,S
E. parvula	--	wet soil	T,R,Tu,S
Lobelia dortmanna	Si,BO,Sa,P	10 - 240	--

[1] Substrate Legend:
Bl = Black mud; BO = brown mud; C = clay; G = gravel; L = loam;
Ma = marl; O = organic; OO = ooze; P = peat; Sa = sand;
Si = silt; SM = sandy mud

[2] Depth Legend: MLW = mean low water

[3] Propagation Legend:
C = cuttings; R = root stocks and rhizomes; S = seeds;
T = transplants; Tu = tubers; W = whole plants and fragments

were pumped from a shallow open-water area within the Bay into a nearby diked containment area where a wetland was constructed. This provided a useful disposal site for the spoils, suitable planting substrates for wetland species, and a seed bank for natural colonization.

Revegetation

There are two general techniques for establishing wetland vegetation at a mitigation site. One is natural colonization from air or waterborne seeds, invasion from adjacent areas, or seed banks within substrates transplanted from other sites. The other technique, artificial establishment, includes seeding and transplanting whole plants, shoots, rhizomes, or tubers (see Table 4). Natural colonization is inexpensive, and the plants that become established are generally ecotypically adapted to the environment of the site. This method, however, has several disadvantages. First, the site may be unvegetated for a period of time during which erosion can occur. Furthermore, the permittee has little control of species composition on the site. The site could become a monotypic stand (e.g., cattail). Therefore, this technique is usually inappropriate when project goals dictate specific target species. Artificial establishment is more expensive and time consuming, however, the permittee has greater control over species composition.

The only method of natural colonization that we will discuss is the seed bank method although other methods may be effective in a given circumstance. The seed bank method uses substrates from a destroyed wetland as the seed source for the wetland being created. This technique is only effective if the substrate has a viable seed bank. Wetland soils can contain 6,405-32,400 seeds m^{-2} in the top 10 cm (Leck and Graveline 1979). Similarly, van der Valk and Davis (1978) found densities ranging from 21,455-42,615 seeds m^{-2} in the top 5 cm of a marsh in Eagle Lake, Iowa. A total of 40 species were germinated under various water level conditions from substrate at Eagle Lake. This diversity of seeds means that a greater potential exists for wetland plants to become established under a variety of environmental conditions. Additionally, seeds can remain viable for up to 30 years (van der Valk, unpublished), this provides resiliency to the wetland. However, the presence of a seed bank in a wetland soil is not guaranteed (van der Valk, unpublished). Van der Valk and Davis (1978) described methods for determining species composition, presence, and viability of seeds.

Artificial establishment is the other method of revegetation. When using this method, the permittee should choose species for planting which are best suited for the environmental conditions found at the site. The Illinois Department of Conservation (1981) published a catalogue of plant species used for habitat restoration in Illinois. This catalogue describes the preferred habitat of some wetland species, including substrates and water depth. It can be used as a guide for species selection throughout the Midwest. Tables 3 and 4 of this publication also list habitat preferences for fringe wetland species.

The permittee's second consideration should be the availability of the planting stock for these species. Types of stock include seeds, rhizomes, tubers, and whole plants. The Illinois Department of Conservation (1981) catalogue also lists nurseries where planting stocks are available.

The Cooperative Wildlife Research Laboratory (CWRL) has demonstrated the importance of using local stock or stocks from a similar environment for revegetation projects (Nawrot 1985, Warburton et al. 1985, and Klimstra and Nawrot 1985). CWRL conducted research on factors which influenced the successful establishment of wetland vegetation in 12 slurry ponds. They found that individuals which naturally colonized the slurry ponds were ecotypically adapted to the conditions found in these ponds (see Nawrot 1985 for hydrogeochemical characteristics). When transplanted to other slurry ponds, these individuals became established more successfully than did individuals of the same species from commercial stocks. For example, rhizomes of hardstem bulrush (Scirpus acutus) collected from slurry ponds had significantly greater survival and produced greater growth and spreading rates than rhizomes from commercial stocks (Warburton et al 1985). CWRL has developed populations of hardstem bulrush, threesquare bulrush (Scirpus americanus) and prairie cordgrass (Spartina pectinata) in a "nursery pond" to provide transplanting stock.

Environments which are less harsh than slurry ponds can also produce ecotypic adaptations in individuals. For example, Keough (1987) described ecotypic variation along a depth gradient in soft-stemmed bulrush. Using individuals which are ecotypically adapted to such site conditions as water depth and quality will increase the success of artificial establishment. Therefore, we recommend taking propagules from the same water body and at the same depth range where they will be transplanted, whenever possible. Using local stocks will also reduce the loss of propagules via long-distance transport (Kadlec and Wentz 1974).

The availability of suitable stock will restrict the permittee's choice of propagule type.

Site characteristcs are another factor. Only certain species types are possible given the life history characteristics of the desired species (see Table 4). The size of the site can also influence propagule choice (Kadlec and Wentz 1974). If the area is large, transplanting may be prohibitively expensive. Environmental conditions at the site must also be considered. For example, seeding may not be as successful as transplanting at a site subject to erosion, siltation, or wave action.

Once the permittee has chosen the desired species and the propagule type, he should outline a planting strategy. The planting method, density, and timing will determine the success of plant establishment. In the discussion below, we describe by propagule type the planting strategies used in several creation projects:

1. The Tennessee Valley Authority has seeded the shorelines of its reservoirs with waterfowl food species (Fowler and Maddox 1974, Fowler and Hammer 1976). The desired species were Japanese millet, common buckwheat, and Italian ryegrass. Seeds of these plants were broadcast onto exposed mud banks and on slopes with 20° to 45° angles using a floating aquaseeder. A commercial fertilizer (N-P-K, 6-12-12) and wood mulch were applied on sloping areas. Seeding was most successful when it occurred immediately after soil exposure. Wood mulch did not affect the success of plant establishment. As an alternative to the aquaseeder, helicopters were also used to broadcast seeds with the same success as was observed with the aquaseeder.

2. Wild rice was seeded in shallow areas of Lake Puckaway, Wisconsin, as a food source for waterfowl (Berge 1987). Berge did not provide planting densities or times. However, he did indicate the importance of water level control to the success of plant establishment. For this project, wild rice seeding was successful until water levels rose during the rice's floating leaf stage, at which time all stands were destroyed.

3. Hoffman (1977) seeded thirteen sites along the shorelines of two South Dakota reservoirs. The sites were plowed and disked in preparation for seeding. Desired species were crested wheatgrass (Agropyron cristatum), tall wheatgrass (A. elongatum), western wheatgrass (A. smithii), Garrison creeping foxtail, turkey brome (Bromus biebersteinii), basin wildrye (Elymus cinereus), switchgrass (Panicum virgatum), and reed canary grass. The sites were seeded in the spring and autumn. Seeds were placed in rows 0.1 m apart which ran perpendicular to the water line. Reed canary grass was successfully planted using these techniques, as measured by a minimum of 10% survival.

4. At the Harbour Project in Sandusky Bay, cells were seeded and mulched in the spring with reed canary grass, rice cut grass (Leersia oryzoides), and manna-grass (Glyceria canadensis). In some areas, natural vegetation had become established from the seed bank, and, therefore, these areas were not seeded (D. Wilson, 1988, ACOE Buffalo District, pers. comm.).

5. Berge (1987) outlines several techniques for planting tubers. He planted tubers of wild celery, wild rice, and sago pondweed (Potamogeton pectinatus) on Lake Puckaway on the leeward side of the breakwaters described under "Hydrology". Tubers placed in peat pot, clay balls, and polyethylene produce bags weighted with gravel were planted on several different sediment types. Single tubers with nails attached for weight were also planted. Berge recommends planting wild celery tubers in late April and early May in densities of 1000-5000 per 0.4 ha (1 acre), depending on wind conditions. The produce bag and nail methods proved to be most successful, whereas the peat pots and clay balls were not productive. Berge (1988, WDNR, pers. comm.) has since found that cotton mesh bags are also successful, and he prefers them. He recommends using bags that will maintain their structure for six to eight weeks to allow the tubers to establish. The best plant growth and survival occurred in areas with a firm bottom and with a sediment of sand and organic matter. Plantings of wild celery were successful despite waves of up to two feet in the lake. The tire breakwater protecting them was removed after three years, at which time the wild celery patch itself could dissipate wave energy itself (D. Berge, 1988, WDNR, pers. comm.).

5. Tubers in weighted mesh bags was also used to plant wild celery at a restoration project on Navigational Pool 8 on the Mississippi River near La Crosse, Wisconsin (Peterson 1987). Seventeen 0.4-ha plots were each planted during the spring with 8,000 tubers of wild celery. These tubers were harvested from nearby Lake Onalaska using a hydraulic dredge. Water depths in the planted plots ranged from one to two meters. The project was successful in establishing wild celery. Seventy to 75% of the tubers sprouted and produced new plants during the summer.

6. Warburton et al. (1985) found that perennial species which produce rhizomes or tubers were most successfully established in slurry

ponds. CWRL's greatest successes occurred when rhizomes were planted in the spring, although threesquare bulrush also did well when planted during summer (Warburton et al. 1985). Rhizomes were successful when planted at a depth of 5 to 13 cm. CWRL recommends spacing rhizomes at 0.3 to 1.5 m intervals, depending on species. Planting at 1.5 m intervals was sufficient for threesquare bulrush, which spread rapidly in one season. Hardstem bulrush was less prolific, and it required a planting density of 0.3 to 0.9 m to be successful.

7. Hoffman (1977) also used transplants to vegetate the South Dakota reservoirs. He used transplants of broad-leaved cattail, giant bulrush, common reed, reed canary grass, Garrison creeping foxtail, and western wheatgrass, placing them in rows 0.1 m apart which ran perpendicular to the water line. Garrison creeping foxtail, reed canary grass, common reed, giant bulrush, and broad-leaved cattail were successfully established, as measured by a minimum of 10% survival. In some areas, garrison creeping foxtail and reed canary grass spread into 1 hectare stands. All the sites flourished from one to three years. After the third year, water inundated the sites long enough to destroy the vegetation.

We feel a permittee should indicate which species are to be established and the methods he will use. These species should be shown to succeed in the environment they will be planted in, and the proposed planting method (including type of propagule and planting method, density, and timing) should be a proven method. Because of the lack of control under natural revegetation, we do not recommend this as a primary revegetation technique. We believe, however, that the seed bank method in particular can be effective when used concurrently with artificial establishment (e.g., transplants), a combination which could provide both immediate cover and a backup source of propagules. These propagules could ensure some type of vegetative cover if the transplants die as a result of changing environmental conditions.

Reintroduction of Fauna

In general, reintroduction of fauna should be passive. Lakes and reservoirs attract wildlife and usually have numerous fish species present. Target organisms will find the created wetland and should remain there if the habitat is suitable. Diked or impounded wetlands, however, exclude fish. To remedy this, fish ladders have been built into dikes to allow fish to move into an impounded wetland (Anderson 1985). On an impoundment on Cass Lake, Minnesota, two fish ladders were constructed to allow northern pike to travel to and from the impounded wetland, however in this case, the pike did not make use of the structure.

Long-term Management

Long-term management plans are a necessity, particularly when water level control structures are used. Drawdown can be used to effectively manage desired plant and animal species. The effect of lake drawdown on wetland plant establishment has been well documented (Salisbury 1970, Richardson 1975, Pederson 1979, Pederson 1981, Knighton 1985, Cooke et al. 1986). The timing of drawdown is also an important consideration. Cooke et al. (1986) provided a table of the response of 74 species to drawdown for an entire year, summer only, and winter only. Animal species are also affected by drawdown, both directly and indirectly due to plant species composition changes (Weller 1981, Jaworski et al. 1980). See Weller (this publication) for a more detailed discussion of drawdown.

Controlled burning is another long-term management method. It is especially useful for controlling nuisance plant species and for waterfowl habitat maintenance (Linde 1985). Burning can also be used to 1) remove the dead canopy of the previous summer's vegetation; 2) control woody vegetation in subimpoundments and dikes; 3) clean impoundment basins after drawdown and prior to reflooding; and 4) produce more nutritious and palatable forage for early spring use by waterfowl. The timing of burning is important, and is best in late summer or early fall where hunting is not a consideration. Otherwise, winter or early spring burns can be used, if done in advance of snows or after snow melt. Burning should never be done in April, May, or June, since this is the nesting period for most waterfowl.

Animal species can also be nuisances, and their population numbers may need to be controlled. For example, Berge (1987) removed carp (*Cyprinus carpio*) from Lake Puckaway to improve water quality. This increased the success of wild celery establishment. An active trapping program initiated on the Sandusky Bay Harbour Project successfully controlled the muskrat population, which had threatened the integrity of the dikes.

We feel that the a long term management plan should include the identification of responsible parties and a financial plan to help ensure that long-term management is initiated and continued at the mitigation site. For example, the permit for the Harbour Project on Sandusky Bay stated that the applicant or a designated party (e.g., the City of Sandusky) was responsible for overseeing and funding water level management.

MONITORING

MONITORING

A notable omission in all the permits we reviewed was a monitoring plan. A monitoring plan is essential to determine the success of a project and to evaluate the need for mid-course corrections. The plan must describe what is monitored, how it is monitored, how long it is monitored, and how the monitoring results will be interpreted.

The plan must first outline those parameters of the wetland which will be monitored. If the goal of the project is to create a wetland to compensate for one that is to be destroyed, then vegetation survival and growth may be the chosen parameters. Specifically, the vegetation should be monitored for species composition, percent survival of plants over time, percent cover relative to the size of the area seeded, the rate of expansion of the planted area, ratio of vegetation to open water, or whether the stand has coalesced.

Next, the plan must describe how the monitoring will be carried out. For example, for vegetation, field measurements can be made using a simple point grid system covering the planted area. To determine percent survival, a count of successful plants per grid could be compared to planting densities for each target species. In addition, the presence of non-target species could also be noted to identify future problems. In large areas where sampling the entire area is unfeasible, random samples could be taken.

Monitoring should occur quarterly for the first year and in early summer and early fall for at least the following two years. Additionally, monitoring should occur immediately after large storm events, particulary in undiked fringe wetlands. When a temporary breakwater is used, monitoring should continue at least one year after the breakwater is removed.

Results from the monitoring program can be compared to criteria established in the permit. These criteria would be based upon the successes exhibited in similar projects. For example, 70-75% of the wild celery tubers planted in one to two feet of water were successfully established in Navigational Pool 8 on the Mississippi River (Peterson 1987). A permittee should be expected to achieve similar success rate given similar circumstances. In harsher environments, a lower success rate may be permissible and, in fact, should be expected.

Additional parameters should be included in the monitoring plan which reflect the goals of the project. If a wetland is being created to provide fish or waterfowl habitat, these organisms should be monitored. This can be done directly by counting individuals or indirectly by counting such surrogates as nests. Sampling schemes such as mark and recapture can be used. As mentioned above, the results should be compared to expected values from other projects or as noted in wetland literature.

MID-COURSE CORRECTIONS

Many things can go awry during the creation of a wetland. Unexpected water level changes is one example. The dikes on the Harbour Project, in Sandusky Bay, were threatened by high lake levels in Lake Erie in 1986-87. To correct this problem, an emergency permit was issued which allowed for reinforcement of the dikes. The wetlands within the impounded area were thus saved. The opposite problem was encountered on an Iowa subimpoundment, which failed to fill with water during a spring with little rainfall (Moore and Pfeiffer 1985). The Iowa DOC pumped water from a nearby river to provide the pool level dictated by management plans. This correction allowed the establishment of the target plant species.

Another example of how plans can go awry is a revegetation plan that was too successful. Berge (1988, WDNR, pers. comm.) suggested that the wild celery became too dense in some of the Wisconsin Department of Natural Resources projects. Although this may be a wetland restorers dream, it can cause problems. Engel (1984, 1985) suggested that a completely vegetated shoreline is not the best fish habitat. He has experimented with harvesting and screening vegetation to open feeding lanes for predatory fish in two Wisconsin lakes. On Halverson Lake, Engel harvested 50-70% of the vegetation to a depth of 1.5 m in June. The vegetation grew rapidly after this harvest and nearly reached preharvest densities by July, when it was harvested again. Vegetation recovery was slower after this July harvest. Macrophyte species composition changed in the harvested areas from the original dominants of Berchtold's pondweed (<u>Potamogeton berchtoldii</u>) and sago pondweed to curly-leaf pondweed (<u>P. crispus</u>), coontail (<u>Ceratophyllum demersum</u>), bushy pondweed (<u>Najas flexilis</u>), and water stargrass (<u>Heteranthera dubia</u>). After harvesting, bluegills (<u>Lepomis macrochirus</u>) continued to feed on aquatic insects in the macrophyte beds but also consumed macroinvertebrates made available by the disruption of the vegetation. Largemouth bass used the harvested channels to cruise for prey, and some individuals were larger than 40 cm. On Cox Hollow Lake, Engel placed screens on the lake bed in May and left them in the lake for one to three years. After one year, some screens began accumulating sediment and supporting vegetation, and required some maintenance. This technique was successful, for few plants grew back after screen removal. Harvesting and screening may also provide a method of maintaining optimum vegetation to open water ratios in waterfowl habitat.

INFORMATION GAPS AND RESEARCH NEEDS

Although a fair amount of information is available about fringe wetlands of the Midwest, further research is needed. For example, relatively few studies of nutrient cycling include a water budget. This precludes the calculation of a true nutrient mass balance. These calculations are necessary to determine whether a wetland functions as a nutrient source or sink. More research is also needed to determine how species (when alive and during decomposition) affect nutrient cycling. For example, the American lotus has spread throughout the Midwest, and little is known of its nutrient cycling functions.

Information is also needed to understand further how water level fluctuations influence species composition and nutrient cycling in fringe wetlands. Although information exists to show how fringe wetlands protect shorelines, this effect needs to be quantified. Research should be conducted to compare shoreline area loss between a vegetated and nonvegetated shoreline and to determine which plant species or communities provide the best protection. Finally, more local nurseries should be developed to provide wetland planting stock which is ecotypically adapted to Midwestern ecoregions.

ACKNOWLEDGEMENTS

We would like to thank M. Beth Levine for her unfathomable patience. Her technical writing skills were much needed and greatly appreciated.

LITERATURE CITED

Allen, H.H. 1978. Role of wetland plants in erosion control of riparian shorelines, p. 403-414. In P.E. Greeson, J.R. Clark, and J.E. Clark (Eds.), Wetland Function and Values: The State of Our Understanding. American Water Resources Association, Minneapolis, Minnesota.

Amundson, R.W. 1970. Nutrient availability of a marsh soil. M.S. Thesis. Department of Water Chemistry, University of Wisconsin, Madison, Wisconsin.

Anderson, G.R. 1985. Design and location of water impoundment structures, p. 126-129. In M.D. Knighton (Ed.), Water Impoundments for Wildlife: A Habitat Management Workshop. U.S. Forest Service, St. Paul, Minnesota.

Banus, M.D., I. Veliela, and J.M. Teal. 1975. Lead, zinc, and cadmium budgets in experimentally enriched salt marsh ecosystems. Estuarine and Coastal Marine Science 3:421-430.

Barnes, M.D. 1980. Appendix B, p. 340-363. In C.E. Herdendorf and S.M. Hartley (Eds.), Fish and Wildlife Resources of the Great Lakes Coastal Wetlands within the United States. Center for Lake Erie Area Research, Technical Report Number 170:1, Columbus, Ohio.

Beam, J.H. 1983. The effect of annual water level management on population trends on white crappie in Elk City Reservoir, Kansas. North Amer. J. Fish. Mgmt. 3:34-40.

Bender, M.E. and D.L. Correll. 1974. The Use of Wetlands as Nutrient Removal Systems. CRC Publication No. 29. Bentley, E.M., III. 1969. The effect of marshes on water quality. Ph.D. Thesis. Department of Water Chemistry, University of Wisconsin, Madison, Wisconsin.

Berge, D.A. 1987. Native plants, fish introduced in a lake restoration effort (Wisconsin). Note #30. Restor. Mgt. Notes 5:1:35.

Boyd, C.E. 1970a. Production, mineral nutrient absorption, and pigment concentrations in Typha latifolia and Scirpus americanus. Ecology 51:285-290.

Boyd, C.E. 1970b. Losses of mineral nutrients during decomposition of Typha latifolia. Arch. Hydrobiol. 66:511-517.

Brady, N.C. 1984. The Nature and Properties of Soils.

Macmillan Publishing Comp., New York.

Braun, E.L. 1950. Deciduous Forests of Eastern North America. Blakiston, Philadelphia, Pennsylvania.

Carbine, W.F. 1943. Egg production of the northern pike, Esox lucius L., and the percentage survival of eggs and young on the spawning grounds. Mich. Acad. Sci. Arts and Letters 20:123-137.

Chamie, J.P.M. 1976. The Effects of Simulated Sewage Effluent upon Decomposition, Nutrient Status, and Litterfall in a Central Michigan Peatland. Ph.D. Thesis, University of Michigan, Ann Arbor, Michigan.

Cooke, G.D., E.B. Welch, S.A. Peterson, and P.R. Newroth. 1986. Lake and Reservoir Restoration. Butterworth Publishers, Boston.

Cowardin, L.M., V. Carter, F.C. Golet, and E.T. LaRoe. 1979. Classification of Wetlands and Deepwater Habitats of the United States. Fish and Wildlife Service OBS-79/31. U.S. Department of the Interior, Washington, D.C.

Davis, C.B. and L.A. Harris. 1978. Marsh plant production and phosphorus flux in Everglades Conservation Area 2, p. 105-131. In M.A. Drew (Ed.), Environmental Quality Through Wetlands utilization. Coordinating Council on the Kissimmee River Valley and Tayler Creek-Nubbin Slough Basin, Tallahasee, Florida.

Davis, C.B. and A.G. van der Valk. 1978. Litter decomposition in prairie glacial marshes, p. 99-113. In R.E. Good, D.F. Whighman, and R.L. Simpson (Eds.), Freshwater Wetlands: Ecological Processes and Management Potential. Academic Press, New York.

Dean, R.G. 1978. Effects of vegetation on shoreline erosional processes, p. 415-426. In P.E. Greeson, J.R. Clark, and J.E. Clark (Eds.), Wetland Function and Values: The State of Our Understanding. American Water Resources Association. Minneapolis, Minnesota.

Edzwald, J.K. 1977. Phosphorus in aquatic systems: The role of the sediments, p 183-214. In I.H. Suffet (Ed.), Fate of Pollutants in the Air and Water Environments, Part 1, Vol. 8. John Wiley and Sons, New York, NY.

Engel, S. 1984. Restructuring littoral zones: a different approach to an old problem, p. 463-466. In J. Taggart (Ed.), Lake and Reservoir Management: Proceedings of the North American Lake Management Society. U.S. EPA Office of Regulations and Standards, Washington, D.C.

Engel, S. 1985. Aquatic community interactions of submerged macrophytes. Technical Bulletin No. 156, Wisconsin Department of Natural Resources, Madison, Wisconsin.

Farmes, R.E. 1985. So you want to build a water impoundment, p. 130-134. In M.D. Knighton (Ed.), Water Impoundments for Wildlife: A Habitat Management Workshop. U.S. Forest Service, St. Paul, Minnesota.

Fenneman, N.M. 1970. Map of the Physical Divisions of the United States, p. 60. In The National Atlas of the United States of America. Department of Interior, Geological Survey, Washington, D.C.

Fowler, D.K. and D.A. Hammer. 1976. Techniques for establishing vegetation on reservoir inundation zones. J. of Soil and Water Conservation. 31:116-118.

Fowler, D.K. and J.B. Maddox. 1974. Habitat improvement along reservoir inundation zones by barge hydroseeding. J. of Soil and Water Conservation 29:263-265.

Garbisch, E.W., Jr. 1977. Marsh development for shore erosion control, p. 77-91. In P.L. Wise (Ed.), Proceedings of the Workshop on the Role of Vegetation in Stabilization of the Great Lakes Shoreline. Great Lakes Basin Commission, Ann Arbor, Michigan.

Great Lakes Basin Commission. 1975a. Wildlife, Appendix 17 of the Great Lakes Basin Framework Study, Ann Arbor, Michigan.

Great Lakes Basin Commission. 1975b. Shore Use and Erosion, Appendix 12 of the Great Lakes Basin Framework Study, Ann Arbor, Michigan.

Hassler, T.J. 1970. Environmental influences on early development and year class strength of northern pike in Lakes Oahe and Sharpe, South Dakota. Trans. Amer. Fish. Soc. 9:369-380.

Hoffman, G.R. 1977. Artificial establishment of vegetation and effects of fertilizer along shorelines of Lakes Oahe and Sakakawea, mainstem Missouri River reservoirs. In P.L. Wise (Ed.), Proceedings of the Workshop on the Role of Vegetation in Stabilization of the Great Lakes Shoreline. Great Lakes Basin Commission, Ann Arbor, Michigan.

Illinois Department of Conservation. 1981. Illinois Plants for Habitat Restoration. Illinois Department of Conservation, Mining Program. Springfield, Illinois.

Jaworski, E., C.N. Raphael. 1978. Fish, Wildlife, and Recreational Values of Michigan's Coastal Wetlands. U.S. Fish and Wildlife Service, Twin Cities, Minnesota.

Jaworski, E. and C.N. Raphael. 1979. Mitigation of fish and wildlife habitat losses in Great Lakes coastal wetlands. In G.A. Swanson (Ed.), Proceedings from the Mitigation Symposium: A National Workshop of Mitigating Losses of Fish and Wildlife Habitats. U.S. Forest Service, Washington, D.C.

Jaworski, E., J.P. McDonald, S.C. McDonald, and C.N. Raphael. 1980. Overview of fish and wildlife resources, p. 54-95. In C.E. Herdendorf and S.M. Hartley (Eds.), Fish and Wildlife Resources of the Great Lakes Coastal Wetlands within the United States. Center for Lake Erie Area Research, Technical Report Number 170:1, Columbus, Ohio.

Kadlec, R.H. and J.A. Kadlec. 1978. Wetlands and water quality, p. 436-456. In P.E. Greeson, J.R. Clark, and J.E. Clark (Eds.), Wetland Function and Values: The State of Our Understanding. American Water Resources Association.

Minneapolis, Minnesota.

Kadlec, J.A. and W.A. Wentz. 1974. State-of-the-Art Survey and Evaluation of Marsh Plant Establishment Techniques: Induced and Natural. Volume 1: Report of Research. U.S. Army Corps of Engineers Research Center, Fort Belvoir, Virginia.

Keith, W.E. 1975. Management by water level manipulation, p. 489-497. In H. Clepper (Ed.), Black Bass Biology and Management. Sport Fishing Institute, Washington, D.C.

Kelley, J.C. 1985. The role of Emergent Macrophytes to Nitrogen and Phosphorus Clycling in a Great Lakes Marsh. Ph.D. Thesis. Department of Fisheries and Wildlife, Michigan State University, East Lansing, Michigan.

Kelley, J.C., T.M. Burton, and W.R. Enslin. 1985. The effects of natural water level fluctuations on N and P cycling in a Great Lakes marsh. Wetlands 4:159-175.

Keough, J.R. 1987. Response by Scirpus validus to the physical environment and consideration of its role in a Great Lakes estuarine system. Ph.D. Thesis, University of Wisconsin, Milwaukee, Wisconsin.

King, D.L. 1985. Nutrient cycling by wetlands and possible effects of water levels, 69-86. In H. H. Prince, and F.M. D'Itri (Eds.), Coastal Wetlands. Lewis Publishers, Inc., Chelsea, Michigan.

Klimstra, W.D. and J.R. Nawrot. 1985. Wetlands as a by-product of surface mining: midwest perspective, p. 107-119. In R.P. Brooks, D.E. Samuel, and J.B. Hill (Eds.), Proceedings of the Wetlands and Water Management on Mined Lands Conference, Penn State University, State College, Pennsylvania.

Klopatek, J.M. 1975. The role of emergent macrophyte in mineral cycling in a freshwater marsh, p. 359-393. In F.G. Howell, J.B. Gentry, and H.M. Smith (Eds.), Mineral Cycling in Southeastern Ecosystems. ERDA Symposium Series (CONF-740513).

Klopatek, J.M. 1978. Nutrient dynamics of freshwater riverine marshes and the role of emergent macrophytes, p. 195-216. In R.E. Good, D.F. Whighman, and R.L. Simpson (Eds.), Freshwater Wetlands: Ecological Processes and Management Potential. Academic Press, New York.

Knighton, M.D. 1985. Vegetation management in water impoundments: water-level control, 39-50. In M.D. Knighton (Ed.), Water Impoundments for Wildlife: A Habitat Management Workshop. U.S. Forest Service, St. Paul, Minnesota.

Landers, D.H. and E. Lottes. 1983. Macrophyte dieback: Effects on nutrient and phytoplankton dynamics, p. 119-122. In J. Taggart (Ed.), Lake Restoration, Protection Management: Proceedings of the North American Lake Management Society. U.S. EPA Office of Water Regulations and Standards, Washington, D.C.

Leck, M.A. and K.J. Graveline. 1979. The seed bank of a freshwater tidal marsh. Amer. J. Bot. 66(9):1006-1015.

Lee, G.F., E. Bentley, and R. Amendson. 1975. Effects of marshes on water quality, p. 105-127. In A.D. Hasler (Ed.), Coupling of Land and Water Systems, Springer-Verlag, New York.

Linde, A.F. 1985. Vegetation management in water impoundments: alternatives and supplements to water-level control, 51-60. In M.D. Knighton (Ed.), Water Impoundments for Wildlife: A Habitat Management Workshop. U.S. Forest Service, St. Paul, Minnesota.

Liston, C.R. and S. Chubb. 1985. Relationships of water level fluctuations and fish, p. 121-140. In H.H. Prince and F.M. D'Itri (Eds.), Coastal Wetlands. Lewis Publishers, Inc., Chelsea, Michigan.

Loucks, O.L. 1981. The littoral zone as a wetland: Its contribution to water quality, p. 125-137. In B. Richardson (Ed.), Selected Proceeding of the Midwest Conference on Wetland Values and Management. Minnesota Water Planning Board, St. Paul, Minnesota.

Lunz, J.D., T. W. Zeigler, R.T. Huffman, R.J. Diaz, E.J. Clarrain, and L.J. Hunt. 1978. Habitat Development Field Investigations, Windmill Point Marsh Development Site, James River, Virginia; Summary Report. U.S. Army Engineer Waterways Experiment Station, Vicksburg, Mississippi.

Martin, A.C. and F.M. Uhler. 1939. Food of Game Ducks of the United States and Canada. U.S.D.A. Technical Bulletin No. 634.

McNicholl, M.K. 1979. Destruction to nesting birds on a marsh bay by a single storm. Prairie Nat. 11:60-62.

McNicholl, M.K. 1985. Avian wetland habitat functions affected by water level fluctuations, p. 87-98. In H.H. Prince and F.M. D'Itri (Eds.), Coastal Wetlands. Lewis Publishers, Inc., Chelsea, Michigan.

Miranda, L.E., W.L. Shelton, and T.D. Bryce. 1984. Effects of water level manipulation on abundance, mortality, and growth of young-of-the-year largemouth bass in West Point Reservoir, Alabama-Georgia. N. Am. J. Fish. Mgmt. 4:314-320.

Mitsch, W.J. and J.G. Gosselink. 1986. Wetlands. Van Nostrand Reinhold Co., New York.

Moore, B.C., H.L. Gibbons, W.H. Funk, T. McKarns, T. Nyznyk, and M.V. Gibbons. 1984. Enhancement of internal cycling of phosphorus by aquatic macrophytes, with implication for lake management, p. 113-117. In J. Taggart (Ed.), Lake and Reservoir Management: Proceedings of the North American Lake Management Society. U.S. EPA Office of Water Regulations and Standards. Washington, D.C.

Moore, R. and D. Pfeiffer. 1985. Subimpoundments: New waterfowl habitat. Iowa Conservationist. 44:11: 2-5.

Nawrot, J.R. 1985. Wetland development on coal mine slurry impoundments: principles, planning, and practices, p. 161-172. In R. P. Brooks, D.E. Samuel, and J.B. Hill (Eds.), Proceedings of the Wetlands and Water Management on Mined Lands Conference. Penn State University, State College, Pennsylvania.

Nichols, D.S. 1983. Capacity of natural wetlands to remove nutrients from wastewater. J. Water Pollut. Control Fed. 55:495-505.

Nute, J.W. 1977. Mountain View Sanitary District Marsh Enhancement Pilot Program Progress Report No. 2. J. Warren Nute, Inc., San Rafael, California.

Pederson, R.L. 1979. Seed Bank Study of the Delta Experimental Ponds. Marsh Ecology Research Program 1979 Annual Report, Delta Waterfowl Research Station, Delta, Manitoba, Canada.

Pederson, R.L. 1981. Seed bank characteristics of the Delta Marsh, Manitoba: applications for wetland management, p. 61-69. In B. Richardson (Ed.), Selected Proceeding of the Midwest Conference on Wetland Values and Management. Minnesota Water Planning Board, St. Paul, Minnesota.

Peterson, L.N. 1987. Wild celery transplanted (Wisconsin). Note #31. Restoration and Management Notes 5:1:35-36.

Prentki, R.T., T.D. Gustafson, and M.S. Adams. 1978. Nutrient movements in lakeshore marshes, p. 169-194. In R.E. Good, D.F. Whighman, and R.L. Simpson (Eds.), Freshwater Wetlands: Ecological Processes and Management Potential. Academic Press, New York.

Prince, H.H. 1985. Avian response to wetland vegetative cycles, p. 99-120. In H.H. Prince and F.M. D'Itri (Eds.), Coastal Wetlands. Lewis Publishers, Inc., Chelsea, Michigan.

Richardson, L.V. 1975. Water level manipulation: a tool for aquatic weed control. Hyacith Control Journal 13:8-11.

Salisbury, E. 1970. The pioneer vegetation of exposed muds and its biological features. Philosophical Transactions of the Royal Society of London Series B. 259:207-255.

Schults, D.W. and K.W. Maleug. 1971. Uptake of radiophosphorus by rooted aquatic plants, p. 417. In D.J. Nelson (Ed.), Radionuclides in Ecosystems. Oak Ridge National Laboratory, Oak Ridge, Tennessee.

Seidel, K. 1976. Macrophytes and water purification, p. 109-121. In J. Tourbier and R.W. Pierson, Jr. (Eds.), Biological Control of Water Pollution. University of Pennsylvania Press, Philadelphia, Pennsylvania.

Semkin, R.G., A.W. McLarty, and D. Craig. 1976. A Water Quality Study of Cootes paradise. Water Resources Assessment, Technical Support Section, Canadian Ministry of the Environment.

Spangler, F., W. Sloey, and C.W. Fetter. 1976. Experimental use of emergent vegetation for the biological treatment of municipal wastewater in Wisconsin, p. 161-171. In J. Tourbier and R.W. Pierson, Jr. (Eds.), Biological Control of Water Pollution. University of Pennsylvania Press, Philadelphia, Pennsylvania.

Sullivan, P. 1976. Versatile wetlands - An endangered species. (reprint from) Conservation News. 42:20:2-5.

Tilton, D.L. and V.R. Schwegler. 1978. The values of wetland habitat in the Great Lakes Basin, p. 267-277. In P.E. Greeson, J.R. Clark, and J.E. Clark (Eds.), Wetland Function and Values: The State of Our Understanding. American Water Resources Association. Minneapolis, Minnesota.

Tiner, R.W., Jr. 1984. Wetlands of the United States: Current Status and Recent Trends. U.S. Department of the Interior, Fish and Wildlife Service, Washington, D.C.

Toth, L. 1972. Reed control eutrophication of Balaton Lake. Water Res. 6:1533-1539.

van der Valk, A.G., and C.B. Davis. 1978. The role of seed banks in the vegetation dynamics of prairie glacial marshes. Ecology 59(2):322-335.

van der Valk, A.G., C.B. Davis, J.L. Baker, and C.E. Beer. 1979. Natural freshwater wetlands as nitrogen and phosphorus traps for land runoff, p. 457-467. In P.E. Greeson, J.R. Clark, and J.E. Clark (Eds.), Wetland Function and Values: The State of Our Understanding. American Water Resources Association. Minneapolis, Minnesota.

Warburton, D.B., W.B. Klimstra, and J.R. Nawrot. 1985. Aquatic macrophyte propagation and planting practices for wetland development, p. 139-152. In R.P. Brooks, D.E. Samuel, and J.B. Hill (Eds.), Proceedings of the Wetlands and Water Management on Mined Lands Conference, Penn State University, State College, Pennsylvania.

Weiler, P.R., E.S. Menges, D.R. Rogers, and O.L. Loucks. 1979. Equations, specifications, and selected results for a nominal simulation of the aquatic ecosystem model WINGRA III. On file at the Institute for Environmental Studies, University of Wisconsin, Madison, Wisconsin.

Weller, M.W. 1981. Freshwater Marshes: Ecology and Wildlife Management. University of Minnesota, Minneapolis, Minnesota.

Willenbring, P.R. 1985. Wetland treatment systems--Why do some work better than others?, p. 234-237. In J. Taggart (Ed.), Lake and Reservoir Management: Practical Applications, Proceedings of the North American Lake Management Society, U.S. EPA Office of Water Regulations and Standards, Washington, D.C.

Wolverton, B.C., R.M. Barlow, and R.C. McDonald. 1976. Application of vascular aquatic plants for pollution removal, energy, and food production in a biological system, p. 141-149. In J. Tourbier and R.W. Pierson, Jr. (Eds.), Biological Control of Water Pollution. University of Pennsylvania Press, Philadelphia, Pennsylvania.

APPENDIX I: ADDITIONAL READINGS

Barten, J. 1983. Nutrient removal from urban stormwater by wetland filtration: the Clear Lake restoration project, p. 23-30. In J. Taggart (Ed.), Lake Restoration: Protection and Management. Proceedings of the North American Lake Management Society. U.S. EPA Office of Water Regulations and Standards, Washington, D.C.

Bedinger, M.S. 1978. Relation between forest species and flooding, p. 427-435. In P.E. Greeson, J.R. Clark, and J.E. Clark (Eds.), Wetland Functions and Values: The State of Our Understanding. American Water Resources Association, Minneapolis, Minnesota.

Branch, W.L. 1988. Design and construction of replacement wetlands on lands mined for sand and gravel, p. 168-172. In J. Zelanzy, and J.S. Feierabend (Eds.), Increasing Our Nations Wetlands Resources. National Wildlife Federation, Washington, D.C.

Burton, T.M., D.L. King, and J.L. Ervin. 1979. Aquatic plant harvesting as a lake restoration technique, p. 177-111. In Lake Restoration: Proceedings of the North American Lake Management Society. U.S. Environmental Protection Agency Office of Water Planning and Standards, Washington, D.C.

Busch, W.D. and L.M. Lewis. 1983. Responses of wetland vegetation to water level variations in Lake Ontario, p. 519-523. In J. Taggart (Ed.), Lake Restoration: Protection and Management. Proceedings of the North American Lake Management Society. U.S. EPA Office of Water Regulations and Standards, Washington, D.C.

Carpenter, S.R. 1983. Submersed macrophyte community structure and internal loading: relationship to lake ecosystem productivity and succession, p. 105-111. In J: Taggart (Ed.), Lake Restoration: Protection and Management. Proceedings of the North American Lake Management Society. U.S. EPA Office of Water Regulations and Standards, Washington, D.C.

Clark, J. 1978. Fresh water wetlands: habitats for aquatic invertebrates, amphibians, reptiles, and fish, p. 330-342. In P.E. Greeson, J.R. Clark, and J.E. Clark (Eds.), Wetland Functions and Values: The State of Our Understanding. American Water Resources Association, Minneapolis, Minnesota.

Crowder, L.B. and W.E. Cooper. 1979. The effects of macrophyte control on the feeding efficiency and growth of sunfishes: evidence from pond studies, p. 251-268. In J.E. Breck, R.T. Prentki, and O.L. Loucks (Eds.), Aquatic Plants, Lake Management and Ecosystem Consequences of Lake Harvesting. Institute of Environmental Studies, University of Wisconsin, Madison.

DeMarte, J.A. and R.T. Hartman. 1974. Studies on absorbtion of ^{32}P, ^{59}Fe, and ^{45}CA by water-milfoil (Myriophyllum exalbescens Fernald). Ecology 55:188-194.

Eckert, J.W., M.L. Giles, and G.M. Smith. 1978. Concepts for In-Water Containment Structures for Marsh Habitat Development. Technical Report D-78-31, Waterways Experiment Station, US ACOE, Vicksburg, Mississippi.

Emerson, F.B., Jr. 1961. Experimental establishment of food and cover plants in marshes created for wildlife in New York state. N.Y. Fish and Game Journal 55:130-144.

Fetter, C.W., Jr., W.E. Sloey, and F.L. Spangler. 1978. Use of a natural marsh for wastewater polishing. J. Water Pollut. Control Fed. 50:290-307.

Frederickson, L.H. and T.S. Taylor. 1982. Managment of Seasonally Flooded Impoundments for Wildlife. U.S. Fish and Wildlife Service Research Publication 148.

Freidman, R.M., C.B. DeWitt. 1978. Wetlands as carbon and nutrient reservoirs: a spatial, historical, and societal perspective, p. 175-185. In P.E. Greeson, J.R. Clark, and J.E. Clark (Eds.), Wetland Functions and Values: The State of Our Understanding. American Water Resources Association, Minneapolis, Minnesota.

Glooschenko, W.A., and I.P. Martini. 1987. Vegetation of river-influenced coastal marshes of the southwestern end of James Bay, Ontario. Wetlands 7:71-84.

Godshalk, G.L. and R.G. Wetzel. 1978. Decomposition in the littoral zone of lakes, 131-143. In R.E. Good, D.F. Whighman, and R.L. Simpson (Eds.), Freshwater Wetlands: Ecological Processes and Management Potential. Academic Press, New York.

Hall, V.L. and J.D. Ludwig. 1975. Evaluation of Potential Use of Vegetation for Erosion Abatement A Long the Great Lakes Shoreline. Miscalaneous Paper 75-7, Coastal Engineering Research Center, US ACOE, Fort Belvoir, Virginia.

Haller, W.T., J.V. Shireman, and F.F. DuRant. 1980. Fish harvest resulting from mechnical control of hydrilla. Trans. Amer. Fisheries Soc. 109:517-520.

Harris, S.W. and W.H. Marshall. 1963. Ecology of water-level manipulations on a northern marsh. Ecology 44(2):331-343.

Huff, D.D. and H.L. Young. 1980. The effect of a marsh on runoff: I. a water-budget model. J. Environ. Qual., 9(4):633-640.

Jaynes, M.L. and S.R. Carpenter. 1985. Effects of submersed macrophytes on phosphorus cycling in surface sediment, p. 370-374. In J. Taggart (Ed.), Lake and Reservoir Management: Practical Applications, North American Lake Management Society. U.S. EPA Office of Water Regulations and Standards, Washington, D.C.

Jewell, W.J. 1971. Aquatic weed decay: dissolved oxygen utilization and nitrogen and phosphorus regeneration. J. Water Pollut. Control Fed. 43:1457-1467.

King, D.L. and T.M. Burton. 1980. The efficiency of weed harvesting for lake restoration, p. 158-161. In Restoration of Lakes and Inland Waters: Proceedings of the North American Lake Management Society. U.S. EPA Office of Water Regulations and Standards, Washington, D.C.

Kistritz, R.U. 1978. Recycling of nutrients in an enclosed aquatic community of decomposing macrophytes (Myriophyllum spicatum). Oikos 30:475-478.

Krull, J.N. 1970. Aquatic plant-macroinvertebrate associations and waterfowl. J. Wildlife Management 34(4):707-718.

Linde, A.F. 1969. Techniques for Wetland Management. Wisconsin Department of Natural Resources Report 45, Madison, Wisconsin.

McMullen, J.M. 1988. Selection of plant species for use in wetland creation and restoration, p. 333-337. In J. Zelanzy and J.S. Feierabend (Eds.), Increasing Our Nations Wetland Resources. National Wildlife Federation, Washington, D.C.

McNabb, C.D., Jr. and D.P. Tierney. 1972. Growth and Mineral Accumulation of Submerged Vascular Hydrophytes in Pleioeutrophic Environs. Institute of Water Resources Technical Report 26, Michigan State University, East Lansing, Michigan.

McRoy, C.P. and R.J. Barsdate. 1970. Phosphate absorbtion in eelgrass. Limnology and Oceanography 15:6-13.

Mickle, A.M. and R.G. Wetzel. 1978. Effectiveness of submersed angiosperm-epiphyte complexes on exchange of nutirents and organic carbon in littoral systems: I. Inorganic nutrients. Aquatic Botany 4:303-316.

Mikol, G.F. 1984. Effects of mechanical control of aquatic vegetation on biomass, regrowth rates and juvenile fish populations at Saratoga Lake, New York, p. 456-462. In J. Taggart (Ed.) Lake and Reservoir Management: Proceedings of the North American Lake Management Society. U.S. EPA Office of Regulations and Standards, Washington, D.C.

Nawrot, J.R., D.B. Warburton, and W.B. Klimstra. 1988. Wetland habitat development techniques for coal mine slurry impoundments, p. 185-193. In J. Zelanzy and J.S. Feierabend (Eds.), Increasing Our Nations Wetlands Resources. National Wildlife Federation, Washington, D.C.

Nichols, D.S. and D.R. Keeney. 1973. Nitrogen and phosphorus release from decaying water-milfoil. Hydrobiologica 42:509-525.

Orberts, G.L. 1981. Impact of wetlands on watershed water quality, p. 213-226. In B. Richardson (Ed.), Selected Proceeding of the Midwest Conference on Wetland Values and Management. Minnesota Water Planning Board, St. Paul, Minnesota.

Perry, J.J., D.E. Armstrong, and D.D. Huff. 1981. Phosphorus fluxes in an urban marsh during runoff, p. 199-211. In B. Richardson (Ed.), Selected Proceeding of the Midwest Conference on Wetland Values and Management. Minnesota Water Planning Board, St. Paul, Minnesota.

Reed, R. B. and D.E. Willard. 1986. Wetland Evolution in Midwestern reservoirs (unpublished), School of Public and Environmental Affairs, Indiana University, Bloomington, Indiana.

Storch, T.A. and J.D. Winter. 1983. Investigation of the Interrelationships Between Aquatic Weed Growth, Fish Communities, and Weed Management Practices in Chautaugua Lake. Environmental Resources Center, State University of New York, Fredonia, New York.

van der Valk, A.G. 1981. Succession in wetlands: a Gleasonian approach. Ecology 62:3:688-696.

Weller, M.W. 1978. Management of freshwater marshes for wildlife, p. 267-284. In R. Good, D. Whigham, and R. Simpson (Eds.), Freshwater Wetlands: Ecological Processes and Management Potential. Academic Press, New York.

Wile, I. 1978. Environmental effects of mechanical harvesting. J. Aquatic Plant Management 16:14-20.

APPENDIX II: PROJECT DESCRIPTIONS

THE HARBOUR PROJECT, SANDUSKY BAY, OHIO

Sandusky Bay on Lake Erie was the site of a large wetland mitigation project permitted by the Buffalo District Army Corps of Engineers (ACOE) (Permit No. 82-475-3 revision A) in 1982. Each step of the project was clearly outlined in the permit. The project goal was to construct a 37.2 ha wetland in Sandusky Bay to mitigate 5-6 ha of emergent wetland lost from dredging channels and constructing a marina and resort.

Complete construction plans were detailed in the permit. Dikes were constructed to impound a 37.2 ha area using 116,000 cubic meters (152,000 cu. yds) of imported material. The bayward face and the tops of the outer dikes were riprapped to ensure the integrity of the dikes during high water events. Over 306,000 cubic meters (400,000 cu. yds) of dredge material were pumped into this impounded area, which was divided into four cells of various sizes. As dictated by the permit, the bottom of each cell was contoured to create a mosaic of islands and channels.

Planting strategies were also outlined in the permit. The cells were mulched and seeded with reed canary grass (Phalaris arundinacea), rice cutgrass (Leersia oryzoides), and manna grass (Glyceria canadensis) in the spring. In some areas, natural vegetation had established from the seed bank, and therefore, these areas were not seeded.

The long-term management plan for the project was primarily for water level management to provide a suitable environment for plant establishment and to allow fish access to the wetland during critical stages in their lifecycles. The permit stated when and how water level was to be managed and identified who was responsible for overseeing and funding the management plan. No monitoring plan was outlined. However, the permit stipulated that research was to be conducted at the site. The research was to 1) document gross and net productivity in the system, 2) identify floral and faunal populations, and 3) determine the impact of carp (Cyprinus carpio) on the establishment of freshwater marshes. The research was to be conducted over a three year period.

The project has been considered a success by the ACOE and local users of the wetland. Wetland plants became established both naturally and as a result of seeding. The wetland provided such a favorable habitat for muskrat (Ondatra zibethica) that the species became a nuisance. Accordingly, an active muskrat trapping program was established. The wetland also provided habitat to bald eagles (Haliaeetus leucocephalus), which use the site for feeding.

This project was not without problems, however. High water levels in 1986-87 undermined the dikes. The ACOE issued emergency permits to allow the dikes to be strengthened. This mid-course correction succeeded in saving the wetlands.

Ironically, the original wetland had not been destroyed at the time of this writing.

Contact: Don Wilson
Army Corps of Engineers Buffalo District
Buffalo, NY 14207
(716) 876-5454

CARLYLE LAKE, ILLINOIS

We understand that Carlyle Lake in Illinois receives substantial pressure for recreation. To relieve this pressure, the St. Louis District of the ACOE has an active dredge-and-fill operation to maintain existing marinas and create new ones. Although they are exempt from a Section 404 permit, the District internally mitigates for wetlands lost due to their operations.

One such project was to create a 0.37 ha lacustrine and palustrine wetland to mitigate the loss of a 0.32 ha lacustrine and palustrine wetland due to marina development at another site on the lake. They modified the site by grading at least 25% of the banks and bottom with a 5:1 slope and the remainder with a 2:1 slope. The rationale for this was not provided. The banks were planted with smooth bromus (Bromus sp.), timothy grass (Phleum pratens), and ladino clover (Trifolium repens). The edge of the banks was planted with Scirpus sp. Vegetation establishment was failing in the summer of 1988 due to a drought.

At another site, tire breakwaters were constructed to accumulate sediment and thereby provide a suitable substrate for planting Scirpus sp. The success of these breakwaters has varied and was not well documented.

The ACOE has diked an incoming stream on Carlyle Lake to create a 4 ha Scirpus sp. nursery. The nursery will provide planting stocks for their numerous mitigation projects.

Contact: Pat McGinnis
Army Corps of Engineers St. Louis District
St. Louis, Missouri
(314) 263-5533

LAKE PUCKAWAY, WISCONSIN

The Wisconsin Department of Natural Resources (WDNR) implemented a plan to restore the natural vegetation and game fish in the lake after they noticed a decline in water quality, aquatic vegetation, and fish populations. A secondary project goal was to provide a food source for canvasback ducks (Aythya valisneria).

To reduce competition with desired fish species and improve water clarity, carp (Cyprinus carpio) were removed using drawdown, chemical treatment, and electric barriers. Drawdown also encouraged the growth of aquatic vegetation.

To provide a low wave-energy environment in which plants could become established, WDNR constructed two types of barriers, tire and geoweb. A tire barrier two meters wide (2 m = 7 tires) and 61 m long was placed in approximately one meter of water. The tires were not flush with the bottom and were slightly exposed at the surface. A 30.5 m geoweb breakwater was also installed. Geoweb is a thick plastic honeycombed wall 6.1 m long, 1.2 m wide, and 20.3 cm thick. Each length is comprised of numerous 10.2 X 20.3 cm rectangular cells. The geoweb barrier was not successful in this application, because it was not heavy enough to remain stationary against two foot waves.

Wild celery (Vallisneria americana), wild rice (Zizania aquatica), and sago pondweed (Potamogeton pectinatus) were planted on the leeward side of the barriers using several planting techniques. Specifically, tubers placed in peat pot, clay balls, and polyethylene produce bags weighted with gravel were planted on several different sediment types. Single tubers with nails attached for weight were also planted. Tubers were planted in late April and early May in densities of 1000-5000 per 0.4 ha (1 acre). Wild rice was also seeded in shallow areas.

The produce bag and nail methods proved to be most successful, whereas the peat pots and clay balls were not. The best growth and survival occurred in areas with a firm bottom and with a sediment of sand and organic matter. Wild rice seeding was successful until water levels rose during the floating leaf stage, which destroyed all stands. Plantings of wild celery behind the tire breakwater were successful despite waves of up to two feet in the lake. The tire barrier was removed after three years, at which time the wild celery patch could dissipate wave energy itself.

The overall project was successful, as measured by an establishment of target vegetation, increase in water clarity, and surviving populations of walleye (Stizostedion vitreum), northern pike (Esox lucius), largemouth bass (Micropterus salmoides), black crappie (Pomoxis nigromaculatus), yellow perch (Perca flavescens), and bluegills (Lepomis macrochirus).

Contact: Dale Berge
Wisconsin Department of Natural Resources
Montello, WI 53949
(608) 297-2888

COOPERATIVE WILDLIFE RESEARCH LABORATORY'S SLURRY POND PROJECTS, ILLINOIS AND INDIANA

The Cooperative Wildlife Research Laboratory (CWRL) has developed planting stock ecotypically adapted to slurry pond environments. CWRL developed this stock based upon their research on factors which influenced the successful natural establishment of wetland vegetation on 12 slurry pond wetlands totaling over 480 acres.

CWRL found that species which naturally colonized the slurry ponds were ecotypically adapted to the conditions found in these ponds. When transplanted to other slurry ponds, individuals of these species became established more successfully than did individuals from commercial stocks. For example, rhizomes of hardstem bulrush (Scirpus acutus) collected from slurry ponds had significantly greater survival and produced greater growth and spreading rates than rhizomes from commercial stocks. CWRL has developed populations of hardstem bulrush, threesquare bulrush (Scirpus americanus) and prairie cordgrass (Spartina pectinata) in a "nursery pond" to provide transplanting stock.

CWRL found that perennial species which produce rhizomes or tubers are most successfully established in these ponds. They have had the greatest success when rhizomes were planted in the spring, although threesquare bulrush also did well when planted during summer. Rhizomes were successful when planted at a depth of 5 to 13 cm and spaced at 0.3 to 1.5 m intervals, depending on species. Planting at 1.5 m intervals was sufficient for threesquare bulrush, which spread rapidly in one season. Hardstem bulrush was less prolific, and it required a planting density of 0.3 to 0.9 m to be successful.

Contact: W.B. Klimstra, J.R. Nawrot,
or D.B. Warburton
Cooperative Wildlife Research Laboratory
Southern Illinois University
Carbondale, Illinois 62901

HOFFMAN'S MISSOURI RIVER RESERVOIR PROJECTS, SOUTH DAKOTA

A shoreline stabilization project was performed on two Missouri River reservoirs in South Dakota. The shorelines of these reservoirs were devoid of permanent vegetation as a result of 3.5 m water level fluctuations. The goal of the project was to establish shoreline vegetation and evaluate different planting techniques and fertilizer application. These planting techniques may be applicable to the Midwest.

Thirteen sites along the shorelines of these reservoirs were plowed and disked in preparation for planting. They were planted with nine grass and two emergent hydrophyte species. Crested wheatgrass (Agropyron cristatum), tall wheatgrass (A. elongatum), western wheatgrass (A. smithii), Garrison creeping foxtail (Alopecurus arundiaceus), turkey brome (Bromus biebersteinii), basin wildrye (Elymus cinereus), switchgrass (Panicum virgatum), and reed canary grass (Phalaris arundinacea) were planted via seeding. Broad-leaved cattail (Typha latifolia), giant bulrush (Scirpus validus), common reed (Phragmites australis), reed canary grass, Garrison creeping foxtail, and western wheatgrass were transplanted. Both seeds and transplants were placed in rows 0.1 m apart which ran perpendicular to the water line. A fertilizer (N-P-K, 50-50-33) was applied to half the area of each plot. Furthermore, each site was surrounded by a barbed-wire fence to prevent cattle grazing, a problem encountered during previous planting attempts.

Garrison creeping foxtail (from transplants), reed canary grass (from both seeds and transplants), common reed, giant bulrush, and broad-leaved cattail were successfully established, as measured by a minimum of 10% survival. In some areas, garrison creeping foxtail and reed canary grass spread into 1 hectare stands. All the sites flourished from one to three years. After the third year, water inundated the sites long enough to destroy the vegetation. The results of fertilization were only preliminary, and no conclusions could be drawn. No attempt was made to quantify the shoreline stabilization provided from these temporary fringe wetlands.

Source: Hoffman 1977.

IOWA DEPARTMENT OF CONSERVATION SUBIMPOUNDMENTS, IOWA

The Iowa Department of Conservation has built numerous subimpoundments around four large reservoirs to provide duck habitat. These subimpoundments are generally located at or above the confluence of the reservoirs and the feeding stream. The area within the dikes was shaped to provide a diverse topography. A habitat combination of 50% palustrine emergent (PEM) and 50% open water (POW), creating a "hemi-marsh" condition, was generally sought as the practical ideal condition for maximizing waterfowl use. Furthermore, maximization of shoreline length was desired to provide the greatest density of breeding territories. Water levels are manipulated to provide water depths appropriate for various species of waterfowl (e.g., giant Canada goose, Branta canadensis maxima) and to allow planting of food stock for waterfowl [e.g., smartweed (Polygonum), sedges (Carex), and millet (Echinochloa)]. Pumping water into these areas was sometimes necessary when natural runoff was insufficient.

Source: Moore and Pfeiffer 1985.

MISSISSIPPI RIVER NAVIGATIONAL POOL 8, WISCONSIN

The U.S. Fish and Wildlife Service created a wetland in Navigational Pool 8 on the Mississippi River near La Crosse, Wisconsin. The goal of the project was to provide a food source for staging canvasbacks. Using techniques described for Lake Puckaway, 17 0.4-ha plots were each planted during the spring with 8,000 tubers of wild celery. These tubers were harvested from nearby Lake Onalaska using a hydraulic dredge. Water depths in the planted plots ranged from one to two meters. The project was successful in establishing wild celery. Seventy to 75% of the tubers sprouted and produced new plants during the summer.

Contact: Leslie N. Peterson
U.S. Fish and Wildlife Service
Upper Mississippi River National Wildlife and Fish Refuge
La Crosse, Wisconsin 54602
(608) 783-6451

TENNESSEE VALLEY AUTHORITY PROJECTS, TENNESSEE

The Tennessee Valley Authority has improved waterfowl habitat in its reservoirs by creating wetlands containing waterfowl food sources. The species planted were Japanese millet (Echinochloa crusgalli), common buckwheat (Fagopyrum esculentum), and Italian ryegrass (Lolium multiflorum). Seeds of these plants were broadcast onto exposed mud banks and on slopes with 20° to 45° angles using a floating aquaseeder. A commercial fertilizer (N-P-K, 6-12-12) and wood mulch were applied on sloping areas. Three to four hectares per hour were seeded using the aquaseeder at a cost of $18.13/acre. Seeding was most successful when it occurred immediately after soil exposure. Wood mulch did not affect the success of plant establishment. As an alternative to the aquaseeder, helicopters were also used to broadcast seeds. They covered 61-81 hectares per hour, cost only $5.59/acre, and had the same success as was observed with the aquaseeder.

Contact: Tennessee Valley Authority
Division of Forestry, Fisheries, and Wildlife Development
Norris, Tennessee 37828

COX HOLLOW AND HALVERSON LAKES, WISCONSIN

The Wisconsin Department of Natural Resources (WDNR) formed a network of open water passageways within two fringe wetlands in Wisconsin to improve fish habitat.

In Cox Hollow Lake, the WDNR placed fiberglass screens n the lake bed in May and left them there for one to three years to inhibit vegetation growth. After one year, some screens began accumulating sediment and supporting vegetation, and some maintenance became necessary. This method was temporarily successful, however, for few plants grew back for one to three months after the screens were removed.

On Halverson Lake, the WDNR harvested 50-70% of the aquatic vegetation to a depth of 1.5 m in June and July. Vegetation grew rapidly after the June harvests and nearly reached preharvest densities. Vegetation recovery was slower, however, after the July harvest. Macrophyte species composition changed in the harvested areas from the original dominants of Berchtold's pondweed (Potamogeton berchtoldii) and sago pondweed (Potamogeton pectinatus) to curly-leaf pondweed (P. crispus), coontail (Ceratophyllum demersum), bushy pondweed (Najas flexilis), and water stargrass (Heteranthera dubia). After harvesting, bluegills continued to feed on aquatic insects in the macrophyte beds, but they also consumed macroinvertebrates made available by the disruption caused by the harvester. Largemouth bass used the harvested channels to cruise for prey. Some individuals were over 400 mm. Both screening and harvesting created a more diverse littoral zone which was beneficial to the fish populations.

Contact: Sandy Engel
Bureau of Research
Nevin Hatchery
Wisconsin Department of Natural Resources
Madison, Wisconsin 53707

CREATION AND RESTORATION OF RIPARIAN WETLANDS IN THE AGRICULTURAL MIDWEST

Daniel E. Willard, Vicki M. Finn, Daniel A. Levine, and John E. Klarquist
School of Public and Environmental Affairs
Indiana University

ABSTRACT. Effective restoration of riparian wetlands in the agricultural midwest demands an early determination of project goals. Established goals will narrow the choices of potential project sites, which can then be evaluated based on hydrology, substrate, seedbank viability, and water quality. Creation and restoration plans should include a realistic timetable that accounts for construction and hydrology constraints, including specifications for revegetation species. Finally, plans should estimate long term vegetation management requirements and establish monitoring schedules to assess project success.

INTRODUCTION

For the purposes of this chapter, the "agricultural midwest" region of the U.S. consists of the upper midwestern states generally associated with major rivers (such as the Mississippi, Missouri, and Wabash rivers). Following this arbitrary (and somewhat shady) distinction, we include Illinois, Indiana, Wisconsin, Missouri, eastern Iowa, southern Michigan, western Kentucky, and western Ohio within this region.

Following Bailey's Ecoregions (Bailey 1978), this region covers portions of both the warm continental division, including portions of the Laurentian mixed and eastern deciduous forest provinces, and the prairie division, including portions of the prairie parkland and tallgrass prairie provinces.

This chapter focuses on riparian ecosystems. The term "riparian zone" is not restricted to riverine systems, and is often applied to meadows, pond margins, etc. We include those communities conceivably affected by periodic flooding (e.g., bottomland hardwood communities) in our discussion of riparian wetlands. Following Cowardin's classification system (Cowardin et al. 1979), we include palustrine emergent, scrub-shrub, and forested systems, in addition to riverine systems. We also discuss some of these systems, as well as lacustrine wetlands, in our paper on fringe wetlands (Levine and Willard, this volume).

Over 70% of the riparian systems in the U.S. have been altered, with natural riparian plant communities reduced 70% overall (95% in some areas). Brinson, et al. (1981) estimate that 10-15 million acres of riparian ecosystems remain in the U.S. (approximately 1.5% of U.S. land area). According to Hey et al. (1982) the rivers in the upper midwest were originally shallow, slow-moving, and meandering. The more extensive floodplain areas were often moist meadows. The more southern floodplains often supported bottomland hardwood communities, many of which have been destroyed or are currently threatened. Virtually all of these systems have been altered to some extent by activities such as channelization or draining.

EXTENT OF CREATION/RESTORATION

TYPICAL PROJECTS AND GOALS

Common goals for creation or restoration projects include stormwater retention, water quality improvement, sediment capture, wastewater treatment, creation of fish and wildlife habitats, and recreational use. Many projects are undertaken simply to perform mitigation required under state and/or federal permits. Although these permits may require specific mitigation activities (eg., habitat replacement), often the mitigator's goal is to complete the project, doing only what is required as quickly and cheaply as possible.

Appendix II contains descriptions of wetland restoration and creation projects that have taken place, or are in progress, in this region. Each description includes information on project types, goals and methods.

SUCCESS IN ACHIEVING GOALS

Goals of wetland creation and restoration projects have rarely been clearly defined, which makes judgement of success in meeting goals difficult. Often projects are considered successful when vegetation is established. However, adequate follow up measurements to gauge long-term success are rare. Some projects, such as the Des Plaines Wetland Demonstration Project, have well-defined goals, but it is too early to judge whether these goals have been achieved (Hey and Philippi 1985).

One ironic example shows the difficulties in determining success or failure in the Midwest. One source (who did not wish to be named) cited an instance in which the highway department of his/her state routed a project through a 200 year old forest to avoid a cattail marsh and, thus, avoid Section 404 mitigation requirements. We have seen several similar situations in which applicants have proposed destroying old upland sites to create emergent marshes. Unfortunately, old upland sites are also valuable habitat and are quite rare over much of the Midwest.

DESIGN CONSIDERATIONS

Wetlands are dynamic systems. It is impossible to isolate specific design criteria as guides to successful projects, since wetland attributes interrelate. Choices of appropriate plant species will depend on hydrology, while choices of water control structures will depend on variables such as desired species and habitat values. Wetland creation and restoration projects require a holistic approach to design considerations to approximate these dynamics. As Chapman and others put it, "knowledge of particular combinations of substrate, microclimate, nutrient and water level regimes, and the dynamism of wetland or riparian plant communities in both time and space is requisite" (Chapman et al. 1982, p. 40).

The Des Plaines River Wetland Demonstration Project can serve as a model for project planning (Hey 1987, Hey and Philippi 1985, Hey et al. 1982). A complete conceptual plan was developed before basic decisions such as construction criteria and species selection were made. This plan is an excellent guide to the kinds of questions and considerations which should be addressed in any project. It is unlikely that our discussion will improve upon it.

DETERMINING PROJECT GOALS

Clearly defined goals are essential to a successful project. **All** design considerations will depend upon these goals. It is not sufficient to simply set goals such as creation of "x acres of wetland".

Several basic questions should be asked, including:

1. Is the wetland being created to replace habitat? What type? Is a similar habitat feasible? What species (plant and animal) are desired?

2. Is the wetland expected to perform specific functions, such as flood control, wastewater treatment, and sediment trapping? What features are required to perform these functions (area, flow rate, etc.)?

All too often sites are selected before goals are established. This severely limits potential project goals, requiring even more careful site evaluation, goal definition, and design tailoring.

PRE-CONSTRUCTION CONSIDERATIONS

Selection of a suitable project site may be the most difficult stage of the planning process. A potential site should be carefully evaluated with particular goals in mind. For example, if wildlife habitat is being created, the proposed site should be in, or adjacent to, an assured corridor.

To create or restore riparian wetlands, adequate room for flood flows is needed. Size requirements can be predicted to some extent by estimating inflow following heavy rains, extent and duration of flooding, and vegetation flood tolerances (Kobriger et al. 1983). Although much of this information will not be available until project planning is completed, educated guesses goals will provide a fair idea of appropriate size.

Past and present land uses also are important considerations. The best sites have historically supported wetlands. However, the hydrology of the area may have been drastically altered since wetlands existed at the sites, hampering attempts to replace them. Planners should also consider the likelihood of future land use changes (eg., development) which might block or contaminate ground and/or surface water supplies (Bell 1981). Verry (1983) provides a "cookbook" approach to selection of suitable impoundment sites.

Potential sites should be carefully evaluated in terms of substrate, seedbank viability, and water quality. Perhaps most importantly, planners should study a potential site's hydrology and historical hydroregime. Is the hydroregime appropriate given project goals? Substrate can be altered and seeded, and water quality can often be improved, but a project cannot significantly change the hydroperiod. The Wetland Evaluation Technique (WET) (Adamus 1988a, Adamus et al. 1987) may be useful in determining whether it is possible to create a wetland with specific functions on a particular site.

Richardson and Nichols (1985) raise several questions that must be answered if project goals include wastewater treatment. Estimates of the hydraulic load and nutrient load, including expected variability, are essential to gauge area, plantings, and other requirements. Reed and Kubiak (1985) present a schematic ecological evaluation procedure useful in the planning stages of wetland wastewater treatment. Their procedure helps determine appropriate impact evaluations for different types of projects.

Practical considerations are also important in site selection. For example, proximity to the fill source may be important. Often mitigators prefer using a portion of the site of the project requiring mitigation for wetland creation since they do not have to acquire additional land. Also, they might be able to work on both projects concurrently, minimizing equipment costs.

Detailed plans cannot be developed without knowledge of pre-construction conditions and consideration of these conditions given project goals. A complete survey of soil, hydrology, water quality, vegetation, and wildlife parameters will be essential data for project planning and future monitoring efforts (see discussion of monitoring methods below). It may be necessary to redefine goals in light of these findings.

CRITICAL ASPECTS OF THE PLAN

Timing

A good timetable for a project should minimize exposure of open ground. Exposure of open ground, especially during rainy seasons, can lead to heavy soil erosion. Also, if revegetation efforts will rely on seedbanks, these must be protected from winter exposure to avoid a frozen seedbank.

The timetable also should allow for planting in the proper season. Different species must be planted at different times of year depending on individual life-histories. A great deal of time and money will be wasted if these requirements are not met.

Include contingency plans. There will be delays and unforeseen problems preventing adherence to the original schedule. Plan for them by developing alternative timetables, including provisions for ground protection, and by budgeting for the added expenses these alternatives might incur.

Construction

Well planned construction will improve the chances of project success. Plans including areas for equipment parking and turning will go a long way towards reducing soil compaction and unnecessary disturbance. If the topsoil will be saved for its seedbed, it must be minimally disturbed and carefully stored. An isolated location should be identified for this purpose.

Controlling erosion during construction is very important. Exposed ground (including topsoil pile, if any) should be covered if possible. Shuldiner et al. (1979) report mixed success with jute mesh and excelsior mat covers. Plastic sheeting and filter cloths are cheap, effective methods of temporary erosion control.

Excavation methods are important

considerations. Linde (1983) discusses several methods of earthmoving including blasting, level ditching and the use of draglines. Weller (1981) advocates blasting or bulldozing over dragging and ditching.

Variable depths are best accomplished by excavating deepest areas first. The depth of excavation is critical since it determines future water depths. Shuldiner et al. (1979) recommend providing water depths of 15-60 cm during the growing season for emergent marshes. For submerged vegetation they recommend deep zones of 0.5-2 m. Scattered pits over 3 m deep will increase aquatic diversity.

A varied bed form adds diversity. Herricks et al. (1982) suggest excavating holes one to three feet deep and using the fill to create riffles. This will provide depth diversity at low flow rates.

Stream channels can be improved by creating meanders. New meanders should follow historical paths if possible. The outside curve of the meander should be the deepest point, with the bed sloping towards the inside curve. Herricks et al. (1982) suggest using riprap to stabilize the banks. Stoplog structures are also commonly constructed.

Extensive floodplains are characteristic of midwestern riparian systems. Bell (1981) recommends creating a gradual slope (under 1%) for riparian shorelines. This will provide gradual vegetation zonation based on water levels, consequently increasing diversity. Kobriger et al. (1983) recommend construction of flat impoundments with slopes less than 1% to slow outflow. Weller (1981) also recommends gradually sloping sidewalls and irregularly shaped basins for wildlife habitat enhancement.

Streambank stabilization is essential to minimize soil erosion. Gray (1977) identified 35° as the maximum slope angle for vegetation establishment, although this varies to some extent with soil type. In addition to establishing stabilizing vegetation, it may be necessary to place stone riprap at the toes of slopes. This is particularly important on undercut banks. Revetments can also be used to stabilize banks (Allen 1978). Although possibly limiting vegetation establishment, rock revetments can promote increased invertebrate populations. (Canter 1985).

Wildlife habitat creation/restoration requires additional considerations. In general, patches of open water and vegetation used to maximize edge effects are needed for adequate habitat diversity. Brinson et al. (1981) suggest creating diverse habitat using live and dead vegetation, areas of standing water, and floating structures. Weller (1981) recommends a 50:50 ratio of cover to open water. Areas of deep open water are necessary for some waterfowl (eg., diving ducks). Other waterfowl species (eg., geese) prefer open water interspersed with islands. Herricks et al. (1982) suggest diverse patches of wetland, herbaceous, and woody vegetation for nongame wildlife species.

Hydrology

Wetland communities are determined by hydrology. Wetland vegetation is adapted to, and often depends upon, water level changes. Managers should strive to maintain hydroregime variability as far as practicable. This practice can have added benefits. For example, a shallow, fluctuating water level that occasionally falls below soil surface is recommended for management of highway pollutant runoff (Kobriger et al. 1983).

Stable water levels can lead to increased nesting success (Weller 1981). This suggests managing water levels during the nesting season, but allowing water level fluctuations during the rest of the year.

Preferably, the project will rely on the natural site hydrology. Of course, this means that managers will have to accept the inevitable fluctuations in water level and flow rates. But, as we discuss elsewhere, these changes will, over time, support a variable dynamic wetland (Willard and Hiller, Volume II).

A second option is to install permanent, low maintenance water control structures. Simple dam structures, such as drop-inlet and whistle stop dams, can be effective impoundments. Several states have established design specifications for dikes, floodgates, and impoundments, so planners should check with state agencies before serious designing begins. (Shuldiner et al. 1979).

Dams are usually placed at the narrowest portion of flow. Typical design parameters include heights 2-3 feet above the expected crest elevation and widths of 10 feet (unless dam will be used as a roadway). Side slopes are determined by the type of material used in construction, with sand dams requiring flatter slopes than clay dams (Anderson 1983).

It is possible to significantly alter the hydrology of a site by installing pumping equipment and drains, creating impoundments, and redirecting water flow. The Des Plaines project is a good example of this type of approach (Hey and Philippi 1985). Project planners hope that by judicious grading and equipment installation they will be able to recreate some of the original floodplain characteristics and provide varied wetlands for research purposes.

However, major alteration of site hydrology through pumps and other approaches requiring continuous mechanical manipulations is very expensive, and often requires a great deal of maintenance. What happens ten years down the line when the money runs out and equipment begins to break down?

Substrate

There are two major substrate considerations important for wetland creation/restoration: will the substrate, in conjunction with the hydroregime, support the desired wetland functions? (consider factors such as soil permeability, water retention); will the substrate support desired vegetation? (consider factors such as nutrients, compaction).

Substrate can be altered through removal or importation of soil materials. Kobriger et al. (1983) suggest adding a layer of clay or other fine material to porous substrates to slow percolation. Using imported topsoil for seedbank revegetation is common and has the advantage of providing suitable substrate for plant establishment.

Liming may be necessary to raise the pH of very acid soils. Kobriger et al. (1983) recommend using fertilizers only on very infertile substrates. In these cases they recommend application of low levels of slow release formulas at the root zone.

Weller (1981) recommends muddy marsh bottoms which have cracked from wetting and drying. These cracked conditions also could be created. Scarification is a useful technique for improving moisture retention and reducing compaction. This is typically done using tractor-pulled disc-harrows or deep chisels. However, site conditions may not be suitable for this type of equipment, suggesting hand hoeing methods (Herricks et al. 1982).

Revegetation

Selection of appropriate species is essential to successful revegetation efforts. Choice of species will depend upon project goals and upon site characteristics such as hydrology, climate, substrate, and grade.

Several factors should be considered when selecting species (Table 1). A short time frame for revegetation may require planting of annuals and biennials for rapid establishment with plantings of perennials for later succession. The availability of ecotypically adapted planting materials (seeds, shoots) is another important consideration. Also it is crucial that species tolerances be matched with soil moisture conditions.

The planners at the Des Plaines project are basing their choices of vegetation on autecological ratings (adapted from Swink and Wilhelm 1979) of the likeliness of invasion and colonization. Another factor being considered is the extent of annual biomass production, since fall burning is planned as a management strategy (Hey et al. 1982).

Different project goals will require different types of vegetation. Herricks et al. (1982) recommend plantings of reed canary grass (Phalaris sp.) and red top (Agrostis stolonifera) for erosion control, although many other grasses are commonly used. Creation of wildlife habitat suggests choices of woody vegetation since woody plant communities increase habitat values (Brinson et al. 1981). Wastewater treatment projects typically use hardy species such as cattails (Typha sp.), bulrushes (Scirpus sp.), and reed grass (Phragmites communis), depending on the specific plant layout. Bell (1981), Kadlec and Wentz (1974), Hey et al. (1982), and many other sources can provide guidance on appropriate species selection for different types of revegetation.

To some extent plant species selection may be superceded by a desire to use the seedbank. At the same time a preference for natural revegetation may give way to a need to establish vegetation quickly (e.g., by transplanting). Natural revegetation is probably the most common method used in the midwest. In many cases projects have relied upon colonization from adjacent wetlands. Nearby natural communities can be used as indicators of likely colonizers. Natural emergent dominants such as cattails, reed grass, bulrushes, burreeds (Sparganium sp.), perennial sedges (Carex sp.), rushes (Juncus sp.) and spike rushes (Eleocharis sp.) can be expected. Woody vegetation would likely include red osier dogwood (Cornus stolonifera), hackberry (Celtis occidentalis), buttonbush (Cephalanthus occidentalis), willows (Salix sp.) and alders (Alnus sp.), cottonwood (Populus deltoides), silver maple (Acer saccharinum), and oaks (Quercus sp.) (Lindsey et al. 1961, Brinson et al. 1981, Kobriger et al. 1983).

Often the seedbank is incidental to the project and may lead to problems of weed invasion. Van der Valk (1981) discusses methods of evaluating seedbanks for seed viability and species composition. Seedbanks are potential sources of diverse native (and introduced) species. However, since seed distribution may not be even and germination rates may be poor, revegetation may yield unpredictable patchy vegetation (Kobriger et al. 1983).

Seedbank relocation is another common technique. Often wetland topsoil is removed from a development site and transferred to the

Table 1. Emergent vegetation types, characteristics, and environmental tolerances. Adapted from Kobriger et al. (1983).

PLANT OCCURRENCES OR TOLERANCES

Vegetation Type	Height	Soil	pH (water)	Salinity	Water Depth	Air Temperature
Cattails <u>Typha</u> spp.	0.7-2.7 spreading mats & thick stands	mud & silt w/ 25 cm detrital & humic layers; organic content up to 32%	wide range 4.7-10.0	0-15 ppt (up to 25 ppt some spp.)	0.2-1 m	9-31 °C seed germ at 18-24 °C
Reeds <u>Phragmites</u> spp.	3-4 m open stands	clay, sand or silt under swampy conditions 6-54%	wide range 2.0-8.5	0-10 ppt optimum; up to 40 ppt some spp.	60-8 cm optimum; ranges from -0.3 to 4 m	11-32 °C seed germ at 10-30 °C
Rushes <u>Juncus</u> spp. over 200 spp.)	0.02-1.5 m dense clumps	organic soils		0-14 ppt; up to 35 ppt some spp.	at or just below soil surface	16-26 °C
Sedges <u>Carex</u> spp.	0.1-1 m in clumps	muck or clay with up to 10 cm detrital layer; up to 45% organic content	4.9-7.4	0-0.4 ppt	-0.05 to 0.95 m	15-21 °C (range of 14-32 °C
Sedges - nutsedges <u>Cyperus</u> spp.	8-50 cm	see <u>Carex</u>	7.1 optimum; range of 3-8	0-0.4 ppt	at or just above ground surface -0.05 to 0.3 m	32 °C optimum (range of 16-45 °C)
Sedges-bulrushes <u>Scirpus</u> spp.	0.3-3 m	see <u>Carex</u>	4-9	4-20 ppt optimum; 0-32 ppt range	min 0-10 cm, max. depth varies w// spp.	17-28 °C
Canary grass <u>Phalaris</u> spp.	30 cm flower spikes 0.6-2 m	silt loam	6.1-7.5	generally freshwater	at or just below soil surface	15-25 °C (up to 30 °C some spp.) seed germ at 18-35 °C
Cordgrass <u>Spartina</u> spp.	short forms 0.3-0.4 m; tall forms 1.2-2 m	sandy substrate	4.7-7.8	9-34 ppt	-.15 to 0.7 m	12-29 °C seed germ 18-35 °C

mitigation site. This may provide a suitable seedbank, but only if topsoil is from a site with species ecotypically adapted to the new project site. Worthington and Helliwell (1987) reported successful transfer of three carefully maintained soil horizons. Revegetation was rapid, and reexaminations six years later found no overall decrease in diversity.

Weller (1981) has found that seedbank revegetation is suitable for sedges, arrowhead (Sagitarria sp.), hardstem and softstem bulrushes (Scirpus acutus and S. validus), cattails, willows, and cottonwood (Populus deltoides), but not for sago pondweed (Potamogeton pectinatus).

Planting vegetation may provide more reliable results than reliance on natural revegetation. Methods of planting will depend upon the species, season, and timeframe of the project (i.e., can you wait several years for seedlings to become established).

Planting success will depend upon species choices and site conditions. Thompson (1984) reports successful plantings of prairie cord grass (Spartina pectinara), common three square (Scirpus americanus), nut sedge (Cyperus sp.), japanese millet (Echinochloa cruzgali), smartweed (Polygonum sp.), switchgrass (Panicum virgatum), reed canary grass (Phalaris arundinacea), rice cutgrass (Leersia oryzoides), hardstem bulrush, and chufa (Cyperus esculentus) in wet areas of an Indiana slurry pond. The area was first fertilized and limed to raise the pH.

Seeding is an alternative to natural revegetation. Seeds must be collected when ripe, and may require stratification, scarification, and/or specific germination conditions (e.g., temperature, photoperiod). These requirements can be identified through the life histories of different species. Many of these conditions can be met by distributing locally collected (ecotypically adapted) seeds at the time of year when they are naturally dispersed (Kobriger et al. 1983). Seeding rates will vary depending on the purity of the seed stock, the expected germination rate, and on factors such as erodibility of the land (Herricks et al. 1982).

There are several methods of seeding which can be used for revegetation. Drill seeding is a good method for fine, fluffy seeds, but requires firm soil conditions to attain correct depths. Commercial drill seeders such as grassland or Truax seeders can be used. Herricks et al. recommend planting depths four times the seed diameter. A hydroseeding truck can be used to spread an aqueous seed suspension on steep slopes. Fertilizer can be incorporated into the suspension if necessary.

Broadcast seeding is the most common and versatile seeding method since seeds can be distributed by hand, or using broadcast seeding equipment, from the ground, a boat, or an aircraft. However, seeding rates are usually higher (up to two times) than for drill seeding.

Aerial seeding can be effective for large areas. Commercial cropdusters often have equipment for aerial seeding. It is also possible to broadcast seeds by hand from light aircrafts (Linde 1983).

Wetseeding has been successful for seeding Japanese millet under flooded conditions. Seeds are soaked until they sink and are then broadcast from a boat. This is only suitable for seeds which absorb water and sink, and which can tolerate standing water (Linde 1983).

Most woody vegetation does not succeed when broadcast seeded. However, Herricks et al. (1982) suggest that this method may be suitable for fall plantings of native species of Cornus and Myrica.

Transplanting of native vegetation is becoming increasingly common in the midwest. Woody vegetation is commonly transplanted from nursery stocks or cuttings, while herbaceous vegetation is often transplanted using rhizomes or plugs. Prairie plant nurseries are also starting to include native wetland plants in their regular stocks (Bell 1981).

Roots and tubers can be collected from nearby wetlands and planted at the appropriate time of year (eg., when dormancy is broken). These materials require protection from damage due to digging, transporting, and planting (Kobriger et al. 1983). Herricks et al. (1982) recommend spring planting to avoid damage from frost heaving and cold temperatures. It is important to make holes big enough for roots to spread and to place plants at the same depth relative to the soil surface as they were previously growing (Herricks et al. 1982).

Kobriger et al. (1983) discuss collection and planting of plugs. This method is most successful with perennials and species adapted to standing water (eg., Typha sp.). Plugs can be collected from local wetlands, hand-grown, or taken from nursery stocks. These plugs can be stored for a few days before planting. Planting intervals of 0.5-1 m for herbaceous perennials have successfully filled areas within one to two growing seasons.

Seagrasses have been successfully planted as sod, with the sod anchored to the substrate surface by spikes. This might be feasible for other grasses, if the sod can be removed so that the sediment is intact without damaging root systems (Kobriger et al. 1983).

Woody vegetation can also be planted with cuttings. Herricks et al. (1982) recommend taking cuttings 12-18 inches long from stems under one inch in diameter, making a clean cut above a vigorous bud. Cuttings should be stored in moist sand. These should be planted in the spring so that short segments below the buds are exposed.

Herricks et al. (1982) recommend planting species such as Populus and Salix in clumps of 6-15 individuals two feet apart at a rate of 10-20 clumps/acre. They also suggest planting single species dense patches of alders, red-osier dogwood, and arrowwood (Viburnum dentatum) near aquatic borders. Allen (1978) suggests planting willows between the expected mean and mean high water levels, and planting alders, poplars (Populus sp.), ashes (Fraxinus sp.), maples (Acer sp.), and elms (Ulmus sp.) at and above the mean high water level. Sigafoos (1964) reminds us that woody seedlings will not survive unless there are extended periods of low flow.

Reintroduction of Fauna

Methods of re-introduction will depend somewhat on which species are desired. Passive reintroduction is by far the most common method, relying on immigration from nearby populations. However, passive methods will work only if there are adequate corridors to allow movement between existing populations and the project site. Construction of nesting structures may help attract desired waterfowl.

More active efforts may be necessary in a variety of situations. If desired species are isolated from the site by manmade or natural barriers (ie., no corridors), or are not very mobile, active stocking will be necessary. In cases of rare and endangered species, planners may not want to rely on passive methods, and instead may endeavor to establish populations through importation, and possibly through supplemental feeding and predator control.

Buffers

As we discuss elsewhere (Willard and Hiller, this volume) buffers are an essential component of wetland systems. In addition to providing protection from outside disturbances, they also act as corridors between sites. There is no formula for appropriate buffer size. The Wild and Scenic Rivers Act (PL 90-542) recommends buffers 400 meters wide. This may not be sufficient since the areas within 200 meters of open water are most frequently used by terrestrial wildlife (Brinson et al. 1981).

Long-term Management

A long-term management strategy is an essential component of the project plan. Strategies will depend upon the goals of the project.

Vegetation management is the most common form of longterm management. Traditional methods include draining of impoundments and mowing. Controlled burning is a common cheap, effective method for control of woody vegetation (fall burns) and removal of dead biomass (winter burns). Care must be taken to minimize wildlife disturbance. In particular, avoid spring burning, which disrupts nesting (Linde 1983).

Warners (1987) studied controlled burns as a means of controlling shrub invasion in sedge meadows. He found no significant reduction in shrub cover, but did find a shift in species composition from bog birch (Betula sandbergii) and poison sumac (Rhus vernix) to red-osier dogwood (Cornus stolonifera). He also found that shrub-invaded areas were wetter than uninvaded areas.

Managers often wish to dredge wetlands. As Novitzki (1978) discusses, wetlands will accumulate sediment. The extent of accumulation will vary with water velocities. Dredging can significantly disturb wetland communities (Darnell 1977). Before dredging, managers should evaluate, and possibly modify, their methods of water control. Alternatively, they can accept the accumulation as a natural part of wetland dynamics.

MONITORING

Monitoring should begin prior to construction. A thorough pre-construction site evaluation will not only provide baseline data for future monitoring efforts, but will also provide valuable information for project planning.

Managers must develop consistent methods. Platts et al. (1987) discuss a broad range of established evaluation methods which could be used for this purpose. A variety of assessment methodologies could also be used.

Frequent evaluations (e.g., monthly) will be necessary immediately following project completion and continuing for the first few years. Regular follow up evaluations (e.g., annually) for subsequent years (a few decades) would be necessary to assess persistence.

Hydrologic characteristics such as streamflow, water levels, dispersion, and stage/discharge balances should be monitored (Kobriger et al. 1983). Platts et al. (1987) and Novitzki (1987) discuss methods suitable for these purposes.

Water quality parameters such as nutrient budgets, suspended solids, and pollutant levels (e.g., metals) should also be monitored above, within, and below site. This can be accomplished using standard sampling and analytic techniques. In addition, sediments should be monitored for excessive sediment and pollutant deposition.

Monitoring programs also should include vegetation characteristics such as planting success, patterns of competition and invasion, and habitat values. Establishing permanent plots will enable repeated evaluations of species composition, diversity, and dominance. The Habitat Evaluation Procedure (HEP) can be used as a way of getting at habitat values (Adamus 1988b).

Wildlife population dynamics and breeding success can be monitored through population sampling (e.g., mark-recapture, seining), and through regular bird counts. Invertebrate and plankton populations should also be monitored.

Assessments of project success will depend on these monitoring efforts, in conjunction with project goals. Traditionally success has been defined in narrow terms, such as vegetation establishment. Allen (1978) states that success of streambank stabilization is usually judged qualitatively, such as by evaluating before and after photographs. Adamus (1988b) discusses more sophisticated methods of evaluating project success. The WET procedure can be used for before/after assessment of a wide range of wetland functions.

CONCLUSIONS/RECOMMENDATIONS

There are no systematic records of the changes and developments that have occurred in and around several hundred large midwestern reservoirs with extensive wetland systems. A few states, such as Michigan and Wisconsin, have well-established wetland programs which can "track" wetland status; while others, such as Kentucky, have only rudimentary programs. Establishing centralized programs capable of monitoring changes in created and restored wetlands, in conjunction with use of historical records, would go a long way towards developing an understanding of the patterns of wetland change in the midwest.

LITERATURE CITED

Adamus, P.R. 1988a. The FHWA/Adamus (WET) method for wetland functional assessment. In D.D. Hook, W.H. McKee Jr., H.K. Smith, J. Gregory, V.G. Burrell Jr., M.R. DeVoe, R.E. Sojka, S. Gilbert, R. Banks, L.H. Stolzy, C. Brooks, T.D. Matthews, and T.H. Shear (Eds.), The Ecology and Management of Wetlands, Volume 2: Management, Use and Value of Wetlands. Timber Press, Portland.

Adamus, P.R. 1988b. Criteria for created or restored wetlands. In D.D. Hook, W.H. McKee Jr., H.K. Smith, J. Gregory, V.G. Burrell Jr., M.R. DeVoe, R.E. Sojka, S. Gilbert, R. Banks, L.H. Stolzy, C. Brooks, T.D. Matthews, and T.H. Shear (Eds.), The Ecology and Management of Wetlands Volume 2: Management, Use and Value of Wetlands. Timber Press, Portland.

Adamus, P.R., E.J. Clairain Jr., R.D. Smith, and R.E. Young. 1987. Wetland Evaluation Technique (WET). U.S. Army Corps of Engineers Technical Report Y-87, Vicksburg, Mississippi.

Allen, H.S. 1978. Role of wetland plants in erosion control of riparian shorelines. In P.E. Greeson, J.R. Clark, and J.E. Clark (Eds.), Wetland Functions and Values: The State of Our Understanding. Proceedings of the National Symposium on Wetland Functions and Values, Nov 7-10, Lake Buena Vista, Florida.

Anderson, G.R. 1983. Design and location of water impoundment structures. In Water Impoundments for Wildlife: A Habitat Workshop. U.S.Dept. of Agric. Forest Service General Technical Report NC-100.

Bailey, R.G. 1978. Ecoregions of the United States. U.S. Forest Service, Intermountain Region, Ogden, Utah.

Bell, H.E. 1981. Illinois Wetlands: Their Value and

Management. State of Illinois Institute of Natural Resources Document Number 81/33.

Brinson, M.M., B.L. Swift, R.C. Plantico, and J.S. Barclay. 1981. Riparian Ecosystems: Their Ecology and Status. U.S. Fish and Wildlife Service FWS/OBS-81/17.

Canter, L.W. 1985. Environmental Impacts of Water Resources Projects. Lewis Publishers, Chelsea, Michigan.

Chapman, R.J., T.M. Hinckley, L.C. Lee, and R.O. Teskey. 1982. Impact of Water Level Changes on Woody Riparian and Wetland Communities, Volume X. U.S. Fish and Wildlife Service FWS/OBS-82/23.

Cowardin, L.M., V. Carter, F.C. Golet, and E.T. LaRoe. 1979. Classifications of Wetlands and Deepwater Habitats of the United States. U.S. Fish and Wildlife Service FWS/OBS-79/31.

Darnell, R.M. 1977. Overview of major development impacts on wetlands. Proceedings of the National Wetland Protection Symposium, June 6-8, Reston, Virginia.

Gray, H.D. 1977. The influence of vegetation on slope processes in the Great Lakes region. In Proceedings of the Workshop on the Role of Vegetation in Stabilization of the Great Lakes Shoreline. Great Lakes Basin Commission, Ann Arbor, Mich. Cited in Allen 1978.

Herricks, E.E., A.J. Krzysik, R.E. Szafoni, and D.J. Tazik. 1982. Best current practices for fish and wildlife on surface-mined lands in the eastern interior coal region. U.S. Fish and Wildlife Service FWS/OBS-80/68.

Hey, D.L. 1987. The Des Plaines River Wetlands Demonstration Project: creating wetlands hydrology. National Wetlands Newsletter 9(2):12-14.

Hey, D.L. and N. Philippi (Eds.). 1985. The Des Plaines River Wetland Demonstration Project, Vol III. Wetlands Research Inc.

Hey, D.L., J.M. Stockdale, D. Kropp, and G. Wilhelm. 1982. Creation of Wetland Habitats in Northeastern Illinois. Illinois Department of Energy and Natural Resources Document Number 82/09.

Kadlec, J.A. and W.A. Wentz. 1974. State of the Art Survey and Evaluation of Marsh Plant Establishment Techniques: Induced and Natural. Vol I: Report of Research. U.S. Army Corps of Engineers, Waterways Experiment Station. Contract Report D-74-9.

Kobriger, N.P., T.V. Dupuis, W.A. Kreutzberger, F. Stearns, G. Guntenspergen, and J.R. Keough. 1983. Guidelines for the Management of Highway Runoff on Wetlands. National Cooperative Highway Research Program Report 264. National Research Council Transportation Research Board, Washington, D.C.

Linde, A.F. 1983. Vegetation management in water impoundments: Alternatives and supplements to water-level control. In Water Impoundments for Wildlife: A Habitat Workshop. U.S. Dept. of Agric. Forest Service General Technical Report NC-100.

Lindsay, A.A., R.O. Petty, D.K. Sterlin, and W. Van Asdall. 1961. Vegetation and environment along the Wabash and Tippecanoe rivers. Ecological Monographs 31(2):105-156.

Novitzki, R.P. 1987. Some observations on our understanding of hydrologic functions. National Wetlands Newsletter 9(2):3-6.

Novitzki, R.P. 1978. Hydrologic characteristics of Wisconsin's wetlands and their influence on floods, stream flow and sediment. In P.E. Greeson, J.R. Clark, and J.E. Clark (Eds.), Wetland Functions and Values: The State of Our Understanding. Proceedings of the National Symposium on Wetland Functions and Values, Nov 7-10, Lake Buena Vista, Florida.

Platts, W.S., C. Armour, G.D. Booth, B. Mason, J.L. Bufford, P. Cuplin, S. Jensen, G.W. Lienkaemper, G.W. Minshall, S.B. Monsen, R.L. Nelson, J.R. Sedell, and J.S. Tuhy. 1987. Methods for evaluating riparian habitats with applications to management. General Technical Report INT-221. U.S. Dept. of Agric. Forest Service, Intermountain Research Station, Ogden, Utah.

Reed, D.M. and T.J. Kubiak. 1985. An ecological evaluation procedure for determining wetland suitability for wastewater treatment and discharge. In P.J. Godfrey, E.R. Kaynor, S. Pelczarski, and J. Benforado (Eds.), Ecological Considerations in Wetlands Treatment of Municipal Wastewaters. Van Nostrand Reinhold Co., New York.

Richardson, C.J. and D.S. Nichols. 1985. Ecological analysis of wastewater management criteria in wetland ecosystems. In P.J. Godfrey, E.R. Kaynor, S. Pelczarski, and J. Benforado (Eds.), Ecological Considerations in Wetlands Treatment of Municipal Wastewaters. Van Nostrand Reinhold Co., New York.

Schuldiner, P.W., D.F. Cope, and R.B. Newton. 1979. Ecological Effects of Highway Runoff on Wetlands. National Cooperative Highway Research Program Report 218B. National Research Council Transportation Research Board, Washington, D.C.

Sigafoos, R.S. 1964. Botanical evidence of floods and floodplain deposition, vegetation and hydrologic phenomena. U.S. Geological Survey Professional Paper 485-A. Cited in Allen 1978.

Swink, F. and G. Wilhelm. 1979. Plants of the Chicago Region. The Morton Arboretum, Lisle, Illinois.

Thompson, C.S. 1984. Experimental practices in surface coal mining--creating wetland habitat. National Wetlands Newsletter 6(2):15-16.

Van der Valk, A.G. 1981. Succession in wetlands: a Gleasonian approach. Ecology 62:688-96.

Verry, E.S. 1983. Selection of water impoundment sites in the lake states. In Water Impoundments for Wildlife: A Habitat Workshop. U.S. Dept. of Agric. Forest Service General Technical Report NC-100.

Warners, D.P. 1987. Effects of burning on sedge meadow studied. Restoration and Management Notes 5(2):90-91.

Weller, M.W. 1981. Freshwater Wetlands: Ecology and Wildlife Management. University of Minnesota Press, Minneapolis.

Worthington, T.R. and D.P. Helliwell. 1987. Transference of semi-natural grassland and marshland onto newly created landfill. Biological Conservation 41: 301-311.

APPENDIX I: RECOMMENDED READINGS

Anon. 1981. Illinois plants for habitat restoration. Illinois Department of Conservation, Mining Program. Springfield, Illinois.

Provides information on habitat requirements, wildlife values, and commercial availability of Illinois native plants.

Hey, D.L. and N. Philippi (Eds.). 1985. The Des Plaines River Wetland Demonstration Project, Vol III. Wetlands Research Inc. AND

Hey, D.L., J.M. Stockdale, D. Kropp, and G. Wilhelm 1982. Creation of wetland habitats in northeastern Illinois. Illinois Department of Energy and Natural Resources Document Number 82/09.

An excellent model for project planning. Thorough discussions of considerations important to a wide variety of project types.

Kadlec, J.A. and W.A. Wentz. 1974. State of the Art Survey and Evaluation of Marsh Plant Establishment Techniques: Induced and Natural. Vol I: Report of Research. U.S. Army Corps of Engineers Waterways Experiment Station Contract Report D-74-9.

Comprehensive survey of values and establishment of wetland species.

Linde, A.F. 1969. Techniques for wetland management. Wisconsin Department of Natural Resources Research Report 45.

A excellent resource for management techniques.

Swink, F. and G. Wilhelm. 1979. Plants of the Chicago Region. The Morton Arboretum, Lisle, Illinois.

Assigns autecological ratings to regional plant species indicating the likelihood that individual species will volunteer at a given site.

Teskey, R.O. and T.M. Hinckley. 1978. Impact of water level changes on woody riparian and wetland communities, Volume III: Central forest region; Volume IV: Eastern deciduous forest region. U.S. Fish and Wildlife Service FWS/OBS-78/87.

Good discussions of bottomland ecology and dominant species. Extensive information on vegetation flood tolerances.

APPENDIX II: PROJECT PROFILES

The following descriptions of projects in the midwest is not a complete survey. However, it does include a broad range of projects somewhat representative of the types that have been attempted.

The project descriptions are organized by state. Project numbers correspond to the sites identified in Figure 1. The Fish and Wildlife Service's Wetland Classification System (Cowardin et al. 1979) is used to classify wetland type (e.g., PEM = palustrine emergent, PFO = palustrine forested). Each description identifies the source(s) used, which could be contacted for further information.

MCHENRY COUNTY FOX RIVER PROJECT, ILLINOIS

Project Number: 1

Part of Required Mitigation: Yes

Parties Involved: McHenry County Highway Department, U.S. Army Corps of Engineers (USAE)--Chicago District.

Wetlands Lost: 1.66 acres of PEM (sedge meadow and wet prairie).

Wetlands Created: 1.75 acres of PEM.

Procedure: Excavation of upland adjacent to wetland, revegetation (seed bank relocation).

Status: Under construction in July 1988.

Source: Public Notice, USAE-Chicago Dist., Feb. 1987, Application #2808703.

Comments: The creation project was proposed as mitigation for PEM habitat destroyed during development. Revegetation will be accomplished by stockpiling soil stripped from sedge meadow and top dressing the excavated site with this soil. Steep slopes will be reseeded with grasses.

DES PLAINES RESTORATION PROJECT, ILLINOIS

Project Number: 2

Part of Required Mitigation: No

Parties Involved: Wetland Research Inc., Illinois Dept. of Energy & Natural Resources, USF&WS, Lake County Forest Preserve District.

Wetlands Created: 250+ acres of PEM/POW and riverine types.

Procedure: Large scale excavation reshaping the original stream and flood plain.

Status: Long term experimental site; should have some initial feedback on vegetation success in late '88.

Source: Creation of wetland habitat in N.E. Ill., Doc. 82/90, Illinois Dept. of Energy & Natural Resources (IDENR), May 1982. Hey, D. and N. Philippi, ed. 1985. The Des Plaines River Wetlands Demonstration Project, Vol. III, Wetlands Research, Inc. (312) 922-0777.

Comments: The goal of this experiment was to restore the drainage characteristics associated with the original creeks and floodplains in N.E. Illinois. To create diverse wetland habitat managers constructed irrigated river terraces, floodway marshes, and aquatic shelves and revegetated with a variety of native wetland species. Although Des Plaines is a large scale project the Wetlands Research, Inc. document discusses a variety of concerns associated with both large and small creation projects.

KLEFSTAD CO., BUFFALO GROVE, ILLINOIS

Project Number: 3

Part of Required Mitigation: Yes

Parties Involved: USAE-Chicago Dist., Klefstad Companies, Inc.

Wetlands Lost: 13.4 acres of PEM and POW.

Wetlands Created: 18.4+ acres of PEM and POW.

Procedure: Excavation and revegetation (seed bank relocation).

Status: Under construction, July 1988.

Source: Public Notice, USAE-Chicago dist., July 1986,

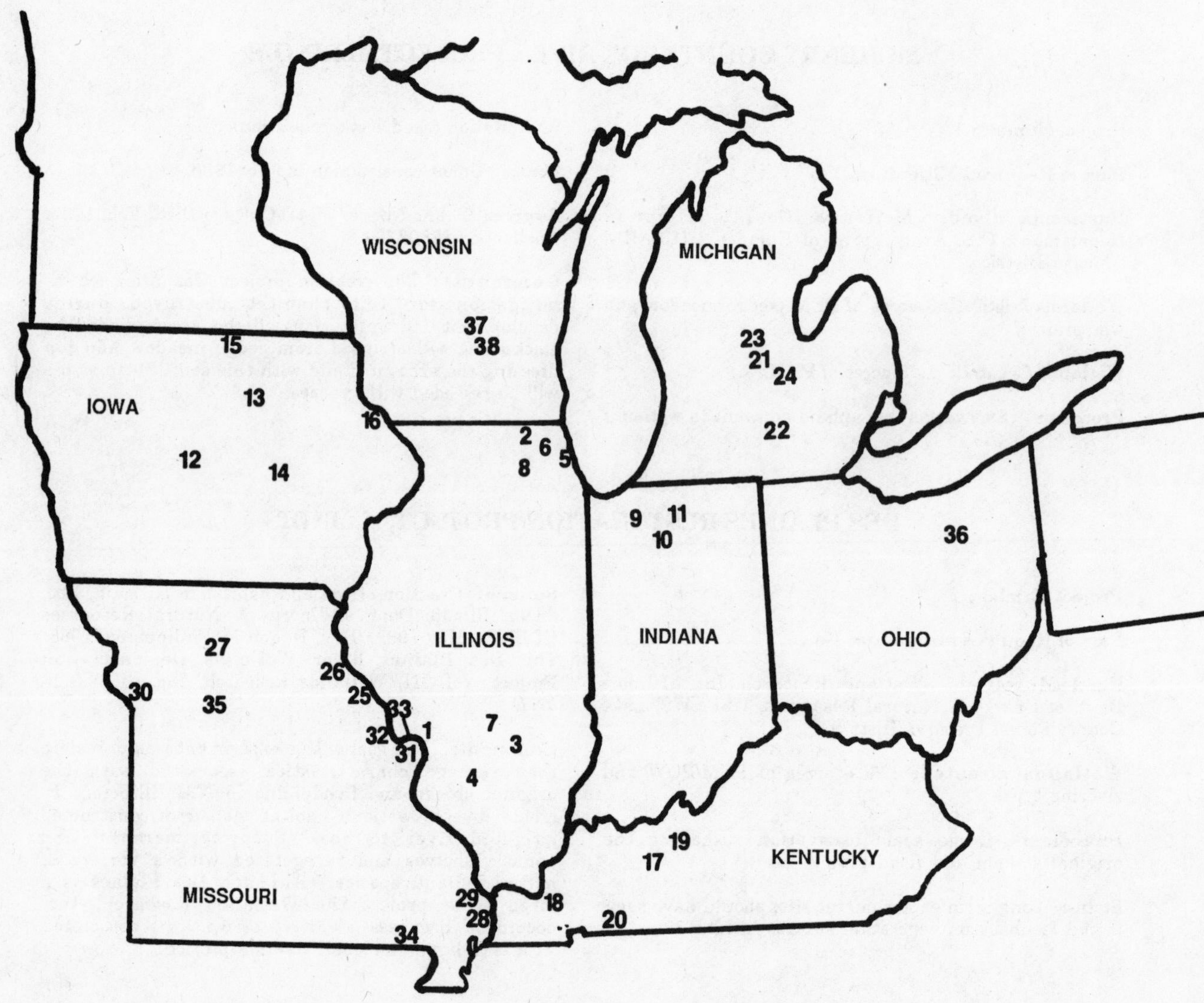

Figure 1. Locations of the projects described in Appendix II. Numbers correspond to the project numbers listed in the appendix.

Permit #NCCCO-R 0148602.

Comments: This project serves as mitigation for filling of portions of a floodway along Aptakisic Creek. The filled area consisted of reed canary grass and cattails with patches of sedges and rushes. The created area will incorporate topsoil removed from the sedge/rush areas as topcover on the site. Additionally seeds of wetland vegetation will be spread on the creation site.

DU PAGE TOLLWAY, ILLINOIS

Project Number: 4

Part of Required Mitigation: Yes

Parties Involved: Ill. State Toll Highway Authority, USAE-Chicago Dist., USF&WS.

Wetlands Lost: 76 ACRES OF PEM wet prairie.

Wetlands Created: 117 acres of PEM wet prairie.

Procedure: Excavation, revegetation (seed bank relocation).

Status: Construction complete.

Source: Mehler, N. du Page fights to save wetlands from dry grave. Chicago Tribune, 9 July 1987, section 2, p.1.

Comments: Revegetation efforts incorporated the transport of excavated wetland soil from the destroyed site several miles to creation sites. On the new sites observers have recorded the occurrence of 60 wetland plant species out of the 200 found on the original site.

WOOD RIVER UPLAND RESERVOIR, ILLINOIS

Project Number: 5

Part of Required Mitigation: Yes

Parties Involved: Illinois Dept. of Transportation (IDOT)

Wetlands Lost: 0.79 acres riparian vegetation.

Wetlands Created: 1.1 acres PEM, 2.5 acres of PSS.

Procedure: Dike and excavation to create a palustrine community from the original smaller riparian wetland, revegetation.

Status: Proposed 1987 but delayed due to property rights questions.

Source: Personal communication, IDOT, Rick Gosch, 6 May 1987 and 19 July 1988.

Comments: This project is proposed to mitigate damage to riparian wetlands caused by the creation of a flood control reservoir. The project replaces the riparian community with a shallow water palustrine community.

Proposed Plantings:

- Submerged areas
  - sago pondweed (Potamogeton pectinatus)
  - duck potato (Saggitaria rigida)
- Seasonally flooded
  - bulrush (Scirpus fluviatilis)
  - pickerel weed (Pontederia cordata)
  - arrowhead (Saggitaria latifolia)
- Scrub/Shrub
  - red osier dogwood (Cornus stolonifera)
  - elderberry (Sambucus canadensis)
- Tree seedlings
  - pin oak (Quercus palustris)
  - sweet gum (Liguidambar styraciflua)
  - green ash (Fraxinus pennsylvanica)

CARLYLE LAKE, BOULDER MARINA AREA, ILLINOIS

Project Number: 6

Part of Required Mitigation: Yes

Parties Involved: USAE-St. Louis Dist.

Wetlands Lost: 0.8 acres of PEM.

Wetlands Created: 0.9 acres of PEM and 10 acres moist soil.

Procedure: Excavation, revegetation disturbed high grade slopes.

Status: Complete.

Source: Personal Communication, P. McGinnis, USAE-St. Louis Dist., 21 July 1988.

Comments: As part of an internal mitigation project the USAE excavated a site to achieve an emergent wetland with diverse elevations. The USAE has created an additional 10 acres of moist soil wetland to act as a nursery area for wetland plants for future Corps' mitigation projects. Despite these efforts wetland revegetation attempts at Carlyle have suffered because of recent droughts. The St. Louis Dist. views the Carlyle Lake site as an opportunity to show how wetland acreage can be created inexpensively. The Dist. plans to continue experimentation with wetland creation around the lake.

GALUM CREEK RESTORATION, ILLINOIS

Project Number: 7

Part of Required Mitigation: Yes

Parties Involved: Consolidation Coal Co.

Wetlands Lost: 137 acres of PFO.

Wetlands Created: 17 to 39 acres of PEM, POW, and PSS.

Wetlands Restored: 100 acres of PFO.

Procedure: Earthmoving and revegetation.

Status: Successful marsh creation, PFO restoration in progress.

Source: Consolidation Coal Company, application for Excellence in Surface Mining Awards, Burning Star #4 mine, 1986.

Comments: The project acts as mitigation for the 137 acres of hardwood bottomland forest destroyed by stream diversion for a coal mine. Instead of attempting to restore the whole site as PFO area with black alder, green ash, river birch, bald cypress, hickory, silver maple, pin oak, sycamore, and sweetgum. The forested areas are still in a scrub/shrub stage of development. The marsh areas are naturally revegetating with cattail, bulrush, pondweed, smartweed, and coontail.

KASKASKIA WETLAND RECLAMATION, ILLINOIS

Project Number: 8

Part of Required Mitigation: Coal mine reclamation.

Parties Involved: Southern Illinois Univ., Peabody Coal Co.

Wetlands Lost: Unknown amount of riverine habitat.

Wetlands Created: 200 acres of PEM and POW.

Procedure: Selective grading of mining spoilbanks, creation of nesting islands, water level control structures.

Status: Initiated Spring 1986.

Source: Coal Research Newsletter, So. Ill. Univ., Vol.8 No.3, 1986.

Comments: This wetland creation is part of a reclamation by Peabody Coal Company. The goal is the development of moist soil and emergent wetland plant communities with the use of water control structures.

CRANE MARSH, INDIANA

Project Number: 9

Part of Required Mitigation: No

Parties Involved: Soil Conservation Service (SCS), Indiana Dept. of Natural Resources (IDNR), Ducks Unlimited.

Wetlands Converted: 8 acres of PEM seasonal converted to 8 acres PEM semipermanent.

Wetlands Created: 10 acres PEM.

Procedure: Flooding, dike, water direction baffles.

Status: Complete 1987.

Source: Personal communication, J. New, 28 July 1987.

Comments: The project had a three part purpose: 1) wildlife habitat creation for wetland species, 2) tertiary treatment marsh for sewage, and 3) brooding pond for muskellunge. IDNR relied on natural revegetation except for the higher gradient dam which was seeded with grasses. The marsh will be monitored once it is established and compared with preimpoundment water testing.

LAKE MAXINKUCKEE, INDIANA

Project Number: 10

Part of Required Mitigation: No

Parties Involved: IDNR, SCS, Dept. of Environmental Mgmt. (DEM), local interests.

Wetlands Created: 6 separate acres. Total: 3 acres PEM saturated, 68 acres PEM intermittently exposed.

Procedure: Dikes, baffles, islands, some revegetation.

Status: 4 structures complete, 2 awaiting funding.

Source: Personal communication, J. New, IDNR, 28 July 1987.

Comments: Project involves the creation of six sites ranging in size from 1 acre to 40 acres. Cattail shoots will be introduced in wetter areas while dikes will be planted with grasses for erosion control. Water quality will be monitored after the project is complete.

WILLOW SLOUGH FISH & WILDLIFE AREA, INDIANA

Project Number: 11

Part of Required Mitigation: No

Parties Involved: IDNR, USAE-Detroit District

Wetlands Created: 700 acres of PEM, 800 acres of POW.

Procedure: Levee damming channel to make reservoir, water level control-ditches and culverts.

Status: Complete.

Source: Permit approval, USAE-Detroit Dist., 19 Oct. 1982, permit #82-156-001.

Comments: The levee construction was performed to stabilize the water level on the management area. The resulting flooding of uplands created a lake bounded by a cattail edge.

ELK CREEK MARSH, IOWA

Project Number: 12

Part of Required Mitigation: No

Parties Involved: ICC

Wetlands Created: 600 acres of PEM.

Procedure: Development of 6 water control structures on Elk Creek.

Status: Complete

Source: Iowa Conservationist, Vol. 38 No.6, June 1979.

Comments: This major creation project was performed on the management area to increase the amount of waterfowl habitat and hunting opportunities.

SWEET MARSH, IOWA

Project Number: 13

Part of Required Mitigation: No

Parties Involved: State agencies.

Wetlands Created: 1098 acres of PEM/POW.

Procedure: 200' dam used to create a shallow water reservoir, 5 additional subimpoundments flooded with reservoir water, drawdowns to stimulate vegetation growth.

Status: Complete.

Source: Zohrer, J. Sweet Marsh. Iowa Conservationist, Feb. 1977.

Comments: First state attempt to create a large marsh area where none had previously existed. Marsh now serves as wildlife habitat and fills hunting/fishing demands.

DUBUQUE PACKING COMPANY, IOWA

Project Number: 14

Part of Required Mitigation: Yes

Parties Involved: USAE-Rock Island Dist., City of Dubuque, USF&WS.

Wetlands Lost: 1.91 acres of PEM.

Wetlands Created: 2.5 acres of PEM/PSS.

Procedure: Excavation/dredging and revegetation.

Status: Completed.

Source: Environmental Assessment, USAE-Rock Island Dist., 1980, permit #NCROD-S-070-0X6-1-079180. Personal communication, N. Johnson, USAE-Rock Island Dist., 29 July 1988.

Comments: This project provides two sites to examine (fill site and creation site). In many cases the fill site will create a new edge which requires grading and revegetation similar to any creation site. The USAE required the fill site to be graded at 3:1 with grass plantings to control erosion. While most plant species at the creation site were expected to revegetate naturally, additional plantings of button bush, deciduous holly, red osier dogwood, and white ash were proposed. Although no specific follow up studies were performed, observations of the site have recorded only about a 50 percent survival rate on the planted woody vegetation. Some willow, cottonwood, and silver maple are beginning to naturally colonize the upland slopes.

SAYLORVILLE WILDLIFE AREA, IOWA

Project Number: 15

Part of Required Mitigation: No

Parties Involved: Iowa Conservation Commission (ICC).

Wetlands Created: 100 acres of PEM.

Procedure: Diking, stop log water control structures.

Status: Completed.

Source: Iowa Conservationist, Vol. 44 No. 11, 1985.

Comments: ICC used four subimpoundments to successfully create waterfowl habitat and control flooding. The Red Rock Reservoir (ID #14, 1000 acres PEM), Hawkeye Wildlife Area (ID #15, 210 acres PEM), and Rathburn Reservoir projects (ID #16, 178 acres PEM) all relied on stop log impoundments for creation. These additional Iowa creation sites are reviewed in the same volume of the Iowa Conservationist.

OTTER CREEK MARSH, IOWA

Project Number: 16

Part of Required Mitigation: No

Parties Involved: ICC

Wetlands Created: 1000 acres of PEM (at maximum water levels).

Procedure: Extensive 8 segment dike system, periodic drawdowns.

Status: Completed.

Source: Iowa Conservationist, Dec. 1977.

Comments: Project met its goals of creating additional wildlife habitat and hunting areas. Drawdowns stimulate the development of beds of smartweed, sedges, duckweed, cattail, and arrowhead.

HARDIN, KENTUCKY

Project Number: 17

Part of Required Mitigation: No

Parties Involved: Tennesee Valley Authority (TVA)

Wetlands Created: 6000 m^2 of PEM (4 cell X 1478 m^2).

Procedure: Excavation, pumping, revegetation (*Phragmites*).

Status: Starting in 8/87.

Source: Steiner et al. 1987. Municipal waste water treatment with artificial wetlands *in* Aquatic plants for waste water treatment and resource recovery (Reddy, K. and W. Smith, ed.), Orlando: Magnolia Publishing Inc., p. 923.

Comments: Similar to the other Steiner/TVA projects but this creation relies on the root zone method for water treatment. A four year evaluation scheme is proposed.

CENTRAL CITY BYPASS, KENTUCKY

Project Number: 18

Part of Required Mitigation: Yes

Parties Involved: Kentucky Dept. of Transportation, USAE-Louisville Dist.

Wetlands Lost: 20 acres of PEM/POW.

Wetlands Created: 20 acres of PEM/POW.

Procedure: Pond excavation with grass seeding on steep slopes.

Status: In progress July 1987.

Source: Adams, J. Bypass wetlands must be replaced. Allen Lake, Messenger-Inquirer, June 1987.

Comments: Four ponds of varying depths are being created to mitigate for wetlands destroyed during highway construction.

BENTON, KENTUCKY

Project Number: 19

Part of Required Mitigation: No

Parties Involved: Tennessee Valley Authority (TVA).

Wetlands Created: 4.4 ha of PEM/POW.

Procedure: Excavation, pumping stations, revegetation.

Status: Operational Jan. 1988.

Source: Steiner et al. 1987. Municipal waste water treatment with artificial wetlands in Aquatic plants for waste water treatment and resource recovery (Reddy, K. and W. Smith, ed.), Orlando: Magnolia Publishing Inc., p. 923.

Comments: This project was created by Steiner/TVA to demonstrate the sewage treatment capability of wetland habitat based on a lagoon method. The creation involves the conversion of a secondary lagoon system to a 3 cell bulrush/cattail waste treatment system. A four year evaluation scheme is proposed to monitor the project's success.

PEMBROKE, KENTUCKY

Project Number: 20

Part of Required Mitigation: No

Parties Involved: Tennesee Valley Authority (TVA)

Wetlands Created: 6000 m^2 of PEM/POW.

Procedure: Excavation, pumping, revegetation.

Status: Construction complete 9/87.

Source: Steiner et al. 1987. Municipal waste water treatment with artificial wetlands in Aquatic plants for waste water treatment and resource recovery (Reddy, K. and W. Smith, ed.), Orlando: Magnolia Publishing Inc., p. 923.

Comments: Similar to the other Steiner/TVA projects in this study but this relies on a 3-part marsh-pond-meadow system for water treatment.

Plantings:

Marsh cell - cattail or bulrush
Pond cell - duckweed
Meadow-Reed Canary grass or sedge/rush

MAPLE RIVER STATE GAME AREA, MICHIGAN

Project Number: 21

Part of Required Mitigation: No

Parties Involved: MWHF, Mich. Chapter of the Mich. Duck Hunter Assoc., Capital Area Audubon Society.

Wetlands Created: 200 acres of PEM.

Procedure: Dikes, flooding with pump (15,000 gal./minute).

Status: Completed 1986.

Source: Michigan Wildlife Habitat News, Vol. 1 No.3, p.2. Personal communication, D. Fijalkowski, MWHF, 28 July 1988.

Comments: This restoration site (200 ac.) abuts the Milli-Ander wetland project (270 ac.) and another 200 ac. wetland. The pump will be used to keep the total 670 acres flooded. MWHF relied on natural colonization for revegetation. No follow up studies performed, but MWHF considers the projects a success because of high wildlife and recreational use of the areas.

MILLI-ANDER WETLANDS, MICHIGAN

Project Number: 22

Part of Required Mitigation: No

Parties Involved: Michigan Wildlife Habitat Foundation (MWHF)

Wetlands Created: 270 acres.

Procedure: Diking, nesting structure construction.

Status: Completed in 1984.

Source: Michigan Wildlife Habitat News, Vol. 1, No. 1, p.3.

Comments: This MWHF restoration project involved the construction of 7000 ft. of diking and four rock spill ways to enhance wildlife habitat. Nesting structures were established for wood ducks, mallards, and osprey.

SHIAWASSEE NATIONAL WILDLIFE REFUGE (AND SATELLITES), MICHIGAN

Project Number: 23

Part of Required Mitigation: No

Parties Involved: U.S. Fish & Wildlife Service.

Wetlands Created: >500 acres made up of several smaller impoundments.

Procedure: Dikes, water control structures (pumps, stop log structures), island creation, drawdowns.

Status: Complete with ongoing management by USF&WS.

Source: Shiawassee Refuge Operations Journal. 1984.

Comments: This is the only USF&WS refuge in the review, but the creation/restoration techniques described indicate that this and other wildlife refuges will prove valuable sources of information of applied creation techniques. Impoundments responded well to drawdowns providing a variety of emergent vegetation (cattails and bulrushes dominant). Stable water levels maintained on other impoundments enhanced muskrat use.

BENGEL MARSH RESTORATION, MICHIGAN

Project Number: 24

Part of Required Mitigation: No

Parties Involved: Michigan Dept. of Natural Resources (MDNR).

Wetlands Created: Restoring 2000 acres.

Procedure: Excavation/dredging, flooding.

Status: Completed in 1985.

Source: Michigan Wildlife Habitat News, Vol. 1 No. 2, p.3.

Comments: Nineteen shallow potholes, 1/6 of an acre each, and 4000 ft. of shallow connecting canals were dug throughout the 400 acre marsh.

FOUNTAIN GROVE WILDLIFE AREA, MISSOURI

Project Number: 25

Part of Required Mitigation: No

Parties Involved: MDOC

Wetlands Created: 2,200+ acres PEM/PFO/POW and upgrade 2,300 acres of PEM/PFO/POW.

Procedure: Acquisition, impoundments, pumping.

Status: In progress or completed.

Source: Fountain Grove Wildlife Area Mgmt. Plan, MDOC, 1983.

Comments: Part of a large plan to both rehabilitate wetlands already present on the refuge and create new wetland acres increasing habitat values and hunter access.

SHOAL CREEK PROJECT - BIRMINGHAM DRAINAGE DISTRICT, MISSOURI

Project Number: 26

Part of Required Mitigation: Yes

Parties Involved: USAE-Kansas City Dist., Birmingham Drainage Dist.

Wetlands Lost: 4 acres of floodplain.

Wetlands Created: 41 acres of PEM/POW (retention basin).

Procedure: Excavation.

Status: Complete and successful.

Source: USAE-Kansas City Dist., May 1981, Permit #2SB OXI 1349.

Comments: Wetland species that became naturally established in the area were broad-leaf arrowhead (Sagittaria latifolia), bulrush (Scripus sp.), and river bulrush (Scripus fluviatilis).

GRAND PASS WILDLIFE AREA, MISSOURI

Project Number: 27

Part of Required Mitigation: No

Parties Involved: MDOC

Wetlands Created: 1600 acres of PEM.

Procedure: Develop 20 moist soil impoundments with independent flood and drainage capabilities, revegetation.

Status: Proposed 1986.

Source: Grand Pass Wildlife Area Mgmt. Plan, MDOC, 1986.

Comments: The goal of this project is to reclaim 1600 acres of wetland from drained cropland creating habitat for migrating dabbling ducks and Canada geese. On 50 - 70% of each unit managers will maintain water depths not exceeding 12 in. to encourage the production of moist soil plants: millet, smartweed, and sedges. On the rest of each unit managers will encourage permanent emergent vegetation: cattails, bulrush, spike rush, water lily, and arrowhead.

BIG ISLAND WETLAND HABITAT REHABILITATION, MISSOURI

Project Number: 28

Part of Required Mitigation: No

Parties Involved: MDOC

Wetlands Created: 500 acres of PEM/POW.

Procedure: Dikes, stop log structures, submersible electric pump (5000 gal./minute).

Status: Proposed Nov. 1986.

Source: Upper Miss. River System Env. Mgmt. Program - General Plan Appendix, MDOC, 1985.

Comments: The proposed area would serve as improved habitat for migratory waterfowl and wintering bald eagles. Based on a 30-year project life expectancy the average annual cost for the project would be $30/acre. Revegetation is proposed only for stabilization of the 3:1 levee slopes.

CLARKSVILLE REFUGE HABITAT REHABILITATION, MISSOURI

Project Number: 29

Part of Required Mitigation: No

Parties Involved: Missouri Dept. of Conservation (MDOC).

Wetlands Created: 325 acres of PEM.

Procedure: Levee (1.5 miles), portable pump (175 h.p.), revegetation of levee slopes with grasses.

Status: Proposed December 1985 (request for funds).

Source: Upper Miss. River System Env. Mgmt. Program - General Plan Appendix, MDOC, 1985.

Comments: This proposal suggests removing soil from other parts of the refuge to build the levee and to simultaneously create additional wetland areas. Based on a thirty year project life expectancy the average annual cost for rehabilitating this wetland would be $14.25/acre.

POOL 25 &26; MOSIER, WESTPORT, DARDENNE, AND BOLTERS ISLAND ON THE MISSISSIPPI RIVER, MISSOURI

Project Number: 30

Part of Required Mitigation: No

Parties Involved: MDOC

Wetlands Created: 830 acres of PEM/POW.

Procedure: Diking, stop log structures, dredging, barge mounted pump.

Status: Proposed 1987.

Source: Upper Miss. River System Env. Mgmt. Program - General Plan Appendix, MDOC, 1987.

Comments: The proposal requests funds for the restoration of wetlands on four separate islands. Pumping facilities are suggested to give managers the ability to maintain wetland water level separate from the river level. The project's goal is to provide habitat for waterfowl and spawning/nursery areas for fisheries.

DRESSER ISLAND WETLAND HABITAT REHABILITATION, MISSOURI

Project Number: 31

Part of Required Mitigation: No

Parties Involved: MDOC

Wetlands Created/Restored: 993 acres PEM/ PFO/ POW.

Procedure: Levee (3.5 miles), dikes with screw gates, stop log structures.

Status: Proposed 1985.

Source: Upper Miss. River System Env. Mgmt. Program - General Plan Appendix, MDOC, 1985.

Comments: The project is proposed to restore wetland acres filled by sedimentation associated with the Miss. River. Based on a 30 year life expectancy the average annual cost is predicted to be $20.44/acre.

RIVERPORT ASSOCIATES, MISSOURI

Project Number: 32

Part of Required Mitigation: Yes

Parties Involved: Riverport Associates, USAE-Kansas City Dist.

Wetlands Lost: PEM

Wetlands Created: 13 acres PEM/POW (retention basin).

Procedure: Excavation, revegetation, water control structure.

Status: Excavation completed, revegetation in progress.

Source: Personal communication, Kathleen Mulder, USAE-Kansas City Dist., 11 Aug. 1988. USAE-Kansas City Dist., 1985, Permit #2SB OXR 2151.

Comments: Wetland plantings were initiated in Spring '88 but have been largely unsuccessful due to the drought. Upland grasses that were planted have survived, but they have begun to move into areas reserved for emergent and moist soil vegetation.

TEN MILE POND WILDLIFE MANAGEMENT AREA, MISSOURI

Project Number: 33

Part of Required Mitigation: No

Parties Involved: MDOC

Wetlands Created: 400 acres of PFO.

Procedure: Seedling planting with periodic flooding when trees are established.

Status: In progress.

Source: Ten Mile Pond Wildlife Mgmt. Area Plan, MCOC, 1984.

Comments: Because of the difficulty in establishing seedlings and the size of the area the MDOC realizes this reforestation project will take several years. Seedling species chosen for reforestation include pin oak, cherry bark oak, overcup oak, cypress, tupelo, ash, maple, sycamore, and willow.

TEN MILE POND WILDLIFE MANAGEMENT AREA, MISSOURI

Project Number: 34

Part of Required Mitigation: No

Parties Involved: MDOC

Wetlands Created: 650 acres of PEM/POW and moist soil.

Procedure: Levee system, water control structures, wells with pumps, and drawdowns.

Status: In progress.

Source: Ten Mile Pond Wildlife Mgmt. Area Plan, MCOC, 1984.

Comments: Goal is to develop multiple wetland units providing "a mosaic of herbaceous vegetation" for use by wetland wildlife species. Managers will rely on agricultural cropping to control succession in moist soil units. This plan shows a heavy reliance on water control structures (pumps & stop log) to modify habitat type and plant distribution.

COON ISLAND WILDLIFE MANAGEMENT PLAN, MISSOURI

Project Number: 35

Part of Required Mitigation: No

Parties Involved: MDOC

Wetlands Created: 2000 PEM, POW, PFO

Procedure: Dikes, water level control.

Status: Proposed 1986.

Source: Coon Island Wildlife Mgmt. Area Plan, MDOC, 1986.

Comments: The 2000 acres will consist of 850 acres of moist soil crop and plant production, 950 acres of seasonally flooded bottomland hardwood forest, and 200 acres of old sloughs and oxbows. Plans provide for 50% open water and 50% emergent vegetation (cattails, pondweed, arrowhead, and sedges). Managers also plan to encourage bald cypress, tupelo, and button bush on the edges of permanent wetlands.

KILLBUCK MARSH, WAYNE COUNTY, OHIO

Project Number: 36

Part of Required Mitigation: No

Parties Involved: Ohio Dept. of Natural Resources (ODNR), Ducks Unlimited.

Wetlands Created: Restores 325 acres of PEM/POW.

Procedure: Repair of damaged dike.

Status: In progress.

Source: Personal communication, R. Whyte, Ohio DNR, 9 July 1987.

Comments: The repair of breaches in the dike will reflood 325 acres enhancing the area's value to migratory waterfowl.

MADISON SOUTH BELTLINE PROJECTS (MONONA, WS), WISCONSIN

Project Number: 37

Part of Required Mitigation: Yes

Parties Involved: Wisconsin Dept. of Transportation (WDOT), USAE-St. Paul Dist.

Wetlands Lost: Unknown

Wetlands Created: 3630 yd^2 of PEM/POW.

Procedure: Excavation, diking, revegetation (seed bank).

Status: Complete.

Source: WDOT project #1206-2-73, Transportation Dist. 1.

Comments: Project was characterized by extensive excavation and revegetation plans.
Plantings:
- High Marsh (0 - 1.5 ft. above water table)
  - Phragmites communis, Spartina pectinata
- Shallow Marsh (0.5 - 1.0 ft. of water)
  - Sparganium eurycarpum, Sagittaria latifolia, Polygonum muhlenbergii, Pontederia cordata, Scirpus fluviatilis, Acorus calamus
- Deep Marsh (1 - 2 ft. of water)
  - Sagittaria rigida, Scirpus acutus, Potamogeton pectinatus, Nymphea tuberosa, Vallisneria spiralis

MADISON SOUTH BELTLINE PROJECTS (MONONA, WS), WISCONSIN

Project Number: 38

Part of Required Mitigation: Yes

Parties Involved: MDOT, USAE-St. Paul Dist.

Wetlands Lost: Unknown

Wetlands Created: 1+ acre

Procedure: Excavation, diking, culverts, sediment basin.

Status: Complete.

Source: WDOT project #1206-2-75, Transportation Dist. 1.

Comments: Like project #40 this creation relied on thorough written plans.
Plantings:
High Marsh (0 - 1.5 ft. above water table)
Sedge mixture, Calamagrostis canadensis, Spartina pectinata, Iris shrevei
Shallow Marsh (0.5 - 1.0 ft. of water)
Scirpus fluviatilis, Sparganium enrycarpum, Scirpus validus, Sagittaria latifolia
Deep Marsh (1 - 2 ft. of water)
Sagittaria rigida, Scirpus acutus

THE CREATION AND RESTORATION OF RIPARIAN HABITAT IN SOUTHWESTERN ARID AND SEMI-ARID REGIONS

Steven W. Carothers and G. Scott Mills
SWCA, Inc., Environmental Consultants

and

R. Roy Johnson
Cooperative National Park Resources Studies Unit
University of Arizona

ABSTRACT. Though the literature on characteristics, values, and functions of riparian habitats in Southwestern arid and semi-arid regions is fairly extensive, few papers that pertain to its creation or restoration are available. Very few creation and restoration projects are more than ten years old and most large projects have been undertaken in the last five years. Because they are so recent, evaluations of successes and failures are based on short-term results; long-term survival and growth rates are, as yet, unknown.

In most cases, creation and restoration projects have involved the planting of vegetation and not the creation of conditions suitable for the natural regeneration of riparian habitats. Many planted riparian forests do not reproduce and their longevity is therefore determined by the lifespan of the individual trees. Mitigation provided by such forests is temporary.

Important considerations for riparian creation or restoration projects in the Southwest include:

1. Depth to water table.

2. Soil salinity and texture.

3. Amount and frequency of irrigation.

4. Effects of rising and dropping water tables on planted trees.

5. Protection from rodent and rabbit predation.

6. Elimination of competing herbaceous weeds.

7. Protection from vandalism, off-road vehicles, and livestock.

8. Monitoring of growth rates as well as survival.

9. Project design flexible enough to allow for major modifications.

Because the creation and restoration of riparian habitats in the Southwest is new and mostly experimental, more information is needed for virtually every aspect of revegetation. Two major questions that need to be answered are whether planted trees survive for more than a few years and reach expected sizes, and what ranges of planting parameters are most cost-effective. Specific information needs include the identification of: the most suitable watering regimes; suitable soil conditions for various tree species; long-term survival and growth rates; and effects of variable water table levels on planted trees.

INTRODUCTION--OVERVIEW OF REGION AND WETLAND TYPES DISCUSSED

CHARACTERISTICS OF REGION

The geographical area discussed in this chapter includes the arid and semi-arid areas of the Southwest, bounded roughly by 27 degrees north and 37 degrees, 30 minutes north latitude and 103 degrees west and 118 degrees west longitude. The climate is warm temperate to subtropical with mild winters and long, hot summers. Rainfall is low throughout most of the region and primarily occurs bi-seasonally (summer and winter) or seasonally (winter). Summer rainfall usually occurs in intense local thundershowers which commonly result in flash floods and ephemeral flows in typically dry washes. Because of rapid evaporation, little of the precipitation is available to plants. In winter, rainfall is generally light and widespread and occurs for longer periods of time. These rains produce fewer flash floods and more of the water that falls is available to plants. Annual precipitation in arid regions varies from 2 to about 15 inches (Benson and Darrow 1981). Topography is varied and ranges from flat plains to rugged mountains.

WETLAND AND RIPARIAN TYPES

Various classification systems and definitions of wetland and riparian habitats exist (Brown 1982, Cowardin et al. 1979, Johnson, et al. 1984, Johnson and Lowe 1985, Ohmart and Anderson 1982, Warner and Hendrix 1984). In the arid Southwest, emphasis has been on riparian habitats because true wetlands as defined by Cowardin et al. (1979) are extremely rare. For the purposes of this chapter, we define riparian habitat as:

> "Environs of freshwater bodies, watercourses, and surface-emergent aquifers (springs, seeps, and oases) whose transported waters provide soil moisture sufficiently in excess of that otherwise available through local precipitation to potentially support the growth of mesic vegetation" (Warner and Hendrix 1984).

Many riparian areas are not wetlands as interpreted by the Army Corps of Engineers under Section 404 of the Clean Water Act, and some, especially the drier ephemeral and possibly intermittent systems, do not even meet the criteria established by Cowardin, et al. (1979). The great majority of riparian vegetation in the Southwest occurs along watercourses, and with few exceptions, all restoration and revegetation efforts summarized in this chapter involve the reestablishment of trees and shrubs along perennial streams.

Riparian habitats of the American Southwest evolved from the major Tertiary Geofloras during the past 100 million years, and are remnants of once-greater biotic communities of an expanded geographic "Southwest" (Brown 1982). Riparian habitats are classified primarily on the basis of vegetation and not on soils or hydrology (Johnson et al. 1984). For example, Brown et al. (1979) classify riparian vegetation in arid regions into four major biomes based on dominant species and plant size. These four major biomes are:

1. Interior Southwestern Riparian Deciduous Forest and Woodland.

2. Sonoran Riparian and Oasis Forest.

3. Interior Southwestern Swamp and Riparian Scrub.

4. Sonoran Deciduous Swamp and Riparian Scrub.

Dominant plants in the Interior Southwestern Riparian Deciduous Forest and Woodland biome include Fremont cottonwood (*Populus fremontii*), willows (*Salix* spp.), Arizona sycamore (*Plantanus wrightii*), ash (*Fraxinus velutina*), and walnut (*Juglans major*). Sonoran Riparian and Oasis Forests are generally dominated by Fremont cottonwood, Goodding's willow (*Salix gooddingii*), and honey mesquite (*Prosopis juliflora*), but may also contain large numbers of net-leaf hackberry (*Celtis reticulata*), ash, blue paloverde (*Cercidium floridum*) or elderberry (*Sambucus mexicana*). Saltcedar (*Tamarix chinensis*) may be dominant in both Interior Southwestern Riparian Scrub and Sonoran Deciduous Swamp and Riparian Scrub biomes. The latter may also support vegetation dominated by screwbean mesquite (*Prosopis pubescens*), honey mesquite, and arrowweed (*Tessaria sericea*).

Habitat creation and restoration has been restricted to forest types which support large trees and consequently the highest densities and diversities of wildlife.

Geographic Range

The riparian habitats considered in this chapter are limited to lowland areas of the Sonoran, Chihuahuan, and Mojave deserts in southern California, Arizona, New Mexico, West Texas and limited areas in Nevada and Utah. The entire area is drained by the Colorado or Rio Grande river systems. Specific drainages where revegetation efforts have been made include the Lower Colorado, Salt, Gila, Verde, and Rio Grande rivers and some of their tributaries.

Key Functions Performed

The many functions and relative values of Southwestern riparian habitats have been reviewed in a number of publications (Johnson and Jones 1977, Warner and Hendrix 1984, Johnson and McCormick 1978). These functions and values can be divided into three major categories:

1. Flora and wildlife.
2. Recreation.
3. Environmental quality.

Southwestern riparian habitats are valued for the high densities and diversities of wildlife they support (Johnson et al. 1985, Stamp 1978, Szaro 1980). Breeding bird densities are among the highest in the United States (Carothers et al. 1974) and have been shown to be proportional to total vegetation volume (Mills and Carothers 1986, Mills et al. 1989). A number of species, such as Bald Eagle (Haliaeetus leucocephalus), Common Black-Hawk (Buteogallus anthracinus), Gray Hawk (Buteo nitidus), and mesquite mouse (Peromyscus merriami), are restricted or functionally dependent on riparian habitats. Riparian vegetation is also important for sustaining healthy fisheries along streams with perennial flow (Platts and Nelson 1985, Platts and Rinne 1985).

Riparian habitats provide a wide range of consumptive and non-consumptive recreational activities (Johnson and Carothers 1982). Studies by Sublette and Martin (1975) in the Salt-Verde River Basin of central Arizona placed a 1972 consumer surplus value of approximately \$50 to \$60 million on recreation in an area comprising only 12 percent of the State's potential recreational area. This unusually large value is probably due in part to the proximity of metropolitan Phoenix. Water-based recreation is in such heavy demand in this desert metropolis that it boasts of having one of the larger concentrations of boats per capita in the United States. More than 20,000 recreationists (Tonto National Forest files) can be found on some weekend days along a stretch of approximately five miles of the Salt River and its riparian environs near Phoenix.

Riparian habitats play a major role in the hydrologic relations of the watershed (Mitsch et al. 1977). Odum believes that the greatest contribution of riparian habitats is a result of their function as a buffer and filter system between man's urban and agricultural developments and the aquatic environment (Johnson and McCormick 1979). Riparian habitats play a major role in improving water quality through the deposition of sediments, assimilation of nutrients and organic matter, degradation of pesticides, and accumulation of heavy metals (McNatt et al. 1980). These unique ecosystems also contribute to flood and erosion control. In healthy riparian habitats, water is stored throughout the floodplain during high flows and released slowly during low flow periods.

EXTENT TO WHICH CREATION AND RESTORATION HAS OCCURRED

GOALS

The goal of most restoration efforts has been to provide mitigation for impacts to wildlife habitat caused by changes in floodplain and streamflow characteristics. Mitigation has usually been out-of-kind. For example, cottonwood and willow trees, because of their disproportionately high wildlife values (apparently due to high foliage volume), have been planted as mitigation for the loss of other riparian species. Wildlife value has rarely been defined, that is, no species or group of species has been identified as a currency upon which to calculate the amount of mitigation required, or to evaluate success or failure of the mitigation attempt. However, in at least one case (James Montgomery Engineers 1986), total breeding bird density was used to quantitatively estimate potential mitigation value for a planted riparian forest.

A goal of some restoration efforts has been the evaluation of various revegetation techniques and their costs. In a few projects, bank stabilization has been at least a secondary goal.

HISTORICAL PERSPECTIVE

Creation and restoration of riparian habitats in the Southwest has occurred on a significant scale only since about 1977. Though a number of projects have been attempted, many have been small (<1 acre) and haphazardly planned and conducted. Only five larger, well-planned projects have been carried out, most in the last three years.

Most revegetation projects have occurred along the Lower Colorado River. Between 18 and 20 "major" projects ranging in size from less than one acre to about 70 acres have been carried out since 1977 (Table 1). A summary of these projects, including an evaluation of their success, is in preparation at the Bureau of Reclamation (Murphy 1988), but is not yet available. Projects elsewhere in the Southwest have been mostly small and casual. A major exception is the one undertaken by the Safford, Arizona District of the Bureau of Land Management, which has planted more than 10,000 dormant cottonwood and willow poles and bareroot walnut and sycamore trees over the past seven years. Unfortunately, no records of planting designs, successes, or costs have been kept (Erick Campbell, BLM Safford District, pers. comm. 1987).

TYPES OF CREATION AND RESTORATION

Unlike most other wetland types, riparian habitat restoration and creation in the arid Southwest has almost exclusively involved planting of trees and rarely has included creating conditions for natural revegetation. Restoration of most other wetland types requires, as a prerequisite, the reestablishment of suitable hydrologic conditions. In the Southwest, natural watercourses have been so impacted by man and are so controlled by dams that it is rarely possible to create conditions for natural revegetation. Riparian plant species that largely rely on floods for successful seed germination, such as cottonwoods, no longer reproduce naturally in many of these areas (Brown 1982). Because of high water demands and residential and agricultural uses of floodplains, restoration of natural flow conditions is unlikely.

As a result of the changes in hydrologic conditions caused by flood control and channelization, most restored riparian forests have not reproduced, and are extremely unlikely to do so. The longevity of most planted forests is therefore determined by the lifespan of the individual trees. Perpetuation of restored riparian forests would require a maintenance program involving periodic plantings. To our knowledge, none of the existing or planned restoration efforts include such a maintenance program.

SUCCESSES AND FAILURES

When defined, mitigation goals usually have been based on long-term survival of trees. Many projects have been obvious failures, with few living trees remaining, however, there are still high percentages of apparently healthy, growing trees at some revegetation sites. Because no revegetation projects are more than 11 years old, none can be considered a long-term success. The oldest revegetation project along the Lower Colorado River (Anderson and Ohmart 1982), which was largely experimental, has trees 10 years old and appears to have been "successful" in demonstrating that revegetation is possible and in identifying some of the key factors necessary for successful plantings. However, a recent inspection of this site determined that all planted willows (46 percent of planted trees) and many of the planted cottonwoods (36 percent of planted trees) are moribund (Murphy 1988). No data are available to evaluate the current wildlife value of this revegetation effort, and almost all other major revegetation projects are less than three years old.

Evaluations of success have been hampered by a lack of monitoring and record keeping. Many projects along the Colorado River have only recently been checked for success and results are not yet fully available. The Safford District of BLM keeps no records of success or failure.

The primary reasons for project failures appear to be lack of proper planning, lack of monitoring, and lack of effective action to solve problems. Though a number of environmental conditions (such as soil types, water table depth, and rabbits, deer, and beavers) make revegetation more difficult or impossible, most of these problems are easily solved or avoided with proper site selection and careful project design and implementation.

DESIGN OF CREATION/RESTORATION PROJECTS

Revegetation projects have included a variety of techniques, such as rooted one-gallon cuttings, dormant pole plantings, and seeding (described in Appendix I), using a variety of upland plant species such as paloverdes and saltbush as well as riparian species such as cottonwood, willow, honey mesquite, and screwbean mesquite.

Table 1: Restoration/Revegetation Attempts in the Arid and Semi-Arid Southwest.

#	LOCATION	DATE	SOURCE	AGENCY*	RIPARIAN SPECIES	AREA	DEPTH TO WATER	IRRIG. NEEDED	APPROXIMATE COST	SUBSTRATE	PLANTING
1	Beal slough	1980/81	Murphy, 1988	BLM/CDGF	Cottonwood, Willow Mesquite, Palo Verde	a) 6 ac. b) 4 ac. c) 4-5 ac.	a) 6- 8 ft. b) 6- 8 ft. c) 11-13 ft.	Yes	$ 20,000	Dredge spoil and fine sand	1-gallon (12-24")
STATUS/COMMENTS: Approximately 40 cottonwood and willow trees (of 650 originally planted) had survived by 1987. The surviving trees range in height from 5 to 13 feet. Floods and lack of irrigation are reasons for mortality.											
2	Topock Marsh	1986	Murphy, 1988	USBR/USFWS	Cottonwood, Willow Mesquite	Not reported	Variable over area (0-10 ft.)	Yes	$ 12,000***	Dredge spoil and fine sand	1-gallon
STATUS/COMMENTS: Present survival over the entire site is estimated at 70 percent. Trees growing closest to water table survived best, some now in excess of 20 feet tall. Mortalities attributed to problems with irrigation, beavers, and soil salinity.											
3	Bankline-Phase I	1986	Mills and Tress, 1988	USBR/ SWCA, INC.	a) 303 Cottonwood and Willows b) Honey Mesquite, and atriplex	2 ac.	25-30 ft.	Yes	$ About 80,000	Sandy loam with a few clay layers	a) 1-gallon b) seeds
STATUS/COMMENTS: Experimental vegetation of riprapped banklines. Most trees planted on sloping riprap died because of bank failure. After two growing seasons, trees averaged > 15 feet tall. Atriplex has done very well. Mesquite seeds were largely unsuccessful.											
4	Bankline-Phase II	1987	Mills and Tress, 1988	USBR/ SWCA, INC.	a) 60 Cottonwood and Willow b) Tree/shrub mix	0.5 ac.	5-12 ft.	No	$ 1,200	Sandy loam	a) dormant poles b) seeds
STATUS/COMMENTS: More than 50 percent of dormant poles were alive more than one year after planting. Mortality greatest for poles planted where water table was deepest. Seed experiments planted directly on riprap were mostly unsuccessful.											
5	No Name Lake	1987	Murphy	USBR/KERR LANDSCAPE	Cottonwood, Willow Mesquite, Wolfberry, Fan Palms	41.5 ac.	4-12 ft.	Yes	$175,000	Sandy loam	1-gallon
STATUS/COMMENTS: Plants presently have a high rate of survival, however sufficient time has not passed to adequately evaluate the project success.											
6	A-10 Backwater, River Mile 115-116	1984-86	Murphy	BLM	Cottonwood, Willow	5.5 ac.	12 ft.	Yes	$ 10,000	Dredge spoil and sand	
STATUS/COMMENTS: Overall, the trees at the site had a 53 percent survival rate through the spring of 1987. Mortality attributed to livestock and deer grazing/browsing and high soil salinity.											

* Bureau of Land Management (BLM)
** Per surviving tree.
*** Some labor costs not included.
**** Soil Conservation Service (SCS)

ALC - Anderson Landscape and Construction Co.
AO - Anderson and Ohmart (ASU)
BIA - Bureau of Indian Affairs
BLM - Bureau of Land Management

CDFF - California Dept. of Fish and Game
CWL - C. W. Cloud
FMTF - Fort McDowell Tribal Farm
MLS - Montana Lawn Service

SCS - Soil Conservation Service
USBR - U.S. Bureau of Reclamation
USFWS - Fish and Wildlife Service

Table 1 (Con't)

#	LOCATION	DATE	SOURCE	AGENCY*	RIPARIAN SPECIES	AREA	DEPTH TO WATER	IRRIG. NEEDED	APPROXIMATE COST	SUBSTRATE	PLANTING
7	River Mile 151	1987	Murphy	USBR	Quailbush, Palo Verde, Mesquite, Wolfberry, Forbs, and Grasses	1 Mile of Bank	2-20 ft.	Yes	$ 3,400	Sandy loam on riprap	Hydromulch
STATUS/COMMENTS: After one month of watering, some quailbush, mesquite, and palo verde had germinated and died from the heat. Irrigation suspended until winter 1988. Project still in process, too early to evaluate.											
8	Palo Verde Oxbow-1	1984-86	Murphy	USBR/CDFG	a) Cottonwood, Willow, Screwbean Mesquite b) Cottonwood, Willow, Honey Mesquite	a) 1 ac. b) 6.5 ac.	1-8 ft.	Yes	a) $ 2,000 b) $17,000	Dredge spoil/sand	1-gallon
STATUS/COMMENTS: There was an initial mortality rate of approximately 40 percent for both sites. Trees on the larger site range in size from 2-12 feet. Fifty (50) percent of the mesquites are infested with damaging insects (psyllids). Gophers killed approximately 25 mesquite trees. No additional data on survivorship available.											
9	Palo Verde Oxbow-2	1983	Murphy	BLM	Cottonwood, Willow Honey Mesquite	4 ac.	405 ft.	Yes	$ 6,800	Dredge spoil/sand	Rooted Starts (12-24")
STATUS/COMMENTS: Thirty (30) percent of the trees were killed immediately after planting due to the unusually high floods of the summer of 1983. After the flood (spring 1984), approximately 200 willow trees germinated naturally on the site. Arrowweed eventually proliferated on the site, dominating the area. Today only a few willows and small cottonwoods remain from the original planting <u>or</u> the natural germination after the 1983 flood.											
10	Palo Verde Oxbow-3 (Experimental seeding)	1986	Murphy	USBR	Screwbean and Honey Mesquite	1.5 ac.	3-5 ft.	Yes	$ 400	Dredge spoil/sand	Seed broadcast
STATUS/COMMENTS: Three one-half acre plots were cleared and seeded. The 30 surviving trees were on the plot where trees were individually fenced from rabbits. Mortality was associated with mammal and insect damage and settling dredge spoil covering and suffocating the plants.											
11	Cibola Refuge (Experimental seeding)	1986	Murphy	USBR	Quailbush, Honey Mesquite, Screwbean Mesquite, Blue Palo Verde	8 20x30 ft. plots	4 ft.	Yes	$ 1,100	Dense clay Salinity range equals 6,000-60,000 ppm	Seed broadcast
STATUS/COMMENTS: Quailbush is growing well. Rabbits have dug under the fences erected on all plots and have subsequently damaged all mesquite trees. Screwbean mesquite is doing better than honey mesquite. There are now six to 24 surviving mesquite trees per plot. None of the palo verde sprouts survived.											
12	Cibola Refuge	1978	Anderson and Ohmart, 1982	USBR/AO	Cottonwood, Honey Mesquite, Willow, Blue Palo Verde	50 ac.	Unknown	Yes	$190,000	Dense soil Salinity range equals 500-90,000 ppm	1-gallon
STATUS/COMMENTS: Surviving cottonwood trees are now in excess of 20 feet and the willow trees are in excess of 20 feet. The quailbush and <u>Suaeda</u> are 15-20 feet tall.											

* Bureau of Land Management (BLM)
** Per surviving tree.
*** Some labor costs not included.
**** Soil Conservation Service (SCS)

ALC - Anderson Landscape and Construction Co.
AO - Anderson and Ohmart (ASU)
BIA - Bureau of Indian Affairs
BLM - Bureau of Land Management
CDFF - California Dept. of Fish and Game
CWL - C. W. Cloud
FMTF - Fort McDowell Tribal Farm
MLS - Montana Lawn Service
SCS - Soil Conservation Service
USBR - U.S. Bureau of Reclamation
USFWS - Fish and Wildlife Service

Table 1 (Con't)

#	LOCATION	DATE	SOURCE	AGENCY*	RIPARIAN SPECIES	AREA	DEPTH TO WATER	IRRIG. NEEDED	APPROXIMATE COST	SUBSTRATE	PLANTING
13	Cibola Refuge Dredge Spoils	1977-78	Anderson and Ohmart, 1982	USBR/AO	Cottonwood, Willow Honey Mesquite	70 ac.	9-14 ft.	Yes	$276,000	Dredge spoil	Rooted cuttings
STATUS/COMMENTS: The initial success of this project was evidenced by trees reaching heights in excess of 40 feet. Presently, the growth has stopped and stabilized and by late 1987 all the willows were dead or dying. Many of the cottonwoods were moribund. Palo verdes, mesquites, and quailbush are apparently still health and growing. The reason for the mortalities is not known.											
14	Mittry Lake	1986	Murphy, 1988	USBR/ALC	Cottonwood, Willow Screwbean Mesquite Honey Mesquite, Quailbush, Blue Palo Verde, Wolf-berry	57 ac.	6-14 ft.	Yes	$474,000	Heterogeneous-sand, rock, gravel, silts, and clay	Trees 1-gallon (12-24") Shrubs
STATUS/COMMENTS: Overall survival of the vegetation was initially over 90 percent. Almost all the paloverde and wolfberry died immediately after planting, however these were replaced. Present growth varies widely between species. Cottonwood have shown the most growth, some having reached eight feet in height. Screwbean mesquite are more vigorous than honey, the former reaching five feet in height. Deer browsing has destroyed most of the willows, damage to cottonwoods was extensive until the beaver was caught, insect (psyllids) damage to some honey mesquites has occurred, and rabbit browsing on the honey mesquite and quailbush is extensive.											
15	Fortuna Wash Fishing Pond	1985	Murphy, 1988	USBR/CWC	Cottonwood, Willow Mesquite	8 ac.	10-11 ft.	Yes	$ 60,000	Sand with underlying clay	Poles and 1-gallon (12-24")
STATUS/COMMENTS: Irrigated trees have grown well and most are in excess of five feet tall; some of the willow cuttings have exceeded ten feet in height. The willow poles have not grown much. Some beaver damage has occurred.											
16	Tacna	1986	Murphy, 1988	USBR/MLS	a) 103 Honey Mesquite/ac., 5 Cottonwood/ac 5 willows/ac. b) Honey Mesquite	a) 109 ac. b) 30 ac.	a) 6- 8 ft. b) 10-14 ft.	Yes	$400,000	Fine sandy soil	1-gallon (6-36")
STATUS/COMMENTS: Survival of sites A and B are approximately 37 and 85 percent, respectively. Mortalities are associated with gopher and insect (psyllid) damage. The mesquites are approximately four to six feet in height, while some cottonwood and willows are in excess of six feet.											
17	Fort McDowell	1986	Murphy, 1988	SCS/FWS/BIA USBR/FMTF	Cottonwood, Willow	N/A	1-5 ft.	No	$ 2,000**	Sand and cobbles	York poles (4-10" diam.)
STATUS/COMMENTS: Of 71 trees originally planted, approximately 26 trees were still alive in late 1987. Mortalities were associated with flooding and/or a dropping water table.											

* Bureau of Land Management (BLM)
** Per surviving tree.
*** Some labor costs not included.
**** Soil Conservation Service (SCS)

ALC - Anderson Landscape and Construction Co.
AO - Anderson and Ohmart (ASU)
BIA - Bureau of Indian Affairs
BLM - Bureau of Land Management

CDFF - California Dept. of Fish and Game
CWL - C. W. Cloud
FMTF - Fort McDowell Tribal Farm
MLS - Montana Lawn Service

SCS - Soil Conservation Service
USBR - U.S. Bureau of Reclamation
USFWS - Fish and Wildlife Service

PRECONSTRUCTION CONSIDERATIONS

A key factor in assuring success of revegetation is matching physiological requirements of the selected plant species to the environmental conditions of the planting site. Considerations in site selection should include soil salinity and type, depth to water table, and probability of prolonged or heavy flooding. For example, in areas where soil moisture availability is limited, drought tolerant species such as mesquite and paloverde will have much greater probability of survival than will species such as cottonwood and willow. Virtually any species can be forced into unsuitable areas if appropriate modifications to soil conditions and water availability are made, but costs are usually very high and such planted trees may be entirely dependent on artificial irrigation.

Availability of water for at least temporary irrigation is necessary for most revegetation projects regardless of species planted.

CRITICAL ASPECTS OF THE PLAN

Timing

Project timing depends upon the revegetation method used. Most one-gallon plantings and dormant pole plantings have been done in late winter or early spring (February to April). Few tree plantings at other times of year have been attempted, but fall plantings may work. Growth rates of Atriplex planted from seed under similar irrigation regimes did not differ between March and September plantings near Parker, Arizona (Mills and Tress 1988). It appears to be advisable to avoid the hot summer months (June to September) because of extremely high transpiration rates.

Construction Considerations: Contouring, Slopes, Elevations, Size

Because typical areas where riparian revegetation efforts occur are usually relatively flat, contouring and slope are generally not considerations. With proper design, it should be possible to establish riparian trees on slopes if other requirements are met. However, in one study (Mills and Tress 1988), a high percentage of trees planted on 1.5:1 riprapped slopes were lost because of bank instability and failure. Elevation should not be a problem as long as climatic conditions are suitable for the species to be planted. Revegetation efforts of any size are possible, though larger projects tend to be more cost-effective. Irrigation labor costs, which often constitutes the most significant portion of a revegetation budget, is usually relatively independent of the number of trees planted.

Hydrology and Irrigation

Depth to water table is a critical factor for most revegetation efforts. For most projects, the idea is to get the roots to the water table so that trees are independent of irrigation or rainfall. The time it takes for trees to reach the water table depends on soil conditions, depth to water table, and amount of irrigation. For most projects so far, the water table has been at 6 to 14 feet and most trees appear to have reached groundwater in the first or second growing season. In one current project (Mills and Tress 1988) where the water table is between 25 and 30 feet, watering has been done for nearly two growing seasons and most trees do not appear to have made it to the water table though most are large and healthy. In some cases, it is easy to determine when a tree has reached the water table simply by observing its growth. Trees become much greener and grow faster after they appear to have reached the water table. It may be necessary to decrease irrigation and closely monitor tree condition to determine when trees are "independent". A neutron probe, which measures soil water content, used in PVC vertical access tubes buried to the water table is a useful device for determining how deep irrigation water and tree roots have gone (see Mills and Tress 1988).

One unresolved issue involving irrigation is its timing and amount. One school of thought, which appears to us the most reasonable, is to give the trees an overabundance of water so that the soil is saturated to the water table nearly constantly. Another school of thought is that trees should be watered less frequently and stressed to some degree to encourage deeper root growth.

In situations where it is not possible or desirable to force roots to the water table, such as trees planted along golf courses or other irrigated areas, rate of tree growth and wildlife value is generally proportional to the amount of irrigation water provided.

Dormant pole cuttings must be placed in wet soil. In most cases, pole cuttings are used where the water table is less than three or four feet below the surface, though poles have been planted in areas with water tables 8 to 12 feet below the surface (Mills and Tress 1988). Twenty-seven of 40 (67.5 percent) cottonwood and willow poles planted where the water table was eight feet survived and grew, but only 6 of 10 cottonwoods and none of 10 willows at 12 feet survived the first four months, though all leafed out initially (Mills and Tress 1980).

Effects of rapidly rising and dropping water tables on planted trees and dormant poles are unknown, but may significantly reduce the success of revegetation efforts. Water levels along parts of the Lower Colorado River have varied by more than 10 feet during the last several years and may vary by more than two feet on a daily basis. The alternate drying and flooding of roots caused by fluctuating water tables may cause stress that results in slow growth rates, susceptibility to disease or insect predation, or death.

Substrate

Along the Colorado River, soil condition and texture have been shown to be critical for successful revegetation (Anderson and Ohmart 1982). High salinity reduces the chances of survival for many species (saltbush and screwbean mesquite appear to be most tolerant). Areas with very high salinities should be avoided. Moderate salinities can be lessened to some degree by leaching before planting and during irrigation. However, long-term effects of moderate salinities, which will increase as infrequent rains wash salts back around planted trees, are unknown. In areas of very low rainfall and especially when trees are planted in upland areas, soils should be tested for salinity before planting. In areas where cottonwoods, willows, and mesquites already occur, soils probably do not need to be tested.

Heavy clay content or clay lenses in the soil also may prevent successful revegetation. Clay prevents irrigated water from reaching the water table; irrigation water moves laterally and not down. Though trees may grow large and appear healthy as long as they are irrigated, they will die if irrigation is stopped. Soil augering has been shown to be critical for successful revegetation to break up clay lenses and insure downward water movement. Vertical holes at least 15 inches in diameter should be augered to the water table before trees are planted. A self-cleaning auger, which mixes soil more thoroughly, works best.

Type of Plant Material, Source, Handling

Most large revegetation efforts have used rooted one-gallon plants grown from cuttings (Table 1). In most cases, these one-gallon plants have been obtained from nurseries in Phoenix, Arizona, such as Mountain State Nurseries. Cottonwoods and willows are grown from plant cuttings from the general planting area (preferably as close as possible) which are taken the previous fall. These plants are usually available only in early spring. Mesquites, paloverdes, and shrubs are grown from seed and, like cuttings, must be ordered in advance. Plants should be transported in closed vehicles to prevent wind damage. When it is necessary to store unplanted trees on site, they should be kept in covered nurseries and watered daily. It is generally better to plant them as soon as possible after reaching the site.

Though we are aware of no cases where transplanted mature trees have been used for mitigation, such trees have been used for revegetation and might be considered for some mitigation projects. These trees are often salvaged from construction sites and are currently available or can be special-ordered from a number of commercial nurseries in Phoenix and Tucson. The advantage of using mature trees is that mitigation for lost vegetation, and much wildlife, is almost instantaneous. It is not necessary to wait several years for planted trees to reach maturity. These trees are very expensive, however, and their use appears to be most warranted in special situations where instant replacement is desirable. Use of salvaged trees is most reasonable for slower-growing species such as mesquite or paloverde; use of salvaged rapidly-growing species such as cottonwood or willow rarely appears to be warranted.

To maximize adaptability to the planting site and reduce possible genetic contamination, dormant pole cuttings (York, in Johnson et al. 1985) should be taken from the nearest available source. Small numbers of poles are usually fairly easy to obtain, but large projects may have difficulty locating large numbers. The ideal source for pole cuttings is from dense even-aged stands three to five years old. These trees are very straight and have few branches; poles 10 to 20 feet long are easily obtained. Taking every fifth or tenth tree in such stands also does negligible damage to the source area; remaining trees quickly fill in gaps. One of the best sources along the Lower Colorado River is the Bill Williams unit of Lake Havasu National Wildlife Refuge. Hundreds of cottonwoods and thousands of willows are available and refuge personnel have allowed some harvesting.

Though dormant poles greater than six inches in diameter are often recommended, our experience has shown that almost any size works if poles are protected from rodents and rabbits. No differences in survival between poles with two and three inch diameters and those with five and six inch diameters was found in one study where 40 poles were planted (Mills and Tress 1988). Non-dormant poles also appear to work if all branches and leaves are stripped off prior to planting. Along the Lower Colorado cottonwoods and willows are "dormant" for only about two weeks in some winters.

Dormant poles can be transported uncovered for short distances, but should be stored in water if kept more than 12 hours before planting. Treating poles with a fungicide or rooting hormone is desirable.

Native seeds are available from a variety of commercial sources in Phoenix and Tucson.

One-gallon cuttings and dormant poles of cottonwoods and willows should be fenced if beavers are present in the planting area. Entire projects have been lost to beavers. A 2x4 inch welded wire mesh fence four feet high around each tree appears to be necessary for complete protection. In some cases a fence around an entire project is adequate, but fencing each tree separately, though sometimes more expensive, appears to be the best long-term strategy. In areas with beavers, trees will need protection for their entire lives; cottonwoods with trunks more than two feet in diameter have been felled by beavers. Because trees do not outgrow their vulnerability to beaver predation, beaver control measures, such as trapping, will provide only a temporary solution to this problem. In some areas, deer are serious willow predators; no entirely successful solution to this problem has been developed.

One-gallon mesquites and paloverdes should be fenced with 18 inch high chicken wire for protection from rodents and rabbits. Screwbean mesquite appears to be largely immune to rabbit attack.

Competing herbaceous plants, especially Bermuda grass (Cynodon dactylon) and Russian thistle (Salsola kali), significantly reduce tree growth and can result in project failure. Weeding of tree wells is necessary, however, only in some cases.

Reintroduction of Fauna

No reintroduction of fauna has been done in conjunction with revegetation efforts and does not appear to be necessary.

Buffers, Protective Structures

Unless planting sites are extremely remote, they should be adequately fenced to keep out off-road vehicles. Fencing may also be necessary to keep livestock from feeding on planted trees. Fences must be monitored and maintained. In some areas it may be wise to leave a cleared fire break around planting sites or plant in patches to prevent total loss from a single wildfire. Irrigation pumps left on site eventually are stolen; it is best to use portable pumps carried to and from the site.

Long-term Management

This appears to be the most critical element for revegetation success. Many past projects have suffered from virtual abandonment after planting. In many cases, the immediate causes of failure could have been easily eliminated if someone had been aware of them and taken corrective action. Agency attitudes on many early projects appeared to assume all that was required was to plant trees; whether or not they survived did not appear to matter. Some projects were "cook-booked"; trees were planted, watered for some predetermined number of months on some specified schedule, and then abandoned. Planting techniques and watering regimes developed at one site were applied to other sites without regard to differences in soils, water tables, beavers, etc. Fixed plans and techniques were drawn up ahead of time that allowed little flexibility in dealing with unanticipated problems. Over the last year, encouraging changes have been occurring. Many agencies have shown more flexibility and a greater commitment to successful projects.

MONITORING

Monitoring is a critical element of revegetation projects that, as pointed out above, has been largely missing from past efforts. Management plans with budgets large enough to handle significant changes, such as an additional year of irrigation, should be an essential part of any revegetation plan.

WHAT AND HOW TO MONITOR

Monitoring should occur throughout the "construction phase" of the revegetation project, and beyond. During the planting and irrigation phase for one-gallon tree plantings, monitoring should concentrate on the health of the trees and the depth of soil saturation and root growth. Health of trees is best assessed by an experienced person who can gain much information simply by inspecting trees and noting leaf color, presence of salt burn, relative growth rate, etc. Frequent measurements of tree growth are not necessary but should probably be made once a year for the first two or three years, and then perhaps every five or so years, as long as no major problems are detected. Because of their apparent relationship to wildlife value, some direct measure or index of vegetation biomass or volume appears to be the most meaningful measurement to make. For even-aged trees, we have found that a simple linear measurement such as height or radius often correlates closely with volume. In these cases, the simpler measurement provides nearly as much information but at a much cheaper cost. Basal diameter has not correlated well with volume in our studies. Depth of water penetration and root depth are best done with a neutron probe.

Pole cuttings simply require inspection and possibly replacement of dead ones.

All planting sites should be monitored for damage caused by rodents and rabbits, and for weed problems. Appropriate actions should be taken when necessary.

Where some "currency" has been chosen to measure "wildlife value", the currency should also be monitored. In some lower Colorado River projects (e.g., James Montgomery Engineers 1986), breeding and winter bird densities have been chosen as a currency. Estimates made during growth of plants can then be compared to conditions before planting to see if predicted increases in wildlife value are being realized. A variety of useful techniques are available for measuring wildlife. Specifics will depend on the exact nature of the site and the experience of the people involved. The most important considerations are an examination of the assumptions upon which sampling techniques are based and use of the same technique throughout the monitoring period.

MONITORING DURATION

Monitoring the site during the first two or three years appears to be critical, but some monitoring should occur throughout the life of the project. In cases where plantings are done for mitigation, perceived values of planted trees are based on the sizes of trees 10 to 20 years old. The extent of actual mitigation provided will not be known unless trees and wildlife are measured during those years. Monitoring in interim years may simply involve casual visits to look for problems and need not necessarily be detailed.

INTERPRETATION OF RESULTS

Tree measurements are fairly easily interpreted. We have found that total vegetation volume is a fairly simple measurement that correlates well with wildlife value as measured by breeding bird density (Mills and Carothers 1986, Mills et al. 1989). If trees are similar in size and shape (fairly likely in even-aged stands), height and radius measurements may provide a simple index of volume.

If a relatively simple currency, such as bird density, is used for wildlife value, interpretations are again straightforward. More complex currencies involving more wildlife types are more difficult to interpret. Interpretation is made much simpler by clearly defining goals at the beginning of the project.

MID-COURSE CORRECTIONS

The major corrections that may be required in revegetation efforts are changes in watering regimes, strengthening or erecting barriers to wildlife around plantings, and weeding around trees. Experience has shown that the amount of water required, and especially the length of time watering is required, cannot be accurately predicted in advance. Though we do not know what watering regimes are best, it is usually obvious when plants are not getting enough water.

Pole plantings should be monitored for animal damage. Once in the ground, little can be done to save poles from dying from lack of water or flooding.

INFORMATION GAPS AND RESEARCH NEEDS

The creation and restoration of riparian habitats in the Southwest is new and still mostly experimental. Though much useful information has been gained from some projects, others have produced little due to a lack of any analysis or monitoring. Because revegetation projects are so recent, no data on their long-term success are available.

More information is needed for virtually every aspect of revegetation. We especially need to know whether planted trees can survive for more than a few years and if they will reach the sizes expected. Secondarily, we need to know the ranges of various parameters that will work and which are most cost-effective. With careful monitoring and project adaptability, it appears that trees can be successfully grown under a wide variety of conditions, but we do not know whether half as much water would have worked as well.

Specific information needs are listed below:

1. Watering regimes. What is the best watering regime for one-gallon trees? Is it best to give them an abundance of water or mildly stress them to encourage root growth? Is it better to water for several days with week intervals or water every other day?

2. Soil conditions. What are the salinity tolerances and limits for each plant species? Will increases in salinity around irrigated trees after irrigation stops prove fatal?

3. What are the long-term survival rates and growth rates of planted trees?

4. Are expected increases in wildlife value ever realized?

5. How deep can rooted one-gallon cuttings and dormant poles be planted? What is the best time of year to plant?

6. Which planting techniques (e.g., rooted one-gallon tree cuttings, dormant poles, or seeding) are most cost-effective and under what conditions?

7. How much variation in water table level can planted trees tolerate?

LITERATURE CITED

Anderson, B.W. and R.D. Ohmart. 1982. Revegetation for Wildlife Enhancement Along the Lower Colorado River. Report to the Bur. Reclamation, Boulder City, Nevada.

Benson, L. and R.A. Darrow. 1981. Trees and Shrubs of the Southwestern Deserts. Univ. Ariz. Press, Tucson.

Brown, D.E. (Ed.). 1982. Biotic communities of the American Southwest--United States and Mexico. Desert Plants 4:1-342.

Brown, D.E., C.H. Lowe, and C.P. Pase. 1979. A digitized classification system for the biotic communities of North America, with community (series) and association examples for the Southwest. J. Ariz.-Nev. Acad. Sci. 14:1-16.

Carothers, S.W., R.R. Johnson, and S.W. Aitchison. 1974. Population structure and social organization in southwestern riparian birds. Amer. Zool. 14:97-108.

Cowardin, L.M., V. Carter, F.C. Golet, and E.T. LaRoe. 1979. Classification of Wetlands and Deepwater Habitats of the United States. U.S. Fish and Wildl. Serv. FWS/OBS-79/31, Washington, D.C.

James M. Montgomery Engineers. 1986. Mittry Lake Revegetation Area 1 Final Master Plan and Design Report. Report to Bur. of Reclamation, Boulder City, Nevada.

Johnson, R.R. and S.W. Carothers. 1982. Riparian Habitat and Recreation: Interrelationships and Impacts in the Southwest and Rocky Mountain Region. Eisenhower Consortium Bull. 12. Rocky Mt. For. and Range Exper. Sta., U.S. Dept. Agric. For. Serv., Ft. Collins, Colorado.

Johnson, R.R., S.W. Carothers, and J.M. Simpson. 1984. A riparian classification system, p. 375-382. In R.E. Warner and K.M. Hendrix (Eds.), California riparian systems. Univ. Calif. Press, Berkeley, Cal.

Johnson, R.R. and D.A. Jones, (Tech. Coords.). 1977. Importance, Preservation and Management of Riparian Habitat. U.S. Dept. Agric. For. Serv. Gen. Tech. Rept. RM-43, Rocky Mtn. For. and Range Exp. Sta., Ft. Collins, Colorado.

Johnson, R.R. and C.H. Lowe. 1985. On the development of riparian ecology, p. 112-116. In R.R. Johnson, C.D. Ziebell, D.R. Patton, P.F. Ffolloitt, and R.H. Hamre, (Tech. Coords.). Riparian Ecosystems and Their Management: Reconciling Conflicting Uses. U.S. Dept. Agric. For. Serv. Gen. Tech. Rep. RM-120. Ft. Collins, Colorado.

Johnson, R.R. and J.F. McCormick, (Tech. Coords.). 1979. Strategies for Protection and Management of Floodplain Wetlands and Other Riparian Ecosystems. Proc. of Symp. U.S. Dept. Agric. For. Serv. Gen. Tech. Rpt. WO-12, Washington, D.C.

Johnson, R.R., C.D. Ziebell, D.R. Patton, P.F. Ffolloitt, and R.H. Hamre, (Tech. Coords.). 1985. Riparian Ecosystems and Their Management: Reconciling Conflicting Uses. U.S. Dept. Agric. For. Serv. Gen. Tech. Rep. RM-120. Ft. Collins, Colorado.

McNatt, R.M., R.J. Hallock, and A.W. Anderson. 1980. RiparianHabitat and Instream Flow Studies. Lower Verde River: Fort McDowell Reservation, Arizona, June 1980. Riparian Habitat Analysis Group. U.S. Fish and Wildlife Service, Region 2. Albuquerque, New Mexico.

Mills, G.S. and S.W. Carothers. 1986. Bird populations in Arizona residential developments, p. 122-127. In K. Stenberg and W.W. Shaw (Eds.), Wildlife Conservation and New Residential Developments; Proceedings of a National Symposium on Urban Wildlife. School of Renewable Natural Resources, Univ. of Ariz., Tucson.

Mills, G.S., J.B. Dunning, and J.M. Bates. 1989. Effects of urbanization on breeding bird community structure in the Southwestern desert. Condor 91:416-428.

Mills, G.S. and J.A. Tress, Jr. 1988. Terrestrial Ecology of Lower Colorado River Bankline Modifications. Draft report to U.S. Bureau of Reclamation, Boulder City, Nevada.

Mitsch, W.J., C.L. Dorge, and J.R. Weimhoff. 1977. Forested Wetlands for Water Resource Management in Southern Illinois. Res. Rep. 132. Illinois Water Res. Center, Univ. of Ill., Urbana.

Murphy, S.K. 1988. Documentation of Revegetation Efforts in the Lower Colorado Region. Draft report, U.S. Bureau of Reclamation, Boulder City, Nevada.

Ohmart, R.D. and B.W. Anderson. 1982. North American desert riparian ecosystems, p. 433-479. In G.L. Bender (Ed.), Reference Handbook on the Deserts of North America. Greenwood Press, Westport, Connecticut.

Platts, W.S. and R.L. Nelson. 1985. Stream habitat and fisheries response to livestock grazing and instream improvement structures, Big Creek, Utah. J. Soil and Water Conser. 40:374-379.

Platts, W.S. and J.N. Rinne. 1985. Riparian and stream enhancement management and research in the Rocky Mountains. North Amer. Jour. of Fish. and Manag. 2:115-125.

Stamp, N.E. 1978. Breeding birds of riparian woodland in south central Arizona. Condor 80:64-71.

Sublette, W.J. and W.E. Martin. 1975. Outdoor Recreation in the Salt-Verde Basin of Central Arizona: Demand and Value. Univ. of Ariz. Exp. Sta. Tech. Bull. 218.

Szaro, R.C. 1980. Factors influencing bird populations in southwestern riparian forests, p. 404-418. In R.M. DeGraf (Tech. Coord.), Workshop Proceedings: Management of Western Forests and Grasslands for Nongame Birds. U.S. For. Serv. Gen. Tech. Rep. INT-86. Ogden, Utah.

Warner, R.E. and K.M. Hendrix (Eds.). 1984. California Riparian Systems. Univ. of Calif. Press, Berkeley, California.

APPENDIX I: RESTORATION TECHNIQUES

ONE-GALLON TREES

Nursery-grown rooted cuttings of cottonwood and willow twigs are planted in holes (preferably 15 inches in diameter) augered to the water table. An irrigation system is set up and holes are prewatered to leach salts and wet the soil profile. Trees are planted in wells of appropriate sizes and watered until roots reach the water table. The amount and timing of watering depends on the soil, weather, and depth to water table. For most projects, watering has been required for 5 to 20 months. Trees should be protected with wire fences and tree wells weeded if necessary.

DORMANT POLES

Cottonwood and willow poles 4 to 20 feet long are cut from dormant living trees. Bases are scored with an axe and dipped in a fungicide/hormone solution (such as Dip-N-Grow). Poles are buried to the water table or wet soil (an auger is best where the water table is deep). Poles should be fenced in areas where beavers are present. This technique also works with non-dormant poles from which all leaves have been removed.

SEEDS

Planting seeds has been tried in plots on flat and sloping ground. Honey mesquite and screwbean mesquite are the primary species used so far. No results are yet available but rabbits have preyed heavily on seedlings in some areas and success on sloping ground appears to be minimal. Seedings of saltbush, especially quailbush (*Atriplex lentiformis*) have been very successful along the Lower Colorado River. Growth rates are proportional to the amount of irrigation.

MATURE PLANT SALVAGE

Mature trees of any size can be boxed and moved. This technique has been used so far to salvage trees from areas to be developed. No creation or restoration projects have been tried but in some cases this technique may be useful. Its main drawback is high cost ($500 to $1000/tree). Before digging, plants are pruned by more than half to reduce transpiration area. Trenches are dug along the sides of the tree and a box is constructed around the root ball. Plants are watered for about two weeks (45 gal/day), after which any taproots are cut and a bottom is placed on the box. Plants can then be moved to a nursery where they can be kept indefinitely (at expense) until they can be replanted at desired locations. Planted trees must be watered for an indefinite period. If the water table is close to the surface and not greatly fluctuating, irrigation can be discontinued after some time period. Success rate averages over 90 percent, regardless of tree size. Species transplanted have included mesquite, paloverde, ironwood, elderberry, ash, desert willow, acacia, hack-berry, and various shrubs.

APPENDIX II: PROJECT PROFILES

The Bureau of Reclamation is currently preparing a summary of revegetation projects along the Lower Colorado River that will include information on project size, numbers of plants, success, costs, etc. For a number of these projects, many details are not available. This summary is currently in draft form (Murphy 1988) and will not be available for general release for several months.

DREDGE SPOILS--ANDERSON AND OHMART 1982

This project was the first major revegetation effort in the Southwest, with trees planted in 1977-79. The project includes two planting sites along the California side of the Colorado River near Parker, Arizona. Trees were planted on 30 ha of dredge spoil that supported no existing vegetation and on a 20 ha area that supported some saltcedars and willows.

These plantings were largely experimental and designed to evaluate effects of soil augering, soil type and irrigation regimes on survival and growth rates of planted trees. Many trees established as controls died. Two thousand trees were planted at each site at an approximate cost per planted tree of $138 and $99 at the two sites, respectively. Water table at the sites ranged from 7 to 15 feet.

Important parameters for tree survival and growth identified during the study included soil salinity and type, effects of augering, and weeding and rodent problems.

Many trees planted in augered holes on dredge spoils are alive today, though most willows and many cottonwoods are moribund. Some evidence suggests that fluctuating water levels may be a problem. When the water table is high for extended periods of time, many roots may drown. When water tables drop roots must regrow to remain in moist soil. Virtually all the trees planted at the 20 ha site have died. Primary cause of death appears to have been lack of water caused by improper irrigation and heavy clay soils.

MITTRY LAKE

A total of 5,594 one-gallon trees (3,376 cottonwood, 408 willow, 733 honey mesquite, 366 screwbean mesquite, 711 paloverdes) were planted on about 57 acres of desert scrub along Mittry Lake, near Yuma, Arizona, in spring 1986. Planting was done in seven areas where the water table ranged from 6 to 14 feet. This project was a substitute for a portion of the mitigation required for the Yuma desalinization plant, which was expected to reduce riverine habitat and riparian vegetation downstream. The cost of planting and the first year of irrigation (March to September) was about $450,000 (about $75 per tree - some shrubs were also planted). This cost does not include project research and design which was done under a separate contract. At present, most trees are alive and appear to be healthy, though some have been lost to rabbits, beavers, and deer. Irrigation was terminated in September of the first year, despite recommendations by design consultants that it be extended through October. Though it was anticipated that irrigation would be required for only one growing season, irrigation was required the next growing season as well. A monitoring program for both tree growth and wildlife use was begun in 1987.

BANKLINE

Three hundred and three one-gallon cottonwoods and willows were planted on about two acres in March 1986 along riprapped banklines of the Lower Colorado River near Parker, Arizona. The water table at this site was 25 feet when trees were planted, but had dropped to about 30 feet by the beginning of the second year. The primary purpose of these plantings was to experimentally develop a low-cost method to revegetate riprapped bankline habitats with the condition that "at least some trees should survive". As the project continued, tree survival became a major goal. One-third of the trees were planted in augered holes behind riprap similar to those at Mittry Lake, one-third were planted on the top edge of riprap, and one-third were planted on the riprap face (half with no soil, half with soil pushed over edge). At the end of the first growing season, all trees behind riprap were alive, 97 of 101 trees on the top edge were alive, and 33 of 101 trees on the riprap face were alive (29 with soil, 1 without). Trees planted behind riprap and on the top edge averaged over 2 m tall with some over 3 m.

Trees were watered through November 1986 and then again from March through December 1987. Most trees did not appear to have reached the water table by the end of 1987, though some appeared to have become independent of irrigation by the end of 1986. Beavers climbed a 20 foot high steep, rocky bank and cut down 26 trees in late summer 1986. Four-foot high welded wire baskets placed around each tree prevented further losses, though beavers continued to prune branches that extended beyond the fences. Most trees that had been cut resprouted, and were only slightly smaller than trees that had not been cut near the end of the second growing season. Cost per tree was about $150.

One hundred 5-foot long willow pole cuttings were placed in water at the base of the riprap in spring 1986. Many leafed out but all died when the river level dropped later in the year. Sixty 12 to 20 foot dormant pole cuttings were planted in augered holes in spring 1987 at a nearby site where the water table was 8 to 14 feet. Almost every pole leafed out and more than 50 percent were alive and appeared healthy in May 1988. Almost all deaths occurred within the first two months after planting. Cost per pole was about $30-40.

Though some honey mesquite seeds planted in irrigated plots behind riprap in 1986 sprouted, all had died by October. Honey and screwbean mesquite seeds were also planted on riprap and in plots behind riprap in the spring of 1987. No germination was observed on riprap and only a few of the plants that grew on plots remained in May of 1988.

TREE SALVAGE

In one recent job 240 trees were salvaged from a 13 acre riparian woodland (ca. one-third of the total trees present) in Tucson, Arizona. Success rate was greater than 90 percent. Most of the trees were planted at other development sites, a few were used for on-site landscaping, and more than 100 trees are being maintained in a nursery more than two years after having been dug. Cost of salvaged trees is high and currently ranges from about \$50 to \$80 per basal diameter inch to box and remove with an additional 40 percent of this cost required to replant and maintain in a nursery. Salvage is limited by soil, which must hold together when trees are boxed, and accessibility by heavy equipment.

BLM SAFFORD DISTRICT

The Safford District of the Bureau of Land Management has planted ca. 10,000 dormant cottonwood and willow poles and "sprigs" and bareroot walnut and sycamore trees in the last seven years. These plantings have been used primarily for habitat enhancement or as general mitigation for long-term habitat losses. Results have varied but no detailed records are available. BLM personnel indicate that success rates are high (> 90 percent) if roots reach the water table. Some trees have been established with drip systems rather than by placing poles or barerooted trees directly in contact with moist soil. Many trees have been planted in cattle exclosures in groups of fewer than 20. Major problems have included predation by rabbits and beavers, salty soils, and floods, which remove trees or drown them. No cost records have been kept but much of the work has been done by volunteer groups.

RESTORATION OF DEGRADED RIVERINE/RIPARIAN HABITAT IN THE GREAT BASIN AND SNAKE RIVER REGIONS

Sherman E. Jensen
White Horse Associates

and

William S. Platts
Platts Consulting

ABSTRACT. Riverine/riparian habitat (RRH) includes interdependent aquatic (riverine) and streamside (riparian) resources that are valuable for fish and wildlife habitat, flood storage and desynchronization, nutrient cycling and water quality, recreation, and heritage values. RRH includes resources both wetter and drier than stipulated for wetlands. Whereas the "natural or achievable state" of a riparian habitat may be wetland, the "existing state" may be non-wetland because of natural or anthropogenically induced changes in the hydrologic character of RRH.

There are many different types of RRH, each with distinctive structure, function and values. Restoration commonly requires:

1. Planning to identify preliminary goals and a general approach.

2. Baseline assessments and inventories.

3. Designs from which the feasibility of accomplishing goals can be assessed.

4. Evaluation to assure compliance with designs.

5. Monitoring of variables important to goals and objectives.

The goals, approach and design of restoration projects must be tailored to each type of RRH. Some general elements important to restoration of degraded RRH are:

1. Establishment of hydrologic conditions compatible with project goals.

2. Efficient handling of soil and substrates in construction.

3. Selection and propagation of plants suited to the site and project goals.

4. Evaluation of features to enhance habitat for target species.

5. Maintenance and control of impacts.

6. Scheduling construction to reflect site constraints and goals.

Perhaps the most universally applicable recommendation is "don't fight the river" but, rather, encourage it to work for you.

INTRODUCTION - OVERVIEW

This chapter addresses restoration of degraded riverine/ riparian habitat (RRH) in the Great Basin Hydrographic Region and the Snake River Subregion of the Columbia River Hydrographic Region. Degraded habitat is that for which the values and beneficial uses have been impaired. The goal of restoration is a less impaired state.

WETLAND TYPES TO BE DISCUSSED

RIVERINE/RIPARIAN ECOSYSTEMS

Riverine/riparian ecosystems are associated with rivers and streams of the western United States. The riverine ecosystem includes nonvegetated substrate and aquatic habitat contained within a channel (Cowardin et al. 1979). The riparian ecosystem is transitional between aquatic and upland habitat and is commonly characterized by hydric vegetation and soil. The riverine/riparian ecosystem (RRE) is the union of these two components. A watershed contains a single RRE that is continuous from headwaters to oceanic or basin sinks.

It is evident that a river is a continuum from headwater to ocean or basin sink. Likewise, the riparian ecosystem associated with a river is a continuum. Changes in the flow of water and sediments at any point along these continua, whether through natural or man-caused impacts, are communicated dynamically throughout the ecosystem. As the ecosystem is contained within a watershed, so must restoration be viewed from a watershed perspective.

The values of riverine and riparian ecosystems are interdependent. Both riverine and riparian ecosystems are essential elements of fish and wildlife habitat; the riparian ecosystem serves to store and desynchronize peak flow conveyed by the riverine ecosystem; the food chain and nutrient cycling of both ecosystems are intertwined; the cultural and heritage values of riverine and riparian ecosystems are intimately linked.

Riverine and riparian ecosystems also function in an integrated fashion. Impoundment, channelization, and diversion in the riverine system can influence the hydrologic qualities of the riparian ecosystem. Similarly, impacts to the riparian ecosystem such as livestock grazing can cause erosion of streambanks and enlargement of channels, thus influencing the functional qualities of the riverine ecosystem. Since the values and function are interdependent, the approach for restoration of riverine and riparian ecosystems must be integrated.

RIVERINE/RIPARIAN HABITAT (RRH)

Riverine/riparian habitat (RRH) is some functional subset of a RRE that includes both riverine and riparian components (see Fig. 1). Given that most streams are less than 2 meters deep, RRH is similar to wetlands as defined by Cowardin et al. (1979) - including permanently flooded aquatic habitat. In contrast to wetland, RRH may include habitat that was historically dominated by hydrophytes but, due to natural or anthropogenically induced impacts, has become colonized by upland vegetation. Upon incision of stream channels, wetland habitat on floodplains may revert to upland vegetation; hydrophytes may recolonize the floodplain if channels aggrade through deposition of sediment. RRH includes not only existing wetland, but also habitats for which the natural or attainable state is wetland. The natural state is that existing prior to or without the influence of man whereas the attainable state is that which could be achieved with best management of the resource.

Riverine habitat includes aquatic habitat and streambars (Fig. 1). Most riparian habitat occurs on floodplains, levees and swales within the valley-bottom. Streambanks are the interface between riverine and riparian habitats. Some riparian habitat is wetland. Wetland can be identified by the prevalence of hydrophytic vegetation, hydric soil, and the frequency and duration of flooding or saturation of the substrate (Sipple 1987). Upland vegetation, sometimes complimented by facultative hydrophytes, is prevalent in other riparian habitat that is not wetland. RRH includes both more hydric and more arid habitats than those defined for wetlands.

Riparian habitat may change between wetland and upland due to natural or anthropogenic causes. For example, incision of stream channels can cause drainage of shallow, alluvial aquifers and subsequent encroachment of upland vegetation onto floodplains that were formerly wetland (Fig. 2). Upon elimination of perturbations, the stream channel may aggrade, causing recharge of shallow aquifers and recolonization of hydrophytic plants. These changes may occur over a few seasons or several decades.

In the manual for identification and delineation of wetlands, Sipple (1987) states that jurisdictional determinations for wetlands which, because of cyclic hydrologic changes, do not have "fixed" boundaries, are an administrative decision beyond the scope of his manual. Similarly, jurisdictional determinations for riparian habitat charac- terized by cyclic hydrologic changes (whether naturally or anthropogenically induced) are beyond the scope of this chapter. We hope that agencies will be influential in restoring not only RRH for which the "existing state" is wetland but also for those which the "natural or attainable" state is wetland.

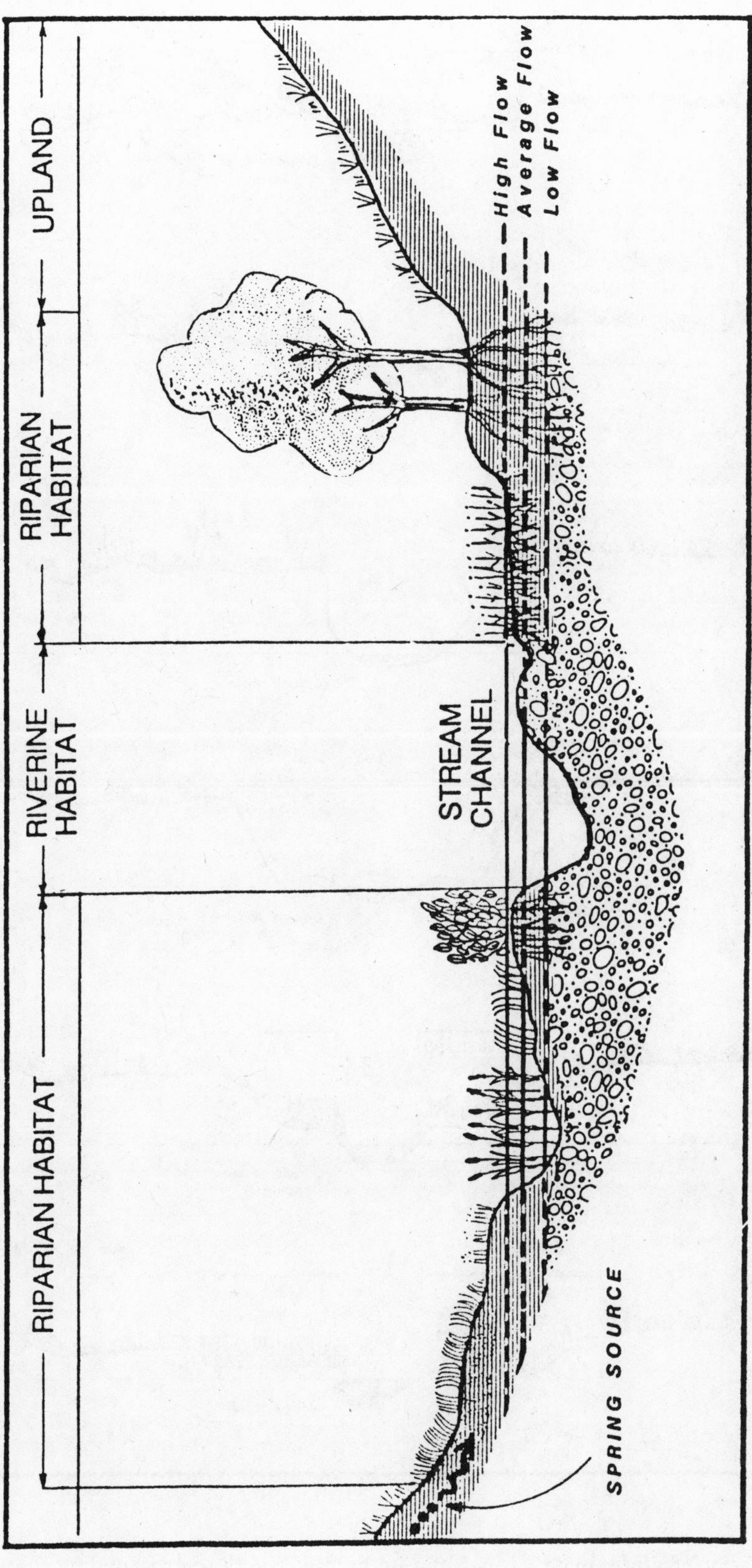

Figure 1. Riverine/riparian habitat.

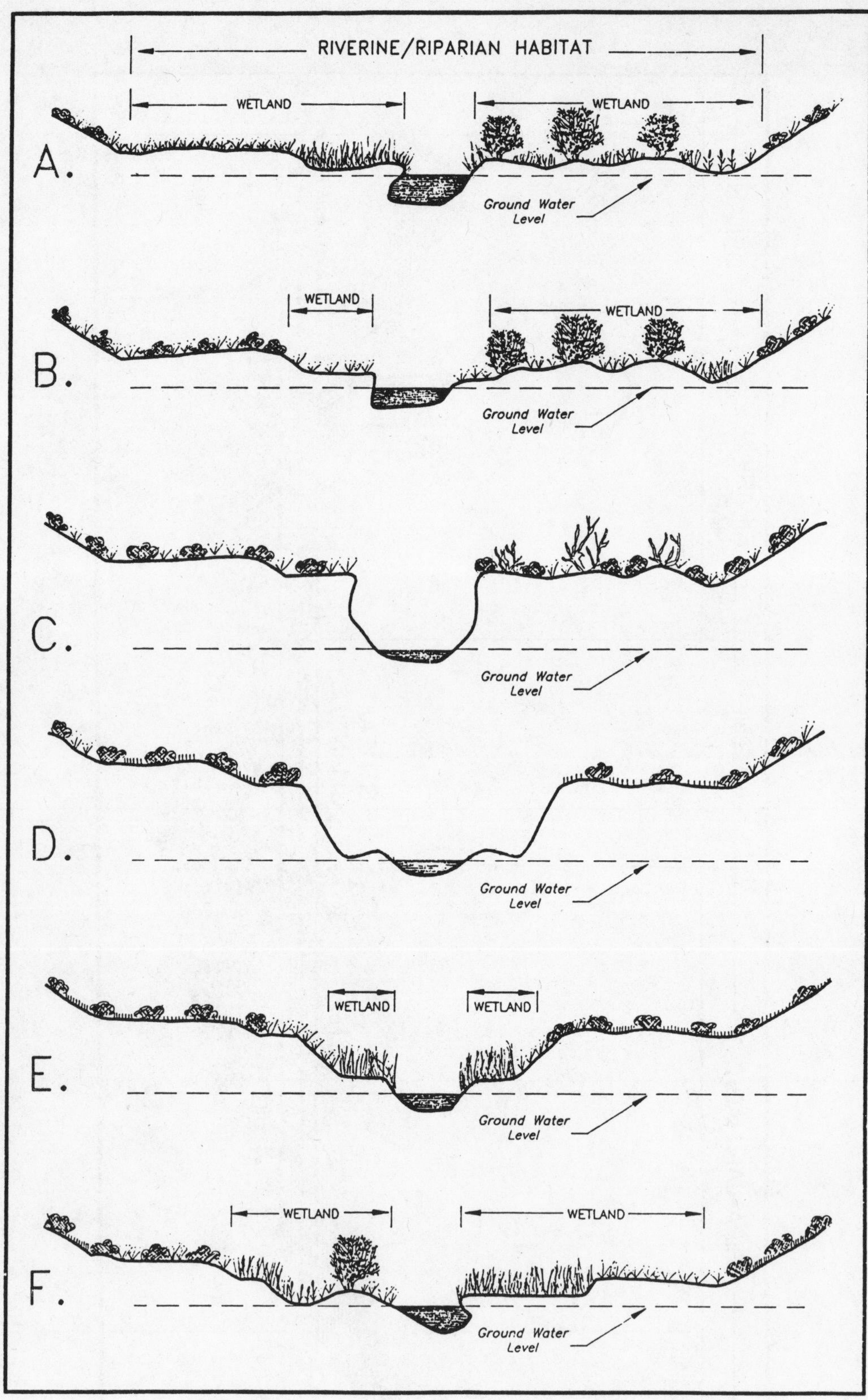

Figure 2. Progression of states in a riverine/riparian habitat and the corresponding changes in the extent of wetland.

CHARACTERIZATION OF RRH

The diversity inherent in RRH precludes any precise description of its characteristics. All RRH functions to transport water and sediments along the elevational gradients of watersheds; however, the manner in which this occurs is influenced by climatic, geologic, geomorphic, soil, and biotic variables, which differ from one location to another. Given that a purpose of restoration is to enhance the functional values and beneficial uses, it is important to recognize the geographical variables that influence the functional characteristics of RRH. As it would be unwise to attempt repair of a car without a basic understanding of how it works, it is unwise to attempt restoration of a RRH without an understanding of its functional characteristics.

The diversity inherent to RRH is apparent at several levels of perspective, ranging from regional to site-specific. A hierarchical format can be used to stratify RRH into successively more homogeneous classes, identified by causal factors influencing their functional characteristics. Bailey (1985, 1983) suggested that such a hierarchical format can be used to recognize several levels of nested ecosystems. At broad levels, ecosystems are identified by regional climatic and structural variables that provide the impetus which drives smaller ecosystems that are identified by geologic and geomorphic variables, which influence still smaller ecosystems identified by biotic attributes. Processes become evident at successively more refined scales that were not apparent at broader levels. Differences in the functional characteristics of RRH will affect the potential, approach and time required for successful restoration. A hierarchical approach can also be used to identify RRH that will behave more-or-less similarly to restoration efforts.

Some hierarchical classes that have been used to identify RRH of more-or-less homogeneous qualities at successively more refined scales are:

1. Ecoregion
2. Geologic District
3. Landtype Association
4. Landtype
5. Valley-bottom Type
6. Landform
7. Riverine/Riparian Types

Broad classes (1-5) include RRH of more-or-less homogeneous "natural or attainable" state (i.e., of similar potential). Refined classes (6-7) are RRH of more-or-less homogeneous "existing state" that may change in response to land and water management.

Ecoregions

Ecoregions are areas of similar land-surface form, potential natural vegetation, land use and soils that are identified at a very broad scale (1:7,500,000) (Omernik 1987). Ecoregions have proved useful in areas of relatively low relief for identifying streams with similar potential to facilitate impact assessments (Hughes and Gammon 1986; Rohm et al. 1987). They have also been used in Ohio for identifying relatively homogeneous stream segments for which attainable water quality can be prescribed (Larsen et al. in review) and for explaining ichthyogeographic, biotic and physiochemical characteristics of streams in Oregon (Hughes et al. 1987; Whittier et al. in review). The variance within ecoregions is not consistent among ecoregions. Functional attributes that influence the manner in which water moves through the terrestrial ecosystem can be used to identify more refined classes of RRH.

Geologic Districts

Geologic districts are distinguished by water-handling characteristics of the rock matrix and weathering products (sediments). Geologic districts can be identified at scales ranging from 1:250,000 to 1:500,000. Differences in the water-handling characteristics may be attributed to the rock matrix, the degree of weathering or the size and character of weathering products. The size of geologic districts varies with the complexity of the geology. In central Idaho, a granitic district covers about 20,000 square km. Extensive areas of basalt cover most of southern Idaho. In northern Nevada, geologic districts are more fragmented and much smaller. Geologic districts yielding an ample supply of relatively fine-textured sediments generally can be restored faster than districts yielding little fine sediment. Differences in drainage density, streamflow regime, morphological character, and fishery values were attributed to geologic districts in the North Fork Humboldt watershed in northern Nevada (Platts et al. 1988b). Geologic districts may also be useful for identifying RRH of relatively homogenous attainable water quality.

Landtype Associations

Landtype associations are areas formed by similar geomorphic processes and can be identified at 1:60,000 to 1:250,000 scale (Werzt and Arnold 1972). Landtype associations have been identified for most Forest Service lands. Some examples of landtype associations are: glaciated lands; fluvial (stream) dissected lands; piedmont (alluvial) valleys; and lacustrine (lake) basins. Landtype associations include RRH of more discrete functional character than Ecoregions or geologic districts. Within a landtype association, RRH is generally restricted to the valley-bottom landtype, which can be identified at 1:24,000 to 1:60,000 scale (ibid.). Valley-bottom landtypes can be further stratified as valley-bottom classes. Valley-bottom classes can be distinguished at 1:12,000 to 1:24,000 scale based on apparent morphological attributes that influence or reflect differences in the manner that water and sediment move through the system. Some examples of valley-bottom classes follow along with discussions of functional attributes influencing restoration.

Glacial Basins--

Glacial basins may include fens or bogs (Fig. 3). The dispersed flow characteristic of these systems favors proliferation of organic material while organic matter decomposition is limited by saturated, anaerobic conditions. The time required for development of deep, organic soils is on a scale of hundreds (or thousands) of years. Organic soils have very high water retention capacity and serve to regulate discharge to streams. Impacts resulting in drainage of organic soils can result in rapid decomposition (mineralization) of organic material and serious impairment of values and functions. Given the length of time required for development of organic soils, it is unlikely that these habitats can be restored in a reasonable timeframe. Plans to mitigate projected impacts to this RRH should be viewed with skepticism.

Glacial (U-shaped) Valleys--

Glacial (U-shaped) valleys (Fig. 4) generally extend considerable distances below cirque basins. The substrate of glacial valleys is commonly very permeable, which allows rapid equilibration of alluvial ground water and streamflow levels (Jensen 1985; Tuhy and Jensen 1982; Jensen and Tuhy 1982). This RRH is important for flood storage and desynchronization, since storage of water high in the basin is essential for reducing flood hazards and for sustaining perennial flow. Impacts tend to be expressed through erosion of streambanks, widening of channels and shallower streams. Extensive streambars and eroding banks are indices of impairment. A drop in stream level may cause a local depression in alluvial groundwater level and succession to more mesic riparian communities on streambanks. The drier riparian communities are less effective in stabilizing streambanks, thus exacerbating erosion. Restoration of this RRH should address impacts to streambanks and channel morphology prior to addressing the condition of riparian habitat. Features to facilitate sediment deposition upon streambars can be used to promote channel narrowing, increasing stream depth, and succession of more hydric communities with greater inherent stability on streambanks.

Erosional (V-shaped) Canyons--

Erosional (V-shaped) canyons (Fig. 5) are associated with steep gradient streams that may be downcutting through consolidated bedrock and/or headcutting toward drainage divides. In areas dominated by hard rocks (e.g., quartzite, limestone, etc.) channel substrate is commonly a jumbled assemblage of angular rock fragments, ranging from gravel to boulder sizes. The lateral extent of the alluvial aquifer is often limited by residual (consolidated) rock forming canyon slopes. Upland habitats may extend down to the edge of stream channels and riparian vegetation may be limited to a narrow band. This RRH conveys surface discharge rapidly and probably is of minimal value for flood control and desynchronization, although it may be of high value to fish and wildlife. Logging of steep canyon slopes commonly results in accelerated erosion and sedimentation to streambeds. Restoration of this RRH may entail slope stabilization and strategies to enhance fish and wildlife habitat.

Depositional (V-shaped) Stream Canyons--

Depositional (V-shaped) stream canyons (Fig. 6) generally occur downstream from erosional (V-shaped) canyons where stream channels are aggrading. Floodplains develop as streams drop their sediment loads and begin to wander across the canyon floors. In depositional stream canyons of northern Nevada, substrates are interbedded layers of gravel and fine (soil-sized) sediment (Platts et al. 1988a). Adverse impacts to this RRH tend to cause incision of streambeds, resulting in drainage of alluvial aquifers and encroachment of upland vegetation onto the floodplain. The higher-and-drier streambanks are more prone to failure. Restoration may entail features to encourage aggradation of the streambed, thus raising alluvial groundwater levels and facilitating stabilization of streambanks by more hydric vegetation. In some areas, streambed aggradation cannot be accomplished unless the impacts causing headcutting and/or reduced flood storage and desynchronization are first addressed.

The preceding examples of RRH are not comprehensive. Preliminary valley-bottom classes have also been identified in intermontaine (piedmont) valleys and lacustrine (lake) basins of the Northern Basin and Range

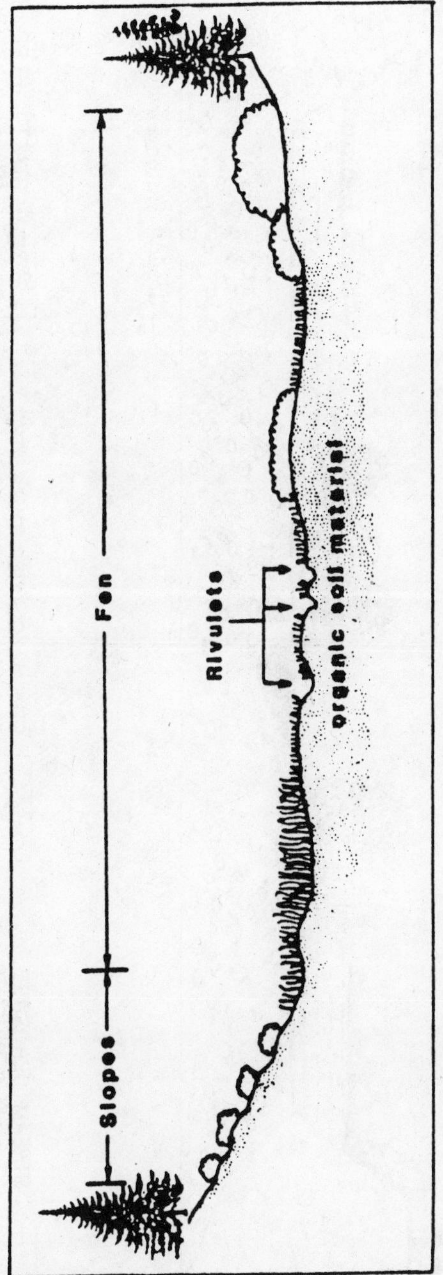

Figure 3. Riverine/riparian habitat in a glacial basin with organic substrate.

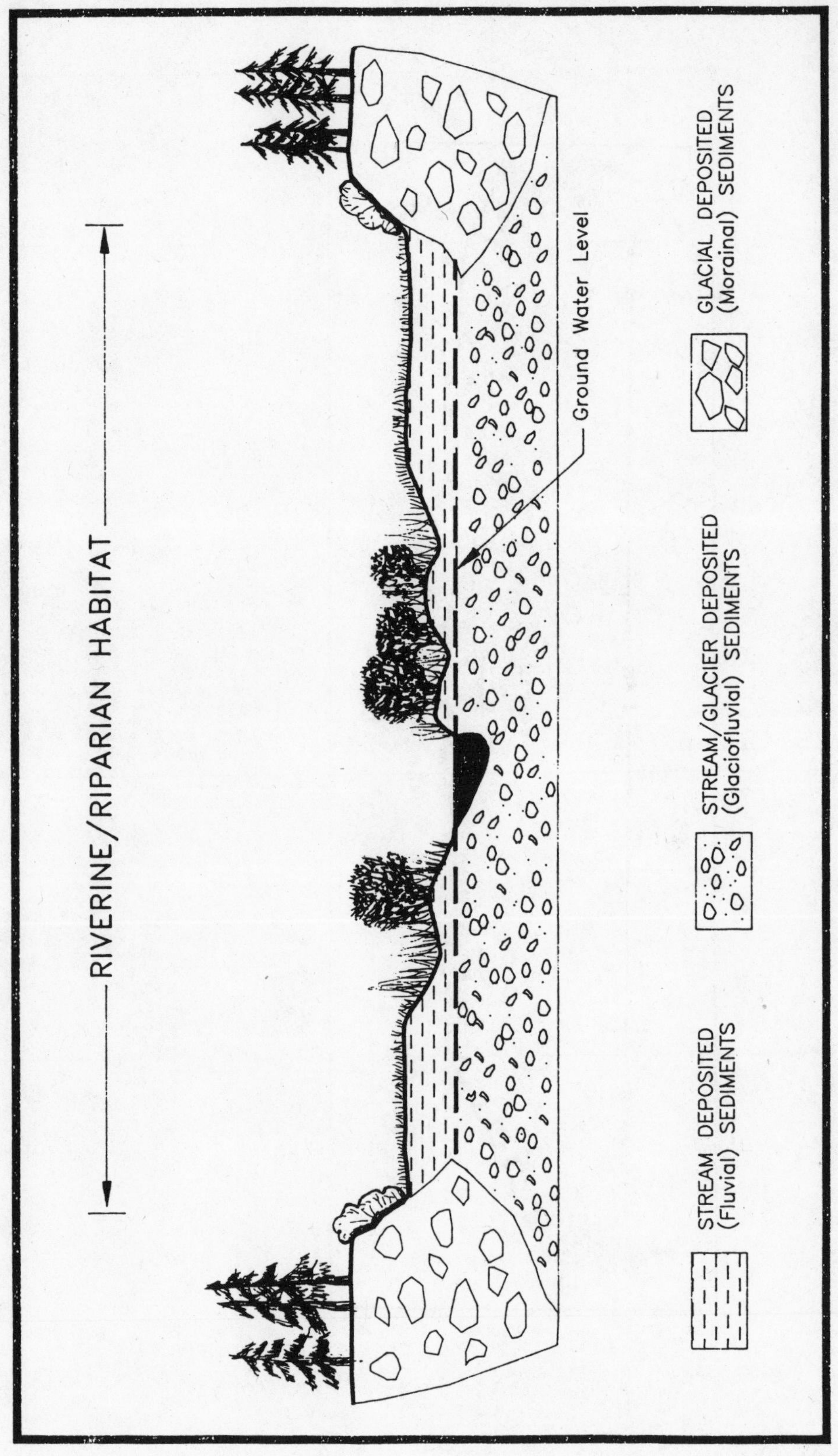

Figure 4. Riverine/riparian habitat in a glacial (U-shaped) valley.

Figure 5. Riverine/riparian habitat in an erosional (V-shaped) stream canyon.

Figure 6. Riverine/riparian habitat in a depositional (V-shaped) stream canyon.

Ecoregion and notch-shaped stream canyons characteristic of the Snake River/High Desert Ecoregion. Harris (in review) found that the distribution of riparian vegetation along the eastern slope of the Sierra-Nevada was associated with geomorphic settings similar to those previously described. Additional landtype associations and valley-bottom types can be identified. RRH of still more discrete functional character can be identified for landforms (e.g., floodplain and levee) that occur within valley-bottom classes. Landforms can usually be identified at 1:2,000 to 1:12,000 scale.

The most homogeneous components of RRH are riverine and riparian types that can usually be identified at scales of 1:2,000 or larger. Rosgen (1985) identified stream types based on gradient, sinuosity, width-depth ratio, substrate, entrenchment, and landform feature. Attributes for identification of stream types are important for assessing fishery habitat. Riparian types can be identified in terms of community physiognomy (e.g., forest, shrub, herbaceous), water regime and/or community types. Water regimes, amended from those developed for description of wetlands (Cowardin et al. 1979), are distinguished by the frequency or duration of flooding and saturation of mineral substrate (Fig. 7). Riparian community types are distinguished by indicator plant species in the overstory and understory strata.

Classifications of riparian community types have been conducted for the Greys River drainage (Norton et al. 1981), the upper Salmon/Middle Fork Salmon Rivers, Idaho (Tuhy and Jensen 1982), the Big Piney Ranger District (Mutz and Graham 1982), the Centennial Mountains and South Fork Salmon River (Mutz and Queiroz 1983), eastern Idaho and western Wyoming (Youngblood et al. 1985), southern Utah (Padgett and Youngblood 1986), Nevada (Padgett and Manning 1988). Similar studies have identified riparian dominance types of Montana (Hansen et al. 1987) and riparian zone associations in southeastern Oregon (Kovalchick 1987). An integrated classification of RRH for the North Fork Humboldt River basin is in preparation (Platts et al. 1988b).

Given the diversity inherent to RRH, the many values for which it is managed, and the myriad of land and water uses that have influenced it to varying degrees, the approach to restoration must be flexible. An understanding of the geographical variables, such as geology and geomorphology, responsible for the distribution of contrasting RRH can be useful for formulating approaches to restoration. This allows identification of RRH that will respond similarly to the same approach and extrapolation of research findings to areas of similar structure and functional character.

GREAT BASINS HYDROGRAPHIC REGION

The Great Basin Hydrographic Region includes most of Nevada, western Utah, and southeastern Oregon, with smaller portions of eastern California and western Wyoming. The region can be further stratified into hydrographic subregions including watersheds of similar hydrographic character (Fig. 8). The hydrographic region includes most of the Northern Basin and Range Ecoregion (Omernik 1987). The watershed also includes portions of the Snake River/High Desert Ecoregion, Wasatch and Uinta Mountains Ecoregion, Eastern Cascades Slopes and Foothills Ecoregion, and the Sierra Nevada Ecoregion.

Contrary to the singular connotation of the term (Great Basin), the region includes over 75 watersheds, each draining to a terminal lake or desert playa. The topography of the Great Basin is characterized by numerous north-south trending ranges separated by broad, nearly level basins. Dutton (1880) likened the pattern of discontinuous subparallel ranges to that of an "army of caterpillars marching to Mexico". These topographic features are important in determining local climate, with higher elevations and windward aspects receiving greater precipitation from the prevailing Pacific storm-tracks, while lee slopes and basins lie in rain shadows. The high mountains of the Sierra-Nevada on the west of the region also produce a rain-shadow extending across the interior of the Great Basin, which is generally semi-arid to arid.

The Great Basin is among the most geologically diverse areas in the United States. The geologic structure was produced by block-faulting of folded and thrust-faulted overlapping geosynclines of Paleozoic and Mesozoic ages (Hunt 1976). Intrusions of igneous rocks and volcanism are also common. Remnants of complex structures are apparent in upthrust block surfaces forming mountain ranges while downthrust block surfaces are covered with alluvial sediments, often to a depth of over 1000 meters. Springs and seeps tend to be most common on the down-dipping flank of mountain ranges.

Most of the major streams of the Great Basin originate in the Sierra-Nevada in the west and

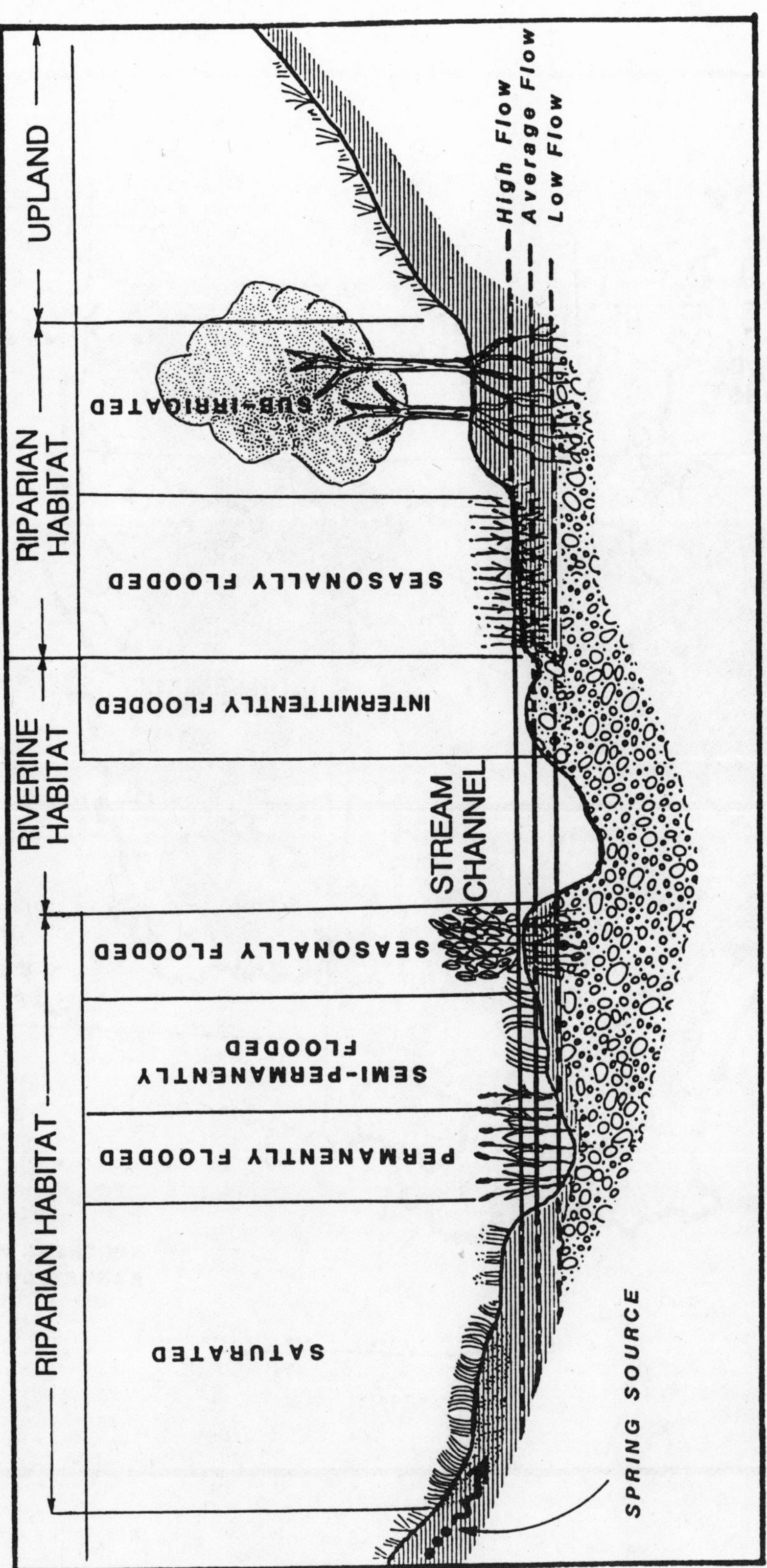

Figure 7. Water regimes associated with riverine/riparian habitat.

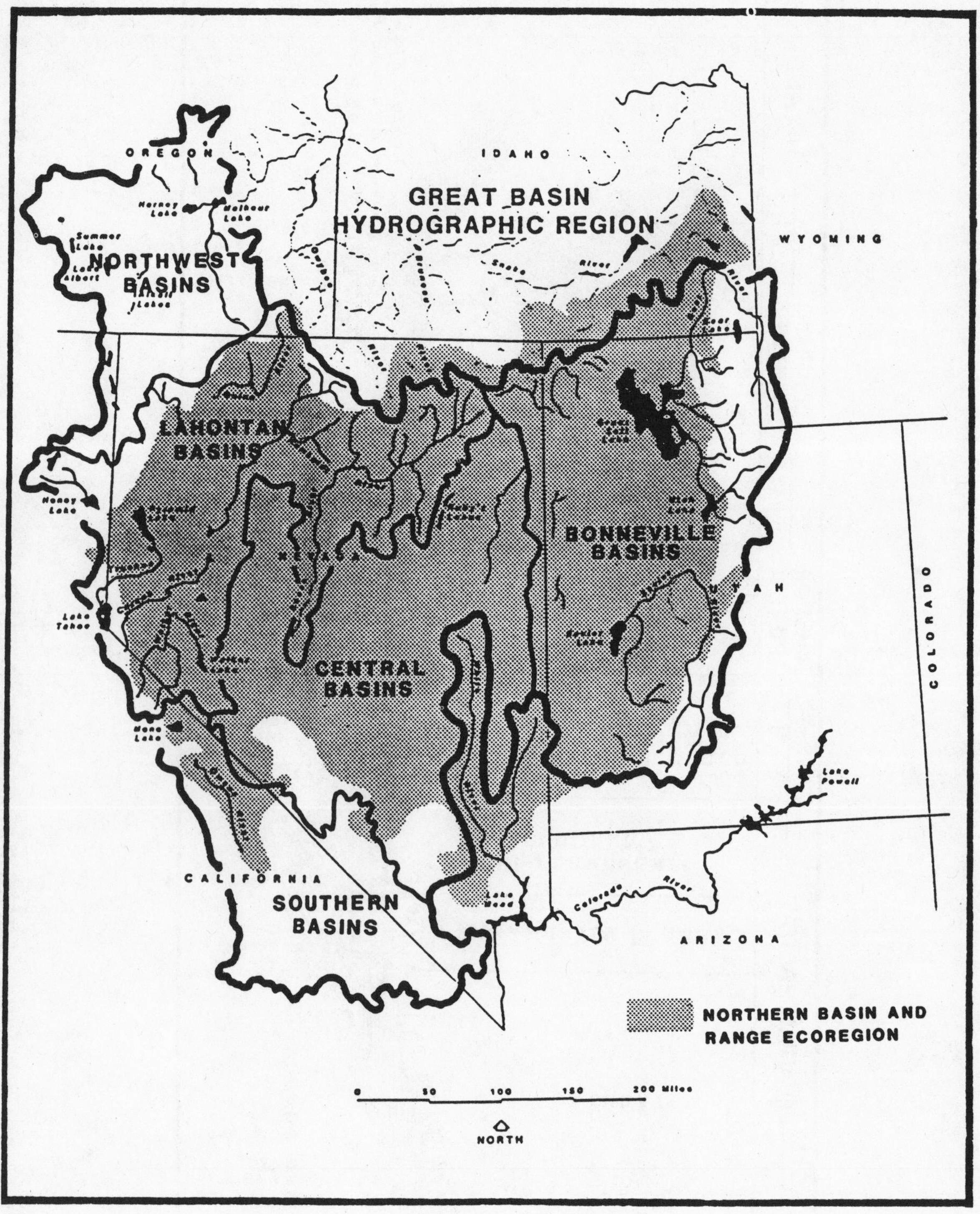

Figure 8. Subregions of the Great Basin Hydrographic Region and their relationship to the Northern Basin and Range Ecoregion (Omernik 1986).

the Wasatch Mountains in the east. The Humboldt River is the only major stream rising within the interior of the Great Basin. A typical major stream arises in glacial basins along watershed divides, flow through glaciated (U-shaped) valleys and stream-cut (V-shaped) canyons along its course before emerging into broad intermontaine valleys, and finally terminates in a lake or playa. Most minor streams draining the interior of the Great Basin originate in stream-cut (V-shaped) canyons draining isolated mountain ranges and are ephemeral along their lower courses through intermontaine basins. Streams are commonly diverted for irrigation of hay meadows soon after leaving the confines of canyons.

The distribution of wetlands generally corresponds with drainage courses and their sinks. Spring-fed wetlands are also prevalent, especially along fault zones. Shaw and Fredine (1956) identified 258,000 hectares of wetlands in Utah and 78,000 hectares in Nevada that were important to waterfowl, most of which were associated with terminal lakes. A National Wetlands Inventory under the direction of the Fish and Wildlife Service is now in progress. Much of the RRH in the Great Basin has been degraded by livestock grazing to the extent that it is no longer wetland (GAO 1988). Large storm events have devastated RRH that were overgrazed in comparison with ungrazed areas (Platts et al. 1985).

Extensive inland seas covered much of the Great Basin during the Pleistocene. With the retreat of the glaciers and erosion of lake outlets, drainage basins became isolated. Further desiccation of the lakes led to the isolation of small, spring-fed pools and streams in which disconnected populations of fishes evolved distinctive genetic characteristics (Minshall et al. in press). Fish species endemic to hydrographically isolated locations of the Great Basin have been reviewed by Sigler and Sigler (1987). Fish populations were greatly reduced (or eliminated) through extensive commercial fishing late in the 19th century, water withdrawal for irrigation and impacts of livestock grazing. Well over half of the RRH in the Great Basin is in only fair or poor condition due to poorly managed livestock grazing (GAO 1988).

SNAKE RIVER SUBREGION

The Snake River Hydrographic Subregion (Fig. 9) is 281,800 square kilometers including most of Idaho and minor portions of Wyoming, Utah, Nevada, Oregon, and Washington. The subregion includes most of the Snake River Basin/High Desert Ecoregion and parts of the Middle Rockies, Northern Rockies, Columbia Basin, and Blue Mountain Ecoregions (Omernik 1986). Climate varies as a function of topography; annual precipitation ranges from greater than 100 cm along the highest ridgelines to less than 25 cm on the Snake River Plain (Hunt 1976).

The geology of the Snake River Subregion is less diverse than the Great Basin to the south. Most of the Columbia Basin is covered with thick, horizontal layers of extrusive igneous rock (basalt) in which streams have cut notch-shaped canyons. The Northern Rockies are dominated by intrusive igneous rock (granite) cut by glaciers to form U-shaped valleys and by streams to form V-shaped canyons.

The Snake, Henry's Fork, and Blackfoot rivers originate in glaciated lands in the vicinity of Yellowstone and Grand Teton National Parks and converge in the eastern portion of the Snake River Plain. The Snake flows through notch-shaped canyons cut in basalt along much of its westwardly course across southern Idaho. Salmon Falls and Owyhee rivers drain to the Snake from the south. The Boise, Payette, Salmon and Clearwater Rivers are major tributaries of the Snake which originate in the Northern Rocky Mountains of central Idaho. The Snake merges with the Columbia River in southeastern Washington. The Columbia, then, flows into the Pacific Ocean.

RRH in the Snake River basin has been impacted by hydroelectric dams and diversions, irrigation withdrawals and return flows, livestock grazing, logging, dredge and placer mining, and industrial waste from widely scattered urban centers. Historically, anadromous fish migrated from the ocean to spawn in the headwaters of Snake River tributaries. But, hydroelectric dams along the Columbia River depleted these runs to a fraction of historical levels. Hell's Canyon Dam, located upstream from the confluence of the Salmon River, stopped salmon and steelhead from passing to upper portions of the Snake River. Only the Salmon and Clearwater Rivers sustain wild anadromous fisheries, albeit impacted by dams on the Columbia River, and by logging, mining, irrigation withdrawal and livestock grazing.

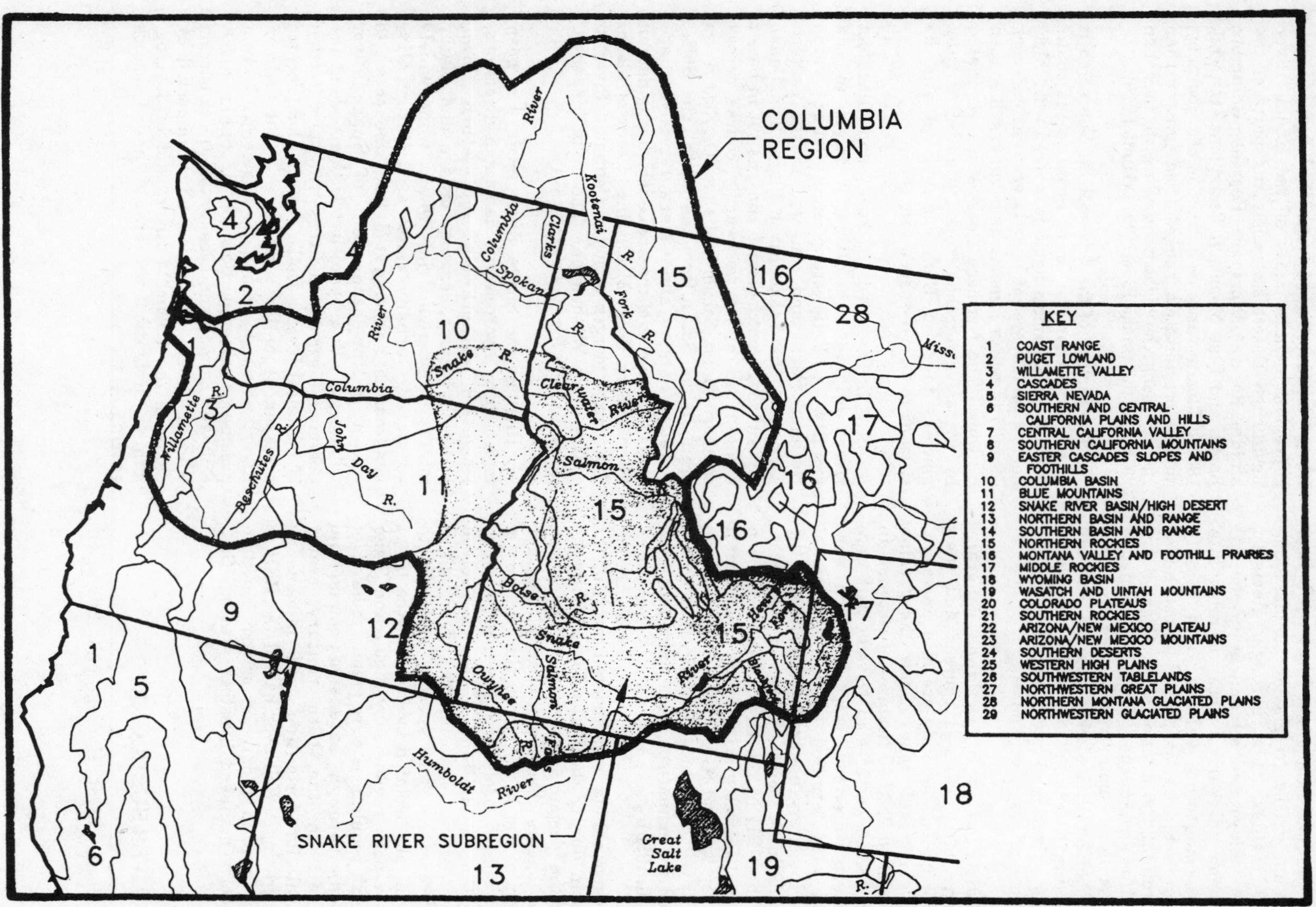

Figure 9. The Snake River Subregion of the Columbia Hydrographic Region and its relationship to the ecoregions of the area (Omernik 1986).

EXTENT TO WHICH RESTORATION HAS OCCURRED

Many projects to enhance and restore aquatic habitat as fisheries have been conducted in the Great Basin and Snake regions. Among the most notable are projects to enhance anadromous fisheries in the Salmon and Clearwater basins of the Snake River subregion. Most of these projects approach restoration only from a fishery perspective. Projects administered by the Bonneville Power Administration, Division of Fish and Wildlife are listed in annual project summaries (Bonneville Power Administration Agency 1986) and documented in numerous project reports.

Many projects to restore RRH are in progress. Several projects to restore RRH devastated by dredge mining in the Salmon and Clearwater basins have been initiated (Konopacky et al. 1985; Richards and Cernera 1987a, b; Hair and Stowell 1986; Platts et al. 1986). A project to create RRH along a diversion leading to a hydroelectric facility in southern Idaho is ongoing (Jensen et al. 1987). Efforts to restore RRH impacted by road construction along the Teton River (Johnson 1987) and the Snake River (Johnson 1986) are in progress. Wetland habitat was created in Cedar Draw, a minor tributary of the Snake River, and is being monitored (Jensen, 1988a). Plans to enhance wetlands at a nature center in Ogden, Utah have been reported (Sempek and Johnson 1987). Creation of RRH is underway in connection with a housing development along the Boise River (Gebhardt 1986). A hydraulic and geomorphic approach to restoration was evaluated for a segment of the Medicine Bow River in Wyoming (Lidstone 1987). These projects are still in their infancy and will require more time for evaluation. A more mature restoration project utilizing dredged material has been reported along the Columbia River (Landin et al. 1987). Some restoration projects are reviewed in Appendix II.

The General Accounting Office (1988) reviewed 22 riparian areas that had been restored by the Bureau of Land Management and the Forest Service. Restoration was accomplished primarily through improving the management of livestock. Only a very small fraction of degraded riparian areas has been restored. In most areas, well over half of the existing RRH is in fair to poor condition.

GOALS OF RESTORATION

The goals of restoration typically address the values and functions attributed to the resource. Enhancement of fish and wildlife habitats, is probably the most common goal of restoration. Other important values and functions are aesthetics, recreation, flood storage and flood desynchronization, streambank stabilization, and water quality. Goals of restoration are typically coincidental with maximizing the potential beneficial uses of the resource. Goals are commonly generic and related to regional and local values and bias (Miller 1987).

SUCCESS IN ACHIEVING GOALS

Success in achieving the goal of enhancing fish and wildlife is often difficult to assess directly. The natural variability in fish populations, removal by sportsmen, seasonal differences in migration, food sources, etc. are difficult to distinguish from changes due to restoration. Similar problems are encountered when monitoring terrestrial wildlife. As a result, use of indirect assessment techniques such as the Habitat Evaluation Procedures (HEP) developed by the U.S. Fish and Wildlife Service (1980), are common.

The success of projects must often be viewed over a long term. Wetlands were created along the eastern shore of the Great Salt Lake through diking of the Bear and Weber Rivers. The Great Salt Lake is terminal; the level of the lake oscillates in response to climatic inputs over its watershed. Most of the facilities were constructed below 4,210 feet elevation, even though the saline lake filled to above this elevation in the late 1800's. These extensive wetlands were managed by the U.S. Fish and Wildlife Service (Bear River Migratory Bird Refuge), the Utah Division of Wildlife Resources (Ogden Bay Waterfowl Management Area) and by private owners who catered to sportsmen. They constituted the most extensive waterfowl habitat in Utah and, for a half-century, were extremely successful. About 85 percent of these wetlands and most of the facilities for managing them were destroyed by a rise of the Great Salt Lake to above 4,211 feet in 1983 and 1984 (Bureau of Land Management 1986). Were these projects successful?

Many wetland restoration projects have been only recently conceived or initiated. The success/failure of these projects cannot be realistically assessed at this time.

FACTORS AFFECTING SUCCESS OR FAILURE

Due to the lack of mature restoration projects in the Great Basin and Snake River regions, factors affecting success or failure can only be hypothetically addressed. Given the importance of hydrologic variables to RRH, they are expected to be the most important factor affecting restoration. Other factors such as climate, topography and soil may influence restoration through their effects on hydrologic variables. Some examples follow:

1. The plugging of a water pipe caused a temporary setback in establishment of wetland habitat in Cedar Draw (Jensen 1986). A pressurized water supply was developed and has been serving well for 2 years.

2. The contour elevation and the size of substrate was limiting to establishment of riparian habitat on a bench constructed primarily from basalt boulders (Johnson 1986). While it was anticipated that fine sediment would be accumulated through flooding of the bench during high flow periods, streamflow levels in 1987 and 1988 remained well below the surface of the bench. Irrigation supplied to the bench has been partially successful in sustaining planted vegetation.

3. The construction of riparian habitat to facilitate establishment of an alluvial aquifer proved successful for establishment of riparian vegetation along a relocated channel for Birch Creek (Jensen et al. 1987). A meandering stream channel served to diversify fishery habitat and to create microsites for a more diverse riparian habitat. Ice "plucked" about half of the willow and birch planted in the riparian zone. Inducement of natural colonization through controlled flooding of the riparian zones during periods when propagules are transported in streamflow is being evaluated.

Richards (Shoshone-Bannock Tribe pers. comm.) suggested that specific procedures for enhancing restoration of one site are usually not applicable to other sites. In restoring two sites in the same general vicinity, both damaged by dredge mining, entirely different approaches were required in order to accomplish different goals. Restoration of a dredged segment of Bear Valley Creek was conducted to reduce sedimentation to streambeds, thus enhancing spawning habitat. Restoration of a dredged segment of the Yankee Fork was conducted to provide sidechannel pools to serve for rearing habitat for young fish. Richard's recommendation for restoration was "don't fight the river".

Improvement of management to restore RRH on public lands may be limited by incomplete inventories and assessments of the resource, lack of commitment by management agencies and conflicts with resource users, especially livestock permittees (General Accounting Office 1988).

DESIGN OF RESTORATION PROJECTS

The diversity of RRH precludes a "boilerplate" approach to restoration. Methods effective for restoration in one ecological setting may not be applicable to another. Different goals may also require entirely different designs for the same ecological setting. Some general steps for design of restoration projects are:

1. Preliminary planning to establish the scope, goals, preliminary objectives and general approach for restoration.

2. Baseline assessments and inventories of project locations to assess the feasibility of preliminary objectives, to refine the approach

to restoration, and to provide input for the project design.

3. Design of restoration projects to reflect objectives and limitations inherent to the project location.

4. Evaluation of construction to identify, correct, or accommodate for inconsistencies with project design.

5. Monitoring of parameters important for assessing goals and objectives of restoration.

Some elements to consider for each of these steps are outlined in Figure 10 and subsequently discussed.

PRECONSTRUCTION CONSIDERATIONS

Preliminary planning should involve developers, contractors, engineers, environmental scientists, and representatives of regulatory agencies. The scope, goals, preliminary objectives and general approach to restoration should be established. Available information pertinent to the project location should be assembled and its relevance and accuracy evaluated. While general information used for area management or broad-scale assessments may be useful for determining a general approach for restoration, it is usually not sufficient for design of restoration projects. More accurate on-site inventories and assessments are generally required.

On-site assessments and inventories commonly require the work of engineers, hydrologists, soil scientists, botanists and wildlife specialists. Depending on the scope and goals of the project, products of on-site investigations might include:

1. Assessment of impacts influencing the existing state and the achievable state of the site.

2. A topographic survey and measurement and/or projection of hydrologic variables.

3. A soil survey.

4. A botanical survey.

5. Fish and wildlife surveys.

6. A baseline report.

Additional or alternative products may be required for projects of more specific nature; priorities must be specific to the project. Inventories and assessments are used to determine the feasibility of restoration, to revise preliminary objectives, and to refine the approach to restoration. Products of on-site studies are inputs to the project design.

ASSESSMENT AND DOCUMENTATION OF SITES

Documentation of the existing state of the site can be used to evaluate the potential for restoration and can serve as a baseline for evaluating the success of restoration. Factors that are limiting to the beneficial uses and values of the site should be identified. These might include measures of stream and channel morphology (e.g., width, depth, substrate, condition of streambanks), fish and wildlife populations, and condition of riparian habitats. Methods for description and evaluation of RRH were assembled by Platts and others (1987). Given that it requires restoration, the existing state is probably different from the "natural or achievable" state for the site.

It is often the case that the "natural" state of RRH is coincidental with maximum beneficial uses and thus may be the ultimate (though probably optimistic) goal of restoration. Given the extent of man's influence upon the environment, it is questionable whether natural states can be identified other than as a theoretical beginning. The achievable state for a site is what can be expected as the end-point of restoration. To project achievable states, it may be necessary to locate minimally impacted reference sites of ecological character similar to the project location. Selection of reference sites should entail evaluation of climatic, geologic, geomorphic, topographic, hydrologic, soil and biotic attributes of both the project-site and potential reference sites.

Restoration usually cannot be accomplished if impacts to the site are not controlled or eliminated. On-site impacts may be attributed to livestock, wildlife, or recreational activities. Fencing and/or closure may be necessary to control on-site impacts. Off-site impacts may influence the site through upstream irrigation withdrawals, regulation of streamflow by upstream reservoirs, headcutting of stream

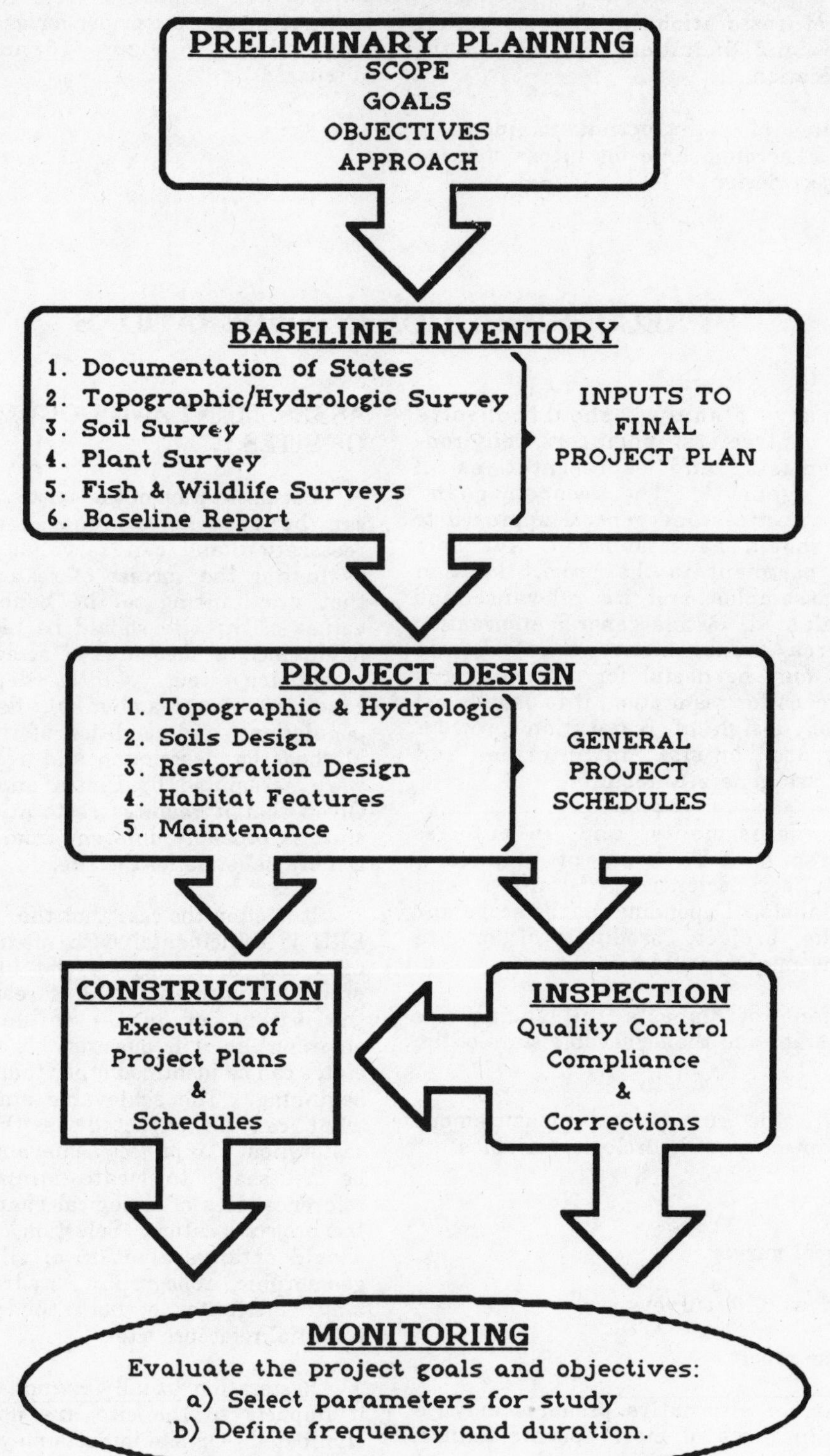

Figure 10. Elements of a restoration plan.

channels, deposition of sediment from upstream sources, and alteration of runoff/retention relationships of contiguous uplands. Methods for dealing with off-site impacts must be addressed in the project design.

TOPOGRAPHIC/HYDROLOGIC SURVEY

A detailed topographic survey is usually needed to determine existing gradients. Miller (1987) suggests a scale of 1:240 to 1:600 and a contour interval of 1-foot for design of freshwater wetlands although 1:1,200 to 1:2,400 surveys with 5-foot contours are used more commonly. The scale and contour interval should be such that hydrologic variables can be accurately assessed.

A reliable source of water is crucial to successful restoration. Water sources might include streamflow, groundwater, spring discharge, and run-on of incident precipitation. Accurate projections of the dynamics of discharge, gradient, velocity, and stage are necessary for sizing stream channels and floodplains, for assessing risk to structures (e.g., check-dams), and for predicting erosion and streambank stability. The effects of anchor ice must also be anticipated. Stream gauging stations monitored by the U.S. Geological Survey are usually the only source of long-term streamflow measurements. The use of computer models developed by the U.S. Army Corps of Engineers (HEC1 and HEC2) was suggested for evaluation of basin hydrology, channel hydraulics and equilibrium analysis for stream channel and wetland construction (Lidstone 1987). Instream Flow Incremental Methodology (IFIM) developed by the Fish and Wildlife Service (Bartholow and Waddle 1986) can be used to evaluate the effects of basin-wide water allocations upon restoration potential.

The relationship between the stream level and alluvial groundwater level is important for designing grades to achieve the proper hydrologic conditions for riparian vegetation. This relationship is influenced by the depth and permeability of substrates and by valley-form. Monitoring wells (perforated plastic pipe inserted in bore-holes to a depth below the minimum groundwater level) can be used to measure spatial and temporal variations in alluvial groundwater level and its relationship to stream levels (Bohn, U.S. Forest Service Intermountain Research Station pers. comm.).

SOIL SURVEY

A soil survey to determine the volume of materials available for use in restoration is needed when contouring is envisioned. The intensity of the survey must be commensurate with the scope and approach to restoration. Where extensive earthwork is envisioned, an order 1 (very detailed) soil survey should be conducted. Soil types (polypedons) are delineated on aerial photos. Representative profiles of each soil type are described (Soil Conservation Service 1982) and classified (Soil Conservation Service 1975). Samples of each soil horizon are obtained for physical and chemical analyses to identify limiting characteristics for use in restoration. Soil properties important for use in restoration are summarized in a document prepared by the U.S. Army Corps of Engineers (Environmental Laboratory 1986). More general (order 3-5) soil survey reports published by the Soil Conservation Service may be adequate for projects not requiring earthwork.

PLANT SURVEY

Botanical surveys can be conducted in conjunction with soil surveys. Plant communities are delineated on aerial photographs. Species composition and cover are described for representative communities of each type. Environmental attributes (e.g., landform, soils, frequency and duration of flooding) are also described. Threatened, endangered and sensitive plant species must be identified. Botanical information can be used to formulate vegetation designs and to determine what plant materials useful for restoration are available at the project location. Brunsfield and Johnson (1985) prepared a guide for identification of willows that has been very useful in the Snake River Basin and the northern portion of the Great Basin. Previously cited classifications of riparian community types should be used where applicable.

FISH AND WILDLIFE SURVEYS

The goals of restoration are often to enhance fish and wildlife resources. Surveys can be used for two purposes:

1. To identify species of concern.

2. To provide baseline information against which restoration can be compared. Inventories can also be used to estimate the potential of the site and to identify factors limiting to the carrying capacity.

Surveys may entail population estimates and/or assessment of habitat.

Populations can be estimated directly or indirectly (U.S. Fish and Wildlife Service 1980). Direct methods entail counting of individuals and are best suited to populations that are sedentary or concentrated in limited areas (e.g., fish in small streams). Indirect estimation of populations involve the use of indices (e.g., pellet count) from which population size is estimated. Population estimates at any point in time may be influenced by sampling errors, cyclic fluctuations, removal by sportsman, migration, etc. Reliable estimates of population potential or carrying capacity may require several years of monitoring and changes attributable to restoration may be difficult to distinguish from variability inherent in population dynamics.

Habitat is that which supplies the space, food, cover, and other requirements for survival of a particular species. Much of the long-term variability in populations can be attributed to changes in the quality and quantity of their habitat (Black and Thomas 1978). Assessment commonly entails measurement of habitat indices (e.g., pool quality, canopy cover, plant composition and density, etc.) thought to be important to fish and wildlife species. Measures of habitat indices are compared with those of optimum habitat to assess condition for a particular species.

The U.S. Fish and Wildlife Service (1980) has developed habitat evaluation procedures (HEP) that have been used in baseline assessments of RRH (Johnson 1986). HEP entails measurement of key habitat indices for a particular species or group of similar species. Measurements are used to calculate a Habitat Suitability Index (HSI) that is linear compared to carrying capacity for optimum habitat conditions. The HSI for a species is calculated using a documented habitat suitability model. The HSI ranges from 0.0 to 1.0; higher scores are closer to optimum habitat. The HSI is multiplied by area of available habitat to obtain Habitat Units (HUs) for individual species. HEP can be used to compare two areas at one point in time or to compare changes in a site over several monitoring periods. Limitations to HEP are:

1. HSI and HUs for different species cannot be aggregated.

2. HEP evaluations are no more reliable than the models used to generate the HSI.

3. Interpretations are specific to the species evaluated and do not relate to other ecosystem components and functions.

4. Habitat suitability models have not been developed for many species.

Given that enhancement of fish and wildlife resources is a prevalent goal of restoration, the application of procedures similar to HEP seems probable. Consequently, selection of species that reflect the goals of restoration is imperative.

Other procedures have been developed for evaluating fishery values. The Index of Biotic Integrity (IBI) was formulated to assess fisheries from measures of species composition, trophic composition, fish abundance, and condition (Karr et al. 1986). This approach has limited application for streams with low species diversity. The General Aquatic Wildlife System (GAWS), developed by the U.S. Forest Service (1985), entails measurement of both fish populations and important habitat parameters. Similar to HEP, GAWS compares existing habitat variables with optimum criteria for fish, not the achievable state of the RRH.

BASELINE REPORT

Baseline information should be assembled as maps and reports. The format for information should allow sorting, summary, and evaluation. The baseline report is used to refine goals, objectives, and the approach to restoration. Information generated may serve in design of restoration.

Geographical Information Systems (GIS) computer software can assemble spatial information in a readily accessible format. Layers of inventory data (e.g., soil, vegetation and hydrology) can be combined and analyzed in terms of ecological relationships. Inventory data can be sorted by attributes (e.g., soil porosity or coarse-fragment composition) for material-summaries. GIS can also be used to prepare maps, rectify scales, measure the dimensions (area, length, and perimeter) of map delineations, and to compile survey information. The U.S. Forest Service has developed MOSS GIS programs that run on Data General computers. The U.S. Corps of Army Engineers has developed GRASS GIS programs to run on most minicomputers. Both MOSS and GRASS are public domain. Commercial GIS software (e.g., ARC-INFO) that can be operated on microcomputers is also available. GIS software is being evaluated as a tool for identifying RRH of similar natural or achievable state in northern Nevada (Platts et al. 1988b).

CRITICAL ASPECTS OF THE PROJECT PLAN

The restoration plan should include all information needed to evaluate the feasibility of successful restoration. Some critical aspects of the project plan are subsequently discussed. Some projects may not require consideration of all of the aspects discussed; other projects may require consideration of additional aspects. Critical aspects are determined by the scope, goals, objectives, and approach, as stipulated in the preliminary planning phase.

TOPOGRAPHIC/HYDRAULIC DESIGN

A topographic design should be provided if restoration entails alteration of the stream course or contouring. Contour maps and cross-section drawings can be used to illustrate layout and relief. The scale and contour interval should enable accurate projection of hydrologic variables such as water depth, extent of flooding and depth to perched aquifers. Successful restoration often hinges on the design and construction of the stream channel. Important perspectives are stream pattern, cross-section, and grade.

Stream pattern is the configuration of the stream as seen from the air. Meanders usually occur in valleys with gentle slope and with soils sufficiently cohesive to provide firm banks (Leopold and Langbein 1966). In these situations, the distance a river is straight is generally less than 10 times its width. Stream pattern is commonly expressed in terms of the sinuosity ratio, defined as the length of the stream divided by the down-valley distance it traverses. A meandering stream pattern was probably the single most beneficial feature for creation of RRH for a relocated channel (Jensen et al. 1987; Vinson 1988). Construction costs are directly proportional to the sinuosity ratio.

The form of meanders is important in determining the rate at which banks are undercut and eroded. In general, bank erosion is proportional to the degree to which the stream channel is bent; erosion tends to be greatest where the stream makes sharp bends. A sine-generated curve tends to minimize sharp bends and total bank erosion (Leopold et al. 1960). Since the upstream and downstream elevations of a valley segment are fixed, the sinuosity of a stream determines its grade, the single most important factor influencing water velocity. Consideration of both stream pattern and grade is necessary to achieve acceptable stream velocities relative to erosional restraints and optimal habitat parameters.

Materials and cross-sectional dimensions for construction of stream channels must take into consideration both channel capacity and the stability of streambeds and banks (Jackson and Van Haveren 1984). Cross-sections should conform to the dimensions of natural channels with similar hydrology. Computer models developed by the Army Corps of Engineers (HEC1 and HEC2) can be used to predict the recurrence interval of flow and the effect of flows upon bed and bank stability.

The design gradient for stream channels should be consistent with that upstream and downstream from the site (Rundquist et al. 1986). Too shallow a gradient may promote headcutting of the downstream segments and aggradation of upstream segments while too high a gradient tends toward the reverse. Where design gradients do not conform to those in the vicinity, grade control structures such as check-dams may be required.

Restoration designed to cause existing stream channels to evolve to a more stable productive state will be appropriate for many situations. All incised channels follow essentially the same evolutionary trend toward more stable conditions (Harvey et al. 1985). Stages of evolution include incision, headward migration, channel widening, channel slope reduction, reduction of bank angles, deposition of sediment, and establishment of vegetation. Restoration should be designed to enhance the evolution of more stable conditions, not to short-cut those evolutionary stages. Restoration of incised channels generally involves control of grade, control of discharge, or a combination of both (Harvey and Watson 1986).

Off-site impacts may affect restoration through influences upon water sources. Methods for control of off-site impacts should be addressed for worst case scenarios. Impacts resulting from high and low flows should be considered. Off-site impacts may also affect restoration through influences upon sediment dynamics (e.g., headcutting of stream channels or deposition of sediment). Grade control structures may be required for these situations.

SOIL DESIGN

Soil resources are identified and evaluated as part of the baseline inventory. When restoration requires contouring, a plan for handling soil/substrate materials should be required. The plan should specify methods for stripping, stockpiling, and redistributing of soil-materials.

A project entailing construction of a channel in coarse alluvium to mitigate for loss of fishery and riparian habitats (Jensen et al. 1987) will be used to illustrate material-handling procedures. The site is characterized by dark-colored, silt loam topsoil about 0.3 m thick subtended by about 100 m of very gravelly glacial outwash (Fig. 11A). Topsoil was stripped and stockpiled along both flanks of the construction area (Fig. 11B). Gravelly substrate was excavated for the stream channel and riparian zones (Fig. 11C). Gravelly substrate was backfilled over a plastic liner that served to eliminate bedloss and to create artificial aquifers in riparian zones (Fig. 11D). Topsoil was used to construct levees and backwashes along straight channel segments (Fig. 12) and to construct concave and convex banks along channel bends (Fig. 13). In this example, the volume of materials needed for construction was equivalent to that available on-site (considering 8% deflation of topsoil). Where materials do not balance, sources or disposal areas should be identified.

In the previous example, two types of soil-materials were identified. In other projects, several types of soil-material may be distinguished by qualities influencing their value for use in restoration. Topsoil can be used for establishment of herbaceous vegetation on floodplains; gravelly-loam substrates are suitable for establishment of willows on channel levees; clean gravel can be used as channel substrate for spawning; sand can be placed along the edge of channels to allow a stream partial freedom to choose its own course; boulders can serve as instream cover for fishes. The mixing of these materials renders each one useless. Segregation and stockpiling of contrasting soil-materials with different values for use in restoration should be carefully evaluated in the light of project goals and cost restraints.

REVEGETATION DESIGN

Revegetation designs must conform with topographic, hydrologic, and soil designs. Plans should identify the types of riparian communities to be established, the area and distribution of each type, the composition of each type, methods for propagation, and sources of propagules. Existing RRH representatives of the achievable state for the site may serve as templates for revegetation design. Cross-sectional designs (Fig. 14) can be useful for illustrating the general layout of vegetation in relation to engineering designs.

Agencies often specify the types and size of RRH to be created as some multiple (e.g., 1.25) of the area impacted. The beneficial use of restored RRH is affected not only by the area of riparian habitats, but also by their location. Tall shrubs and trees can be placed to shade the stream channel for fish. Wetlands can serve as "moats" to restrict access to critical wildlife areas or to enhance the quality of wastewater effluents.

Selection of species for revegetation should consider project goals, location, climate, microclimate, tolerance, soil, plant growth habits, availability of propagules, maintenance requirements and costs (Army Corps of Engineers 1987). If a project goal is to establish habitat for target wildlife species, any plant known to be of value for cover, food, resting, or nesting of those species should be considered. If streambank stabilization is a goal, other species may be considered. Lists of wetland plants, their regional distribution and wetland status (e.g., obligate hydrophyte, facultative hydrophyte, etc.) have been assembled for the Northwest Region (Reed 1986a) and the Intermountain Region (Reed 1986b). Whitlow and Harris (1979) and Allen and Klimas (1986) review flood tolerance, values, and habitat requirements of plant species for different regions of the United States. Kadlec and Wentz (1979) and the Environmental Laboratory (1978) tabulate soil and moisture conditions, geographic regions, morphological characteristics, potential uses, and propagation techniques for numerous plant species. Several other references pertinent to hydrologic requirements of vegetation have been recommended (Walters et al. 1980; Teskey and Hinkley 1977; Stephenson 1980).

Alan and Klimas (1986) suggest that the adaptations of pest plants (e.g., rapid dispersal, fast growth, and hardiness) such as canary grass are the very characteristics that favor their growth on new or bare substrates; advantages/disadvantages to their use merit consideration, especially where they are dominant in surrounding RRH. Baseline descriptions of reference RRH in the vicinity of the site can also be useful for selection of plant species for revegetation.

Techniques for propagation must be selected based on site constraints, scheduling, species, and costs. Methods for establishing herbaceous vegetation include seeding (broadcast, drill, hydroseeding and aerial seeding) and transplanting (springs, rootstocks, rhizomes and tubers). Trees and shrubs are propagated from cuttings, bare-root, or containerized stock. Alan and Klimas (1986) discuss methods, costs and relative advantages of these techniques and suggest some special techniques applicable to unstable situations. Doer and Landin (1983) generally suggest 180 to 270 pure live seeds per square meter for drill seeding and double this rate for broadcast seeding. Jensen (1988a) found

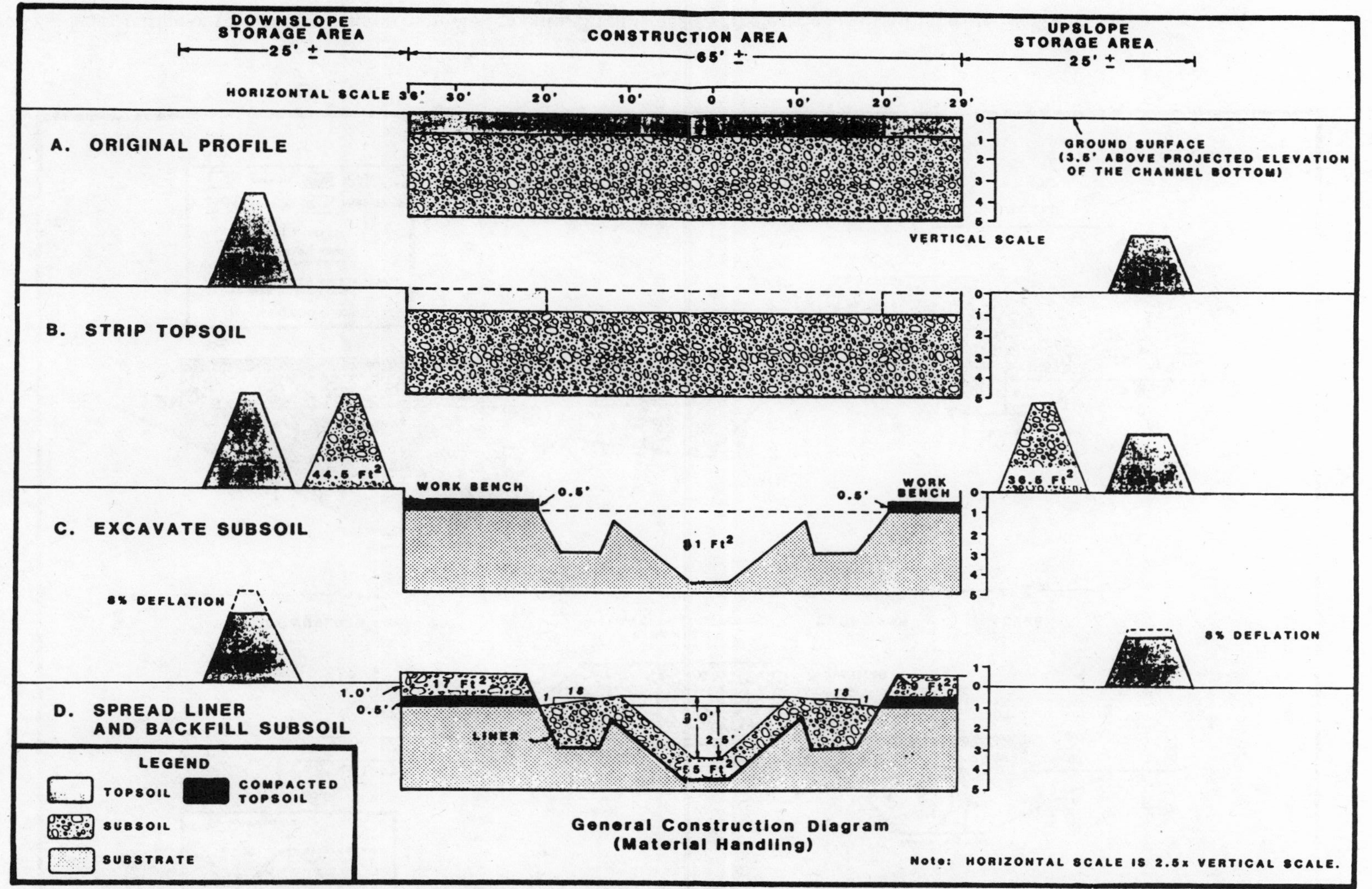

Figure 11. Material handling for creation of fish and riparian habitats, Birch Creek hydroelectric facility, Idaho (Jensen et al. 1987).

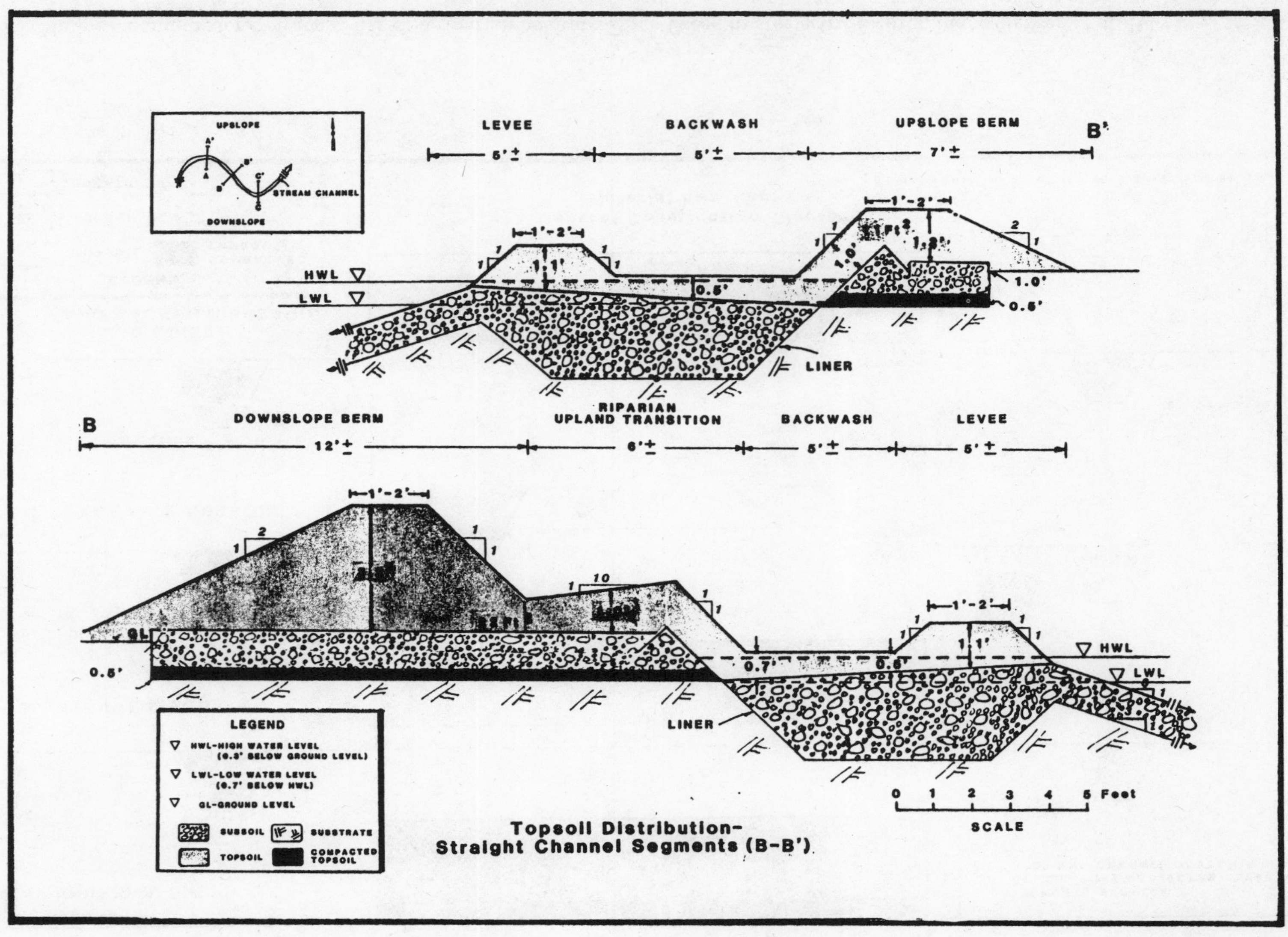

Figure 12. Distribution of topsoil on straight channel segments, Birch Creek hydroelectric facility, Idaho (Jensen et al. 1987).

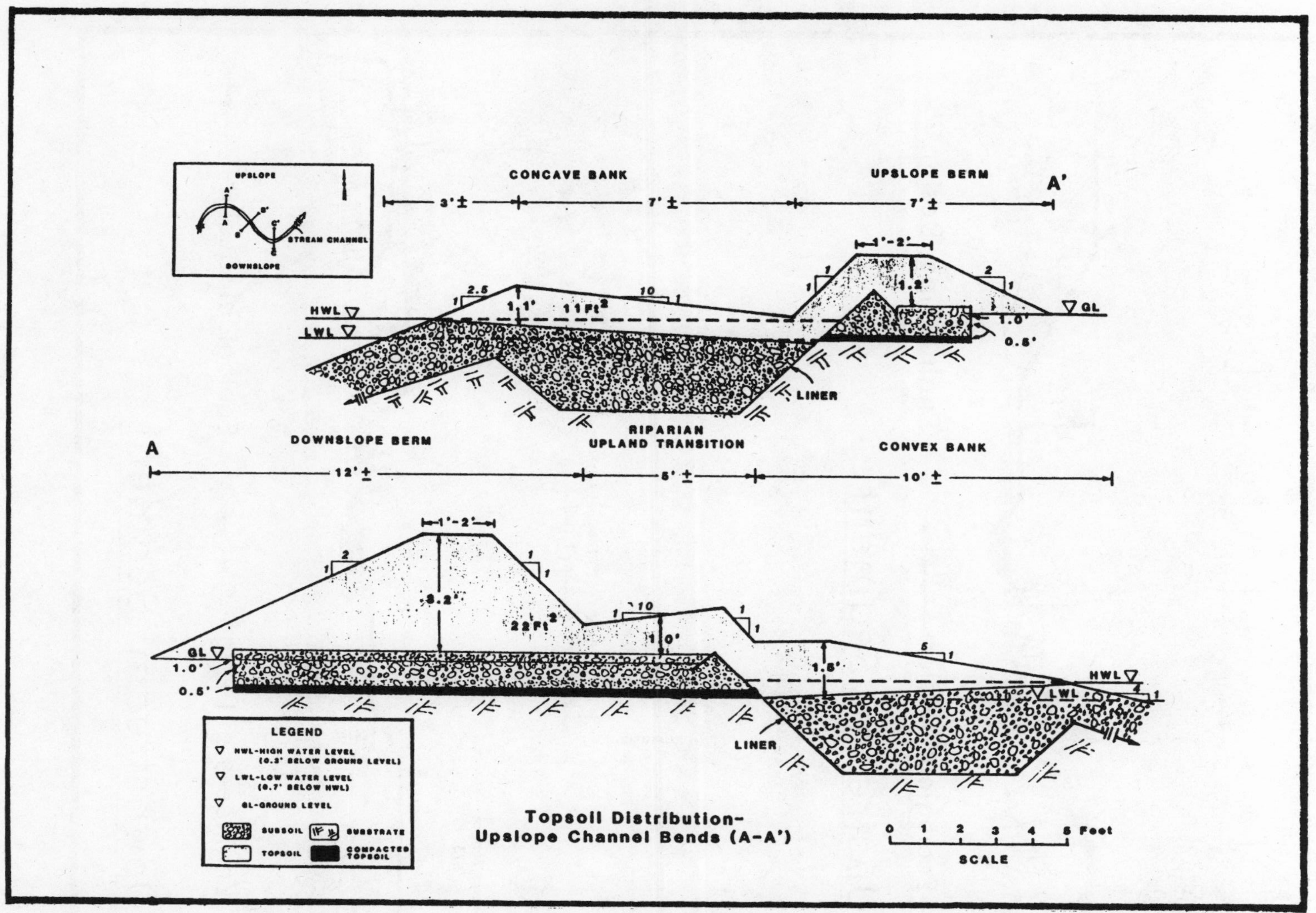

Figure 13. Distribution of topsoil on channel bends, Birch Creek hydroelectric facility, Idaho (Jensen et al. 1987)

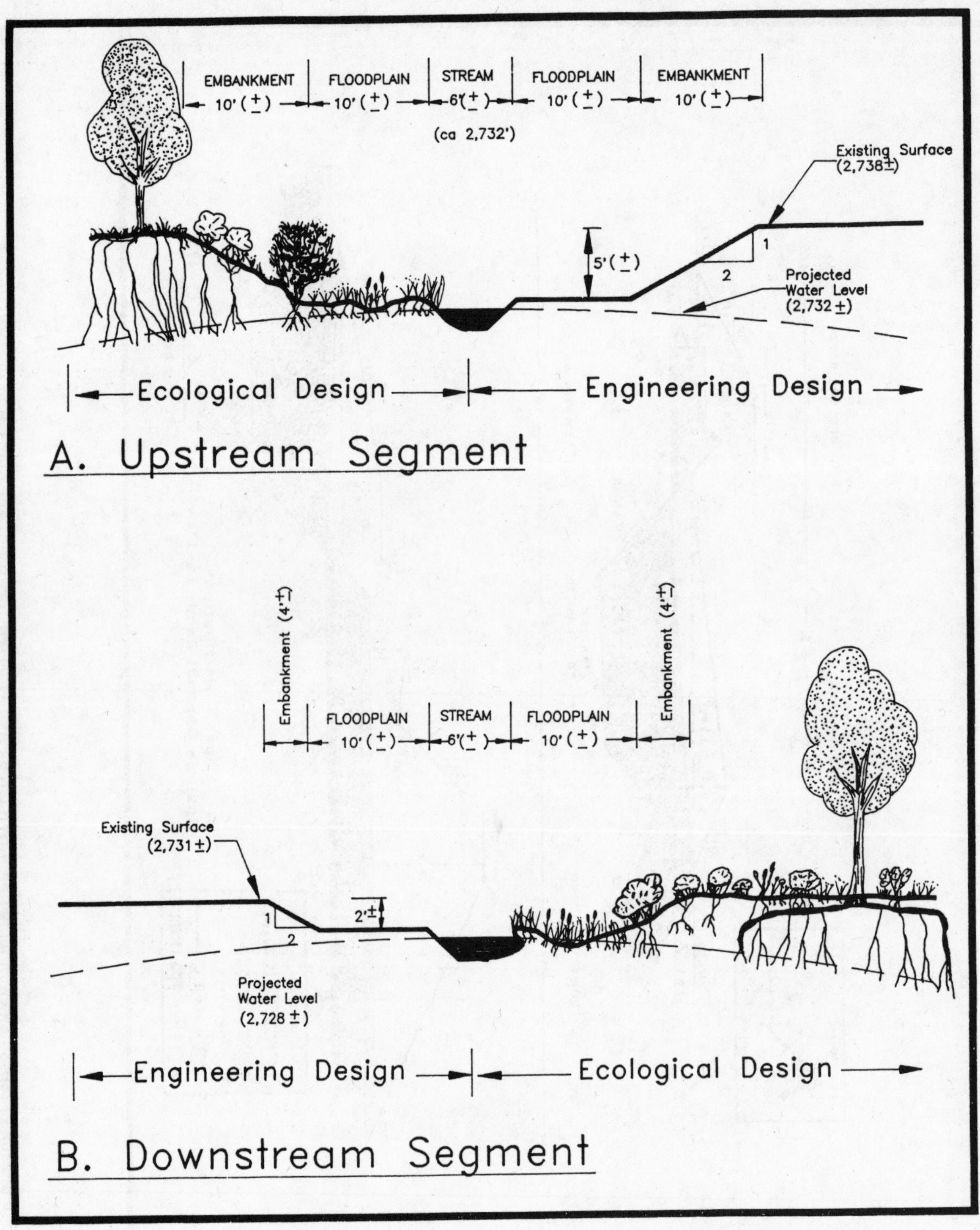

Figure 14. General vegetation layout relative to engineering design, River Run development, Boise, Idaho (Jensen 1988c).

that natural colonization from surrounding wetlands far outpaced his transplanting efforts. Preliminary evaluations of controlled flooding during critical periods when streams are transporting catkins of willow and birch are encouraging (Jensen 1988b). Controlled flooding can also be effective for diversifying the composition of the herbaceous stratum, but should be timed to preclude dissemination of weedy species.

Sources of propagules for revegetation may be identified in baseline inventories of the project site and vicinity. Doer and Landin (1983) list seed and nursery stock sources in the western United States. Propagules from commercial sources should be acquired from areas environmentally similar to the project site. The percent pure live seed (weight), germination percentage, origin, percent impurities (weight) and seed class influence application rates for commercial seeds (Army Corps of Engineers 1987).

HABITAT FEATURES

Designs may include features to enhance specific values of restored RRH. Artificial bank covers and instream boulder placements were used to enhance fish habitat in a relocated segment of Birch Creek (Vinson 1988). Fish screens may be needed to prevent fertilization of agricultural fields. Perches and nesting platforms may be used to enhance habitat for birds. Instream structures can be beneficial for enhancing fish habitat (Binns 1986). Rosgen and Fittante (1986) discuss selection of fish habitat structures suitable for specific stream types. Hubert (1986) assessed longevity and costs for maintenance of stream improvement structures.

Successful restoration is often dependent upon the control of on-site and off-site impacts. On-site impacts may include livestock grazing and recreation. Fences and/or signed closures may be required to facilitate or maintain restoration.

MAINTENANCE

Responsibility for maintenance should be clearly specified in the design plan. Maintenance may include weeding, irrigation, fixing fences, and repair of habitat features. The period and frequency for maintenance should also be specified.

SCHEDULE

A schedule for completion of critical aspects of the design plan is usually required. The schedule should generally conform to seasonal variations inherent to the site. Construction and stabilization of sites must often be completed within a single growing season to preclude serious impacts from spring runoff.

INSPECTION/MONITORING

The purpose of inspection is to verify compliance with restoration designs and to evaluate construction according to project goals. Mid-course corrections may also be suggested. It is recommended that inspection personnel be identified before starting construction. Inspection personnel should have the authority to stop work when it is out of compliance. Inspection personnel, usually from regulatory agencies, are especially important in projects requiring workers who are responding to orders from a contractor who may not be constantly available on-site. Education of workers in the purpose and goals of the project may contribute to the success of restoration. Frequent inspections should be conducted during critical aspects of construction and less frequently once it is clear that construction is proceeding correctly.

Monitoring is used to evaluate the success or failure of restoration. Parameters monitored derive from project goals and objectives, often addressing both habitat and population measurements. Habitat parameters may include: frequency and duration of flooding; groundwater dynamics; channel morphology; streambank stability; streamflow characteristics; water quality; vegetative composition, cover and production; stream shading; etc. Population measurements may include fish and wildlife counts. Monitoring may entail periodic remeasurement of variables identified in the baseline inventory (e.g., HEP, IBI, GAWS) in addition to other variables important for assessing objectives of restoration.

The frequency of monitoring is determined by project goals and deadlines. Monitoring should be conducted frequently early in the project. Frequent monitoring can be used to identify variables that limit restoration and to support mid-course corrections. Once it is clear that restoration is proceeding at an acceptable rate, monitoring may be conducted less frequently. Projects are commonly monitored monthly during the initial period of vegetation establishment and biannually thereafter. Observations should also be conducted during winter months. The period that a project is monitored must be based on that required for accomplishment of objectives, commonly 3 to 5 years. Results of monitoring must be documented.

INFORMATION GAPS AND RESEARCH NEEDS

Differences in the structure and dynamics of RRH require correspondingly different approaches to restoration. Differences may be attributed to geologic, geomorphic, hydrologic, and biotic characteristics that vary at different scales. A procedure for identifying more-or-less homogeneous classes of RRH could be useful for assessing methods for restoration. Participants in a workshop on the enhancement of stream habitat concluded that stratification of the resource into hierarchical classes of similar function and values was the only acceptable way to extrapolate results from one project to another with some rational basis (Buel 1986). A hierarchical structure is also needed to facilitate integration of restoration results.

Inventories of the amount and condition of RRH on BLM and Forest Service lands are incomplete (GAO 1988). Given that the values of a RRH may be related to its area (quantified in terms of either spatial distribution or uniqueness) and its condition, inventories are needed to prioritize restoration projects to maximize the beneficial uses of the resource.

The values of a RRE are an integrated expression of the many different types of RRH it contains. The restoration of a RRH may be limited by the condition of other RRH contained in the same watershed or by land and water uses in distal portions of the same watershed. Information pertinent to restoration in a watershed context is lacking (Platts and Rinne 1985). Such an approach could be useful for assessing the potential for restoration and for prioritizing restoration projects.

Methods for assessing the functional attributes of RRH and the functional relationships between RRH of a watershed have not been developed; these could be useful for evaluating restoration. Methods have been developed for functional assessment of wetlands (Adamus 1983).

Reference sites representing the natural or attainable state of RRH are scarce throughout much of the region. Several demonstration and research sites established by management agencies have been valuable for illustrating that restoration can be achieved through proper management (GAO 1988). Lands owned and protected by the Nature Conservancy and Research Natural Areas may also serve as reference sites. Identification and description of reference sites could help to determine the endpoints of successful restoration. Establishment of additional reference sites will probably be necessary to represent the diversity of RRH in the region.

A critical aspect of restoration projects is cost. Fiscal evaluations are complicated by the difficulty in placing monetary values on resources and by the uncertainty of restoration success. Methods for evaluating cost/benefit of restoration projects should be developed. Given that most RRH is impaired to some degree, fiscal evaluations could be used to prioritize restoration to achieve maximum beneficial use of the resource with available funds.

ACKNOWLEDGEMENTS

Cynthia K. Johnson gathered information about restoration projects. M. Christy Donaldson assisted in the literature review and C.L. Rawlins edited this chapter.

LITERATURE CITED

Adamus, P.R. 1983. A Method for Wetland Functional Assessment: Volume I and II. Federal Highway Administration Report FHWA-IP-82-23. National Technical Information Service, Springfield, Virginia.

Allen, H.H. and C.V. Klimas. 1986. Reservoir Shoreline Revegetation Guidelines. U.S. Army Engineer Waterways Experiment Station, Vicksburg, Mississippi.

Army Corps of Engineers. 1987. Beneficial Uses of Dredged Material-Engineer Manual. EM1110-2-5026. U.S. Government Printing Office, Washington D.C.

Bailey, R.G. 1985. The factor of scale in ecosystem mapping. Environmental Management 9(4):271-276.

Bailey, R.G. 1983. Delineation of ecosystem regions. Environmental Management 7(4):365-373.

Bartholow, J.M. and T.J. Waddle. 1986. Introduction to Stream Network Habitat Analysis, Instream Flow Information Paper 22. Fish and Wildlife Service, Instream Flow and Aquatic Systems Group, National Ecology Center, Fort Collins, Colorado.

Binns, N.A. 1986. Habitat, macroinvertebrate and fishery response to stream improvement efforts in the Thomas Fork Bear River drainage, Wyoming, p. 105-116. In J.G. Miller, J.A. Arway and R.F. Carline (Eds.), Fifth Trout Stream Improvement Workshop. Pennsylvania Fish Commission, Harrisburg, Pennsylvania.

Black, H., Jr., and J.W. Thomas. 1978. Forest and range wildlife habitat management: ecological principals and management systems, p. 47-55. In R.M. DeGraaf (Tech. Coord.), Proceedings of the Workshop on Nongame Bird Habitat Management in the Coniferous Forests of the Western United States. Forest Service Gen. Tech. Rep. PNW-64. Portland, Oregon.

Bonneville Power Administration. 1986. Fish and Wildlife Annual Project Summary. Dept. of Energy, Bonneville Power Administration, Division of Fish and Wildlife. Portland, Oregon.

Brunsfield, S.J. and F.D. Johnson. 1985. Field Guide to the Willows of East-Central Idaho. Forest, Wildlife and Range Experiment Station, University of Idaho, Moscow, Idaho.

Buel, J.W. 1986. Stream Habitat Enhancement Evaluation Workshop: a Synthesis of Views. Bonneville Power Administration, Division of Fish and Wildlife, Portland, Oregon.

Bureau of Land Management. 1986. West Desert Pumping Project, Final EIS. Salt Lake District Office, Salt Lake City Utah.

Cowardin, L.M., V. Carter, F.C. Golet, and E.T. Laroe. 1979. Classification of Wetlands and Deepwater Habitats of the United States. FWS/OBS 79/31. U.S. Government Printing Office, Washington, D.C.

Doer, T.B. and M.C. Landin. 1983. Vegetation Stabilization of Training Areas of Selected Western United States Military Reservations. U.S. Army Environmental Laboratory, Vicksburg, Mississippi.

Dutton, C.E. 1880. Geology of the High Plateaus of Utah. U.S. Government Printing Office. Washington, D.C.

Environmental Laboratory. 1986. Field Guide for Low-Maintenance Vegetation Establishment and Management. Instruction Report R-86-2. U.S. Army Engineer Waterways Experiment Station, Vicksburg, Mississippi.

Environmental Laboratory. 1978. Wetland Habitat Development with Dredged Material: Engineering and Plant Propagation. Technical Report DS-78-16. U.S. Army Engineer Waterways Experiment Station, Vicksburg, Mississippi.

Gebardt, K. 1986. Environmental Assessment for the River Run South Development. Resource Systems, Boise, Idaho.

General Accounting Office. 1988. Public Rangelands--Some Riparian Areas Restored but Widespread Improvement will be Slow. GAO/RCED-88-105. Gaithersburg, Maryland.

Hair, D. and R. Stowell. 1986. South Fork Clearwater River Habitat Enhancement--Annual Report. In Natural Propagation and Habitat Improvement, Vol. II. Department of Energy, Bonneville Power Administration, Division of Fish and Wildlife.

Hansen, P.L., S.W. Chadde, and R.D. Pfister. 1987. Riparian Dominance Types in Montana. Montana Riparian Association, School of Forestry, University of Montana, Missoula, Montana.

Harris, R.R. In review. Associations between Stream Valley Geomorphology and Riparian Vegetation as a Basis for Landscape Analysis in the Eastern Sierra Nevada, California. Dept. of Forest Science, Oregon State University, Corvallis, Oregon.

Harvey, M.D. and C.C. Watson. 1986. Fluvial processes and morphological thresholds in incised channel restoration. Water Resources Bulletin 22(3):359-368.

Harvey, M.D., C.C. Watson, and S.A. Schumm. 1985. Gully Erosion. Bureau of Land Management Tech. Note 366. U.S. Government Printing Office, Washington D.C.

Hubert, P.J. 1986. Longevity and maintenance requirements of stream improvement structures in New York waters. p. 199-207. In J.G. Miller, J.A. Arway, and R.F. Carline (Eds.), Fifth Trout Stream Improvement Workshop. Pennsylvania Fish Commission, Harrisburg, Pennsylvania.

Hughes, R.M. and J.R. Gammon. 1986. Longitudinal changes in fish assemblages and water quality in the Willamette River, Oregon. Trans. of the Amer. Fish. Soc. 116(2):196-209.

Hughes, R.M., E. Rextad, and C.E. Bond. 1987. The relationship of aquatic ecoregions, river basins and physiographic provinces to the ichthyogeographic regions of Oregon. Coperia 1987 (2):423-432.

Hunt, C.B. 1976. Natural Regions of the United States and Canada. W.H. Freeman and Company, San Francisco, California.

Jackson, W.L. and B.P. Van Haveren. 1984. Design for stable channel in coarse alluvium for riparian zone restoration. Water Resources Bulletin 20(5):695-703.

Jensen, S.E. 1988a. Monitoring Report-Establishment of Wetland Habitat, Cedar Draw. White Horse Associates, Smithfield, Utah.

Jensen, S.E. 1988b. Establishment of Riparian Habitat, Birch Creek, Monitoring Report. White Horse Associates, Smithfield, Utah.

Jensen, S.E. 1988c. Preliminary Planning, River Run Project, Boise, Idaho. White Horse Associates, Smithfield, Utah.

Jensen, S.E. 1987. Progress Report, Wetland Establishment, Cedar Draw Creek, Idaho. White Horse Associates, Smithfield, Utah.

Jensen, S.E. 1986. Wetland Establishment, Cedar Draw Creek, Twin Falls County, Idaho. White Horse Associates, Smithfield, Utah.

Jensen, S.E. 1985. Effects of Altered Bear River Streamflow upon associated Palustrine Wetlands. In Environmental Evaluation of Smiths Fork Reservoir Project. Ecosystems Research Institute, Logan, Utah.

Jensen, S.E. and J.S. Tuhy. 1982. Soils Investigation of Riparian Communities of East Smiths Fork and Henrys Fork Drainages, North Slope Uinta Mountains, Utah. White Horse Associates, Smithfield, Utah.

Jensen, S., M. Vinson, and J. Griffith. 1987. Creation of riparian and fish habitats, Birch Creek Hydroelectric Facility, Clark County, Idaho, p. 144-149. In K.M. Mutz and L.C. Lee (Eds.), Proceedings of the Society of Wetland Scientists Eighth Annual Meeting. Seattle, Washington.

Johnson, C.K. 1987. Final Mitigation Plan for the Felt Hydroelectric Project. FERC No. 5089. Ecosystems Research Institute, Logan, Utah.

Johnson, C.K. 1986. Final Off-Site Mitigation Plan for LQ/LS Drain (Pigeon Cove) Hydroelectric Project. Ecosystems Research Institute, Logan, Utah.

Kadlec, J.A. and W.A. Wentz. 1979. State-of-the-Art Survey and Evaluation of Marsh Plant Establishment Techniques: Induced and Natural; Vol. I: Report of Research. D-74-9. U.S. Army Engineer Waterways Experiment Station, Vicksburg, Mississippi.

Karr, J.R., K.D. Fausch, P.L. Angermeier, P.R. Yant, and I.J. Schlosser. 1986. Assessing Biological Integrity in Running Waters--A Method and Its Rationale. Illinois Natural History Survey, Special Publication 5.

Konopacky, R.C., E.C. Bowles, and P.J. Cernera. 1985. Salmon River Habitat Enhancement. U.S. Department of Energy Bonneville Power Administration Division of Fish and Wildlife, 83-359. Corvallis, Oregon.

Kovalchick, B.L. 1987. Riparian Zone Associations, Deschutes, Ochoco, Freemont, and Winema National Forests. R6 ECOL TP-279-87. U.S. Forest Service, Pacific Northwest Region.

Landin, M.C., C.J. Newling, and E.J. Clarion Jr. 1987. Miller Sands Island: a dredged material wetland in the Columbia River, Oregon, p. 150-155. In K.M. Mutz and L.C. Lee (Eds.), Proceedings of the Society of Wetland Scientists Eighth Annual Meeting. Seattle,Washington.

Larsen, D.P., D.R. Dudley, and R.M. Hughes. In Review. A regional approach for assessing attainable surface water quality: Ohio as a case study. Jour. of Soil and Water Cons.

Leopold, L.B., R.A. Bangold, M.G. Wolman, and L.M. Brush. 1960. Flow Resistance in Sinuous or Irregular Channels. Geological Survey Professional Paper 282-D. U.S. Government Printing Office, Washington D.C.

Leopold, L.B. and W.B. Langbein. 1966. River meanders. Scientific American. 214(6):60-70.

Lidstone, C.D. 1987. Stream channel and wetland reconstruction, p. 131-135. In K.M. Mutz and L.C. Lee (Eds.), Proceedings of the Society of Wetland Scientists Eighth Annual Meeting. Seattle, Washington.

May, B.E. and R.W. Rose. 1986. Camas Creek (Meyers Cove) Anadromous Species Habitat Improvement Plant Final Report. In Natural Propagation and Habitat Improvement; Vol II--Idaho. Annual and Final Reports. Dept. of Energy, Bonneville Power Administration, Division of Fish and Wildlife.

Miller, R. and J.P. Olivarez. 1986. Joseph Creek Drainage Off-Site Enhancement Project. Annual Report FY 1985. In K.M. Mutz and C.L. Lee (Eds.), Natural Propagation and Habitat Enhancement; Vol. I--Oregon. Annual and Final Reports 1985. Dept. of Energy, Bonneville Power Administration, Division of Fish and Wildlife.

Miller, T.S. 1987. Techniques used to enhance, restore, or create freshwater wetlands in the Pacific Northwest, p. 116-121. In K.M. Mutz and L.C. Lee (Eds.), Proceedings of the Society of Wetland Scientists Eighth Annual Meeting. Seattle, Washington.

Minshall, G.W., S.E. Jensen, and W.S. Platts. In press. The Ecology of Great Basin Streams and Riparian Habitats: A Community Profile. U.S. Fish and Wildlife Service Report.

Murphy, W. and A. Espinosa, Jr. 1985. Eldorado Creek Fish Passage Final Report, Modification M001 to Agreement DE-A179-54BP16535. In Natural Propagation and Habitat Improvement; Volume II Idaho. Annual and Final Reports. Dept. of Energy, Bonneville Power Administration, Division of Fish and Wildlife.

Mutz, K.M. and R. Graham. 1982. Riparian Community Type Classification - Big Piney Ranger District, Wyoming. Unpublished U.S. Forest Service document. Ogden, Utah.

Mutz, K.M. and J. Queiroz. 1983. Riparian Community Classification for the Centennial Mountains and South Fork Salmon River, Idaho. Unpublished U.S. Forest Service Document. Ogden, Utah.

Noll, W.T. 1986. Grande Ronde Habitat Improvement Project: Joseph Creek and Upper Grande Ronde River Drainages. Annual Report. In Natural Propagation and Habitat Enhancement; Vol. I Oregon. Annual and Final Reports 1985. Dept. of Energy, Bonneville Power Administration, Division of Fish and Wildlife.

Norton, B.E., J.S. Tuhy, and S.E. Jensen. 1981.

Riparian Community Classification for the Greys River, Wyoming. Unpublished U.S. Forest Service document. Ogden, Utah.

Omernik, J.M. 1987. Ecoregions of the conterminous United States. Annals Assoc. of Amer. Geog. 77:118-125.

Omernik, J.M. 1986. Ecoregions of the United States (1:7,500,000 scale map). U.S. Environmental Protection Agency.

Padgett, W.G. and M.E. Manning. 1988. Preliminary Riparian Community Type Classification for Nevada. Draft Report. U.S. Forest Service, Intermountain Region, Ogden, Utah.

Padgett, W.G. and A.P Youngblood. 1986. Riparian Community Type Classification of Southern Utah. U.S. Forest Service. Ogden, Utah.

Platts, W.S., C. Armour, G.D. Booth, M. Bryant, J.L. Bufford, P. Cuplin, S.E. Jensen, G.W. Lienkaemper, G.W. Minshall, S.B. Monsen, R.L. Nelson, J.R. Sedell, and J.S. Tuhy. 1987. Methods for Evaluating Riparian Habitats with Applications to Management. U.S. Dept. Agric., Forest Service, Intermountain Research Station, Gen. Tech. Rep. INT-221. Boise, Idaho.

Platts, W.S., K.A. Gebhardt, and W.L. Jackson. 1985. The effects of large storm events on Basin-Range riparian stream habitats, p. 30-34. In Riparian Ecosystems and their Management. U.S. Dept. Agric., Forest Service Gen. Tech. Rep. RM-120. Tuscon, Arizona.

Platts, W.S., S.E. Jensen, and F. Smith. 1988a. Preliminary Classification and Inventory of Riparian Communities, Livestock/Fisheries Study Areas, Nevada. Unpublished report. U.S. Forest Service Intermountain Research Station, Boise, Idaho.

Platts, W.S., S.E. Jensen, and R. Ryel. 1988b. Classification of Riverine/Riparian Habitat and Assessment of Nonpoint Source Impacts, North Fork Humboldt River, Nevada. Draft report to U.S. Environmental Protection Agency, Denver, Colorado.

Platts, W.S., M. McHenry, and R. Torquemada. 1986. Evaluation of fish habitat enhancement projects in Crooked River, Red River, and Bear Valley Creek; Progress Report II. In: Idaho Habitat Evaluation for Off-Site Mitigation Record. U.S. Department of Energy, Bonneville Power Administration, Div. of Fish and Wildlife.

Platts, W.S. and J.N. Rinne. 1985. Riparian and stream enhancement management and research in the Rocky Mountains. North American Journal of Fisheries Management 5(2A): 115-125.

Reed, P.B. 1986a. Wetland Plant List--Northwest Region. U.S. Fish and Wildlife Service, WELUT-86/W13.09. St. Petersburg, Florida.

Reed, P.B. 1986b. Wetland Plant List--Intermountain Region. U.S. Fish and Wildlife Service, WELUT-86/W13.09. St. Petersburg, Florida.

Richards, C. and P.J. Cernera. 1987a. Salmon River Enhancement. U.S. Department of Energy Bonneville Power Administration Division of Fish and Wildlife, 83-359. Corvallis, Oregon.

Richards, C. and P.J. Cernera. 1987b. Yankee Fork of the Salmon River: Inventory, Problem Identification and Enhancement Feasibility. U.S. Department of Energy Bonneville Power Administration Division of Fish and Wildlife, 83-359. Corvallis, Oregon.

Rohm, C.M., J.W. Giese, and C.C. Bennet. 1987. Evaluation of an aquatic ecoregion classification of streams in Arkansas. Journal of Freshwater Ecology 4(1):127-140.

Rosgen, D.L. 1985. A stream classification system, p. 91-95. In Riparian Ecosystems and Their Management. U.S. Forest Service Gen. Tech. Report RM-120.

Rosgen, D.L. and B.L. Fittante. 1986. Fish habitat structures--a selection guide using stream classification, p. 163-179. In J.G. Miller, J.A. Arway, and R.F. Carline (Eds.), Fifth Trout Stream Improvement Workshop. Pennsylvania Fish Commission, Harrisburg, Pennsylvania.

Rundquist, L.A., N.E. Bradley, and T.R. Jennings. 1986. Planning and design of fish stream rehabilitation, p. 119-132. In J.G. Miller, J.A. Arway, and R.F. Carline (Eds.), Fifth Trout Stream Habitat Improvement Workshop. Pennsylvania Fish Commission, Harrisburg, Pennsylvania.

Sempek, J.E. and C.W. Johnson. 1987. Wetlands enhancement at the Ogden Nature Center in Ogden, Utah, p. 161-165. In K.M. Mutz and L.C. Lee (Eds.), Proceedings of the Society of Wetland Scientists Eighth Annual Meeting. Seattle, Washington.

Shaw, S.P. and C.G. Fredine. 1956. Wetlands of the United States. U.S. Fish and Wildlife Service Circ. 39.

Sigler, W. and J. Sigler. 1987. Fishes of the Great Basin--A Natural History. University of Nevada Press, Reno, Nevada.

Sipple, W.S. 1987. Wetland Identification and Delineation Manual. U.S. Environmental Protection Agency, Washington D.C.

Soil Conservation Service. 1982. Soil Survey Manual. U.S. Government Printing Office, Washington, D.C.

Soil Conservation Service. 1975. Soil Classification--Handbook No. 436. U.S. Government Printing Office, Washington, D.C.

Stevenson, M. 1980. The Environmental Requirements of Aquatic Plants. Appendix A to Publication No. 65. The California State Water Resources Control Board, Sacramento, California.

Tesky, R.O. and T.M. Hinkley. 1977. Impact of Water Level Changes on Woody Riparian and Wetland Plant Communities; Vol. I: Plant and Soil Responses to Flooding. U.S. Fish and Wildlife

Service FWS/OBS-77-58.

Thomas, T. and M. Collette. 1986. BPA Project 84-9: Grande Ronde River Habitat Enhancement Project. Annual Report FY 1985. In Natural Propagation and Habitat Enhancement. Vol. I--Oregon. Annual and Final Reports 1985. Dept. of Energy, Bonneville Power Administration, Division of Fish and Wildlife.

Tuhy, J.S. and S.E. Jensen. 1982. Riparian Classification for the Upper Salmon/Middle Fork Salmon Rivers, Idaho. Unpublished U.S. Forest Service Report, Ogden, Utah.

U.S. Fish and Wildlife Service. 1980. Habitat Evaluation Procedures. Division of Ecological Services, Washington D.C.

U.S. Forest Service. 1985. Fisheries Habitat Surveys Handbook. Region 4, FSH 2609.23.

Vinson, M.R. 1988. An ecological and sedimentological evaluation of a relocated cold desert stream. M.S. Thesis. Idaho State University, Pocatello, Idaho.

Walters, A.M., R.O. Tesky, and T.M. Hinkley. 1980. Impact of Water Level Changes on Woody Riparian and Wetland Communities; Volume VIII: Pacific Northwest and Rocky Mountain Regions. U.S. Fish and Wildlife Service FWS/OBS 78/94.

Welling, K. 1986. Final On-Site Revegetation Plan. LQ/LS Drain (Pigeon Cove). FERC No. 5767. Hosey and Associates.

Wertz, W.A. and J.F. Arnold. 1972. Land Systems Inventory. U.S. Forest Service, Ogden, Utah.

Whitlow, T.H. and R.W. Harris. 1979. Flood Tolerance in Plants: A State-of-the-Art Review. Technical Report E-79-2, U.S. Army Engineer Waterways Experiment Station, Vicksburg, Mississippi.

Whittier, T.R., R.M. Hughes, and D.P. Larsen. 1988. The correspondence between ecoregions and spatial patterns in stream ecosystems in Oregon. Can. Jour. Fish. Aquat. Sci. 45:1-15.

Youngblood, A.P., W.G. Padgett, and A.H. Winward. 1985. Riparian Community Type Classification of Eastern Idaho--Western Wyoming. U.S. Forest Service, R4-Ecol-85-01.

APPENDIX I: RECOMMENDED READING

Adamus, P. R. and L. T. Stockwell. 1983. A Method for Wetland Functional Assessment; Volume 1: Critical Review and Evaluation Concepts. FHWA, U.S. Dept. of Transportation, Report No. FHWA-IP-82-23.

This manual reviews wetland functions and the potential impacts of highways on these functions.

Harvey, M.D., C.C. Watson, and S.A. Schumm. 1985. Gully Erosion. Bureau of Land Management Tech. Note 366. U.S. Government Printing Office, Washington D.C.

Reviews literature on incised channels, discusses their causes and evolution, and provides resource specialists a conceptual framework for dealing with them.

Heede, B. H. 1980. Stream Dynamics: An Overview for Land Managers. U.S. Depart. Agric., A Forest Service, Rocky Mountain Research Station, General Technical Report RM-72, Fort Collins, Colorado.

Covers the general concepts of fluvial dynamics and how management actions affect streams and vegetative bank cover.

Jackson, W. L. and B. P. Van Haveren. 1984. Design for a stable channel in coarse alluvium for riparian zone restoration. Water Resources Bulletin 20(5):695-703.

Geomorphic, hydraulic, and hydrologic principles are applied to the design of a stable stream channel and riparian habitat.

Mutz, K.M. and L.C. Lee (Eds.). 1987. Wetland and Riparian Ecosystems of the American West. Eighth Annual Meeting of the Society of Wetland Scientists. Seattle, Washington.

A compendium of articles relating to wetland/ riparian habitats including several papers on creation and restoration.

Platts, W.S., C. Armour, G.D. Booth, M. Bryant, J.L. Bufford, P. Cuplin, S. Jensen, G.W. Lienkaemper, G.W. Minshall, S.B. Monsen, R.L. Nelson, J.R. Sedell, and J.S. Tuhy. 1987. Methods for Evaluating Riparian Habitats with Applications to Management. U.S. Dept. Agric., Forest Service, Intermountain Research Station, Gen. Tech. Rep. INT-221. Boise, Idaho.

A compilation of methods for managing, evaluating, and monitoring riverine/riparian habitat.

Platts, W.S. and J.N. Rinne. 1985. Riparian and stream enhancement management and research in the Rocky Mountains. North Amer. Jour. Fish. Manag. 5(2A):115-125.

Reviews stream enhancement research in the Rocky Mountains, its adequacy, and research needs to improve the effectiveness of stream enhancement projects.

United States General Accounting Office. 1988. Public Rangelands--Some Riparian Areas Restored But Widespread Improvement Will Be Slow. GAO/RCED-88-105. Gaithersburg, Maryland.

Evaluates the condition of riparian habitats in the western United States, gives examples of some restored riparian areas, discusses barriers to restoration and makes recommendations to the Committee of Interior and Insular Affairs. This document doesn't cut the Forest Service and BLM any slack!

Warner, R.E. and K.M. Hendrix (Eds.). 1984. California Riparian Systems--Ecology, Conservation, and Productive Management. University of California Press, Berkeley, California.

A compendium of papers addressing a wide range of riparian topics.

APPENDIX II

BIRCH CREEK, IDAHO

Riparian and fish habitats were created in association with a feeder canal leading to a hydroelectric generating facility in southeastern Idaho. The goal of the project was to mitigate for losses of fish and riparian habitat downstream from the diversion. Due to extremely permeable substrate, an innovative design was necessary.

The 2 km channel was constructed to include 12 meanders (sinuosity ratio = 1.25) and a 6 m wide riparian zone on each side. Substrate was excavated and a plastic liner was installed to eliminate bedloss from the channel and riparian zones. Gravelly substrate was backfilled over the liner to facilitate establishment of an artificial aquifer for sustenance of riparian habitat. The substrate was judged suitable for creation of fish habitat in the constructed channel. Topsoil was used to create topographic features on the riparian zones (e.g., levees and backwashes). A structure for controlling stream stage and a fish screen were placed at the downstream end of the project area. Artificial bank overhangs and boulder placements were used to enhance fish habitat.

Preliminary evaluations indicate that the meandering pattern of the constructed channel is probably the most important factor in the success of the project. Fine sediments are accumulating on convex banks (point bars), creating favorable conditions for vegetation. A pool-riffle sequence similar to that of a natural stream is developing. Fish populations about 200 percent of the stipulated goal were measured only 18 months after construction.

An experimental approach was used to establish riparian vegetation. The stream reach was divided into 24 experimental units, each consisting of a convex bank, concave bank, and adjacent straight channel segments. Variables tested for establishment of riparian vegetation were:

1. Controlled flooding during periods when propagules of desirable species were being transported in the stream.

2. Cuttings versus rooted propagules for several shrub species.

3. Transplanting versus seeding of sedges. The entire riparian zone was broadcast with native grass seeds.

Survival of unrooted willow cuttings was about the same as for rooted stock after one season. Site conditions (i.e. available moisture) appeared to be more important to survival than method of propagation. Ten to twenty unrooted cuttings could be planted with the same effort (and expense) of one rooted cutting. Unrooted cuttings of red-osier dogwood were less successful; cuttings of choke cherry all died. Controlled flooding of the riparian zone was very successful for diversifying the composition of riparian habitat and resulted in numbers of shrub seedlings many times greater than were planted. Seeding of sedges was not successful; transplanted sedges are spreading rapidly; natural colonization of sedges is also occurring. Native grasses cover the most riparian zone (80 to 90 percent cover) and have stabilized streambanks, which are evolving to enhance cover for fish. Monitoring of revegetation is continuing.

Some factors limiting the success of this project were:

1. Failure of the contractor to follow procedures for stripping and stockpiling of topsoil. Mixing of topsoil with substrates resulted in shortage for construction of design features.

2. Failure of the contractor to adhere to the design plan. Some mid-course design changes were required to compensate for errors in construction.

3. Failure to anticipate the effects of ice, which covered the riparian zone to depths of several feet during winter months. Many young shrubs were "plucked" out of the ground by ice.

Cost to enhance fish and riparian habitats was about $100,000.

Reports:

Jensen, S.E., J. Griffith, and M. Vinson. 1987. Creation of riparian and fish habitat, Birch Creek hydroelectric facility, Clark County, Idaho, p. 144-149. In Mutz, K.M. and C.L. Lee (Eds.) Wetland and Riparian Ecosystems of the American West. Proceedings of the Eight Annual Meeting of the Society of Wetland Scientists, May 26-29, 1987, Seattle, Washington.

Vinson, M.R. 1988. An ecological and sedimentological evaluation of a relocated cold desert stream. M.S. Thesis, Department of Biology, Idaho State University, Pocatello, Idaho.

CEDAR DRAW, SNAKE RIVER, IDAHO

Wetland habitat was created as mitigation for that destroyed by construction of a road and penstock along a minor tributary of the Snake River, near Twin Falls. The goal of the project was to create about 0.3 hectares of emergent wetland habitat to compensate for that filled by construction.

A series of 12 hydrologically linked basins were

constructed and flooded to a maximum depth of about 1.03 meters. Wetland vegetation was planted around the periphery of each pond. Soon after completion, the inlet pipe to the ponds clogged, the ponds drained and emergent vegetation went dormant or died. A more reliable water source was constructed and the basins re-flooded. In September, 1987, Less than 4 weeks following reflooding, about 11 percent of the pond surfaces were vegetated with emergent plants. Propagules are thought to have originated from an existing wetland in the immediate vicinity. Emergent plant cover varied as a function of distance from the existing wetland. In June, 1988 emergent vegetation covered about 19 percent of the ponded area. The rate of emergent vegetation establishment in a pond varied directly with the ratio of its circumference and surface area. Vegetation establishment proceeded from the edges of the ponds towards the centers. The smaller the pond, the more rapidly it was vegetated.

Cost for construction of about 0.3 hectares of wetland habitat was about $30,000.

Reports:

Jensen, S.E. 1986. Wetland Establishment, Cedar Draw Creek, Twin Falls County, Idaho. White Horse Associates, Smithfield, Utah.

Jensen, S.E. 1987. Progress Report, Wetland Establishment, Cedar Draw Creek, Idaho. White Horse Associates, Smithfield, Utah.

Jensen, S.E. 1988a. Monitoring Report, Wetland Establishment, Cedar Draw Creek, Idaho. White Horse Associates, Smithfield, Utah.

BEAR VALLEY CREEK, IDAHO

Bear Valley Creek is located in a glaciated portion of a granite batholith in central Idaho. It joins Marsh Creek to form the Middle Fork of the Salmon River. Historically, Bear Valley Creek provided spawning and rearing habitat for chinook salmon, but these habitats were seriously degraded by dredge mining in the 1950's.

A project to restore about 3.2 km of Bear Valley Creek was implemented in 1985 and 1986. Objectives of restoration were:

1. To reduce sediment recruitment from dredge-mined areas and sediment transport to lower stream segments.
2. To stabilize the channel and streambanks.
3. To improve water quality through minimizing turbidity.
4. To improve the aesthetic qualities of the mined areas.
5. To create or improve habitats for spawning and rearing of chinook salmon.

The approach to restoration entailed realignment of the channel, construction of a floodplain about 100 m wide, stabilization, and revegetation.

Contouring was conducted to create surfaces similar to those existing prior to mining activity. A combination of geotextile fabric, erosion control blankets, vegetation, and riprap was used to stabilize recontoured surfaces. The realigned segment of Bear Valley Creek was designed to allow the river freedom to form meanders in the constructed floodplain. Design criteria approximated the natural stream pattern and grade. Shallow sinks for collection of sediment were also created. Construction of this project was a major engineering endeavor.

Seeding and rooted willow cuttings were used to facilitate revegetation. Willow cuttings obtained from the vicinity of the site were propagated in a greenhouse and treated with an anti-transpirant to reduce dehydration before transplanting to the site. The success of one, two, and three year-old willow plantings was 97%, 88% and 82%, respectively. Redd counts have increased from 80 prior to restoration to 230 in October, 1988. The increase in redds may be attributed to factors other than restoration of the site.

Cost of the project was about $2.5 million.

Reports:

Konopacky, R.C., E.C. Bowles, and P.J. Cernera. 1985. Salmon River habitat enhancement. Annual Report FY 1984, Part 1. In Natural Propagation and Habitat Improvement. Idaho: Salmon River Habitat Enhancement. Annual Report 1984. U.S. Dept. of Energy, Bonneville Power Administration, Div. of Fish and Wildlife.

Richards, C. and P.J. Cernera. 1987. Salmon River Enhancement. U.S. Department of Energy Bonneville Power Administration Division of Fish and Wildlife, 83-359. Corvallis, Oregon.

FELT, TETON RIVER, IDAHO

The site is located in a notch-shaped canyon incised in basalt. During construction of a road into the Teton River Canyon, fill that was side-cast down steep slopes blocked fish passage to upstream spawning habitat and covered about 0.15 hectares of riparian habitat. The purpose of the project was to mitigate for lost wetland values and functions.

The objectives were to restore fish passage, to re-establish riparian vegetation similar to that existing prior to disturbance, and to stabilize the site. Fish passage was restored by removal of rock from the stream channel. The goal of revegetation was to establish communities similar to those existing prior to disturbance. The percent shrub canopy cover and species composition were stipulated by the regulatory agency. Riparian shrubs were planted on 0.15 hectares.

Preliminary results indicate that fish passage has been restored. Permanent vegetation quadrants have been established in each community type and monitoring will take place every June and September. Revegetation has been largely unsuccessful in the riparian zone due to lack of adequate moisture and porous substrates. The incomplete status of riparian restoration precludes any judgement of success/failure at this time.

To provide compensation for temporary and permanent losses of wildlife habitat a plot of land encompassing 730 meters of Teton Creek that was impacted by livestock grazing was purchased and will be donated to The Nature Conservancy. Plans to restore this section of Teton Creek were prepared. Restoration plans call for planting of riparian shrubs and sedges, tree revetments to enhance streambank stability, and fencing to eliminate livestock grazing. The restored habitat will compensate for that damaged by road construction.

Reports:

Johnson, C.K. 1987. Final Mitigation Plan for the Felt Hydroelectric Project. FERC No. 5089. Ecosystems Research Institute, Logan, Utah.

PIGEON COVE/SNAKE RIVER, IDAHO

Several hundred feet of riparian vegetation were destroyed during construction of a road into the Snake River Canyon near Twin Falls. A rock slide initiated by construction activities blocked much of the river, causing increased streamflow velocities that eroded streambanks. Wetland habitat supported by seeps along the canyon walls was also destroyed. A total of about 7.6 hectares of wetland/riparian habitat was impacted. Baseline assessments indicated that about 3.8 hectares could be restored.

Rock was removed from the river and used to create a bench. Sediment dredged from the river was spread over the rock to enhance vegetative establishment. Specific revegetation goals were based on Habitat Evaluation Procedures (HEP) conducted in the vicinity. Vegetation was planted in 1987.

Plans called for the surface of the bench to be 2-3 feet below the high water line to ensure seasonal flooding. Abnormally low runoff in 1987 and 1988 resulted in peak flow levels well below the elevation of the bench, resulting in minimally successful revegetation. The coarse nature of substrates used to construct the bench limited its water retention and its capacity for establishment and sustenance of vegetation. Seeps from the canyon walls rapidly percolated through the porous substrate and were thus unavailable to plants. It is anticipated that future flooding will deposit fine sediments on the bench and improve conditions for revegetation.

Monitoring of revegetation indicated that shrub/tree plantings were 80% successful in seep areas along the canyon wall. Revegetation of riparian areas were only 30 to 50 percent successful due to droughty conditions. Irrigation will be required for revegetation of the riparian areas. Transplanted xeric shrubs were mostly unsuccessful due to drought.

Important lessons learned during restoration at Pigeon Cove include:

1. Seasonal flooding of rivers that are heavily committed for irrigation may not be counted on for restoration.

2. A reliable source of water may make the difference between nearly complete transplant survival (>95%) and low transplant survival (<50%).

3. Plant protectors intended to prevent wildlife browsing may damage transplants by falling over and bending or breaking the plants.

Revegetation efforts and monitoring of this site are continuing.

To mitigate for the 3.8 hectares of habitat that were permanently lost as well as for the temporary loss of habitat, approximately 20 hectares of relatively pristine habitat in Gooding County, Idaho, were purchased and donated to The Nature Conservancy.

Construction costs for this project were about $500,000.

Reports:

Johnson, C.K. 1986. Final Off-Site Mitigation Plan for LQ/LS Drain (Pigeon Cove) Hydroelectric Project. Ecosystems Research Institute, Logan, Utah.

Welling, K. 1986. Final On-Site Revegetation Plan. LQ/LS Drain (Pigeon Cove). FERC No. 5767. Hosey and Associates.

RED RIVER AND CROOKED RIVER, IDAHO

The Red River project area consists of approximately 30 km of stream, including meandering meadow reaches and timbered valley bottoms. The area has been impacted by dredge mining and livestock grazing. The Crooked River project area consists of 16 km of stream devastated by dredge mining.

The project areas were divided into reaches where problems and potential solutions were identified. In 1983-1985, habitat improvement was implemented, including contouring, weirs, deflectors, bank overhangs, bank stabilization structures, boulder placements, fencing of riparian zones and planting of riparian vegetation.

Monitoring of aquatic habitat condition for two years revealed several significant differences in habitat quality following stream enhancement at the lower Red River study site. Stream depth, percent pool width, bank water depth and pool quality improved in the treatment site as compared to the control site. Increases in fine sediments were evident in the treatment section, and were probably associated with the increased pool-riffle ratio. The riparian habitat variables exhibited no significant changes.

Evaluation of Crooked River habitat enhancement activities has been postponed until those activities are completed. No data are available with which to judge success.

Problems encountered included inability to obtain easements from land owners to implement plans. Low survival of vegetation was attributed to planting during the hottest time of the year.

Reports:

Hair, D. and R. Stowell. 1986. South Fork Clearwater River Habitat Enhancement. Annual Report-1985. In Natural Propagation and Habitat Improvement, Volume II-Idaho. Annual and Final Reports. Dept. of Energy, Bonneville Power Administration, Division of Fish and Wildlife.

Platts, W.S., M. McHenry, and R. Torquemada. 1986. Evaluation of Fish Habitat Enhancement Projects in Crooked River, Red River, and Bear Valley Creek. Progress Report II. In Idaho Habitat Evaluation for Off-Site Mitigation Record. Annual Report 1985. U.S. Dept. of Energy, Bonneville Power Administration, Div. of Fish and Wildlife.

ELDORADO CREEK, IDAHO

Eldorado Creek is a sixth order tributary of Lolo Creek, an important anadromous fishery. Construction of a road which parallels Eldorado Creek substantially altered the configuration of the channel and blocked upstream migration of steelhead trout.

The goals of the project were to enhance streambank stability, increase pool frequency and quality, reduce substrate embeddedness, and to increase in-stream and streambank cover for three stream reaches. The approach included construction of instream structures (Log and boulder weirs, boulder placements, root wads, tree revetments), partial removal of debris dams, and planting of riparian vegetation.

Measures were implemented in 1984, 1985, and 1986. At this time, all physical stream enhancements have been completed and fish have been stocked in the improved reaches. No judgement of results in terms of fisheries or revegetation success was given in the final report.

Reports:

Murphy, W. and A. Espinosa, Jr., 1985. Eldorado Creek Fish Passage Final Report, Modification M001 to Agreement DE-A179-54BP16535, In Natural Propagation and Habitat Improvement, Volume II Idaho. Annual and Final Reports. Dept. of Energy, Bonneville Power Administration. Division of Fish and Wildlife.

GRANDE RONDE RIVER, OREGON/WASHINGTON

The Joseph Creek and Upper Grande Ronde River drainages were examined as part of a study to identify, evaluate, prioritize, and recommend solutions to problems influencing anadromous fisheries. These two rivers have historically been excellent producers of anadromous fish but recent censuses indicate a decline in populations from those observed in the late 1960's and early 1970's. Declines were attributed to passage problems at mainstem Columbia and Snake River dams, user demands for the fishery resource, and impacts of logging, agriculture, roads, placer mining, and stream channelization. The goal of this project was to provide optimum spawning and rearing habitats for summer steelhead and spring chinook in selected portions of the Grande Ronde River Basin. Summaries of work proposed and implemented on these portions of streams follow.

New pools were excavated at numerous locations on Elk Creek. A number of existing pools were enlarged. Instream structures were installed to stabilize pools and to improve instream spawning and rearing habitat. Willow, alder, Siberian crabapple, and dogwood were planted along unstable streambanks. Planted areas were seeded with a grass mixture and fenced to prevent grazing by livestock.

Trees and shrubs were planted along 3.9 km of Swamp Creek. Species included white willow, Siberian crabapple, Midwest crabapple, wavey-leaf alder, river birch, and red-osier dogwood. Planted areas were seeded with a grass mixture and fenced to exclude livestock.

A habitat inventory was performed and instream structures were installed to improve pools along Fly and Sheep Creeks. Structures include wing deflectors, sills, weirs, revetments, and digger logs, rootwads, boulder placements, and bank-covers. Planting of riparian vegetation and fencing to protect plantings were also anticipated.

Monitoring of stream and riparian enhancement projects has been scheduled but little data is available at this time. Observation of instream structures indicates that creation of pool habitat has generally accomplished project goals. Structures show little damage since their installation. No judgment of success in terms of revegetation success or fishery enhancement is possible with the information presently available.

Reports:

Miller, R. and J.P. Olivarez. 1986. Joseph Creek Drainage Off-Site Enhancement Project. Annual Report FY 1985. (July 1, 1985 through June 30, 1986). In Natural Propagation and Habitat Volume I - Oregon. Annual and Final Reports 1985. Dept. of Energy, Bonneville Power Administration, Division of Fish and Wildlife.

Noll, W.T. 1986. Grande Ronde Habitat Improvement Project: Joseph Creek and Upper Grande Ronde River Drainages, Annual Report. In Natural Propagation and Habitat Enhancement. Volume I - Oregon. Annual and Final Reports 1985. Dept. of Energy, Bonneville Power Administration, Division of Fish and Wildlife.

Thomas, T. and M. Collette. 1986. BPA Project 84-9: Grande Ronde River Habitat Enhancement. Effective Period: April 1, 1985 to June 30, 1986. In Natural Propagation and Habitat Enhancement. Volume I - Oregon. Annual and Final Reports 1985. Dept. of Energy, Bonneville Power Administration, Division of Fish and Wildlife.

MEYERS COVE/CAMAS CREEK, IDAHO

Meyers Cove is located on Camas Creek, a tributary of the Middle Fork of the Salmon River, which is a significant producer of wild anadromous fish. The site was impacted by livestock grazing and agricultural practices. The goal of restoration was to improve habitat for spring chinook and steelhead trout spawning and rearing.

An enhancement plan was developed based on evaluations of streambank and channel morphology. Livestock management was evaluated to determine solutions to multiple use resource conflicts.

The riparian zone will be fenced and alternate water sources will be developed for livestock. Upland meadows will be reseeded with more productive grasses to compensate for loss of forage in riparian areas. The riparian zone will be contoured to eliminate high streambanks and to reduce sediment recruitment. Revegetation will consist of transplanting of willow, cottonwood, and alder seedlings and seeding of herbaceous plants. Unstable banks will be stabilized using deflectors and revetments. Instream cover will be enhanced with boulder placements. It is anticipated that streambank cover will be enhanced as the riparian zone recovers.

No data are available regarding success of implementation of these plans.

Report:

May, B.E. and R.W. Rose. 1986. Camas Creek (Meyers Cove) Anadromous Species Habitat Improvement Plant Final Report. In Natural Propagation and Habitat Improvement. Volume II--Idaho. Annual and Final Reports. Dept. of Energy, Bonneville Power Administration, Division of Fish and Wildlife.

OVERVIEW AND FUTURE DIRECTIONS

Joy B. Zedler
Pacific Estuarine Research Laboratory
San Diego State University

and

Milton W. Weller
Department of Wildlife and Fisheries Sciences
Texas A & M University

INTRODUCTION

Despite loss of over 50% of the wetlands in the contiguous United States, there is continuing pressure to use wetlands for immediate economic gain. Functional values that are nearly perpetual (self-maintaining) are rarely considered, and the complexity, integrity, and uniqueness of natural wetlands are undervalued. It is commonly assumed that wetland losses can be mitigated by restoring or creating wetlands of equal value. Some feel that replication is not always necessary if certain functions are replaced; others, including most wetland scientists, recognize that duplication is impossible and simulation is improbable. All would agree that we need substantially more information about what functions are being lost and how to replace them. This overview highlights the topics for which information needs are greatest and provides a research strategy to: a) improve wetland restoration/creation efforts, b) determine the degree to which constructed systems can replace lost functions, and c) determine the potential for persistence (resilience) of restored and constructed wetlands. We rely on authors of papers in this volume (hereafter called reviewers), our own experience, other wetland restoration literature, and discussions with other members of the National Wetlands Technical Council.

REASONS FOR RESTORING/CONSTRUCTING WETLANDS

While mitigation policies are the stimulus for most wetland restoration/creation projects today, there have been many efforts to alter or create wetlands in the past. Replacement of lost values has not always been required; rather, wetlands have been created or modified specifically to: 1) provide waterfowl and other wetland wildlife habitats; 2) minimize flood damage through increased flood-storage capacity; 3) store rainwater for livestock use or crop irrigation; 4) create agricultural basins for rice, cranberries, fish or crayfish; 5) improve water quality by trapping sediments; and 6) confine acid mine wastes, chemical contaminants, or fertilizer waste products. In addition, wetlands have been created inadvertently where topographic changes, such as roadway construction, have impounded water, or where gravel extraction or other mining has created suitable basins for wetland development. The marshes adjacent to the Salton Sea are perhaps the most extensive inadvertent wetlands in North America; the dry basin became a sea in 1904 when an irrigation canal overflowed with floodwater from the Colorado River.

Today, wetland restoration and creation projects involve a variety of ecosystems and many kinds of target species (e.g., fishes, birds, plants). Most of the freshwater projects have been in palustrine or open marsh wetlands that are aesthetically pleasing and have high, or at least more visible, wildlife values. Intentional conversion of shrub-dominated wetlands to herbaceous marshes has been common in the northeast, where it is easy to do and where wildlife become easier to view than in forests. The creation of open-water areas for agricultural water supplies has been extensive, involving tens of thousands of ponds in Missouri, Kansas, Oklahoma, and Texas. Their functions are limited by pond depth, turbidity and watershed characteristics. Nevertheless, these wetlands are new to the landscape. In the southwest, riparian restoration projects have been implemented to improve bird habitat. Throughout the country, streams have been restored for fisheries through structural modifications and the reduction of disturbance from livestock. Bogs and fens, which are extremely mature and complex systems, seem not to have been the target of restoration/creation attempts.

Forested palustrine wetlands are being lost at up to five times the rate of upland forests, according to a recent review of U.S. Forest Service data (Abernethy and Turner 1987). These habitats have been studied for plant-water relationships, and the water purification values of trees such as cypress are now well understood. Still, little has been done to create such forests beyond tree planting in agricultural settings, and much needs to be learned about the management of woody succession, the creation of natural understory communities, and the impoundment of waters to create bottomland habitat for wildlife.

Along the coasts, restoration efforts have been widespread (but not extensive) in intertidal marshes dominated by cordgrass, in mangrove forests, and in subtidal seagrass beds. Compared to inland restoration projects, coastal systems involve fewer plant species. Although providing monotypic vegetation (e.g., cordgrass) is often the immediate restoration objective, the ultimate goal is to provide habitat for a diverse coastal animal community that includes benthic infauna, fishes, terrestrial and aquatic insects, and birds. In California, mitigation agreements often require enhancement of habitat for target species, such as commercially valuable fishes or mammals, waterfowl, or plants and animals that are threatened with extinction and protected by federal and state laws.

Most projects involve modifications of former or degraded wetlands. Because many freshwater wetlands have been degraded by sedimentation, water turbidity, eutrophication, or altered water levels, and because many coastal wetlands have been filled or diked, there are numerous opportunities to restore or enhance existing sites. Some mitigation projects attempt to compensate for lost area by changing one wetland type (e.g., shallow water) to another (e.g., deeper water), or vice versa. Few projects replace lost wetland by converting upland to new wetland. A few attempts have been made to create vernal pools from upland habitat for the purpose of conserving endangered plant species. However, upland conversions or deeper water conversions are complex and expensive, and less likely to achieve agreed-upon goals than enhancement of degraded sites. Most former wetlands sites retain at least some vegetation and/or seed banks, and thus have potential for recovery. This is especially true of wildlife habitat development using water control structures and of the construction of farm ponds in the southern Great Plains, which often occurs in small drainages.

SETTING GOALS AND EVALUATING SUCCESS

Evaluating whether an attempt to create or restore a wetland has been successful is always controversial, largely because criteria for success differ. As Josselyn et al. (this volume) reiterate, success may be viewed as replacement of natural functional values or as compliance with a specific contract. The former is often the mitigation requirement, but natural functions are too complex to be identified in detail in a construction contract. The reviewers agreed that goals are too often vague or unrealistic. On the other hand, objectives listed in a contract may be too narrow in scope or too few in number.

The overall recommendation is to improve goal setting by beginning the mitigation permit process with a thorough evaluation of functions that will be lost when the wetland is destroyed or modified. In order to set goals for in-kind replacement, it is essential to know how the wetland is functioning; then, mitigation can be planned, and compliance of the resulting project evaluated. In discussing restoration/creation projects, the term "success" is best avoided or used only with qualifiers that identify the measurement criteria. Scientifically defensible standards are needed, based on research to develop suitable sampling protocols and identify appropriate reference (comparison) data sets.

Follow-up efforts to evaluate progress toward some goal have been almost totally lacking in the mitigation process. Legal mandates to do so vary by project and often are not enforced. A few studies have incorporated vegetation assessments of the mitigation sites, but quantification is often lacking and the timing of the assessment may be inadequate (e.g., two weeks after planting is insufficient to distinguish live from dead cordgrass). It is imperative that constructed wetlands be persistent. Reviewers agreed that more detailed and longer-term monitoring programs are required to determine if wetland functions are being replaced in perpetuity. Elsewhere in this volume (cf. paper by Erwin), specific suggestions are made for evaluating wetland ecosystem development.

Long-term commitments to maintain restored/created wetlands seem rare except in government agency projects, where weed control and maintenance of dikes and water-level control structures are often included in the long-range plan or forced by local advocates. Coastal wetlands are dynamic, and mitigation wetlands won't necessarily develop according to engineering or landscape plans. Initial seedings or plantings may not perform as expected due to weather, substrate type, elevation, or other unpredictable conditions. Water regimes may not meet expectations. Thus, many reviewers recommended that projects provide for "mid-course corrections"-- recognizing that something unplanned may happen. The conscientious evaluator sees the need for a change of direction to meet the original or changing goals, but contracts may not allow it. Such readjustments are more common in wildlife or flood control projects where goals are more general and perhaps less restrictive. Ongoing monitoring and evaluation of data can and should allow for readjustments. If an unplanned event is beneficial to the region, it can be capitalized upon; if it is a problem, plans can be made to correct it.

THE ROLE OF RESEARCH

The process of restoring wetlands involves at least three steps that can benefit from scientific research.

Step 1: Setting general (large-scale, region-wide) goals. Usually these are loosely stated as "maintain regional biodiversity" and/or solve societal problems, such as "improve water quality" or "enhance fisheries" or "reduce shoreline erosion". For this step, the kinds of information needed include: a) broad surveys of species distributions and knowledge of the relationships of species and their biotic and abiotic habitats and; b) general models of wetland functions for filtering materials from flowing water, supporting fisheries, and reducing hydrologic hazards. The fields of study range from taxonomy, biogeography, and landscape ecology, to water and soil chemistry and hydrology.

Step 2: Specifying project objectives and implementation procedures. The targets here are usually biological ones – with waterfowl, fisheries, endangered species, and/or selected vegetation types to be enhanced or exotic and pest species to be removed. Altering the topography or changing channel or shoreline morphometry is often prescribed. The information required includes detailed knowledge of species-habitat relationships and existing hydrology; in many cases, hydrologic models of existing and future changes in water circulation and sediment distribution will also be needed. Plant and animal population ecology, autecology, sedimentology, and hydrology are fields of study that can contribute understanding at this stage of restoration planning.

Step 3: Assessing how well the project matches its goals. Monitoring plans and on-site sampling are needed to characterize the effectiveness of restored, enhanced, or created wetlands. Controversies over whether projects are successful or not have already developed, and it is not clear which aspects of the project must be measured (contract compliance versus wetland functional value). In the future, it is likely that the methods of sampling the site and the data analysis required to interpret the results will also become more controversial, especially if penalties are established for inadequate mitigation. Sampling design and statistics become important.

INFORMATION NEEDS: SIMILARITIES AMONG REGIONS

The reviewers agree on one general principle---that mitigation efforts cannot yet claim to have duplicated lost wetland functional values. It has not been shown that restored or constructed wetlands maintain regional biodiversity and recreate functional ecosystems---there is some evidence that constructed wetlands can look like natural ones; there are few data to show they behave like natural ones. We can plant cordgrass gardens, but we don't know how well they resemble native cordgrass ecosystems. The restoration planning process does not include baseline studies of wetland ecosystem functioning, i.e. the dynamics of the wetland. Permits may require inventories, but these are rarely more than "snapshots" of the ecosystem, i.e., one-time characterizations of structure.

The greatest need is to understand wetland hydrology. As one hydrologist concluded, the interaction between wetlands, surface water, and ground water is still poorly understood or documented for most wetland types. The interactions that occur in wetlands among hydrologic elements, soils, water quality, and plant communities are rarely studied long enough, or in sufficient detail or precision, to provide complete understanding. We need to understand wetland hydrology before we can effectively understand any of the other interactions or processes that occur in wetlands. The range of hydrologic conditions typical for each wetland type, and the extremes that stress or destroy each wetland type, must be determined (R. Novitzki, U.S.G.S., pers. comm.).

A widespread need is to understand the relationships between wetland species and wetland topography. Grading must be planned and implemented with high precision---a few centimeters too high or too low will prevent the desired community from developing. While the concept seems simple, one of the most common errors in site construction is incorrect elevation.

Closely tied to hydrology and topography is the need to understand the importance of substrate type to wetland species. It is not yet clear how particle size and organic matter content interact to influence the establishment, distribution and growth of plants and animals, or the chemistry of the soil (e.g., pH, nutrients, oxygen, sulfides).

Several reviewers identified the need for more information about the gene pools that will be affected by wetland alterations. There is concern both about loss of local diversity and introduction of "alien" genes during transplantation of stock from distant sources. Because of ecotypic variation, foreign plants may lack necessary adaptations for environmental conditions at the transplant site. Such shortcomings may not be obvious until several years after transplantation, if the gene lacking is one that allows tolerance to rare extremes in temperature, salinity, or other conditions. It is also possible that foreign stock might outcompete local material in the short run and eliminate important genes from remnant populations. Because there are so many unknowns, several reviewers recommended: that transplant stock be taken from the nearest possible source, that nursery grown stock be labeled to specify the location where seeds or sprigs were obtained, and that contracts require local stock.

More information is needed to predict which methods of planting will achieve vegetative cover most rapidly across a variety of site conditions. Existing techniques include: 1) reliance on natural germination, often enhanced by water level management or irrigation; 2) seeding; 3) planting of tubers, rhizomes or whole emergent plants; and 4) planting of trees from cuttings, dormant poles, or potted plants.

Individual reviewers identified additional information needs for problems that are widespread among regions of the U.S.: 1) We need to understand what conditions favor and discourage nuisance species. Weedy exotics are more likely to invade disturbed sites; they may outcompete the restoration target species. Both preventative and control measures are needed. 2) We need to understand the dispersal of plants and the movements of animals among wetlands. Linkages with adjacent wetlands may be cut, so that the native species may not have ready access to the mitigation site. 3) We need to know more about the content of toxic substances at mitigation sites. Old landfills may be exposed by grading, heavy metals and organic toxins may be present in concentrations high enough to stress transplants and reduce their survival. The mobility and the potential for trophic concentration of exposed materials need to be predictable so that hazards can be identified. 4) We need to understand wetland soil chemistry. Nutrient dynamics are influenced by inundation and exposure. Aeration of some wetland soils (cat clays) may lead to very low pH and soils that retain their acidity for many years.

INFORMATION NEEDS: DIFFERENCES AMONG REGIONS

The concerns of reviewers differed widely by region and wetland type:

1. The scale of restoration problems differs enormously from region to region. In Louisiana, 40 square miles of wetland are being lost each year. Along the Florida coast, mangrove forests are being chipped away a few hectares at a time.

2. Sedimentation problems differ drastically by region: In Louisiana, river flows need to be redirected so that sediment-starved marshes will accrete and keep up with subsidence and rising sea level. In southern California, sediment-choked lagoons are filling too fast. The tidal prism declines, and the ocean inlet becomes blocked to tidal flow. Extreme environmental conditions eliminate species with narrow ranges of tolerance.

3. Dredge spoil islands can provide an

important substrate for marsh creation where the coastal shelf is flat and wide (e.g., Atlantic and Gulf Coasts); they compete for rare subtidal habitat where the shelf is steep and narrow (Pacific Coast). In most regions, bay disposal is not permitted.

4. The functions that are most valued differ for each region.

a. Hydrologic benefits that are most valued range from shoreline anchoring along the Atlantic Coast, to flood storage along major rivers, to providing wet habitat for wetland plants and animals in southern California.

b. Water quality problems may focus on sediment load in Louisiana and industrial wastes in the northeast.

c. The dominant vegetation differs, and research on transplanting cordgrass in North Carolina doesn't necessarily transfer to problems concerning mangroves in Florida or sedge-dominated marshes in Oregon. Where endangered plants are present (e.g., Mesa mint in vernal pools, salt marsh bird's beak in coastal marshes), very specific vegetation plantings are called for. One might expect that a single transplanting protocol could be developed, at least for sea grasses, which occur nearly everywhere along the nation's coast, including Hawaii and Alaska. But, among the ecological regions, the species grow differently and transplanting recommendations do not transfer from one region to the next (cf. Fonseca, this volume).

d. The animal uses of different wetland types vary by region. In the Pacific Northwest, and the Atlantic and Gulf of Mexico, wetlands may be managed for various fisheries, while some in San Francisco Bay may be designed for the endangered salt marsh harvest mouse, and many in southern California and Florida are managed for endangered birds.

5. Finally, regions differ in information availability, which is a function of the time span over which scientists have been involved in restoration research (10+ years in N. Carolina), the number of projects that have been done in a region (e.g., Florida mangrove work), the type of restoration problem (e.g., U.S. Army Corps of Engineers Dredged Material Research Program) and whether or not anyone has taken the initiative to review past projects (e.g., Shisler's work in New Jersey).

Because of these different information needs, individual papers provided research recommendations specific to their region or a single wetland type.

RESEARCH STRATEGIES

To meet the wide variety of information needs, we propose a research program (Figure 1) that is based on improved understanding of wetland functioning (a national wetland ecosystem initiative), that will characterize similarities and differences in the functioning of natural and constructed wetlands (comparisons within regions), and that will respond to specific needs for local projects and regional regulatory personnel (demonstration sites, each with an experimental component). Of these three goals, the first is broad and open-ended, requiring a long-term funding commitment and many investigations; yet a commitment to basic understanding underlies all wetland science. The second and third goals focus on comparisons within biogeographic regions and these studies would provide the most rapid and direct input to regulatory personnel. The rationale for each research goal is discussed, with priorities for the types of studies that should be undertaken:

I. NATIONAL WETLAND ECOSYSTEM RESEARCH PROGRAM

A national wetland ecosystem research program is needed to provide better understanding of wetland ecosystem functioning, wherein science is applied to wetland restoration and construction. We need to move from descriptive to hypothesis generating and testing studies. Long-term efforts of 5 to 15 years and multidisciplinary involvement is required. Studies of soils, chemistry, microbiology, and physiology of wetland components must be included. A national program encompassing many wetland types would allow an interactive and creative setting for basic science, as well as problem solving. Where one branch of science has provided predictive powers (e.g., non-point source pollution models and lake restoration models), technology transfers should be included in the research programs (O. Loucks, Butler University, pers.

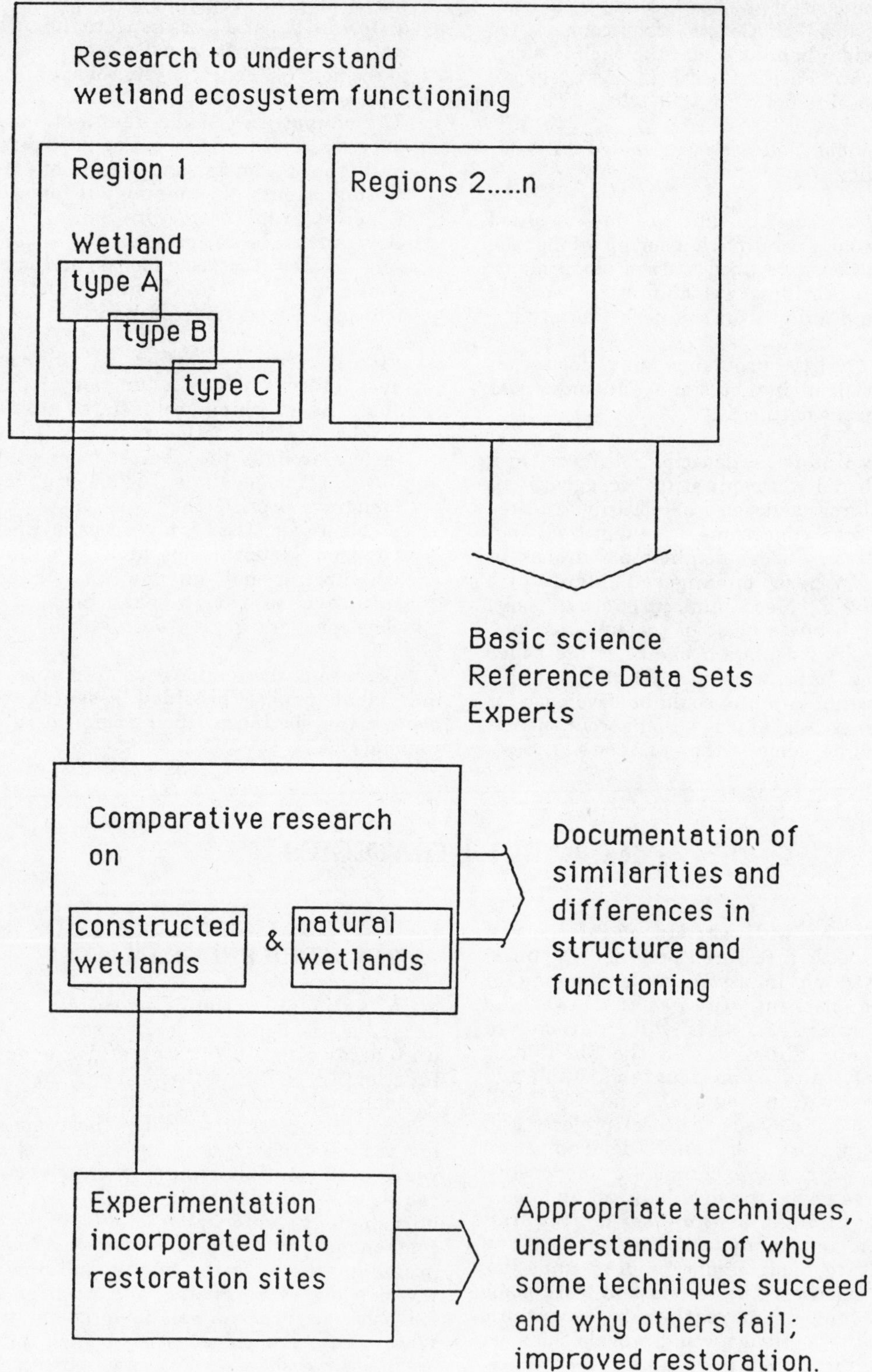

Figure 1. Research strategy to improve wetland restoration programs. Elements should be implemented simultaneously, i.e., research need not proceed from top to bottom.

comm.). Studies of vegetation should recognize the dynamic, often stochastic, nature of wetlands that are subject to catastrophic events and rare environmental extremes (W. Niering, Connecticut College, pers. comm.).

Current ecosystem research programs (e.g., NSF-LTER program) focus on a few sites chosen to represent major biomes. We recommend a broader approach for wetland ecosystem research that focuses on wetland types---not just a single site for each type but an understanding of wetland types with a biogeographic region---to provide reference data sets that characterize the spatial and temporal dynamics of ecosystem structure and functioning. Specific studies should address:

1. The history of species composition, microstratigraphy, development of organic soil deposits and other soil characteristics. While considerable work of this type has been done in peat beds and bogs, goals have focused on describing climate and species change, and have been less oriented to processes and changes in functional components of the system.

2. The importance and functional significance of organic soil and of the chemical and biological processes in wetlands.

3. The geochemical and geomorphological processes that influence soil-water holding capacity, movement and quality.

4. Relationships between substrate type and geomorphology to wetland autonomy, e.g., perched wetlands, water-table wetlands, and flow-through basins.

5. Biochemical processes of nutrient turnover at various water regimes. The relationship of macroinvertebrates to nutrient dynamics.

6. Life-history patterns of plants at various water regimes, especially minima and maxima. Plant responses to fixed-water regimes.

7. Persistence of communities and resilience to extreme events.

8. The relative values of wetlands and uplands in performing hydrologic functions such as flood desynchronization and ground water recharge.

II. LANDSCAPE-LEVEL COMPARISONS

Landscape-level comparisons of natural and constructed wetlands are needed to understand controls on short and long-term achievement of restoration goals. Within regions and within wetland types, we need to identify the rates with which constructed wetlands approach the functional characteristics of more natural systems, the factors that slow wetland development, and the techniques for accelerating the achievement of restoration goals. Pilot studies are underway in Florida, the Pacific Northwest, and Louisiana, with funding from EPA. Their focus is on structural similarities and differences between constructed and "reference" wetlands (M. Kentula, NSI Technology Services, Corp., pers. comm.). These studies need to be extended to identify beneficial attributes and shortcomings of wetland mitigation projects and to identify causes of successes and failures in developing the proper hydrologic and biologic functions. They also need to be extended to all regions of the country.

Reviewers provided a number of specific suggestions for research. These include:

1. Determination of the range of hydrologic conditions typical for each wetland type, and identification of the extremes that stress or destroy each wetland type. Important questions are: what changes in groundwater are tolerable; what degree of streamflow alteration is permissible; what degree of salinity change will alter ecosystem structure; what degree of sediment input can the ecosystem absorb?

2. Physiological requirements and tolerance of plants to soil and water chemistry (including salinity) and temperature. Important questions for natural and constructed wetlands are: is nitrogen fixation similar; is the nutrient trapping function present; do toxic materials have similar impacts; is productivity similar; are the species groupings likely to persist?

3. Genetic make-up of hydrophytes across their distributional range. It is important to know if the vegetation in constructed wetlands has the potential to withstand extremes in physical and biological (e.g., insect irruptions) conditions.

4. Cumulative impacts of adjacent projects on hydrology and water quality of the developed site. Can the constructed wetlands persist within regions where multiple impacts occur?

5. Ecological merits of mitigating damages on-site vs. off-site and with restoration vs. construction efforts. Can regional biodiversity and functional values be maintained with off-site mitigation; are newly constructed systems less likely to succeed than restored systems?

6. Weed problems and control methods. It is hypothesized that constructed wetlands are less resilient than natural systems, and that considerable effort will be needed to control weedy species.

7. Studies of various "buffer concepts", including protective zones (bands around wetlands) that might reduce negative impacts at the upland-wetland interface. How broad must buffers be, and what qualities must they possess to protect functional values? Are requirements similar for natural and constructed systems?

The comparative research programs should include interaction with regulatory agencies to provide rapid information transfer. The research should analyze monitoring data from regional restoration sites to identify problems and evaluate the need for mid-course corrections. The biological goals of ongoing and future restoration projects should be updated continuously as information on regionally significant species becomes available.

Finally, we recommend that the regional approach initiated in this review book be augmented with preparation of <u>regional</u> resource information banks that list high-priority restoration goals, characterize additional restoration projects by describing their problems and achievements, and continue to update the literature on restoration within the region and nationwide.

III. WETLAND RESTORATION DEMONSTRATION SITES

Wetland restoration demonstration sites with manipulative experimentation and rapid information transfer are needed to improve restoration projects and to foster a close working relationship between regulatory personnel and scientists---beyond the publication of research findings. Wetland scientists can offer new data, but there should be incentive and opportunity to contribute more---those scientists interested in wetland restoration can identify questions that should be asked of permit applicants and can offer opinions based on unique experiences. In many situations, this interaction already exists and should be fostered. In other situations, it needs to be developed. Therefore, we suggest that EPA fund at least one restoration research program in each of its regions.

The priority goals of the research program should be to develop best restoration methods, to understand the causes of successes and failures, to transfer information to regulatory personnel, and to provide expert consultation on highly controversial permits. The criteria used to identify suitable research programs should include the following:

1. The mitigation project(s) should be "representative" of the region---preferably, it will include more than one type of habitat type, including the one most commonly mitigated in the region. At least one reference site should also be selected.

2. The project should include an experimental component, so that cause-effect relationships can be tested. At least some of the region's important questions should be answerable through manipulative experimentation at the restoration site.

3. There should be a designated mechanism for rapid information transfer to regulatory personnel, including field site visits and video or slide shows. Where multiple responses to the requests for proposals are obtained, on-site reviews that include regulatory personnel should be implemented. This review process alone would identify specific problems, sites, and available expertise.

4. There should be opportunities for several experts in the region to interact with agency personnel---in reviewing the research program and in evaluating successful and unsuccessful mitigation programs.

5. The research program should begin with 2-3 years of intensive study, and include a long-term component, if only reevaluation at 5, 10 and 15-year intervals. Continued commitment by funding agencies will allow the most basic restoration question to be answered, i.e., are we creating gardens or functional, resilient ecosystems? Long-term projects will insure that advice continues even after regulatory personnel change.

Specific objectives should be geared to each region according to recommendations in this volume; they include studies of each of the research objectives recommended above, plus detailed, site-specific studies identifying and investigating:

1. Dependencies between animals, plants and substrate, including hydroperiods required by individual plant species, host specificity of herbivorous insects, dependence of infauna on substrate texture, nesting and feeding requirements of birds. Special, detailed studies are needed for each region's rare and endangered species.

2. Merits of using natural seed banks or seed-bearing mulch from other wetlands vs.

adding seeds and the merits of various planting techniques.

3. Required watering regimes (irrigation requirements, if any; elevation and inundation tolerances), fertilizer needs, and salinity tolerances for species planted at the site. The value of (need for) organic soil for successful plant establishment.

4. Control regimes to reduce or eradicate local nuisance species.

5. Indicator species that can help quantify wetland functions, e.g., use of invertebrates as measures of food chain support; abundance of blue-green algae or legumes as indicators of nitrogen fixation potential; presence of clean-water species of fishes or invertebrates that indicate filtering functions.

IV. PRIORITIES FOR FUNDING

If EPA research budgets are limited, every attempt should be made to identify other federal, regional, and local sources of research support, so that efforts can be coordinated. Immediate funding of regional demonstration sites will provide the most direct contribution of scientific understanding to improved restoration (Strategy III). A long-term commitment to systematic data gathering (monitoring) of selected sites will be necessary to understand if restored wetlands have the resilience to persist well into the future. Expansion of research to compare a diverse selection of wetland types within and among regions (Strategy II) would broaden the basis for sound management. Finally, the commitment to understanding wetland functioning at the broadest geographic scale (Strategy I) is basic to wetland science. There will be progress toward this goal if Strategies III and II are adopted. The demonstration sites and regional comparisons are important components of the research program needed to insure successful restoration and management of the nation's existing wetland resources.

CONCLUSION

There is a national need to understand wetland functioning at its most basic and detailed levels, to compare natural and constructed wetlands within biogeographic regions, and to test methods of restoring wetlands using manipulative experimentation. Wetland research should be directed toward the proximate goal of preserving wetland values, and ultimately, of understanding the details of how wetlands function. Problems and objectives involve different research levels (e.g., size of areas, time scales, precision and complexity). Some problems require immediate solutions for crisis situations; with answers based on research that uses a less precise level or a smaller size scale. The research program must build short-term goals into a long-term design, both to gain perspective and to prioritize projects and funding.

Interaction between researchers and regulatory personnel will help solve practical problems and crisis decision-making and help develop technology for restoring degraded wetlands or constructing new ones. Transfer of information between programs will enhance effectiveness and aid in research problem identification. A national program of wetland ecosystem research will move the process of wetland construction, restoration and enhancement from trial-and-error to a sound management program.

LITERATURE CITED

Abernethy, Y., and R.E. Turner. 1987. U.S. Forested Wetlands: 1940-1980. BioScience 37:721-727.

PART 2

PERSPECTIVES

WETLANDS RESTORATION/CREATION/ENHANCEMENT TERMINOLOGY: SUGGESTIONS FOR STANDARDIZATION

Roy R. Lewis III
Lewis Environmental Services, Inc.

INTRODUCTION

This document includes a glossary that was prepared after review by all the authors. Four versions of the manuscript have been circulated for reviewers' comments, and each version was an improvement on the previous one. The specific definitions in the glossary represent an attempt to bring some order to the terminology applied to the topic of wetland creation and restoration. It has been our collective experience that much confusion exists about specific terms, and they are used in different ways by different authors in different parts of the country. Unfortunately, much of the existing confusion is becoming formalized as states, counties, and municipalities develop their own regulations related to wetland creation and restoration. This discussion of terminology is meant to highlight the major problem areas.

HISTORICAL CONTEXT

In looking for a starting point we were able to find only three existing glossaries applicable to the topic. These were contained in the U.S. Army Corps of Engineers Wetlands Delineation Manual prepared by the Environmental Laboratory Waterways Experiment Station, Vicksburg (Environmental Laboratory 1987), the U.S. Fish and Wildlife Service's classification of wetlands and deepwater habitats of the United States (Cowardin et al. 1979), and the proceedings of a conference titled Wetland Functions, Rehabilitation and Creation in the Pacific Northwest: The State of Our Understanding, prepared by the Washington State Department of Ecology (Strickland 1986). Three additional glossaries (Helm 1985, Rawlins 1986, and Soil Survey Staff 1975) were recommended by reviewers and have been used to improve this section. To these combined glossaries were added definitions from individual authors of published papers or proceedings, for example Zedler (1984) and Schaller and Sutton (1978), and regulatory or review agency rule promulgation, such as U.S. Fish and Wildlife Service (1981). Where the existing definitions were checked against dictionary definitions, Websters Unabridged Dictionary, Second Edition (McKechnie 1983) was used as the reference dictionary. Some geological terms were taken from Bates and Jackson (1984) and Gary et al. (1972) as recommended by reviewers.

DISCUSSION

The five key definitions are: mitigation, restoration, creation, enhancement, and success. Briefly, McKechnie (1983) defines these terms as follows:

MITIGATION alleviation; abatement or diminution, as of anything painful, harsh, severe, afflictive, or calamitous (p. 1152);

RESTORATION a putting or bringing back into a former, normal, or unimpaired state or condition (p. 1544);

CREATION the act of bringing into existence (p. 427);

ENHANCEMENT the state or quality of being enhanced; rise, increase, augmentation (p. 603);

SUCCESS favorable or satisfactory outcome or result (p. 1819).

For the purposes of this document, we are defining these terms so that there is as little ambiguity and overlap as possible. The glossary definition and an explanation of each of the key terms is provided below.

MITIGATION - For the purposes of this document, the actual restoration, creation, or enhancement of wetlands to compensate for permitted wetland losses. The use of the word mitigation here is limited to the above cases and is not used in the general manner as outlined in the President's Council on Environmental Quality National Environmental Policy Act regulations (40 CFR 1508.20).

MITIGATION BANKING - Wetland restoration, creation, or enhancement undertaken expressly for the purpose of providing compensation for wetland losses from future development activities. It includes only actual wetland restoration, creation, or enhancement occurring prior to elimination of another wetland as part of a credit program. Credits may then be withdrawn from the bank to compensate for an individual wetland destruction. Each bank will probably have its own unique credit system based upon the functional values of the wetlands unique to the area. As defined here, mitigation banking does not involve any exchange of money for permits. However, some mitigation programs, such as those in California, do accept money in lieu of actual wetland restoration, creation or enhancement.

RESTORATION - Returned from a disturbed or totally altered condition to a previously existing natural, or altered condition by some action of man. Restoration refers to the return to a pre-existing condition. It is not necessary to have complete knowledge of what those pre-existing conditions were; it is enough to know a wetland of whatever type was there and have as a goal the return to that same wetland type. Restoration also occurs if an altered wetland is further damaged and is then returned to its previous, though altered condition. That is, for restoration to occur it is not necessary that a system be returned to a pristine condition. It is, therefore, important to define the goals of a restoration project in order to properly measure the success.

In contrast with restoration, creation (defined below) involves the conversion of a non-wetland habitat type into wetlands where wetlands never existed (at least within the recent past, 100-200 years). The term re-creation is not recommended here due to confusion over its meanings. Schaller and Sutton (1978) define restoration as a return to the exact pre-existing conditions, as does Zedler (1984). Both believe restoration is therefore seldom, if ever, possible. Schaller and Sutton (1978) use the term rehabilitation equivalent to our restoration. For our purposes, "rehabilitation" refers to the conversion of uplands to wetlands where wetlands previously existed. It differs from restoration in that the goal is not a return to previously existing conditions but conversion to a new or altered wetland that has been determined to be "better" for the system as a whole. Reclamation is also used to mean the same thing by some, but "wetland reclamation" often means filling and conversion to uplands, therefore its use is not recommended.

CREATION - The conversion of a persistent non-wetland area into a wetland through some activity of man. This definition presumes the site has not been a wetland within recent times (100-200 years) and thus restoration is not occurring. Created wetlands are subdivided into two types: artificial and man-induced. An artificial created wetland exists only as long as some continuous or persistent activity of man (i.e., irrigation, weeding) continues. Without attention from man, artificial wetlands revert to their original habitat type. Man-induced created wetlands generally result from a one-time action of man and persist on their own. The one-time action might be intentional (i.e., earthmoving to lower elevations) or unintentional (i.e., dam building). Wetlands created as a result of dredged material deposition may have subsequent periods during which additional deposits occur. Man-initiated is an acceptable synonym.

ENHANCEMENT - The increase in one or more values of all or a portion of an existing wetland by man's activities, often with the accompanying decline in other wetland values. Enhancement and restoration are often confused. For our purposes, the intentional alteration of an existing wetland to provide conditions which previously did not exist and which by consensus increase one or more values is enhancement. The diking of emergent wetlands to create persistent open-water duck habitat is an example; the creation of a littoral shelf from open water habitat is another example. Some of the value of the emergent marsh may be lost as a result (i.e., brown shrimp nursery habitat).

SUCCESS - Achieving established goals. Unlike the dictionary definition, success in wetlands restoration, creation, and enhancement ideally requires that criteria, preferably measurable as quantitative values, be established prior to commencement of these activities. However, it is important to note that a project may not succeed in achieving its goals yet provide some other values deemed acceptable when evaluated. In other words, the project failed but the wetland was a "success". This may result in changing the success criteria for future projects. It is important, however, to acknowledge the non-attainment of previously established goals (the unsuccessful project) in order to improve goal setting. In situations where poor or nonexistent goal setting occurred, functional equivalency may be determined by comparison with a reference wetland, and success defined by this comparison. In reality, this is easier said than done.

LITERATURE CITED

Bates, R.L. and J.A. Jackson (Eds.). 1984. Dictionary of Geologic Terms. American Geological Institute. Anchor Books, Garden City, New York.

Cowardin, L.M., V. Carter, F.G. Golet, and E.T. LaRoe. 1979. Classification of Wetlands and Deepwater Habitats of the United States. U.S. Fish & Wildlife Service. FWS/OBS-79/31.

Environmental Laboratory. 1987. Corps of Engineers. Wetlands Delineation Manual, Technical Report Y-87-1. U.S. Army Engineer Waterways Experiment Station, Vicksburg, Mississippi.

Gary, M., R. McAfee, Jr., and C.L. Wolf (Eds.). 1972. Glossary of Geology. American Geological Institute, Washington, D.C.

Helm, W.T. (Ed.). 1985. Aquatic Habitat Inventory: Glossary and Standard Methods. Western Division, American Fisheries Society, Utah State University, Logan, Utah.

McKechnie, J.L. (Ed.). 1983. Websters New Universal Unabridged Dictionary. Simon and Schuster, Cleveland, Ohio.

Rawlins, C.L. 1986. Glossary. In S. Jensen, An Approach to Classification of Riparian Ecosystems. White Horse Associates, Smithfield, Utah. [mimeo]

Schaller, F.W. and P. Sutton. 1978. Reclamation of Drastically Disturbed Lands. American Society of Agronomy, Madison, Wisconsin.

Soil Survey Staff. 1975. Soil Taxonomy, A Basic System of Soil Classification for Making and Interpreting Soil Surveys. Agriculture Handbook No. 436. U.S. Dept. of Agriculture, Soil Conservation Service. U.S. Government Printing Office, Washington, D.C.

Strickland, R. (Ed.). 1986. Wetland Functions, Rehabilitation, and Creation in the Pacific Northwest. Washington State Department of Ecology, Olympia, Washington.

U.S. Fish and Wildlife Service. 1981. U.S. Fish and Wildlife Service Mitigation Policy. Federal Register 46(15):7644-7663.

Zedler, J.B. 1984. Salt Marsh Restoration--A Guidebook for Southern California. California Sea Grant Report No. T-CSGC P-009.

GLOSSARY

AREAL COVER - A measure of dominance that defines the degree to which above-ground portions of plants (not limited to those rooted in a sample plot) cover the ground surface. It is possible for the total areal cover in a community to exceed 100% because (a) many plant communities consist of two or more vegetative strata (overstory, understory, ground cover, undergrowth); (b) areal cover is estimated by vegetative layer; and (c) foliage within a single layer may overlap.

ARTIFICIAL WETLAND - A created wetland requiring constant application of water or maintenance to provide wetland values.

BASAL AREA - The cross-sectional area of a tree trunk measured in square inches, square centimeters, etc. Basal area is normally measured at 4.5 feet (1.4 m) above ground level or just above the buttress if the buttress exceeds that height and is used as a measure of dominance. The most easily used tool for measuring basal area is a tape marked in units of area (i.e., square inches). When plotless methods are used, an angle gauge or prism will provide a means of rapidly determining basal area. This term is also applicable to the cross-sectional area of a clumped herbaceous plant, measured at 1.0 inch (2.54 cm) above the soil surface.

BASELINE STUDY - An inventory of a natural community or environment that may serve as a model for planning or establishing goals for success criteria. Synonym: reference study.

BENCH MARK - A fixed, more or less permanent reference point or object, the elevation and horizontal location of which is known. The U.S. Geological Survey [USGS] installs brass caps in bridge abutments or otherwise permanently sets bench marks at convenient locations nationwide. The elevations on these marks are referenced to the National Geodetic Vertical Datum [NGVD], also commonly known as Mean Sea Level [MSL] although they may not be exactly the same. For most purposes of wetland mitigation, they can be assumed to be equivalent although a local surveyor should be consulted for final determination. Locations of these bench marks on USGS quadrangle maps are shown as small triangles. The existence of any bench mark should be field verified before planning work that relies on a particular reference point. The USGS, local state surveyor's office, or city or town engineer can provide information on the existence, exact location and exact elevation of bench marks, and the equivalency of NGVD and MSL.

CANOPY LAYER - The uppermost layer of vegetation in a plant community. In forested areas, mature trees comprise the canopy layer, while the tallest herbaceous species constitute the canopy layer in a marsh.

CONTROL PLOT - An area of land used for measuring or observing existing undisturbed conditions.

CONTOUR - An imaginary line of constant elevation on the ground surface. The corresponding line on a map is called a "contour line".

CREATED WETLAND - The conversion of a persistent upland or shallow water area into a wetland through some activity of man.

DEGRADED WETLAND - A wetland altered by man through impairment of some physical or chemical property which results in a reduction of habitat value or other reduction of functions (i.e., flood storage).

DENSITY - The number of individuals per unit area.

DIAMETER AT BREAST HEIGHT [DBH] - The width of a plant stem as measured at 4.5 feet (1.4 m) above the ground surface or just above the buttress if over 4.5 feet (1.4 m).

DISTURBED WETLAND - A wetland directly or indirectly altered from a natural condition, yet retaining some natural characteristics; includes natural perturbations.

DOMINANCE - As used herein, a descriptor of vegetation that is related to the standing crop of a species in an area, usually measured by height, areal cover, density, or basal area (for trees), or a combination of parameters.

DOMINANT PLANT SPECIES - A plant species that exerts a controlling influence on or defines the character of a community.

DRAINED - A condition in which the level or volume of ground or surface water has been reduced or eliminated from an area by artificial means.

DRIFT LINE - An accumulation of debris along a contour (parallel to the water flow) that represents the height of an inundation event.

EMERGENT PLANT - A rooted plant that has parts extending above a water surface, at least during portions of the year but does not tolerate prolonged inundation.

ENHANCED WETLAND - An existing wetland where some activity of man increases one or more values, often with the accompanying decline in other wetland values.

EXOTIC - Not indigenous to a region; intentionally or accidentally introduced and often persisting.

EXPERIMENTAL PLOT - An area of land used for measuring or observing conditions resulting from a treatment (i.e., an installation of particular plants).

FILL MATERIAL - Any material placed in an area to increase surface elevation.

FREQUENCY (vegetation) - The distribution of individuals of a species in an area. It is quantitatively expressed as:

$$\frac{\text{Number of samples containing species A}}{\text{Total number of samples}} \times 100$$

FUNCTIONAL VALUES - Values determined by abiotic and biotic interactions as opposed to static measurements (e.g., biomass).

HABITAT - The environment occupied by individuals of a particular species, population, or community.

HABITAT VALUE - The suitability of an area to support a given evaluation species.

HEADWATER FLOODING - A situation in which an area becomes inundated primarily by surface runoff from upland areas.

HERB - A nonwoody individual of a macrophytic species.

HERBACEOUS LAYER - Any vegetative stratum of a plant community that is composed predominantly of herbs.

HYDRIC SOIL - A soil that is saturated, flooded, or ponded long enough during the growing season to develop anaerobic conditions that favor the growth and regeneration of hydrophytic vegetation. Hydric soils that occur in areas having positive indicators of hydrophytic vegetation and wetland hydrology are wetland soils.

HYDROLOGIC REGIME - The distribution and circulation of water in an area on average during a given period including normal fluctuations and periodicity.

HYDROLOGY - The science dealing with the properties, distribution, and circulation of water both on the surface and under the earth.

HYDROPHYTE - Any macrophyte that grows in water or on a substrate that is at least periodically deficient in oxygen as a result of excessive water content; plants typically found in wet habitats. Obligate hydrophytes require water and cannot survive in dry areas. Facultative hydrophytes may invade upland areas.

HYDROPHYTIC VEGETATION - The sum total of macrophytic plant life growing in water or on a substrate that is at least periodically deficient in oxygen as a result of excessive water content. When hydrophytic vegetation comprises a community where indicators of hydric soils and wetland hydrology also occur, the area has wetland vegetation.

IMPORTANCE VALUE - A quantitative term describing the relative influence of a plant species in a plant community, obtained by summing any combination of relative frequency, relative density, and relative dominance.

INDIGENOUS SPECIES - Native to a region.

IN-KIND REPLACEMENT - Providing or managing substitute resources to replace the functional values of the resources lost, where such substitute resources are also physically and biologically the same or closely approximate those lost.

INUNDATION - A condition in which water from any source temporarily or permanently covers a land surface.

MACROPHYTE - Any plant species that can be readily observed without the aid of optical magnification. This includes all vascular plant species and mosses (e.g., Sphagnum spp.), as well as large algae (e.g., Chara spp., kelp).

MAINTENANCE - Any activities required to assure successful restoration after a project has begun (i.e., erosion control, water level manipulations).

MAN-INDUCED WETLAND - Any area of created wetlands that develops wetland characteristics due to some discrete non-continuous activity of man.

MEAN SEA LEVEL - A datum, or "plane of zero elevation", established by averaging hourly tidal elevations over a 19-year tidal cycle or "epoch". This plane is corrected for curvature of the earth and is the standard reference for elevations on the earth's surface. The National Geodetic Vertical Datum [NGVD] is a fixed reference relative to Mean Sea Level in 1929. The relationship between MSL and NGVD is site-specific.

MESOPHYTIC - Any plant species growing where soil moisture and aeration conditions lie between extremes. These species are typically found in habitats with average moisture conditions, neither very dry nor very wet.

MITIGATION - The President's Council on Environmental Quality defined the term "mitigation" in the National Environmental Policy Act regulations to include "(a) avoiding the impact altogether by not taking a certain action or parts of an action; (b) minimizing impacts by limiting the degree or magnitude of the action and its implementation; (c) rectifying the impact by repairing, rehabilitating, or restoring the affected environment; (d) reducing or eliminating the impact over time by preservation and maintenance operations during the life of the action; and (e) compensating for the impact by replacing or providing substitute resources or environments" (40 CFR Part 1508.20(a-e)). For the purposes of this document, mitigation refers only to restoration, creation, or enhancement of wetlands to compensate for permitted wetland losses.

MITIGATION BANKING - Wetland restoration, creation or enhancement undertaken expressly for the purpose of providing compensation credits for wetland losses from future development activities.

MONITORING - Periodic evaluation of a mitigation site to determine success in attaining goals. Typical monitoring periods for wetland mitigation sites are three to five years.

NATURAL - Dominated by native biota and occurring within a physical system which has developed through natural processes (without human intervention), in which natural processes continue to take place.

NUISANCE SPECIES - Species of plants that detract from or interfere with a mitigation project, such as most exotic species and those indigenous species whose populations proliferate to abnormal proportions. Nuisance species may require removal through maintenance programs.

OUT-OF-KIND REPLACEMENT - Providing or managing substitute resources to replace the functional values of the resources lost, where such substitute resources are physically or biologically different from those lost.

PHYSIOGNOMY - A term used to describe a plant community based on community stratification and growth habit (e.g., trees, herbs, lianas) of the dominant species.

PLANT COMMUNITY - All of the plant populations occurring in a shared habitat or environment.

PLANT COVER - see AREAL COVER.

PONDED - A condition in which water stands in a closed depression. Water may be naturally removed only by percolation, evaporation, and/or transpiration.

POORLY DRAINED - Soils that are commonly wet at or near the surface during a sufficient part of the year that field crops cannot be grown under natural conditions. Poorly drained conditions are caused by a saturated zone, a layer with low hydraulic conductivity, seepage, or a combination of these conditions.

PRODUCTIVITY - Net annual primary productivity; the amount of plant biomass that is generated per unit area per year.

QUANTITATIVE - A precise measurement or determination expressed numerically.

RECLAIMED WETLANDS - Same as restored wetland, but often used in other parts of the world to refer to wetland destruction due to filling or draining.

REHABILITATION - Conversion of an upland area that was previously a wetland into another wetland type deemed to be better for the overall ecology of the system.

RELATIVE DENSITY - A quantitative descriptor, expressed as a percent, of the relative number of individuals in an area; it is calculated by:

$$\frac{\text{Number of individuals of species A}}{\text{Total number of individuals of all species}} \times 100$$

RELATIVE DOMINANCE - A quantitative descriptor, expressed as a percent, of the relative amount of individuals of a species in an area; it is calculated by:

$$\frac{\text{Amount of species A}}{\text{Total amount of all species}} \times 100$$

The amount of a species may be based on percent areal cover, basal area, or height.

RELATIVE FREQUENCY - A quantitative descriptor, expressed as a percent, of the relative distribution of individuals in an area; it is calculated by:

$$\frac{\text{Frequency of species A}}{\text{Total frequency of all species}} \times 100$$

RELIEF - The change in elevation of a land surface between two points; collectively, the configuration of the earth's surface, including such features as hills and valleys. See also TOPOGRAPHY.

RESTORED WETLAND - A wetland returned from a disturbed or altered condition to a previously existing natural or altered condition by some action of man (i.e., fill removal).

SAMPLE PLOT - An area of land used for measuring or observing existing conditions.

SOIL - The collection of natural bodies on the earth's surface containing living matter and supporting or capable of supporting plants out-of-doors. Places modified or even made by man of earthy materials are included. The upper limit of soil is air or shallow water and at its margins it grades to deep water or to barren areas of rock or ice. Soil includes the horizons that differ from the parent material as a result of interaction through time of climate, living organisms, parent materials and relief.

SLOPE - A piece of ground that is not flat or level.

SUBSTRATE - The base or substance on which an attached species is growing.

TIDAL - A situation in which the water level periodically fluctuates due to the action of lunar and solar forces upon the rotating earth.

TOPOGRAPHY - The configuration of a surface, including its relief and the position of its natural and man-made features.

TRANSECT - As used here, a line on the ground along which observations are made at some interval.

TRANSITION ZONE - The area in which a change from wetlands to nonwetlands occurs. The transition zone may be narrow or broad.

TREE - A woody plant >3.0 inches in diameter at breast height, regardless of height (exclusive of woody vines).

UPLAND - As used herein, any area that does not qualify as a wetland because the associated hydrologic regime is not sufficiently wet to elicit development of vegetation, soils, and/or hydrologic characteristics associated with wetlands. Such areas occurring within floodplains are more appropriately termed non wetlands.

WATER TABLE - The upper surface of groundwater or that level below which the soil is saturated with water. The saturated zone must be at least 6 inches thick and persist in the soil for more than a few weeks.

WETLANDS - Those areas that are inundated or saturated by surface or groundwater at a frequency and duration sufficient to support, and that under normal circumstances do support, a prevalence of vegetation typically adapted for life in saturated soil conditions. Wetlands generally include swamps, marshes, bogs, and similar areas.

INFORMATION NEEDS IN THE PLANNING PROCESS FOR WETLAND CREATION AND RESTORATION

Edgar W. Garbisch
Environmental Concern Inc.

ABSTRACT. This chapter addresses both the factors which should be considered at various stages of the permitting process and those which should be contained in plans to create or restore wetlands. The information wetland regulatory agencies need to critically evaluate proposed mitigation projects in terms of acceptability, feasibility, and soundness is presented. If the process suggested is followed, the permit conditions should contain (1) details of construction and landscape plans, (2) specifications to facilitate verification by the regulatory agencies that the wetland creation/restoration project has been constructed according to the plans, and (3) the criteria by which to determine if the project has been maintained and monitored during the life of the permit.

INTRODUCTION

State wetland regulations and policies vary widely and many are still under development. Marked variances also occur in the administration of federal regulations in the east and probably nationwide. Consequently, it is not possible to recommend a planning process for wetland creation and restoration that will be uniformly acceptable. This chapter reflects the opinions of the author whose experiences have been limited largely to the eastern United States. Hopefully, the recommendations provided here will prove useful and applicable to some regions of the United States.

Much of the wetland creation and restoration work conducted throughout the United States results from regulatory requirements that compensation (mitigation) take place for permitted wetland impacts and losses. Prior to issuing permits, regulatory agencies review the applicants' mitigation plans to ensure that disturbed wetlands are restored or appropriate compensation is provided through compensation. If proposed mitigation plans are found acceptable, they generally become part of the permits together with stipulations or conditions relating to criteria for success and acceptability, timetables for wetland creation/restoration, monitoring, and reporting--especially when these stipulations or conditions have not been specifically addressed in the mitigation plans.

Most of the wetland creation/restoration plans published in the Corps of Engineers Public Notices lack the details necessary to evaluate their potential for successful execution. Such plans are often the only ones available for review by interested members of the general public and by the state and federal regulatory agencies. Moreover, they are often of insufficient detail for the agencies to verify that the "as built" project will compare acceptably to the one conceptually proposed. Without such details, the regulatory agencies must place full responsibility on the applicant, and indirectly on the applicant's mitigation consultant, to design and acceptably construct the wetland creation/restoration as called for in the permits.

If a regulatory agency is charged with the preservation of wetlands by legislation and policy, and if the agency permits a given wetland to be destroyed provided there is adequate compensation, then the regulatory agency's responsibility is to (1) engage in all aspects of review and evaluation of the wetland construction details, as well as other matters of planning, and (2) help ensure that such compensation is constructed successfully. Requiring "successful compensation" as a permit condition does not, in itself, ensure success. The compensation site and/or the plans and specifications may be inappropriate and success may not be possible. The regulatory agencies should have complete confidence in the compensation site and in the plans and specifications before making success a permit condition.

The objectives of this chapter are to define the information needed by the regulatory agencies at pre-permit application, at permit application, and during the life of the permit in order for proposed wetland creation/restoration mitigation projects to be fully and critically evaluated in terms of (1) acceptability, feasibility, and soundness of the proposed plan, (2) the construction performance as related to the project being constructed in accordance with the plans and specifications, and (3) the post construction performance as related to maintenance of hydrological requirements and vegetation establishment. All of the above combine to provide the regulatory agencies' planning process for wetland creation and restoration.

If a regulatory agency feels that it does not have in-house the qualified staff to conduct the necessary critical review and evaluation of detailed mitigation plans involving wetland creation/restoration, a qualified consultant might be retained to perform this service.

PRE-PERMIT APPLICATION

The development of detailed construction plans and specifications for wetland creation/restoration mitigations is usually time consuming and expensive. Their submittal should not be required until such time that (1) the regulatory agencies have agreed that the proposed project will be permitted pending review and acceptance of a final mitigation plan or (2) it is determined to be in the best public interest that they be submitted (e.g., for complicated or controversial projects).

Generally, the regulatory agencies will not discuss wetland creation/restoration as a compensatory measure for proposed wetland losses and impacts at early stages of pre-permit application meetings. They must be assured first that all other measures to mitigate such losses and impacts have been explored. However, after this is done, the applicant should be prepared to discuss wetland creation/restoration when certain wetland losses and impacts are unavoidable. In this event, the applicant should have available for distribution and discussion a preliminary mitigation plan that contains the basic information provided below.

PRELIMINARY MITIGATION PLAN

The Preliminary Mitigation Plan should contain the following:

1. **Text, 8.5" X 11" plans, and photographs describing the existing conditions at the project site** and particularly the wetlands on site and the portion(s) of these wetlands where disturbance and/or loss is unavoidable. Accurate areas of all wetlands to be disturbed and/or lost should be provided according to wetland type, if more than one type is involved.

2. **An evaluation of all wetlands that are proposed to be disturbed and/or lost** including their apparent stabilities, their dominant vegetative compositions, and their prevailing functions. An objective evaluation such as provided by WET (Wetland Evaluation Technique) to level-1 is suggested.

3. **Text, 8.5" X 11" plans, and photographs briefly describing the existing conditions at the wetland creation site.**

4. **Text and 8.5" X 11" plans that describe conceptually the proposed wetland creation** together with arguments, data, and calculations that demonstrate that the necessary hydrological requirements will be realized. Accurate areas of all wetlands to be created should be provided according to wetland type, if more than one type is proposed to be created.

5. **An evaluation of the proposed created wetland(s)**, as in Section 2, with an emphasis on functional replacement and enhancement relative to those functions provided by the existing wetland(s) to be lost.

6. **Text providing methods of any wetland restoration that is proposed** together with discussion of any possible enhancement of functional values that may be provided as part of the restoration. Issues related to the impact of soils compaction and other wetland disturbances on the success of the restoration should be addressed. If such impact(s) may limit the success of the restoration, approaches to circumvent the problem(s) should be discussed.

The Preliminary Mitigation Plan should not be a voluminous submittal. It should be brief and to the point. It is intended to be the precursor to the Draft Mitigation Plan which should be submitted later with the permit application, following reviews and comments by the regulatory agencies.

If the necessary hydrological requirements for the created wetland cannot be verified, monitoring of stream flows, ground water levels, etc. will be necessary for up to one year before the Draft Mitigation Plan can be prepared. Detailed soil borings throughout the proposed wetland creation site should be completed prior to preparing the Draft Mitigation Plan to verify that the soil characteristics will support the desired hydrology and functions of the created wetland.

PERMIT APPLICATION

DRAFT MITIGATION PLAN

Following any necessary monitoring and testing, and receipt and consideration of comments by the reviewing agencies, the Draft Mitigation Plan can be prepared. The Draft Mitigation Plan is a revised Preliminary Mitigation Plan and will include changes primarily in Sections 4-6.

The 8.5" X 11" plans provided in the Draft Mitigation Plan should be sufficient to be included in the Corps of Engineers Public Notice. Larger plans that are reduced to 8.5" X 11" are not recommended, as details and letterings may be reduced beyond recognition. After receipt of the comments on the Public Notice and review of comments derived from any public meetings, the regulatory agencies will come to a decision regarding issuance of permits. If the decision is to issue such permits pending receipt and acceptance of the construction and landscape plans and specifications for any created and restored wetlands, these materials must be provided. These plans and specifications together with the Draft Mitigation Plan constitutes the Final Mitigation Plan. The Final Mitigation Plan may have been requested by the regulatory agencies at an earlier time or it may have been provided voluntarily by the applicant.

FINAL MITIGATION PLAN

Draft Mitigation Plan + Construction and Landscape Plans and Specifications

The construction and landscape plans and specifications should be sufficiently detailed for bidding purposes, engineering and biological review, and verification of the "as built" condition. All monitoring, inspections, reporting, and maintenance during the life of the permit or during the required period of time should be detailed on the plans and specifications. The extent and duration of all landscape guarantees should be specified. It is recommended that the plans and specifications submitted as part of the Final Mitigation Plan include but not necessarily be limited to the following items:

1. **All plans should be scaled at 1" = 100' or larger** i.e., 1" = 50') and show 1.0' contours or less, if important.

2. **All slopes should be designed to be stable** in the absence of vegetation.

3. **Sufficient cross-sections of land and structures should be provided** so as to clarify all typical and atypical conditions.

4. **In addition to wetlands, all land** (e.g., transition and buffer zones and upland) included in the proposed mitigation should be shown.

5. **A summary of the sizes and types of wetlands** lost and created should be given.

6. **A summary of the sizes and types of non-wetland** habitats created as enhancement features should be given.

7. **The site hydrology should be clearly shown.** For example:

   Pool Elevation: if water level is static and non-fluctuating.

   Seasonal Pool Elevations: spring, summer, fall/winter if water level fluctuates.

   Tidal Elevations: when flooding water is tidal. Mean high water (MHW), and mean low water (MLW) should be indicated. Corrections to National Geodetic Vertical Datum (NGVD) or other local datum should be provided in the NOTES.

   Ground Water Levels: when flooding is temporary during times of storms and spring thaws. The expected seasonal ground water levels should be provided in the NOTES.

8. **Verification of hydrology should be detailed in the NOTES:** e.g., stream flow year-round and weir controls pool level; groundwater given in soil boring logs; stream/river water level data and analyses; calculations if stormwater is the only source of water; vegetation zonation of existing nearby wetlands sharing the same hydrology as the proposed vegetated wetlands; etc.

9. **The construction timetable should be provided** together with notations of any elements whose timing may be critical to biological success; e.g., coordination of completion of earthwork with the installation of certain species of plants to minimize the impact of salt buildup in soils; timing of plant installation to minimize the impact of waterfowl and drought; specify time windows for seeding to ensure vegetation establishment.

10. **The locations and elevations of all bench**

marks on site should be shown on the plans.

11. **A Summary** of the volume of earthwork and total tonnage of stonework should be given.

12. **The proposed disposition of any excavated materials** should be given.

13. **The elevations and elevation ranges for the planting and seeding** of all plant species should be shown. Plant spacings and seeding rates should be given.

14. **Landscape lists, notes, and specifications** should include the following:

    a. Plant lists for seeding and planting that provide total quantities, plant sizes, and plant conditions (e.g., bare root, can, peat pot, etc.). Acceptable substitutes should be indicated if the availability of some species might be limited.

    b. Because of the variable quality of nursery-produced wetland plant materials, acceptable plant conditions should be clearly specified. Some examples follow: Container grown nursery stock shall have been grown in a container long enough for the root system to have developed sufficiently to hold its soil together. Peat-potted nursery stock shall have been grown in 1.50" to 1.75" square peat pots long enough and under proper conditions for the root systems to be sufficiently well-developed through the sides and bottoms of the pots to prevent easy removal of the plants from the pots. Each pot shall contain a minimum of (specify) stems. Container grown nursery stock to be transplanted to wet areas year-round shall have been grown under hydric soil conditions for at least one growing season. The nursery providing these materials must certify that these growing conditions were met.

    c. Fertilization requirements that include rates and fertilizer formulations.

    d. Any special conditioning of the plant materials that may be required. For example, conditioning plant materials to specified water salinities or conditioning facultative/facultative wet species to hydric soil cultivation.

    e. Any geographical constraints regarding the origin of the plant materials.

    f. The names and addresses of all acceptable commercial sources of plant materials.

    g. A requirement that the supplier of seeds specified provide the purity and the current germination percentages of the seeds.

    h. What plant materials, if any, may be field collected and from where they will be taken.

    i. Construction details and timetable for any required controls against wildlife depredation.

    j. Details and definitions of any landscape guarantees, including the guarantee periods.

15. **Maintenance program during the guarantee period, the life of the permit, or other required period** should be detailed. Such maintenance may include invasive weed control of algae, common reed, purple loosestrife, etc.; removal of deposited litter and debris; watering; replanting; repair of water control structures; clearing of culverts; etc.

16. **Any critical elements and possible problems (with solutions)** that may influence the success of the project should be described, even if these items have been addressed in other sections; i.e., 9, 14i, 15. For example, a watering program for vegetation establishment in a floodplain wetland construction may be critical for success and should be restated, even though such a program was included in Section 15. In many instances, wildlife management will be critical for success and should be restated, even though Section 14i describes the item.

17. **Reporting timetable** during the life of the permit or until final approval should be included. The regulatory agencies will want to be informed periodically regarding the wetland construction progress. For example, is construction on schedule and, if not, why and what is being done to get it back on schedule. Are the criteria for success being realized (i.e., has the project been constructed according to the plans and specifications) and if not what corrective action is being taken. Reporting of the results of monitoring should be included in the reporting timetable. Generally, it would seem appropriate to report quarterly during the construction phase of the wetland and annually thereafter during the life of the permit or until final

inspection and approval. Photographs that are keyed on the site plan and that show the existing conditions should be included with all reports to facilitate verifications by the regulatory agencies.

18. **The monitoring program for the life of the permit** should be provided in the specifications or on the plans.

It is the author's opinion that if the wetland is constructed or restored according to the detailed construction and landscape plans and specifications that are part of the Final Mitigation Plan, the project must be considered successful. If the "as built" project is according to plans and specifications, then the wetland functional replacement and enhancement, as determined in Section 5 of the Draft Mitigation Plan, have been realized. Consequently, the monitoring program should be one of inspection and verification of the "as built" project according to hydrological performance, vegetation establishment, and other key elements in the plans and specifications. Scientific studies should not be part of a monitoring program sanctioned by the regulatory agencies. While such studies often will be important and should be encouraged, they should not be part of the required mitigation process.

CONCLUSION: NEED FOR CERTIFICATION OF MITIGATION CONSULTANTS

To a very large degree, the success of wetland creation/restoration projects will depend on the correctness of the plans and specifications and the execution of the construction according to these plans and specifications. Consequently, it is important that people with a background in both wetland creation/restoration design and the practicalities of construction become associated with such projects. To ensure that future wetland creation/restoration projects are planned and directed by qualified people, it is suggested that the Preliminary, Draft, and Final Mitigation Plans be signed and stamped by an individual who has been certified as a qualified wetland creation/restoration scientist. It is further suggested that such a certification program be undertaken by an organization such as The Society of Wetland Scientists.

WETLAND EVALUATION FOR RESTORATION AND CREATION

Kevin L. Erwin
Kevin L. Erwin Consulting Ecologist, Inc.

ABSTRACT. One of the principal questions that must be addressed when evaluating the success of a created, restored, or enhanced wetland is, to what extent does the wetland provide biological and hydrological functions similar to those of the original or desired "reference" wetland. Wetland evaluation methods are widely discussed throughout the literature. However, many would not be appropriate to evaluate a created or restored wetland, particularly given the time and financial limitations often placed upon the investigator and reviewer. The selected method must adequately characterize and evaluate the functions of the created and reference wetlands given the limitations of time, budget, type of wetland, size of wetland, context, degree of alteration from original wetland, location, and expertise of investigator. A qualitative wetland evaluation plan should include: a baseline vegetation survey, annual reporting of post construction monitoring conducted for a minimum of five years, fixed point panoramic photographs, rainfall and water level data, a plan view showing all sampling and recording station locations, wildlife utilization observations, fish and macroinvertebrate data, a maintenance plan, and a qualified individual to conduct monitoring. Quantitative evaluation is recommended when the proposed construction technique is unproven, where the ability to successfully create or restore the habitat is unproven, or when success criteria are related to obtaining specific thresholds of plant cover, diversity, and wildlife utilization. Quantitative evaluation should include: surface and groundwater hydrological monitoring, and vegetation analysis. The methods will often require some site specific fine tuning to prevent the over simplification of the wetlands complexity.

A rapidly accessible, easily understood, and cost effective database on wetland creation and restoration projects is needed to support environmental regulatory agency review, decision making, and action on specific projects. Any comprehensive wetland evaluation effort must be proceeded by the establishment of criteria which the investigator and regulator believe to be fundamental to the existence, functions, and contributions of the wetland system and its surrounding landscape. Failure to address the wetlands system's surrounding landscape leads to an inaccurate characterization of the wetland. Additional research is needed to establish the inter-relationships between wetlands, transitional areas, and adjacent uplands.

INTRODUCTION

Wetland evaluation is needed prior to a project to set goals and develop a plan, as a component of the monitoring program, and as a means for ultimately determining compliance. Although the timing differs for each of these evaluations, the factors to be considered and the general needs and approaches are much the same.

The following chapter has been prepared to assist consultants, client/permit applicants, and regulatory personnel in evaluating restored and created wetlands. It does not exhaustively review potential evaluation approaches, but presents a general framework and discusses selected topics. The chapter draws heavily upon the author's own experience and his many discussions with colleagues. An extensive bibliography of publications dealing with wetland evaluation is provided to assist the reader. As can be seen from the bibliography, a number of efforts have developed and assessed methods for evaluating wetlands. The author draws your attention to: Golet 1973, Winchester and Harris 1979, Reppert et al. 1979, U.S. Army Engineer Division 1980, U.S. Fish and Wildlife Service 1980, Lonard et al. 1981, Adamus and Stockwell 1983, Adamus 1983, Euler et al. 1983, Lonard et al. 1984, and Marble and Gross 1984.

EVALUATION NEEDS IN RESTORATION/CREATION

Evaluation may be needed for any or all of the following purposes:

1. **Assessing the Original Wetland.** The investigation must obtain, if possible, baseline data which evaluates the reference wetland's form and functions. The reference wetland may be the wetland to be impacted or another wetland chosen as a model for the mitigation project. This baseline data should be used to aid in the establishment of selected success criteria and the design of the wetland project.

2. **Setting goals for the enhancement, restoration or creation of a wetland required as mitigation.** Prior to designing the wetland required as mitigation, if possible, the wetland to be restored or enhanced, or an acceptable reference wetland should be evaluated. This information should be used to set goals for the mitigation project. It also should be used as a baseline from which to design the mitigation project and measure its success. In the case of wetland enhancement, the pre-enhancement baseline evaluation data will be compared with post-enhancement data.

3. **Assessing Project During Maturation.** Monitoring a project periodically during maturation will determine the need for corrections in design or maintenance to get the project back on course.

4. **Determining Post-Project Compliance.** At this stage the evaluation is used to establish compliance with goals or success criteria and to obtain the regulatory agencies' approval.

5. **Describing the Long Term Status.** Information about the wetland's responses to changes in site conditions (i.e., increased water levels, decreased hydroperiod, or colonization by problematic exotic vegetation) is obtained. This will indicate the ability of the system to persist.

The following is a general discussion of factors and considerations in wetland evaluation. It is intended as an overview of the choices available and not as an instructional guide to performing detailed data collection and analysis. These methods, when used individually or in some combination, will provide a varied database: qualitative or quantitative, inexpensive or costly, and relatively quick or lengthy.

PRACTICAL CONSIDERATIONS

What is the practical approach to wetland evaluation in a particular restoration/creation context or at a specific stage of a project? The answer depends upon a variety of factors. Economic and spacial constraints must be considered for each project evaluation. The investigator should evaluate the available methods and select or develop the method best suited to the situation given its limitations. In many instances the limitations placed upon the investigator have a greater influence on the methods finally selected than the objectives of the study.

TIME

Time may be a factor when conducting baseline monitoring of a wetland area because of the constraints of the permit application review procedure. In most cases, the regulatory agency should require baseline monitoring to be presented with the permit application to aid in the evaluation of existing conditions and in the establishment of success criteria.

In many instances the investigator may not have the time necessary to conduct a thorough study over the desired number of wet/dry or growing seasons. In such cases, the investigator should choose a time which will provide the greatest amount of information about the site. This information should be easily gathered over time. The most satisfactory time for a limited event evaluation is during the late phase of the growing season and, if possible, when the site is inundated to allow for the collection of fish and macroinvertebrate samples.

BUDGET

In some instances (i.e., where the project is small or where a public agency is involved) only limited funds will be available for evaluation. In cases where budget constraints exist, some compromises will inevitably be necessary. The investigator must choose an evaluation method and monitoring plan which is the most efficient and provides the greatest amount of desired information at the least cost ("the most bang for the buck").

TYPE OF WETLAND

Certain methods are more appropriate for one type of wetland than another (i.e., line intercept for nonforested wetlands and line strip or belt transects for forested wetlands). In addition, the fact that certain wetland habitats have proven to be less difficult than others to restore or create should have a bearing on the evaluation method used, and the scope of the baseline monitoring of the reference wetland and the post-construction monitoring.

SIZE OF WETLAND

The size of the wetland, the number and types of habitats to be evaluated, and the parameters to be examined will place constraints on the method selected.

CONTEXT

The selection of an evaluation method should depend upon whether single or multiple parameters are chosen as success criteria for the enhanced, created, or restored wetland. If wildlife utilization is the major goal, then a detailed vegetative analysis could be replaced by a more simple floral characterization with greater emphasis on monitoring for wildlife utilization. The science of creating certain marsh habitats is more advanced than for most forested wetland habitats (e.g., bottomland hardwoods), therefore, the monitoring of a marsh restoration project may need to be intensive for a shorter period of time. Many marsh restoration projects can be successfully completed and agency approval received within three growing seasons following construction, whereas a forested wetland project may take one to three decades.

DEGREE OF ALTERATION FROM ORIGINAL WETLAND

The greater the deviation of the proposed restoration or enhancement project from the original wetland, the more comprehensive the baseline and post wetland construction evaluation methods should be.

LOCATION

Wetlands are "open" systems with strong links to their adjacent ecosystems. A major factor determining the ecological value of the wetland is its relationship with other ecosystems. These relationships make the wetland an integral part of the landscape of a region or watershed.

EXPERTISE

The biases, objectives, and the expertise of an investigator will influence the choice of a method, therefore care must be taken to objectively select a method of evaluation that can successfully be used. This caution also holds true for the reviewer who must have an adequate understanding of the method and presentation of data.

WETLAND FUNCTIONS NEEDING ASSESSMENT

At each stage, the wetland evaluation should be geared toward evaluating particular functions of the created or reference wetland. Excellent references on wetland functions are in the proceedings of a national symposium on wetlands held in 1978 (Greeson, Clark and Clark, 1979), in Reppert et al. (1979), Larson (1982), Adamus (1983), Sather and Smith (1984), Gosselink (1984), Mitsch and Gosselink (1986), and Kusler and Riexinger (1985).

The Federal Highway Administration's Wetland Functional Assessment Method recognizes eleven functions (Adamus 1983) which form a good checklist. These functions are:

GROUNDWATER RECHARGE

Groundwater recharge by wetlands is generally poorly understood. The majority of hydrologists believe that while some wetlands do recharge groundwater systems, most wetlands do not (Sather and Smith 1984). The soils underlying most wetlands are impermeable which is why there is standing water during the annual cycle (Larson 1982). In the few studies available, recharge was related to the edge:volume ratio of the wetland. Recharge appears to be relatively more important in small wetlands such as prairie potholes than in large wetlands (Mitsch and Gosselink 1986). These small wetlands can contribute significantly to recharge of regional groundwater (Weller 1981). Heimberg (1984) found significant radial infiltration from cypress domes in Florida, with the rate of infiltration relative to the area of the wetland and the depth of the surficial water table. If groundwater recharge is a goal of the restoration or creation project, the design should emphasize the wetland edge to maximize potential for groundwater recharge.

GROUNDWATER DISCHARGE

Wetlands are generally considered by hydrologists to be a discharge area in terms of total water budget, however, recharge and discharge may be occurring at the same time in some wetlands. The recharge/discharge relationship of a wetland is a function of groundwater piezometric surface ("head") relationships and antecedent conditions (Hollands 1985). Water may be recharging an aquifer and/or discharging to a down gradient wetland, attenuating flows, and possibly providing baseline water flows to the down gradient wetland.

FLOOD STORAGE

Wetlands may intercept and store stormwater runoff, and hence change sharp runoff peaks to slower discharges of longer duration. Since it is usually the peak flows that produce flood damage, wetlands can reduce the danger of flooding (Novitzki 1979, Verry and Boelter 1979). A study undertaken by Ogawa and Male (1983) found that for floods with a 100-year recurrence, interval or greater, the increase in peak stream flow was very significant for all sizes of streams when the wetlands within the watershed were removed.

Ogawa and Male (1983) summarized that the usefulness of wetlands in reducing downstream flooding increases with: (a) an increase in wetland area, (b) the seriousness of the flooding downstream of the wetland, (c) the size of the flood, (d) the closeness to the upstream wetland, and (e) the lack of other storage areas such as reservoirs. These factors should be considered if the proposed restoration or creation project is within a flood prone area where some improvement to these conditions is desirable.

SHORELINE ANCHORING

Wetlands such as tropical mangrove forests and temperate Spartina-Juncus saltmarshes, bind shoreline sediments with their root systems, thus anchoring the substrate. The aboveground biomass provides friction to overland sheetflow, wave energy, and storm surges, providing a degree of stabilization to the shoreline under natural conditions.

SEDIMENT TRAPPING

Wetlands can serve as sinks for particular inorganic nutrients. Many marshes are nutrient traps that purify the water flooding them. Wetlands have several attributes that cause them to have major influences on chemical materials that flow through them (Sather and Smith 1984). Mitsch and Gosselink (1986) describe these attributes in the following manner:

A. A reduction and velocity of streams entering wetlands, causes sediments and chemicals to drop into the wetland.

B. A variety of anaerobic and aerobic processes such as denitrification and chemical precipitation remove certain kinds of chemicals from the water.

C. The high rate of productivity of many wetlands can lead to high rates of mineral uptake by vegetation and subsequent burial in sediments when the plants die.

D. A diversity of decomposers and decomposition processes occur in wetland sediments.

E. A high amount of contact of water with sediments, because of the shallow depths, lead to significant sediment-water exchange.

F. The accumulation of organic peat in many wetlands causes the permanent burial of chemicals.

FOOD CHAIN SUPPORT

Wetlands possess an inherent ability to trap nutrients. They often store nutrients when there is an abundance, then frequently release them when they are most needed (Niering 1985). In mature wetlands, food chains are elaborate, species diversity is high, the space is well-organized into many different niches, organisms are larger than in immature systems, and life cycles tend to be long and complex. Approximately 60% of the fish and shellfish species that are harvested commercially are associated with wetlands. For example, many fish species utilize wetlands as spawning and/or nursery areas. Some important species are permanent residents and others are transients that periodically feed in the wetlands. Virtually all freshwater species are somewhat dependent upon wetlands, often spawning in marshes bordering lakes or in riparian forests during spring flooding. Saltwater species tend to spawn offshore, moving into the coastal marshes during their juvenile stages, then migrating offshore as they mature. The importance of wetlands to the sport and commercial fishery harvest is well documented in the literature (Peters et al. 1979).

WILDLIFE HABITAT

It has been estimated that within North America 150 kinds of birds and some 200 kinds of animals are wetland-dependent. Other animals including deer, bear, and racoon also use wetlands (Niering 1985). In addition, wetland habitats are necessary for the survival of a disproportionately high percentage of endangered and threatened species.

ACTIVE RECREATION, PASSIVE RECREATION, HERITAGE, AND EDUCATION

Wetlands are living museums, where the dynamics of ecological systems can be taught. The high productivity of wetlands is related to the efficient functioning of both the grazing and detritus food chains. In many wetlands there are two major energy flow patterns: (1) the grazing food chain, which involves the direct consumption of green plants, and (2) the detrital food chain, composed of those organisms that depend primarily on detritus or organic debris as their food source. Often the two patterns are interrelated. In lake and pond ecosystems, submerged aquatic plants and floating algae serve as the basis of the food chain. Zooplankton feed on the algae and aquatic insects eat the zooplankton. These are eaten by small fish, which in turn are consumed by larger fish, which in turn may end up on a fisherman's dinner table. In streams, the main sources of organic input, or food for stream organisms, include partly decomposed leaves or other organic material flowing down stream. This debris, or detritus, may be caught in nets set by the larvae of caddis flies. Stone flies also glean the rocks for algae. These insects are in turn consumed by fish, many of which are commercially important.

Activities such as sport fishing along a wetland edge of a lake and canoeing through a hardwood swamp are pursued by thousands of people on a regular basis. Wetlands are an important national heritage providing the sites and experiences many of us attribute to our country's heritage.

FISHERY HABITAT

Wetlands have been documented as important sources of food and habitat for sport and commercial fisheries. These outdoor laboratories can demonstrate such basic ecological principles as energy flow, recycling, and limiting carrying capacity (Niering 1985).

The Federal Highway Administration (FHWA) assessment procedures are among several which can by used for wetland evaluation for restoration/creation purposes. There are limitations, however, with this approach. Manual implementation of the FHWA assessment procedure (Adamus 1983) is cumbersome and time consuming. The U.S. Army Corps of Engineers Waterways Experiment Station (WES) developed a wetland evaluation technique (WET) that can reliably assess and partially quantify wetland functions and values for Corps of Engineers use. The main structural reorganization of the FHWA technique was to computerize the analytical portion. A discussion of WET is provided in Clarian (1985) and more detailed information on WET is contained in Winchester (1981a and 1981b).

LEVEL OF DETAIL--DEGREE OF QUANTIFICATION

Having determined the stages in a restoration/creation project at which evaluation should take place and the functions that need assessment, the next major decision relates to the level of detail needed. In general, quantitative evaluation is much more expensive and time-consuming than qualitative approaches. However, quantitative approaches are essential in some instances.

To determine whether qualitative or quantitative evaluation methods are appropriate, the investigator should consider the established history of success in creating the type of wetland proposed for mitigation. In general, much more quantitative and detailed analyses are needed for wetlands with no history of success in creation or restoration.

Choosing the appropriate level and detail ofevaluation and the factors to be evaluated in a particular instance is a process that must be thoroughly considered by each party involved in the evaluation process including the investigator, the reviewer, and the client/applicant. Each should consider the appropriateness of the selected method to provide an adequate characterization of the wetland and the ability to produce the required data and analysis within a realistic time frame. The client or project manager must also assess his/her ability to provide the required budget for the expected duration of monitoring and reporting.

Assuming appropriate funding, enough time, and an attempt to create a wetland with no or little history of success, what should the proponent of a wetland restoration/creation project evaluate? The author suggests the "quantitative" evaluation described in the following section.

QUANTITATIVE EVALUATION

When the investigator requires quantitative data, both detailed field studies and office evaluation are required. A detailed evaluation often involves hydrologic analysis, studies on plant and animal population dynamics, water quality sampling, soils analysis, topographic mapping, wildlife counts, and a regional watershed analysis. These studies are time consuming, labor intensive, costly, and subject to producing biased results when not properly conducted.

Quantitative evaluation is particularly needed when (1) the proposed construction technique is unproven, (2) where the ability to successfully create or restore the habitat has not been established, or (3) when success criteria are related to attaining specific thresholds of plant cover, diversity, wildlife utilization, etc. Properly applied quantitative evaluation may often be replaced by less intensive evaluation methods after a sufficient period of study (i.e., the latter stages of a restoration/creation project).

Investigators need rapidly accessible, easily understandable, and cost effective data in support of environmental regulatory agency review, decision making, and action on specific projects pursuant to local, state, and federal policies and regulations. A variety of systematic and quantified approaches for evaluating either individual or the full range of wetland functions have been developed by agencies and researchers (see Appendix I). These assessment models vary from very simple to quite sophisticated in the types of factors considered. Their outputs range from a qualitative to a quantitative evaluation of a particular wetland's ability to provide a particular service or function. Some models produce a single numerical rating for the wetlands, while others provide a rating for each function.

HYDROLOGY

Hydrology is the single most important factor to consider in designing and implementing restoration/creation projects for specific types of wetland systems and their related functions.

GROUNDWATER

Gathering actual wetland groundwater data is time consuming and expensive; extrapolating data from one wetland to another can be problematic. No quick, accurate, and inexpensive groundwater function predictors are available. Even hydrogeologists experienced in wetland hydrology cannot consistently predict the hydrogeologic functions of specific wetlands (Hollands 1985). Data requirements for understanding the groundwater function of a specific wetland include:

1) Geologic history, including an understanding of the current theories relative to the geologic processes that created the topographic and hydrologic setting in which the wetland is located, (e.g., bedrock and surficial geology).

2) Stratigraphy of the geologic units underlying the wetland and their physical properties, such as permeability.

3) History, stratigraphy, and physical properties of the wetland's organic or mineral soils.

4) Description of the wetland vegetative community.

5) Groundwater and surface water hydrology, including a water budget for the wetland based on items 1 through 4 above.

The recharge/discharge relationship of a wetland is a function of groundwater (head) relationships and antecedent conditions. To determine head relationships, nested water table observation wells (piezometers) are required. These permit simultaneous measurements of head at various levels within the aquifer.

Measurements for at least one year should be required to establish a complete record of recharge/discharge functions. This is normally a costly process.

Perched wetlands, water table wetlands, and other hydrogeologic classifications such as artisan and water table/artisan wetlands (Motts and O'Brien 1980) also require nested wells for identification. Hydrogeologic classifications of wetlands are important in understanding a wetland's water balance and the effect of hydrology on other wetland functions (Hollands 1985). The wetland hydrogeologic classification that appears to be most used by non-hydrogeologist wetland regulators is that of Novitziky (1978). Novitziky classified wetlands in Wisconsin as "surface water depression", "groundwater depression", "surface water slope" or "groundwater slope". This classification combines topography, surface water, and groundwater parameters. However, without wetland specific hydrogeologic data, it is doubtful if this method can be accurately applied by non-hydrogeologists (Hollands 1985).

SURFACE WATER

A hydrological model should be developed to determine the watershed dynamics which affect the subject wetland system. Usually a very simple model can at least establish the extent of the watershed, timing and volume of input to the wetland, depth and duration of flooding, and discharge from the wetland. Post-construction monitoring of the created wetland should establish where fine tuning is required in order to provide the desired levels of inundation and hydroperiod. As noted above, the wetland's relationship to the surrounding groundwater system should be identified when constructing the hydrological model. Water quality analysis is also recommended at upstream and downstream locations as well as within the wetland itself to determine inputs to the wetland and its present ability to handle pollutants. Riparian wetland systems will require evaluation of stream flow and the sedimentation process.

Ideally, monitoring of ground and surface water quantity and quality should be done in the reference wetland area for at least one annual cycle, and if possible, including two wet seasons and one dry season. Similar monitoring for the created wetland should be done until the project goals are met and possibly longer where this information is of value to the long term management of the system. Factors critical to the maintenance of the wetlands's hydrology and that of surrounding lands should be used to assist in future land use decisions and to prevent adverse impacts from taking place.

VEGETATION

Analysis of the vegetation in a wetland system is usually second only to understanding the hydrology of the area when characterizing the wetland and evaluating its functions. The method of monitoring/evaluation will depend on the type and size of the wetland. The methods discussed below are "goal oriented", that is, they will provide sufficient data to adequately characterize the reference and created wetland systems for quantitative measurement of success criteria.

In order to adequately characterize reference wetland vegetation within the scope of most mitigation related evaluations, three methods are recommended and described below: (1) belt transects for forested wetlands, (2) replicate quadrats for herbaceous wetlands, and (3) multiple quadrats for shrub wetlands.

BELT TRANSECT

This method, also called the modified line intercept method (Bauer 1943), consists of observation of plant species occurring along a belt transect extending through the study area. A single belt transect 6.10 meters in width divided into 15.25 meter intervals is established through each forested wetland. The belt transect is positioned so that each vegetation zone of the wetland is sampled. Belt transects should extend into adjacent upland in order to characterize the wetland-upland ecotone as well as the upland habitat. Each interval of the belt transect (quadrat) covers 93.025 square meters. Within each quadrat canopy, midstory, and groundcover taxa are recorded. The diameter at breast height (DBH) of all canopy trees (DBH > 25.4 millimeters) are measured to the nearest 30.48 millimeters. Trees with multiple stems originating from a common trunk are recorded as individual trees. The percent cover of midstory taxa (DBH < 25.4 millimeters and 457.2 millimeters or greater in height) should be estimated for the entire quadrat. Percent cover of ground cover taxa (less than 0.91 meters in height) is estimated for the entire quadrat. Water depth and percent cover of bare ground should also recorded.

REPLICATE QUADRATS

The vegetation within individual reference or created herbaceous wetlands should be delineated into major macrophyte zones. Seven 1 m^2 quadrats are established in each zone. The vegetation within the quadrat is divided into as many as three strata based on relative height. The percent cover of each taxa within each strata is estimated and the average height recorded. Water depth and percent cover of non-vegetated areas should also be recorded.

MULTIPLE QUADRATS

The ground cover in a shrub dominated wetland is recorded for seven replicate 1 m^2 quadrats as described above. Two 3.0 meter x 3.0 meter quadrats should be established to describe the shrub strata. Within both quadrats the number and average height of individuals from all non-herbaceous taxa is recorded and the DBH of the five largest individual's of each taxa is recorded.

These sampling methods were developed and used for several reasons. First and most important, these methods have been modified and refined to develop a standardized method for establishing an absolute measure of species occurrence by using defined frequency intervals and cover estimates. Use of small continuous frequency intervals allows increases or decreases of colonizing vegetation to be accurately mapped and subsequent changes easily followed with time. In addition, since frequency data is based on species presence or absence it is absolute. Therefore, no error is introduced as is the possibility when using ocular estimates. Frequency is needed in determining cover percentages. Although cover estimates are not absolute (and may be somewhat variable when performed by different people), they serve as comparative indices for evaluating cover between different treatments and/or wetlands.

Occurrence of non-vegetated areas (bare ground) throughout the transects were given the same consideration as plant species cover. Bare ground or non-vegetated surfaces are present in all systems and as such are not necessarily a definitive characteristic of newly created wetland areas. Bare ground is defined as all ground area not covered by some form of vegetative structure as viewed from above. Analyses of bare ground allows for determining vegetation stratification. With bare ground considered, vegetation coverage of an area will seldom be greater than 100% cover. Analyses may indicate that a great degree of plant stratification occurs; however, areas are most often not 100% covered by vegetation. The bare ground method is recommended because coverages based totally upon species occurrence (which often total much greater than 100%) may no longer be an acceptable method of reclamation success determination.

CREATED OR RESTORED WETLANDS

The methods recommended to characterize the plant community within created or restored wetlands overlap in scope with the reference wetland evaluation methods. The differences are those modifications required to monitor survival and growth of planted woody species in a created or restored wetland. The line-strip (elongated quadrat) technique (Lindsey 1955, Woodin and Lindsey 1954) has been used to facilitate an intensive, accurate, and repeatable sampling program. Permanent quadrats are established at a constant width to allow for a maximum sampling of trees concomitant with planting density such that generally four to five parallel planting rows (average 1.525-3.05 meter centers) can be monitored within each quadrat. Elongated quadrats can be extended parallel to the slope of the wetland to allow for survival and growth comparisons to be made on a gradient from flooded through moist to dry conditions. Best and Erwin (1984) and Erwin (1987) used this method to evaluate the effects of hydroperiod on survival and growth of tree seedlings in a phosphate surface-mined reclaimed wetland.

MEASUREMENT PARAMETERS AND CRITERIA FOR PLANT CONDITION ASSESSMENT

All trees occurring within the sample quadrat should be measured during the growing season for height. Water depth in the quadrat should also be measured. Qualitative observations should be made concerning the individuals' overall appearance. Generally seven different categories are suggested for condition assessment. Categories and descriptive criteria are:

Live

Tree appears in apparently good condition--leaves green, no symptoms of wilting, die back, or chlorotic appearance of leaves.

Stressed

Tree appears to be in a generally poor condition--chlorotic leaves, wilting, and leaf drop.

Tip Die Back

The main stem is in good condition, but the most apical portions are in very poor condition exhibiting wilting and die back symptoms.

Basal Sprouts

The main stem is dead but new growth is initiated from the stem base or the root stock.

Not Found

In some cases seedlings are not found during a particular sampling period. If a seedling is not found on two successive sampling periods, the seedling is counted as dead.

Apparently Dead

The general appearance of the stem is dry and brittle with no live wood observed and there is no observable green foliage growth.

Dead

A decision as to whether a tree is dead is generally made only following a sampling period in which the tree was classified as "apparently dead". Only if initial observation indicated that the stem was in such poor condition that survival was unlikely should a tree be listed as dead.

To completely evaluate the potential for "forest" development in a created or restored wetland, crown cover should be recorded for species above the herbaceous stratum. In addition, trees producing seed should be noted. Table 1 is a summary of planted tree survival (total of all species), change in height and crown size from a created wetland in central Florida (Erwin 1987).

These methods for evaluating a created forested wetland have provided data which established trends of survival and growth for certain tree species after four years of monitoring (Erwin 1987). Forested wetland creation projects should be monitored using this method for a minimum of five years. The established trends will dictate whether further intensive monitoring is required or if a reduced periodic evaluation is appropriate to maintain conditions required for maturation of the system.

BIOMASS

Biomass of vegetation per unit area may be an important parameter to be measured in a restored wetland when standing crop of the restored wetland is a criteria for its success. For small quadrats and herbaceous vegetation, the biomass can be measured by cutting all above ground matter, drying it in an oven, and weighing it. Ideally roots are also excavated, but they are often ignored and consequently most of the biomass data represents only the above ground plant matter. Quadrat size and shape are important. The significantly limiting factor is generally man-hours and the cost to perform the analysis.

Productivity can be determined from these measures as the rate change and biomass per unit area over the course of a growing season, or a year or several years during the maturation of a restored or created wetland. This process is described below.

PRODUCTIVITY

Productivity may be a criteria for evaluating the success of wetland to be created or restored.

The most accurate means of measuring primary productivity is to measure the net photosynthetic rates of photosynthetic tissues, and extrapolate to the community level, using the net production per gram of biomass of each species in a community. Obviously, this assessment of net primary productivity is not possible under the circumstances associated with the typical wetland creation or restoration project. Consequently, the most practical measurement of net primary productivity (NPP) in frequently conducted by calculating the change in biomass through time where $NPP = (W_{t+1} - W_t) + D + H$ where $W_{t+1} - W_t$ is the difference in standing crop biomass between two harvest times, D is the biomass lost to decomposition and H is the biomass consumed by herbivores during the period between harvests.

Above ground biomass may be measured with little error in herbaceous vegetation by replicate samples harvested randomly from a grid. This technique is most effective with annual vegetation, where little biomass is lost to decomposition during the growing season. If herbivore activity is significant, comparisons between replicate samples taken inside and outside herbivore enclosures are often employed (Barbour et al. 1987).

Table 1. Summary of planted tree survival, change in height and crown size during 1987 growing season.

SURVIVAL

	SPRING		SUMMER		FALL	
	Number/Acre	%	Number/Acre	%	Number/Acre	%
Live	757	71	696	65	679	63
Dead	315	29	382	35	406	37

GROWTH (HEIGHT cm)

SPRING	FALL	
Height	Height	Change in Crown
123	151	+28

CROWN (DIAMETER cm)

SPRING	FALL	
Height	Height	Change in Crown
38	46	+8

MACROINVERTEBRATE MONITORING

One of the least known facets of freshwater wetland systems is the role of the macroinvertebrate fauna. These communities are very important, forming an intermediate level in the wetland's food chain, providing a primary food source for higher organisms such as fish and wading birds. Certain macroinvertebrate species are excellent indicators of water quality.

Comparisons from one marsh to the next, even when they may closely resemble one another with regard to hydrology and physiognomy, may yield different species composition and diversity. Species composition, richness, and diversity will depend on the season in which the monitoring was conducted, the macrophyte community from which the samples were taken and the method of collection.

The lack of baseline data for wetland habitats makes it impossible to detail the exact degree of macroinvertebrate utilization in a wetland creation project. However, monitoring should be required to determine whether, in fact, a project is being utilized by at least some of the desired species. The Macroinvertebrate monitoring program should be designed to complement the wetland community and water quality monitoring of the project. The objective is to develop a predictive model of success for the long term trends in biological community development in a created wetland. In riparian systems where flowing water is present, the use of Hester-Dendy multi-plate artificial substrate samplers seasonally, in combination with some of the methods described below, may be appropriate. A number of monitoring plans included in recently issued permits have required the use of Hester-Dendy plate samplers in freshwater marsh systems where no flowing water is present. The author's research indicates that the use of Hester-Dendy plate samplers in static water situations provides unrepresentative data when compared with other methods. Thus, they should not be used in these situations.

Substrate coring and leaf and/or stem scraping are usually satisfactory methods for obtaining quantitative data as long as the substrate areas are measured and computed for each sample. This method has not been widely used in many wetland systems, probably due to the great expense in time and labor. It is often common to find stem samples harboring few organisms. The author has used the method on many reference and created wetlands and has observed that proper qualitative sampling, usually consisting of dipnet samples, can provide a more accurate characterization of the wetland. Sample size should be as large as possible with collections made within each niche of each macrophyte community.

Seasonal data collection for at least one annual cycle is recommended for the natural wetland to be altered. Monitoring of the created wetland should be continued until the vegetation-related goals are met. Freshwater marsh and some salt water marsh creation projects generally will require less intensive monitoring over a shorter period of time (two to five years) than forested wetland projects.

WILDLIFE UTILIZATION

The most widely used generally successful method of evaluating wildlife utilization of a natural or created site is reliable observation. The observations are made during the correct season, time of day, and over a satisfactory number of events for the type of wildlife anticipated by qualified personnel. Once again, where more specific goals have been established with regard to particular species utilization, more intense monitoring may be required which may involve quantitative surveys to determine, for example, the number of nests per acre or breeding pairs per season. Wildlife utilization of a wetland creation project is almost always one of the specified or inferred goals, but actual monitoring or observation of wildlife utilization is often lacking in the permit conditions. Special consideration should be given to endangered, threatened, or listed species (of special concern). The reference wetland should always be evaluated with respect to current or possible future utilization by and suitability for listed species. Habitat characteristics that are necessary for use by listed species should be thoroughly documented. In addition, the wetland's proximity to other wetlands or specific types of upland habitat may dictate its degree of utilization by particular wildlife species. Thus, it is important to describe the location and type of connecting corridors.

There are cases when the proposed biological structure of a created wetland focuses on preserving those species threatened with extinction. Managing a created habitat to favor these endangered species will inevitably hamper the growth of some more abundant species. This approach is appropriate if the criterion is to preserve a full diversity of species regardless of relative abundance. However, management is often intensive.

RARITY

An often overlooked aspect of wetland

evaluation is the rarity or uniqueness of one or more components of the habitats within a region. While fish, wildlife, and vegetation are usually evaluated, other aspects are often overlooked. Items of importance include the rarity of the specific habitat in that particular stage of succession; geomorphology; water quality; and other characteristics such as stream flow and cultural criteria such as archaeological, scientific, and public/recreation significance.

SOCIAL ECONOMIC VALUES

Sather and Smith (1984) state that nonconsumptive values up to this time have been given secondary status to other wetland values for which scientific criteria can be developed or direct economic gain can be realized (Sather and Smith 1984, Gosselink et al. 1974, and Niering 1985). This is partly related to the fact that aesthetic or cultural values are more difficult to measure since they involve a more personal approach as well as value judgments (Niering 1985). There is a need to further develop a method for assessing nonconsumptive values of wetlands (Niering and Palmisano 1979).

ECOLOGICAL WATERSHED CONTEXT

The author believes that wherever possible, wetland evaluation should be broadened to assess original or restored/created wetlands in their broader ecological and hydrologic context, including regional wetland functions. Any comprehensive wetland evaluation effort should be preceded by the establishment of certain criteria or goals which the investigator believes to be fundamental to the existence, functions, and contributions of the wetland system to its surrounding landscape and vice versa. Failure to address the wetland system's surrounding landscape leads to inaccurate characterization of the wetland. Additional work is needed to develop definitive techniques and models for better assessing the importance of individual wetlands in a broader watershed context for flood control, flood conveyance, pollution control, food chain support, habitat, and other purposes. However, approximations can be made with existing approaches. Inter-relationships between wetlands, transitional areas, and immediate uplands need to be studied to determine the importance, not only of adjacent lands, but of the functioning of wetlands in adjacent areas as total systems.

QUALITATIVE EVALUATION

As discussed above, quantitative and detailed evaluations are rarely possible for all wetland functions and all aspects of a wetland at any stage in a wetland restoration/creation project due to cost or time limitations. In some instances, they are simply not needed, as, for example, with proven designs or in the later stages of a restoration/creation project where the goal is to determine compliance rather than to design a system.

The author's experience with wetland reclamation in Florida suggests that successful freshwater marsh creation (see Erwin this volume), and mangrove and saltmarsh creation, can take place with more generalized qualitative evaluations. Proper planning, design, and management can result in the creation of a functioning freshwater marsh within three full growing seasons (Erwin 1986). In this case, qualitative baseline monitoring of the reference wetland and post-construction monitoring of the created wetland for a minimum of three or four years is usually adequate.

The topics needing attention in a quantitative evaluation remain much the same in a qualitative evaluation, but the approaches differ.

The author offers several suggestions with regard to specific aspects of qualitative approaches in the following sections.

VEGETATION MAPPING

Vegetation mapping of wetlands can provide both physiognomic and floristic information, and may be useful in several different ways. If the wetland area to be evaluated is too large to evaluate by the methods previously described due to short time allotted for analysis and/or a restricted budget, vegetation mapping on an aerial photograph verified by groundtruthing may be the answer. The investigator should try to distinguish as many different communities or vegetation types as possible and outline the boundaries of each on the aerial photograph. Species richness and assorted observations on topography, water depth, and wildlife utilization should be recorded for each vegetative type. Acreages for each type can then be computed from the vegetation map.

Vegetation mapping may also be used in a final, but long term phase, following short term intensive data collection using the previously described methods. A vegetation map could be prepared on a currently studied wetland and correlated with data for each mapped vegetation type (Figure 1). The author has satisfactorily used this method for evaluating large landscapes with wetlands. Vegetation mapping on high quality aerial photographs (black and white, color, and infrared sensitive film) taken on an annual cycle can confirm the continuity of trends or changes in habitat types and vegetative cover. Figure 2 is a vegetation map of a portion of a wetland reclamation study site where the quantitative data collected will be correlated with the detailed mapping (Erwin 1987 and Erwin this volume). While this method does not produce the detailed data previously produced by belt transects, it will provide a relatively inexpensive and reliable confirmation and description of the habitat. Maps should be regularly groundtruthed to confirm the reliability of the habitat types and boundary definitions. Recently developed computer aided drawing (CAD) allows for great flexibility in generating scaled vegetative maps and habitat acreage figures.

In addition to generalized vegetation maps, a more specific qualitative baseline vegetation survey should be performed in most cases. Transects should be established through wetlands to be preserved, impacted, and created so that each major vegetation zone, including adjacent upland habitat, will be represented. Each major vegetation zone should be identified and a sampling station (3.0 meter x 3.0 meter quadrat) located within each zone. All plant species should be recorded. Relative abundance and percent cover of species should be noted. The same transects and stations should be used for all future post-construction monitoring.

An example of the results obtained using this method is shown in Table 2, which represents baseline vegetation monitoring data collected in four quadrats from one of several transects (Transect D, shown in Figure 3) established in selected preserved and constructed wetlands of a proposed surface water management system. Each major vegetation zone is represented.

Transect D (Figure 3) was aligned to intersect each major vegetation zone in which a 3.0 meter x 3.0 meter quadrat (sampling station) was established. All plant species were recorded. Percent cover was estimated for each species. The results are presented in Figure 4. Figure 4 also illustrates the basic structural differences in major species composition and cover between the three macrophyte zones of this marsh and the adjacent uplands. In some of the quadrats the vegetated areas were structurally complex and the stratified vegetation layers yielded cover values greater than 100% (Table 2). The same transects and quadrats will be used for all future monitoring.

In some cases where multiple wetlands are present (i.e., _Cladium_, _Pontederia_, and _Spartina_ marshes), it may be possible to group wetlands of similar physiognomy and monitor a subset of each group using this method as a _minimum_ requirement. This will still provide the required data as long as the investigator selects a representative wetland from each group and all vegetation zones are monitored. If there are adjacent upland habitats that are recognized as an important component of the wetland system, the transects and quadrats should extend into the adjacent upland habitat.

POST CONSTRUCTION MONITORING

Project evaluation is essential both during and after construction to determine compliance with project goals and permit mid-course corrections. Certain aspects of such evaluation may need to be quantitative (depending upon the project goals and measures of success), but much of it can be qualitative. Based upon the author's experience, the following are suggested as key components of a monitoring plan for post project evaluation.

MAINTENANCE PLAN

The methods to be used for maintaining the wetland after construction should be submitted with the original baseline survey. The plan should address removal of nuisance species, i.e., _Melaleuca_, Brazilian pepper, purple loosestrife, and cattails, and assure an 80% survival rate for planted or recruited species. An evaluation of the success of the maintenance effort should be discussed in annual reports.

ANNUAL REPORTS

Post-construction monitoring should be conducted annually for five years (minimum three years) at the end of the wet season (October-

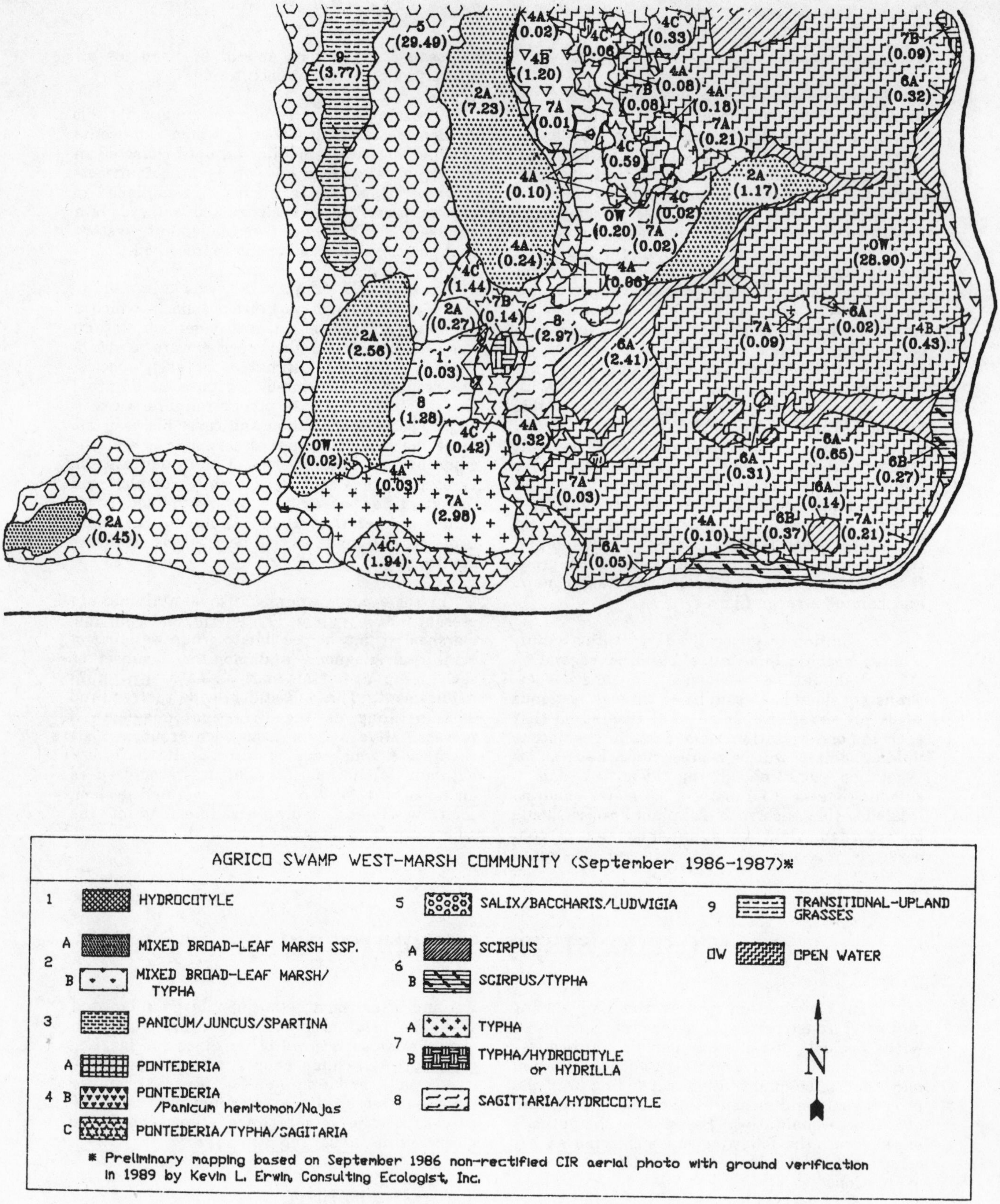

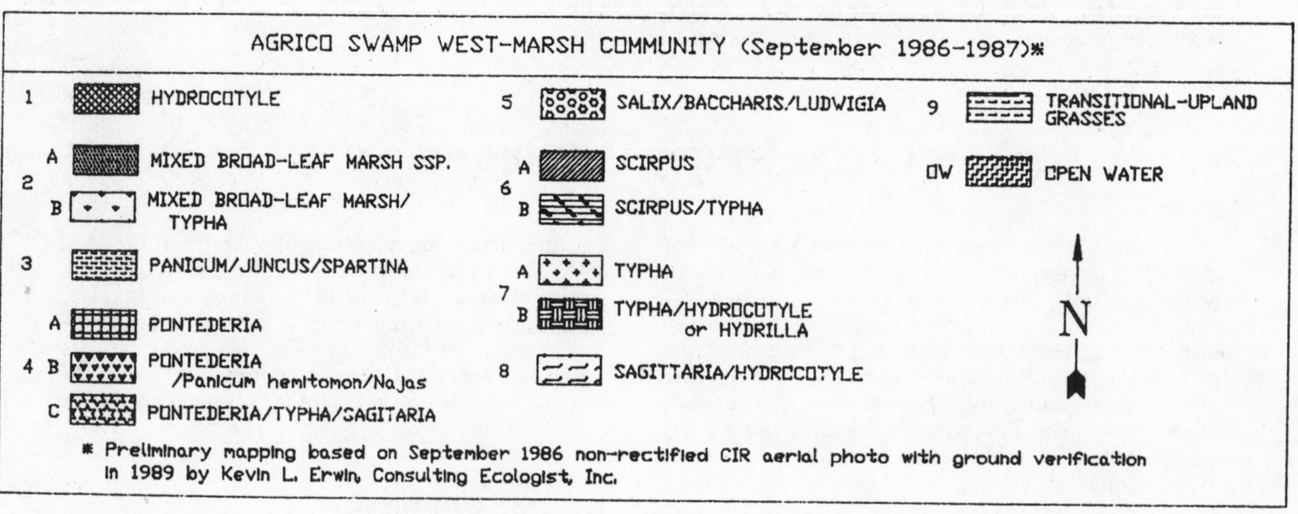

Figure 1. Vegetation map of plant communities, acreages, and percent cover on a 106.94 acre parcel in southwest Florida produced by Computer Aided Drawing (CAD).

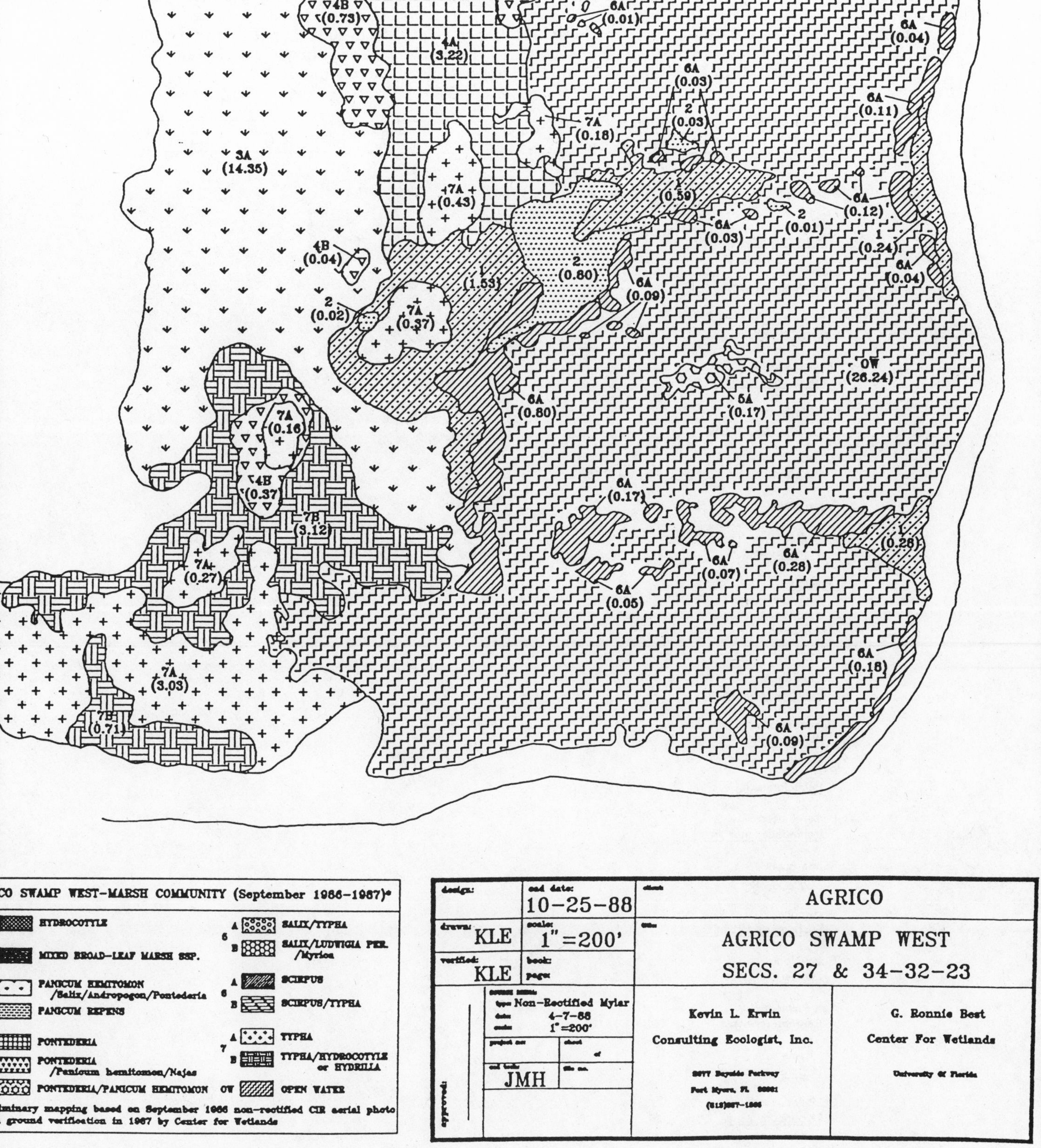

Figure 2. A portion of a vegetation map of plant communities and acreages for a wetland reclamation study site (Agrico Swamp West) in central Florida produced by computer Aided Drawing (CAD).

Table 2. Baseline vegetation data (%-Cover, Ht.-Water Depth) from four 3 m x 3 m quadrats along a wetland/upland transect (Transect D) in a surface water management system. M: Species present within Macrophyte community, but outside of sample quadrat.

	QUAD D1		QUAD D2		QUAD D3		QUAD D4	
SPECIES	%	Ht (cm)	%	Ht (cm)	%	Ht (cm)	%	Ht (cm)
Background	15		5		65			
Water depth		dry		moist		7-8		
Beak rush Rhynchospora spp.					1	30-60		
Broom sedge Andropogon spp.					<1	8		
Cyperus sedge Cyperus spp.					M			
Floating orchid Habenaria repens							<1	30
Floating heart Nymphoides cordata					1	--		
Gallberry Ilex glabra	10	122-183						
Green algae					5	--		
Joint grass Manisuris spp.			70	30-60				
Ludwigia Ludwigia repens			<1	8				
Marsh aster Aster spp.					M			
Maiden cane Panicum hemitomon			M		30	30-60		
Marsh fleabane Pluchea rosea			1	30				
Paspalum Paspalum spp.			<1	25				
Pickerel weed Pontederia lanceolata							90	30-75
Plume grass Erianthus spp.					M			
Red root Lachnathes caroliniana			M		<1	0.25	15	60-90
Rusty lyonia Lyonia ferruginea	1	30						
San Palmetto Serenoa repens	80	90-120						
St. Johns wort Hypericum spp. A			25	60-90				
St. Peter's wort Hypericum stans	M							
Sundew Drosera spp.	M		<1	8				
Hire grass Aristida stricta	5	15-30						
Yellow eyed grass Xyris spp.			1	30-90				

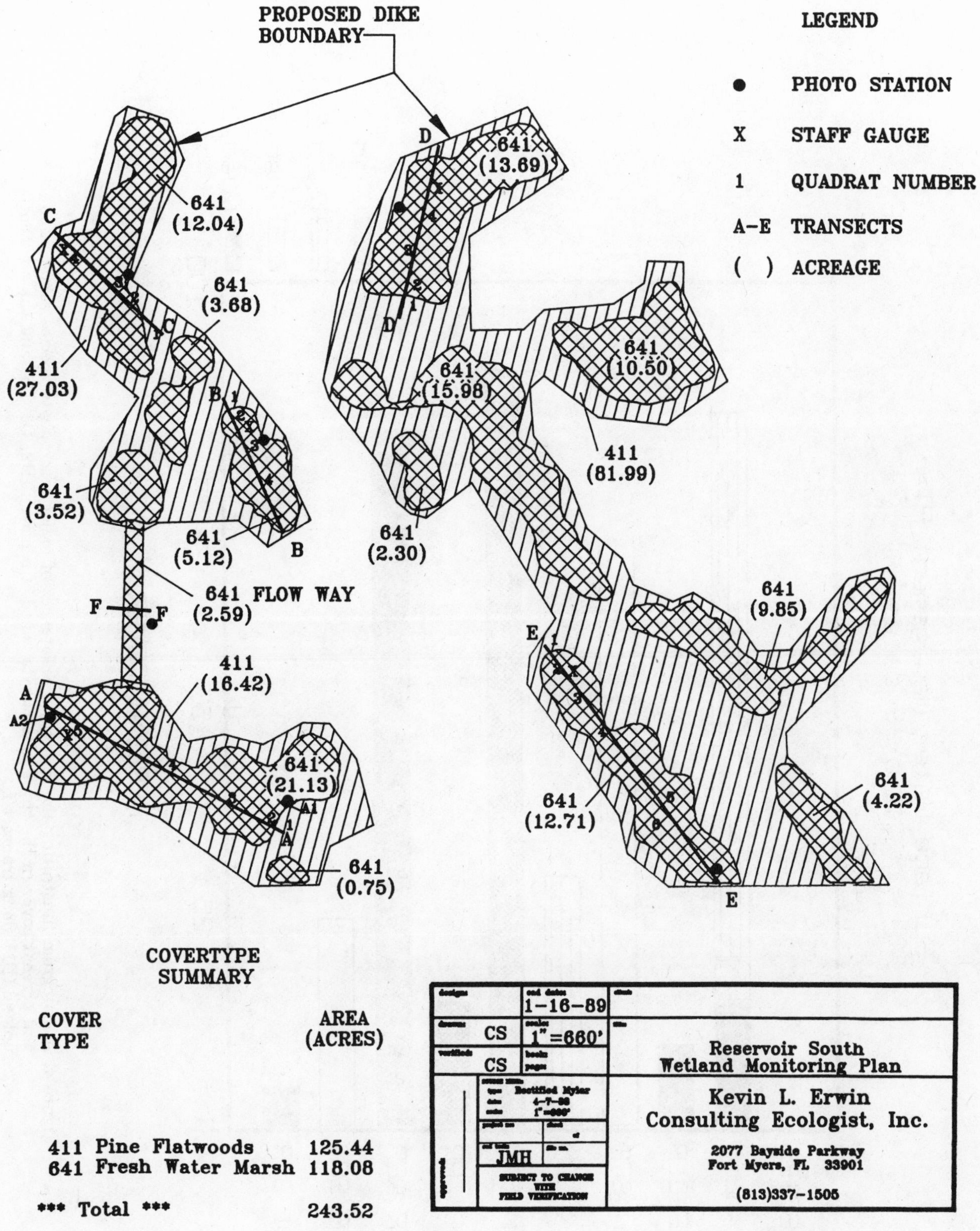

Figure 3. Locations of transects, quadrats, water level recording stations, and photo location stations in preserved and created marshes within a surface water management-reservoir system.

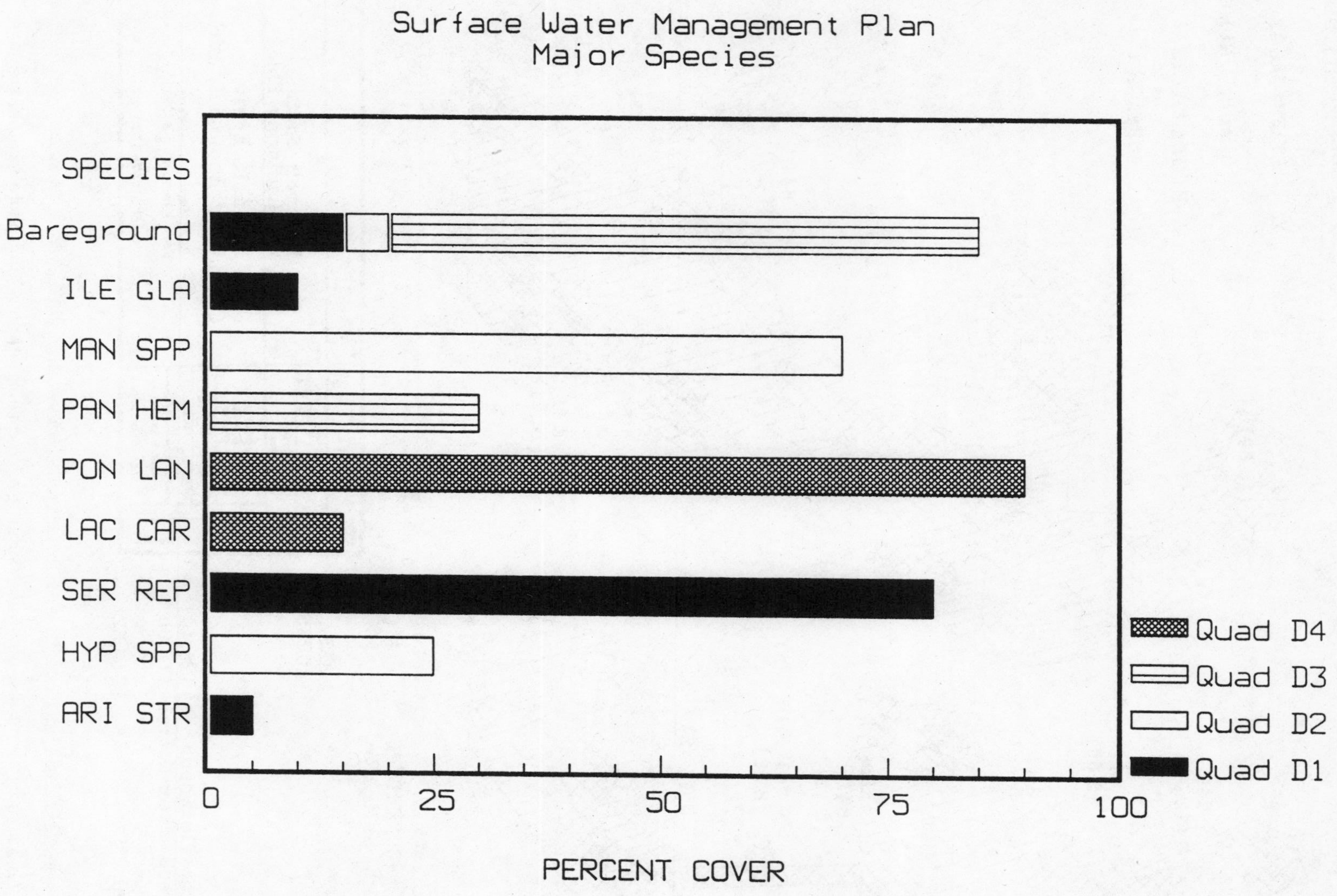

Figure 4. A histogram illustrating the structural differences of major macrophyte species composition and percent cover in the three macrophyte zones of a marsh (D2, D3, D4) and the adjacent upland (D1) along transect D.

November). Reports should be submitted within 90 days following sampling, and should document any vegetation changes including percent survival and cover of planted and/or recruited species (trees and herbs). Issues related to water levels, water quality, sedimentation, etc., should be addressed and recommendations or changes for improving the degree of success discussed.

FIXED POINT PANORAMIC PHOTOGRAPHS

Establish locations for fixed point photos in each wetland area to be monitored by providing a range pole in each section of photo panorama for scaling purposes. Photos should provide physical documentation of the condition of the wetland and any changes taking place within it. They should accompany the baseline vegetation survey and each annual report. Photo points and range pole locations are to remain the same throughout the duration of the monitoring program.

RAIN GAUGE

A rain gauge in a conveniently located area of the project site should, in general, be provided. It is unnecessary for some projects. Rainfall should be recorded daily. A summary of rainfall should accompany annual reports.

STAFF GAUGE

Staff gauges should be provided in each wetland area to be monitored. Staff gauges should be located in the deepest portion of the wetland and set to National Geodetic Vertical Datum (NGVD) elevation. Water levels should be recorded monthly and summarized with annual reports.

PLAN VIEW

A plan view showing locations of transects through wetland areas with rain gauge, staff gauge(s), and photo points should be provided with the baseline vegetation survey.

OBSERVED WILDLIFE UTILIZATION

Qualitative observations of wildlife utilization of the created/restored wetland should be recorded during all visits and annual surveys. If wildlife utilization is a major success criteria, extensive observations should be taken at least monthly.

FISH AND MACROINVERTEBRATES

Qualitative macroinvertebrate and fish should in many instances, be collected within each macrophyte zone containing standing water. Samples should be collected utilizing a D-frame dip net for a period of at least 20 minutes per station. All samples should be field sorted, preserved in 70% ethanol, and identified to the lowest possible taxon. A checklist of species collected should be compiled for submittal with annual reports.

CONCLUDING REMARKS

Wetland evaluation approaches including post project monitoring must, of course, be tailored to the specifics of the site and project goals. This requires expertise and creativity on the parts of the project designers and reviewers. Without such expertise and creativity, huge amounts of money may be spent on gathering useless data while no attempt is made to gather essential data. Qualified wetland scientists with knowledge of wetland ecology, hydrology, wildlife, and an appreciation of practical considerations in restoration or creation must be involved in the design and execution of evaluation efforts.

LITERATURE CITED

Adamus, P.R. 1983. A Method for Wetland Functional Assessment. Volume II. The Method. U.S. Department of Transportation, Federal Highway Administration. Office of Research, Environmental Division. Washington, D.C. (No. FHWA-IP-82-24).

Adamus, P.R. and Stockwell, L.R. 1983. A Method for Wetland Functional Assessment. Volume 1. Critical Review and Evaluation of Concepts. U.S. Department of Transportation. Federal Highway Administration. Office Research, Environmental Division. Washington, D.C. (No. FHWA-IP-82-23).

Barbour, M.G., J.H. Burk, and W.D. Pitts. 1987. Terrestrial Plant Ecology. The Benjamin/Cummings Publishing Company, Inc., Menlo Park, California.

Bauer, H.L. 1943. The statistical analysis of chaparral and other plant communities by means of transect samples. Ecology 24:45-60.

Best, G.R. and K.L. Erwin. 1984. Effects of hydroperiod on survival and growth of tree seedlings in a phosphate surface-mined reclaimed wetland, p. 221-225. In National Symposium on Surface Mining, Hydrology, Sedimentology, and Reclamation, University of Kentucky, Lexington, Kentucky.

Clairain, E.J., Jr. 1985. National wetlands functions and values study plan, p. 994-1009. In Proceedings of the 50th North American Wildlife and Natural Resources Conference. The Wildlife Management Institute, Washington, D.C.

Erwin, K.L. 1986. Agrico Fort Green Reclamation Project, Fourth Annual Report. Agrico Mining Company, Mulberry, Florida.

Erwin, K.L. 1987. Agrico Fort Green Reclamation Project, Fifth Annual Report. Agrico Mining Company, Mulberry, Florida.

Euler, D.L., F.T. Carreiro, G.B. McCullough, E.A Snell, V. Glooschenko, and R.H. Spurr. 1983. An Evaluation System for Wetlands of Ontario South of the Precambrian Shield, First Edition. Ontario Ministry of Natural Resources and Canadian Wildlife Service, Ontario Region.

Golet, F.C. 1973. Classification and evaluation of freshwater wetlands as wildlife habitat in the glaciated Northeast, p. 257-279. In Transactions of the Northeast Fish and Wildlife Conference, Vol. 30.

Gosselink, J.G. 1984. The Ecology of Delta Marshes of Coastal Louisiana: A Community Profile. U.S. Fish and Wildlife Service, Biological Services FWS/OBS-84/09, Washington, D.C.

Gosselink, J.G., E.P. Odum, and R.M. Pope. 1974. The Value of the Tidal Marsh. Center for Wetlands Resources Pub. LSU-SG-74-03. Louisiana State University, Baton Rouge.

Greeson, P.E., J.R. Clark, and J.E. Clark (Eds.). 1979. Wetland Functions and Values: The State of Our Understanding. Proceedings of the National Symposium on Wetlands, Lake Buena Vista, Florida, American Water Resources Association Tech. Publ. TPS 79-2, Minneapolis, Minnesota.

Heimberg, K. 1984. Hydrology of north-central Florida cypress domes, p. 72-82. In K.C. Ewel and H.T. Odum (Eds.), Cypress Swamps. University Presses of Florida, Gainesville.

Hollands, G.G. 1985. Assessing the relationship of groundwater and wetlands, p. 55-57. In J.A. Kusler and P. Riexinger (Eds.), Proceedings: National Wetlands Assessment Symposium. Association of State Wetland Managers, Berne, New York.

Kusler, J.A. and P. Riexinger (Eds.). 1985. Proceedings of the National Wetland Assessment Symposium. Association of State Wetland Managers, Berne, New York.

Larson, J.S. 1982. Understanding the ecological values of wetlands, p. 108-118. In Research on Fish and Wildlife Habitat. EPA-600/8-82-002. U.S. Environmental Protection Agency, Washington, D.C.

Lindsey, A.A. 1955. Testing the line strip method against full tallies in diverse forest types. Ecology 36:485-495.

Lonard, R.I., E.J. Clairain, Jr., R. T. Huffman, J.W. Hardy, L.D. Brown, P.E. Ballard, and J.W. Watts. 1981. Analysis of Methodologies Used for the Assessment of Wetlands Values. U.S. Water Resources Council, Washington, D.C.

Lonard, R.I., E.J. Clairain, Jr., R.T. Huffman, J.W. Hardy, L.D. Brown, P.E. Ballard, and J.W. Watts. 1984. Wetlands Function and Values Study Plan, Appendix A: Analysis of Methodologies for Assessing Wetlands Values. Technical Report Y-83-2, U.S. Army Engineer Waterways Experiment Station, Vicksburg, Mississippi.

Marble, A.D. and M. Gross. 1984. A method for assessing wetland characteristics and values, Landscape Planning II 1-17.

Mitsch, W.J. and J.G. Gosselink. 1986. Wetlands. Van Nostrand Reinhold Company Inc., New York.

Motts, W. and A. O'Brien. 1980. Hydrogeologic evaluation of wetland basins for land use planning. Water Resources Bulletin 16(5):

Niering, W.A. 1985. Wetlands. Alfred A. Knopf, Inc. New York.

Niering, W.A. and A.W. Palmisano. 1979. Use values: harvest and heritage, p 100-113. In J.R. Clark and J.E. Clark (Eds.), Scientists Report, The National Symposium on Wetlands. National Wetlands Tech. Council, Washington, D.C.

Novitzki, R.P. 1978. Hydrology of the Nevin Wetland near Madison, Wisconsin. U.S. Geol. Sur. Water Resources Investigation. No. 78-48.

Novitzki, R.P. 1979. Hydrologic characteristics of Wisconsin's wetlands and their influence on floods, stream flow, and sediment, p. 377-388. In P.E. Greeson, J.R. Clark, and J.E. Clark (Eds.), Wetland Functions and Values: The State of Our Understanding. American Water Resources Association, Minneapolis, Minnesota.

Ogawa, H. and J.W. Male. 1983. The Flood Mitigation Potential of Inland Wetlands. Water Resources Research Center Publication No. 138, University of Massachusetts, Amherst.

Peters, D.S., D.W. Ahrenholz, and T.R. Rice. 1979. Harvest and value of wetland associated fish and shellfish, p. 606-617. In P.E. Greeson, J.R. Clark, and J.E. Clark (Eds.), Wetland Functions and Values: The State of Our Understanding. American Water Resources Association, Minneapolis, Minnesota.

Reppert, R.T., G. Sigleo, E. Stakniv, L. Messman, and C. Myer. 1979. Wetlands Values: Concepts and Methods for Wetlands Evaluation. IWR Research Report 79-R-1, U.S. Army Engineer Institute for Water Resources, Fort Belvoir, Virginia.

Sather, J.H. and R.D. Smith. 1984. An Overview of Major Wetland Functions and Values. NWS/OBS-84/18. U.S. Department of the Interior, Fish and Wildlife Service, Washington, D.C.

U.S. Army Engineer Division, Lower Mississippi Valley. 1980. A Habitat Evaluation System (HES) for Water Resources Planning. U.S. Army Engineer Division, Lower Mississippi Valley, Vicksburg, Mississippi.

U.S. Fish and Wildlife Service. 1980. Habitat Evaluation Procedures (HEP) manual. 102 ESM, Washington, D.C.

Verry, E.S. and D.H. Boelter. 1979. Peatland hydrology, p. 389-402. In P.E. Greeson, J.R. Clark,and J.E. Clark (Eds.), Wetland Functions and Values: The State of Our Understanding. American Water Resources Association, Minneapolis, Minnesota.

Weller, M.W. 1981. Freshwater Marshes. University of Minnesota Press, Minneapolis, Minnesota.

Winchester, B.H. and L.D. Harris. 1979. An approach to valuation of Florida freshwater wetlands. In Proceedings of the Sixth Annual Conference on Wetlands Restoration and Creation. Hillsborough Community College, Tampa, Florida.

Winchester, B.H. 1981a. Assessing ecological value of central Florida wetlands: a case study, p. 25-38. In D. Cole (Ed.), Proceedings of the Eighth Annual Conference on Wetlands Restoration and Creation. Hillsborough Community College, Tampa, Florida.

Winchester, B.H. 1981b. Valuation of coastal plain wetlands in the southeastern United States, p. 285-298. In Symposium on Progress in Wetlands Utilization and Management, Orlando, Florida.

Woodin, H.E. and A.A. Lindsey. 1954. Juniper-Pinyon east of the continental divide as analyzed by the line strip method. Ecology 35:474-489.

APPENDIX I: A SELECTED BIBLIOGRAPHY ON WETLAND EVALUATION

Abele, L.G. 1974. Species diversity of decapod crustaceans in marine habitats. Ecology 55:156-161.

Adamus, P.R. 1983. A Method for Wetland Functional Assessment. Volume II. The Method. U.S. Department of Transportation, Federal Highway Administration. Office of Research, Environmental Division. Washington, D.C. (No. FHWA-IP-82-24).

Adamus, P.R. and Stockwell, L. R. 1983. A Method for Wetland Functional Assessment. Volume 1. Critical Review and Evaluation of Concepts. U.S. Department of Transportation. Federal Highway Administration. Office Research, Environmental Division. Washington, D.C. (No. FHWA-IP-82-23).

American Public Health Association. 1971. Standard Methods for the Examination of Water and Wastewater. 13th ed. Amer. Public Health Assn., New York.

Anderson, B.W., R.D. Ohmart, and J.D. Disano. 1978. Revegetating the riparian floodplain for wildlife, p. 318-331. In R.R. Johnson and J.F. McCormick (Tech. Coord.), Strategies for Protection and Management of Floodplain Wetlands and Other Riparian Ecosystems. General Technical Report WO-12, U.S. Dept. Agric., Forest Service, Washington, D.C.

Bailey, N.T.J. 1951. On estimating the size of mobile populations from recapture data. Biometrika 38:293-306.

Bailey, R.G. 1976. Ecoregions of the United States. U.S. Forest Service, Ogden, Utah.

Bailey, R.G. 1978. Description of the Ecoregions of the United States. U.S. Forest Service, Ogden, Utah.

Bakelaar, R.G. and E.P. Odum. 1978. Community and population level responses to fertilization in an old-field ecosystem. Ecology 59:660-665.

Ball, M.E. 1974. Floristic changes on grasslands and heaths on the Isle of Rhum after a reduction or exclusion of grazing. J. Environ. Manage. 2:299-318.

Barbour, M.G., J.H. Burk, and W.D. Pitts. 1987. Terrestrial Plant Ecology. The Benjamin/Cummings Publishing Company, Inc., Menlo Park, California.

Bauer, H.L. 1943. The statistical analysis of chaparral and other plant communities by means of transect samples. Ecology 24:45-60.

Beauvis, T.W. 1984. Evaluating wetlands in federal land exchanges. Wetlands 4:19-28.

Bell, III, H.E. 1981. Illinois Wetlands: Their Value in Management. Illinois Institute of Natural Resources Report 81/33, Chicago, Illinois.

Beschel, R.E. and P.J. Webber. 1962. Gradient analysis in swamp forests. Nature 194:207-209.

Best, G.R. and K.L. Erwin. 1984. Effects of hydroperiod on survival and growth of tree seedlings in a phosphate surface-mined reclaimed wetland, p. 221-225. In National Symposium on Surface Mining, Hydrology, Sedimentology, and Reclamation. University of Kentucky, Lexington, Kentucky.

Bormann, F.H. and G.E. Likens. 1979. Pattern and Process in a Forested Ecosystem. Springer-Verlag, New York.

Botkin, D.B., et al. 1982. Ecological characteristics of ecosystems. In W.R. Siegfried and B.R. Davies (Eds.), Conservation of Ecosystems: Theory and Practice. South African Natl. Sci. Prog. Rep. 61.CSIR, Pretoria.

Bowman, K.O., K. Hutcheson, E.P. Odum, and L.R. Shenton. 1969. Comments on the distribution of indices of diversity, p. 315-359. In G.P. Patil, E.C. Pielou, and W.E. Waters (Eds.), Statistical Ecology. Vol. 3. Pennsylvania State Univ. Press, University Park, Pennsylvania.

Boyd, C.E. 1972. A bibliography of interest in the utilization of vascular aquatic plants. Economic Botany 26:74-84.

Boyd, M. 1982. Salt marsh faunas: colonization and monitoring, p. 75-82. In M. Josselyn (Ed.), Wetland Restoration and Enhancement in California. Report T-CSGCP-007, California Sea Grant College Program, University of California, LaJolla, California.

Bradshaw, A.D. and M.J. Chadwick. 1980. The Restoration of Land: The Ecology and Reclamation of Derelict and Degraded Land. Univ. California Press, Berkeley, California.

Bray, J.R. and J. T. Curtis. 1957. An ordination of the upland forest communities of southern Wisconsin. Ecol. Monogr. 27:325-349.

Briand, F. 1983. Environmental control of food web structure. Ecology 64:253-263.

Brown, M.T. and H.T. Odum. 1985. Studies of a Method of Wetland Reconstruction Following Phosphate Mining. Final Report. Florida Institute of Phosphate Research, Publication #03-022-032.

Buikema, Jr., A.L. and J. Cairns, Jr. (Eds.). 1980. Aquatic Invertebrate Bioassays. STP 715. Amer. Soc. Testing and Materials, Philadelphia, Pennsylvania.

Butler, G.C. 1976. Principles of Ecotoxicology. SCOPE 12. Wiley, New York.

Butler, G.C. 1978. Principles of Ecotoxicology. SCOPE 12. Wiley, New York.

Cain, S.A. and G. M. de Oliveira Castro. 1959. Manual of Vegetation Analysis. Harper, New York.

Cairns, Jr., J. 1982. Artificial Substrates. Ann Arbor Sci., Ann Arbor, Michigan.

Cairns, J., Jr. and Dickson, K.L., eds. 1973. Biological Methods for the Assessment of Water Quality. STP 528. Amer. Soc. Testing and Material, Philadelphia, Pennsylvania.

Cairns, Jr., J. and K.L. Dickson. 1978. Field and laboratory protocols for evaluating the effects of chemical substances on aquatic life. J. Test Eval. 6:81-90.

Cairns, Jr., J. and K.L. Dickson. 1980. Risk analysis for aquatic ecosystems, p. 73-83. In Biological Evaluation of Environmental Impacts. Rep. FWS/OBS-80/26. U.S. Dept. Interior, Fish and Wildlife Service, and Council Environ. Quality, Washington, D.C.

Cairns, Jr., J., K.L. Dickson, and E.E. Herricks (Eds.). 1977. Recovery and Restoration of Damaged Ecosystems. Univ. Virginia Press, Charlottesville, Virginia.

Cairns, Jr., J. and W.H. van der Schalie. 1980. Biological monitoring, Part I. Early warning systems. Water Res. 14:1179-1196.

Cairns, Jr., J., K.L. Dickson, and G. F. Westlake (Eds.). 1977. Biological Monitoring of Water and Effluent Quality. STP 607. Amer. Soc. Testing and Materials, Philadelphia, Pennsylvania.

Cairns, Jr., J., J.R. Stauffer, Jr., and C.H. Hocutt. 1978. Opportunities for maintenance and rehabilitation of riparian habitats: eastern United States, p. 304-317. In R.R. Johnson and J.F. McCormick (Tech. Coord.), Strategies for Protection and Management of Floodplain Wetlands and Other Riparian Ecosystems. General Technical Report WO-12, U.S. Dept. of Agric., Forest Service, Washington, D.C.

Calkins, H.W., and R. F. Tomlinson. 1977. Geographic Information Systems: Methods and Equipment for Land Use Planning. U.S. Geological Survey, Reston, Virginia.

Canter, L.W. 1977. Environmental Impact Assessment. McGraw Hill, New York.

Canter, L.W. 1979. Water Resources Assessment Methodology and Technology Source Book. Ann Arbor Sci., Ann Arbor, Michigan.

Carpenter, S.R. and J. E. Chaney. 1983. Scale of spatial pattern: four methods compared. Vegetatio 53:153-160.

Chabreck, R.H. 1979. Wildlife harvest and wetlands of the United States, p. 618-631. In P.E. Greeson, J.R. Clark, and J.E. Clark (Eds.), Wetland Function and Values: The State of Our Understanding. American Water Resources Association, Minneapolis, Minnesota.

Chamberlain, R. 1982. Methods used to evaluate fish utilization of a salt marsh restoration site in Humboldt Bay, California, p. 97. In M. Josselyn (Ed.), Wetland Restoration and Enhancement in California. Report T-CSGCP-007, California Sea Grant College Program, University of California, LaJolla, California.

Christenson, J.W. 1979. Environmental Assessment Using Remotely Sensed Data. Gov. Print. Off., Washington, D.C.

Christian, C.S. and G. A. Stewart. 1968. Methodology of integrated surveys, p. 233-280. In Aerial Surveys and Integrated Studies. UNESCO, Paris.

Clairain, Jr., E.J. 1985. National wetlands functions and values study plan, p. 994-1009. In Proceedings of the 50th North American Wildlife and Natural Resources Conference. The Wildlife Management Institute, Washington, D.C.

Clairain, Jr., E.J., R.A. Cole, R.J. Diaz, A.W. Ford, R.T. Huffman, L.J. Hunt, and B.R. Wells. 1978. Habitat Development Field Investigations, Miller Sands Marsh and Upland Habitat Development Site, Columbia River, Oregon. Technical Report D-77-38, Waterways Experiment Station, U.S. Army Corps of Engineers, Vicksburg, Mississippi.

Clark, B.D., R. Bisset, and P. Wathern. 1980. Environmental Impact Assessment: A Bibliography with Abstracts. Mansell, London.

Clark, B.D., K. Chapman, R. Bisset, and P. Wathern. 1978. Methods of environmental impact analysis. Built Environ. 4:111-121.

Clark, J.E. 1979. Freshwater wetlands: habitats for aquatic invertebrates, amphibians, reptiles, and fish, p. 330-343. In P.E. Greeson, J.R. Clark, and J.E. Clark (Eds.), Wetland Functions and Values: The State of Our Understanding. American Water Resources Association, Minneapolis, Minnesota.

Clifford, H.T. and W. Stephenson. 1975. An Introduction to Numerical Classification. Academic Press, New York.

Cole, D. (Ed.). 1979. Proceedings of the 6th Annual Conference on Wetlands Restoration and Creation. Hillsborough Community College, Tampa, Florida.

Cole, D. (Ed.). 1980. Proceedings of the 7th Annual Conference on Wetlands Restoration and Creation. Hillsborough Community College, Tampa, Florida.

Cole, D. (Ed.). 1981. Proceedings of the 8th Annual Conference on Wetlands Restoration and Creation. Hillsborough Community College, Tampa, Florida.

Coleman, B.D., M.A. Mares, M.R. Willig, and Y.H. Hsieh. 1982. Randomness, area and species richness. Ecology 63:112-1133.

Collins, B. and L. Maltby. 1984. A Statistical Analysis of "An Evaluation System for Wetlands of Ontario, First Edition 1983". Canadian Wildlife Service Publication.

Connor, E.F. and E.D. McCoy. 1979. The statistics and biology of the species-area relationship. Am. Nat. 113:791-833.

Costanza, R. 1984. Natural resource evaluation and management: Tollward Ecological Economics, p. 7-18. In A.M. Jansen (Ed.), Integration of Economy and Ecology--An Outlook for the Eighties. Univ. of Stockholm Press, Stockholm, Sweden.

Costanza, R. and S.C. Farber. 1985. Theories and methods evaluation of natural systems: a comparison of willingness to pay and energy analysis based approaches. Man. Environment.

Space and Time 4:1-38.

Cottam, G. 1947. A point method for making rapid surveys of woodlands. Bull. Ecol. Soc. Amer. 28:60.

CRC. 1981. Handbook of Chemistry and Physics: A Ready-Reference Book of Chemical and Physical Data. 62nd ed. CRC Press, Boca Raton, Florida.

Daubenmire, R. 1966. Vegetation: Identification of typal communities. Science 151:291-298.

Denslow, J.S. 1980. Patterns of plant species diversity during succession under different disturbance regimes. Oecologia 46:18-21.

Dillon, T.M. and M.P. Lynch. 1981. Physiological responses as determinants of stress in marine and estuarine organisms, 227-241. In G.W. Barrett and R. Rosenberg (Eds.), Stress Effects on Natural Ecosystems. Wiley, New York.

Dunn, W.J. and G.R. Best. 1983. Enhancing ecological succession: 5. seed bank survey of some Florida marshes and the role of seed banks in marsh reclamation. In Proceedings, National Symposium on Surface Mining, Hydrology, Sedimentology and Reclamation. Office of Continuing Education, University of Kentucky, Lexington, Kentucky.

Dunne, T. and L. B. Leopold. 1978. Water in Environmental Planning. Freeman, San Francisco, California.

Ecologistics Limited. 1981. A Wetland Evaluation System for Southern Ontario. Prepared for the Canada/Ontario Steering Committee on Wetland Evaluation and the Canadian Wildlife Service, Environment Canada.

Egler, F.E. 1954. Vegetation science concepts: I. Initial floristic composition, a factor in old-field vegetation development. Vegetatio 4:412-417.

Errington, J.C. 1976. The effect of regular and random distributions on the analysis of pattern. J. Ecol. 61:99-105.

Erwin, K.L. 1983. Agrico Fort Green Reclamation Project, First Annual Report. Agrico Mining Company, Mulberry, Florida.

Erwin, K.L. 1984. Agrico Fort Green Reclamation Project, Second Annual Report. Agrico Mining Company, Mulberry, Florida.

Erwin, K.L. 1985. Agrico Fort Green Reclamation Project, Third Annual Report. Agrico Mining Company, Mulberry, Florida.

Erwin, K.L. 1986. Agrico Fort Green Reclamation Project, Fourth Annual Report. Agrico Mining Company, Mulberry, Florida.

Erwin, K.L. 1987. Agrico Fort Green Reclamation Project, Fifth Annual Report. Agrico Mining Company, Mulberry, Florida.

Erwin, K.L. 1988. Agrico Fort Green Reclamation Project, Sixth Annual Report. Agrico Mining Company, Mulberry, Florida.

Erwin, K.L. and F. D. Bartleson. 1985. Water quality within a central Florida phosphate surface mined reclaimed wetland, p. 84-95. In F.J. Webb, Jr. (Ed.), Proceedings of the 12th Annual Conference on Wetland Restoration and Creation. Hillsborough Community College Environmental Studies Center, Tampa, Florida.

Erwin, K.L. and G.R. Best. 1985. Marsh community development in a central Florida phosphate surface-mined reclaimed wetland. Wetlands 5:155-166.

Erwin, K.L., G.R. Best, W.J. Dunn, and P.M. Wallace. 1984. Marsh and forested wetland reclamation of a central Florida phosphate mine, p. 87-103. Wetlands 4:87-103.

Euler, D.L., F.T. Carreiro, G.B. McCullough, E.A Snell, V. Glooschenko, and R.H. Spurr. 1983. An Evaluation System for Wetlands of Ontario South of the Precambrian Shield. First Edition. Ontario Ministry of Natural Resources and Canadian Wildlife Service, Ontario Region.

Ewel, K.C. 1976. Effects of sewage effluent on ecosystem dynamics in cypress domes, p. 169-195. In D.L. Tilton, R.H. Kadlec, and C.J. Richardson (Eds.), Freshwater Wetlands and Sewage Effluent Disposal. University of Michigan, Ann Arbor, Michigan.

Ewel, K.C. and H.T. Odum. 1978. Cypress swamps for nutrient removal and wastewater recycling, p. 181-198. In M.P. Wanielista and W.W. Eckenfelder, Jr. (Eds.), Advances in Water and Wastewater Treatment Biological Nutrient Removal. Ann Arbor Sci. Publ., Inc., Ann Arbor, Michigan,

Ewel, K.C. and H.T. Odum. 1979. Cypress domes: nature's tertiary treatment filter, p. 103-114. In W.E. Sopper and S.N. Kerr (Eds.), Utilization of Municipal Sewage Effluent and Sludge on Forest and Disturbed Land. The Pennsylvania State University Press, University Park, Pennsylvania.

Ewel K.C. and H.T. Odum (Eds.). 1984. Cypress Swamps. University Presses of Florida, Gainesville, Florida.

Fager, E.W. 1972. Diversity: a sampling study. Am. Nat. 106:293-310.

Fetter, Jr., C.W., W.E. Sloey, and F.L. Spangler. 1978. Use of a natural marsh for wastewater polishing. J. Water Pollution Control Fed. 50:290-307.

Forman, R.T.T. 1979. The pine barrens of New Jersey: an ecological mosaic, p. 569-585. In R.T.T. Forman (Ed.), Pine Barrens: Ecosystem and Landscape. Academic Press, New York.

Forman, R.T.T. 1982. Interactions Among Landscape Elements: A Core Landscape Ecology, p. 35-48. In S.P. Tjallingii and A.A. de Veer (Eds.), Perspectives in Landscape Ecology. PUDOC, Wageningen, The Netherlands.

Forman, R.T.T. 1983. Corridors in a landscape: their ecological structure and function. Ekologia (USSR).

Forman, R.T.T., A.E. Galli, and C.F. Leck. 1976. Forest size and avian diversity in New Jersey wood lots with some land use implications. Oecologia 26:1-8.

Forman, R.T.T. and M. Godron. 1981. Patches and structural components for a landscape ecology. BioScience 31:733-740.

Fox, J.E.D. 1976. Constraints on the natural regeneration of tropical moist forest. For. Ecol. Manage. 1:37-65.

Frankel, O.H., and M.E. Soule. 1981. Conservation and Evolution. Cambridge Univ. Press, Cambridge.

Game, M. 1980. Best shape for nature reserves. Nature 287:630-632.

Game, M. and G.F. Peterken. 1980. Nature reserve selection in central Lincolnshire Woodlands. In M.D. Hooper (Ed.), Proceedings of the Symposium on Area and Isolation. Institute of Terrestrial Ecol., Cambridge, U.K.

Gauch, Jr., H.G. 1982. Multivariate Analysis in Community Ecology. Cambridge Univ. Press, Cambridge.

Gauch, Jr., H.G. and R.H. Whittaker. 1981. Hierarchical classification of community data. J. Ecol. 69:135-152.

Gehlbach, F.R. 1975. Investigation, evaluation, and priority ranking of natural areas. Biol. Conserv. 8:79-88.

Gillison, A.N. and D.J. Anderson (Eds.). 1981. Vegetation Classification in Australia. Austr. Natl. Univ. Press, Canberra.

Glooschenko, V. 1983. Development of an evaluation system wetlands in southern Ontario. Wetlands 3:192-200.

Goff, F.G., G.A. Dawson, and J.J. Rochow. 1982. Site Examination for Threatened and Endangered Plant Species. Environ. Manage. 6:307-316.

Goh, B.S. 1980. Management and Analysis of Biological Population. Elsevier, Amsterdam.

Goldstein, R.A. and D.F. Grigall. 1972. Computer Programs for the Ordination and Classification of Ecosystems. Rep. ORNL-IBP-71-10. Oak Ridge Natl. Lab., Oak Ridge, Tennessee.

Golet, F.C. 1973. Classification and evaluation of freshwater wetlands as wildlife habitat in the glaciated northeast, p. 257-279. In Transactions of the Northeast Fish and Wildlife Conference, Vol. 30.

Goodall, D.W. 1974. A new method for the analysis of spatial pattern by random pairing of quadrats. Vegetatio 29:135-146.

Goodall, D.W. 1978. Sample similarity and species correlation, p. 99-149. In R.H. Whittaker (Ed.), Ordination of Plant Communities. 2nd ed. Junk, The Hague.

Goodman, G.T. and S.A. Bray. 1975. Ecological Aspects of the Reclamation of Derelict and Disturbed Land. GEO Abstract, Norwich, U.K.

Gosselink, J.G. 1984. The Ecology of Delta Marshes of Coastal Louisiana: A Community Profile. U.S. Fish and Wildlife Service, Biological Services FWS/OBS-84/09, Washington, D.C.

Gosselink, J.G., E.P. Odum, and R.M. Pope. 1974. The value of the tidal marsh. Center for Wetlands Resources Pub. LSU-SG-74-03. Louisiana State University, Baton Rouge, Louisiana.

Gosselink, J.G. and R.E. Turner. 1987. The role of hydrology in freshwater ecosystems, p. 63-78. In R.E. Good, D.F. Whigham, and R.L. Simpson (Eds.), Freshwater Wetlands: Ecological Processes and Management Potential. Academic Press, New York.

Green, R.H. 1979. Sampling Design and Statistical Methods for Environmental Biologists. Wiley-Interscience, New York.

Greeson, P.E., J.R. Clark, and J.E. Clark (Eds.). 1979. Wetland Functions and Values: The State of Our Understanding, Proceedings of the National Symposium on Wetlands, Lake Buena Vista, Florida. American Water Resources Association Tech. Publ. TPS 79-2, Minneapolis, Minnesota.

Greig-Smith, P. 1952. The use of random and contiguous quadrats in the study of the structure of plant communities. Ann. Bot. Lond. NS 16:293-316.

Greig-Smith, P. 1961. Data on pattern within plant communities: I. the analysis of pattern. J. Ecol. 49:695-702.

Greig-Smith, P. 1964. Quantitative Plant Ecology. 2nd ed. Butterworths, London.

Greig-Smith, P. 1983. Quantitative Plant Ecology. 3rd ed. University of California Press, Berkeley, California.

Grime, J.P. 1974. Vegetation classification by reference to strategies. Nature 250:26-31.

Grime, J.P. 1979. Plant Strategies and Vegetation Processes. Wiley, New York.

Harman, W.N. 1972. Benthic substrates: their effect on freshwater mollusca. Ecology 53:271-277.

Harper, J.L. 1969. The roles of predation in vegetational diversity. Brookhaven Symp. Biol. 22:48-62.

Harrison, G.W. 1979. Stability under environmental stress: resistance, resilience, persistence and variability. Am. Nat. 113:659-669.

Hartshorn, G.S. 1975. A matrix model of tree population dynamics, p. 41-51. In F.B. Golley and E. Medina (Eds.), Tropical Ecological Systems. Springer-Verlag, New York.

Hefner, J.M. 1982. The national wetlands inventory: tools for wetland creation and restoration, p. 265-277. In F.J. Webb, Jr. (Ed.), Proceedings of the Ninth Annual Conference on Wetlands Restoration and Creation. Hillsborough Community College, Tampa, Florida.

Heimburg, K. 1984. Hydrology of north-central Florida cypress domes, p. 72-82. In K.C. Ewel and H.T. Odum (Eds.), Cypress Swamps. University Presses of Florida, Gainesville, Florida.

Helliwell, D.R. 1969. Valuation of Wildlife Resources. Regional Studies 3:41-47.

Higgs, A.J. and M.B. Usher. 1980. Should nature reserves be large or small? Nature 285:568-569.

Hill, A.R. 1975. Ecosystem stability in relation to stresses caused by human activities. Can. Geogr. 19:206-220.

Holdgate, M.W. and M.J. Woodman (Eds.). 1976. The Breakdown and Restoration of Ecosystems. Plenum, New York.

Hollands, G.G. 1985. Assessing the relationship of groundwater and wetlands, p. 55-57. In J.A. Kusler and P. Riexinger (Eds.), Proceedings: National Wetlands Assessment Symposium. Association of State Wetland Managers, Berne, New York.

Hollick, M. 1981. The role of quantitative decision-making methods in environmental impact assessment. J. Environ. Manage. 12:65-78.

Holling, C.S. 1973. Resilience and stability of ecological systems. Ann. Rev. Ecol. Syst. 4:1-24.

Horn, H.S. 1975. Forest Succession. Sci. Amer. 232:90-98.

Horn, H.S. 1976. Succession, p. 187-204. In R.M. May (Ed.), Theoretical Ecology. Principles and Applications. Saunders, Philadelphia.

Howell, E.A. 1981. Landscape design, planning, and management: an approach to the analysis of vegetation. Environ. Manag. 5:207-212.

Humphreys, W.F. and D.J. Kitchener. 1982. The effect of habitat utilization on species-area curves: implications for optimal reserve area. J. Biogeogr. 9:391-396.

Hutcheson, K. 1970. A test for comparing diversities based on the Shannon Formula. J. Theor. Biol. 29:151-154.

Innis, G.S. and R.V. O'Neill (Eds.). 1979. Systems Analysis of Ecosystems. Intl. Coop. Publ. House, Fairland, Maryland.

Ivanovici, A.M. and W.J. Wiebe. 1981. Towards a working "definition" of stress: a review and critique, p. 13-47. In G.W. Barrett and R. Rosenberg (Eds.), Stress Effects on Natural Ecosystems. Wiley, New York.

Johnson, M.P., L.G. Mason, and P.H. Raven. 1968. Ecological parameters and plant species diversity. Am. Nat. 102:297-306.

Johnson, R.A. 1981. Application of the guild concept to environmental impact analysis of terrestrial vegetation. J. Environ. Manage. 13:205-222.

Kadlec, R.H. 1979. Wetlands for tertiary treatment, p. 490-540. In P.E. Greeson, J.R. Clark, and J.E. Clark (Eds.), Wetland Functions and Values: The State of Our Understanding. American Water Resources Association, Minneapolis, Minnesota.

Kadlec, R.H. and D.L. Tilton. 1979. The use of freshwater wetlands as a wastewater treatment alternative. CRC Crit. Rev. Environ. Control 9:185-212.

Karr, J.R. 1981. Assessment of biotic integrity using fish communities. Fisheries 6:21-27.

Keammerer, W.R., W.C. Johnson, and R.L. Burgess. 1975. Floristic analysis of the Missouri River bottomland forest in North Dakota. Can. Field Net. 98: 5-19.

Kershaw, K.A. 1973. Quantitative and Dynamic Plant Ecology. 2nd ed. Elsevier, New York.

Kessell, S.R. 1981. Application of gradient analysis concepts to resource management modeling. Proc. Ecol. Soc. Aust. 11:163-173.

Klopatek, J.M., J.T. Kitchings, R.J. Olson, K.D. Kumar, and L.K. Mann. 1981. A hierarchical system for evaluating regional ecological resources. Biol. Conserv. 20:271-290.

Krebs, C.J. 1972. Ecology: The Experimental Analysis of Distribution and Abundance. Harper & Row, New York.

Krenkel, P.A. and V. Novotny. 1980. Water Quality Management. Academic Press, New York.

Kusler, J.A., and P. Riexinger. 1986. Proceedings of the National Wetlands Assessment Symposium. Association of State Wetland Managers, Berne, New York.

Lacasse, N.L. and W.J. Moroz (Eds.). 1969. Handbook of Effects Assessment: Vegetation Damage. Center for Air Environ. Stud., Pennsylvania State Univ., University Park, Pennsylvania.

Larson, J.S. (Ed.). 1976. Models for Evaluation of Freshwater Wetlands. Water Resources Centre. University of Massachusetts, Amherst, Massachusetts.

Larson, J.S. 1982. Wetland Value Assessment-State-of-the Art, p. 417-424. In B. Gopal, R.E. Turner, R.G. Wetzel, and D.F. Whigham (Eds.), Wetlands: Ecology and Management. Natural Institute of Ecology and International Scientific Publications, Jaipur, India.

Larson, J.S. 1982. Understanding the ecological values of wetlands, p. 108-118. In Research on Fish and Wildlife Habitat. EPA-600/8-82-002. U.S. Environmental Protection Agency, Washington, D.C.

Legendre, L. and P. Legendre. 1983. Development in Environmental Modeling. Vol 3: Numerical Ecology. Elsevier, Amsterdam.

Levin, S.A. and R.T. Paine. 1974. Disturbance, patch formation, and community structure. Proc. Natl. Acad. Sci. 71:2744-2747.

Levin, S.A. 1976. Population dynamic models in

heterogeneous environments. Ann. Rev. Ecol. Syst. 7:287-310.

Lewis, III, R.R. (Ed.). 1982. Creation and Restoration of Coastal Plant Communities. CRC Press, Boca Raton, Florida.

Lindsey, A.A. 1955. Testing the line strip method against full tallies in diverse forest types. Ecology 36:485-495.

Lintz, Jr., J. and D.S. Simonett (Eds.). 1976. Remote Sensing of Environment. Addison-Wesley, Reading, Massachusetts.

Llyod, M. and R.J. Ghelardi, R.J. 1964. A table for calculating the "equitability" component of species diversity. J. Annim. Ecol. 33:217-225.

Lonard, R.I., E.J. Clairain, Jr., R.T. Hoffman, J.W. Hardy, L.D. Brown, P.E. Ballard, and J.W. Watts. 1981. Analysis of Methodologies Used for the Assessment of Wetland Values. U.S. Water Resources Council, Washington, D.C.

Lonard, R.I., E.J. Clairain, Jr., R.T. Huffman, J.W. Hardy, L.D.Brown, P.E. Ballard, and J.W. Watts. 1984. Wetlands Function and Values Study Plan, Appendix A: Analysis of Methodologies for Assessing Wetlands Values. Technical Report Y-83-2, U.S. Army Engineer Waterways Experiment Station, Vicksburg, Mississippi.

MacArthur, R.H. 1955. Fluctuations of animal populations, and a measure of community stability. Ecology 36:533-536.

Marble, A.D. and M. Gross. 1984. A method for assessing wetland characteristics and values, Landscape Planning II 1-17.

Marsh, W.M. and Dozier, J. 1981. Landscape: An Introduction to Physical Geography. Addison-Wesley, Reading, Massachusetts.

May, R.M. 1973. Stability and Complexity in Model Ecosystems. Princeton Univ. Press, Princeton, New Jersey.

Merks, R.L. 1968. The accumulation of 36 CL ring-labelled DDT in a freshwater marsh. J. Wildlife Manage. 32:376-398.

McEntyre, J.G. 1978. Land Survey Systems. Wiley, New York.

McIntosh, R.P. 1980. The relationship between succession and the recovery process in ecosystems, p. 1-62. In J. Cairns, Jr. (Ed.), The Recovery Process in Damaged Ecosystems. Ann Arbor Sci., Ann Arbor, Michigan.

McIntyre, G.A. 1953. Estimation of plant density using line transects. J. Ecol. 41:319-330.

Mitsch, W.J. and J.G. Gosselink. 1986. Wetlands. Van Nostrand Reinhold Company, Inc., New York.

Montanari, J.H. and J.A. Kusler. 1978. Proceedings of the National Wetlands Protection Symposium, June 6-8, 1977, Reston, Virginia. FWS/OBS-78/97, Office of Biological Services, Fish and Wildlife Service, Washington, D.C.

Mooney, H.A. and M. Gordon (Eds.). 1983. Disturbance and Ecosystems-Components of Response. Springer-Verlag, New York.

Motts, W. and A. O'Brien. 1980. Hydrogeologic evaluation of wetland basins for land use planning. Water Res. Bull. 16(5)

Mueller-Dombois, D. and H. Ellenberg. 1974. Aims and Methods of Vegetation Ecology. Wiley, New York.

Naiman, R.J. 1983. The annual pattern and spatial distribution of aquatic oxygen metabolism in Boreal Forest watersheds. Ecol. Monogr. 53:73-94.

Naveh, Z. 1982. Landscape ecology as an emerging branch of human ecosystem science. Adv. Ecol. Res. 12:189-237.

Naveh, Z. and A.S. Lieberman. 1984. Landscape Ecology. Theory and Application. Springer-Verlag, New York.

Nelder, J.A. and R.W.M. Wedderburn. 1972. Generalized linear models. J. Roy. Statist. Soc. 135:370-384.

Nichols, R. and E. Hyman. 1980. A review and analysis of fifteen methodologies for environmental assessment. Water Research & Technology Report. U.S. Dept. Interior, Washington, D.C.

Niering, W.A. 1985. Wetlands. Alfred A. Knopf, Inc., New York.

Niering, W.A. and A.W. Palmisano. 1979. Use values: harvest and heritage, p 100-113. In J.R. Clark and J.E. Clark (Eds.), Scientists Report, The National Symposium on Wetlands. National Wetlands Tech. Council, Washington, D.C.

Nixon, S.W. and V. Lee. 1985. Wetlands and Water Quality--A Regional Review of Recent Research in the United State on the Role of Fresh and Salt Water Wetlands as Sources, Sinks, and Transformers of Nitrogen, Phosphorus, and Various Heavy Metals. Report to the Waterways Experiment Station, U.S. Army Corps. of Engineers, Vicksburg, Mississippi.

Novitzki, R.P. 1979. Hydrologic characteristics of Wisconsin's wetlands and their influence on floods, stream flow, and sediment, p. 377-388. In P.E. Greeson, J.R. Clark, and J.E. Clark (Eds.), Wetland Functions and Values: The State of Our Understanding. American Water Resources Association, Minneapolis, Minnesota.

Novitzki, R.P. 1978. Hydrology of the Nevin Wetland near Madison, Wisconsin. U.S. Geol. Sur. Water Resources Investigation. No. 78-48.

O'Banion, K. 1980. Use of Value Functions in Environmental Decisions. Environ. Manage. 4:3-6.

Odum, E.P. 1969. The Strategy of Ecosystem Development. Science 164:262-270.

Odum, E.P. 1979. The value of wetlands: a hierarchical approach, p. 16-25. In P.E. Gresson, J.R. Clark, and J.E. Clark (Eds.), Wetland Functions and Values: The State of Our Understanding. American Water Resources Association, Minneapolis, Minnesota.

Ontario Ministry of Natural Resources and Environment Canada 1983. An Evaluation System for Wetlands of Ontario South of the Precambrian Shield. 1st ed. Toronto, Ontario, Canada.

Ontario Ministry of Natural Resources and Environment Canada. 1984. An Evaluation System for Wetlands of Ontario South of the Precambrian Shield. 2nd ed. Toronto, Ontario, Canada.

Orians, G.H. 1975. Diversity, Stability and Maturity in Natural Ecosystem, p. 64-65. In W.H. van Dobben and R.H. Lowe-McConnell (Eds.), Unifying Concepts in Ecology. Junk, The Hague.

Orloci, L. 1967. An agglomerative method for classification of plant communities. J. Ecol. 55:193-205.

Orloci, L. 1978. Multivariate Analysis in Vegetation Research. 2nd ed. Junk, The Hague.

Paine, R.T. 1966. Food web complexity and species diversity. Amer. Nat. 100:65-75.

Paine, R.T. 1974. Intertidal community structure: experimental studies on the relationship between a dominant competitor and its principal predator. Oecologia 15:93-120.

Paine, R.T. 1977. Controlled manipulations in the marine intertidal zone and their contributions to ecological theory. Acad. Nat. Sci. Phila. Spec. Pub. 12:245-270.

Paine, R.T. 1980. Food webs: linking interaction strength and community infrastructure. J. Anim. Ecol. 49:667-685.

Patton, D.R. 1975. A diversity index for quantifying habitat edge. Wildl. Soc. Bull. 394:171-173.

Paysen, T.E., J.A. Derby, H. Black, V.C. Bleich, and J.W. Mincks. 1981. A vegetation classification system applied to southern California. U.S. Forest Service, Gen. Tech. Rep. PSW-45. Berkeley, California.

Peet, R.K. 1974. The measurement of species diversity. Ann.Rev. Ecol. Syst. 5:285-307.

Peet, R.K. and N.L. Christensen. 1980. Succession: a population process. Vegetatio 43:131-140.

Peters, D.S., D.W. Ahrenholz, and T.R. Rice. 1979. Harvest and value of wetland associated fish and shellfish, p. 606-617. In P.E. Greeson, J.R. Clark, and J.E. Clark (Eds.), Wetland Functions and Values: The State of Our Understanding. American Water Resources Association, Minneapolis, Minnesota.

Phillips, E.A. 1959. Methods of Vegetation Study. Holt, Rinehart and Winston, Inc., New York.

Pielou, E.C. 1966. Species-diversity and pattern-diversity in the study of ecological succession. J. Theor. Biol. 10:370-383.

Pielou, E.C. 1975. Ecological Diversity. Wiley-Interscience, New York.

Pielou, E.C. 1977. Mathematical Ecology. 2nd ed. Wiley-Interscience, New York.

Pimm, S.L. 1979. The structure of food webs. Theor. Pop. Biol. 16:144-158.

Pimm, S.L. 1982. Food Webs. Chapman & Hall, London.

Poole, R.W. 1974. An Introduction to Quantitative Ecology. McGraw-Hill, New York.

Preston, F.W. 1948. The commonness and rarity of species. Ecology 29:254-283.

Preston, F.W. 1960. Time and space and the variation of Species. Ecology 41:611-627.

Quammen, M.L. 1986. Measuring the success of wetlands mitigation. In J.A. Kusler, M.L. Quammen, and G.Brooks (Eds.), Proceedings of the National Wetlands Symposium, Mitigation of Impacts and Losses. Association of State Wetland Managers, Berne, New York.

Rau, J.G. 1980. Summarization of environmental impact. In J.G. Rau and D.C. Wooten (Eds.), Environmental Impact Analysis Handbook. McGraw-Hill, New York.

Rau, J.G. and D.C. Wooten (Eds.). 1980. Environmental Impact Analysis Handbook. McGraw-Hill, New York.

Reid, R.A., N. Patterson, L. Armour, and A. Champagne. 1980. A Wetlands Evaluation Model for Southern Ontario. The Federation of Ontario Naturalists.

Reppert, R.T. and W.R. Sigleo. 1979. Concepts and methods for wetlands evaluation under development by the U.S. Army Corps of Engineers, p. 57-62. In P.E. Greeson, J.R. Clark, and J.E. Clark (Eds.), Wetlands Function and Values: The State of Our Understanding. Technical Publication TPS79-2, Water Resources Association, Minneapolis, Minnesota.

Reppert, R.T., W. Sigleo, E. Stakhiv, L. Messman, and C.D.Meyers. 1979. Wetland Value: Concepts and Methods for Wetland Evaluation. Research Report 79-Rl, Institute for Water Resources, U.S. Army Corps of Engineers, Fort Belvoir, Virginia.

Rowe, J.S. and J.W. Sheard. 1981. Ecological land classification: a survey approach. Environ. Manage. 5:451-464.

Sabins, F.F. 1978. Remote Sensing: Principles and Interpretation. Freeman, San Francisco, California.

Sather, J.H. and R.D. Smith. 1984. An Overview of Major Wetland Functions and Values. NWS/OBS-84/18. U.S. Fish and Wildlife Service, Washington, D.C.

Schaller, F.W. and P. Sutton (Eds.). 1979. Reclamation of Drastically Disturbed Lands. Amer. Soc. Agron., Madison, Wisconsin.

Schamberger, M.L., C. Short, and A. Farmer. 1979. Evaluation Wetlands as a Wildlife Habitat, p. 74-83. In P.E. Greeson, J.R. Clark, and J.E. Clark (Eds.), Wetland Function and Values: The State of Our Understanding. American Water Resources Association, Minneapolis, Minnesota.

Schanda, E. (Ed.). 1976. Remote Sensing for Environmental Sciences: Ecological Studies, Analysis and Synthesis, Vol. 18. Springer-Verlag, Berlin.

Shaffer, M.L. 1981. Minimum population size for species conservation. BioScience 31:131-134.

Shannon, C.E. and W. Weaver. 1949. The Mathematical Theory of Communication. Univ. Illinois Press, Urbana, Illinois.

Sharitz, R.R. 1982. Plant community structure and processes, p. II-1-58. In M.H. Smith, R.R. Sharitz, and J.B. Gladden (Eds.), An Evaluation of the Steel Creek Ecosystem in Relation to the Proposed Start of L-Reactor. SREL-12,ITCH-66e. NTIS, Springfield, Virginia.

Shugart, H.H., T.R. Crow, and J.M. Hett. 1973. Forest succession models: a rationale and methodology for modeling forest succession over large regions. For. Sci. 19:203-212.

Shugart, H.H., J.M. Klopatek, and W.R. Emanuel. 1981. Ecosystems analysis and land-use planning, p. 665-699. In E.J. Kormondy and J.F. McCormick (Eds.), Handbook of Contemporary Development in World Ecology. Greenwood, Westport, Connecticut.

Simpson, E.H. 1949. Measurement of diversity. Nature 163:688.

Sinden, J.A. and G.K. Windsor. 1981. Estimating the value of wildlife for preservation: a comparison of approaches. J. Environ. Manage. 12:111-125.

Smardon, R.C. 1979. Visual-cultural values of wetlands, p. 535-544. In P.E. Greeson, J.R. Clark, and J.E. Clark (Eds.), Wetland Functions and Values: The State of Our Understanding. American Water Resources Association, Minneapolis, Minnesota.

Sondheim, M.W. 1978. A comprehensive methodology for assessing environmental impact. J. Environ. Manage. 6:27-42.

Soule, M.E. 1980. Thresholds for survival: maintaining fitness and evolutionary potential, p. 1-8. In M.E. Soule and B.A. Wilcox (Eds.), Conservation Biology: An Evolutionary-Ecological Perspective. Sinauer, Sunderland, Massachusetts.

Sousa, W.P. 1980. The responses of a community to disturbance: the importance of successional age and species life histories. Oecologia 45:72-81.

Southwood, T.R.E. 1978. Ecological Methods, with Particular Reference to the Study of Insect Population. Wiley, New York.

Strahler, A.H. 1978. Binary discriminant analysis: a new method for investigating species-environment relationships. Ecology 59:108-116.

Strahler, A.N. 1957. Quantitative analysis of watershed geomorphology. Trans. Am. Geophys. Union 38:913-920.

Stumm, W. and J.J. Morgan. 1981. Aquatic Chemistry: An Introduction Emphasizing Chemical Equilibria in Natural Waters. 2nd ed. Wiley-Interscience, New York.

Suffling, R. 1980. An index of ecological sensitivity to disturbance, based on ecosystem age, and related to landscape diversity. J. Environ. Manage. 10:253-262.

Sutherland, J.P. 1974. Multiple stable points in natural Communities. Am. Nat. 108:859-873.

Swartz, R.C. 1980. Application of diversity indices in marine pollution investigations, p. 230-237. In Biological Evaluation of Environmental Impacts. Rep. FWS/OBS-80/26. U.S. Fish and Wildlife Service, and Council on Environ. Quality, Washington, D.C.

Terborgh, J. 1974. Preservation of natural diversity: the problem of extinctionprone species. BioScience 24:715-722.

Tjallingii, S.P. and A.A. de Veer (Eds.). 1982. Perspectives in Landscape Ecology. PUDOC, Wageningen, The Netherlands.

U.S. Army Corps of Engineers, Lower Mississippi Valley. 1980. A Habitat Evaluation System (HES) for Water Resources Planning. Vicksburg, Mississippi.

U.S. Environmental Protection Agency. 1977. Quality Criteria for Water. Off. Water Hazardous Materials, Washington, D.C.

U.S. Fish and Wildlife Service. 1980. Habitat Evaluation Procedures (HEP) Manual. Washington, D.C.

U.S. Fish and Wildlife Service. 1980. Habitat as a Basis Environmental Assessment. 101 ESM. Div. Ecol. Serv., Washington, D.C.

U.S. Fish and Wildlife Service. 1980. Habitat Evaluation Procedures (HEP). 102 ESM. Div. Ecol. Serv., Washington, D.C.

U.S. Fish and Wildlife Service. 1980. Human Use and Economic Evaluation. 104 ESM. Div. Ecol. Serv., Washington, D.C.

U.S. Fish and Wildlife Service. 1981. Standards for the Development of Habitat Suitability Index Models. 103 ESM. Div. Ecol. Serv., Washington, D.C.

U.S. Geological Survey. 1970. Potential Natural Vegetation of the United States (map). In The National Atlas of the United States of America. Washington, D.C.

Usher, M.B. 1975. Analysis of pattern in real and artificial plant populations. J. Ecol. 63:569-586.

Verry, E.S. and D.H. Boelter. 1979. Peatland hydrology, p. 389-402. In P.E. Greeson, J.R. Clark, and J.E. Clark (Eds.), Wetland Functions and Values: The State of Our Understanding. American Water Resources Association, Minneapolis, Minnesota.

Warner, M.L., J.L. Moore, S. Chatterjee, D.C. Copper, C. Ifeadi, W.T. Lawhon, and R.S. Reimers. 1974. An Assessment Methodology for the Environmental Impact of Water Resources Projects. EPA Rep. 600/5-74-016. Washington, D.C.

Weller, N.W. 1981. Freshwater Marshes. University of Minnesota Press, Minneapolis, Minnesota.

Weitzel, R.L. (Ed.). 1979. Methods and Measurements of Attached Microcommunities. Amer. Soc. Testing and Materials, Philadelphia, Pennsylvania.

Westman, W.E. 1970. Mathematical models of contagion and their relation to density and basal area sampling techniques, p. 515-536. In G.P. Patil, E.C. Pielou, and W.E. Waters (Eds.), Statistical Ecology, Vol I., Spatial Patterns and Statistical Distributions. Pennsylvania State Univ. Press, University Park, Pennsylvania.

Westman, W.E. 1980. Gaussian analysis: identifying environmental factors influencing bell-shaped species distributions. Ecology 61:733-739.

Whitlatch, E.E. 1976. Systematic approaches to environmental impact assessment. Water Res. Bull. 12:123-138.

Whittaker, R.H. 1956. Vegetation of the Great Smokey Mountains. Ecol. Monogr. 26:1-80.

Whittaker, R.H. 1960. Vegetation of the Siskiyou Mountains, Oregon and California. Ecol. Monogr. 30:279-338.

Whittaker, R.H. 1965. Dominance and diversity in land plant communities. Science 147:250-260.

Whittaker, R.H. 1967. Gradient analysis of vegetation. Biol. Rev. 42:207-264.

Whittaker, R.H. 1970. Communities and Ecosystems. 1st ed. Macmillan, New York.

Whittaker, R.H. 1972. Evolution and Measurement of Species Diversity. Taxon. 21:213-251.

Whittaker, R.H. 1975. Communities and Ecosystems. 2nd ed. Macmillan, New York.

Whittaker, R.H. 1975. The design and stability of plant communities, p. 169-181. In W.H. van Dobben and R.H. Lowe-McConnell (Eds.), Unifying Concepts in Ecology. Junk, The Hague.

Whittaker, R.H. 1978. Direct gradient analysis, p. 7-50. In R.H. Whittaker (Ed.), Ordination of Plant Communities. 2nd ed. Junk, The Hague.

Whittaker, R.H. (Ed.). 1978. Ordination of Plant Communities. 2nd ed. Junk, The Hague.

Whittaker, R.H. and S.A. Levin. 1977. The role of mosaic phenomena in natural communities. Theor. Popul. Biol. 12:117-139.

Whittaker, R.H. and H.G. Gauch, Jr. 1978. Evaluation of ordination techniques, p. 227-336. In R.H. Whittaker (Ed.), Ordination of Plant Communities. 2nd ed. Junk, The Hague.

Whittaker, R.H. and G.M. Woodwell. 1978. Retrogression and coenocline distance, p. 51.70. In R.H. Whittaker (Ed.), Ordination of Plant Communities. 2nd. ed. Junk, The Hague.

Whyte, R.O. 1976. Land and Land Appraisal. Junk, The Hague.

Williams, J.D. and C.K. Dodd, Jr. 1979. Importance of wetlands to endangered and threatened species, p. 565-575. In P.E. Greeson, J.R. Clark, and J.E. Clark (Eds.), Wetlands Functions and Values: The State of Our Understanding. American Water Resources Association, Minneapolis, Minnesota.

Wilhm, J.L. and T.C. Dorris. 1968. Biological parameters for water quality criteria. BioScience 18:477-481.

Williams, W.T., G.N. Lance, L.J. Webb, and J.G. Tracey. 1973. Studies in the numerical analysis of complex rain-forest communities; VI: models for the classification of quantitative data. J. Ecol. 61:47-70.

Winchester, B.H. 1981. Assessing ecological value of central Florida wetlands: a case study, p. 25-38. In R.H. Stovall (Ed.), Proceedings of the Eighth Annual Conference on Wetlands Restoration and Creation. Hillsborough Community College, Tampa, Florida.

Winchester, B.H. 1981. Valuation of coastal plain wetlands in the southeastern United States. In Progress in Wetlands Utilization and Management, Coordinating Council on the Restoration of the Kissimee River Valley and Taylor Creek-Nubbin Slough Basin. Orlando, Florida.

Winchester, B.H and L.D. Harris. 1979. An approach to valuation of Florida freshwater wetlands, p. 1-26. In D. Cole (Ed.) Proceedings of the Sixth Annual Conference on Wetlands Restoration and Creation. Hillsborough Community College, Tampa, Florida.

Woodin, H.E. and A.A. Lindsey. 1954. Juniper-pinyon East of the Continental Divide as Analyzed by the Line Strip Method. Ecology 35:474-489.

Woodwell, G.M. 1970. Effects of pollution on the structure and physiology of ecosystems. Science 168:429-433.

Word, D.L. (Ed.). 1980. Biological monitoring for environmental effects. Lexington Books, Lexington, Massachusetts.

WETLAND DYNAMICS: CONSIDERATIONS FOR RESTORED AND CREATED WETLANDS

Daniel E. Willard and Amanda K. Hiller
School of Public and Environmental Affairs
Indiana University

"Who can explain why one species ranges widely and is very numerous, and why another allied species has a narrow range and is rare? Yet these relations are of the highest importance, or they determine the present welfare, and, as I believe, the future success and modifications of every inhabitant of this world." (Darwin, 1859).

ABSTRACT. Wetlands are affected by intrinsic and extrinsic forces. Managers cannot always predict or control the extrinsic forces leading to wetland changes, nor can they predict the range of species adaptations to those changes. Managers should plan for extreme circumstances by including mechanisms for wetland adjustment and persistence and by maintaining multiple sites as refugia to spread the risk of catastrophe. Our recommendations include:

- creating buffer zones and corridors

- recreating spatial and temporal habitat variability

- maintaining marginal wetlands as reserve sites

- planning for worst case scenarios (cumulative impacts)

- suggestions for managing for uncertainty and risk in wetland restoration and creation.

INTRODUCTION

For the purpose of this chapter, we consider wetlands as ecosystems with water. As such they respond to biological, chemical and physical change and they demonstrate properties more or less common to other ecosystems. Many of these properties have been discussed at length in many textbooks of ecology. Here we will review several concepts that seem to have applicability to restoration and creation of wetlands.

This discussion focuses on the methods by which wetland ecosystems change and the methods by which plant and animal species adapt to those changes. To introduce the discussion we will briefly review succession, which we call "change by intrinsic elements" and perturbations, which we call "change by extrinsic forces". Because many wetlands are open ecosystems, these kinds of changes often operate together in ways that make them difficult to distinguish. For restoration and creation purposes, this separation helps managers understand what parts of the ecosystem they can manipulate, what parts change in fairly predictable ways and what parts remain uncertain.

Planning and regulation must consider landscape consequences. Often when we cannot understand the consequences of change at the local level, we can gain insight by looking at the change regionally, or vice versa. The role of a wetland in the landscape depends on its change pattern, its proximity to other habitat and its persistence. These considerations become particularly important as wetlands become stressed. That a wetland has become entangled in the permit process may provide sufficient evidence for stress.

EQUILIBRIUM, INTRINSIC ELEMENTS AND EXTRINSIC FORCES

Wetlands are dynamic systems. While some wetlands appear constant on a human-relevant time scale (e.g., bogs and fens), others change much more frequently. Generally, newly restored or created wetlands are not in equilibrium and may undergo a period of adjustment before functioning in socially and ecologically desirable ways. Some ecologists consider internal vegetation readjustments not as change, but rather as the maintenance of ecosystem equilibrium through homeostatic balancing. Urban et al. (1987) show that our perception of the significance of wetland responses changes as the scale from which we view changes. For them equilibrium is a function of geographic scale and may emerge if that scale is large enough. DeAngelis and Waterhouse (1987) state that wetland changes may occur regularly or irregularly and may result from internal or external causes. Each of these views attempts to understand the relation between the internal dynamics of ecosystems and external events that affect ecosystems.

Managers, restorers, and creators of wetlands act through extrinsic forces on wetlands. These extrinsic forces can cause modifications in the intrinsic elements so that the wetland provides functions which society values. Unfortunately the application of management to wetlands is not yet an absolute science, so outcomes vary. Unpredictable natural extrinsic forces such as drought, fire, or insect infestation further complicate the wetland manager's ability to predict the outcome of his or her project with confidence. Those who work in environments affected by anthropogenic sources of extrinsic forces, such as cities or heavily farmed land, should consider the sorts of wetland ecosystems which can tolerate these forces.

HIERARCHICAL SYSTEMS APPROACH

We discuss wetland creation and restoration considerations from the perspective of wetland dynamics. We take a hierarchical systems approach (which looks at a hierarchy of individualistic events) with particular emphasis on the landscape ramifications of wetland dynamics and management activities. Urban et al. (1987) defines "landscape as a mosaic of patches, each a component of a pattern". Landscape ecology focuses "on the wide range of natural phenomena by considering the apparent complexity of landscape dynamics and illustrating how a hierarchical paradigm lends itself to simplifying such complexity".

Random stochastic events play an extremely important role in the creation and development of wetlands. Habitat can only change in certain ways at certain rates. The dominant variables controlling a wetland frame the possibilities for wetland response to a stochastic event. Community responses to intrinsic elements limit the possible responses to extrinsic forces. In other words, if the wetland is in state X then a, b, and c are possible responses, but if it is Y then d, e, and f are possible. This ecological "roulette" confuses the outcome of any wetland change sequence. Managers are stuck with statements like, "If x doesn't happen and y happens the way it usually does, this wetland will look like this in 10 years." Annoyingly, x and y are uncontrollable.

LOCAL WETLAND CHANGE IN A REGIONAL CONTEXT

Andrewartha and Birch (1984) describe the differential effects of local and regional extrinsic events on populations and species. Local populations of plants and animals prosper and wane in response to the interaction between their population dynamics and the local environmental conditions. In wetlands subject to change, local populations may vary considerably from season to season and year to year. Contrarily, regional populations may remain relatively constant. This regional constancy results from the asynchrony of the separate local populations: when some have low levels, others are high.

Bertness and Ellison (1987) studied change patterns in New England salt marshes. Local small scale extrinsic events caused differential plant mortality. These short term disturbances changed the relationship between species, allowing those with greater colonizing ability to recolonize disturbances. This scrambled the areal distribution of the species and provided new patterns.

Most plants and animals which have evolved to exploit wetlands have adapted to the change patterns of wetlands. Plants and animals have considerable physiological and genetic adaptability. Clausen et al. (1940) showed with altitudinally separate populations of Phlox along an environmental gradient that ecotypic variability allows many plant species to use a variety of habitats. McNaughton (1966) showed that Typha latifolia varies somewhat within populations and broadly across its range. Typha can adjust its position along a gradient quickly through vegetative growth.

Animals adapt to changing wetland conditions through a variety of behavioral strategies in addition to the genetic and physiological mechanisms suggested above. These strategies include movement, dispersal, environmental change, or no action. Weller (1981) describes a variety of behavioral adaptations for freshwater species (e.g., muskrats, Ondatra; black terns, Chlidonias niger). Many species have adapted to wetland environments which combine both spatial and temporal heterogeneity. This heterogeneity allows internal adaptation.

Most species do not occupy all of their environment at any given time. Skeate (1987) demonstrated regular periodic movement by 22 bird species to exploit ripening fruit in northern Florida hammocks. These birds opportunistically actively foraged on only a small segment of their habitat at any one time. The portion of the habitat which the birds were not using at any time was relieved of predation until it had replenished the supply of fruit. Wiens (1985) shows the effect of local rainfall on establishing varied patches in the desert. Areas that receive rain at a particular time "reset" themselves and begin new intrinsic change sequences. In a single square kilometer several different sequences operate simultaneously. Switching from patch to patch allows more species of animal to exploit these local areas. This opportunistic habitat switching is a behavioral adaptation to changing conditions in the landscape. The survival of the animals depends to some extent on the continuing pattern of changes in the various patches included in their range.

Opportunistic habitat switching may lead to fluctuating population levels on a given wetland. Willard (1980) describes the resetting of small ephemeral farm ponds in Wisconsin and their subsequent use by migratory shorebirds. Generally, farmers drained the ponds on their farms every four or five years, but patterns varied. The shorebirds preferred ponds on the first and second wet year because the opportunistic invertebrates on which the birds prey had higher population densities during those years. The shorebirds exploited fairly specific environments by finding which pond provided that environment at a given time. Myers et al. (1987) discusses the importance of these changing patches of foraging habitats to transcontinental migratory shorebirds. They argue that even though the ruddy turnstones (Arenaria interpres) and red knots (Calidris canutus) used the wetlands of Delaware Bay only as stopover sites, these sites were essential to their survival. "On these stopovers they [the birds] doubled their weight in preparation for the last stage of their migration from South America to the Arctic."

For many of the species we have just discussed, the various patches of habitat are isolated from one another. Each small bit of habitat may be of essential importance to a species which uses it only irregularly, yet short-term observation of these wetland habitats may reveal little habitat value. The loss of these apparently low-value isolated sites may appear as a small loss of acreage but may instead constitute a large loss for the survival potential of some desirable species. This critical, but ephemeral habitat contributes to the difficulty of measuring the cumulative impacts of threats to these otherwise unimportant appearing wetlands. When Gosselink and Lee (1987) analyzed cumulative impacts in bottomland hardwoods, they expressed these impacts as a special case of hierarchical organization of the landscape. They considered that an area contains a pattern of optional patches that a population can use. The areas they described often contained watersheds. By analogy they described the collection of disjunct sites used by a duck as a "duckshed". This analogy helps understand the movement and flow of the population as it exploits available habitat within an area.

Lewin (1986) describes the necessary movement of African elephants (Loxodonta africana) within their large home range. Elephant foraging behavior destroys trees so that elephants must move continually. However, as they eat the canopy trees they open up the forest for the rapidly increased growth of new forage plants. The "elephantshed" must be big enough to allow this cycle to take place before the elephants revisit. The foraging activity of the elephants causes faster regrowth. In the absence of elephants, the forage plants become replaced by non-elephant food plants. The interaction of the elephant and its food plants maintains a persistent community which is composed of a fairly rapidly changing array of habitat patches. In this case the animal itself causes the local inconstancy of habitat, but is adapted to achieve a balanced ecosystem which allows both the animal and its food to survive. From this Lewin suggests that change can provide persistence for organisms and that if we attempt to manage for constancy we may ultimately fail to conserve the very resources we intended to save. He summarizes by urging that we manage for persistence, not constancy.

All change may not necessarily help wetland species. Many impacts, both natural and human-induced, are environmental changes of great magnitude and size. These extrinsic forces sometimes cause environmental change beyond adaptability of the species. Andrewartha and Birch (1984) explain the effect of local and regional environmental events on the range and success of species. They suggest that populations succeed best when local habitat patches change asynchronously. By the same reasoning, not all of the habitat patches change in a manner, or at a rate, beyond the ability of the species to adapt at the same time. In essence, each population distributes risk by using independently varying habitats, so that as some of these habitats become depleted others will improve. Plants and animals maintain themselves by using each site to its limit. Some regional extrinsic forces can affect all of the habitat sites of a species, but most affect only a portion of the species range. Those sites which remain habitable act as reserves to "spread the risk". As human activities remove portions of habitat, even though these parts seem marginal at the time, we may have critically reduced the species' ability to react to other perturbations, by removing these "reserve" or "emergency" habitats, and increasing the risk of population extinction.

An example may help tie these ideas together. The Kesterson (California), Stillwater (Nevada), and Malheur (Oregon) National Wildlife Refuges all provide wetland habitat for a variety of water birds. All occur in essentially desert habitats where they experience irregular cycles of drought and flood. Each refuge cycles differently. A population of White Pelicans (Pelecanus erythrorhynchos) uses several of these refuges alternatively. The number of individuals on each refuge varies from year to year as the population tries to exploit the best conditions. All these refuges (and probably the Great Salt Lake) constitute the "pelicanshed". All the refuges act to allow pelicans to "spread the risk".

In contrast the sandhill crane (Grus canadensis) populations, which winter on the gulf coast and nest in eastern Canada, all concentrate on the Jasper-Pulaski Wildlife Refuge in northern Indiana during spring and fall migration. Because wetland habitat in the midwest is so reduced, the cranes have few other choices. For several weeks each year the entire population is at risk on the same site. If that small refuge fails to provide resting and forage for the cranes because of drought, fire, or contamination the crane population will suffer severely. Wildlife managers have attempted to adapt other sites with some success, but have difficulty convincing politicians that it is wise to conserve habitat on the speculation that a species might need it briefly sometime.

The sandhill crane population during migration has only a single locality which all of the subpopulations use. An impact to this site affects the entire crane population. Since the pelican population is comprised of several scattered subpopulations, there is a greater likelihood that the population as a whole will survive environmental changes at a particular refuge. These wetlands taken together comprise a single pelican habitat unit. Should one of the several refuges suffer an impact which reduces its habitat value, the other refuges will still support viable populations. The removal of any one option lowers the survival ability of the pelicans by reducing their ability to distribute risk, but they will survive. However, for the cranes the Jasper-Pulaski Refuge and surrounding fields constitute the only option for migratory stopover. The loss of this site would cause the crane population considerable distress. The problem is real because currently each refuge has a water quality problem induced by agriculture, with the problems differing in kind and severity.

RESTORATION AND CREATION CONSIDERATIONS AND RECOMMENDATIONS

PERSISTENCE VS. CONSTANCY

In some cases the wetland complex survives because various portions of the system continually change from one type to another, but the sum of each habitat type more or less balances. This dynamic balancing, which may destroy a particular type of a subunit, also creates that type elsewhere in the wetland system. This principle of dynamic balancing is not new, but merely adds a temporal dimension to the concept of spatial heterogeneity. Simply stated, some wetlands persist by balanced change over time and space.

Resilience is a measure of the ability to persist in the presence of perturbations arising from weather, physiochemical factors, other organisms, and human activities. (Krebs 1978). Persistence is "continuous existence" (Webster 1970).

A conceptual difficulty arises until we clarify whether we wish to maintain any wetland at a specific locality, or whether we wish to maintain a specific kind of wetland at a specific locality. In many cases the goal requires that the site support a persistent bundle of wetland functions, not necessarily a wetland that looks a particular way. Because form and function are related, some wetland types do perform some functions better than other types. Operationally, the regulator has to balance form, function, and persistence.

When planning and managing a wetland system it is important to examine the goals of the project. A balance must be found between form, function, and persistence. For example, wetlands are often created for stormwater retention. These need to be persistent functional wetlands. In other cases maintaining specific habitat characteristics requires management practices emphasizing constancy.

Managed habitats can fail by becoming too constant, changing the wrong way, or changing too fast. Jacobson and Kushlan (1986) and Kushlan (1987) illustrate habitat sharing on an alternate time basis for endangered species in the Everglades National Park. These species require quite different seasonal water levels. Natural flood and drought cycles alternatively allowed manatees, crocodiles, woodstorks and other significant species high reproductive success. Each got a turn often enough to maintain viable populations. This was at least the situation before Florida irrigation and drainage projects limited the upper end of the gradient. Now animals using it can find no appropriate sites during high water years. Current regulated water use must recreate these variable sequences. Many organisms exist through their ability to adapt quickly. Habitat must vary if these species are to survive.

FREEBOARD, BUFFERS, AND CORRIDORS

Our management practices have hurt the habitat value of scattered wetlands by limiting the adaptability of individual wetlands. Individual wetlands have considerable but limited powers of recovery. Managers often place dikes or other structures in the flood plain so that wetlands cannot expand up gradient during high water periods, leading to limited adaptability. Artificial wetland boundaries often cut off the 100 or 50 year high water level of the wetland (e.g., Great Salt Lake). Wetland creation, restoration and other management plans must contain considerable "freeboard" for extra protection.

Unconfined wetlands allow vegetation to grow up and down the bank and adjust itself to changing hydroregimes. Steep landward banks may occur as a result of bulkheads on bars or beaches, levees on stream channels, or seawalls on lakes. The steep sides of confined systems remove the potential for adjustment and therefore force the loss of plant species, animals and habitat. Buffers are needed to avoid such losses. Buffers allow for the expansion and contraction of the plant communities to respond to variations in hydroperiod, and thus increase the probability of persistence.

Planning for environmental corridors helps reduce the isolation of the wetlands in or adjacent to the corridors, which in turn facilitates movement and recolonization of these wetlands. The corridors themselves can provide a variety of beneficial functions, as well as acting as buffers.

An increase in the area of a created wetland may be as simple as adding a modest buffer area and yet may have a large improvement in persistence. The buffer will allow the biota in the wetland to adapt to changing water levels. The buffer can also become additional refugia for the resident animals.

Little formal information exists on the definition and construction of buffer zones. But the rationale and need for buffers seem clear. Buffers provide an area of refuge for plants and animals between their normal or preferred habitat and human activities.

SPATIAL AND TEMPORAL HETEROGENEITY AND SIZE

Increased freeboard, corridors, and buffer areas add internal spatial and temporal heterogeneity to a wetland construction and restoration project. Because these forms of habitat may provide water connections between potential portions of the wetland, they can become reserve sites and refugia for aquatic plants and animals which disperse through water. All of these techniques increase the "effective" size of the site.

The effective size of a wetland includes the area of the wetland available and accessible to the plant and animal species of concern. In practice, the effective area for a species is measured by determining the interconnections available with the dispersal mechanisms used by that species. For example, if a manager wished to use topminnows (Poecilidae) for mosquito control, he or she should provide water connections for at least some portion of the year between the portions of the site that might allow mosquitos to breed. The best season for dispersal will vary depending on the climate and the natural history of the species.

Given the heterogeneity discussed above, larger wetlands invariably provide greater habitat value (Andrewartha and Birch, 1984). This occurs for two reasons. First, greater area provides additional spatial heterogeneity and consequent opportunity to spread the risk. Second, additional spatial heterogeneity over a larger area creates a more complex pattern. This pattern complexity allows for an increase in habitat diversity which supports and encourages greater species diversity (MacArthur 1972). In terms of genetic resilience, more species present on the site help guarantee that some species will adapt and survive if severe impacts occur. In this sense, species diversity spreads the risk.

MANAGING UNCERTAINTY

Two kinds of uncertainty simultaneously plague and bless careful managers. The first of these includes our inability to predict with certainty both natural and anthropogenic extrinsic impacts. The second includes the unpredictable behavioral adaptations that plants and animals sometimes make.

The poet Robert Burns once said that the best laid plans of mice and men oft times go astray. For those who attempt to manage natural resources this regularly applies. Things happen. In the midwest within the last year we have experienced a 17-year cicada (Magicicada septendecim) outbreak, record high wind conditions, and a prolonged summer drought. All of these have increased the danger from fire. While scientists predicted the cicada emergence, the event happens so infrequently that little was known about the effects of the cicadas on natural ecosystems. Thus, managed habitats may combine natural events which occur simultaneously through chance. This stochastic summing of independently varying impacts can over stress the adaptability of the species and the ecosystem even when the same impacts taken singly would cause changes well within the homeostatic limits of the ecosystem.

A third subtle uncertainty frustrates wetland restoration efforts. Wetlands often act as sinks for waterborne contaminants. Many accumulate potential pollutants while they act to clean water. These contaminants can build up to levels which harm wetland plants and animals.

Our land use management practices have hurt the habitat value of many scattered wetlands by creating habitats which contain threats animals cannot detect. Animals lack the ability to detect many fatal pollutants such as DDT, selenium, mercury, and PCBs. A wetland with valued habitat characteristics, but accumulated contaminants may lure animals to their death. For example, wetlands which receive high Habitat Evaluation Procedure (HEP) ratings may have contaminants present. These contaminants may occur at levels of concern while the site is otherwise physically and biologically attractive to the important species. Our study of the Kesterson National Wildlife Refuge may help us understand this problem.

Wetland evaluation systems seldom include detailed chemical analysis and even if evaluation considers the presence of chemical contaminants, the data are often not available to predict impact on particular species. For many contaminants, we know little about their effect on plants and animals. Chemical analyses are also costly (up to $1000/sample) and often raise questions which defy easy solution.

In some parts of the United States, wetlands have become so reduced that little habitat remains and plant and animals quickly colonize even marginally appropriate sites. Thus, managers accidentally create or restore wetlands which can become attractive nuisances.

We know of no easy answer to the contaminant problem. If the situation indicates that such a problem might exist (e.g., an abandoned landfill, an agricultural wastewater sump, an area which receives stormwater runoff from urban or industrial sites, the prudent manager should get some samples analyzed for potential pollution problems. Site restoration may require pollutant removal or abatement. Unfortunately, this may involve hazardous waste control agencies and years of cost and delay. Due to the different responsibilities and training of agency personnel, habitat assessments rarely consider pollutant hazards and chemical analyses seldom contain wildlife risk estimates. Improved agency coordination and an interdisciplinary approach to wetland evaluation could help identify these problem areas.

To counter the potential for adverse and unpredictable impacts, wetland restoration and creation projects could be designed and planned for a variety of unknown worst cases. No simple recipe exists to assure the health and survival of the wetland, but applying a few guidelines taken from our discussion may help:

1. Understand the natural history of the individual species of interest.

2. Provide extra space and diverse conditions to improve the odds for survival of a variety of species.

3. Provide spatial heterogeneity which usually increases with size and gradient.

4. Plan reserve sites as refugia when possible. Many species of plants and animals require reserve sites. HEP analysis can often recognize candidate reserve sites even though the population levels of the target species may be low at the time.

5. Imagine the consequences of potential adverse impacts. Try to design safeguards against these.

6. When the potential for chemical contamination exists, attempt to get an analysis of the risk to the plants and animals that may use the site.

DISCUSSION OF RECOMMENDATIONS

Assume that all wetlands naturally change in size, in community structure and in locality. That is, they get bigger and smaller; they become different sorts of habitats and, from time to time, wetlands appear on the landscape and disappear. Wetlands probably maintain some regional dynamic balance, but this balance is not precise.

It is difficult to design for precise wetland types and boundaries. They both change regularly. Design for a general type which contains elements for self regulation and maintenance. Regulators should set standards for long term monitoring that demand "persistence" not "constancy" for restored and created wetlands.

Don't get concerned if the restored or created wetland does what it will, as long as it does something (Lewis, pers. comm. 1987). Remember that the homeostatic strength of any wetland or habitat cluster causes the system to adjust to some external perturbations and not others. Most landscape units develop as a result of probabilistic external events which no one can control.

Encourage designs which include room for change in size and type. Hydraulic gradients allow greater spatial heterogeneity to develop. Climatic variation may cause vegetation to self-adjust higher or lower on the gradient. Rare species of plants and animals may require more rigid designs. Species achieve rarity in several ways. Some exploit specialized local niches which occur infrequently in the landscape (Yucca Night Lizards, Xantusia vigilis, Limpkins, Aramus guarauna; Southern Cougar, Felis concolor). Occasionally these species become locally abundant. Other rare species scatter widely and exploit ephemeral circumstances that provide food, reduced competition, or predation (White Pelicans). The former depend on persistent local conditions which wetland builders have difficulty constructing. These local habitats may represent the species only chance. The latter fall prey to incremental habitat destruction because each part of their habitat is scattered over a large region and is ephemeral.

We recommend that managers plan for the worst combination of events the wetland they are restoring or creating may encounter. Natural causes may force animals into marginal habitats as refuges. These then become essential. The current population of animals may be at a low and will use this habitat later. An animal population may use a currently marginal or unused site as a reserve under different unpredictable circumstances.

All wetlands have some public value, though sometimes these values are not readily apparent. Some disturbed wetlands have reduced functional value. Undisturbed naturally functioning wetlands are essentially irreplaceable in the short run.

Expect change in size, wetness and ecosystem type in created and restored wetlands. Therefore, they should be designed well "oversize" compared to the wetlands for which they compensate.

Unfortunately, the natural variability of ecosystems, the unpredictability of extrinsic forces, and the perversity of complex systems make any absolute guarantee of success impossible. Therefore, in addition to worst case and oversize design, most projects should include a monitoring program geared to some explicit but flexible and realistic goals. To protect against the potential (if improbable) failure of the project, the project sponsors should be asked to provide, in every permit, some financial guarantee such as a long term bond. This bond should cover the cost of trying to repair the original effort or if necessary, trying again.

LITERATURE CITED

Andrewartha, H.G. and L.C. Birch. 1984. The Ecological Web: More on the Distribution and Abundance of Animals. University of Chicago Press.

Bertness, M.D. and A.M. Ellison. 1987. Determinants of patterns in a New England salt marsh plant community. Ecological Monographs 57(2):129-147.

Clausen, J., D.D. Deck, and W.M. Heisey. 1940. Experimental Studies on the Nature of Species. I. The Effect of Varied Environments on Western North American Plants. Carnegie Institution of Washington, Pub. 520, Washington, D.C.

Darwin, C. 1859. On the Origin of Species. Harvard University Press, Cambridge, Massachusetts (1964 ed.).

DeAngelis, D.L. and J.C. Waterhouse. 1987. Equilibrium and nonequilibrium concepts in ecological models. Ecological Monographs 57(1):1-21.

Gosselink, J.G. and L.C. Lee. 1987. Cumulative Impact Assessment in Bottomland Hardwood Forests. Center for Wetland Resources, Louisiana State University, Baton Rouge. LSU GEI-86-09.

Jacobson, T. and J.A. Kushlan. 1986. Alligators in natural areas: Choosing conservation policies consistent with local objectives. Biological Conservation 36:181-196.

Krebs, C.J. 1978. Ecology: The Experimental Analysis of Distribution and Abundance. Second Edition. Harper & Row, New York.

Kushlan, J.A. 1987. External threats and internal management: the hydrologic regulation of the Everglades, Florida, USA. Environmental Management 11(1):109-119.

Lewin, R. 1986. In ecology, change brings stability. Science 234:1071-1074.

MacArthur, R.H. 1972. Geographical Ecology. Harper & Row, New York.

McNaughton, S.S. 1966. Ecotype function in the Typha community type. Ecological Monographs 36:297-325.

Myers, J.P., R.I.G. Morrison, P.Z. Antas, B.A. Harrington, T.E. Lovejoy, M. Sallaberry, S.E. Senner, and A. Tarak 1987. Conservation strategy for migratory species. American Scientist 75:19-26.

Skeate, S.T. 1987. Interactions between birds and fruits in a northern Florida hammock community. Ecology 68(2):297-309.

Urban, D.L., R.V. O'Neill, and H.H. Shugart, Jr. 1987. Landscape ecology. BioScience 37:119-127.

Webster, N. 1970. Webster's New Twentieth Century Dictionary. World Pub. Cleveland.

Weller, M.W. 1981. Freshwater Marshes: Ecology and Wildlife Management. University of Minnesota Press.

Wiens, J.A. 1985. Vertebrate responses to environmental patchiness in arid and semiarid ecosystems. In S.T.A. Pickett and P.S. White (Eds.), The Ecology of Natural Disturbance and Patch Dynamics. Academic Press, New York.

Willard, D.E. 1980. Vertebrate use of wetlands. Proceedings of Indiana Water Resources Association Meeting.

RESTORATION OF THE PULSE CONTROL FUNCTION OF WETLANDS AND ITS RELATIONSHIP TO WATER QUALITY OBJECTIVES

Orie L. Loucks
Holcomb Research Institute
Butler University

ABSTRACT. Many wetlands and wetland restoration opportunities occur in the poorly drained headwaters of streams, along the stream floodplains, and at discharge points to larger water bodies. All of these are greatly changed by upland development that accelerates flows and increases the runoff pulse from headwater areas. In turn, the runoff increases scouring and transport of sediments, and subsequent deposition in or erosion of downstream wetland types. Successful restoration must consider how the hydrologic pulse may have been changed and whether pulse control measures can bring stream flows within a range consistent with historical development of downstream wetlands.

Comparison of spring versus summer loadings pulses indicates major differences in the seasonality of transport of excess nutrients into wetlands and downstream water bodies. The annual average loading is misleading, indicating that statements on wetland functions which ignore their role during pulsed events probably understate their significance in the landscape. Modelling studies of runoff and sediment transport suggest the combination of reduced soil exposures and restoration of wetland cover in temporary detention areas can produce major benefits in stream water quality. With additional parameters, quantitative estimates could be made of the cumulative impact of wetland restoration toward mitigation of flood peaks and the transport of sediment and toxic substances into adjacent aquatic systems.

At present, the general physical relationships between land use and hydrology provide only a guide to the prospective benefits we associate with investment in wetland restoration. They suggest how to evaluate tradeoffs in benefits and costs between a lower cost investment higher in the watershed (carried out over large areas), versus investment in a higher risk but potentially higher quality wetland restoration along the main stem or outlet of a drainage system. Although limited predictive capabilities exist for assessing the efficacy of restored wetlands, they have not been subjected to quantitative testing within the environment of wetland restoration technology. There is a need for more complete treatment-response modeling and model testing if the predictive capability needed to improve wetland restoration is to become available.

INTRODUCTION

Many studies have shown that impairment of rural (and suburban) surface water quality occurs during infrequent but unusually large storm events. These events produce pulses in runoff, leading to flood peaks, unusual levels of sediment transport or "washouts", and mobilization of contaminants from agricultural lands or industrial sites. Studies also have shown that hydrologic detention in wetlands, or other means of delaying the peak hydrologic response, provides important buffering in the transport of sediment and chemicals by reducing the size of the pulsed events. This chapter examines the basic principles of hydrologic response and material transport, emphasizing pulsed events, and considers the potential for improving water quality and aquatic ecosystem functions through restoration of pulse-control processes in wetlands.

To meet these objectives the paper reviews the relevant literature on wetland hydrology, sediment transport and nutrient detention. Historical or "reference ranges" in hydrologic or detention functions are emphasized as well as modern conditions, as a synthesis of both will contribute to understanding the potential for restoration. In landscapes where widespread alteration of hydrologic and related wetland functions has occurred, little possibility exists for consensus on a normal or reference state for wetland functions, particularly for processes that ameliorate unusual pulsed events. One important reference point, however, is the buffering associated with the original vegetation and wetlands of a watershed. In the long-run, the success of a newly created or restored wetland needs to be evaluated in relation to the capacity of even the most natural of remaining wetlands to function within present-day hydrologically altered watersheds.

The question of how to further reduce the effects of unusual hydrologic events, including the transport of foreign substances, is significant now because the Agricultural Stabilization Act of 1985 could withdraw up to 50 million acres of land from agricultural use. Implementation of this program is already being seen as significant for further wetland restorations. Although the set-aside programs (sodbuster and swampbuster) emphasize highly erodible soils, some of the alternate uses will permit the restoration of wetlands, or more likely restoration of the buffering capacity of vegetation and wetlands in the headwater areas of watersheds. Indeed, a potential exists to focus certain of the set-aside programs so that they greatly enhance the water quality improvement potential of wetlands in agricultural areas, thereby enhancing the recovery of wetland functions and downstream water quality.

PRINCIPLES EVIDENT FROM EXISTING STUDIES

DISTRIBUTION OF WETLANDS ON A LANDSCAPE

Although wetland types vary widely from one area to another across the landscapes of the United States, a very common wetland type, making up by far the largest part of wetland acreages in the central regions, are depressions that provide temporary detention of water. These wetlands are characterized simply by poorly drained soils (gleysols) and the presence of a number of wet-habitat indicator species. This wetland type is best illustrated by the shallow basin type described by Novitzki (1979), Figure 1. Frequently these shallow wetlands are drained for agriculture, and no longer function as wetlands. They are not usually protected by state or federal wetland programs, and most often are not subject to permitting or mitigation requirements. Indeed, although the classification by Cowardin et al. provides for the inclusion of such temporarily flooded wetland types within the palustrine wetland systems, and they can be mapped by following the distribution of gleysol soils, they are not well illustrated in the cross-sectional diagrams shown by Cowardin et al. (1979).

Despite the very large area of temporary detention in many watersheds, the withdrawal of shallow basin wetlands through ditching and drainage has greatly increased the rate at which water is discharged from the upland landscape into the remaining wetlands, streams and floodplains. The importance of this relationship is shown in Figure 2, where the stream hydrograph with good retention has a much lower flood peak and a longer discharge time than the stream without detention (Reppert et al. 1979). The presence of the drainage channel system increases the flood peak and greatly increases the potential for transport of substances into the aquatic environment (Carter et al. 1979, Novitzki 1979).

DETENTION IN NATURAL DRAINAGES

Studies in the Lake Wingra Basin around Madison, Wisconsin from 1969 to 1974 (Loucks and Watson 1978, Loucks 1981, Loucks 1986) have illustrated the hydrologic and nutrient detention potential of wetlands in natural drainage systems. One part of this study focused on a comparison of the modern hydrology of the Lake Wingra basin with the pre-settlement hydrology reconstructed from measurements on a subwatershed within the University of Wisconsin Arboretum. Data summarized by Prentki et al. (1977) have shown that the natural subwatershed (Marshland Creek, a large natural area of forest, prairie and shallow basin wetlands), yielded much less runoff than that characterizing watersheds of similar size in the suburban areas around the Arboretum. Runoff from the natural watershed occurred only during snowmelt when the soils were still partially frozen. Runoff occurred from the other watersheds during almost every significant rain event throughout the study period. The hydrology for the entire Lake Wingra basin, for both present and presettlement conditions, is summarized in Table 1. The runoff under current conditions, where shallow basin wetlands have been filled or drained, is about twice that estimated for the presettlement watershed. A related decrease of almost ten percent is shown for the inflow from springs and groundwater to the lake.

Because the concentrations of nitrogen and phosphorus are higher in runoff waters compared to groundwater, the hydrologic changes have resulted in large differences in the nutrient

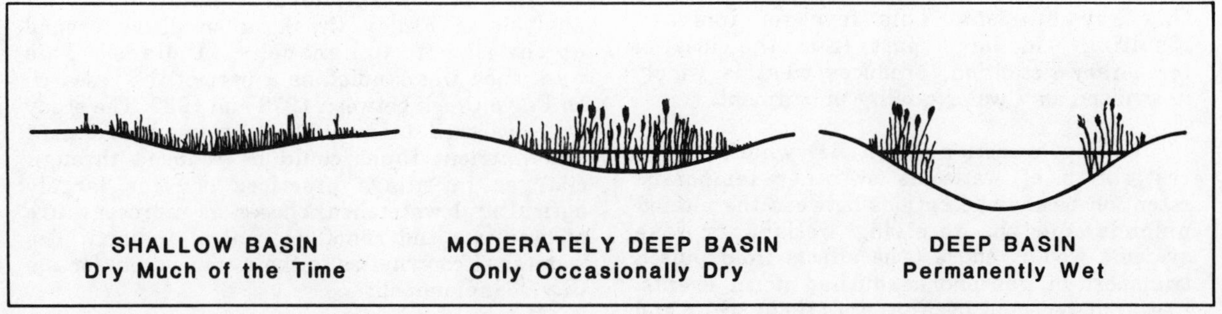

Figure 1. Different plant communities in surface water depression wetlands related to water permanence and basin depth (Novitzki 1979).

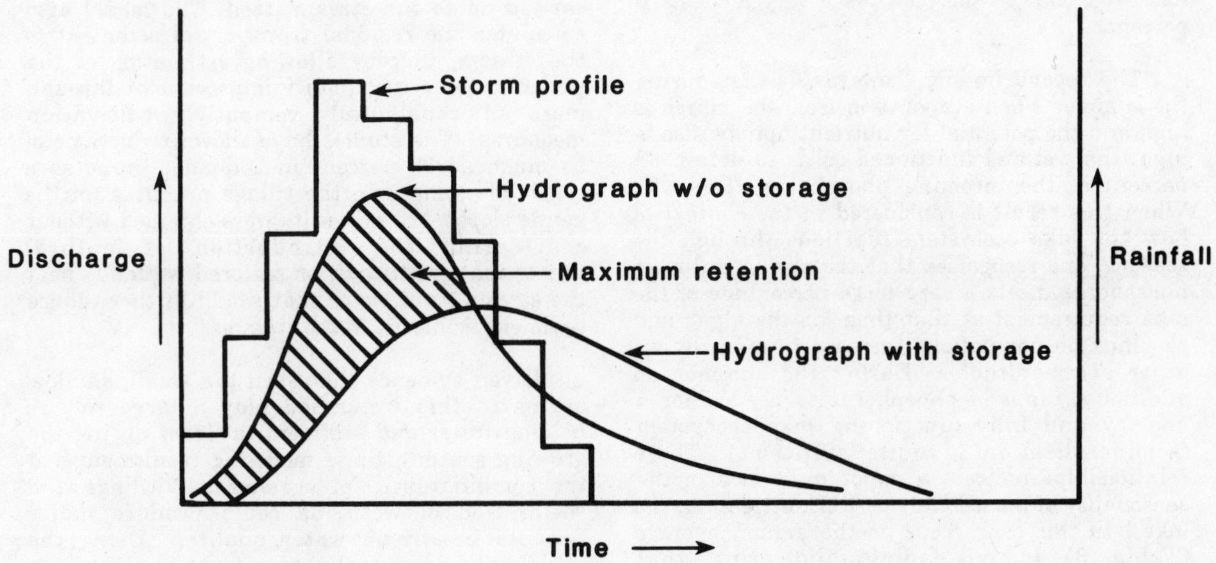

Figure 2. Flood hydrographs for watersheds with and without wetland storage capacity (from Reppert et al. 1979).

loadings to this lake, as summarized in Table 2. Nitrogen input has increased slightly due to the increased runoff. Phosphorus has much higher concentrations in runoff, and along with the increased proportion of surface runoff, yields an increase of almost two-thirds in the loading of this key nutrient. This level of increase, resulting in large part from the loss of temporary detention, produces what is widely recognized as a water quality impairment.

A key question then is whether the restoration of wetlands or other temporary retention basins at locations between the altered uplands and the receiving wetland or lake systems would reduce the effects from pulsed transport of phosphorus during storm events. Research reported by Flatness (1980), Huff and Young (1980), and Perry et al. (1981) provides an opportunity to answer this question. West Marsh, a small area remaining at the west end of Lake Wingra, is the receiving site for storm drainage from a golf course and urban area on the west side of the City of Madison, Wisconsin. Measurements of water inflow, and the net detention of water and nutrients, were carried out over a two year period (Flatness 1980). Two findings are significant: first, the largest part of the net flow through of water, sediment, and nutrients across the marsh and into the lake occurred during the spring runoff when the marsh was frozen (as was seen in the natural watershed in the Arboretum). The transport of water and nutrients is large enough during this time that the annual average detention achieved by the wetland, expressed on an annual basis (as has been done in the past), is a very modest 10 percent.

The second finding, however, is that during the summer when evaporation from the marsh is high and the potential for nutrient uptake also is high, the wetland functioned so as to detain 83 percent of the incoming phosphorus (Table 3). When this result is considered in the context of how the lake ecosystem functions through the seasons, one recognizes that the spring influx of phosphorus meets a very large percentage of the lake requirement at that time (as the algae and zooplankton populations increase in mass by an order of magnitude). During the summer no additional input of phosphorus is needed for a healthy and fully functioning lake ecosystem (remineralization is quite sufficient). These relationships indicate a major difference in the seasonality of pulsed transport of nutrients to the lake. In this case, focus on the annual average (Table 3) is misleading, indicating that statements on wetland functions that pass over their role during pulsed events will also understate their significance in the landscape.

MODELING AREAWIDE SEDIMENT SOURCES AND DETENTION

Consider another study of detention of sediment and water within the shallow basin systems of an agricultural landscape: an analysis of Finley Creek, a small watershed northwest of Indianapolis, Indiana. This watershed was studied as a part of the research on Eagle Creek between 1978 and 1982. The study was designed to evaluate how much sediment and nutrient input could be reduced through changes in tillage practices on this largely agricultural watershed, chosen as representative of land use and runoff in central Indiana. The watershed contributes to the water supply for the City of Indianapolis.

A capability for evaluating runoff, transport, and deposition of sediment within a small watershed was developed by Beasley and others at Purdue University (Beasley et al. 1985, Huggins et al. 1982, Lee et al. 1985.) Their work led to the ANSWRS model (Areal Nonpoint Source Watershed Response Simulation). The model is, in effect, a fine grid geo-information system that allows two-dimensional calculation of the proportion of runoff from each area in the watershed. Given the volumes of water from each area and the rates of flow (from the runoff and slopes), the amount of sediment mobilized from the eroding surfaces can be calculated. These calculations then allow estimation of the sediment subsequently deposited in the temporary detention depressions (shallow basin wetlands) in the watershed (see Figure 3), or carried on to the stream itself. The model also calculates the residual transport of sediment to the stream, thereby allowing estimation of the potential for water quality improvement through more environmentally compatible cultivation measures. The studies have shown reductions of as much as 36 percent in sediment input as a result of changes in the tillage practices on the gentle slopes. These results were obtained without considering the introduction of natural vegetation, greenways, or restored wetlands near the stream itself, steps that would further reduce sediment transport to the streams.

Given evidence that even the small, shallow basins in this watershed play a large role in holding water and sediment (at least during the growing season), these modeling results suggest the combination of reduced tillage and restoration of wetlands could produce major benefits in stream water quality. Using the model with additional parameters (for the effects of increased permanent green cover), an estimate can be made of the areawide <u>cumulative impact</u> of wetland restoration toward the mitigation of flood peaks and transported sediment and associated chemicals.

Table 1. A comparison of hydrologic inputs for a modern and predisturbance watershed, in volumetric and percent-of-total forms. (From Loucks and Watson 1978).

	Hydrologic Input (10^3 m^3/yr)			
Source	Present	Percent	Presettlement	Percent
Rainfall	920	15	920	15
Runoff	990	16	450	7
Springs and Groundwater	430	69	4500-4700	78
TOTALS	6200	100	5900-6100	100

Table 2. Estimated present and presettlement loadings, Lake Wingra. (From Loucks and Watson 1978).

	Nitrogen Loading (kg yr^{-1})		Phosphorus Loading (kg yr^{-1})	
Source	Present	Presettlement	Present	Presettlement
Rainfall	1200	1200	34	34
Runoff	5200	3000	710	320
Springs and Groundwater	18000	18000-19000	160	170-180
Dryfall	1900	1800-1900	95	92-95
N Fixation	3800	(assume 3800)	--	--
TOTALS	30000	28000-29000	1000	620-630

Table 3. Phosphorus mass balance for Wingra runoff water, August 1975-August 1976 (From Livingston and Loucks 1978).

	Dissolved Reactive Phosphorus			
Season	Input (kg/day)	Output (kg/day)	Retention (kg/day)	Percent
Autumn (14 August - 13 December)	0.380	0.318	0.061	16
Winter (14 December - 10 February)	1.232	1.231	0.002	1
Spring (11 February - 7 May)	3.862	3.553	0.309	8
Summer*	0.145	0.024	0.120	83
Annual*	1.282	1.155	0.127	10

*Assumes 100 percent infiltration for summer except 15 May and 24 June runoff event.

RESULTS FROM EXPERIENCE IN LAKE RESTORATION

Research on lake restoration (Cooke et al. 1986), related to the lake/watershed studies cited previously, also should be considered for its potential relevance in a synthesis of mitigative measures for wetlands. The principles of lake restoration have matured over the past fifteen years, and are beginning to be viewed as a predictive science, despite our recognition that virtually no two lakes are alike or will respond similarly to the same treatment. The investigations used to design and evaluate the prospective response of lakes to restoration measures show that, before mitigation can be expected to meet goals set for a given lake, key characteristics of that system need to be understood. These include knowing the entire hydrology of the lake and watershed, particularly flushing time (i.e., the time required for the incoming water volume to equal the volume of water in the lake or wetland), the annual and seasonal net loading of sediment and nutrients (expressed in terms of equivalent weight per unit area of lake), and seasonal hydrologic fluctuation, shoreline aeration, and related questions of sediment toxicities. Let us consider how the lake restoration procedures and principles would apply to evaluating the potential for success in wetland restoration.

Prior to restoration, the essential principles for wetland evaluation need to address the hydrology of the wetland, just as it would be addressed in evaluating the restoration of a small lake. Included are questions of the total magnitude of the hydrologic inflows and outflows, from which can be calculated an apparent flushing time. The flushing-time term already is recognized as the dominant factor in coastal wetlands where tidal processes produce daily flushing (de la Cruz 1978). Freshwater wetlands, on the other hand, experience flushing principally during unusual runoff events, although not all freshwater wetlands experience flushing. In addition, just as for lakes, the relative importance of surface water input as opposed to groundwater input, and how the balance of these inputs controls the resultant chemistry of the wetland, are essential to understanding the biological community that can be sustained on a restored wetland.

Related to these questions are the seasonal and longer-period hydroperiods for the sites being considered for restoration. Although we are accustomed to dealing with annual averages, the key characteristics of the wetland and the associated water courses are often dominated by pulsed events only partially mediated by the system as a whole--as the examples cited earlier in this chapter illustrate.

For the annual hydroperiod of wetlands, saturation is expected during the winter or spring, followed by a major drying down in most parts of the country during the summer. Here, as we saw earlier, one needs to consider a reference

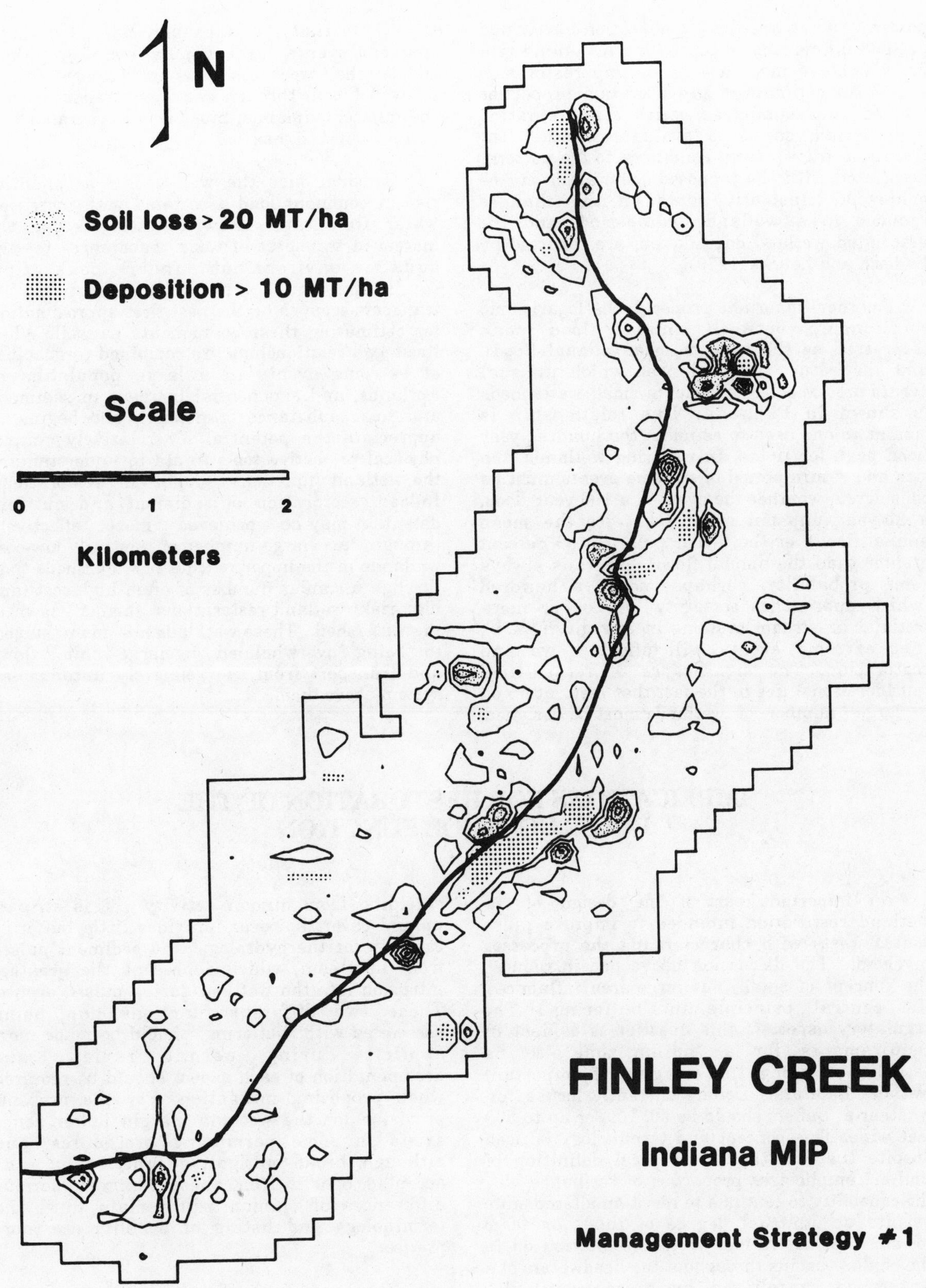

Figure 3. Sediment yield contour map of Finley Creek (from Huggins et al. 1982).

pattern: What was the pre-alteration fluctuation pattern influencing or controlling a wetland type for which we may now be setting restoration goals? An estimate of pre-alteration properties should be compared with the existing hydroperiod, so as to evaluate whether the transition from present conditions to the patterns associated with the proposed restoration can be achieved. Illustrative data on the changes induced in a wetland hydroperiod, and the associated wetland degradation, are reported by Bedford and Loucks (1979).

Another important property, the logarithmic relationship between annual flood peak (expressed as the ratio to mean annual flood), and the return interval over which unusual events are expected to recur in small watersheds is shown in Figure 4. This relationship is similar to one used to estimate the hundred-year flood peak for rivers. In restoring wetlands, the size and return-period of extreme events must be considered, whether the event is a 100-year flood, a 20-year unusual event (at twice the mean annual flood), or the 5-year pulse (at 30 percent greater than the annual flood). There is always some probability, perhaps one in a hundred (which approaches certainty for 100 or more wetland or stream systems in any given year), that extreme events will affect a wetland restoration. The significance of return-time considerations lies in the fact that restoration on a large number of wetlands must be designed for events that are unusual locally, but fairly frequent over a large population of wetlands. Indeed, the threat of serious erosion of wetlands, or even burial through sediment transport from the adjacent uplands, must be incorporated into restoration designs.

Consider also the well-known exponential rise in sediment load associated with increased water flow rates. Few studies discuss the increased transport of toxic substances to the aquatic environment under peak flow conditions, but exponential increases in transport seem to be the best first approximation for estimating these components as well. When these two relationships are combined (probability of extreme events in a large population of wetlands, and exponential increase in sediment and toxic substance transport), one begins to appreciate the potential for relatively simple, physical predictive tools to aid in understanding the wetland functions sought in restorations. Indeed, restoration of sediment and nutrient detention may be achieved more effectively through a large number of relatively low-cost wetlands in the upper reaches of watersheds than through a similar number of often high-cost (and high-risk) wetland restorations farther down in the watershed. These wetlands are more subject to being overwhelmed by large-volume flows and transport from the relatively uncontrolled system above them.

IMPLICATIONS FOR RESTORATION OF THE PULSE CONTROL FUNCTION

An important part of the design of a wetland restoration intended to improve pulse control rests with characterizing the processes involved. The discussion above has introduced the concept of small, but large-area influences, the central principle in "buffering". The regulatory aspect of this question is evident in requirements for a "buffer zone" to be established along the margins of important wetland habitats. Debate currently focuses on whether a "buffer" should be 50, 100, or up to 300 feet wide. This concept and terminology is used despite the fact that a technical definition of "buffer" emphasizes properties of resilience, i.e., the capability to continue to resist an altered state despite an unusual degree of input or force toward that alteration. Wetlands distributed in the shallow basins throughout the headwaters of a watershed, unrelieved by present-day tile-drainage and channeling, provide buffering capability in the original technical sense. On the other hand, a strip of land retained around the edge of a wetland provides important habitat for species that may utilize the wetland border as shielding from human activity. This strip of upland cover, however, provides little buffering capacity for the hydrologic and sediment pulses from upstream, which represent the greatest intrusion into the wetland during pulsed events. These two different functions now being associated with "buffering" should be made more explicit during permit review, and documentation of each aspect should be required when proposing mitigation. For the present, provision for the wetland margin buffer zones exists in some permitting procedures, and although broad mitigative functions are an accepted benefit from these reserves, serious differences of opinion will remain until the terminology and listing of benefits are more precise.

Given that unbuffered events impacting wetlands can be many times larger than average peak flow conditions, designs for an optimal buffering capacity to control man-induced peaks should seek a severalfold reduction in the size of these fluctuations. A reduction in the size and frequency of these events, expressed in clearly measurable standards (e.g., as in Figure 4), should be articulated as a performance criterion for wetland restoration. Such measures should then be incorporated into design criteria for restoration projects. In the examples considered here, one reference point for performance criteria is the magnitude of the extreme-event pulses in natural systems, while the magnitude of the extreme-event pulses characteristic of urban watersheds is an opposing reference point. Major pulse reduction can be expected downstream when restoration is carried out close to the headwaters of small drainage systems, but water quality benefits will be expressed locally as well as downstream. However, as more and more of the area of a drainage basin loses this "headwater" buffering capacity, the more extreme are the pulsed events midway or lower in the watershed and the more difficult is restoration or protection of wetland functions in those areas.

These general physical relationships can provide only a guide to the expectations likely to be associated with investment in wetland restoration. However, they also are a means for guiding estimates of the long-term survivability of a restoration project, at least in relation to the return-time of extreme events that could negatively impact the restoration. These physical relationships also suggest how we might evaluate the tradeoff in benefits and costs. In its simplest form, the tradeoff is between a lower-cost investment higher in the watershed (little more than small area set-asides), versus investment in a higher risk (but potentially higher quality) wetland restoration along the main stem or outlet of a drainage system. Additional research is needed before fully quantitative expressions of these tradeoffs can be formulated.

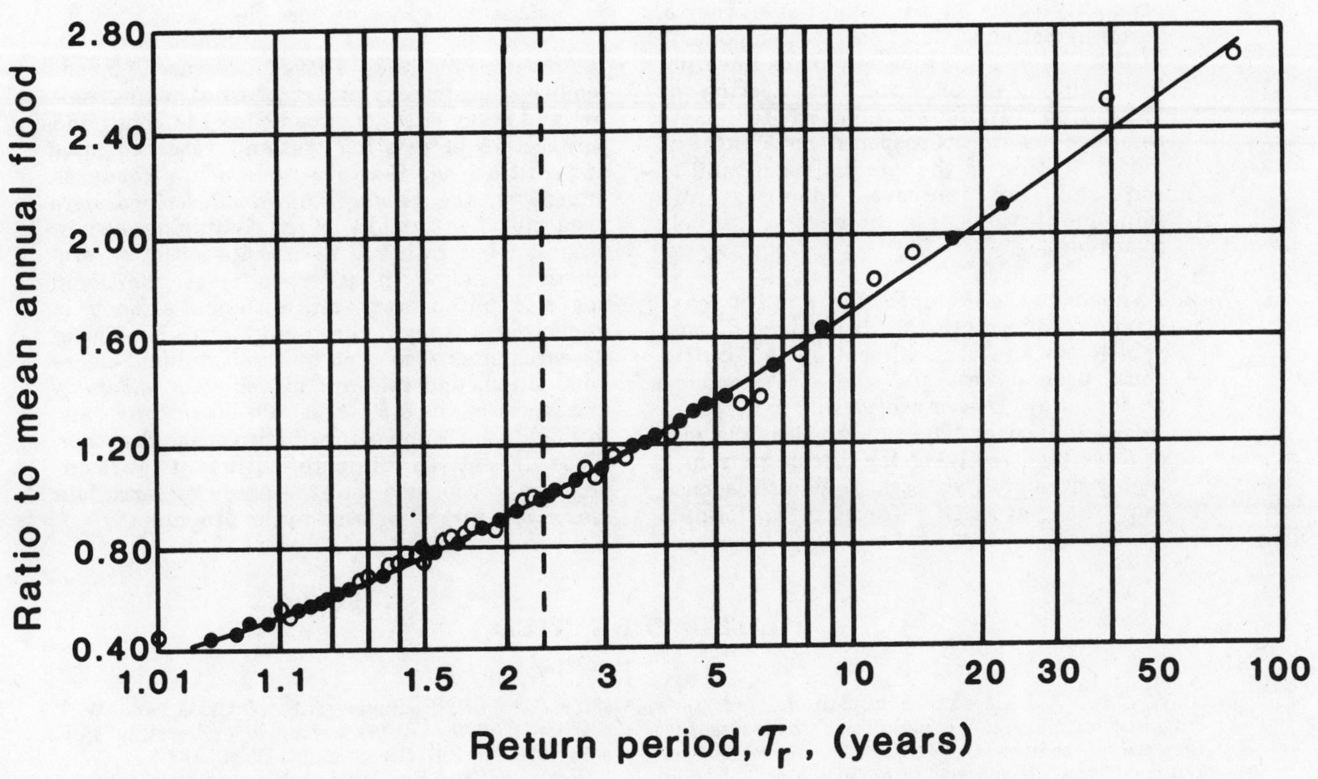

Figure 4. Regional-flood-frequency curve for selected stations in the Youghiogheny and Kiskiminetas River basins (Pennsylvania and Maryland), showing the return time for pulsed events expressed as a ratio to mean annual flood. A reduction in the hydrologic detention of wetlands raises the slope of this curve (U.S. Geological Survey, from Linsley et al. 1975).

RESEARCH NEEDS

While wetlands function in many important ways, and while technology exists for proceeding with confidence on a moderate number of restoration goals, great limitations exist in our understanding of the variability in response among sites. These limitations restrict our ability to predict the outcomes of restoration measures. Priority research needs for improving our understanding of risks and benefits in wetland restoration, particularly during pulsed events, can be addressed under the following four headings:

1. A need exists for additional case studies of hydrologic and water quality responses to the distribution of wetland buffering capacity under a wide variety of weather conditions, soils, and topography. These studies should include an increased emphasis on the role of wetlands in intercepting and retaining nonpoint source pollutants, including toxic substances mobilized from adjacent uplands.

2. Although some quantitative predictive capability exists for assessing the efficacy of restored wetlands, or of specific components in the restoration processes, these have not been subjected to significant testing or validation. There is a need for more complete treatment-response modeling and model testing if the predictive capability required for improved designs and implementations is to become widely available.

3. A need has developed for specific case studies regarding the benefit and cost relationships from allocating a "buffer zone" around wetlands where development is underway. How much benefit is achieved from a 100-foot or 300-foot strip along the edge of a wetland receiving the discharge from a highly developed watershed? What response can be expected from a restoration and rehabilitation under such conditions? A tradeoff may exist between values in the 300-foot strip, some of which may not be of great benefit to the wetland, and a comparable investment at strategic locations in headwater areas. This research requires predictive capabilities for both the border or edge benefits of the buffer zone, as well as the pulse-control processes.

4. Finally, research should evaluate the hierarchical relationships among hydrologic and nutrient pulse control functions, as one proceeds from the small watersheds that we are beginning to understand to the larger watersheds and landscapes where water quality questions are paramount. Do new principles come to bear? Can a geoinformation system such as the ANSWRS model be extended to adjacent watersheds to assess cumulative impacts from the restoration of a pattern of wetlands across the landscape? The tools appear to be available to address these questions, and they should be evaluated as soon as possible.

Some of the above research needs require a strategy either of establishing new wetland study sites, or broadening the depth and perspective of existing wetland research sites and initiatives. Because so much already is underway, the research needs described here need not be thought of as requiring a new program, but rather a re-articulation of several existing studies in order to meet additional needs. Limited work on wetlands already is underway through the Long-Term Ecological Research sites sponsored by the National Science Foundation, and relevant modeling is underway at several of the EPA research laboratories and through their supporting institutions. Together, these initiatives indicate sufficient work-in-progress to conclude that the above research is a reasonable target for a five-year program.

LITERATURE CITED

Beasley, D.B., E.J. Monke, E.R. Miller, and L.F. Huggins. 1985. Using simulation to assess the impacts of conservation tillage on movement of sediment and phosphorus into Lake Erie. J. Soil Water Conserv. 40(2):233-237.

Bedford, B.L. and O.L. Loucks. 1979. Changes in the Structure, Function, and Stability of a Wetland Ecosystem Following a Sustained Perturbation. Progress report on 1978 research. Water Resources Center, Univ. of Wisconsin-Madison.

Carter, V., M.S. Bedinger, R.P. Novitzki, and W.O. Wilen. 1979. Water resources and wetlands, p. 344-376. In P.E. Greeson, J.R. Clark, and J.E. Clark (Eds.), Wetland Functions and Values: The State of Our Understanding. Proceedings of the National Symposium on Wetlands, American Water Resources Association, Minneapolis, Minnesota.

Cooke, G.D., E.B. Welch, S.A. Peterson, and P.R. Newroth. 1986. Limnology, lake diagnosis, and selection of restoration methods, p. 9-51. In G.D.

Cooke, E.B. Welch, S.A. Peterson, and P.R. Newroth (Eds.), Lake and Reservoir Restoration, Butterworth Publishers, Boston, Massachusetts.

Cowardin, L.M., V. Carter, F.C. Golet, and E.T. LaRoe. 1979. Classification of Wetlands and Deepwater Habitats of the United States. U.S. Department of the Interior, Fish and Wildlife Service, Washington, D.C.

de la Cruz, Armando A. 1978. Production and transport of detritus in wetlands. In P.E. Greeson, J.R. Clark, and J.E. Clark (Eds.), Wetland Functions and Values: The State of our Understanding. Proceedings of the National Symposium on Wetlands, American Water Resources Association, Minneapolis, Minnesota.

Flatness, D.E. 1980. Particulate Phosphorus-Availability in Tributaries of the Great Lakes and Removal in a Marsh System. M.S. Thesis, Water Chemistry Program, University of Wisconsin, Madison, Wisconsin.

Huff, D.D. and H.L. Young. 1980. The effect of a marsh on runoff: I. A water budget model. J. Environ. Qual. 9:633-640.

Huggins, L.F., D.B. Beasley, D.W. Nelson, T.A. Dillaha, III, D.L. Thomas, C. Heatwole, E.J. Monke, R.A. Dorich, and L.A. Houston. 1982. NPS pollution: evaluating alternative controls by simulation and monitoring, p. 5-1 - 5-80. In A.H. Preston (Ed.), Insights into Water Quality--Final Report. Indiana Heartland Model Implementation Project.

Lee, J. Gary, S.B. Lovejoy, and D.B. Beasley. 1985. Soil loss reduction in Finley Creek, Indiana: An economic analysis of alternative policies. J. Soil Water Conserv., January-February 132-135.

Linsley, R.K., M.A. Kohler, and J.L.H. Paulhus. 1975. Hydrology for Engineers. McGraw-Hill Series in Water Resources and Environmental Engineering. McGraw-Hill Book Company.

Livingston, R.J. and O.L. Loucks. 1978. Productivity, trophic interactions, food-web relationships in wetlands and associated systems, p. 101-119. In P.E. Greeson, J.R. Clark, and J.E. Clark (Eds.), Wetland Functions and Values: The State of Our Understanding. Proceedings of the National Symposium on Wetlands, American Water Resources Association, Minneapolis, Minnesota.

Loucks, O.L. 1981. The littoral zone as a wetland: Its contribution to water quality, p. 125-138. In B. Richardson (Ed.), Selected Proceedings of the Midwest Conference on Wetland Values and Management, June 17-19, 1981.

Loucks, O.L. 1986. Role of basic ecological knowledge in the mitigation of impacts from complex technological systems: agriculture, transportation and urban. In O.L. Loucks (Ed.), Proceedings of a Conference on Long Term Environmental Research and Development. Council on Environmental Quality (CEQ), Washington, D.C.

Loucks, O.L. and V. Watson. 1978. The use of models to study wetland regulation of nutrient loading to Lake Mendota, p. 242-252. In C.B. DeWitt and E. Soloway (Eds.), Wetlands, Ecology, Values, and Impacts. Proceedings of the Waubesa Conference on Wetlands. University of Wisconsin-Madison, Inst. for Env. Studies, Madison, Wisconsin.

Novitzki, R.P. 1979. Hydrologic characteristics of Wisconsin's wetlands and their influence on floods, stream flow, and sediment, p. 337-388. In P.E. Greeson, J.R. Clark, and J.E. Clark (Eds.), Wetland Functions and Values: The State of Our Understanding. Proceedings of the National Symposium on Wetlands, American Water Resources Association, Minneapolis, Minnesota.

Perry, J.J., D.E. Armstrong, and D.D. Huff. 1981. Phosphorus fluxes in an urban marsh during runoff, p. 199-211. In B. Richardson (Ed.), Selected Proceedings of the Midwest Conference on Wetland Values and Management, Freshwater Society, Navarre, Minnesota.

Prentki, R.T., D.S. Rogers, V.J. Watson, P.R. Weiler, and O.L. Loucks. 1977. Summary tables of Lake Wingra Basin data, p. 89. In Univ. of Wisc. Institute for Env. Studies, Report 85, December 1977. Madison, Wisconsin.

Reppert, R.T., W. Sigleo, F. Stakhiv, L. Messuram, and C. Meyers. 1979. Wetland Values: Concepts and Methods for Wetlands Evaluation. Institute for Water Resources, U.S. Army Corps of Engineers, Ft. Belvoir, Virginia.

VEGETATION DYNAMICS IN RELATION TO WETLAND CREATION

William A. Niering
Connecticut College and Connecticut College Arboretum

ABSTRACT. An understanding of the ecological processes involved in wetland vegetation development is essential to wetland managers concerned with wetland creation. Ascertaining a sound hydrologic system is basic in any attempt to re-create a wetland system since the vegetation and associated fauna are dependent upon a consistent but usually fluctuating hydrologic regime. Any hydrologic manipulations can also greatly modify what species will become established in a given site or those that may decline in abundance. Traditional succession-climax dogma has limited usefulness in interpreting vegetation change. Thus an understanding of the complex of factors involved in the process, including chance and coincidence, makes the task of the manager even more challenging.

Since vegetation change is not necessarily predictable and orderly, as is sometimes thought, it is often difficult to predict the ultimate vegetation in a given created site. Some wetland communities once created may be relatively stable; others may undergo directional or cyclic change, thus adding to the complexity of the ultimate vegetation.

Therefore, one of the major goals in wetland creation should be the persistence of the wetland as a self-perpetuating oscillating system. This can be achieved by assuring a sound hydrologic regime.

INTRODUCTION

In wetland creation it is important to understand the patterns and processes involved in vegetation or biotic change. Such ecological concepts as succession and climax usually come to mind in this regard. However, depending upon one's interpretation, they can actually hamper rather than aid in understanding wetland dynamics. How do wetlands change? Is it an orderly, predictable, directional pattern as suggested by traditional succession? Will a given wetland reach a so-called climax state? Does it succeed to upland communities? Although most ecologists have modified their views concerning these concepts, there is still much debate concerning their interpretation as well as their continued use (Egler 1947; Heinselman 1970; Drury & Nisbet 1973; Niering & Goodwin 1974; Pickett 1976; McIntosh 1980, 1981; MacMahon 1980, 1981; Zedler 1981; Patterson 1986; Niering 1987).

The purpose of this chapter is to review wetland vegetation dynamics occurring in natural systems in order to provide a background for evaluating wetland change in created systems.

WETLAND VEGETATION DYNAMICS

The concept of succession has been most closely associated with vegetation dynamics. As traditionally conceived by Clements (1916), it set forth a rather orderly, predictable and directional process for vegetation change in which one set of communities replaced another until a relatively stable system (climax) was established. It was primarily community controlled and, in the case of wetlands, the ultimate vegetation was believed to be an upland climax.

These traditional concepts have been considerably modified during the last half century. Yet, this does not diminish Clements' contribution to plant ecology since his six basic processes involved in vegetation change (nudation, migration, ecesis, competition, reaction and stabilization) are still relevant. However, it is now recognized that autogenic (soil development, competition for light, mineral depletion or accumulation) as well as allogenic factors (flooding, drought, fire, wind, and anthropogenic influences) are important in the process. In fact, the latter often have an overriding effect on the former. For example, a major disturbance that periodically interrupts a wetland ecosystem can produce changes that may last for decades or centuries. This introduces uniformitarian vs. catastrophic ecology (Egler 1977) in which the latter, representing allogenic change, may be so infrequent that it can be overlooked by short-term studies or by researchers who fail to recognize the role of historical factors. Disturbance is a natural and normal part of ecosystem dynamics (White 1979; Pickett & White 1985) to which ecological systems have become adjusted, especially in terms of their recovery (Marks 1974; Bormann & Likens 1979).

Gleason (1926) was one of the first to challenge Clements, especially his idea of the plant association and his organismic approach to vegetation. Instead, Gleason promoted the Individualistic Concept in which the differential establishment and survival of the various species in a given site were critical to the composition of any resulting unit of vegetation. In essence, the genetic characteristics of each species limit its ecological tolerance and every environment has its own biotic potential. For example, on an Alaskan floodplain, life history processes or species longevity, not facilitative interactions, explained forest development (Walker et al. 1986). This concept has been recently developed by van der Valk (1981, 1982) and applied to the prairie pothole wetlands (van der Valk & Davis 1978). He found 12 basic life history types in which the life forms of the plants, propagule longevity, and propagule establishment requirements are critical. Thus the plants which develop in a given situation will be dictated primarily by the life history requirements of each species and its presence or absence in the seed bank. No two sites, even though similar, will support exactly the same plant association.

An extension of the Gleasonian approach has resulted in the continuum concept in which discrete communities are not thought to exist, but rather continua, since species tolerances overlap along environmental gradients. In studying shoreline vegetation Raup (1975) was unable to recognize sufficient integration of species populations to identify community types and thus proposed the term "assemblages" which is logical in certain wetlands. It is also important to recognize that vegetation is highly variable and that a community is really a relative continuum between two discontinua (Egler 1977).

Chance and coincidence are especially relevant in wetland vegetation development (McCune & Cottam 1985; Egler 1987). The role of these stochastic factors is often difficult to measure and quantify and therefore sometimes overlooked. The three successional models (Facilitation, Tolerance and Inhibition) proposed by Connell and Slatyer (1977) are also relevant in understanding wetland dynamics. The Facilitation model relates to autogenic processes resulting in vegetation change in which the existing biota so influence the environment as to induce replacement of one set of species by another. Accumulation of solely organic sediments may lead to such changes, but evidence of autogenic development at the community level is limited. This model is the traditional concept of vegetation change as conceived by Clements. It can occur but it is only one of several possible processes or models that can be involved in vegetation change. The Tolerance model implies that various species may continue to become established over time, and that differential tolerance to light, and other limiting factors, will determine those species which will ultimately dominate. The Inhibition model suggests that those species which get established initially following a disturbance may well inhibit others from taking over. This may be relevant in wetland situations where relatively pure stands of cattail (Typha spp.) or reed grass (Phragmites australis) initially get established and essentially exclude other species. This parallels Egler's (1954) Initial Floristic Composition factor which sets forth the same idea. For example, in intertidal rocky shore wetland communities where new sites are exposed following scouring, those species that get established first frequently exclude others (Lubchenco & Menge 1978; Sousa 1979). In fact, this results in the patch pattern dynamics so typical in intertidal systems (Paine & Levin 1981; Dethier 1984). The concept of inhibition is somewhat counter to the concept of succession, since the vegetation development may be arrested at a phase that is not considered "climax". The role of these models, and possibly other factors, should be carefully evaluated when interpreting the causal factors contributing to biotic change.

WETLAND ZONATION AND VEGETATION CHANGE

Wetlands have long been regarded as transitional communities with a trend toward terrestrialization or upland communities (Gates 1926). In fact, this general misconception is still portrayed in certain current biology and environmental texts using successional diagrams which show a series of belts from open water to upland forest and suggesting that one vegetation belt is replacing another centripetally. This zonation pattern often represents a set of species populations that has found its optimum ecological requirements or tolerance in terms of water depth and frequency and duration of flooding. Thus these belts may not be a successional sequence of one community replacing another, but may represent relatively stable vegetation types possibly in a state of flux, depending upon changing hydrologic conditions

(Gallagher 1977). Yet, there may be situations where purely autogenic processes are involved in this developmental sequence.

In Michigan, Daubenmire (1968) indicates that upland trees cannot grow on peat soils and that a transition to upland will not occur. In Connecticut, Nichols (1915) also indicates that replacement of forested wetlands, such as red maple swamps, by upland oak forest is highly unlikely. More recently Walker (1970) states that, "Although certain sequences of transition are 'preferred' in certain site types, variety is the keynote of the hydroseral succession. In spite of this the data clearly indicate that bog is the natural 'climax' of autogenic hydroseres throughout the British Isles and the transition from fen to oak wood is unsubstantiated."

Bogs frequently exhibit distinctive belts but fail to represent a successional sequence. In Connecticut, Egler (personal communication) found trees the same age in all belts, indicating the effect of post-fire or cutting. At Cedar Creek Bog in Minnesota, Buell et al. (1968) followed vegetation change over three decades. They found that the width of the bog mat did not change in position over this time. However, the width of the various vegetation belts did change. The larch-shrub zone expanded outward into the floating sedge mat, greatly reducing its width. An earlier droughty period apparently favored sedge development but as the water level rose over the 33 years of observation, shrubs and larch trees expanded outward. Yet Lindeman (1941) observing this bog over part of the same period (1937-1947) found that the mat had advanced one meter, which emphasizes the limitations of short-term observations. At Bryants Bog in Michigan, studied since 1917, the bog mat has advanced into the bog pool in an irregular manner averaging 2.1 cm per year (Schwintzer & Williams 1974). In 1972 the open water was 76% of its extent in 1926. The mat vegetation has also changed over more than five decades. The advancing leatherleaf belt which was dominant in 1917 was succeeded by a high bog shrub community in the drier years and eventually by a bog forest by the late 1960's. Then in the early 1970's the trees died as the water level rose and leatherleaf was reestablished. This emphasizes the dynamic non-predictable cyclic pattern of vegetation change. In Connecticut, I have observed the mortality of a mature spruce bog forest due to extreme prolonged flooding. In fact, beaver activities play a major role in altering bog vegetation (Rebertus 1986) and vegetation along boreal forest streams (Naiman et al. 1986). In the Minnesota peatlands, Heinselman (1970) found no consistent trend toward mesophytism, terrestrialization, or uniformity but rather a swamping of the landscape, rise in water tables, deterioration of the growth and a diversification of landscape types. In the northern Canadian peatlands both water level fluctuations and fire are primary factors governing vegetation change. Here long-term records show that the spatial-temporal approach does not accurately describe the dynamics of peatlands (Jasieniuk & Johnson 1982).

In the bottomland hardwood forests six vegetation zones can often be recognized, based on soil moisture and hydrology. However, as Wharton et al. (1982) point out, the term "zone" is somewhat misleading since many of these plant communities are arranged in a mosaic pattern, depending upon the hydrologic conditions. Along the Northeast riverine systems the vegetation pattern is also related to the frequency and duration of flooding (Metzler & Damman 1985).

In the marine environment, mangrove and salt marshes also exhibit distinct belting patterns. In south Florida three mangroves - the red (Rhizophora mangle [most oceanward]), black (Avicennia germinans), and white (Laguncularia racemosa [most landward]) - frequently form a belting pattern which has been interpreted as a succession oceanward (Davis 1940). Egler (1948) questioned this interpretation and, more recently, Ball (1980) found that interspecific competition was an important factor in controlling the zonation. In Panama, Rabinowitz (1975) found that reciprocal transplants can grow well in either zone. She also found that a primary mechanism controlling zonation is tidal sorting of propagules due to their size, rather than habitat adaptation. It appears that, once a given zone reaches equilibrium, it is unlikely to change unless disturbance occurs (Odum et al. 1982). In fact, major storms such as hurricanes are prime site builders for the establishment of mangrove (Craighead & Gilbert 1962; Craighead 1964). However, in Australia, biotic influences such as seed predation by crabs can also have a significant influence on these intertidal mangrove forests (Smith 1987).

Tidal wetlands of the Northeast offer still another example of close integration between vegetation change and hydrology. With coastal submergence peat cores document a vegetation development from the intertidal salt water cordgrass (Spartina alterniflora) to salt meadow cordgrass (Spartina patens). There is also a tendency for spike grass (Distichlis spicata) and switch grass (Panicum virgatum) to replace upland species as the marsh advances landward with sea level rise. However, once the high marsh has developed, oscillations in the vegetation patterns are primarily hydrologically induced. Waterlogged, poorly drained sites favor the short form of S. alterniflora. This condition can be induced by mosquito ditching, with the levees along the ditches preventing the flooded high marsh from draining.

Once the high marsh has developed, myriad vegetation changes can occur as documented by the hundreds of peat cores taken by Orson (1982) in tracing the ontogeny of the Pataguanset marshes in Connecticut, and by the author along the coastline of Connecticut. For example, a single core 1 meter in length, representing 500-1000 years, may show five or six vegetation types or changes based on the preserved rhizomes in the core (Niering et al. 1977). There appears to be no predictable unidirectional pattern. Hydroperiod, changes in micro-relief, accretion rates, salinity, redox potential, storms, and other factors make these systems too complex to be orderly or predictable (Niering & Warren 1980). As stated by Miller and Egler (1950), "...the present mosaic may be thought of as a momentary expression, different in the past, destined to be different in the future, and yet as typical as would be a photograph of moving clouds."

WETLANDS AS PULSED SYSTEMS

Odum (1971) set forth the concept of pulsed stability as related to wetland systems. Subjected to more or less regular but acute physical disturbance imposed from without, they are often maintained at an intermediate state in development. This may further reflect why traditional successional concepts frequently have limited application in wetland systems. Tidal wetlands, for example, may be maintained in a relatively fertile state by a "tidal energy subsidy" which provides rapid nutrient cycling and favors substrate aeration. Among the freshwater systems the prairie potholes are pulsed in an even more striking manner, often completely drying up in droughty periods and then reappearing with the advent of adequate precipitation. During droughts, aerobic breakdown of the organic matter replenishes the nutrient supply to favor future productivity. In south Florida, the Everglades is another fluctuating system. The importance of this pulsing is dramatically correlated with the successful breeding of the wood stork (Kahl 1964) since lower water levels are necessary to concentrate small fish to feed nestlings. Yet, an "energy subsidy" from agricultural runoff, with its nutrient enrichment, can actually be deleterious to the low nutrient demanding sawgrass, resulting in its demise and favoring a dramatic increase in cattail. The sawgrass (Cladium jamaicensis) can also be destroyed by peat fires which recently have increased due to drainage. Under the natural fluctuating water regime the Glades burned when flooded and viable sawgrass marsh was maintained (Egler 1952). Other wetlands that are pulsed by drought and fire are the evergreen shrub pocosins along the southeast coastal plain (Richardson 1981) and the Okefenokee Swamp where fire and drought have set the pattern for vegetation change for decades (Schlesinger 1978; Hamilton 1984). As previously mentioned, dry periods favor rapid decomposition and peat fires can aid in maintaining more hydric conditions. Some species like the bald cypress, which normally grows under flooded conditions, actually require bare soil conditions for seedling establishment.

The pulsing concept is especially relevant in wetland creation. Will there be fluctuating hydrologic conditions in the newly created wetland? Fixed water levels are not the rule in nature. Continuous flooding or the absence of pulsing are deleterious to most trees. Pulsed, not static water regimes, should be one of the major objectives in any mitigation project, especially those in inland waterways and lake systems.

WETLAND DEVELOPMENT AND ECOSYSTEM PROCESSES

In respect to ecosystem development, Odum (1969) set forth a series of ecosystem processes--community structure and energetics--as related to the stage of maturity in the process. Mitsch and Gosselink (1986, see Table 7-1, p. 160-1) illustrate how this scheme relates to some of the major wetlands in the United States. It is obvious that wetlands are highly variable with respect to these criteria, exhibiting aspects of both immature and mature systems, an attribute to be expected in pulsed systems. For example, in most wetlands, except bogs, mineral cycles are open and life cycles are short, typical of immature systems; food chains, however, are often complex, characteristic of mature systems. Figure 1 attempts to more holistically integrate the multiplicity of factors involved in biotic change for both terrestrial and wetland ecosystems without using traditional succession-climax dogma. Relatively stable states can occur only to

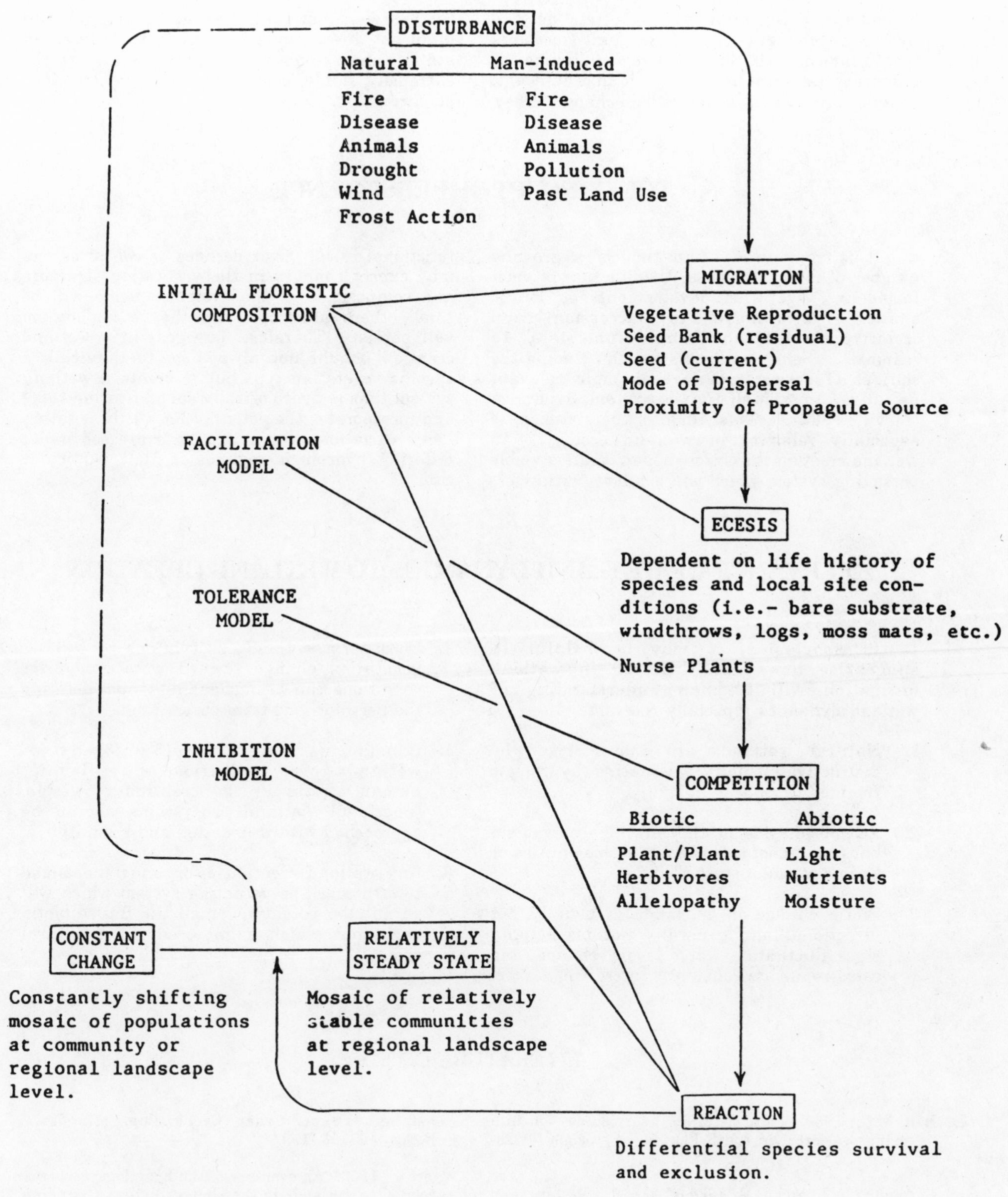

Figure 1. Holistic view of some of the factors and processes involved in vegetation change. Following disturbance a given system may eventually reach a relatively stable state or be in a continuous state of flux (Niering 1987).

be modified by disturbance, often due to hydrologic changes. As Mitsch and Gosselink (1986) indicate, the idea of a regional climax is inappropriate since both allogenic and autogenic factors are operative in wetland change. They further point out that changes in wetlands are often not directional and "...in fact, wetlands in stable environmental regimes seem to be extremely stable, contravening the central idea of succession."

THE CONCEPT OF PERSISTENCE

It is my opinion, and that of a growing number of other ecologists, that the use of such terms as "vegetation development" or "biotic change" is preferred to "succession," and "relative stability" or "equilibrium state" to "climax." Some ecologists are now finding the concept of persistence an even more relevant paradigm in visualizing ecosystem dynamics (Lewin 1986). This idea of persistence is especially relevant in wetland ecology. In wetland creation the objective is to create a viable persisting system which will exhibit a variety of functional roles. Over decades or centuries one may expect changes in the vegetation structure and composition of the system. Some will be small, others catastrophic, but the wetland system will persist. Therefore, our goal in wetland creation should not always be to duplicate a specific vegetation type but to create a wetland system that is hydrologically sound (Carter 1986) and incorporates the potential for all those future biotic variations that might be expressed under differing hydrologic regimes in that particular site.

RELEVANCE OF WETLAND DYNAMICS TO WETLAND CREATION

In conclusion, it may be helpful to summarize how those involved in wetland mitigation will find an understanding of wetland dynamics especially relevant.

1. Natural wetlands are characterized by distinctive, usually fluctuating hydrologic regimes.

2. As pulsed systems, they are highly dynamic but can persist as relatively stable entities or be in a constant state of flux.

3. Biotic change in wetlands is usually not directional and generally not predictable since fluctuating water levels, chance, and catastrophe are constantly interacting.

4. Short-term wetland observations concerning vegetation change toward wetter or drier conditions can be misleading, thus dictating the need for long-term observations.

5. Considering the natural ontogeny of wetlands over centuries or millennia, human efforts in the creation of viable, functional wetland ecosystems should be approached with trepidation and humility.

6. Any wetland creation effort must be aimed toward a self-perpetuating system which will permit the potential for all the future biotic variations which might occur in a natural system.

LITERATURE CITED

Ball, M.C. 1980. Patterns of secondary succession in a mangrove forest in south Florida. Oecologia 44:226-235.

Bormann, F.H. and G.E. Likens. 1979. Pattern and Process in a Forested Ecosystem: Disturbance, Development and the Steady State Based on the Hubbard Brook Ecosystem Study. Springer-Verlag, New York.

Buell, M.F., H.F. Buell, and W.A. Reiners. 1968. Radial mat growth on Cedar Creek Bog, Minnesota. Ecology 49:1198-1199.

Carter, V. 1986. An overview of the hydrologic concerns related to wetlands in the United States. Canadian Jour. Bot. 64:364-374.

Clements, F.E. 1916. Plant Succession: An Analysis of the Development of Vegetation. Carnegie Institution of Washington Publ. 242. Washington, D.C.

Connell, J.H. and R.O. Slatyer. 1977. Mechanisms of succession in natural communities and their role in community stability and organization. Am. Naturalist 3(982):1120-1144.

Craighead, F.C. 1964. Land mangroves and hurricanes. Fairchild Trop. Gard. Bull. 19:5-32.

Craighead, F.C. and V.C. Gilbert. 1962. The effects of Hurricane Donna on the vegetation of southern Florida. Q.J. Florida Acad. Sci. 25:1-28.

Daubenmire, R.F. 1968. Plant Communities. Harper and Row, New York.

Davis, J.H., Jr. 1940. The ecology and geologic role of mangroves in Florida. Publ. 527. Portugas Lab. Paper 32:303-412. Carnegie Inst., Washington, D.C.

Dethier, M.N. 1984. Disturbance and recovery in intertidal pools: maintenance of mosaic patterns. Ecol. Monog. 54:99-118.

Drury, W.H. and I.C.T. Nisbet. 1973. Succession. J. Arnold Arb. 54:331-368.

Egler, F.E. 1947. Arid southwest Oahu vegetation, Hawaii. Ecol. Monog. 17:383-435.

Egler, F.E. 1948. The dispersal and establishment of the red mangrove, Rhizophora, in Florida. Caribbean Forester 9:299-310.

Egler, F.E. 1952. Southeast saline Everglades vegetation, Florida, and its management. Vegetatio 3(4-5):213-265.

Egler, F.E. 1954. Vegetation science concepts. I. Initial floristic composition, a factor in old-field vegetation development. Vegetatio 4:412-417.

Egler, F.E. 1977. The Nature of Vegetation, Its Management and Mismanagement: An Introduction to Vegetation Science. Aton Forest, Norfolk, Connecticut.

Egler, F.E. 1987. Trail Wood, Hampton, Connecticut. Vegetation of the Edwin Way Teale Memorial Sanctuary. Connecticut Conservation Assoc.

Gallagher, J.L. 1977. Zonation of wetland vegetation, p. 752-758. In J. Clark (Ed.), Ecosystem Management: a Technical Manual for Conservation of Coastal Zone Resources. The Conservation Foundation, Washington, D.C.

Gates, F.C. 1926. Plant succession about Douglas Lake, Cheboygan County, Michigan. Bot. Gaz. 82:170-182.

Gleason, H.A. 1926. The individualistic concept of the plant association. Bull. Torrey Bot. Club 53:7-16.

Hamilton, D.B. 1984. Plant succession and the influence of disturbance in Okefenokee Swamp, p. 86-111. In A.D. Cohen, D.J.Casagrande, M.J. Andrejko, and G.R. Best (Eds.), Okefenokee Swamp: Its Natural History, Geology, Geochemistry. Wetlands Surveys, Los Alamos, New Mexico.

Heinselman, M.L. 1970. Landscape evolution, peatland types, and the environment in the Lake Agassiz Peatlands Natural Area, Minnesota. Ecol. Monog. 40:235-261.

Jasieniuk, M.A. and E.A. Johnson. 1982. Peatland vegetation organization and dynamics in the western subarctic, Northwest Territories, Canada. Canadian J. Bot. 60:2581-2593.

Kahl, M.P. 1964. The food ecology of the wood stork. Ecol. Monog. 34:97-117.

Lewin, R. 1986. In ecology, change brings stability. Science 234:1071-1074.

Lindeman, R.I. 1941. The developmental history of Cedar Creek Bog, Minnesota. Am. Midland Nat. 25:101-112.

Lubchenco, J. and B.A. Menge. 1978. Community development in a low rocky intertidal zone. Ecol. Monog. 48:67-94.

MacMahon, J.A. 1980. Ecosystems over time: succession and other types of change, p. 27-58. In R.H. Waring (Ed.) Forests: Fresh Perspectives from Ecosystem Analysis. Proc. 40th Annual Biology Colloquium, Oregon State Univ. Press. Corvallis, Oregon.

MacMahon, J.A. 1981. Successional processes: comparisons among biomes with special reference to probable roles of and influences on animals, p. 277-304. In D.C. West, H.H. Shugart, and D.B. Botkin (Eds.), Forest Succession: Concepts and Application. Springer-Verlag, New York.

McCune, B. and G. Cottam. 1985. The successional status of a southern Wisconsin oak woods. Ecology 66:1270-1278.

McIntosh, R.P. 1980. The relationship between succession and the recovery process in ecosystems, p. 11-62. In J. Cairns, Jr. (Ed.), The Recovery Process in Damaged Ecosystems, Ann Arbor Science Publ.

McIntosh, R.P. 1981. Succession and ecological theory, p. 10-23. In D.C. West, H.H. Shugart, and D.B. Botkin (Eds.), Forest Succession: Concepts and Application. Springer-Verlag, New York.

Marks, P.L. 1974. The role of pin cherry (Prunus pensylvanica L.) in the maintenance of stability in northern hardwood ecosystems. Ecol. Monog. 44:73-88.

Metzler, K.J. and A.W.H. Damman. 1985. Vegetation patterns in the Connecticut River flood plain in relation to frequency and duration of flooding. Nat. Canad. 112:535-547.

Miller, W.R. and F.E. Egler. 1950. Vegetation of the Wequetequock-Pawcatuck tidal marshes, Connecticut. Ecol. Monog. 20:144-172.

Mitsch, W.J. and J.G. Gosselink. 1986. Wetlands. Van Nostrand Reinhold, New York.

Naiman, R.J., J.M. Melillo, and J. E. Hobbie. 1986. Ecosystem alteration of boreal forest streams by beaver (Castor canadensis). Ecology 67:1254-1269.

Nichols, G.E. 1915. The vegetation of Connecticut IV. Plant societies in lowlands. Bull. Torrey Bot. Club 42:168-217.

Niering, W.A. 1987. Vegetation dynamics ("Succession" and "Climax") in relation to plant-community management. Conservation Biology 1:287-295.

Niering, W.A. and R.H. Goodwin. 1974. Creation of relatively stable shrublands with herbicides: arresting "succession" on rights-of-way and pastureland. Ecology 55:784-795.

Niering, W.A. and R.S. Warren. 1980. Vegetation patterns and processes in New England salt marshes. BioScience 30:301-307.

Niering, W.A., R.S. Warren, and C.G. Weymouth. 1977. Our dynamic tidal marshes: vegetation changes as revealed by peat analysis. Conn. Arboretum Bull. 22:1-12.

Odum, E.P. 1969. The strategy of ecosystem development. Science 164:262-270.

Odum, E.P. 1971. Fundamentals of Ecology. W.B. Saunders Co., Philadelphia, Pennsylvania.

Odum, W.E., C.C. McIvor, and T.J. Smith III. 1982. The Ecology of the Mangroves of South Florida: A Community Profile. U.S. Fish and Wildlife Service FWS/OBS-81-24, Office of Biological Services, Washington, D.C.

Orson, R.A. 1982. Development of the lower Pataguanset estuarine tidal marshes, Niantic, Connecticut. Masters Thesis, Connecticut College, New London.

Paine, R.T. and S.A. Levin. 1981. Intertidal landscapes: disturbance and the dynamics of pattern. Ecol. Monog. 51:145-178.

Patterson, W.H. III. 1986. A "new ecology" - implications of modern forest management. "Let's strike 'climax' from forest terminology." J. Forestry 84:73.

Pickett, S.T.A. 1976. Succession: an evolutionary interpretation. Am. Naturalist 110:107-119.

Pickett, S.T.A. and P.S. White (Eds.). 1985. The Ecology of Natural Disturbance and Patch Dynamics. Academic Press, Inc., New York.

Rabinowitz, D. 1975. Planting experiments in mangrove swamps of Panama, p. 355-393. In G. Walsh, S. Snedaker, and H. Teas (Eds.), Proc. of the International Symposium on Biology and Management of Mangroves. Univ. of Florida, Gainesville.

Raup, H.M. 1975. Species versatility in shore habitats. J. Arnold Arboretum 55:126-165.

Rebertus, A.J. 1986. Bogs as beaver habitat in north-central Minnesota. Am. Midl. Nat. 116:240-245.

Richardson, C.J. (Ed.). 1981. Pocosin Wetlands. Hutchinson Ross Pub. Co., Stroudsburg, Pennsylvania.

Schlesinger, W.H. 1978. Community structure, dynamics, and nutrient cycling in the Okefenokee cypress forest. Ecol. Monog. 48:43-65.

Schwintzer, C.G. and G. Williams. 1974. Vegetation changes in a small bog from 1917 to 1972. Am. Midl. Nat. 12:447-459.

Smith, T.G. III. 1987. Seed predation in relation to tree dominance and distribution in mangrove forests. Ecology 68:266-273.

Sousa, W.P. 1979. Experimental investigations of disturbance and ecological succession in a rocky intertidal algal community. Ecol. Monog. 49:227-254.

van der Valk, A.G. 1981. Succession in wetlands: a Gleasonian approach. Ecology 62:688-696.

van der Valk, A.G. 1982. Succession in temperate North American wetlands, p. 169-179. In B. Gopal, R.E. Turner, R.G. Wetzel, and D.F. Whigham (Eds.), Wetlands: Ecology and Management. National Inst. of Ecology and International Scientific Publications, Jaipur, India.

van der Valk, A.G. and C.B. Davis. 1978. The role of seed banks in the vegetation dynamics of prairie glacial marshes. Ecology 59:322-335.

Walker, D. 1970. Direction and rate in some British post-glacial hydroseres, p. 117-139. In D. Walker and R.G. West (Eds.), The Vegetation History of the British Isles. Cambridge Univ. Press.

Walker, L.R., J.C. Zasada, and F.S. Chapin III. 1986. The role of life history processes in primary succession on an Alaskan floodplain. Ecology 67:1243-1253.

Wharton, C.H., W.M. Kitchens, E.C. Pendleton, and T.W. Sipe. 1982. The Ecology of Bottomland Hardwood Swamps of the Southeast: A Community Profile. U.S. Fish and Wildlife Services FWS/OBS-81/37, Biological Services Program.

White, P.S. 1979. Pattern, process and natural disturbance in vegetation. Bot. Rev. 45:229-299.

Zedler, P.H. 1981. Vegetation change in chaparral and desert communities in San Diego County, California, p. 431-447. In D.C. West, H.H. Shugart, and D.B. Botkin (Eds.), Forest Succession: Concepts and Application. Springer-Verlag, New York.

LONG-TERM EVALUATION OF WETLAND CREATION PROJECTS

Charlene D'Avanzo
School of Natural Science
Hampshire College

ABSTRACT. Long-term success of wetland restoration and creation projects may be quite different from short-term success. In this chapter six criteria are used to evaluate the long-term success of more than 100 artificial wetland projects reported in the literature. Results from numerous U.S. Army Corps of Engineers' dredged material stabilization projects demonstrate the importance of long-term monitoring and increasing long-term as well as short-term success. Several studies reviewing wetland creations are also used to demonstrate problems with projects in both the short and long-term.

The long-term evaluation of artificial wetlands is very difficult because wetlands are created for a variety of purposes. We know little about basic aspects of many wetland systems, "succession" in wetlands is less straightforward than previously assumed, and it is difficult to generalize from one wetland type to another. There is a striking range of opinions about the success of wetlands that have been created. On the one hand, the U.S. Army Corps of Engineers' dredged material stabilization program exemplifies artificial wetland projects that appear successful over a decade or more. Several types of criteria including vegetation characteristics, soil chemistry, and animal studies suggest that several dredged material wetlands are becoming similar to reference wetlands with time. But, some wetlands characteristics (soil carbon) may require many years to reach natural levels.

In contrast, a great many other artificial wetland projects are problematic or failures. Reasons for failures include improper hydrology, erosion, herbivory, and invasion by upland plants. Many projects have never been evaluated so their permanence is not known, and a disturbing number of required projects have never been created.

In evaluating projects with regard to persistence (long-term success) of the created wetlands, the following points are especially important: 1) 1-2 years of monitoring is too short; evaluations over as long a period of time as possible (10-20 years) are desirable; 2) vegetation characteristics are useful but do not necessarily indicate function; at a minimum, several parameters should be used (e.g., belowground/aboveground biomass comparisons); 3) chemical/physical aspects of wetland soils are also useful in evaluating trends in created sites; 4) local reference wetlands are critical for comparative purposes; and 5) some wetlands should be created with great caution because they have failed in the past (e.g., high salt marsh in the northeast) or because we know little about these wetland types (e.g., forested wetlands).

INTRODUCTION: A CHALLENGING TASK

This chapter is a review and evaluation of changes that have occurred over time in wetland creation projects. The results of more than 100 artificial wetland studies are discussed; the sites range from large-acreage federal projects to small private plantings. The main questions addressed are: 1) how have artificial wetlands evolved over time, and 2) what can we learn from these effects concerning the feasibility of creating wetlands with long-term functions?

Evaluating the long-term "success" of artificial wetlands is very difficult for a number of reasons. First, wetlands are created for a wide variety of purposes--some are created as mitigation for destroyed wetlands, some are experimental plantings, and still others are aimed at stabilization of dredged material. Methods of evaluation are also not standardized. Access to publications of the studies differ as well; publications in refereed journals are more accessible while some federal studies, such as those by the U.S. Army Corps of Engineers (USACE), are more difficult to obtain.

A second reason why the long-term

evaluation of wetland creation projects is especially challenging was pointed out by Larson and Loucks (1978) a decade ago. We know little about many basic aspects of ecosystem-level processes in wetlands. For example, a knowledge of wetland seed bank dynamics is important for creation and evaluation of human-made wetlands. However, we are particularly ignorant about this aspect of wetland ecosystems. The most basic of questions--how water level influences seedling recruitment (Keddy and Ellis 1985) and how seed dispersal and on-site hydrology influences plant community composition (Schneider and Sharitz 1986)--are just being addressed for some wetlands. A wetland-creation evaluator who lacks such community/ecosystem information will find it difficult to understand why the outcome of one project differs from another.

A third reason particularly relevant to long-term wetland studies is the great difficulty in predicting vegetation change over time ("succession") in many wetland types. The classical Clementsian view of wetlands developing towards an upland climax is not held today by most wetland ecologists (Niering 1987; Guntenspergen and Stearns 1985). For example, bogs can become more hydric with time and plants typical of low marsh or brackish areas can grow in the high salt marsh zone (Niering 1987; Odum 1988). The salt marsh example is particularly important because these are our best studied wetlands. However, papers are now being published that discuss factors (e.g., flooding, disturbance, and herbivory) influencing the distribution of plants in the high marsh of northeast US coastal marshes (Valiela 1984). Once again, without this kind of information long-term changes in high salt marsh vegetation will be difficult to predict.

A fourth reason why wetland creation projects are challenging to evaluate over the long term, is that wetlands are exceedingly varied, highly dynamic systems. They are dynamic because they exist at the interface between terrestrial and aquatic systems and are unusually sensitive to variations in hydrologic regime (Guntenspergen and Stearns 1985). As a result, it is difficult to generalize from the response of creation projects in one wetland type to creations in different locales and habitats.

We do know more about some wetland types than about others. As an example of this difference, Mitsch (1988) summarizes the state of art of wetland modelling; he points out that freshwater marsh models are "primitive" while coastal marsh models are "well developed". Therefore, generalization about saltmarsh creations may be more accurate than generalizations about freshwater marsh creations.

DIVERGENT VIEWS ABOUT THE LONG-TERM SUCCESS OF WETLAND CREATIONS

There is a striking range of opinions about the success of wetland creation projects. The reported outcomes of USACE Dredged Material projects are generally positive (e.g., Newling and Landin 1985). Other researchers also conclude from relatively long-term studies that under proper hydrologic regimes created salt marshes appear similar to natural ones (e.g., Seneca et al. 1976).

On the other hand, the success of many other wetland creations and mitigations is less certain. For example, several summaries of wetland restoration projects in California point to problems. Josselyn and Buchholz (1984) concluded from a statewide analysis that most California sites were not carefully monitored after project completion; therefore, long-term success or failure of these creations is not known. Race (1985) says that "...it is debatable whether any sites in San Francisco Bay can be described as completed, active or successful restoration projects at present". Eliot (1985) evaluated permits of 58 projects in San Francisco Bay that required wetland restoration. She states that "...the 58 projects are diverse, frequently unsuccessful, and do not adhere to established mitigation policies. Many projects have not been completed. Of those that have been, many do not resemble the existing remnant marshes in San Francisco Bay".

Difficulties with mitigation projects are not limited to California. In Washington state Kunz et al. (1988) reviewed Section 404 projects and concluded that 1) mitigations resulted in a net wetland loss of 33% in 6 years (1980-1986), 2) some wetlands (forested) were not replicated at all, 3) time lags between project initiation and mitigation completion resulted in a loss of at least 1-3 growing seasons per project, 4) there was no routine procedure for tracking compliance, and 5) 5 of the 35 projects were never restored or negotiated.

Why do these conclusions differ so greatly? The people involved in the USACE Dredged Materials Program partially answer this question. They have a great deal of experience which they can apply to each project, particularly with regard to hydrologic design. For example, Newling (USACE, pers. comm.) states that someone familiar with the project design must be on-site when dredged material is applied to a site because elevation above water is critical to project success. Such experience may be lacking in other creations.

THE SIX CRITERIA USED FOR WETLAND ASSESSMENT

In this chapter I address wetland creations, as opposed to restoration or mitigation, unless these other activities have taken place. I define wetland creations as artificial wetland habitats established in new locations. Manipulations of already existing wetlands--flooding marshes to enhance wildlife use, for example--are not described here. Terminology is debatable in wetland creation studies. For example, Harvey and Josselyn (1986) criticized Race (1985) for her use of the term "wetland restoration" in describing various experimental plantings in San Francisco Bay.

The projects I discuss differ greatly in 1) age of the human-made wetland when evaluated (14 months to 40 years); 2) use of natural controls for comparisons; 3) use of quantitative methods in contrast to qualitative ones; and 4) reasons for the creations. Nevertheless, the wetland creation projects discussed provide considerable information to address questions concerning change of these artificial wetlands with time.

Although the focus of this chapter is permanence of created wetlands and their evolution with time, most artificial wetlands are too young to provide information for long-term studies. Many of the projects described in this chapter were evaluated 1-2 years after they were created. Exceptions are USACE dredged material projects; some of these have been studied for 14 years.

The criteria used in this chapter to evaluate success and describe how the human-made wetlands change with time are:

1. Comparison of vegetation growth characteristics (for example, biomass or density) in artificial and natural wetlands after two or more growing seasons;

2. Habitat requirements (for example, upland vs wetland) of plants invading the created site;

3. Success of planted species;

4. Comparison of animal species composition and biomass in human-made and natural sites;

5. Chemical analyses of artificial wetland soils compared to natural wetlands; and

6. Evidence of geologic or hydrologic changes with time.

These criteria are typically used in wetland ecosystem studies (e.g., Valiela 1984). In addition, plants are emphasized as wetland indicators because they reflect the hydrologic regime and perform numerous important functions (D'Avanzo 1987).

COMPARISON OF VEGETATION GROWTH CHARACTERISTICS IN HUMAN-MADE AND NATURAL WETLANDS

Using vegetation criteria, USACE scientists have generally judged successful marine wetland creations on dredged materials. But these studies also note the importance of long-term monitoring. For example, Hardisky (1978) found that in Buttermilk Sound, Georgia, aerial biomass of saltwater cordgrass, _Spartina alterniflora_, planted in dredged spoil was 1.3-5.5 times less after four growing seasons than that of cordgrass in natural sites. Of the 16 comparisons of aboveground biomass of various plants listed in this study, in 13 cases the biomass was greater in natural saltmarshes. Belowground biomass for Buttermilk Sound _S. alterniflora_ was also 2.1-12.4 times less than that of comparison marshes. Hardisky expressed concern about erosion, herbivory, and competition between saltmarsh and invading plants. By 1982 Newling and Landing (1985) were more positive about this site because aboveground biomass was more similar to that in reference marshes. Belowground biomass still lagged behind.

In contrast, in a different dredged spoil stabilization project in North Carolina, aboveground _S. alterniflora_ production measured in the third through fifth growing seasons was within the range of that seen in similar natural marshes; belowground production exceeded natural controls (Hardisky 1978).

At the Bolivar Peninsula dredged materials site in Texas the marsh grasses, _Spartina alterniflora_ and _Spartina patens_, dominated plots after 3 years although erosion was noted there (Webb et al. 1986). Newling and Landin (1985) concluded from preliminary analysis of this site after 4 years that stem height and aboveground biomass equaled or exceeded reference locales while root biomass was less. In addition, at the Apalachicola Bay salt marsh site, Newling et al. (1983) monitored stem density, height, occurrence and flowering in 8 quadrats 6 years after the site was planted with _S. alterniflora_ and _S. patens_; these parameters were similar to those in reference marshes.

Not all Corps salt marsh projects have been

entirely successful. Stedman Island in Aransas Bay, Texas, was almost entirely vegetated after two years but after 39 months S. alterniflora at low elevations died (Landin and Webb 1986). The reason for the die-off is not known, but this study shows the importance of long term monitoring.

Some freshwater marshes created within the Corps dredged materials program are also judged successful over a relatively short period of time when vegetation criteria were used. Windmill Point on the James River, Virginia, is a freshwater tidal marsh established in 1974-5. Emergent plants similar to those in comparison marshes were observed in 1978-82 (Newling and Landin 1985). Another freshwater intertidal locale, Miller Sands on the Columbia River in Oregon, was planted in 1976. Vegetation development has been slower here than at other Army Corps sites and some plantings failed. However, vegetation began to invade bare areas after several years. Since freshwater tidal marshes have only been studied in the last decade or so, experience and information which could help explain the results here are limited.

Other projects in which vegetation was used to evaluate the long-term success of created salt marshes show mixed results. In two artificial Spartina foliosa marshes in San Francisco Bay, plant density in experimental plots was less than a third of that in nearby natural stands, but the site was only evaluated after two growing seasons (Morris et al. 1978). Plantings in a salt marsh mitigation for a marina in Bourne, Massachusetts failed because the marsh grass was planted too low in the intertidal zone and because the grasses were eaten by waterfowl (Reimold and Cobler 1986). Josselyn and Bùchholz (1984) analyzed 3 wetland restoration projects in Marin County, California; plantings failed in 2 sites because proper elevation was difficult to achieve and contaminated dredged spoil used to raise elevation may have hindered plant growth.

Shisler and Charette (1984) compared eight artificial salt marshes to eight adjacent natural marshes in New Jersey. Overall live biomass after 2-6 years was similar in created sites and natural marshes and, not surprisingly, total biomass (including dead litter) was significantly lower in the human-made marshes. However, density and number of reproductive grass heads in the artificial wetlands were also lower than in controls. These vegetation differences, invasion by plants not characteristic of salt marshes, and significantly different soil chemical parameters (described below) led Shisler and Charette to recommend no further construction of high marsh habitat in New Jersey.

Restoration of areas previously vegetated by marine plants appears less problematic. Thorhaug (1979) planted the seagrass Thalassia in Biscayne Bay, Florida in areas that had been denuded by thermal effluent. After four years, she measured similar grass densities in planted areas compared to controls; for these planted seagrasses, flowering and fruiting compared well with controls. It is important to note that these were sites that had previously supported Thalassia and, therefore, the success of revegetation after thermal emissions ceased was promising.

The vegetation characteristics described above--above and belowground biomass, plant density, and number of reproductive stalks--are among the most commonly used quantitative measures of plant growth. Using these characteristics as measures of success, it is difficult to make long-term generalizations about wetland studies. In New Jersey for example, only low S. alterniflora has been successfully established and S. patens exhibited very limited success; therefore, one cannot predict replacement of low marsh by high marsh, a change that can occur in natural marsh ontogeny. We can say that growth of plants in the artificial habitats is sometimes different from that in controls even after 4-6 years. It is difficult to determine temporal trends, however, since many sites were not analyzed over time. In several dredged material sites that have been evaluated over time, vegetation becomes more similar to reference sites. This is not so in the California examples. Therefore, it is impossible from many existing descriptions to determine whether these created wetlands are becoming more like the natural controls as they age.

What generalizations about wetland creation have been drawn by others from vegetation studies in artificial wetlands? Again, opinions greatly differ. On one hand, Zedler et al. (1982) conclude: "Regardless of the techniques used, the examples are too few, and their period of existence too short to provide an instructional guide for marsh restoration projects in California. At present restoration must be viewed as experimental". In contrast, the Corps appears more confident about the information base. Landin and Webb (1986) state that: "The Corps has strived for development of viable wetland sites and will continue to do so. When problems have arisen on sites, or failure noted ...lessons were learned ... and these mistakes were not repeated on later sites".

COMPARISON OF ANIMAL SPECIES COMPOSITION AND BIOMASS IN ARTIFICIAL AND NATURAL WETLANDS

While we emphasize vegetation in this chapter, it is instructive to evaluate animal responses in created wetlands. One set of studies again demonstrates the value of long term analyses. In a North Carolina dredged material site, Cammen (1976a) found significantly more macroinvertebrates and 10 fold greater biomass in natural plots compared to those in the 2 year old human-made marsh. In addition, in the natural marsh, isopods, polychaetes, and mussels were the dominant fauna while amphipods and flies dominated the artificial plots. In one sampling location, less than 40% overall faunal similarity was seen in the natural/created comparison and in another site the similarity was less than 10% after three growing seasons. Sacco et al. (1988) studied the macrofauna in this marsh 15 years after it was created. The macrofauna had greatly changed and was mainly composed oligochaetes (56%) and polychaetes (36%). Therefore, Sacco et al. concluded that the macrofauna in the human-initiated marsh began to resemble natural marshes within 15 years, although fauna in reference sites were not listed in this abstract.

In a different comparison of North Carolina dredged spoil restoration, Cammen (1976b) also found significantly lower animal density and biomass as well as different animal populations in several year old created systems. In this case, insect larvae dominated the created wetland while polychaetes accounted for most of the biomass in the natural marsh. In contrast to Cammen's findings, in the New Jersey mitigation sites evaluated by Shisler and Charette (1984), many species of macroinvertebrates were common to natural and artificial marshes and populations were highly variable in each.

HABITAT REQUIREMENTS OF INVADING PLANTS

The persistence of obligate wetland plants with time--either planted or naturally colonizing--and their successful dominance over other vegetation, is one good measure of creation failure or success. Data on species changes of wetland versus nonwetland plants over time indicate mixed success of creation projects.

Kruczynski and Huffman (1978) studied marsh and dune vegetation on dredged material in Apalachicola Bay, Florida. One island supported no plant growth after 17 months because of erosion. Dikes stabilized another island where 42 plants--many upland indicators such as morning glory (Ipomoea sp.) and cudweed (Gifola germanica)--were already growing alongside planted Spartina after 14 months. Shisler and Charette (1984) describe plant species characteristic of upland/marsh ecotones in numerous artificial marsh projects in New Jersey. In Creekside Park, a San Francisco Bay restoration site, upland species and bare ground occupied as much of the marsh surface area as marsh vegetation after eight years. High marsh in particular was not vegetated due to high salt concentrations (Josselyn and Buchholz 1984).

On 40 Florida coastal islands of various ages composed of dredged material studied by Lewis and Lewis (1978), exotic upland plants were common invaders, while these plants were unusual on natural islands; "the predominance of the exotic Australian pine and Brazilian pepper in the later seral stages is unique to dredged material islands in Florida. The maritime forest climax is rare ..." Birds may have influenced plant invasion and success on these islands.

Invasion by upland plants into artificial wetlands was not seen in many studies. However, development of non-wetland flora was not an unusual occurrence and is cause for concern (Odum 1988). In addition, many artificial wetlands were observed only after 2 or 3 growing seasons, which may be insufficient time for establishment and growth of upland vegetation or to determine if wetland flora will persist.

SUCCESS OF PLANTED VEGETATION

Particular types of vegetation are planted in artificial wetlands to temporarily stabilize soil, provide wildlife habitat, or for aesthetic reasons. Disappearance of these plants over time is not unusual (Hardisky 1978; Shisler and Charette 1984; Dial and Deis 1986; Odum 1988). Certainly, plants in artificial wetlands, as in natural ones, will likely change with time as, for example, seeds in the soil or imported seeds germinate. If these newly observed plants are obligate wetland plants, we may, by definition, call the creation site a wetland. However, it is much more difficult to decide whether the new wetland is a success if unanticipated plants invade a project. What if freshwater marsh plants grow in a saltmarsh project? Is the creation a failure? The answer to this question largely depends on the specific functions the artificial wetland is designed to serve and these must be outlined in detail in the management plan. In any case, it is useful to document unanticipated results.

Examples of unplanned vegetation communities in artificial wetlands are common. Hardisky (1978) noted after four growing seasons an influx of fresh and brackish water plants overtopping planted spikegrass (Distichlis spicata) in a Georgia estuary dredged material site. Hardisky predicted that the salt tolerant D. spicata would be outcompeted. The growth of freshwater plants in this location indicated a complete inability to predict the hydrology of the area. Similarly, freshwater wetland plants such as the royal fern (Osmunda regalis) and a rush (Juncus sp.) dominated the high marsh in a West Florida dredged material project where Spartina patens had originally been planted. In the New Jersey artificial marshes, Spartina patens was planted in some locations but Spartina alterniflora was dominant during subsequent samplings (Shisler and Charette 1984). Spartina foliosa was planted in a 95 acre (38.4 ha) dredged material site in San Francisco Bay; here, as in other locations in California where S. foliosa has been planted, Salicornia has invaded the area (Race 1985). Dial and Deis (1986) reviewed 10 mitigation or restoration projects in Tampa Bay, Florida. The survival of Spartina alterniflora ranged from 10-93% and the number of plants per square meter ranged from 0 to 230. Dial and Deis (1986) attribute plant deaths to erosion, competition by upland plants, and poor planting techniques. Finally, Savage (1978) photographed the same mangrove plantings in Tampa Bay, Florida over a six year period; Rhizophora and Laguncularia did not survive while Avicennia did.

Odum (1988) points out that invasion by unwanted plants is common in freshwater artificial wetlands. Typha spp. often crowd out more valuable planted species, leading to the "cattailization of America".

Other factors influencing the success of plantings of created wetlands include predictions of hydrologic conditions and proximity to seed source. High water levels and lack of control of water level resulted in death of trees planted on the shore of Missouri River reservoirs (Hoffman 1978). Eastern cottonwood (Populus deltoides) and green ash (Fraxinus pennsylvanica) did not survive inundation, while broadleaf cattail (Typha latifolia) and white willow (Salix alba) did. Cattle grazing also influenced vegetation success on the banks of these dams. Reimold and Cobler (1986) evaluated five mitigation projects in the northeast U.S.; they rated one freshwater site after two growing seasons as "marginally successful" because banks were too steep and water too deep for emergent vegetation. Two other freshwater mitigations rated "ineffective" by Reimold and Cobler were only seen after one year (D'Avanzo 1987). Gilbert et al. (1981) studied a 49-acre tract in Florida that has been mined for phosphate and noted invasion by 50 wetland plant species after 3 years. However, plantings had failed because the hydrology of the site had been incorrectly predicted. Gilbert et al. (1981) concluded that species potentially invading the approximately 27,000 acres in Central Florida used for phosphate mining were site-specific; invading types depend on source material and type of habitats close to the restoration. In the case studied, unmined wetlands supporting a diverse native flora were adjacent to the mitigation project.

Race (1985) reviewed 15 experimental plantings in San Francisco Bay. She concluded that many problems--high soil salinities, incorrect slope and tidal elevations, erosion and sedimentation, and poor water circulation--accounted for numerous failures of the plantings. For example, in the Bay Bridge site, 10% of the Spartina foliosa and 20% of the Distichlis spicata transplants survived one year; Salicornia transplanting was more successful. S. foliosa spread well from plugs at the Marin County Day School location after two years, but the stated objective of the project--erosion control--was not met. All S. foliosa experimental plantings in the Anza Pacifica lagoon failed within 2-3 years; the mitigation site was replanted and, again, after three years only remnants of the planted plugs remained. In three USACE erosion control projects, neither seedlings, sprigs, nor plugs survived longer than eight months (Race 1985); survival of marsh plugs was good only in unexposed areas of marsh and creeks. Experimental plantings of cordgrass seedlings in Muzzi Marsh prior to mitigation were dead after one year. The 125 acre (50.6 ha) Muzzi Marsh project is becoming naturally colonized by Salicornia and Spartina.

The highly experimental nature of marsh creation is clear from Race's critical review of these projects. (See Harvey and Josselyn, 1986 for a critique of this review and Race, 1986 for a reply). Since saltmarsh restoration is a new technology and one with a relatively poor science base, failure of experimental plantings is not surprising. However, it is disturbing when projects that are largely unvegetated or that support exotic vegetation are called successful restorations (Race 1985).

CHEMICAL ANALYSES OF SOILS IN CREATED AND NATURAL WETLANDS

Little data exist on sediment characteristics of human-made wetlands or of comparisons between these sites and natural controls (Race and Christie 1982), although this data base is growing. Several studies do show that nitrogen, phosphorous, and organic matter increase with age of the created site (Reimold et al. 1978, Lindau and Hossner 1981, Craft et al. 1988a). While organic carbon at various depths was considerably less in human-made marshes in

North Carolina, Cammen et al. (1974) estimated that organic content of soils in these creation projects would reach reference concentrations in 4-26 years. Studies by Craft et al. (1988b) with natural isotopes support this trend since marsh plants were the main source of organic carbon in both natural and transplanted marshes. After 2 years, organic matter concentrations, total nitrogen, and ammonium-nitrogen levels in experimental marsh soils from Texas dredged spoil projects were on average 2-3 times lower than those in natural marshes (Lindau and Hossner 1981). Concentrations of these parameters increased with time and Lindau and Hossner concluded that, assuming a linear rate of increase, concentrations would be equal to those in surrounding marshes in 2-5 years.

Despite such predictions, Race and Christie (1982) are cautious in their analysis of these findings; "no man-made marsh to date has shown the stabilization of physical and chemical properties in the range of values for natural marshes". Their caution is supported in the findings of Craft et al. (1988a) who compared natural and planted soil in 5 sites; they concluded that organic matter pools develop in 15-30 years but development of soil C, N, and P pools take much longer.

In some cases, the substrate in the created wetland differs greatly from that in genuine wetlands. Shisler and Charette (1984) found that sand was the substrate most often used in eight artificial marshes studied and this resulted in distinct edaphic differences. Artificial marsh sediment was lower in organic matter, nitrogen, phosphorous and salinity when compared to nearby reference marshes. Only pH was the same.

Chemical/physical analyses of artificial wetland soils are particularly useful indicators of project progress and success with regard to changes with time. It is possible to predict trends (increasing organic carbon concentration, for example) and to determine rates of change of these parameters. The few studies in which this approach was used show that created sites become more like natural ones with time. An important question for mitigation projects is: how much time? The time scale of these projects is several years and sediment in genuine wetlands has developed during hundreds and thousands of years.

EVIDENCE OF GEOLOGIC OR HYDROLOGIC CHANGES WITH TIME IN ARTIFICIAL WETLANDS

Much of this information has been described above but it deserves reemphasis because the geologic/hydrologic setting is so critical in wetlands. Clearly, dramatic geologic or hydrologic changes--including sediment erosion or deposition, or groundwater seeps--will alter creation projects as planned.

Some creations, such as the Panacea Island project in Florida (Kruczynski and Huffman 1978), have entirely eroded away. Waves killed planted mangroves in a Tampa Bay creation (Savage 1978). Wave erosion and sediment inundation was also a problem in some New Jersey mitigations (Shisler and Charette 1984). In a freshwater bank stabilization project, high water killed numerous planted floodplain trees (Hoffman 1978). Finally, Gilbert et al. (1981) noted that plantings failed in a phosphate mine revegetation project because the hydrology of the site was poorly understood.

Some of these events could have easily been prevented. Dikes can be better constructed and creations should be not attempted in areas where erosive forces may negate the project. However, storm damage is impossible to predict in many locations, including the coast and floodplains of rivers. Therefore, it is not surprising that some creations fail.

CONCLUSION

What conclusions can be drawn concerning the long-term evaluation of wetland creation projects discussed in this chapter? Using six criteria as measures of success, there is a striking contrast in the 2-15 year success of different projects. On one hand, for a decade or more the U.S. Army Corps of Engineers has evaluated a large number of wetlands created with dredged materials. When vegetation parameters are used, many of these projects become structurally similar to reference sites with time. In addition, one 15-year old animal study showed a similar trend. Several evaluations of soil chemistry also indicate that these wetlands become more like natural ones with time. USACE researchers evaluate their experiments and use this information in new projects; a large data base about similar types of projects is communicated within the program.

It is important to recognize that even the "old" USACE artificial wetlands are not identical to reference wetlands; for example, soil carbon and belowground plant biomass are developing slowly. Therefore, when an artificial wetland is built as a mitigation for a lost wetland, decades may pass before the created project assumes the structure and function of the lost habitat. During this time, the important functions that the destroyed wetland may have served (Larson and Neill 1987) may be lost to society.

In contrast to these USACE projects, many other artificial wetlands--mitigation projects and experimental plantings--are judged problematic or partial failures in studies of up to several years. Reasons for failures include contamination of soils, herbivory, erosion, and inappropriate hydrologic regime. In addition, many created wetlands have never been evaluated and, therefore, their success in not known. Studies also indicate that a small but disturbing number of required projects were never even initiated.

Many creation projects fail because of improper hydrology. Basic to the entire concept of wetland creation is the existence of a functional hydrologic regime appropriate for the establishment and development of the specific wetland species. For example, water level depth, seed bank potential, and sloping marginal contours are crucial to the development of emergent aquatic plants (Niering 1987). Some types of artificial wetlands do appear stable after several years; perhaps the hydrology of these habitats is less challenging to predict than in other locales. For example, the establishment of low Spartina alterniflora salt marsh has met with considerable success while the creation of high Spartina patens salt marsh has been problematic (Shisler and Charette 1984). Most high marsh sites are adjacent to upland and therefore the hydrology of the high marsh is more unpredictable than that of the low marsh and more difficult to reproduce.

Some created wetlands systems will remain relatively stable over time while others can be expected to change. Hydrology is an important factor determining wetland community changes with time. The basic goal is to create persistent functional wetland systems. In some situations this may be more important than creation of specific wetland types because the present structure of a wetland may be a momentary expression of the wetland of the future.

LITERATURE CITED

Cammen, L.M. 1976a. Macroinvertebrate colonization of Spartina marshes artificially established in dredge spoil. Est. Coastal Mar. Sci. 4: 357-372.

Cammen, L.M. 1976b. Abundance and production of macroinvertebrates from natural and artificially established salt marshes in North Carolina. Amer. Midl. Nat. 96:487-493.

Cammen, L.M. 1976c. Accumulation rate and turnover time of organic carbon in salt marsh sediments. Limnology and Oceanography. 20:1012-1015.

Craft, C.B., S.W. Broome, and E.D. Seneca. 1988a. Soil nitrogen, phosphorus and organic carbon in transplanted estuarine marshes, p. 351-358. In D.D. Hook (Ed.), The Ecology and Management of Wetlands, Vol. I: Ecology of Wetlands. Timber Press, Portland Oregon.

Craft, C.B., S.W. Broome, E.D. Seneca, and W.J. Shower. 1988b. Estimating sources of soil organic matter in natural and transplanted estuarine marshes using stable isotopes of carbon and nitrogen. Est. Coast. Shelf Sci. 26:633-641.

D'Avanzo, C. 1987. Vegetation in freshwater replacement wetlands in the northeast, p. 53-81. In J.S. Larson and C. Neill (Eds.), Mitigating Freshwater Wetlands Alterations in the Glaciated Northeastern U.S.: An Assessment of the Science Base, Publ. No. 87-1, The Environmental Institute, University of Massachusetts, Massachusetts.

Dial. R.S. and D.R. Deis. 1986. Mitigation Options for Fish and Wildlife Resources Affected by Port and Other Water Dependent Developments in Tampa Bay, Florida. Fish and Wildlife Service Biological Report 86(6).

Eliot, W. 1985. Implementing Mitigation Policies in San Francisco Bay: A Critique. State Coastal Conservancy, Oakland, California.

Gilbert, T., T. King, and B. Barnett. 1981. An Assessment of Wetland Habitat Establishment at a Central Florida Phosphate Mine. Fish and Wildlife Service, U.S. Department of Interior, FWS/OBS-81/38.

Guntenspergen, G.R. and F. Stearns. 1985. Ecological perspective on wetland systems, p. 69-97. In P.J. Godfrey, E.R. Kaynor, S. Pelczarski, and J. Benforado (Eds.), Ecological Considerations in Wetlands Treatment of Municipal Wastewaters. Van Nostrand and Reinhold Co., New York.

Harvey, H.T. and M.N. Josselyn. 1986. Wetlands restoration and mitigation policies: comment. Environ. Manag. 10:567-569.

Hoffman, G.R. 1978. Shore Vegetation of Lakes Oahe and Sakakawea, Mainstem Missouri River Reservoirs. Tech. Report U.S. Army Engineer, Waterways Experiment Station, Vicksburg, Mississippi.

Hardisky, M. 1978. Marsh restoration on dredged material, Buttermilk Sound, Georgia, p. 136-151. In D.P. Cole (Ed.), Proceedings of the Fifth Annual

Conference of the Restoration and Creation of Wetlands, Hillsborough Community College, Tampa, Florida.

Josselyn, M. and J. Buchholz. 1984. Marsh Restoration in San Francisco Bay: A Guide to Design and Planning. Tech. Report #3, Tiburon Center for Environmental Studies, San Francisco State University.

Keddy, P.A. and T.H. Ellis. 1985. Seedling recruitment of 11 wetland plants along a water level gradient: shared or distinct responses? Can. J. Bot. 63:1876-1879.

Kruczynski, W.L. and R.T. Huffman. 1978. Use of selected marsh and dune plants in stabilizing dredged materials at Panacea and Apalachicola Bay, Florida, p. 99-135. In D.P. Cole (Ed.), Proceedings of the Fifth Annual Conference on the Restoration and Creation of Wetlands, Hillsborough Community College, Tampa, Florida.

Kunz, K., M. Rylko, and E. Somers. 1988. An assessment of wetland mitigation practices in Washington state. National Wetlands Newsletter 10:2-4.

Landin, M.C. and J.W. Webb. 1988. Wetland development and restoration as part of Corps of Engineer programs: case studies, p. 388-391. In J. Kusler, M.L. Quammen, and G. Brooks (Eds.), Proceedings of the National Wetlands Symposium: Mitigation of Impacts and Losses. Assoc. of State Wetland Mgrs. Berne, New York.

Larson, J.S. and O.L. Loucks. 1978. Research Priorities for Wetlands Ecosystem Analysis. Workshop report by the National Wetlands Technical Council to the National Science Foundation.

Larson, J.S. and C. Neill. 1987. Mitigating Freshwater Wetland Alterations in the Glaciated Northeast: An Assessment of the Science Base. Publ. No. 87-1, The Environmental Institute, University of Massachusetts, Amherst.

Lewis, R.R. and C.S. Lewis. 1978. Colonial Bird Use and Plant Succession on Dredged Material Islands in Florida: Patterns of Plant Succession. Tech. Report D-78-14. U.S. Army Engineer Waterways Experiment Station, CE, Vicksburg, Mississippi.

Lindau, C.W. and L.R. Hossner. 1981. Substrate characterization of an experimental marsh and three natural marshes. Soil Sci. Soc. Am. J. 45:1171-1176.

Mitsch, W.J. 1988. Wetland modelling, p. 1-10. In W.J. Mitsch, M. Straskraba, and S.E. Jrgensen (Eds.), Wetland Modelling. Elsevier. New York.

Morris, J.H., C.L. Newcombe, R.T. Huffman, and J.S. Wilson. 1978. Habitat Development Field Investigations, Salt Pond No. 3 Marsh Development Site, South San Francisco Bay, California. Tech. Report D-78-57. U.S. Army Engineer Waterways Experiment Station, CE, Vicksburg, Mississippi.

Niering, W.A. 1987. Wetlands hydrology and vegetation dynamics. National Wetlands Newsletter 9:9-11.

Newling, C.J., M.C. Landin, and S.D. Parris. 1983. Long-term monitoring of the Apalachicola Bay wetland habitat development site, p. 164-186. In F.J. Webb (Ed.), Proceedings of the Tenth Annual Conference of Wetland Restoration and Creation, Hillsborough Community College, Tampa, Florida.

Newling, C.J. and M.C. Landin. 1985. Long-Term Monitoring of Habitat Development at Upland and Wetland Dredged Material Disposal Sites, 1974-1982. Tech. Report D-85-5. U.S. Army Engineer Waterways Experiment Station, CE, Vicksburg, Mississippi.

Odum, W.E. 1988. Predicting ecosystem development following creation and restoration of wetlands, p. 67-70. In J. Zelazny and J.S. Feierabend (Eds.), Increasing our Wetland Resources, National Wildlife Federation Conference Proceedings. Washington, D.C., October 4-7.

Race, M.S. and D.R. Christie. 1982. Coastal zone development: mitigation, marsh creation, and decision making. Environ. Manag. 6:317-328.

Race. M.S. 1985. Critique of present wetlands mitigation policies in the United States based on an analysis of past restoration projects in San Francisco. Envir. Manag. 9:71-82.

Race, M.S. 1986. Wetlands restoration and mitigation policies: reply. Environ. Manag. 10:571-572.

Reimold, R.J., M.A. Hardisky, and P.C. Adams. 1978. Habitat Development Field Investigations, Buttermilk Sound Marsh Development Site, Atlantic Intracoastal Waterway, Georgia. Technical Report D-78-26, U.S. Army Engineer Waterway Exp. Station, Vicksburg, Mississippi.

Reimold, R.J. and S.A. Cobler. 1986. Wetlands Mitigation Effectiveness. A Report to the Environmental Protection Agency Region I, Contract No. 68-04-0015.

Sacco, J.N., S.L. Booker, and E.D. Seneca. 1988. Comparison of the macrofaunal communities of a human-initiated salt marsh at two and fifteen years of age, abstract. In Benthic Ecology Meeting, Portland, Maine.

Savage, T. 1978. The 1972 experimental mangrove planting--an update with comments on continued research needs, p. 43-71. In D.P. Cole (Ed.), Proceedings of the Fifth Annual Conference on Restoration of Coastal Vegetation in Florida, Hillsborough Community College, Tampa, Florida.

Schneider, R.L. and R.R. Sharitz. 1986. Seed bank dynamics in a southeastern riverine swamp. Amer. J. Bot. 73:1022-1030.

Seneca, E.D., S.W. Broome, W.W. Woodhouse, L.M. Cammen, and J.T. Lyon. 1976. Establishing Spartina alterniflora marsh in North Carolina. Environ. Conservation 3:185-188.

Shisler, J.K. and D.J. Charette. 1984. Evaluation of Artificial Salt Marshes in New Jersey. New Jersey Agri. Exp. Station Publ. No. P-40502-01-84.

Thorhaug, A. 1979. The flowering and fruiting of restored Thalassia beds, a preliminary note. Aquat. Bot. 6:189-192.

Thorhaug, A. 1980. Environmental management of a highly impacted, urbanized tropical estuary: rehabilitation and restoration. Helgolander Meeresunters 33:614-623.

Valiela, I. 1984. Marine Ecological Processes. Springer-Verlag, New York.

Zedler, J., M. Josselyn and C. Onuf. 1982. Restoration techniques, research and monitoring vegetation, p. 63-74. In M. Josselyn (Ed.), Wetland Restoration and Enhancement, Report No. T-CSGCP-007, Tiburon Center for Environmental Studies, Tiburon, California.

REGIONAL ASPECTS OF WETLANDS RESTORATION AND ENHANCEMENT IN THE URBAN WATERFRONT ENVIRONMENT

John R. Clark
Rosenstiel School of Marine and Atmospheric Science
University of Miami

ABSTRACT. In urban settings, wetland resources are typically degraded and often seriously dysfunctional. Loss of wetland function in this manner reduces the productivity of the larger aquatic ecosystems of which the wetlands are a component. Therefore, in urban settings a high priority must be given to restoration and enhancement of aquatic ecosystems and to their component wetlands. Success in system-wide restoration requires formulation of a regional strategy with goals, objectives, methodologies, and predesignated restoration sites. Such strategies must be locally generated and cannot be substituted by existing one-agency programs. All levels of government and private interests must be involved. Moreover, the existing system of site-by-site permit review must be altered to ensure that permit decisions are oriented toward the regional restoration strategy. It is particularly important to recognize that the developers' resources will be the main source of restoration project funds through voluntary or mitigative restoration and enhancement. Therefore, mitigation has to be given a role at the front end of the review process and not held until the end as a "last resort".

INTRODUCTION

The urban setting presents distinctive problems for waterfront development as well as special opportunities for restoration and enhancement. For several reasons mitigation has the potential to become a positive tool for restoration and enhancement rather than just an obstacle to developers (Wessel and Hershman, in press). Currently, the limited waterfront property that exists in most urban areas is being aggressively sought for residential and commercial development. The pressures are great, front-foot prices are astronomical, investment funds are abundant, and profits are assured, providing that permits can be obtained with reasonable effort. This pressure has resulted in renewal projects for much of the old, degraded, urban waterfront area in coastal cities (Figure 1). For example, the buildout cost of renewal projects now underway or proposed for the New Jersey side of the Hudson River opposite New York City is estimated at $10 to $12 billion.

In urban waterfront settings, there are virtually no original or unaltered wetlands or intertidal flats left. Wilderness is not found here. Most of the urban shoreline edge has been "hardened" and dredged or otherwise altered, changing its ecological character to the detriment of fish and wildlife. The result is an overall reduction in biological diversity and carrying capacity for desirable species within the adjacent estuary, river, or lake. This habitat needs to be repaired as much as to be protected. Much of the urban wetland we try to protect is so damaged that to be worth saving it should be repaired. Therefore, in urban settings, restoration and rehabilitation should be given maximum attention.

It was predicted in 1980 that restoration would reach the top of the wetland agenda during the decade of the 1980's (Clark and McCreary 1980). This has certainly occurred for urban aquatic ecosystems but in most regions it has happened de facto , not as the result of policy decisions or program commitments (one exception is the California State Coastal Program). Because funding possibilities are so limited in the urban setting, rehabilitation should be primarily focused on biological needs, such as restoring habitat for endangered or commercially valuable species or enhancing critical processes of the wider wetland ecosystem (Clark 1979). As put by Batha and Pendleton (1987): "Lack of suitable enhancement sites at reasonable cost and conflicts among agencies as to what type of habitats are of greatest importance to the Bay system greatly concern all people interested in the future of San Francisco Bay.... Rapid urbanization will make the possibility of adding wetlands to the Bay increasingly difficult in the future".

One example of the multiobjective approach

Figure 1. Typical waterfront scene on the Hudson River in Manhattan. Original wetlands and tideflats were replaced with piers and channels which are now deteriorated and dysfunctional. Some interests want to retain these piers, others want them removed, and still others want ecological rehabilitation in combination with residential development.

that can be used in an urban situation is the Pt. Liberte canal side residential project in Jersey City, N.J. (see following section: "Case Study Port Liberte"). Here the development site itself was small, but peripheral areas and edges were used in a voluntary program of multiple enhancements (Figure 2) to: rehabilitate and reroute a seriously degraded stream, rehabilitate a degraded tideflat/ slough system, enhance the beach-dune system, create a least tern nesting site, build an artificial reef, enhance a small peninsula owned by Liberty State Park, and protect the state-designated Caven Point Natural Area (tideflats and shallow waters) adjacent to the site from boater damage.

It is unfortunate that aquatic habitat restoration has not been given the same priority as water quality restoration which has received great attention and lavish budgets in the past 15 years. Direct appropriations for physical habitat repair and ecosystem restoration by either Federal or state governments have been minuscule, regardless of how compelling the need may have been (Clark 1985). Progress in habitat restoration has for the most part, been left

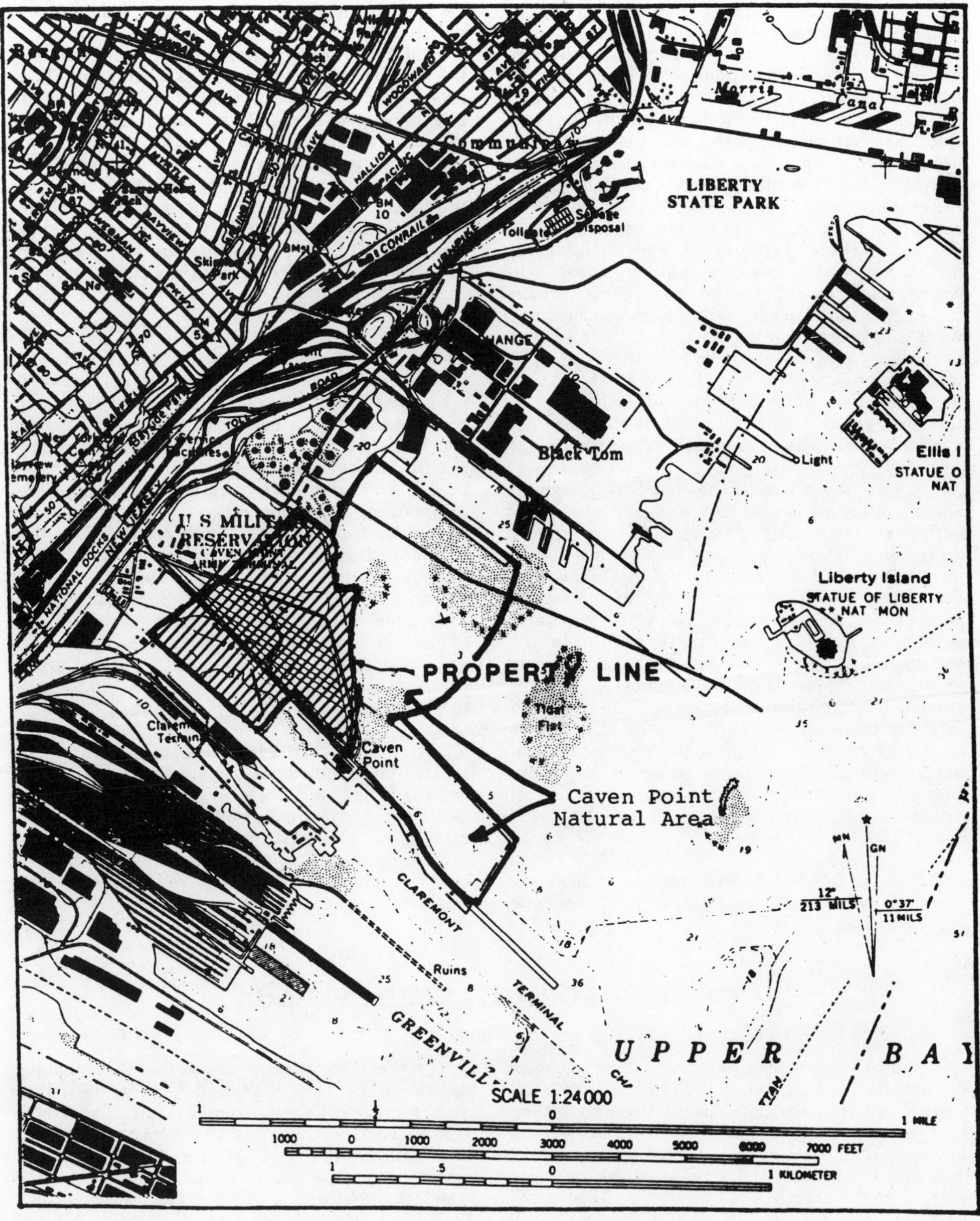

Figure 2: Jersey City, N.J., waterfront now in active redevelopment from industrial to residential/commercial. Shaded section is the site of the new Port Liberte project which is occupying an abandoned military site.

to casual and secondary mechanisms. The most promising of these mechanisms is mitigation in exchange for development permits, whereby physical habitat restoration is, in effect, exacted from developers by regulators as a quid pro quo for obtaining waterfront development permits. But this approach has been less than successful because the mitigation process under most permit programs has been poorly organized and ad hoc. Since long term future goals have not been established for particular ecosystems or regions, nor guidelines formulated for developers, mitigation has been a case-by-case, uncoordinated activity. Often this seems to have been deliberate, because regulatory agencies may want to relegate mitigation to a "last resort" basis. This may be commendable in rural areas where more natural conditions exist, but in urban settings mitigation may often have to be the "first resort" where no other mechanisms are available for repair of damaged aquatic habitats.

A mitigation goal of "no net loss of habitat value", often advocated for rural settings is not sufficient for highly damaged urban aquatic ecosystems. Here the goal should be "a net gain in habitat value" if we are to regain the losses of the past, reach higher levels of biological productivity, and accomplish recovery of depleted populations of economically valuable or endangered species. This goal could be stated in another way, specifically as a policy "to achieve net positive cumulative impacts". This approach to ecosystem recovery through strategic, rather than reactive mitigation has promise for certain urban aquatic ecosystems where a basis for cooperation among regulators, developers, scientists, and environmentalists can be found (Clark 1985).

It is far too easy to drift into an attitude, or approach, where only potential negative impacts are addressed. I believe we should try hard to prevent future losses of wetland, but we also should work to regain lost functions. To be workable, the restoration approach must address the total individual wetlands ecosystem (lake, river, estuary, marsh, etc.); that is, the whole aquatic system of which the wetland is a part. We should recognize that certain functions are being lost in an aquatic system, see which are dependent upon wetlands, establish priorities for the functions of greatest value (e.g., bird habitat, flood storage, productivity) and enhance these functions. This would reverse the serious decline of productivity and diversity in aquatic habitats in the urban setting.

It is fair to say that most mitigation consultants find the typical permit-by-permit approach of regulatory agencies ineffective in advancing long-term goals for aquatic ecosystem conservation and a deterrent to strategic system restoration. Further, individual permit reviews should be evaluated wherever and whenever possible through a regional strategy for restoration. I recommend that goals and targets should be determined in advance according to a regional strategy and used to guide subsequent permit actions involving restoration and enhancement. For example, Sorensen (1982) concludes that "...the relative scarcity and abundance of the resource needs to be determined on a region-wide basis in order to set priorities on the types and locations of habitats that should be provided in a restoration site plan".

The ideas expressed in this chapter are particularly applicable to the urban settings of coastal cities and their surroundings. But the principal recommendations involving regional strategies, goal-setting, ecosystem focus, and predetermination of restoration needs and sites could apply to less urbanized seacoast and freshwater areas.

THE NEED FOR ADVANCED CRITERIA

If urban developers (whether public agencies or private corporations) are to cooperate in restoration of aquatic habitats through either mitigation or voluntary enhancement, they must have some guidance. They should know what is expected of them in the context of the regional ecosystem in which the project is located. If voluntary enhancement is to be encouraged, they should know what specific opportunities exist and what public interests they can best serve. Moreover,they should know this information at the time they are planning their projects, not after the application has been submitted and the Section 10 or 404 Public Notice has been circulated by the Army Corps of Engineers (COE).

In the mitigation process, the COE usually defers to the U.S. Fish and Wildlife (FWS) to assess mitigation requirements and expects to receive FWS advice after the developer's permit is submitted (COE 1985) and the developer is already committed to a certain plan. If the COE does not agree with FWS or other commentators, including EPA and the National Marine Fisheries Service (NMFS), a prolonged and expensive delay will often result (months or even years). For example, Zagata (1985) states, "The mitigation requirement in 404 of the Clean Water Act has been a source of controversy between the regulating agencies and permit applicants. The need to mitigate is considered at the end, rather than the beginning, of the permit process, after other alternatives have been examined. Thus, industry frequently perceives mitigation as an additional source of delay and money...an add-on-cost since it is considered after completion of the proposed project's normal budgeting and planning process". All of which may be welcome if you're only trying to stop a project but not if you're trying to promote restoration of urban aquatic ecosystems.

In order to succeed, effective restorative mitigation must be a cooperative venture between developer and agencies. As of now, the developers' opinion of mitigation is typically one of frustration (Wilmar 1986):

> "The Corps usually recommends that the applicant embark upon a series of negotiations with the various commenting agencies. This is generally a frustrating exercise because there are few rules, and commenting agencies have broad discretion to interpret those standards that do exist. Moreover, the applicant has usually already obtained approvals from the local and state agencies, each of which has extracted concessions as the price of project approval. Thus, the unsophisticated, generous, or inexperienced applicant often has no more to give, and the federal agencies have little other incentive to reach an agreement."

By the time the application has been submitted, positions have hardened and options have been closed for both the developer and the regulator. Aware of this situation, EPA and other principal agencies including state agencies will often meet with the developer in "pre-application" conferences that FWS mitigation policy presumably encourages (FWS 1981). This can be beneficial if various agency staff can come to agreement among themselves and give unambiguous advice to the developer (Is high marsh lower priority than low marsh? Are bird breeding islands a beneficial substitute for open water surface, water column, and bottom? Are piers over bare bottom beneficial or detrimental? Should mudflats be converted to marshland? Are ducks more important than fish?). However, all too often an agency's staff has not had a chance to come to consensus in advance on the issues of habitat option preferences. The developer is too often left to gamble on which mitigation approach might be best to get him through the permit process. It is a major problem for the permit process that a formal method for prior consensus on regional mitigation priorities does not exist. It is also a major problem that current mitigation manuals or guidebooks on mitigation methods and preferences are not available to development planners to consult throughout the siting and design process.

What can an individual EPA or other agency permit reviewer do to improve this situation in a region where no organized mitigation policies and programs are in place? The answer is to work to clarify and reach consensus among agency colleagues on mitigation procedures and priorities, and to assist in conveying the results to the applicant at the earliest possible time. An appreciation of the urgent necessity for restoring urban wetland-related ecosystems using mitigation is, of course, the precursor to agreement on mitigation targets. While FWS and COE are the main Federal agency actors in early permit skirmishes, EPA has a strong influence because of the agency's ultimate "veto" power.

The advantages of advance criteria and early coordination in project planning are stated by Dial et al. (1985): "To be most effective in preserving habitat, mitigation activities should begin during the planning phase of a project. It is usually only at this phase that the avoidance or minimization of the impacts is possible, and mitigation in the literal sense of the word occurs".

THE ADVANTAGE OF THE REGIONAL STRATEGY

A major challenge to EPA and other agencies is to make the fundamental shift from a site-by-site focus to a regional focus for urban areas. Because permit actions are typically confined to a project site, often there may be little knowledge or concern about the relationship between that site and the regional ecosystem incorporating the project site. Ecosystem thinking is engendered by the regional approach. One can't think of each individual wetland as a unit of landscape in isolation, but rather in terms of the whole system.

A major advantage of taking the regional ecosystem view by thinking beyond the immediate project site to consider the whole, is that most wetland functions that we value involve a related aquatic system that is larger than the affected wetland itself. The wetland unit often depends on the larger aquatic system to actually realize the potential of a particular wetland function. For example, detrital output is a value only if there is a living community beyond the wetland to utilize it. Likewise, if a wetland is to serve as a nursery for fish it must be accessible to an adjacent healthy, functioning, major aquatic system which depends upon many components other than wetlands. For example, a snook nursery area needs to have a shallow intertidal area with mangrove edge and an admixture of fresh water and, outside, a productive feeding and cryptic habitat of seagrass beds. After this period (1/2 year or so) the snook move into deeper waters and utilize a variety of habitats in enclosed waters of estuaries and around channels, where good water quality becomes important. Obviously, we need to be concerned about more than acres of wetlands (mangrove) if we want to improve the snook's lot (now greatly depleted). We want to reverse the degradation of the whole aquatic system and achieve a positive direction in various permit review and mitigation activities. Consequently, in restoring or creating wetland units in mitigation, we have to decide what is the optimum balance of, say, high wetlands, low wetlands, flats, channels, and open shallow waters (Clark 1986b).

When the potential for mitigation or voluntary enhancement arises, the question follows of what specific restoration projects should be recommended. With private developers, the question is most often fielded by a consultant; with public projects, a staff professional usually provides advice. Interaction of these professionals with regulatory agency personnel is most often the key to efficiency in subsequent review of the permit and in approval of the mitigation/restoration program. This is all that would be necessary in a perfect world characterized by mutuality, omniscience, and altruism. However, in the real world of permitting, the process is typically an adversarial one, each permit is handled de nouveau, advance goals are absent, and agency reviewers are often unsympathetic to development and reluctant to commit to specifics and foreclose their post-submittal options. Thus, the official pre-application conference (as advocated, for example, by FWS mitigation guidelines, Fed. Reg., Jan. 23, 1981, Vol. 46, No. 15, p. 7644 et seq.) may fall short of developer needs and may discourage restoration initiatives. This confirms the strong need for directive guidelines to make the process predictable to developers and to make available to their environmental experts advance mitigation criteria for use at early planning stages. This can only be accomplished effectively in a regional context.

The shift from the reactive to the strategic approach would bring a shift from "supply-side" to "demand-side" thinking about wetlands. That is, assessing the condition of a wetlands system begs the strategic question "What are the societal demands for natural goods and services from this system and how well are they being met?". This replaces the reactive question "What natural goods and services does this wetland supply?".

Effectiveness in aquatic habitat restoration requires understanding the regional ecosystem, its present condition (how far degraded), and what values are most important and should be given priority for rehabilitation (plant productivity? bird habitat? nursery area?). Given this, one can formulate goals and advance criteria for permit review and mitigation and even reverse the trend of negative cumulative impacts and bring about a positive cumulative impact sequence.

REGIONAL ORGANIZATION

It is within the purview of the EPA or other agency reviewers to consider each wetland unit as part of a greater ecological and hydrologic system when dealing with restoration and reversal of cumulative impacts. Thinking discriminately is important, including considering the variety of configurations, functions, and social needs that restoration projects can meet. From the ecological engineering point of view, given the money, one can do almost anything to a wetland. It can be regraded, reshaped, rewatered or replanted. The substrate can be changed, the elevation, topography, or the supply of water (Clark 1986a).

But beyond the technical issues, there are judgmental questions to be answered: What wetland design would yield the highest socioeconomic benefit, considering regional needs for natural goods and services? If waterfowl habitat is critical, then a relatively shallow open water area would be most appropriate (Figure 3). If shorebird habitat, shoreline stabilization, or run-off water purification are the priority needs, then different designs are indicated. The strategic approach to mitigative restoration requires that someone other than the project developer or permit reviewer--preferably a regional entity--select the regional priorities for aquatic ecosystem outputs of natural goods and services. Once that is accomplished, an environmental professional can convert these priorities to functional criteria and engineers can convert the criteria to design specifications and construction (Clark 1986a).

Figure 3. Degraded streambed on the Port Liberte site scheduled for rehabilitation to enhance its value to waterfowl, shorebirds, and nursery size fishes and to remove high concentrations of toxics in the streambed.

Because of the variety of public interest questions involved in mitigative restoration, individual agency permit reviewers would benefit from advance formulation of regional goals and restoration priorities by a recognized entity charged with balancing the variety of private and public interests. Where regional entities have been established to deal with aquatic habitats and permits and have formulated guidelines and criteria, the results seem to have been helpful.

Federal/state programs that can be used to explore regional possibilities include the following:

1. EPA's authority for "advance identification" or "predesignation" of wetlands (under Sect. 404{c} or Sect. 230.80 of the Guidelines) which enables EPA to list those that are off limits to dredging and filling in a particular region (e.g. the Hackensack Meadows). The procedures for advance identification provide for input from a variety of interests, scientific evaluation of wetlands values, and a plan for priority protection of critical habitats. But the program may or may not deal effectively with whole ecosystems or with restoration needs (Studt 1987).

2. The COE's long-term management strategy (LTMS) for dredging activities on a regional basis strongly encourages and assists Districts in developing LTMS's within their boundaries (Klesh 1987). Districts with LTMS's in place or in planning include St. Paul, Rock Island, Seattle, and Portland.

3. States' authority under the Federal Coastal Zone Management Act (1980 Amendments) to do regional Special Area Management Plans (SAMP's), whereby all aspects can be considered in a regional planning context. SAMP's can effectively establish regional strategies for aquatic

habitat restoration, including restoration criteria, identification of mitigation sites, preparation of guidelines, and mechanisms for incorporating mitigation into a restoration master plan. Federal agencies must be consistent with approved state SAMP's (Studt 1987). A major example of a successful SAMP is that for Rhode Island's salt pond region (Olsen and Lee 1985) whereby strong guidelines for development and aquatic system restoration in the salt ponds were formulated with the participation of a wide spectrum of agencies and environmental and private interests. The main example of the difficulty of the SAMP approach is Gray's Harbor, Washington, in which Federal and state agencies and the local planning entity and port authority have spent more than 8 years trying to agree to a plan.

4. "Area wide" advance Environmental Impact Statements developed by the COE, and often advocated by FWS, provide an opportunity to review the condition and restoration needs for regional ecosystems. These need to be set up for wide consensus, public interest balancing, formulation of policy, and implementation of positive programs.

5. Use of the Estuarine Reserves program, authorized by the Federal Coastal Zone Management Act to organize regional aquatic ecosystem conservation programs. Examples of strong local estuarine reserve programs which have enhanced restoration and provided an advance framework for permit decisions include Apalachicola Bay (Florida) and Tijuana Estuary and Elkhorn Slough (California) National Estuarine Reserves (Clark and McCreary 1987).

6. State-organized advance designation of mitigation sites. This approach, currently operating in California and proposed by New Jersey, is especially applicable to the urban setting where on-site mitigation opportunities are limited (however, they do imply advance acceptance of the idea of offsite mitigation, which is not viewed favorably by some agencies). Such programs are an excellent way to pinpoint the need for restoration and to take definitive steps to establish priorities for restorative mitigation.

To the extent possible, regional needs and opportunities for restoration should be included in any initiatives under the above programs.

Locally organized programs designed for specific regional aquatic ecosystems have been successful in many coastal urban areas. Although not usually designed specifically for restorations, these locally organized programs can be most helpful in generating a consensus on restoration needs and guiding regulatory agencies toward restoration priorities in permit decisions involving mitigation and voluntary enhancement. Examples of such programs in urban coastal areas are:

1. The Bay Conservation and Development Commission (BCDC), the original regional organization for aquatic ecosystem conservation, was founded in 1965. All development around the shoreline of San Francisco Bay must be permitted by BCDC which has worked to find broad consensus on aquatic ecosystem conservation and to formulate mitigation guidelines. A major restoration mitigation goal is to require opening of diked wetlands in compensation for any filling allowed.

2. The Environmental Enhancement Plan for Baltimore Harbor (1982) by the Regional Planning Council for the Baltimore Metropolitan Area (a Maryland state body) broke a 10-year deadlock over fill, dredging, and dredge spoil (dredged material) disposal when it was modified to be acceptable to Federal agencies (by eliminating a mitigation bank). The Plan includes rehabilitation of aquatic habitats and creation of wetlands. Five sites were selected in advance for mitigation activity. This approach made mitigation more rational and expedited the permit process.

3. The Tampa Bay Regional Planning Council initiated aquatic habitat management planning action that resulted in a cooperative agreement with FWS to, among other things, identify mitigation options and select mitigation sites. This action has resulted in enhanced cooperation among various interests and has expedited permit approvals.

4. The Ports of San Pedro and Los Angeles jointly developed the "2020 Plan" for port expansion which includes specific mitigation commitments according to a Memorandum of Understanding signed by the interested agencies. Mitigation requirements will be met by off-site restoration (there being extreme limits on available mitigation sites in the Los Angeles harbor area). Specifically, as a first goal the entire Bataquitos Estuary ecosystem will be restored to a prescribed level of function.

5. The Bataquitos Lagoon restoration project is a regional effort, organized cooperatively with several Federal and state agencies and local institutions with a goal to restore the entire aquatic ecosystem of the lagoon (ca. 1,000 acres) by off-site, out-of-kind (mostly) mitigation. Included is sediment removal,

building of least tern nesting sites, beach nourishment, inlet stability, creation of freshwater marsh, etc. This project, funded as mitigation for dredge-and-fill by the Port of Los Angeles, is seen as the first in a series of restorations under a long-term cooperative integrated regional plan (Marcus 1987).

6. Fraser River Estuary Management Programs (British Columbia, Canada) initiated in 1977, resulted in a plan in 1982 and 1985 for the joint management and restoration of the estuary by a network of national and provincial government authorities. The plan was preceded by a thorough inventory, consensusing, goal setting, and criteria formulation process. The management plan makes decision making predictable. Under the coordinated Project Review System, developers get a 30-day response (the interagency committee meets bi-weekly). The North Fraser Harbour component has a mitigation bank with preselected littoral sites for restoration and a system to intercalibrate relative values of different wetland types and other shallow aquatic habitats (Williams and Colquhoun 1987).

7. The Biscayne Bay Management Project (Florida) has integrated a variety of authorities and actions toward a master plan approach to water quality and aquatic habitat restoration and conservation. While management is less centralized than examples above, the integrated consensus formation and networking have enhanced restoration and made development constraints more predictable.

REGIONAL GOALS AND MITIGATION TARGETS

Any regional strategy for aquatic habitat restoration requires formulation of goals, often followed by objectives, guidelines, and criteria for project evaluation. The process of goal setting should incorporate the policies of agencies and the views of the full spectrum of private and public interests involved. Even when completed, the strategy will most likely be advisory and not a substitute for existing agency authorities and prerogatives.

The following is recommended as a conceptual approach to the goal setting process (Josselyn and Buchholz 1984, quoted in Quamman 1986):

> "The government agencies involved in managing and regulating natural resources need to identify restoration goals which state the habitats and functions deemed to be important within each ecoregion. This will result in improved project coordination within an ecoregion, and also allow for identification of the cumulative effects of piecemeal alterations in the region. The initial step in identifying restoration goals involves determining the types and area of the different habitats present, as well as their rates of losses and gains. This determination, coupled with knowledge about the importance of each habitat to the ecoregion's key species, will provide the information needed to decide which habitat types should be restored or replaced."

An example of Regional Restoration Goals generated by the California Coastal Commission is the following for the South Region of the California Coast (Calif. Coastal Comm. 1987):

> "The predominant restoration goal for this region should emphasize the creation of open circulation, low intertidal habitat interspersed with salt marsh patches to enhance shorebird, diving duck, and marsh and wading bird populations. The open circulation pattern will enhance local fish and invertebrate populations and keep mosquitos and flood control activities relatively easy. The salt marsh areas should be sufficient in size to maintain endangered species populations."

Such goals provide a good starting point for the technical realization of restoration but obviously need to be extended with detailed criteria.

If you are fortunate enough to be reviewing permits for an aquatic ecosystem that is covered by a regional strategy with goals and criteria for restorative mitigation where previous analytic steps have been taken, you have only to match the development project with identified-in-advance restoration targets and procedures. If not, you may nevertheless be able to analyze the regional ecosystem involved and identify targets that would be acceptable to your colleagues in the other agencies, environmentalists, and the developer. One major issue is to evaluate the mitigation or voluntary restoration in the context of the needs of the wider ecosystem. Most wetlands do not exist in isolation; they are coupled to wider aquatic ecosystems (Figure 4). Another major issue is to recommend mitigation targets that have high priority in terms of your perception of regional ecosystem needs.

This approach is different from the familiar acre-for-acre compensation requirements because of its urban orientation. Urban aquatic systems are always in need of repair. These needs can be diagnosed and specific treatments prescribed that will be of much greater value than formula acre-for-acre replacement of a particular marsh, mudflat, or beach habitat type. For example, ducks may be more in need of protected shallow water area than of emergent marsh, or terns more in need of sandy nesting islands than mudflats. Critical habitat needs such as these can be identified in most urban aquatic ecosystems. In another chapter of this volume, Erwin suggests giving priority to defining mitigation goals specifically in terms of fish and wildlife targets and in fulfilling those goals, allowing the maximum flexibility and creativity.

Trying to restore an existing wetland to its original, pristine condition may not always produce the most appropriate result in terms of meeting the region's critical need for wetland output. For example, creating a coastal high marsh area of sea daisy and saltwort, although a close replication of the original wetland, may be of far less value than a replacement low marsh of mixed cordgrass and mangrove with open channels, which would both provide a nursery area for fishes and an export of detritus to the estuary. Often there is a current regional demand caused by shortages of particular types of wetland function that are recognized for a particular region (Clark 1986b). Whether the shortage has occurred because of wetlands conversion or wetland dysfunction, the demand can be at least partially provided through repair of dysfunctional ecosystem units in many circumstances (Figure 5).

Any regional strategy can be organized to respond to "cumulative impacts" and to provide "offsets" for any environmental damage in degraded ecosystems (as for air quality "non-attainment" areas). Under a regional strategy, a priority goal would be to reverse the accumulation of negative impacts, and begin a trend of positive cumulative impacts for the regional ecosystem. The regional authority would determine a baseline condition, or threshold level, for attainment by examining historical trends of resource losses for the ecosystem. Future restorative mitigation would have an overall target to return the system, via positive cumulative impacts, to an earlier designated level of productivity (e.g., for the Chesapeake Bay, return to the status of the year 1950 seems to be favored).

MULTIPLE IMPACT PROJECTS

In urban waterfront projects, several different impact types can often be identified; some positive and some negative. In this situation, the balance of net benefits and losses must be determined in some fashion based on qualitative and quantitative factors. If no predetermined scheme is available to convert "apples to oranges", the process may have to be more judgmental than analytical. In the previously cited official FWS mitigation policy (p. 765f2) it is stated that: "... the net biological impact of a project proposal is the difference in predicted habitat value between the future with the action and the future without the action". In effect this encourages the developer to present an actual "balance sheet" in support of his application (including voluntary enhancements) which shows for each of the important functional categories the extent to which the project will benefit or harm the ecosystem in the "without project" and "with project" scenarios. Table 1 illustrates a very simplified example of a summary sheet.

Sophistications that can easily be introduced into such comparisons include FWS "resource categories", HEP analysis, and relative value calibrations.

This approach encourages the developer (with his consultant's advice) to incorporate a variety of voluntary restorative enhancements into project design in the early planning stages. However, ambiguity is introduced by the FWS interest in holding to itself the determination of "...whether these positive effects can be applied towards mitigation" (FWS mitigation policy, p. 7652). If such interpretation is actually delayed until permit application is submitted, no improvement in predictability is achieved and developer-supported restorative mitigation is frustrated. Where restorative mitigation will be in the offing, it behooves EPA and other agency reviewers to provide secure advice to developers as early in the pre-application process as possible.

Figure 4. This Spartina patens marsh on the Port Liberte site appears isolated but it is strongly linked to the lower Hudson Estuary through runoff drainage, tide action, detrital outflow, animal movements, and other factors

Figure 5. Repair of this shallow, polluted, marsh/tideflat/slough is part of the Port Liberte enhancement program.

Table 1. A very simplified example of a summary sheet that could be used to review the possible impacts of a proposed project.

Category	Without Project	With Project	Net Impact	Comments
Channel circulation	Restricted, stagnation, eutrophic	Free flow non-stagnant, non-eutrophic, improved access, fewer mosquitos	Highly positive	Depth of 4' specified; range of 3-5' would be acceptable
Macrophytes	Unimpeded insolation	Piers, boardwalks will shadow 3-1/2 acres	Moderately negative	Plank spacing of 1-1/4" will reduce shading effect

MITIGATION BANKS AND ALTERNATIVES

In the urban setting it is most difficult to avoid going off site with mitigation where it is required of a development. Waterfront is so scarce and valuable that land parcels for development are small and use is intensive. Therefore, land available for mitigation is at a premium. On the other hand, waterfront redevelopment creates extensive benefits through value added to the land, taxes, jobs, and particularly, through ridding the waterfront of the blight of decaying warehouses, collapsing docks, and health and security nuisances. Most communities will vigorously support waterfront renewal. This must be strongly considered by COE in its public interest review.

These two factors, motivation for intensive use of waterfront parcels and the limited options for onsite mitigation, create strong pressure to find offsite solutions for mitigation demands, often through some type of "mitigation bank". One solu- tion is a mitigation bank whereby developers are "taxed" for impacts and the proceeds "deposited" in a habitat creation/restoration account similar to the "impact fees" for infrastructure often charged to dryland developers. A second solution is a different kind of bank whereby areas in need of habitat restoration or suitable for habitat creation are "banked" so as to be available in the future for mitigation requirements levied against developers.

A third solution is a regional cooperative restoration plan, whereby mitigation for various projects is done at predesignated sites. It would be as though, at Bataquitos Lagoon for example, several developers had participated sequentially in the restoration project. The major requirement is advance designation of sites for restoration as the COE does now for dredge spoil (dredged material) disposal areas. Designation can be done by a regional body (e.g., North Fraser River), a state (e.g., California coast, or New Jersey's proposed advance site program), a Federal agency, or a special agency like the California Conservancy (McCreary and Zentner 1983). This approach avoids the appearance of a developer "buying" a permit, as in the first solution (which is not looked on favorably by most agencies; e.g., the Baltimore Harbor plan was rejected by the COE for this reason).

CASE STUDY PORT LIBERTE

This case study describes aspects of an innovative residential project in Jersey City, N.J. along the shores of the Hudson River. A 125-acre parcel of previously filled land (by the U.S. Army in the early 1940's) was excavated to provide 2 linear miles of canals for waterside housing (Figure 6). The developer engaged a panel of experts to prepare a series of enhancements (to be paid for by the developer) in advance of formal permit review.

Port Liberte, bounded by the Caven Point Natural Area and Liberty State Park is a canal side residential marine community currently under construction with 1690 condominium units, commercial space and marina. Caven Point Natural Area represents one of the few remaining remnants of the natural estuary with a Spartina salt marsh and tidal mudflat (Burger and Clark 1987). The genesis of the Port Liberte project was and continues to be one of sound environmental planning whereby appropriate geological and biological expertise guided the development, architectural design, and construction schedule of the project and the monitoring programs. With the acquisition of the permits, major construction began in 1986 but careful monitoring of water quality, fish populations (Figure 7) and avian use (the three critical resources on the site) continues and will continue during the project. This is one of a few projects to involve a year of monitoring prior to permitting, with continued monitoring during buildout. Monitoring data obtained before construction was used to physically design the marina, channels, canals, and boardwalks as well as the timing of particular construction schedules. One notable and unusual aspect of the Port Liberte project was the cooperative nature of the interactions between project personnel, government personnel, scientists, and environmentalists, rather than the usual more adversarial approach (Burger and Clark 1987).

That the complex project received its New Jersey permit in less than 1 1/2 years is owed to the collaborative spirit in which the project evolved. Another important factor was that, due to confidence in the process, the state permitting authority was willing to extensively "condition" the permit rather than wait until the multitude of details were settled and the COE was agreeable. By this means, the project could get underway and the issues could be worked out simultaneously, requiring an extended in-process dialogue with state and federal agencies (still ongoing) and continuing ecological baseline and monitoring studies. Also, full use was made of the opportunity for pre-application conferences and informal dialogue with state and Federal agency personnel (Burger and Clark 1987).

The Port Liberte Restoration Design Panel, an interdisciplinary group of ecological experts, (academics and consultants) was formed to review Port Liberte development plans and to provide advice on ecological enhancement. At this point no specific mitigation requirements had been mandated, but permit review authorities did expect a good and sufficient enhancement effort which was strongly supported by the developer, the Port Liberte Partners. Consequently, the Panel was encouraged to brainstorm freely and to formulate an optimum variety of ecological enhancements for the project.

The panel was concerned with using the opportunity to fulfill current ecological demands. That is, rather than simply planning to supply so many acres of habitat, the panel wanted to meet responsible local demand for ecological goods and services. For example, it was recognized that the endangered Least Tern needed safe nesting sites and that the profusion of aquatic birds using the littoral zone of the project area needed both an adjacent source of fresh to slightly brackish water and continued access to low cover on the beach berms at high tide, as well as protection from disturbance. A second priority goal was to see that adversely impacted aquatic areas were rehabilitated. While accomplishing the above, the panel recognized that many constraints were operating and attempted to stay within the limits of practicality imposed by permit conditions and project requirements.

The Port Liberte Restoration Design Panel met in September, 1985 to develop criteria for ecosystem enhancements, including restoration, rehabilitation, and creation of aquatic subsystems. The enhancement concepts had been reviewed in advance by the State of New Jersey (Department of Environmental Protection, Division of Coastal Resources) and conditions had been imposed. Consequently, the Design Panel was simultaneously considering the developer's proposals, the state's reactions and requests, and the individual ideas of panel members.

The mandate was specifically to advise the developer on restoration and enhancement of natural systems within and adjacent to the project area, particularly in regard to the following:

1. Rehabilitation and rerouting of Caven Creek, a drainage channel that transects the

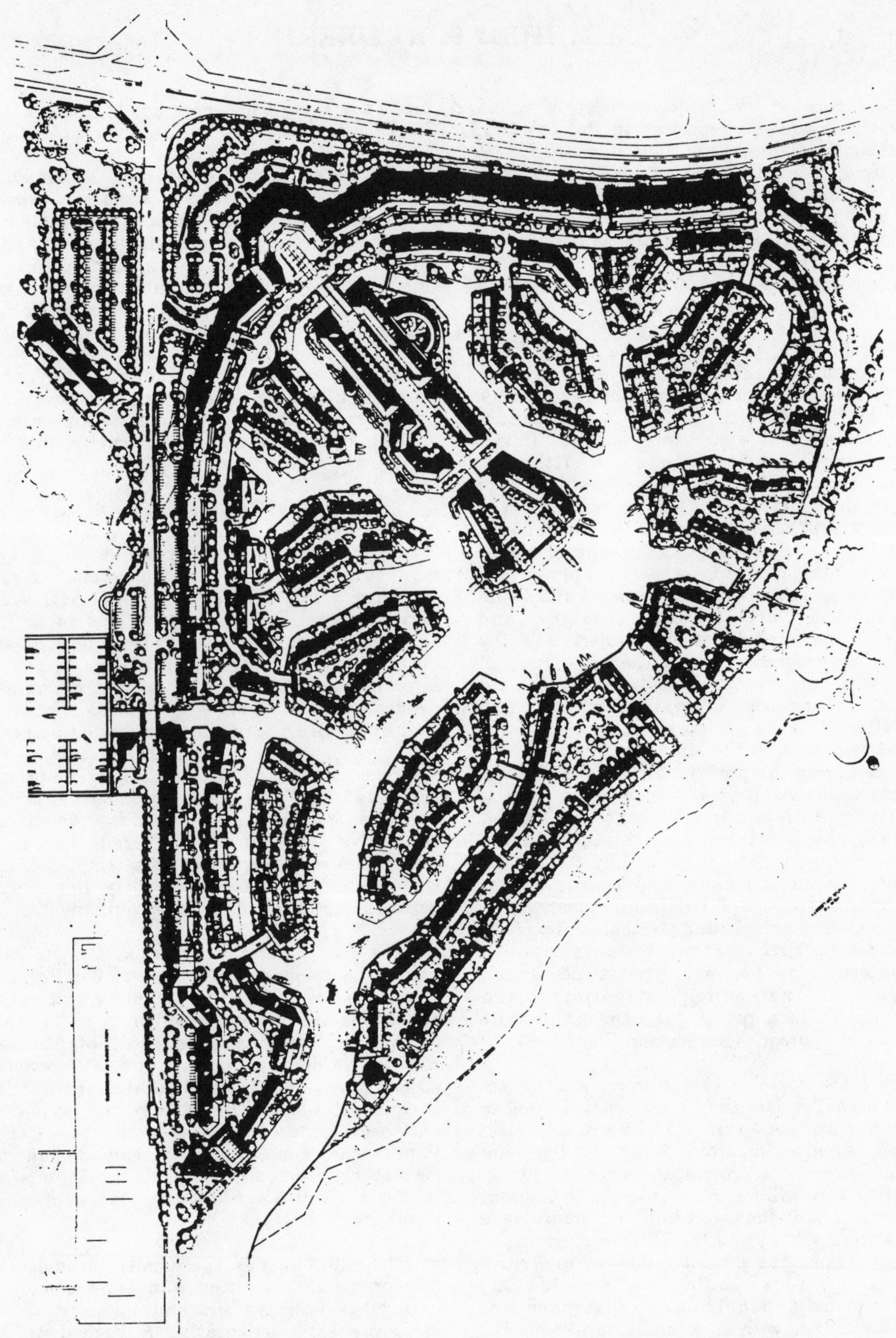

Figure 6. General plan for the Port Liberte waterfront community. The artificial canal system is open to tidal flow on the north, south, and at the main boat entrance at the southeast. The canals shallow from the central trunk to the laterals. Deadends have been virtually eliminated.

Figure 7: Biological baseline and monitoring activities include seining in the shallow waters of the east beach at Port Liberte along with offshore trawls, benthic and water quality samples, and intensive bird censusing.

property near its northern boundary along with ecological improvements of the Caven Point Peninsula.

2. Rehabilitation of the North Slough, a tidal embayment lying west of Caven Peninsula that has been adversely affected by pollution.

3. Enhancement and maintenance of the Spartina-beach-dune system that fronts the east side of the project along Caven Cove on the Hudson Estuary (Figure 8).

4. Creation of a special nesting habitat for the endangered (New Jersey state list) Least Tern.

5. General ecological enhancement of Caven Point Peninsula with considerations for public access and education.

The Panel's charge was to generate recommendations with sufficient detail to enable project planners to draw detailed plans and write specifications for the work (Figure 9). After receiving its mandate, the six-person panel worked with full independence, generating some recommendations that neither the state nor the developer might have favored, but which the panel was obliged to offer by virture of its knowledge or the principles involved. While it is still too early to determine the panel's success, all recommendations were accepted and acted upon by developers and regulators.

SUMMARY

In dense urban settings, restoration is a priority goal for mitigation or voluntary enhancement of aquatic habitats. Therefore the needs of regional ecosystems must be considered, not just project sites or single wetland units. In expanding urban areas a combination of protection, set-aside, and restoration may be required. Because strategies for aquatic ecosystem restoration are planning programs, they conflict strongly with the ad hoc nature of regulatory programs. Adjustments are necessary to enable permit evaluations to respond to the goals of regional restoration strategies. Effective restorative mitigation depends upon cooperation from private and public development entities; this means that unambiguous advice can be given to developers in project planning phases.

LITERATURE CITED

Batha, R. and A. Pendleton. 1987. Mitigation: A good tool that needs sharpening. Calif. Waterfront Age 3(2):15-17.

Burger, J. and J. Clark. 1987. Port Liberte. An example of collaborative planning for a coastal development on the lower Hudson River. Paper presented at Conference on the Impacts of New York Harbor Development on Aquatic Resources, Hudson River Foundation.

California Coastal Commission. 1987. Draft working paper on wetland restoration goals.

Clark, J. 1979. Mitigation and grassroots conservation of wetlands urban issues, p. 141-151. In The Mitigation Symposium: A National Workshop on Mitigating Losses of Fish and Wildlife Habitats. Genl. Tech. Rept. RM65, U.S. Forest Service, Fort Collins, Colorado.

Clark, J. 1985. A perspective on wetland rehabilitation, p. 342-349. In J. Kusler, R. Hamaan (Eds.), Wetland Protection: Strengthening the Role of the States. Center for Government Responsibility, U. of Florida, Gainesville.

Clark, J. 1986a (in press). Assessment for wetlands restoration, p. 250-253. In J. Kusler and P. Riexinger (Eds.), Proceedings: National Wetlands Assessment Symposium. (Portland, Maine). Assoc. of State Wetland Managers, Berne, New York.

Clark, J. 1986b. Setting the agenda for new research, regulations, and policy, p. 309-318. In E.D. Estevez, J. Miller, J. Morris and J. Hamman (Eds.), Managing Cumulative Effects in Florida Wetlands Conference Proceedings. Mote Marine Lab., E.S.P. Publ. 38.

Clark, J. and S. McCreary. 1980. Prospects for coastal conservation in the 1980's. Oceanus 23(4): 22-31.

Clark, J. and S. McCreary. 1987. Special area management at estuarine reserves, p. 49-93. In D.J. Brower and D.S. Carol (Eds.), Managing Land-Use Conflicts, Duke Univ. Press.

COE. 1985. Regulatory Guidance Letter, Nov. 8, 1985. U.S. Army Corps of Engineers, Office of Chief of Engineers.

Dial, S., M. Quamman, D. Deis and J. Johnston. 1985. Estuary-wide mitigation options for port development in Tampa Bay, Florida, p. 1332-1344. In O. T. Magoon and H. Converse (Eds.), Coastal Zone '85, Vol. 2, American Society of Civil Engineers.

Figure 8. Restoration of the east beach is a major enhancement activity at Port Liberte. The beach is screened and protected from landside disturbance by a buffer zone of Phragmites.

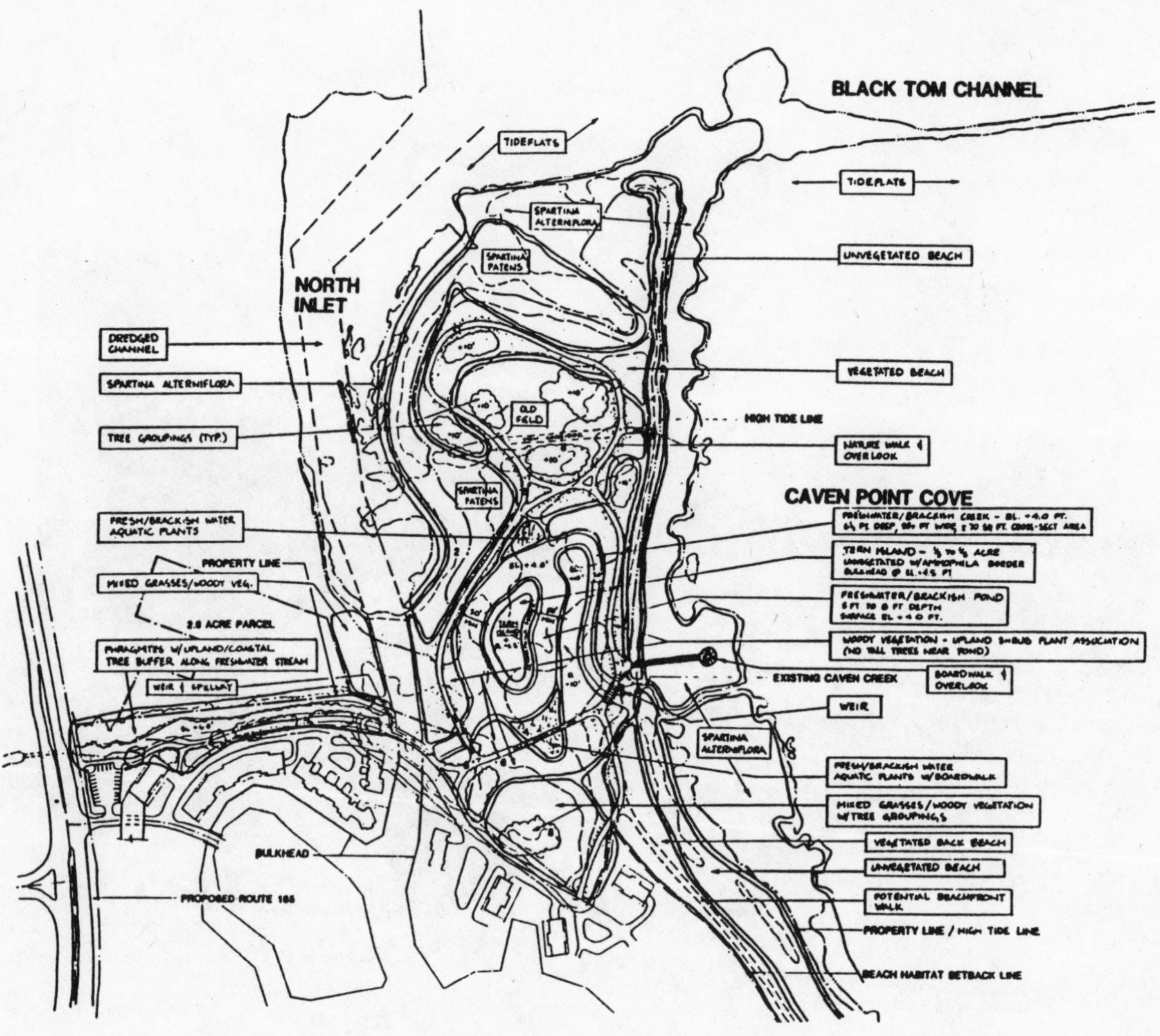

Figure 9. General enhancement plan for the natural areas lying east and north of the Port Liberte project site.

Vol. 2, American Society of Civil Engineers.

Josselyn, M.N. and J.W. Buchholz. 1984. Marsh Restoration in San Francisco Bay: A Guide to Design and Planning. Technical Report #3. Tiburon Center for Environmental Studies, San Francisco State University.

Klesh, W.L. 1987. Long-term management strategy for the disposal of dredged material: Corps-wide implementation. In Proc. North Atlantic Regional Conf. on the Beneficial Uses of Dredged Material, 12-14 May 1987, Baltimore, Md. p. 185-192.

Marcus, L. 1987. Wetland restoration and port development: the Bataquitos Lagoon case. p. 4152-4165. In O. T. Magoon, H. Converse, D. Miner, L. T. Tobin, D. Clark, and G. Domurat (Eds.), Coastal Zone '87, Vol. 4, Am. Soc. of Civ. Eng.

McCreary, S. and T. Zentner. 1983. Innovative estuarine restoration and management, p. 2527-2551. In O. T. Magoon and H. Converse (Eds.), Coastal Zone '83, Vol. 3. Am. Soc. of Civ. Eng.

Olsen, Stephen and V. Lee. 1985. Rhode Island's Salt Pond Region: a special area management plan. Coastal Resources Management Council, Providence, Rhode Island.

Quamman, M.L. 1986. Measuring the success of wetlands mitigation. National Wetlands Newsletter, Sept.-Oct.: 6-8.

Sorensen, J. 1982. Towards an overall strategy in designing wetland restoration, p. 85-96. In M. Josselyn (Ed.), Wetland Restoration and Enhancement in California. California Sea Grant, U. of California, La Jolla.

Studt, J.F. 1987. Special area management plans in the Army Corps of Engineers regulatory program. National Wetlands Newsletter, May-June: 8-10.

United States Fish and Wildlife Service. 1981. United States Fish and Wildlife Service Mitigation Policy. Federal Register 46(15):7644-7655.

Wessel, A.E. and M.J. Hershman. (In press). Mitigation: Compensating the Environment for Unavoidable Harm. In M.J. Hershman (Ed.), Urban Ports and Harbor Management: Changing Environments Along the U.S. Waterfront. Taylor and Francis, N.Y.

Williams, G.L. and G.W. Colquhoun. 1987. North Fraser Harbour environmental plan, p. 4179-4192. In Coastal Zone '87, Vol. 4, Am. Soc. of Civ. Eng.

Wilmar, M. 1986. Mitigation: the applicant's perspective. National Wetlands Newsletter, Sept.-Oct.: 16-17.

Zagata, M.D. 1985. Mitigation by "banking" credits a Louisiana pilot project. National Wetlands Newsletter, 7(3):9-11.

WATERFOWL MANAGEMENT TECHNIQUES FOR WETLAND ENHANCEMENT, RESTORATION AND CREATION USEFUL IN MITIGATION PROCEDURES

Milton W. Weller
Caesar Kleberg Professor in Wildlife Ecology
Department of Wildlife & Fisheries Sciences
Texas A & M University

ABSTRACT. Waterfowl and other wetland wildlife managers have long been involved in wetland restoration and enhancement, and have developed functional techniques for management of certain wetland types in various geographic regions. These procedures can serve other wetland managers in many useful ways, and are worthy of experimentation for other purposes. Most use natural processes to tap natural seed banks, modify cover-water ratios, or control weeds via water level control and herbivores. Wetland types where procedures have been standardized include those dominated by palustrine persistent emergents, moist-soil nonpersistent emergents, estuarine emergents, and forested palustrine communities.

This chapter presents some general concepts based on a selection of the extensive literature designed to facilitate adaptation of these strategies to the special situations of restoring, enhancing or creating wetlands to meet mitigation requirements. The procedures described also can be used to enhance various wetland functions such as water quality, shoreline protection, and esthetic values.

Major information gaps include: long-term ecological effects of management processes; methods for speeding natural events that aid in wetland restoration; and lack of quantitative information on other groups of wildlife such as fish, herptiles, and even nongame birds and mammals.

INTRODUCTION

Mitigation efforts requiring enhancement of established wetlands, restoration of former wetlands, or the creation of new wetlands where none existed, often are viewed as new efforts that are untested and uncertain. In fact, wildlife managers and some fisheries managers have been involved in such efforts for many years. Local wildlife or fisheries managers often are the best source of wetland restoration techniques that have been tried and work in that region. Most of the habitat management techniques are based on natural processes in wetland systems, and thus also influence other wetland values and functions. However, few of these practices have been subjected to long-term experimental testing and evaluation. Much of this material has been published but it is not available in a single document that covers all wetland types and their regional variations.

This chapter brings together some generalizations, a selection of the extensive literature, and a few examples of wetland types and procedures that have been standardized in certain areas. Such efforts have emphasized waterfowl and muskrats and occasionally other furbearers; very little work has dealt with other groups although some work indicates that successful efforts for game species also may favor nongame species. These procedures can be used equally well to enhance other wetland functions, such as reducing turbidity, protecting shorelines from erosion, and esthetic values. Thus, a review of the backgrounds of this management for waterfowl will facilitate consideration of all available strategies to the special situations of restoring, enhancing or creating wetlands to meet mitigation requirements.

DEFINITIONS OF MANAGEMENT STRATEGIES AND MANIPULATIONS

Although a fairly complex and somewhat inconsistent terminology has evolved among wetland managers and consultants, wildlife managers generally include a variety of situations and problems under the terms "habitat development" and "management" (e.g., Sanderson and Bellrose 1969, Atlantic Flyway Council 1972). However, there have been several typical patterns: Restoration of drained wetlands has been common where there was an opportunity to reclaim a major wetland of high wildlife value that has been degraded. The usual situation involves a failed drainage project where land values have declined, so that federal or state agencies can purchase the land and attempt restoration. Numerous National Wildlife Refuges (NWR) such as Tule Lake NWR in California and Aggasiz Lake NWR in Minnesota fall into this category (Laycock 1965). Still older examples are available in the European literature (Fog 1980).

Enhancement of wetlands involves an attempt to improve the wildlife values of a wetland that has not been drastically perturbed, but one that managers believe could be producing wildlife at a higher level more of the time. Periodic manipulations of water levels to enhance nesting conditions or modify plant succession rates would be typical examples.

If we define creation of wetlands as establishment of wetland communities and functions where none existed, this has been a less common practice among wildlife managers. However, this is partly a matter of terminology as procedures have not been categorized in relation to the nature and status of the wetland. The typical pattern has been the conversion of terrestrial communities to wetlands via identification of a natural drainage or basin that has had some history of moist-soil vegetation resulting from periodically wetter years. This conversion is usually done by creating an impoundment or diverting water to provide more regular flooding and encourage wetland plants and associated game species. Large impoundments created for other purposes such as water supply or flood control also may result in wetland development at the margins or upper reaches of the pool. Additionally, mining operations such as gravel removal may result in suitable wildlife habitats and considerable information is available (Svedarsky and Crawford 1982). The reverse of this, the conversion of aquatic areas to wetland, has been less common but has been accomplished by diking to keep out water and reduce continual deep flooding--as has been common at refuges and other wildlife management areas along coastlines of large lakes or oceans.

GOALS AND OBJECTIVES

Most conservation agencies have long-range planing processes for management areas that involve 3-to 10-year plans, often with annual updates. By their legal charge (e.g., migratory bird treaty or Federal-Aid funding) or by policies of guiding groups such as commissions, wetland managers in charge of specific areas or regions usually attempt to maximize wildlife attractiveness and wildlife production. This target could be in conflict with other wetland values and functions, but this is not always the case.

Philosophically, many of the goals of wildlife managers overlap with those of other wetland managers: There is a desire to preserve natural landscape units and functional values. Managers differ, however, in whether they prefer to use artificial methods that may give more immediate results, versus the use of more natural processes that tend to take longer but are less expensive and longer lasting (Weller 1978). Benefit-cost ratio strongly influences the choice of strategy as some functional but very expensive techniques such as hand transplants can be more easily justified in mitigation procedures than in conservation efforts for wildlife. Short-term goals often have been the driving force in habitat management of wetlands for wildlife: 1) increased production of game for hunting via creation of more and better nesting sites, increased food supplies, resting and roosting condition, or reduced predation; 2) improved conditions for hunters in pursuit of game, such as cover patches or blind sites; or 3) improved access to wetland areas for hunting by means of roads and boat channels.

Many wetland wildlife managers cherish the rich natural values and aesthetic aspects of the natural system. They expound a natural management philosophy as a code of ethics (Errington 1957), and are dedicated to preserving a naturally functioning ecosytem in perpetuity (Weller 1978, 1987) and maintaining or adding diversity (Sanderson 1974, Mathisen 1985). Others place first priority on maximal production of the targeted game species, but not intentionally in opposition to other natural values and functions. Ideally, these approaches should both be incorporated into a unified plan that involves other wetland functions as well. The consideration of multiple interests and approaches has been enhanced by the use of public hearings for local residents, laypersons and experts so that greater unity of goals is possible.

REVIEW OF PRODUCTION, SUCCESSION AND VEGETATION STRUCTURE

Research biologists working with marsh and other wetland wildlife have long recognized the importance of natural ecological processes such as seed germination, plant growth and production, succession, flooding, and water quality in regulating the plant community that ultimately dictates the diversity and populations of wildlife. As a result, there have been numerous vegetation studies oriented toward understanding habitats for waterfowl, including: vegetation community, structure and dynamics (Dane 1959, Kadlec 1962, Weller and Spatcher 1965, Weller and Fredrickson 1974); germination and plant growth (Weller 1975, Beule 1979); effects of drawdowns or dewatering (Kadlec 1962, and others); hydroperiod and other water-driven vegetation patterns (Meeks 1969, Knighton 1985); wetland wildlife responses to changes in structure and availability of vegetation (Weller and Fredrickson 1974, Ortega et al. 1976); influence of vegetation on predation rates (Keith 1961, Duebbert and Lokemoen 1976); and competition between wildlife species (Weller and Spatcher 1965).

Much of this work has been done on inland freshwater marshes, but there has been some excellent work on interior saline (Bolen 1964, Christiansen and Low 1970) or alkaline wetlands (Stewart and Kantrud 1972). Tidal regime, turbidity and salinity influences on plant growth and survival in coastal marshes have been the subject of numerous wildlife-oriented plant succession studies (Chabreck 1972, Palmisano 1972, and others), and have provided a basis for sound management. Some studies by botanists and plant ecologists have preceded and supplemented these studies by wildlife biologists, but their research questions often were directed toward different goals. There are many opportunities to improve upon this foundation with management applications in mind, but the effectiveness of such studies will depend upon the questions asked.

PRECONSTRUCTION CONSIDERATIONS

GEOMORPHOLOGY, LANDSCAPE, PATCH SIZE AND PATTERNS

It has become increasingly apparent that waterfowl and other migratory birds, and some wetland fish, amphibians, reptiles and mammals, do not satisfy all their needs in one wetland. This is particularly true of the breeding period when very specific requirements for foods or nest sites may exist in comparison to post-breeding activities such as migration or wintering. Thus, wetland diversity and density may figure prominently in satisfying these needs, a consideration that may not be met with the restoration or creation of only one wetland when a complex may be necessary. Wetland complexes have been recognized as important by many workers (Swanson et al. 1979, Weller 1981) and data presented by Brown and Dinsmore (1986) suggest that complexes increase species richness over solitary wetlands of similar size. General studies of wetland density in Prairie Pothole habitats, where water cycles influence wetland numbers from year to year, show general correlations between wetlands and waterfowl abundance, as one would expect (Sugden 1978, Leitch and Kaminski 1985). The diversity of wetland types also is deemed vital as the loss of small wetlands in drought years has a particularly great impact on certain species (Evans and Black 1956). Patterns of vegetation (cover-water ratios or cover-cover interspersion patterns) also influence bird use and are important designs for management of wetlands intended to enhance bird use (Weller and Spatcher 1965, Patterson 1976, Kaminski and Prince 1984).

Larger wetlands are known to provide greater numbers of habitats, and therefore are more likely to attract greater number of species (i.e., species richness) (Brown & Dinsmore 1986). There is a tendency to acquire and restore large units for these reasons, but it is also recognized that small areas also may provide specialized habitats for certain species, and management for those may require size considerations (Evans and Black 1956).

Configuration (e.g., length of shoreline in

relation to area of the wetland) seems to be an important influence on numbers of territorial species that an area can support, but there are little substantiating data (Mack and Flake 1980, Kaminski and Prince 1984). Some workers have suggested that other vegetation patterns are more influential than shoreline length (Patterson 1976). Managers often successfully create artificial wetlands with complex configurations to provide isolation for breeding pairs. Contiguity between wetlands is especially important to fish (e.g., Herke 1978) and some other vertebrates, and one obvious conclusion is that it provides habitat diversity that meets various needs throughout life.

PRECONSTRUCTION DESIGNING FOR WATER DEPTH

Preconstruction planning involving detailed contour mapping of prospective sites is essential (Verry 1985). Site observations during natural flooding periods also are useful because contour maps may not provide the precision essential in water level regulation. For example, large-scale contour intervals are unusual on construction maps, when the ultimate precision required for water level control may be a matter of centimeters.

Contouring with earth-moving equipment is commonplace, and should be used to create water depths associated with the desired plant community. Islands, bays and other structural features can be created during construction if soil character and shoreline protection are considered. Where such work is done on areas with a rich seed bank, soil should be moved off-site and returned as topsoil both for the merits of its organic content and as a seed bank. This will reduce invasions by exotics where they are an issue and the necessity of seeding with cultivated varieties that result in low natural diversity.

REGULATING RUNOFF-EVAPORATION RATIO AND FLOW-THROUGH RATES

Rainfall-evaporation patterns and watershed-wetland size ratios are of special importance in impoundment site selection (Verry 1985). Special considerations are necessary where uplands have been modified by land-use practices. Farming may increase the mean annual silt load and eutrophication, and intensive grazing on the waterway slopes can increase peak surges of water entering the wetland during storm events, washing out vegetation and water control structures. Urban development may increase runoff due to parking lots, roadways and roofs. In wetland restoration or enhancement projects, water control structures must be carefully engineered to handle these added burdens. Additionally, up-slope protection can be achieved through water diversions to streams, grass plantings to absorb more rainfall, or smaller impoundments that serve as catch basins for both water and silt.

These solutions also are relevant where wetlands are created from terrestrial sites, in which case site selection is extremely crucial and requires knowledge of the slope, area, and rainfall data of the watershed (Verry 1985). Storm events always seem to exceed runoff projections and are particularly damaging in the early stages of wetland development. In coastal areas, tidal action and wind fetch are important influences that must be considered, and considerable work has been done on shoreline protection.

CONTROLLING EROSION AND TURBIDITY

In any wetland management program, modified water levels, exposed banks and unvegetated bottom are vulnerable to wave and current action. Decreased wetland productivity may result from erosion of shorelines, elimination of vegetation, and increased water turbidity. Most wildlife managers have dealt less successfully with these problems than have other wetland designers, due in part to factors of need and cost. Importing firmer soil, gravel or rock may be necessary, and prepared rip-rap can be used in extreme cases of erosion. Several steps can be taken to prevent erosion: 1) exposing the shoreline by dewatering until vegetation has become established, and 2) delaying flooding until the bottom has been stabilized with emergent and preferably submergent vegetation. These measures will reduce turbidity that may prevent vegetation establishment when reflooding occurs.

CONSIDERING SPECIAL WILDLIFE NEEDS

Waterfowl and other wildlife are highly selective in their choice of habitat: it must supply vegetation cover, nesting sites, protection from predators, reduced disturbance and food supplies. On a worldwide basis, systems have been developed to manage for these needs (e.g., Scott 1982). In recent years, it has become clear that habitat quality is very important, and providing a place to live must also include nutritional food at the proper time of the reproductive cycle (Fredrickson 1985). Work on invertebrates of wetlands still is inadequate but several workers have demonstrated how patterns vary and how wildlife seem to respond (Voigts 1976, Swanson et al. 1979), which managers more and more take into account in their management strategies (Whitman 1976, Reid 1985).

PROCEDURES BY PROCESS AND PROXIMATE OBJECTIVE

MANAGING THE SEED BANK

Although the longevity and abundance of seeds of wetland plants has been known at least since the 1930's (Billington 1938), and early marsh managers advised location and utilization of sites with a natural seed bank (Addy and MacNamara 1948, Crail 1955, Singleton 1951), managers have varied in their utilization of this general principle. Seeding of Japanese millet or use of native millets (Echinochloa crus-galli) and smartweeds (Polygonum spp.) for seed sources were widespread among National Wildlife Refuges in the 1940's and 1950's (Linduska 1964). Planting of millets, smartweeds, and agricultural crops was used to feed large numbers of migrant and wintering waterfowl in preference to less predictable processes that capitalized on natural sources (Givens and Atkeson 1957). More recently, moist-soil management and other types of strategies that use natural seed banks have been viewed more positively by a generation of managers who favor natural diversity and minimal expenditures on machinery and manpower. Management of both water depth and hydroperiod is the major strategy employed (see for example studies by Fredrickson and Taylor 1982) and the techniques are now more widely recognized and appreciated.

Another aspect of seed banks generally recognized by wetland ecologists but unknown to many others is the importance of preserving the seed bank in a newly created wetland. Dams and levees often are built with borrowed soil taken from the water side to create some deep-water sites or to facilitate installation of water control structures. When basin shape is modified by scraping, a barren substrate may be created (Kelting and Penfound 1950) that must be reseeded or await the natural processes of seed transport, germination and local seed production. Marsh hay cut at seeding times (as occurs with seeding of wildflowers on highway right-of-ways) may be a good seed source, but I know of no experiments in wetlands to demonstrate this.

Despite their common longevity, seed banks may be limited in diversity or even non-existent in situations of long inundation where wetland emergents have not existed for many years; aquatic plants or long-lived terrestrial plant seeds may survive, but species of value to the wetland restoration may not occur (Pederson and van der Valk 1984). Where available, use of soil from local wetlands may be useful in resolving this problem.

MANAGING PLANT SUCCESSION THROUGH WATER LEVEL MANIPULATION

Because it has long been recognized that hydroperiod and water depth dictate plant species composition, density, and growth, waterfowl managers have studied the availability and germination characteristics of seed banks (Crail 1955), the physical and physiological requirements of the seed for germination, germination rates and rates of growth, tuber production, plant productivity, and plant life-history strategies (Weller 1975, Beule 1979) that influence wetland succession (Dane 1959).

To provide the water conditions that induce germination seed and plant growth, most wetlands created or modified for wildlife make provisions for complete dewatering via control structures (Atlantic Flyway Council 1972). The latter is a vital consideration if any influence over plant community is desired, and the engineer should be alerted to the need for total dewatering. Subsequent modifications of dike height must consider water control structures and the potential of dam failure or erosion. Strategies for achieving the desired goals will be discussed under examples of several wetland types that have been commonly and successfully managed.

Although the terminology of succession may not be very useful in many wetlands (Weller and Spatcher 1965, van der Valk 1981), manipulations involve dewatering to return the plant community to mudflat annuals and seeding perennials that are characteristic of more shallow marsh (Bednarik 1963, Linde 1969, and others). Plant growth rates or germination also can be influenced by extreme flooding or drying, by enhancing nutrients with fire or fertilizers, and by weed control.

MODIFYING WATER DEPTH AFTER WETLAND FORMATION

The most common method of deepening a natural or created wetland for waterfowl has been to install a dam or dike to impound more water, and to incorporate a water-control structure that limits the pressure on the dam and allows dewatering. Increasing water depths in basins that do not have an adequate supply can best be accomplished most economically by gravity flow systems such as stream diversions and up-slope reservoirs. A more reliable but expensive alternative is a well and pump, but this is a perpetual expense, and may be a source of disagreement among different user groups.

Modifying shallow wetlands to create open pools and deeper water can be done by dewatering and a bulldozer, by drag-line removal of basin substrate, or by blasting with dynamite (Strohmeyer and Fredrickson 1967) or ammonium nitrate fertilizer charged with a more volatile explosive (Mathiak 1965).

REGULATING VEGETATION VIA HERBIVORES AND FIRE

Wetlands, especially those dominated by persistent, perennial emergent plants, are renowned for their attractiveness to native muskrats (Ondatra zibethicus) and the introduced nutria (Myocaster coypu). Overpopulations resulting in "eat-outs" are well documented in northern and midwestern marshes (Errington et al. 1963), and dramatic population changes also exist in eastern brackish marshes (Dozier 1947), and southern deltaic and chenier marshes (Lynch et al. 1947, O'Neil 1949, Palmisano 1972). Muskrat populations in particular expand rapidly because of their high reproductive potential and adaptability to new food resources. Ultimately, the area is denuded and populations of other wildlife are drastically impacted (Weller and Spatcher 1965). Managers often are unprepared for this event and may lack methods for control because of harvest regulations. In large areas, control may be impossible due to size and logistics of trapping. The resultant open water may persist for many years unless dewatering is used to induce revegetation.

Beavers (Castor canadensis) likewise can impact on willows (Salix spp.), cottonwood (Populus deltoides) and other highly palatable plants, whether the plant are used only for food or for lodges and dams as well (Beard 1953). Flooding by beavers of other terrestrial or wetland vegetation often is regarded as serious by managers not only because of plant mortality but because water level stability within a wetland may not be conducive to maximal community diversity and productivity of many wetland species.

Another group of herbivorous animals are fish such as the introduced common carp (Cyprinus carpio), and more recently, white amur or grass carp (Ctenopharyngodon idella). Whereas sterile hybrids of grass carp are being used to reduce the chances of reproduction in the wild, the common carp reproduces readily and is spread by fishermen. It also moves upstream into shallow wetlands. Invading carp can be extremely serious because of their direct consumption of submergent plants and invertebrates, and indirectly because of the turbid waters they induce which reduces light penetration and therefore plant production (Robel 1961).

Livestock such as cattle, sheep and goats can be useful management tools, but such grazing may be difficult to regulate because of public pressure for grazing rights or due to our lack of understanding of the carrying capacity of wetlands under variable and often uncontrollable conditions. Grazing exposes tubers then utilized by grubbing geese such as snow geese (Anser caerulescens) in both southern and eastern coastal marshes (Glazener 1946), but can also can eliminate favored duck food plants (Whyte and Silvy 1981). Grazing in northern areas is generally regarded as detrimental to nesting waterfowl (Kirsch 1969), but obviously can be beneficial to species like upland sandpipers (Bartramia longicauda) that prefer short vegetation. Fencing is the simple tool used by managers all over the world to change the character of overgrazed wetlands (Fog 1980), but better management may require regulation of the grazing level and not merely total exclusion.

Fire has been used by farmers and ranchers for years to increase forage and hay crops, but it can be devastating to nesting ducks and other birds (Cartwright 1942). The response of wintering geese to fire has resulted in a policy of periodic burning on refuges to reduce vegetation and expose tubers for grubbing geese (Lynch 1941). Fire also has been used in northern marshes with peat bottoms to create deep water openings; however, control of the fire is often difficult (Linde 1985). Considerable work has been done recently to explore the role of fire in marsh succession (Smith 1985a, 1985b), and clearly much more of this type of work is needed in all vegetation types (Kantrud 1986).

SEEDING AND TRANSPLANTING

Both seeding and transplanting were used extensively in the 1940's for marsh edge plants, and sometimes for emergent plants along the shallow water's edge (Linduska 1964). Planting Japanese millet and other seeds available from farm suppliers and wildlife nurseries is still widely done on small areas, especially by private landowners, but is less widespread as an operational procedure on refuges and waterfowl management areas because of cost. Upland farming is more common because equipment is available or sharecropping is a cost-effective way to produce wildlife foods while appeasing local farmers over local land lost to agriculture (Givens and Akeson 1957). Planting in wetland areas used for waterfowl hunting may involve legal issues because of "baiting" laws, and must be done in counsel with local wildlife officers.

Planting natural rootstocks or runners and other plant parts has been used with cattail (Bedish 1967) and other perennials, but is labor-intensive and of questionable success--not just because of plant growth potential but because of losses due to their attractiveness to muskrats and other herbivores. Erosion on wave-swept shores is an equally serious problem, and floating boards have been used to protect as well as facilitate wetland establishment and survival.

CONTROLLING WEEDS

Part of maintaining a wetland that is attractive to wildlife requires a balance of open water and vegetation. Because nesting birds and migratory waterfowl are especially influenced by these conditions, considerable effort goes into creating suitable habitat. In addition to the control of succession and cover-water ratios through water-level management (Weller 1978) or fire (Kantrud 1986), more direct (artificial) and immediate methods have been utilized such as chemical sprays (Martin et al. 1957, Beule 1979); mechanical destruction by roller or cutting (in or out of water) (Nelson and Dietz 1966, Linde 1969); and grazing (Kirsch 1969).

EXAMPLES OF COMMONLY MANAGED WETLAND TYPES BY OBJECTIVE

PALUSTRINE PERSISTENT EMERGENT WETLANDS

For nesting waterfowl and other marsh wildlife, most managers prefer sturdy water level control structures and a reliable source of water for use in modifying water depths to control plants for nest sites (birds) and food sources (muskrats as well as birds). A diversity of plants of various life forms is preferred to serve various animals. Marginal nonpersistent emergent plants produce large seed crops; deeper water persistent emergents are excellent for nest sites and provide tuberous bases useful to herbivores; and submergent plants often provide food directly or serve as substrates for invertebrates.

Water depths dictate dynamic vegetation patterns in wetlands subject to seasonal and annual variation in water supply. These may vary from lake-like aquatic conditions to near-terrestrial vegetation due to hydrologic perturbations. Wildlife respond directly and vary greatly in species richness and population abundance (Weller and Spatcher 1965).

Examples of plant succession following drawdowns to re-establish vegetation were provided by Kadlec (1962), Harris and Marshall (1963), and Weller and Fredrickson (1974). Subsequent changes in vegetation due to the influence of water level and muskrats (Errington et al. 1963, Weller and Fredrickson 1974) and common carp (Robel 1961) demonstrate that: 1) the natural short-term water dymamics of midwestern wetlands must be dealt with in restoration projects, and 2) the inherent adaptability of wetland plants provides them with great powers of recovery and repair.

A well-established technique to reestablish vegetation after it has been eliminated by high water or muskrat "eat-out" is to dewater ("drawdown") the area by use of a water control structure or pumping. Germination from the natural seed bank (van der Valk and Davis 1978) provides most of the source of plants but enhanced production of persistent plants via tubers and rhizomes also results (Weller 1975). Re-establishment may result in excessively dense vegetation not suitable for the intended wildlife, whereupon the natural herbivores may move in and create suitable openings.

Modification of this pattern occurs with various methods of creating artificial openings described elsewhere, and these may be useful in intensive management or restoration projects where time is vital. The drawdown-revegetation cycle is commonly practiced by conservation agencies in many midwestern states (Linde 1969), and in situations where time is less important relative to costs. This method seems to simulate natural processes and events without lasting damage. Even fish populations of those areas seem responsive and pioneering.

MOIST-SOIL IMPOUNDMENTS FOR NONPERSISTENT EMERGENTS

Migratory waterfowl feed less on high protein animal foods in fall and winter and instead use seeds or foliage. Seed production is especially enhanced by creating or maintaining shallow water areas where nonpersistent emergents such as millets, smartweeds, spikerushes (Eleocharis spp.) and other marsh edge plants germinate and produce seed (Fredrickson and Taylor 1982). This requires mud flat conditions, typical of any marsh drawdown, except that such wetlands are normally not breeding marshes for waterfowl (although they may be suitable for some rails and songbirds). Drawdowns are performed early in the year to allow sufficient drying so that annual seeds will be produced. Because of the climatic regime in southerly areas, both spring and late summer crops may be produced where water is available and levels can be controlled. The usual procedure is to construct low dams or dikes to impound water equipped with a simple structure for water level control. In rice areas, low-level terraces that are opened and closed by machinery work quite well, but the intended crop of annuals must determine the structure design and water depth.

Areas flooded in spring and dewatered gradually create mudflat conditions attractive to migrant shorebirds and ducks (e.g., green-winged teal, Anas crecca), but also induce germination from the seed bank (Fredrickson and Taylor 1982, Rundel and Fredrickson 1981). Too rapid drying produces more terrestrial species (Harris and Marshall 1963), so soil conditions and water level regulation are extremely important. The retention of some moisture on the flat is essential to ensure full maturity of seeds and the germination of late maturing plants like smartweeds. Here, as in any drawdown, a thunderstorm can produce flooding and the loss of a year's crop. Typically, such areas are reflooded in the fall prior to the arrival of waterfowl and other migrants. Flooding to make food available to migrant waterfowl involves regulation of the water control structure (except in cases of high rainfall and flow-through rates) to maintain depths of 6 to 18 inches so that birds can swim but still dabble and tip up for food. Dabbling ducks like green-winged and blue-winged teal (A. discors), mallards (Anas platyrhynchos) and pintails (A. acuta) find ideal food and water conditions in such situations. Deeper areas may be utilized by shallow divers like the ring-necked duck (Aythya collaris) (Fredrickson and Taylor 1982),

Undesirable plants, particularly willow and cattail can be extremely troublesome as they outcompete annuals and eliminate openings favored by ducks. The capability to dry out the area, or to flood it to excessive depths, may allow control of some nuisance species (Fredrickson and Taylor 1982).

TIDAL ESTUARINE (BRACKISH AND SALT) WETLANDS

Many of the processes that occur in fresh marshes also occur in brackish marshes, and can be influenced by muskrats or nutria populations. However, sites under a tidal regime usually cannot be dewatered for revegetation, and serious marsh loss can occur (as considered elsewhere by other authors in this volume). Herbivore control is one of the most important and effective methods of preventing this loss as it is in freshwater wetlands lacking water source and level control. Such marshes also may serve as nesting and feeding areas for waterbirds, although no long-term studies seem available.

The usual management strategy for attracting breeding, migrant and wintering waterfowl has been to build freshwater impoundments as catch basins that exclude salt water. This approach developed in part from the adaptation of rice impoundments for use by waterfowl along the eastern U.S. coast (MacNamara 1949). It is still commonplace on many coastal refuges, but is now strongly discouraged by the National Marine Fisheries Services to ensure free access of marine organisms such as finfish and shellfish to brackish and fresh tidal marshes that are vital for feeding, breeding and nursery areas. Hence, weirs are often used in place of dikes, and these semi-impoundments reduce turbidity but may not affect salinity markedly (Chabreck and Hoffpauir 1962, Chabreck et al. 1979). In this way, submergent vegetation attractive to waterfowl and also to marine organisms that frequent these shallow areas is enhanced. There has been intensive study of this type of impoundment on fish and shrimp populations in Louisiana (Herke 1978). Undoubtedly there are changes in the species composition and diversity of benthic invertebrate populations caused by this more continuous flooding of areas that once were periodic mudflats. Impounding areas for private waterfowl hunting areas has been legally challenged by federal agencies on the east coast, and this practice may demand extensive evaluation before it is allowed to continue. Nevertheless, it is widely regarded as the best way to enhance waterfowl habitat in saline coastal areas. In South Carolina coastal impoundments, brackish rather than fresh water has been used in the management of such areas (DeVoe and Baughman 1986). Waterfowl capitalize on brackish food species like widgeongrass or musk grass. Dewatering by the use of dikes and water-control structures can also be used to produce the mudflat plant, Sesuvisum (Swiderek et al. 1988) which has small but highly palatable seeds. Waterfowl response in these situations has been impressive, but the species composition may shift. Additionally, such areas may be attractive to shorebirds at low water stages.

Burning has previously been mentioned as a tool for seasonally opening up vegetation for geese, but this more commonly occurs in higher marshes that are only periodically flooded.

GREEN-TREE IMPOUNDMENTS (FORESTED PALUSTRINE WETLANDS)

The natural winter flooding of flood plain oxbows, sloughs and backwaters, especially in the southeastern United States, provides superb wintering habitat for mallards and other ducks that eat acorns and other large tree seeds and fruits (Allen 1980). Additionally, beaver ponds have been important to waterfowl, but tree mortality results from such water stabilization (Beard 1953). Flooding is erratic and dependent on the timing and the rate of rainfall. Moreover, flooding that occurs during the growing season may kill trees that are not tolerant to prolonged inundation at that time. To enhance the reliability of water and food supplies for migrants, low level impoundments have been used with water control structures to flood mast-producing trees during the dormant season (Cowardin 1965, Schnick et al. 1982, Mitchell and Newling 1986). Typically, a stream is diverted to fill the area, and some flood prevention plan often is necessary. Some areas use low level dikes that will withstand overflowing water but, as in any wetland impoundment, the potential for complete drawdown is essential. If a water control structure is not used, the impounding dam may be cut with a front-end scoop, and then repaired after the drying has occurred.

To take advantage of natural seed crops, site selection for impoundments is crucial and must include those mast producing species such as willow oak (Quercus Phellos) and water oak (Q. nigra) that tolerate prolonged flooding, but also produce acorns of a size suitable for ducks (Allen 1980). Depending upon the forest species composition and flood duration, some tree mortality may occur due to the reduced drainage capacity of the area. However, these openings are attractive to waterfowl and will produce moist-soil annual seed crops if the areas are naturally or artificially exposed during late summer and early fall. Some managers are creating openings by clear-cutting small patches of less desirable species with the intent of combining the moist-soil management technique with the mast production of the green-tree impoundment strategy (Harrison and Chabreck 1988).

CONCLUSION

Waterfowl and other wetland wildlife managers have been involved in wetland enhancement and modification for many years, and have developed a series of techniques that are fairly standard for management of various wetland types and geographic regions. These can serve other wetland managers in many useful ways, and are worthy of exploration and experimentation. Some of these techniques could result in highly significant cost reduction where time is available for the use of natural processes. However, because of the Gleasonian nature of succession in many wetlands (van der Valk 1981), plant communities are difficult to predict and the desired or original ecotype may not develop; therefore some range in specifications is essential.

Understanding natural patterns and processes of wetlands is a vital first step to proper and lasting restoration, enhancement or creation. Additionally, one must tap the expertise of the many disciplines that can contribute to such wetland preservation processes. As long-term evolutionary products of extremely dynamic systems, wetland plants and animals are quickly responsive to the availability of resources in newly created and enhanced areas. But we must not become over-confident that we can "create" a normally functioning and naturally diverse system. In most situations, we can provide the environmental needs to allow dominant wetland plants and animals to succeed, and the product will satisfy many if not most viewers. We cannot, however, expect to replace the complex and diverse natural systems that are a product of many centuries of evolution and randomness, and we should not let the ease of creating the structure and simple features of a wetland for mitigation lead us to accept unnecessary and perhaps unsatisfactory substitutes.

LITERATURE CITED

Addy, C.E. and L.G. MacNamara. 1948. Waterfowl Management on Small Areas. Wildl. Manage. Inst. Washington, D.C.

Allen, C.E. 1980. Feeding habits of ducks in a green-tree reservoir in eastern Texas. J. Wildl. Manage. 44:232-236.

Atlantic Flyway Council. 1972. Techniques Handbook of the Waterfowl Habitat Development and Management Committee, 2nd ed. Atlantic Flyway Council, Boston, Massachusetts.

Beard, E.B. 1953. The importance of beaver ponds in waterfowl management at the Seney National Wildlife Refuge. J. Wildl. Manage. 17:398-436.

Bedish, J.W. 1967. Cattail moisture requirements and their significance to marsh management. Am. Midl. Natur. 78:288-300.

Bednarik, K.E. 1963. Marsh management techniques, 1960. Ohio Dept. Nat. Resour. Game Res. Ohio 2:132-144.

Beule, J.D. 1979. Control and Management of Cattails in Southeastern Wisconsin Wetlands. Wisc. Dept. Nat. Resour. Tech. Publ. No. 112.

Billington, C. 1938. The vegetation of the Cranbrook Lake bottom. Cranbrook Inst. Sci. Bull. No. 11.

Bolen, E.C. 1964. Plant ecology of spring-fed salt marshes in western Utah. Ecol. Monogr. 34:143-166.

Brown, M. and J.D. Dinsmore. 1986. Implications of marsh size and isolation for marsh management. J. Wildl. Manage. 50:392-397.

Cartwright, B.W. 1942. Regulated burning as a marsh management technique. Trans. N. Am. Wildl. Nat. Resour. Conf. 7:257-263.

Chabreck, R.H. 1972. Vegetation, Water and Soil Characteristics of the Louisiana Coastal Region. La. St. Univ. Agric. Exp. Sta. Bull. No. 644.

Chabreck, R.H., R.J. Hoar, and W.D. Larrick, Jr. 1979. Soil and water characteristics of coastal marshes influenced by weirs, p. 129-146. In J.W. Day Jr., D.D. Culley Jr., R.E. Turner, and A.J. Mumphrey Jr. (Eds.), Proc. Third Coastal Marsh and Estuary Management Symposium. Louisiana State Univ. Div. of Continuing Ed., Baton Rouge, Louisiana.

Chabreck, R.H. and C.M. Hoffpauir. 1962. The use of weirs in coastal marsh management in Louisiana, p. 103-112. In Proc. 16th Ann. Conf. Southeastern Assoc. of Game and Fish Comm. Charleston, South Carolina.

Christiansen, J.E. and J.B. Low. 1970. Water Requirements of Waterfowl Marshlands in Northern Utah. Utah Div. Fish Game Publ. No. 69-12. Salt Lake City.

Cowardin, L.M. 1965. Flooded Timber as Waterfowl Habitat at the Montezuma National Wildlife Refuge. New York Coop. Wildl. Research Unit, Cornell University, Ithaca.

Crail, L.R. 1955. Viability of Smartweed and Millet Seeds in Relation to Marsh management in Missouri. Mo. Conserv. Comm. PR Project Rep. 13-R-5.

Dane, C.W. 1959. Succession of aquatic plants in small artificial marshes in New York state. N.Y. Fish & Game J. 6:57-76.

DeVoe, M.R. and D.S. Baughman (Eds.). 1986. South Carolina Coastal Impoundments: Ecological Characterization, Management, Status and Use. Vol. II. Public. No. SC-SG-TR-82-2. So. Car. Sea Grant Consortium, Charleston, South Carolina.

Dozier, H.L. 1947. Salinity, as a factor in Atlantic Coast tidewater muskrat production. Trans. No. Amer. Wild. Conf. 12:398-420.

Duebbert, H.F. and J.T. Lokemoen. 1976. Duck nesting in fields of undisturbed grass-legume cover. J. Wildl. Manage. 40:39-49.

Errington, P.L. 1957. Of Men and Marshes. Macmillan Co., New York.

Errington, P.L., R. Siglin, and R. Clark. 1963. The decline of a muskrat population. J. Wildl. Manage. 27:1-8.

Evans, C.D. and K.E. Black. 1956. Duck production studies on the prairie potholes of South Dakota. Fish & Wildl. Serv. Spec. Sci. Rept (Wildl.) No. 32.

Fog, J. 1980. Methods and results of wetland management for waterfowl. Acta Ornithologica XVII (12):147-160.

Fredrickson, L.H. 1985. Managed wetland habitats for wildlife: why are they important?, p. 1-8. In M.D. Knighton (Ed.), Water Impoundments for Wildlife: A Habitat Management Workshop. North Central Forest Exper. Sta. Tech. Rep. NC-100. St. Paul, Minnesota.

Fredrickson, L.H. and T.S. Taylor. 1982. Management of Seasonally Flooded Impoundments for Wildlife. U.S. Fish & Wildl. Serv. Resour. Pub.148.

Givens, L.S. and T.Z. Atkeson. 1957. The use of dewatered land in southeastern waterfowl management. J. Wildl. Manage. 21:465-467.

Glazener, W.C. 1946. Food habits of wild geese on the Gulf Coast of Texas. J. Wildl. Manage. 10:322-329.

Harris, S.W. and W.H. Marshall. 1963. Ecology of water-level manipulations on a northern marsh. Ecology 44:331-343.

Harrison, A.J. and R.H. Chabreck. 1988. Duck food production in openings in forested wetlands, p. 339-351. In M.W. Weller (Ed.), Waterfowl in

Winter. Univ. Minn. Press, Minneapolis, Minnesota.

Herke, W.H. 1978. Some effects of semi-impoundment on coastal Louisana fish and crustracean nursery usage, p. 325-346. In J.W. Day (Ed.), Proc. Third Coastal Marsh and Estuary Management Symposium. La. St. Univ. Cont. Ed. Div., Baton Rouge, Louisiana.

Kadlec, J.A. 1962. Effects of a drawdown on a waterfowl impoundment. Ecology 43:267-281.

Kaminski, R.M. and H.H. Prince. 1984. Dabbling duck--habitat associations during spring in the Delta Marsh, Manitoba. J. Wildl. Manage. 48:37-50.

Kantrud, H. 1986. Effects of Vegetation Manipulation on Breeding Waterfowl in Prairie Wetland--A Literature Review. U.S. Fish & Wildl. Serv. Tech. Rep. 3.

Keith, L.B. 1961. A study of waterfowl ecology on small impoundments in southeastern Alberta. Wildl. Mongr. 6.

Kelting, R.W. and W.T. Penfound. 1950. The vegetation of stock pond dams in Central Oklahoma. Am. Midl. Nat. 44:69-75.

Kirsch, L.M. 1969. Waterfowl production in relation to grazing. J. Wildl. Manage. 33:821-828.

Knighton, M.D. 1985. Vegetation management in water impoundments: water level control, p. 39-50. In M.D. Knighton (Ed.), Water Impoundments for Wildlife: A Habitat Management Workshop. North Central Forest Exper. Sta. Gen. Tech. Rept. NC-100. St. Paul, Minnesota.

Laycock, G. 1965. The Sign of the Flying Goose. Natural History Press, Garden City, New York.

Leitch, W.G. and R.M. Kaminski. 1985. Long-term wetland-waterfowl trends in Saskatchewan grassland. J. Wildl. Manage. 49:212-222.

Linde, A.F. 1969. Techniques for Wetland Management. Wisc. Dept. Nat. Resour. Rep. No. 45.

Linde, A.F. 1985. Vegetation management in water impoundments: alternatives and supplements to water-level control, p. 51-60. In M.D. Knighton (Ed.), Water Impoundments for Wildlife; A Habitat Management Workshop. North Central Forest Exper. Sta. Tech. Rep. NC-100, St. Paul, Minnesota.

Linduska, J.P. (Ed.). 1964. Waterfowl Tomorrow. Fish and Wildlife Service, Washington, D.C.

Lynch, J.J. 1941. The place of burning in the management of Gulf Coast wildlife refuges. J. Wildl. Manage. 5:454-459.

Lynch, J.J., T.O. O'Neil, and D.W. Lay. 1947. Management significance of damage by geese and muskrats to Gulf Coast marshes. J. Wildl. Manage. 11:50-76.

Mack, G.D. and L.D. Flake. 1980. Habitat relationships of waterfowl broods in South Dakota stock ponds. J. Wildl. Manage. 44:695-700.

MacNamara, L.G. 1949. Salt-marsh development at Tuckahoe, New Jersey. Trans. No. Amer. Wildl. Conf. 14:100-117.

Martin, A.C., R.C. Erickson, and J.H. Steenis. 1957. Improving Duck Marshes by Weed Control. Fish and Wildl. Serv. Circ. 19-rev. Washington D.C.

Mathisen, J.E. 1985. Wildlife impoundments in the North-Central states: why do we need them?, p. 23-30. In M.D. Knighton (Ed.), Water Impoundments for Wildlife; A Habitat Management Workshop. North Central Forest Exper. Sta.Tech. Rep. NC-100, St. Paul, Minnesota.

Mathiak, H. 1965. Pothole Blasting for Wildlife. Public. 352, Wisc. Cons. Dept. Madison, Wisconsin.

Meeks, R.L. 1969. The effect of drawdown date on wetland plant succession. J. Wildl. Manage. 33:817-821.

Mitchell, W.A. and C.J. Newling. 1986. Greentree Reservoirs. TR EL-86-9. U.S. Army Engineers, Waterways Exp. Sta., Vicksburg, Mississippi.

Nelson, N.F. and R.F. Dietz. 1966. Cattail Control Methods in Utah. Utah State Dept. Fish & Game Publ. No. 66-2.

O'Neil, T. 1949. The Muskrat in the Louisiana Coastal Marshes. La. Dept. Wildl. & Fish., New Orleans, Louisiana.

Ortega, B., R.B. Hamilton, and R.E. Noble. 1976. Bird usage by habitat types in a large freshwater lake. Proc. S.E. Fish & Game Conf. 13:627-633.

Palmisano, A.W. 1972. Habitat preference of waterfowl and fur animals in the northern Gulf Coast marshes, p. 163-190. In R.H. Chabreck (Ed.), Proc. Coastal Marsh and Estuary Management Symposium. La. St. Univ. Cont. Ed. Div., Baton Rouge, Louisiana.

Patterson, J.H. 1976. The role of environmental heterogeneity in the regulation of duck populations. J. Wildl. Mange. 40:22-32.

Pederson, R.L. and A.G. van der Valk. 1984. Vegetation change and seed banks in marshes: ecological and management implications. Trans. N. Am. Wildl. Nat. Resour. Conf. 49:271-280.

Reid, F.A. 1985. Wetland invertebrates in relation to hydrology and water chemistry, p. 72-79. In M.D. Knighton (Ed.), Water Impoundments for Wildlife; A Habitat Management Workshop. North Central Forest Exper. Sta.Tech. Rep. NC-100, St. Paul, Minnesota.

Robel, R.J. 1961. Water depth and turbidity in relation to growth of sago pondweed. J. Wildl. Manage. 25:436-438.

Rundel, W.D. and L.H. Fredrickson. 1981. Managing seasonally flooded impoundments for migrant rails and shorebirds. Wildl. Soc. Bull. 9:80-87.

Sanderson, G.C. 1974. Habitat--key to wildlife perpetuation; aquatic areas, p. 21-40. In How Do We Achieve and Maintain Variety and Optimum

Numbers of Wildlife? Symposium, Nat. Wildl. Fed., Washington, D.C.

Sanderson, G.C. and F.C. Bellrose. 1969. Wildlife habitat management of wetlands. An. Acad. Brasil. Cienc., 41:153-204 (supplement).

Schnick, R.A., J.M. Morton, J.C. Mochalski, and J.T. Beall. 1982. Mitigation and Enhancement Techniques for the Upper Mississippi River System and Other Large River Systems. Fish & Wildl. Serv. Resour. Publ. 149. Washington, D.C.

Scott, D.A. (Ed.). 1982. Managing Wetlands and Their Birds. Int. Waterfowl Res. Bur., Slimbridge, England.

Singleton, J.R. 1951. Production and utilization of waterfowl food plants on the East Texas Gulf Coast. J. Wildl. Manage. 15:46-56.

Smith, L.M. 1985a. Fire and herbivory in a Great Salt Lake marsh. Ecology 66:259-265.

Smith, L.M. 1985b. Predictions of vegetation change following fire in a Great Salt Lake marsh. Aquatic Botany 21:43-51.

Stewart, R.E. and H.A. Kantrud. 1972. Vegetation of Prairie Potholes, North Dakota, in Relation to Quality of Water and Other Environmental Factors. Geol. Surv. Prof. Paper 585-D. Washington, D.C.

Strohmeyer, D.S. and L.H. Fredrickson. 1967. An evaluation of dynamited potholes in northwest Iowa. J. Wildl. Manage. 31:525-532.

Sugden, L.G. 1978. Canvasback habitat use and production in Saskatchewan parklands. Can. Wildl. Serv. No. 34.

Svedarsky, D. and R.D. Crawford (Eds.). 1982. Wildlife Values of Gravel Pits. Miscellaneous Publ. 17-1982, University of Minnesota Agricultural Experiment Station, St. Paul, Minnesota.

Swanson, G.A., G.L. Krapu, and J.R. Serie. 1979. Foods of laying female dabbling ducks on the breeding grounds, p. 47-55. In T.A. Bookhout (Ed.), Waterfowl and Wetlands--An Integrated Review. NC Section Wildl. Soc., Madison, Wisconsin.

Swiderek, P.K., A.S. Johnson, P.E. Hale, and R.L. Joyner. 1987. Production, management and waterfowl use of sea purslane, Gulf Coast muskgrass, and widgeongrass in brackish impoundments, p. 441-457. In M.W. Weller (Ed.), Waterfowl in Winter. Univ. Minn. Press, Minneapolis, Minnesota.

van der Valk, A.G. 1981. Succession in wetlands: a Gleasonian approach. Ecology 62:688-696.

van der Valk, A.G. and C.B. Davis. 1978. The role of seed banks in the vegetation dynamics of prairie glacial marshes. Ecology 59:322-335.

Verry, E.S. 1985. Selection of water impoundment sites in the Lake States, p. 31-38. In M.D. Knighton (Ed.), Water Impoundments for Wildlife; A Habitat Management Workshop. North Central Forest Exper. Sta. Tech. Rpt. NC-100, St. Paul, Minnesota.

Voigts, D.K. 1976. Aquatic invertebrate abundance in relation to changing marsh vegetation. Amer. Midl. Nat. 95:312-322.

Weller, M.W. 1975. Studies of cattail in relation to management for marsh wildlife. Iowa State J. of Res. 49:383-412.

Weller, M.W. 1978. Management of freshwater marshes for wildlife, p. 267-284. In R.E. Good, D.F. Whigham, and R.L. Simpson (Eds.), Freshwater Wetlands, Ecological Processes and Management Potential. Academic Press, New York.

Weller, M.W. 1981. Estimating wildlife and wetland losses due to drainage and other perturbations, p. 337-346. In B. Richardson (Ed.), Selected Proceedings of the Midwest Conference on Wetland Values and Functions. Minn. Water Planning Bd., St. Paul, Minnesota.

Weller, M.W. 1987. Freshwater Marshes, Ecology and Wildlife Management. 2nd ed., Univ. Minn. Press, Minneapolis, Minnesota.

Weller, M.W. and L.H. Fredrickson. 1974. Avian ecology of a managed glacial marsh. Living Bird 12:269-291.

Weller, M.W. and C.E. Spatcher. 1965. The Role of Habitat in the Distribution and Abundance of Marsh Birds. Iowa St. Univ. Agric. & Home Econ. Experiment Sta. Spec. Sci. Rep. N0. 43.

Whitman, W.R. 1976. Impoundments for Waterfowl. Canadian Wildl. Serv. Occas. Paper No.22.

Whyte, R.J. and N.J. Silvy. 1981. Effects of cattle on duck food plants in southern Texas. J. Wildl. Manage. 45:512-515.

WETLAND AND WATERBODY RESTORATION AND CREATION ASSOCIATED WITH MINING

Robert P. Brooks
School of Forest Resources
Forest Resources Laboratory
Pennsylvania State University

ABSTRACT. A review of published and unpublished reports was combined with personal experience to produce a summary of the strategies and techniques used to facilitate the establishment of wetlands and waterbodies during mine reclamation. Although the emphasis is on coal, phosphate, and sand and gravel operations, the methods are relevant to other types of mining and mitigation activities. Practical suggestions are emphasized in lieu of either extensive justification or historical review of wetlands mitigation on mined lands.

The following key points should receive attention during planning and mitigation processes:

* Develop site specific objectives that are related to regional wetland trends. Check for potential conflicts among the proposed objectives.

* Wetland mitigation plans should be integrated with mining operations and reclamation at the beginning of any project.

* Designs for wetlands should mimic natural systems and provide flexibility for unforeseen events.

* Ensure that basin morphometry and control of the hydrologic regime are properly addressed before considering other aspects of a project.

* Mandatory monitoring (a minimum of three years is recommended) should be identified as a known cost. Rely on standard methods whenever possible.

Well-designed studies that use comparative approaches (e.g., pre- vs. post-mining, natural vs. restored systems) are needed to increase the database on wetland restoration technology. Meanwhile, regional success criteria for different classes of wetlands need to be developed by consensus agreement among professionals. The rationale for a particular mitigation strategy must have a sound, scientific basis if the needs of mining industries are to be balanced against the necessity of wetland protection.

OVERVIEW

The protection of wetlands is an issue of national concern. Of primary concern is how to mitigate for wetland losses. Few cost-effective opportunities exist to restore and create wetlands, thereby helping to reverse the trend in wetlands loss and perhaps create an increase. Surface mining, which historically has had substantial negative impacts upon the landscape, may offer some realistic and inexpensive mitigation options, if mitigation plans are integrated into mine reclamation plans at an early stage. To help guide those who make day-to-day decisions about wetland mitigation, this review provides a summary of the methods used to create and restore wetlands and waterbodies during mine reclamation. The recommendations presented at the end of this review can serve as a checklist to help ensure that constructed wetland systems function properly.

This review focuses on surface mining for coal, phosphate, and sand and gravel. It must be recognized that mining of these materials will continue in the U.S. into the foreseeable future. Coal is an essential component of electrical energy production, the fertilizer and chemical industries depend heavily on phosphate rock, and the construction industry requires continued access to sand and gravel reserves. Therefore, mitigation decisions should be based on consensus agreement among knowledgeable individuals who are familiar with both mining operations and wetland trends in the particular physiographic region in question.

The principles pertaining to these three extractable materials will, of course, have a bearing on other types of mining and severe landscape disturbances (e.g., placer mining; hydraulic mining and in-stream dredge mining; open pit mining and sand mining for metal ores, limestone and other rocks; peat extraction). Important literature on wetland mitigation from other types of mining is cited where it is relevant to the discussion. The extraction of peat differs markedly from mineral mining, and is beyond the scope of this review. Readers are referred to the following publications regarding peatland values, impacts and mitigation (Carpenter and Farmer 1981, Minnesota Department of Natural Resources 1981, Damman and French 1987).

The impacts of surface mining activities on wetlands and waterbodies have been well-documented (Darnell 1976, Cardamone et al. 1984). They differ depending on the material extracted, the methods used, and regional differences in topography, geology, soils, and climate. Even if it is assumed that reclamation is performed according to current regulations, mining will have significant effects upon the environment. In addition to the direct removal and filling of wetlands, the removal of soil and overburden severely alters local topography. This in turn disrupts local and regional groundwater and surface water flow patterns. Mining activities typically result in a decrease in groundwater tables and an increase in surface water runoff, both of which significantly affect the restoration and creation of aquatic systems. The removal of vegetation and disturbance of land surfaces increases sedimentation rates, with resultant increases in water turbidity. Access roads cause erosion in steep terrain, and can block the flow of water in areas of low relief, resulting in the formation of ponds. Exposed coal mine spoils readily oxidize, causing pollution problems such as acidic mine drainage. There are increases in the formation and deposition of materials, such as iron, manganese, aluminum, and sulfur, sometimes in amounts toxic to biota. Tailings from metal mines also can produce biologically-toxic discharges. Sedimentation rather than metal toxicity is a major problem associated with phosphate and sand and gravel mining.

In summary, habitat loss, chronic environmental stress, and toxic levels of pollution can occur during the mining process, especially if reclamation practices are poorly implemented (Darnell 1976). Any efforts to encourage wetland and waterbody creation on mined lands, whether to mitigate for losses attributable directly to mining, or as a means of increasing wetland area, should be cognizant of mining impacts on surficial and groundwater hydrology as outlined above.

Ten years have passed since the Surface Mining Control and Reclamation Act of 1977 (SMCRA, P.L. 95-87) was enacted. This federal act, coupled with the appropriate state statutes, has halted many of the past environmental abuses associated with surface mining, particularly with respect to the mining of coal. Although viewed as among the most encompassing and detailed pieces of environmental legislation, the Act often relies on vague notions, such as "higher and better uses" to guide decision-makers about reclamation strategies (Wyngaard 1985). Thus, the overall success of this law must be tempered by an examination of the relatively sterile landscapes that are often created under the guise of reclamation. Wetlands and waterbodies are allowed under existing regulations, but provisions are strict, and anything but encouraging. Decades of pre-SMCRA experience with polluted waters have resulted in cautious approaches to managing water on mined lands. Permanent impoundments are allowed under SMCRA guidelines, but "are prohibited unless authorized by the regulatory authority" (Sec. 816.49a). Thus, unless variances are sought, it is often viewed as less expensive to remove an impoundment or wet depression rather than to develop plans to leave it in place (Grandt 1981).

The Experimental Practices section of SMCRA (Sec. 711) produces the same result. Virtually any innovative reclamation technique can be tried if the operator is willing to justify the practice to state and federal regulatory authorities (Thompson 1984). This additional effort is often perceived as adding expense to a project, but overall costs may actually be reduced if permanent wetlands or waterbodies are left in place (Fowler and Turner 1981). The net result of these regulatory stumbling blocks has been to discourage the intentional creation of wetlands on surface mined lands (e.g., Gleich 1985) unless they are either demanded by an informed landowner, or based on in-kind replacement of a wetland that has been lost or degraded during mining. The vast majority of wetlands and waterbodies on mined lands exist not because of astute planning, but by accident.

Mining and related activities have disturbed less than 0.2% of the land mass of the U.S. (Schaller and Sutton 1978), yet in mining regions, disturbances can exceed 20% of a given land area (e.g., coal mining in Clearfield County, Pennsylvania; phosphate mining in Polk County, Florida). Coal reserves occupy large areas in selected regions of the U.S. (Fig. 1). Phosphate deposits, although large in area, occur only in a few regions. Wetland mitigation

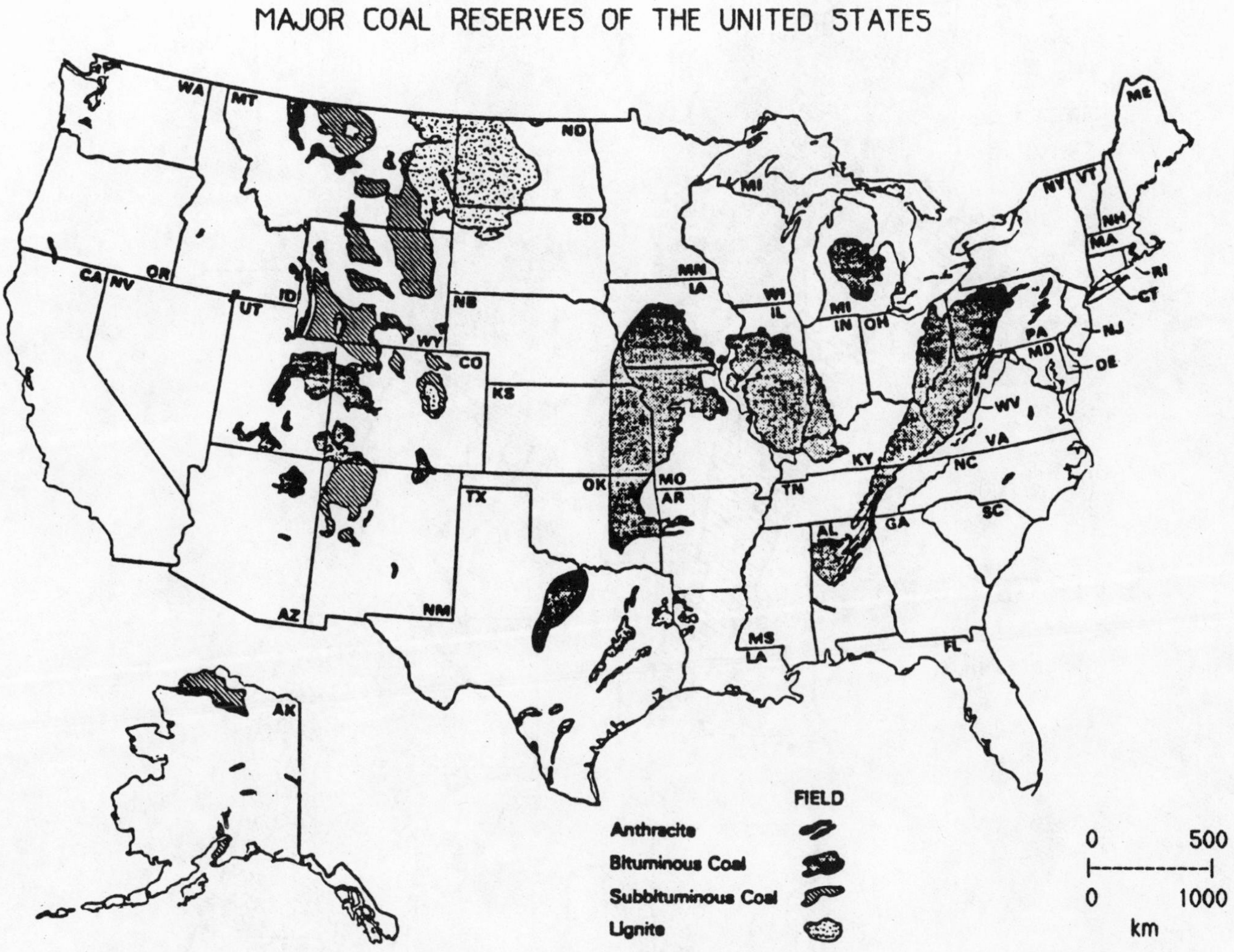

MAJOR COAL RESERVES OF THE UNITED STATES

Figure 1. Major coal reserves of the continental United States and Alaska (modified from Energy Information Service 1984).

issues related to phosphate mining occur primarily in Florida (Fig. 2). Sand and gravel deposits are dispersed throughout the U.S.

Mining activity has not always resulted in a net decrease in wetlands. The gain in open-water, palustrine wetlands nationwide during the 1950's to the 1970's (Frayer et al. 1983) is apparent in land use surveys of mined lands. Brooks and Hill (1987) reported that mined lands in Pennsylvania supported 18% more palustrine wetlands than unmined lands, primarily because of a 270% gain in permanent, open water wetlands in the glaciated coal region of the state. Conversely, Hayes et al. (1984) observed a reduction in the number of impoundments, particularly shallow, vegetated waterbodies, following passage of SMCRA in 1977. Palustrine vegetated wetlands are often converted to open-water wetlands, which may result in a significant change in regional wetland types (Tiner and Finn 1986, Brooks and Hill 1987).

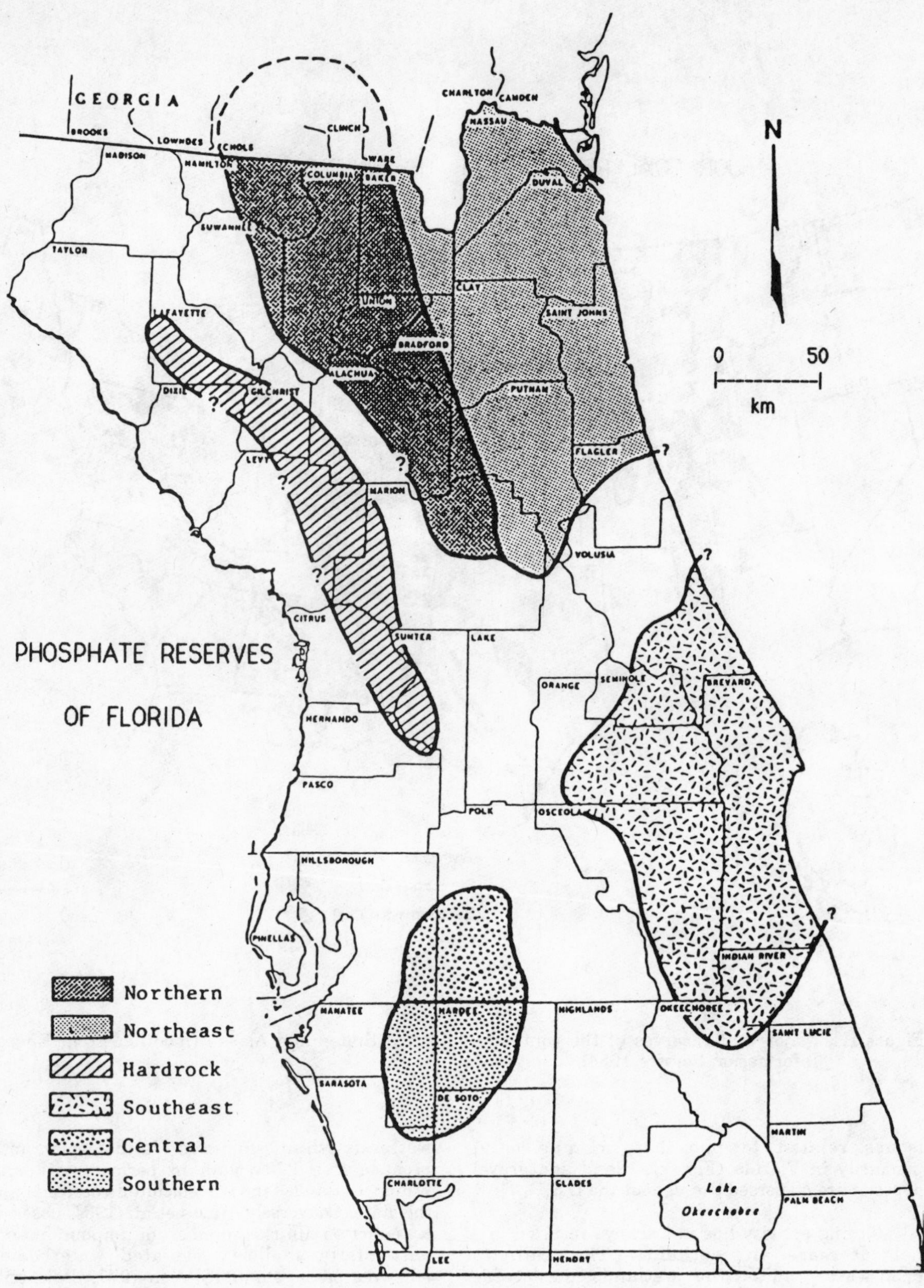

Figure 2. Major phosphate reserves of Florida (modified from Florida Defenders of the Environment 1984).

Before examining specifically how wetlands and waterbodies have been restored and created on mined lands, it is useful to discuss the characteristics and functions of volunteer wetlands, which are much more abundant than those purposefully designed. As mining practices differ, so do the types of aquatic environments left behind. The following will profile the inadvertent creation of aquatic environments by past mining and reclamation practices.

SURFACE MINING FOR COAL

Based on a survey of the literature, the following types of wetlands (listed approximately in order of declining numerical abundance) are commonly found on coal-mined lands: 1) sediment basins; 2) shallow wet depressions and emergent marshes; 3) moss-dominated springs and seeps; 4) final-cut and other deep lakes; 5) intermittent streams; and 6) slurry ponds and other coal refuse disposal areas.

During surface mine reclamation for coal, wetlands and waterbodies are created intentionally for erosion and sedimentation control as sediment basins. Basin size is determined by the anticipated runoff for a mine site and thus, is a function of the area of land disturbed. Sediment basins are usually <0.5 ha in size (Brooks and Hill 1987); often only 0.1 ha (Fowler and Turner 1981). They are geometrically shaped (i.e., circular, oval, rectangular), and have steep slopes (usually >30°), and flat bottoms. Volunteer palustrine wetlands also occur as a function of local changes in hydrology following reclamation. These include moss-dominated springs and seeps, persistent and non-persistent emergent marshes, and shallow wet depressions (similar to the prairie potholes of the northcentral U.S. (Cole 1986).

Before the advent of reclamation legislation, final-cut lakes were left inadvertently when the final excavation was not back-filled. Characteristics of these lacustrine waterbodies vary considerably, but they are often linear in shape, large (1-50 ha), deep (2-30 m), and have poorly developed littoral zones (Jones et al 1985a, Hill 1986). Other deep water bodies are formed after pits are excavated in regions with water tables near the surface. Lakes of varying shape and size have formed in this manner in glaciated regions of Pennsylvania (Brooks and Hill 1987), and several midwestern states (Jones et al. 1985b, Klimstra and Nawrot 1982); 6,000 ha of these lakes exist in Illinois (Coss et al. 1985) and 3,600 ha in Ohio (Glesne and Suprenant 1979) (Fig. 1).

After coal is processed, coal fines and associated particles are discharged into basins known as slurry ponds. The usual reclamation procedure for these typically acidic disposal areas is to cover them with at least 1.3 m of topsoil. However, a variety of vascular hydrophytes will colonize slurry ponds, thus establishing emergent wetlands (Nawrot 1985) (Fig. 3).

Although a discussion of streams and rivers is beyond the scope of this review, intermittent streams containing emergent vegetation are also fairly common, and therefore, constitute another wetland type. Relocation and restoration of major streams is discussed in another chapter of this document (see Jensen and Platts).

A variety of ecological functions and economic uses have been documented for the types of wetlands listed above, including wildlife and fisheries habitat, agricultural and recreational activities, sediment retention, treatment of acidic mine drainage, and public water supplies.

Sediment basins provide for uses beyond their intended purpose. They provide habitat for a variety of vertebrate taxa, including birds (Burley and Hopkins 1984, Sponsler et al. 1984, Brooks et al. 1985a), mammals (Brooks et al 1985a), and herpetofauna (Fowler and Turner 1981, Brooks et al. 1985a). A diverse macroinvertebrate community also has been identified with sediment basins (Hepp 1987).

Mine lakes are known to produce excellent fisheries (Jones et al. 1985b, Mannz 1985), in part due to adequate primary production (Brenner et al. 1985) and macroinvertebrate production (Jones et al. 1985a). They can serve as foci for recreational activities such as fishing, boating, and waterfowl hunting (Klimstra et al. 1985). Other uses, particularly in the Midwest, include lake-side housing and community open spaces, crop irrigation and livestock watering, and water supplies for homes, fire protection, and industrial purposes (Glazier et al. 1981).

Vegetated wetlands dominated by either vascular or non-vascular species can effectively sequester some of the constituent of mine drainage. Observations on the removal of metals by naturally occurring Sphagnum moss (e.g., Wieder and Lang 1986) has led to further investigations of how mosses, algae and macrophytes with their associated bacteria, can be used to ameliorate the effects of mine drainage (see Girts and Kleinmann 1986 for a review).

Figure 3. Seasonally inundated zone of a wetland created on coal slurry in Indiana showing four years of growth. (Courtesy of Jack Nawrot, Cooperative Wildlife Research Laboratory, Southern Illinois University.)

SURFACE MINING FOR PHOSPHATE

Primary phosphate deposits in the U.S. occur in Florida, Tennessee, South Carolina, and the Phosphoria Formation of Montana, Wyoming, Idaho and Utah. However, surface mining activities affecting wetlands occur almost exclusively in Florida, where wetlands typically comprise 8-17% of the landscape area (Florida Defenders of the Environment 1984) (Fig. 2). The principal focus of reclamation in Florida has been on mitigating for wetland losses. Substantial progress in developing restoration and creation technology has been made through the combined efforts of the phosphate industry and state regulatory agencies. Florida provides an example of how regulatory pressures can accelerate a desired technology if the pressures are firmly and reasonably applied.

Reclamation of phosphate mines in Florida was voluntary until 1975, when the Department of Natural Resources developed regulations (Florida Administrative Code Section 16C-16, 16C-17) in response to legislation (Florida Statute 211.32, 370.021). Due to extraordinary residential and commercial development, there have been increasing efforts to push reclamation technology toward the goal of replicating original wetland conditions as a provision of mining. Public pressure combined with a flat topography suitable for wetland establishment and awareness of wetland functions and values has led to sophisticated efforts to restore and create complex wetland systems.

Historically, the removal of phosphate ore by draglines produced mounds of sand tailings interspersed among waterbodies and clay settling ponds. Many of these waterbodies were used by wintering waterfowl that were attracted by volunteering hydrophytes (e.g., Najas spp.). Gradual filling of these waterbodies with clay produced a successional trend from submergent species, to emergents (e.g., cattail, Typha spp.), and finally to shrubs (e.g., willow, Salix spp.) (Clewell 1981). Wildlife use diminished during this process (King et al. 1980). After 1975, reclamation required regrading of the sand tailings, and planting them to pasture. Depressions left from the removal of phosphate ore fill with water, producing a mosaic of pastures and lakes. Marion and O'Meara (1983) reported that the reclamation laws had produced both positive and negative effects on wildlife. One of the positive impacts, was an increase in wetland edge following reclamation.

Boody (1983) studied 12 reclaimed lakes, 6 classified as deep, and six that were considered shallow. Deep lakes had a mean area or 59 ±113 ha (range = 2-287 ha) with depths of 3-5 m. Shallow lakes had a mean area of 9 ± 17 ha (range = 2-30 ha) with depths of 2-3 m. The pH was typically 5-7, but ranged from 4.2-9.3. Most reclaimed lakes supported fewer species of fish (mean = 10 ± 6 species, range = 4-22) than the 4 natural lakes that were studied (mean = 18 ± 2 species, range = 16-20). Of the 29 fish species collected, 27 were native and 2 were introduced (Brice and Boody 1983).

Recent reclamation plans have included littoral zones and periodically flooded areas as part of the lake ecosystem. Freshwater emergent marshes have been successfully established, and to a lesser extent, forested wetlands have been created (Haynes 1984).

The clay settling ponds, which usually occupy >50% of a mine site, continue to pose problems and are perceived negatively by the public. Waste clays from the mining process are suspended in a slurry and pumped into settling ponds. Attention has been focused on de-watering these ponds as rapidly as possible; typically within 10 years. Although the ponds support fewer plant species than natural wetlands, site management in conjunction with planned species introductions can create a heterogeneous mix of wetland vegetation and open water (Robertson 1983). Montalbano et al. (1978) discussed the value of clay settling ponds as wintering habitat for waterfowl (7 species reported), and suggested that water level manipulation would help create high interspersion of emergents and open water. Haynes (1984) believes that these settling ponds may have substantial positive values, and thus should be manipulated and managed as productive wetlands. The settling ponds are large (81-405 ha) and are currently increasing at a rate of 1,000 ha/yr beyond the 30,000+ ha already present. Several authors advocate a drainage basin approach for mitigating wetland losses in the phosphate region, so that areas beyond the individual mining unit are considered in the planning process (Breedlove and Dennis 1983), although Fletcher (1986) believes that the current knowledge is only suitable for restoration of small drainage basins.

SURFACE MINING FOR SAND AND GRAVEL

Sand and gravel is defined as unconsolidated mineral and rock particles. Generally the particles have been transported by water, and therefore, many deposits still occur in and around waterbodies. Inland deposits are typically classified as fluvial, glacial or alluvial depending on their origin.

Sand and gravel is vital to the construction industry. Excavation is the chief method of removal, and 18% of the lands disturbed by mining are for sand and gravel. Sand and gravel operations are regulated primarily through state laws. Demand continues to increase, with deposits near urban areas in highest demand. More excavation can be expected in nearly every state in the U.S. (National Research Council 1980). Concurrently, opportunities for restoration and creation of wetlands on reclaimed sand and gravel sites will also increase.

Although waterbodies that remain following sand and gravel mining have been used for a variety of purposes, reclamation plans have often lacked advanced planning and imagination (McRae 1986). Fishing, boating, and wildlife observation commonly take place in water-filled sand and gravel pits in both the U.S. and Great Britain (Koopman 1982, McRae 1986, respectively).

Lomax (1982) found that reclaimed lakes in the southern coastal plain of New Jersey were colonized first by emergents (e.g., _Typha_ spp., _Cyperus_ spp., _Juncus_ spp., _Scirpus_ spp.), and then by woody vegetation such as black willow (_Salix nigra_), red maple (_Acer rubrum_), and black tupelo (_Nyssa sylvatica_). Occasionally, bog-like communities developed in pits <1 ha in area. Lomax recorded use by 194 species of vertebrates over a 16-year period. Gallagher (1982) found that the Delta Ponds of Eugene, Oregon became completely revegetated through volunteer colonization. Species found included cattail, pondweed (_Potamageton_ spp.), willow, and alder (_Alnus_ spp.). Birds (78 species), mammals, and fish were observed using the 65-ha area.

Street (1982) reported on the gravel-pits of Great Britain. Pits ranged in size from 1-100 ha, and were 3-30 m deep. Most had steep sides and flat bottoms. A restoration project was initiated at the gravel-pit complex of the A.R.C. Wildfowl Centre in Great Britain in 1972, and has developed into a highly productive 37-ha wetland system. Waterfowl density within the managed site ranged between 2.4 to 38.7 birds/ha, whereas avian density on unmanaged gravel pits did not exceed 2.8 birds/ha. By manipulating basin morphometry, plant communities, and the availability of organic matter, both vertebrate and invertebrate numbers and diversity were increased.

RECOMMENDATIONS FOR RESTORATION AND CREATION WETLANDS AND WATERBODIES ON MINED LANDS

There have been recent efforts to stimulate the restoration and creation of wetlands on mined lands (e.g., Klimstra and Nawrot 1982, Brooks 1984, Brenner 1986, Brooks 1986, Haynes 1986, McRae 1986). Sufficient recommendations exist to provide guidance for wetland mitigation on mined lands. Due to procedural and regional differences in mining coal, phosphate, and sand and gravel, the recommendations will be discussed under separate headings below.

One cannot incorporate all possible mitigation into a single wetland project. It is best to work within clearly stated objectives that are tied to specific wetland functions. Only then can the wetland be designed optimally with respect to the desired objectives. There may be conflicts among objectives which should be resolved in the planning process (e.g., public water supply vs. ecological productivity, recreational fishery vs. habitat for diverse amphibian community). Whenever possible, pre-mining conditions and regional reference wetlands should be used as guides to how a wetland ought to be created or restored.

WETLANDS AND WATERBODIES ON COAL MINED LANDS

Basin Morphometry

Area--

The area covered by a wetland or waterbody is constrained both by objective and site location. Peltz and Maughan (1978) suggested that several ponds of small size (0.25-1.0 ha) were preferred to a few large ones with regard to fish production; 0.1 ha being the minimum recommended size. Sandusky (1978) found that some species of waterfowl (e.g., blue-winged teal, Anas discors) nested on ponds as small as 0.04-0.2 ha. Hudson (1983) recommended pond areas >0.5 ha for waterfowl production in stock ponds, as did Allaire (1979) for wildlife in general. Based on an inventory of 35 existing wetlands on mined lands in Pennsylvania, Hill (1986) recommended areas of 1-3 ha to maximize wildlife diversity. For natural habitats, bird species richness has been found to increase with wetland area, but to level off for areas >4 ha (Williams 1985). As sediment ponds are typically less than 0.5 ha in area, a slight increase in pond size, while still maintaining a diversity of sizes, would seem to meet multiple objectives.

Mine lakes can be as large as the remaining mined pit or depression. Lakes exceeding 10 ha are not uncommon (Jones et al. 1985a). Glazier et al. (1981), Nelson (1982), and Doxtater (1985) provide scenarios for planning multiple-uses of large, deep-water lakes.

Depth–

The need for water permanence will determine the appropriate depth of a given wetland or waterbody. Again, project objectives coupled with mining operations will influence the eventual depth characteristics of the wetland. It is important to remember that deep water lakes will tend to be either mesotrophic or oligotrophic, whereas shallow waterbodies and vegetated wetlands usually have higher primary productivity, tending toward eutrophic conditions.

To enhance year round survival for fish, Peltz and Maughan (1978) recommended depths of 2-3 m for ponds with groundwater sources, and >5 m for ponds supplied by surface runoff alone. Although water depth in excess of 3 m may be desirable for fish survival, retention of flood waters, as a water supply reservoir, and for some recreational activities, most investigators have stressed the need for construction of an extensive littoral zone. Many species of sport fish require depths of 0.5-2.0 m for spawning (Peltz and Maughan 1978, Leedy 1981). Some submergent hydrophytes grow better in depths >0.5 m (e.g., Potomageton spp., Chara spp.). The regulatory guidelines of SMCRA require stability of water levels in impoundments, but this may not be feasible or desirable for many wetlands. Colonization by emergent hydrophytes requires fluctuating water levels. Cole (1986) found that water volume, and hence depth, varied by >40% in five ponds on mined land in Illinois. These changes in water level exposed the littoral zone much like the wetlands of the Prairie Pothole region further west. Fluctuating water depths of <0.5 m are recommended to promote the growth of emergent hydrophytes, which in turn encourage macroinvertebrate production in the littoral zone.

Slope–

Whereas a shelf 1 m in depth may benefit aquatic species such as fish, other species benefit from slopes that grade gently from upland to wetland. A wetland basin that has a variety of slopes, ranging from <5° to almost 90° will benefit a diversity of wildlife species and provide visual variety. The majority of the shoreline should have gentle slopes. Amphibians, reptiles, and some fishes require gentle slopes, typically <15°. Sand and mud flats used by foraging shorebirds should have slopes of <5°. Access areas for swimming and boating also require gently sloping terrain. Some species will benefit from steeply sloped or overhung banks, including burrow-dwelling muskrats (Ondatra zibethicus, Brooks and Dodge 1986), belted kingfisher (Cerle alcyon, Brooks and Davis 1987), swallows, cliff-nesting raptors, and some fishes.

Shape–

The shorelines of wetlands and waterbodies should be convoluted to produce an irregular shape (Brooks 1984). Basins with a high shoreline development index (i.e., length of shoreline divided by the circumference of a circle of equal area, Wetzel 1975) provide more edge for wildlife, and reduce wind and wave action on larger waterbodies (Coss et al. 1985). Coves, peninsulas, and islands contribute substantially to shoreline development (Leedy and Franklin 1981, Brooks 1984). Islands (>3 m in diameter, Emerick 1985) and even large rocks (0.5-1.5 m in diameter, O'Leary et al. 1984) provide nesting and resting places for many species of waterfowl and shorebirds. Irregularities in basin shape tend to disperse water flows thus helping to maximize retention time in the basin if flood control or water treatment are desirable characteristics of the wetland.

Soils--

It is important to consider both hydric soils within a wetland and the soils of adjacent uplands. Proper management of upland soils will protect aquatic systems from unnecessary sediment, chemical, and thermal pollution (Rogowski 1978, Leedy 1981).

Upland Soils--

During reclamation of mined land it is preferable to have topsoil cover the overburden to protect and conserve the available water and provide a better medium for plant growth. Vegetated topsoil will reduce evaporation, allow more infiltration into groundwater supplies, produce temporary ponds in depressions, and reduce peak infiltration rates that lead to abnormal fluctuations in the water table and droughty surface conditions (Rogowski 1978). The ability of a soil to retain water is dependent on its texture (i.e., sand is more droughty than clay), depth, the content of organic matter, and the distribution of pore size (Schaller and Sutton 1978). Thus, a careful study of soil conditions will enhance the probability of successful restoration and creation of wetlands, and the reclamation of upland areas.

Sediment yields from exposed soils are typically highest in the first 6 months after regrading, and are halved in subsequent 6-month periods as the site progressively revegetates (Schaller and Sutton 1978). Silts and sediments that enter waterbodies tend to reduce light transmission (and hence, photosynthesis), raise water temperature, and cover sensitive organisms (Leedy 1981). Thus, it is important to revegetate exposed soils as rapidly as possible to avoid interfering with wetland establishment.

Based on the reclamation literature for upland portions of mined sites, several recommendations can be made to protect aquatic systems from upland runoff. Exposed soils must be seeded, fertilized, and mulched as soon as possible. Rafaill and Vogel (1978) recommended 60 lbs/acre (67 kg/ha) of nitrogen and 100 lbs/acre (112 kg/ha) of phosphorus, but no potassium for reclaiming mined land in Appalachia. In acidic areas, soil should be limed to a pH of at least 5.5. Use of high quality seeds is advised (i.e., high germination and purity percentages, McGee and Harper 1986). Seeds and fertilizer should be applied first, followed by an appropriate mulch to avoid perching seeds above the ground's surface (Schaller and Sutton 1978). Straw and hay were suggested as the best mulch to use, particularly if applied by a mulch blower that cuts, shreds, and evenly spreads the material. Estimated costs for purchase and application of straw were \$100-200/ton (909 kg) with an application rate of 1-2 tons/acre (2,245-4,490 kg/ha), whereas nets and mats may cost > \$1,500/acre (\$3,700/ha), especially on steep slopes (Mining and Reclamation Council of America and Hess and Fisher Engineers 1985). Advice for selecting the appropriate plant species, and seeding, fertilizing, and liming rates for a given soil type are generally available from mining agencies (e.g., Office of Surface Mining, U.S. Bureau of Mines, state mining agencies) and county offices of the Soil Conservation Service.

Vegetative buffers should be installed around wetland basins. Although recommendations for buffer widths range from 1-300 m, vegetated strips as narrow as 15-20 m can remove 50-75% of the sediments (Barfield and Albrecht 1982). Whenever appropriate for a given region, buffers should include shrub and forested zones. Gilliam (1985) studied unmined agricultural areas in North Carolina and found that wooded buffers about 100 m wide removed >50% of the sediment, including much of the nitrogen and phosphorus. In addition to serving a water quality function, vegetative buffers can act as travel corridors and refugia for wildlife, thereby reducing the isolation of the wetland. If desired, wetland edges can be shaded by planting properly oriented tree species that will grow to a height of twice the distance to the water (Leedy 1981).

In severe cases of upland runoff, structural diversions may be necessary to divert sediment-laden waters. Diversion ditches, concave depressions, and sediment traps are some of the techniques available. Mining agencies, the Soil Conservation Service, or experienced consultants can provide the expertise needed to design these systems.

Hydric Soils--

Hydric soils, or those previously saturated, usually must be constructed, unless soil is available from a wetland scheduled to be altered or removed. Routine cleaning of roadside ditches or other wet depressions can also act as a source of hydric soil, although pollutants such as road salt, oil, or lead may be present in substantial quantities. Hydric soils should be stockpiled, preferably for less than one month, and then spread to the desired thickness in newly constructed basins. These soils typically have a relatively high organic matter content, and often act as a seed source or seed bank. Longer storage periods will result in desiccation of plant materials, and possibly re-oxidation of metals and other potentially damaging materials.

When existing hydric soils are not available, they can be constructed by using a relatively fertile topsoil. Good plant survivorship and seed germination rates have been obtained by mixing about 30% (by volume) livestock manure in with the topsoil to act as a source or organic matter and nitrogen (Brooks, unpublished). Small quantities of superphosphate were added to the soil around each planted propagule. Chemical fertilizers have been recommended as an additive to ponds and lakes designed for fish production; 12-12-12 or 8-8-2 (nitrogen-phosphorous-potassium; Glesne and Surprenant 1979, Leedy 1981, respectively). Leedy (1981) suggested that no more than 200kg/ha of 8-8-2 fertilizer be added at one time, although application rates for infertile waters might exceed 1,500 kg/ha/yr.

Whenever possible, soil tests should be made to provide more accurate estimates of fertilizer and lime additions for both hydric, and adjacent upland soils depending on the plant species desired and the intended use of the wetland or waterbody. Basins constructed on mined lands often contain acidic soils. Assuming that a circumneutral pH is desired (although some wetland plants require acidic or alkaline conditions), the pH of the bottom soils should be raised to about 6.5 using lime (Peltz and Maughan 1978). If the pH of the soil is less than 5.5, then at least 1,000 kg/ha of lime is probably needed (Leedy 1981). Slurry ponds with acidic soils require more alkaline additions to promote growth of hydrophytes; >20,000 kg/ha of limestone (Nawrot 1985). Warburton et al. (1985) reported improved growth rates for bulrushes (Scirpus spp.) with addition of slow-release fertilizer tablets (22-8-2) to each propagule. If acidophilic plants occur naturally on the site or their presence is a desired objective of reclamation, then it may not be necessary to adjust pH. The presence of certain species of moss and algae in springs and seeps is a good indicator of waters with low pH and usually low concentrations of nutrients (Brooks, unpublished).

Basins constructed below the water table rarely need to be sealed, whereas perched wetlands need a water-conserving layer of material on the bottom and sides of the basin. Clay is commonly used in this manner and should be compacted to a thickness of about 30 cm (Soil Conservation Service 1979). Bentonite, and synthetic membranes can also serve as sealants. Specifications for a specific soil type and climate are generally available from county offices of the Soil Conservation Service or mining agencies.

Vegetation

Studies of existing wetlands have shown that a diversity of hydrophytes will volunteer over time. Cattail (Typha latifolia) is by far the most successful vascular hydrophyte on mined lands. Cattail, soft rush (Juncus effusus), and woolgrass (Scirpus cyperinus) were the first invaders of four wetland basins in central Pennsylvania; all were present within 1-1.5 years of regrading (Hepp 1987). Twelve species of vascular plants had volunteered on one site after 6 years. Fowler et al. (1985) found that cattail, soft rush, and spike rush (Eleocharis obtusa) rapidly invaded sediment ponds in Tennessee; 10 species were eventually present. Coss et al. (1985) found 14 species of vascular hydrophytes growing in four lake complexes in Illinois. The lake with the greatest hydrophyte diversity had 7 species.

After volunteer macrophytes were observed on slurry ponds in southern Illinois (e.g., reedgrass (Phragmites australis), a planting program was started in that has led to revegetation of more than 200 ha of wetlands on 12 sites (Nawrot 1985). Investigators found that perennial rootstocks of hardstem bulrush (Scirpus acutus), three-square (S. americana), and prairie cordgrass (Spartina pectinata) were more dependable than seed because sub-surface conditions were more amenable to plant establishment than surface conditions. Rootstocks were collected at a rate of 75-100 propagules/man-hour, and hand-planted with bars and shovels. Collection of propagules in early spring is preferred over autumn collection. Spacing was on 0.3-1.5 m centers, and each propagule was planted in 5-13 cm of soil, depending on the species. They recommended a water-level control structure to assure adequate control over seasonally variable water levels. Plants collected locally under similar conditions had better survival rates than commercially available stock. Whenever possible, local planting stock should be used.

We have constructed smaller wetlands, designed specifically for treating acidic mine drainage (Brooks, unpublished). Cattail rhizomes were collected from existing sediment basins at a rate of 50-100/man-hour, and planted at a rate >100/man-hour. Multi-stemmed clumps of sedges (Carex gyandra) and soft rush (20-cm dia. plugs) were also collected. Both were planted on 0.5-1.0 m centers. We had 75-80% survival of these plants after one year. Costs for constructing wetlands designed to treat mine drainage are slightly less than \$10/m^2 (\$1/ft^2) (Girts and Kleinmann 1986, Rightnour, pers. comm.), including all planting and basin construction costs.

There are several ways to enhance the proliferation of aquatic vegetation if a decision is made to encourage volunteer colonization. The morphometry of the basin, as previously discussed, must be suitable with respect to depth and slope. The desired zones of vegetation can be controlled by manipulating morphometric variables. Emergent species will colonize the littoral zone up to about 1 m in depth. Shrubs will be restricted to very shallow or seasonally flooded zones. By creating topographic diversity within a site, there will be more opportunities for a variety of species to successfully colonize. Mitigated wetlands that have hydrologic connections with natural wetlands or other mitigated sites will be more likely to receive plant propagules, either through wind, water, or animal dispersal.

Fauna

Diverse vertebrate and invertebrate communities have been found in wetlands and waterbodies on coal surface mines. Brooks et al. (1985a) reported that 125 vertebrate species were observed on 35 wetlands studied in western Pennsylvania (86 birds, 19 mammals, 11 reptiles, and 9 amphibians). The mean number of vertebrates per wetland was 23 ± 12 (1 SE), with a range of 7-60 species/wetland (Hill 1986). Hepp (1987), in a more intensive study on four wetlands in the same region, reported use by 90 vertebrate species (64 birds, 15 mammals, 3 reptiles, and 8 amphibians) and more than 39 invertebrate taxa. O'Leary et al. (1984) observed 76 avian species, including 18 species of waterfowl, and 10 mammalian species on 47 wetlands in southwestern Illinois. Also in Illinois, Coss et al. (1985) studied four mined lake complexes and found 89 vertebrate species. In nine sediment ponds in Tennessee, Fowler et al. (1985) reported use by 61 invertebrate taxa, 6 fish species, and 12 amphibian species. Jones et al. (1985a), in a study of 33 mine lakes in Illinois and Missouri, identified almost 200 invertebrate taxa and 33 fish species. Nineteen species of fish were collected from one 86-ha mine lake in Illinois (Jones et al. 1985b).

With the exception of fish species and some invertebrate taxa (e.g., Mollusca), most species voluntarily colonize surface mine wetlands. Some invertebrates can be introduced by water birds as larvae attached to feet and feathers. Fish are also introduced by local anglers. If wetlands are hydrologically connected with other reclaimed or natural systems, the opportunities for rapid colonization are greatly improved. If wetlands are juxtaposed to a variety of upland habitats that provide shelter and travel corridors, colonization rates and numbers probably will be greater. Hepp (1987) reported rapid colonization rates (within 3 years of final grading) for both invertebrate (e.g., dipterans, coleopterans, hemipterans) and vertebrate (e.g., amphibians, some small mammals, and many birds) species, followed by a period of stabilization in community structure. Pentecost and Stupka (1979) found that common amphibian species invaded sediment ponds within one month of formation; founder populations were located 100 m away.

Artificial structures and substrates can be introduced to supplement existing shelter, such as nest boxes for cavity nesters and artificial reefs for fish and invertebrates. As with flora, wetlands designed with specific objectives will provide suitable habitat for the desired species, whether fish, waterfowl, or a diversity of faunal groups.

WETLANDS AND WATERBODIES ON PHOSPHATE MINED LANDS

Basin Morphometry

The same parameters discussed for coal mined lands are equally important for phosphate areas (e.g., area, depth, slope, shape). A major difference exists for the Florida landscape, however, because of its low relief. To accommodate the sheet flow of water over flat lands and to match the adaptions of plant species to subtle changes in elevation, the slopes of basin banks and bottoms need to be carefully established. In addition, hydroperiod variations between wet and dry seasons must be considered in project design. Wetland systems must be capable of storing large quantities of water during the wet season. This can be accomplished by designing wetlands of sufficient size. During the dry season, when the water table drops below the land surface, there must be enough deep depressions to harbor aquatic organisms, such as fish, amphibians, and invertebrates (King et al. 1985). Depressions should be at least 2 m below the high water marks to maintain aquatic habitat during droughts (King et al. 1985). Reduced slopes (<3%) will allow the development of wide soil moisture zones. This will provide a wider tolerance zone for many species of wetland vegetation and compensate for environmental disturbances, such as drought, fluctuating water tables, and fire. Conversely, steeper slopes result in narrow moisture zones that leave little room for error in predicting the eventual composition of the floral community.

Soils

Soils of the central phosphate region in Florida are typically circumneutral and quite fertile due to the abundance of phosphorus and calcium, although potassium may be limiting (Clewell 1981). Phosphate deposits further north may be more acidic and less fertile. Soils being prepared for the establishment of wetlands are usually regraded to the proper elevation and conformation using the sand tailings. Additional overburden, if available, can then be added up to a depth of 30 cm (Erwin 1985). Numerous studies have shown that the addition of wetland topsoil (i.e., mulch, organic muck) from natural wetlands scheduled for mining greatly enhances the chances for successful reclamation (Clewell 1981, Dunn and Best 1983, Erwin 1985). "Topsoiling" or "mulching" can provide a variety of propagules (e.g., seeds, roots, rhizomes) from native plant species that result in a more natural vegetative community at the exclusion of weed species.

Vegetation

Wetland restoration efforts in the Florida phosphate region have focused on the establishment of three major types of wetland communities: open water, emergent marshes, and forested wetlands. Open water areas are primarily a function of water depth and need not be discussed further. The two types of vegetative communities will be discussed separately. The techniques developed have been influenced by regulations that require rapid revegetation (within one year) and the creation of a self-sustaining community.

Emergent Marshes--

Techniques used to establish herbaceous hydrophytes include: 1) transplanting from natural wetlands; 2) application of hydric topsoil from natural wetlands; and 3) reliance on voluntary establishment. Florida allows licensed individuals to remove native species from natural wetlands for the purpose of mitigation. In addition, plants from wetlands scheduled for mining can be transplanted to newly prepared sites. However, the availability of plants from natural wetlands may not always match the timing of mitigation projects.

For the Agrico Swamp project in central Florida (restoration of 61 ha of wetland on a 148 ha project site) Erwin (1985) and Erwin and Best (1985) reported that the application of wetland topsoils resulted in the establishment of 41 plant species in a restored marsh, whereas overburden alone resulted in the establishment of only 26 species. The common species present included, cattail (Typha latifolia), pickerelweed (Pontederia cordata), rushes (Scirpus californicus), dog fennel (Eupatorium capillifolum), and arrowhead (Sagittaria lanceolata). The topsoil areas contained both perennial and annual species, whereas the overburden areas contained primarily annuals. The rapid establishment of late-successional species, such as many perennials, either through "topsoiling" or transplanting may help to eliminate undesirable species such as cattail (Erwin and Best 1985).

Volunteer plants contribute substantially to restoration efforts in Florida. Certain factors can increase the role of volunteers. When natural communities are proximal to restored sites, the likelihood of propagule dispersal is enhanced. Hydrologic connections with streams can also distribute the seeds and propagules of desirable species. Self-sustaining seed banks with their inherent benefits can become established within 3 years if artificial planting is done (Erwin 1985), and within 4-5 years if based solely on volunteer species (Dunn and Best 1983).

Restoration of emergent marshes in Florida is further enhanced if good quality planting stock is used and if specific site preparation and planting methods are properly applied (Haynes 1984). Several long-term monitoring studies of wetland mitigation projects are underway in Florida (e.g., Erwin 1985) which should help determine how closely created match natural conditions.

Forested Wetlands--

The slow growth of woody species prevents rapid assessment of the success of creating forested wetlands, but there are indications that the techniques applied in Florida will be successful. As part of the 61 ha of wetlands created on the Agrico Swamp project site, 66,000 tree seedlings were planted. Twelve species were represented. The most abundant species included, cypress (Taxodium distichum), Florida red maple (Acer floridium), loblolly bay (Gordonia lasianthus), black gum (Nyssa sylvatica), sweetgum (Liquidambar styraciflua), and Carolina ash (Fraxinus carolinia). Seedlings were planted on about 2-m centers by hand in the summer and fall of 1982. Survivorship was 72% in 1982, 77% in 1983, 72% in 1984, and dropped to 58% in 1985 following a drought (Erwin 1985). Growth of some species was apparently enhanced when water levels during the wet season did not exceed 20 cm (Best and Erwin 1984).

Gilbert et al. (1980) reported on the success of planting 10,400 seedlings representing 16 species. After the first year, survival was 85% for cypress and green ash (Fraxinus pennsylvanica), 72% for sweetgum, and 62% for red maple (Acer rubrum).

Clewell (1981, 1983) also had tree seedling survival in excess of 70% while creating a riverine forested wetland in central Florida. Mechanical planting of potted, nursery-grown seedlings increased efficiency and enhanced survival, but may not be feasible, depending upon the substrate. Direct seeding may also be possible, but germination and survival rates are lower. Clewell (1983) suggested that enclaves of saplings could be established through transplanting to provide shade for shade-tolerant species. A combination of seedlings, saplings, topsoil, and natural colonization were recommended.

Fauna

Studies of faunal communities on reclaimed phosphate mines have been less common than studies of vegetation. Erwin (1985) reported that 56-62 taxa of macroinvertebrates were collected seasonally in open water, submergent, and emergent wetland communities for an annual total of 107 taxa. A total of 83 bird species were recorded on the same 148 ha site. Use of clay settling ponds by waterfowl and shorebirds has also been reported (Montalbano et al. 1978). King et al. (1985) provide extensive recommendations for enhancing fish and wildlife habitat on both wetland and upland mine sites in the phosphate region. To maximize fish and wildlife diversity, they suggest the creation of heterogeneous physical and vegetative habitats among a diversity of aquatic systems.

WETLANDS AND WATERBODIES ON SAND AND GRAVEL MINES

As many of the recommendations discussed for coal and phosphate mining apply to sand and gravel, only those techniques that differ will be included in this section. Mining of mineral sands for rare metals (e.g., rutile, zircon) is a unique type of sand mining. Although most prevalent in Australian coastal zones, the wetland restoration techniques developed by this industry also warrant inclusion in this section (e.g., Brooks 1987, 1988).

Basin Morphometry

Most authors recommend increasing the area of wetland basins, and having a heterogeneous shoreline. Slopes with a horizontal to vertical ratio as gentle as 10:1 or 20:1 are recommended to increase the zone widths of plant communities (Street 1982, Crawford and Rossiter 1982). Water-level control devices are encouraged to allow optimal management of vegetation.

Soils

The addition of topsoil to pit floors was recommended where plant colonization is desired (Crawford and Rossiter 1982). Leaving some areas bare will meet the foraging requirements of wading birds and shorebirds (Lomax 1982), and the spawning needs of fish (Herricks 1982). The bare zones should have a variety of particle sizes as substrate to meet the specific needs of various species. Compaction of bottom material is an effective means to discourage volunteer plant species. In newly reclaimed sites, organic matter is often lacking, so straw or hay can be added as food and substrate for both plants and invertebrates; 1 kg/m^2 of straw was suggested by Street (1982). Nutrients are often lacking as well, so fertilizing may be necessary. Stabilization of upland banks and surrounds is also emphasized to reduce erosion and sedimentation (Branch 1985).

Brooks (1987, 1988) identified three major factors that enhanced the recovery of both herbaceous and woody vegetation after mining for mineral sands. First, basin morphometry must be reclaimed properly to provide suitable drainage patterns and water-level control. Second, the use of drains before and after mining under saturated conditions facilitated the establishment of seedlings by avoiding excessive drying or flooding. Drains were removed once the plants adapted to the variable water regime. Third, careful manipulation of existing topsoil enhanced the survival of propagules of native species. A "double-stripping" method was used. The upper 20-25 cm was removed and stockpiled in large lumps. A second layer that was 10-15 cm deep was stockpiled separately. Topsoil layers were returned in their original order after mining. Storage time was usually 1-3 months. This additional care later reduced planting costs during reclamation. Other recommendations for enhancing restoration of vegetation and fauna were similar to those discussed for coal and phosphate.

CONCLUSIONS AND RESEARCH NEEDS

There are examples throughout the U.S. and other countries of innovative approaches to successful restoration and creation of wetlands during mine reclamation, however, specific guidance applicable to different physiographic regions is still needed. The recommendations presented in this chapter should provide the basis on which to build a mitigation plan for a specific project. We are still in a rapid learning phase in restoration technology, and thus, must be open to new ideas and willing to experiment with innovative methods.

Managers need to move away from easily constructed geometric shapes and must attempt to create landforms that mimic natural systems. Overly engineered designs with specifications that are difficult to meet are not appropriate given the nature of biological systems and the current level of understanding in restoration technology. Designs and plans must be flexible to allow room for error and unpredictable events. Regulatory reform may be necessary to allow this flexibility to occur.

One way to ensure that the proper information is collected is to require mandatory monitoring programs of all wetland mitigation projects. It is suggested that a 3-year monitoring period be part of the known costs to a permittee before a project gets underway (e.g., Brooks and Hughes 1987). Short-term monitoring of individual sites coupled with a few long-term research projects will enhance our ability to predict the outcome of mitigation policies. As there is some scientific evidence for the stabilization of emergent marsh systems after 3 years (e.g., Erwin 1985, Hepp 1987), a 3-year period will allow evaluation of the project's success after three growing seasons. Also, some annual variability in climatic and growing conditions can be assessed during this time period. Finally, a modest 3-year monitoring plan does not put an unbearable economic burden on the permittee.

Long-term studies should seek to improve our predictive capabilities regarding the seasonal, annual, and successional variation inherent in most wetland systems. How do fluctuating water levels influence the composition and abundance of floral and faunal communities? What is an acceptable load of pollutants for a wetland to absorb before significant changes are observed in food webs and the health of individual organisms?

A number of studies have used comparative approaches to gain insight into how to replicate the functions of natural wetlands. Pre-mining and post-mining studies are valuable, as are comparisons between natural and restored systems, the latter being quite scarce (e.g., Brooks and Hughes 1987, Brooks 1988). Land managers need to establish their mitigation policies in the context of what changes are occurring in wetland types throughout a given physiographic region, not just on a particular mine site. In some regions (e.g., glaciated) wetland restoration has a greater chance for success because of inherent water and soil characteristics. Thus, what may work well for one area, may fail in another.

Based on this survey of the literature, it appears that the techniques appropriate to restoration and creation of simple open water and emergent marsh wetlands are fairly well established. The success of shrub and forested wetland projects, because of their slower rates of succession, has been more difficult to assess, and therefore, needs more attention. Questions remain with regard to plant materials: Are the proper propagules available? What is the best mix of native and exotic species to use? How do we balance the variable success of different planting methods against economic realities? How adept are we at predicting the successional outcome of a newly restored wetland system?

Success criteria for wetland mitigation need to be established. I do not believe, however, that satisfactory criteria can be developed on a national scale. Criteria necessarily vary with the type of wetland being established (e.g., tidal mud flats vs. freshwater emergent marshes vs. evergreen forested swamps). They also vary with the differential pressures placed on wetland resources within a region. For some wetland types, we may not yet know how to characterize their hydrology or biotic diversity, let alone satisfactorily replicate them.

At the current level of knowledge, it is ludicrous to demand 100% replication of species richness and abundance for all projects, but what are the minimum standards for replacing equivalent functions? Allowances must be made for variable growth patterns among floral species and for seasonally and annually fluctuating hydrologic regimes. Naturally occurring changes in wetland characteristics are commonplace. How will these changes be assessed, and then applied to a mitigation project? As with many environmental regulations, success criteria must evolve incrementally as new information becomes available. In time, a broad criterion such as "establish locally-occurring plant species" will be replaced by quantitative specifications for designated species arranged in suitable patterns on the landscape.

An interim solution is to establish regional success criteria for major wetland classes through consensus agreement among knowledgeable individuals (e.g., academics, regulatory scientists, industrial researchers, professional consultants). These criteria should be compatible with regional mitigation policies that are established by even broader representation from the community (i.e., planners, administrators, politicians, citizen's groups, business and industry leaders). Dames and Moore (1983) used a questionnaire sent to phosphate mining companies to gather opinions regarding success criteria for wetland restoration projects. Combined with information from regulatory authorities, this type of survey could form the basis for establishing success criteria for any physiographic region.

More attention must be placed on how to decide among multiple objectives for a given mitigation project. When are the utilitarian functions of wetlands (e.g., water supply, water treatment) to be substituted for in-kind replications of natural systems? Numerous authors suggested that planning and decision-making by consensus among scientific, industrial, regulatory, and citizen's groups is the appropriate strategy for establishing mitigation policy.

It needs to be mentioned that as wetland restoration technology improves, the mining industries will demand access to additional reserves. Therefore, the rationale for a particular mitigation strategy must have a sound, scientific basis if we are to successfully balance the needs of industry with the necessity of wetland protection.

RECOMMENDATIONS

PLANNING

1. Develop site-specific objectives that are related to regional wetland trends. Check for potential conflicts among the proposed objectives.

2. Wetland mitigation should be integrated with mining operations and reclamation plans at the beginning of any project, especially with regard to hydrologic plans for the site.

3. Project planning and evaluation should include input from trained professionals and local constituencies.

4. Mitigation plans for single wetlands should be related directly to the adjacent waterbodies and uplands. Be cognizant of regional trends and needs.

IMPLEMENTATION

1. Designs for wetlands should mimic natural systems and provide flexibility for unforeseen events.

2. The key elements to successful wetland restoration and creation are basin morphometry and hydrologic control. Assess these parameters first before specifying requirements for soil preparation or establishment of floral and faunal communities.

3. Varying the areas of the wetlands and waterbodies constructed between 0.5-10 ha will meet the needs of many species, as well as human users.

4. Bank slopes and basin bottoms should be varied with emphasis on gentle slopes and irregular bottoms unless dictated otherwise by project objectives.

5. A heterogeneous shoreline is recommended to increase habitat diversity. Extensive littoral zones should be encouraged.

6. A capability to regulate the hydroperiod using water-level control structures is highly recommended.

7. The addition of upland or hydric topsoil provides a good substrate for plant growth, serves as a source for seeds and propagules, and reduces moisture loss of exposed substrates.

8. An integrated approach to establishing vegetation that incorporates direct seeding, transplanting, "topsoiling or mulching", and natural colonization can increase plant diversity and survivorship at a reasonable cost.

9. Revegetate exposed substrates rapidly, preferably with native species. Vegetative buffers around wetlands and waterbodies are essential.

10. Diverse vertebrate and invertebrate communities will colonize newly restored wetlands if basin morphometry and vegetative communities are suitable.

LITERATURE CITED

Allaire, P.N. 1979. Coal mining reclamation in Appalachia: low cost recommendations to improve bird/wildlife habitat, p. 245-251. In G.A. Swanson (Tech. Coord.), The Mitigation Symposium: A National Workshop on Mitigating Losses of Fish and Wildlife Habitats. Gen. Tech. Rep. RM-65, U.S. Dep. Agric., For. Serv., Rocky Mt. For. Range Exp. Stn., Fort Collins, Colorado.

Barfield, B.J. and S.C. Albrecht. 1982. Use of a vegetative filter zone to control fine-grained sediments from surface mines, p. 481-490. In D.H. Graves (Ed.), Proc. Symp. on Surface Mining Hydrol., Sedimentol., and Reclam. University of Kentucky, Lexington.

Best, G.R. and K.L. Erwin. 1984. Effects of hydroperiod on survival and growth of tree seedlings in a phosphate surface-mined reclaimed wetland, p. 221-225. In D.H. Graves (Ed.), Proc. Symp. on Surface Mining Hydrol., Sedimentol., and Reclam. University of Kentucky, Lexington.

Boody, O.C., IV. 1983. Physico-chemical analysis of reclaimed and natural lakes in central Florida's phosphate region, p. 339-358. In D.J. Robertson (Ed.), Reclamation and the Phosphate Industry. Proc. Symp. of Florida Inst. of Phosphate Res., Bartow, Florida.

Branch, W.L. 1985. Design and construction of replacement wetlands on land mined for sand and gravel, p. 173-179. In R.P. Brooks, D.E. Samuel and J.B. Hill (Eds.), Proc. Conf. Wetlands and Water Management on Mined Lands. Pennsylvania State Univ. University Park, Pennsylvania.

Breedlove, B.W. and W.M. Dennis. 1983. Wetland reclamation: a drainage basin approach, p. 90-99. In D.J. Robertson (Ed.), Reclamation and the Phosphate Industry. Proc. Symp. of Florida Inst. of Phosphate Res., Bartow, Florida.

Brenner, F.J., W. Snyder, J.F. Schalles, J.P. Miller and C. Miller. 1985. Primary productivity of deep-water habitats on reclaimed mined lands, p. 199-209. In R.P. Brooks, D.E. Samuel, and J.B. Hill (Eds.), Proc. Conf. Wetlands and Water Management on Mined Lands. Pennsylvania State Univ. University Park, Pennsylvania.

Brenner, F.J. 1986. Evaluation and mitigation of wetland habitats on mined lands, p. 181-184. In D.H. Graves (Ed.), Proc. Symp. on Surface Mining Hydrol., Sedimentol., and Reclam. University of Kentucky, Lexington.

Brice, J.R., and O.C. Boody, IV. 1983. Fish populations in reclaimed and natural lakes in central Florida's phosphate region: a preliminary report, p. 359-372. In D.J. Robertson (Ed.), Reclamation and the Phosphate Industry. Proc. Symp. of Florida Inst. of Phosphate Res., Bartow, Florida.

Brooks, D.R. 1987. Rehabilitation following mineral sands mining on North Stradbroke Island, p. 24-34. In T. Farrell (Ed.), Australian Mining Industry Council, Canberra.

Brooks, D.R. 1988. Wetland rehabilitation following mineral sands mining in Australia. Paper presented at Mine Drainage and Surface Mine Reclamation Conf., U.S. Dep. Interior Bur. of Mines, 17-22 April 1988, Pittsburgh.

Brooks, R.P. 1984. Optimal designs for restored wetlands, p. 19-29. In J.E. Burris (Ed.), Treatment of Mine Drainage by Wetlands. Contrib. No. 264, Dep. Biology, Pennsylvania State Univ., University Park, PA.

Brooks, R.P. 1986. Wetlands as a compatible land use on coal surface mines. National Wetlands Newsletter 8(2):4-6.

Brooks, R.P., D.E. Samuel, and J.B. Hill (Eds.). 1985a. Proc. Conf. Wetlands and Water Management on Mined Lands. Pennsylvania State Univ., University Park, Pennsylvania.

Brooks, R.P., J.B. Hill, F.J. Brenner, and S. Capets. 1985b. Wildlife use of wetlands on coal surface mines in western Pennsylvania, p. 337-352. In R.P. Brooks, D.E. Samuel, and J.B. Hill (Eds.), Proc. Conf. Wetlands and Water Management on Mined Lands. Pennsylvania State Univ. University Park, Pennsylvania.

Brooks, R.P., and W.E. Dodge. 1986. Estimation of habitat quality and summer population density for muskrats on a watershed basis. J. Wildl. Manage. 40:269-273.

Brooks, R.P. and W.J. Davis. 1987. Habitat selection by breeding belted kingfishers (Ceryle alcyon). Am. Midl. Nat. 117:63-70.

Brooks, R.P. and J.B. Hill. 1987. Status and trends of freshwater wetlands in the coal-mining region of Pennsylvania. Environ. Manage. 11(1):29-34.

Brooks, R.P. and R.M. Hughes. 1987. Guidelines for assessing the biotic communities of freshwater wetlands, p. 278-282. In J.A. Kusler, M.L. Quammen, and G.Brooks (Eds.), National Wetland Symposium: Mitigation of Impacts and Losses. Association of State Wetland Managers, Berne, New York.

Burley, J.B., and R.B. Hopkins. 1984. Potential for enhancing nongame bird habitat values on abandoned mine lands of western North Dakota, p. 333-343. In D.H. Graves (Ed.), Proc. Symp. on Surface Mining Hydrol., Sedimentol., and Reclam. University of Kentucky, Lexington.

Burris, J.E. (Ed.). 1984. Treatment of Mine Drainage by Wetlands. Contrib. No. 264, Dep. Biology, Pennsylvania State Univ., University Park, Pennsylvania.

Cardamone, M.A., J.R. Taylor, and W.J. Mitsch. 1984. Wetlands and Coal Surface Mining: A Management Handbook. Water Resour. Res. Inst., Univ. of Kentucky, Lexington.

Carpenter, J.M., and G.T. Farmer. 1981. Peat Mining: An Initial Assessment of Wetland Impacts and Measures to Mitigate Adverse Effects. Final Report. U.S. Environ. Prot. Agency, Washington, D.C.

Clewell, A.F. 1981. Vegetational restoration techniques on reclaimed phosphate strip mines in Florida. Wetlands 1:158-170.

Clewell, A.F. 1983. Riverine forest restoration on reclaimed mines at Brewster Phosphates, central Florida, p. 122-133. In D.J. Robertson (Ed.), Reclamation and the Phosphate Industry. Proc. Symp. of Florida Inst. of Phosphate Res., Bartow, Florida.

Cole, C.A. 1986. Morphometry, hydrology, and some associated water quality fluctuations in a surface mine wetland complex in southern Illinois, p. 157-163. In D.H. Graves (Ed). Proc. Symp. on Surface Mining Hydrol., Sedimentol., and Reclam. University of Kentucky, Lexington.

Coss, R.D., J.R. Nawrot, and W.D. Klimstra. 1985. Wildlife habitats provided by aquatic plant communities of surface mine lakes, p. 29-39. In D.H. Graves (Ed.), Proc. Symp. on Surface Mining Hydrol., Sedimentol., and Reclam. University of Kentucky, Lexington.

Crawford, R.D., and J.A. Rossiter. 1982. General design considerations in creating artificial wetlands for wildlife, p. 44-47. In D. Svedarsky and R.D. Crawford. (Eds.), Wildlife Values of Gravel Pits. Misc. Pub. 17-1982 Minnesota Agric. Exp. Stn., University of Minnesota, St. Paul.

Dames and Moore. 1983. A Survey of Wetland Reclamation Projects in the Florida Phosphate Industry. Publ. No. 03-019-011. Florida Institute of Phosphate Research, Bartow.

Damman, A.W.H., and T.W. French. 1987. The Ecology of Peat Bogs of the Glaciated Northeastern United States: A Community Profile. U.S. Fish Wildl. Serv. Biol. Rep. 87-38.

Darnell, R.M. 1976. Impacts of Construction Activities in Wetlands of the United States. U.S. Environ. Prot. Agency, Ecol. Res. Ser. EPA-600/3-76-045.

Doxtater, G.D. 1985. Potential of future mine-cut lakes, p. 1-6. In L.B. Starnes (Ed.), Fish and Wildlife Relationships to Mining. Proc. Symp. 113th Ann. Meet. Am. Fish. Soc., Milwaukee, Wisconsin.

Dunn, W.J., and G.R. Best. 1983. Enhancing ecological succession: 5. seed bank survey of some Florida marshes and role of seed banks in marsh reclamation, p. 365-370. In D.H. Graves (Ed.), Proc. Symp. on Surface Mining Hydrol., Sedimentol., and Reclam. University of Kentucky, Lexington.

Emerick, N.R. 1985. Nesting islands for giant Canada geese on west-central Illinois strip-mine lands, p. 381. In R.P. Brooks, D.E. Samuel and J.B. Hill (Eds.), Proc. Conf. Wetlands and Water Management on Mined Lands. Pennsylvania State Univ. University Park, Pennsylvania.

Erwin, K.L. 1985. Fort Green reclamation project. 3rd Ann. Rep. Agrico Chemical Co., Mulberry, Florida.

Erwin, K.L., and G.R. Best. 1985. Marsh community development in a central Florida phosphate surface-mined reclaimed wetland. Wetlands 5:155-166.

Fletcher, S.W. 1986. Planning and evaluation techniques for replacement of complex stream and wetland drainage systems, p. 195-200. In J. Harper and B. Plass (Eds.), New Horizons for Mined Land Reclamation. Proc. Nat. Meet. Amer. Soc. for Surface Mining and Reclamation, Princeton, West Virginia.

Florida Defenders of the Environment. 1984. Phosphate Mining in Florida: A Source Book. Environ. Serv. Cen., Gainesville, Florida.

Fowler, D.K., and L.J. Turner. 1981. Surface Mine Reclamation for Wildlife. U.S. Dep. Interior, Fish Wildl. Serv., FWS/OBS-81/09, Washington, D.C.

Fowler, D.K., D.M. Hill, and L.J. Fowler. 1985. Colonization of coal surface mine sediment ponds in southern Appalachia by aquatic organisms and breeding amphibians, p. 261-280. In R.P. Brooks, D.E. Samuel and J.B. Hill (Eds.), Proc. Conf. Wetlands and Water Management on Mined Lands. Pennsylvania State Univ. University Park, Pennsylvania.

Frayer, W.E., T.J. Monahan, D.C. Bowden and F.A. Graybill. 1983. Status and Trends of Wetlands and Deepwater Habitats in the Conterminous United States, 1950's to 1970's. Dep. of Forest and Wood Sciences, Colorado State University, Fort Collins Colorado.

Gallagher, T.J. 1982. Eugene's urban wildlife area: Delta Ponds, p. 122-126. In W.D. Svedarsky and R.D. Crawford. (Eds.), Wildlife Values of Gravel Pits. Misc. Pub. 17-1982 Minnesota Agric. Exp. Stn., University of Minnesota, St. Paul.

Gilbert, T., T. King, L. Hord, and J.N. Allen, Jr. 1980. An assessment of wetlands establishment techniques at a Florida phosphate mine site. Proc. Ann. Conf. Restoration and Creation of Wetlands. Hillsborough Community College, Tampa, Florida. 7:245-263.

Gilliam, J.W. 1985. Management of agricultural drainage water for water quality, p. 208-215. In H.A. Groman, T.R. Henderson, E.J. Meyers, D.M. Burke and J.A. Kusler (Eds.), Proc. Conf. Wetlands of the Chesapeake. Environmental Law Institute, Washington, D.C.

Girts, M.A., and R.L.P. Kleinmann. 1986. Constructed wetlands for treatment of acid mine drainage: a preliminary review, p. 165-171. In D.H. Graves (Ed.), Proc. Symp. on Surface Mining Hydrol., Sedimentol., and Reclam. University of Kentucky, Lexington.

Glazier, R.C., R.W. Nelson, and W.J. Logan. 1981. Planning for mine cut lakes, p. 533-540. In D.H. Graves (Ed.), Proc. Symp. on Surface Mining Hydrol., Sedimentol., and Reclam. University of Kentucky, Lexington.

Gleich, J.G. 1985. Why don't coal companies build wetlands?, p. 191- 194. In R.P. Brooks, D.E. Samuel and J.B. Hill (Eds.), Proc. Conf. Wetlands and Water Management on Mined Lands. Pennsylvania State Univ. University Park, Pennsylvania.

Glesne, R.S., and C.J. Surprenant. 1979. Summary of Central States Fishery Station strip mine lake project reports 1973-1978, p. 123-125. In D.E. Samuel, J.R. Stauffer, Jr., C.H. Hocutt and W.T. Mason, Jr. (Eds.), Surface Mining and Fish/Wildlife Needs in the Eastern United States. Addendum. U.S. Dep. Interior, Fish Wildl. Serv. FWS/OBS-78/81, Washington, D.C.

Grandt, A.F. 1981. Permanent water impoundments, p. 123-136. In Surface Coal Mining and Reclamation Symp. McGraw-Hill, Inc., New York.

Haynes, R.J. 1984. Summary of wetlands reestablishment on surface-mined lands in Florida, p. 357-362. In D.H. Graves (Ed.), Proc. Symp. on Surface Mining Hydrol., Sedimentol., and Reclam. University of Kentucky, Lexington.

Haynes, R.J. 1986. Surface mining and wetland reclamation strategies, p. 209-213. In J. Harper and B. Plass (Eds.), New Horizons for Mined Land Reclamation. Proc. Nat. Meet. Amer. Soc. for Surface Mining and Reclamation, Princeton, West Virginia.

Hayes, L.A., J.R. Nawrot, and W.D. Klimstra. 1984. Habitat diversity change associated with reclamation in Illinois, p. 363-368. In D.H. Graves (Ed.), Proc. Symp. on Surface Mining Hydrol., Sedimentol., and Reclam. University of Kentucky, Lexington.

Hepp, J.P. 1987. An ecological survey of four newly created surface-mine wetlands in central Pennsylvania. M.S. Thesis. Pennsylvania State Univ., University Park.

Herricks, E.E. 1982. Development of aquatic habitat potential of gravel pits, p. 196-207. In W.D. Svedarsky and R.D. Crawford (Eds.), Wildlife Values of Gravel Pits. Misc. Pub. 17-1982 Minnesota Agric. Exp. Stn., University of Minnesota, St. Paul.

Hill, J.B. 1986. Wildlife use of wetlands on coal surface mines in western Pennsylvania. M.S. Thesis. Pennsylvania State Univ., University Park, Pennsylvania.

Hudson, M.S. 1983. Waterfowl production of three age-classes of stock ponds in Montana. J. Wildl. Manage. 47:112-117.

Jones, D.W., M.J. McElligott, and R.H. Mannz. 1985a. Biological, Chemical, and Morphological Characterization of 33 Surface Mine Lakes in Illinois and Missouri. Peabody Coal Co., Freeburg, Illinois.

Jones, D.W., R.H. Mannz, M.J. McElligott, and B. Imboden. 1985b. A rotenone survey to determine the standing crop of fishes in a 21-acre surface-mine lake in St. Clair County, Illinois, p. 7-13. In L.B. Starnes (Ed.), Fish and Wildlife Relationships to Mining. Proc. Symp. 113th Ann. Meet. Am. Fish. Soc., Milwaukee, Wisconsin.

King, T., L. Hord, T. Gilbert, F. Montalbano, III, and J.A. Allen, Jr. 1980. An evaluation of wetland habitat establishment and wildlife utilization of phosphate clay settling ponds. Proc. Ann. Conf. Restoration and Creation of Wetlands. Hillsborough Community College, Tampa, Florida. 7:245-263.

King, T., R. Stout, and T. Gilbert. 1985. Habitat Reclamation Guidelines. Off. Environ. Serv., Florida Game and Fresh Water Fish Comm., Barstow, Florida.

Klimstra, W.D. and J.R. Nawrot. 1982. Water as a reclamation alternative: an assessment of values, p. 39-44. In D.H. Graves (Ed.), Proc. Symp. on Surface Mining Hydrol., Sedimentol., and Reclam. University of Kentucky, Lexington.

Klimstra, W.D., J.R. Nawrot, and M.R. Santner. 1985. Recreation utilization of surface-mined lands, p. 403-408. In D.H. Graves (Ed.), Proc. Symp. on Surface Mining Hydrol., Sedimentol., and Reclam. University of Kentucky, Lexington.

Koopman, R.W. 1982. Pits, ponds, and people: reclamation and public use, p. 127-131. In W.D. Svedarsky and R.D. Crawford (Eds.), Wildlife Values of Gravel Pits. Misc. Pub. 17-1982, Minnesota Agric. Exp. Stn., University of Minnesota, St. Paul.

Leedy, D.L. 1981. Coal Surface Mining Reclamation and Fish and Wildlife Relationships in the Eastern United States. Vol. 1. U.S. Dep. Interior, Fish Wildl. Serv. FWS/OBS-80/24, Washington, D.C.

Leedy, D.L., and T.L Franklin. 1981. Coal Surface Mining Reclamation and Fish and Wildlife Relationships in the Eastern United States. Vol. 2. U.S. Dep. Interior, Fish Wildl. Serv. FWS/OBS-80/25, Washington, D.C.

Lomax, J.L. 1982. Wildlife use of mineral extraction industry sites in and coastal plains of New Jersey, p. 115-121. In W.D. Svedarsky and R.D. Crawford (Eds.), Wildlife Values of Gravel Pits. Misc. Pub. 17-1982, Minnesota Agric. Exp. Stn., University of Minnesota, St. Paul.

Mannz, R.H. 1985. Recreational fishing in surface mine lakes - a case study in St. Clair County, Illinois, p. 409-415. In D.H. Graves (Ed.), Proc. Symp. on Surface Mining Hydrol., Sedimentol., and Reclam. University of Kentucky, Lexington.

Marion, W.R., and T.E. O'Meara. 1983. Phosphate reclamation plans and changes in wildlife habitat diversity, p. 498-509. In D.J. Robertson (Ed.), Reclamation and the Phosphate Industry. Proc. Symp. of Florida Inst. of Phosphate Res., Bartow, Florida.

McGee, G.W., and J.C. Harper, II. 1986. Guidelines for Reclamation of Severely Disturbed Areas. ST-7 Pennsylvania State University Extension Service, University Park.

McRae, S.G. 1986. Opportunities for creative reclamation following sand and gravel extraction, p. 51-53. In J. Harper and B. Plass (Eds.), New Horizons for mined Land Reclamation. Proc. Nat. Meet. Amer. Soc. for Surface Mining and Reclamation, Princeton, West Virginia.

Mining and Reclamation Council of America and Hess & Fisher Engineers, Inc. 1985. Handbook of Alternative Sediment Control Methodologies for Mined Lands. U.S. Dep. Interior, Off. Surface Mining, Washington, D.C.

Minnesota Department of Natural Resources. 1981. Minnesota Peat Program. Final Report. St. Paul, Minnesota.

Montalbano, F., W.M. Hetrick, and T.C. Hines. 1978. Duck foods in central Florida phosphate settling ponds, p. 247-255. In D.E. Samuel, J.R. Stauffer, Jr., C.H. Hocutt and W.T. Mason, Jr. (Eds.), Surface Mining and Fish/Wildlife Needs in the Eastern United States. U.S. Dep. Interior, Fish Wildl. Serv. FWS/OBS-78/81, Washington, D.C.

National Research Council. 1980. Surface Mining of Non-Coal Minerals. Appendix I: Sand and gravel mining and quarrying and blasting for crushed stone and other construction materials. National Academy of Sciences, Washington, D.C.

Nawrot, J.R. 1985. Wetland development on coal mine slurry impoundments: principles, planning, and practices, p. 161-172. In R.P. Brooks, D.E. Samuel and J.B. Hill (Eds.), Proc. Conf. Wetlands and Water Management on Mined Lands. Pennsylvania State Univ. University Park, Pennsylvania.

Nelson, R.W. and Associates. 1982. Planning and Management of Mine-Cut Lakes at Surface Coal Mines. U.S. Dep. Interior, Off. Surface Mining. OSM/TR-82/1, Washington, D.C.

O'Leary, W.G., W.D. Klimstra, and J.R. Nawrot. 1984. Waterfowl habitats on reclaimed surface mined lands in southwestern Illinois, p. 377-382. In D.H. Graves (Ed.), Proc. Symp. on Surface Mining Hydrol., Sedimentol., and Reclam. University of Kentucky, Lexington.

Peltz, L.R. and O.E. Maughan. 1978. Analysis of fish populations and selected physical and chemical parameters of five strip mine ponds in Wise County, Virginia with implications for management, p. 171-176. In D.E. Samuel, J.R. Stauffer, Jr., C.H.

Hocutt and W.T. Mason, Jr. (Eds.), Surface Mining and Fish/Wildlife Needs in the Eastern United States. U.S. Dep. Interior, Fish Wildl. Serv. FWS/OBS-78/81, Washington, D.C.

Pentecost, E.D. and R.C. Stupka. 1979. Wildlife investigations at a coal refuse reclamation site in southern Illinois, p. 107-118. In D.E. Samuel, J.R. Stauffer, Jr., C.H. Hocutt and W.T. Mason, Jr. (Eds.), Surface Mining and Fish/Wildlife Needs in the Eastern United States. U.S. Dep. Interior, Fish Wildl. Serv. FWS/OBS-78/81, Washington, D.C.

Rafaill, B.L. and W.G. Vogel. 1978. A Guide for Vegetating Surface-Mined Lands for Wildlife in Eastern Kentucky and West Virginia. U.S. Dep. Interior, Fish Wildl. Serv. FWS/OBS-78/84, Washington, D.C.

Robertson, D.J. (Ed). 1983. New directions for phosphate mine reclamation, p. 510-516. In D.J. Robertson (Ed.), Reclamation and the Phosphate Industry. Proc. Symp. of Florida Inst. of Phosphate Res., Bartow, Florida.

Rogowski, A.S. 1978. Water regime in strip mine spoil, p. 137-145. In D.E. Samuel, J.R. Stauffer, Jr., C.H. Hocutt and W.T. Mason, Jr. (Eds.), Surface Mining and Fish/Wildlife Needs in the Eastern United States. U.S. Dep. Interior, Fish Wildl. Serv. FWS/OBS-78/81, Washington, D.C.

Sandusky, J.E. 1978. The potential for management of waterfowl nesting habitat on reclaimed mined land, p. 325-327. In D.E. Samuel, J.R. Stauffer, Jr., C.H. Hocutt and W.T. Mason, Jr. (Eds.), Surface Mining and Fish/Wildlife Needs in the Eastern United States. U.S. Dep. Interior, Fish Wildl. Serv. FWS/OBS-78/81, Washington, D.C.

Schaller, F.W. and P. Sutton (Eds.). 1978. Reclamation of Drastically Disturbed Lands: Proceedings of a Symposium. Amer. Soc. Agronomy, Madison, Wisconsin.

Soil Conservation Service. 1979. Engineering FieldManual for Conservation Practices. U.S. Dep. Agric., Washington, D.C.

Sponsler, M., W.D. Klimstra and J.R. Nawrot. 1984. Comparison of avian populations on unmined and reclaimed lands in Illinois, p. 369-376. In D.H. Graves (Ed.), Symp. Surface Mining, Hydrology, Sedimentology, and Reclamation. University of Kentucky, Lexington.

Street, M. 1982. The Great Linford Wildfowl Research Project - a case history, p. 21-33. In Wildlife on Man-Made Wetlands. Proc. Symp. A.R.C. Wildfowl Centre, Great Linford, England.

Thompson, C.S. 1984. Experimental practices in surface mining coal--creating wetland habitat. National Wetlands Newsletter 6(2):15-16.

Tiner, R.W., Jr. and J.T. Finn. 1986. Status and Recent Trends of Wetlands in Five Mid-Atlantic States. U.S. Fish Wildl. Serv. National Wetlands Inventory, Newton Corner, Massachusetts/U.S. Environ. Prot. Agency Region III, Philadelphia.

Warburton, D.B., W.B. Klimstra, and J.R. Nawrot. 1985. Aquatic macrophyte propagation and planting practices for wetland establishment, p. 139-152. In R.P. Brooks, D.E. Samuel, and J.B. Hill (Eds.), Proc. Conf. Wetlands and Water Management on Mined Lands. Pennsylvania State Univ. University Park, Pennsylvania.

Wetzel, R.G. 1975. Limnology. W.B. Saunders Co., Philadelphia, Pennsylvania.

Wieder, R.K. and G.E. Lang. 1986. Fe, Al, Mn, and S chemistry of Sphagnum peat in four peatlands with different metal and sulfur input. Water, Air, and Soil Poll. 29:309-320.

Williams, G.L. 1985. Classifying wetlands according to relative wildlife value: application to water impoundments, p. 110-119. In M.D. Knighton (comp.), Proc. Water Impoundments for Wildlife: A Habitat Management Workshop. U.S. Dep. Agric., For. Serv., North Central For. Exp. Stn., St. Paul, Minnesota.

Wyngaard, G.A. 1985. Ethical and ecological concerns in land reclamation policy: an analysis of the Surface Mining Control and Reclamation Act of 1977, p. 75-81. In R.P. Brooks, D.E. Samuel, and J.B. Hill (Eds.), Proc. Conf. Wetlands and Water Management on Mined Lands. Pennsylvania State Univ. University Park, Pennsylvania.

MITIGATION AND THE SECTION 404 PROGRAM: A PERSPECTIVE

William L. Kruczynski[1]
U.S. Environmental Protection Agency
Region IV

ABSTRACT. Although the basic language of Section 404 of the Federal Water Pollution Control Act Amendments of 1972 has not changed substantially since the Program's inception, the Program has evolved through revisions in U.S. Army Corps of Engineers (Corps) Regulations and Environmental Protection Agency (EPA) Guidelines, and the judicial history of wetland case law. Compensatory replacement mitigation appeared early in the program as an attempt to replace loss of wetlands, at least on paper. It appeared in projects for which federal commenting agencies chose not to dispute issuance of a Corps permit.

The EPA and the Corps are currently negotiating a joint mitigation policy, but there remains a difference of opinion between the agencies concerning how mitigation should be considered in the permitting process. It is EPA's position that the presumption that there are alternatives to the destruction of wetlands cannot be overcome by the applicant's promise to create new wetlands. However, compensatory replacement mitigation may be appropriate for projects for which there are no practicable alternatives and all appropriate and practicable minimization has been required. There are three categories of proposed projects, those for which impacts are: (1) significant regardless of proposed mitigation, (2) significant unless sufficiently offset by mitigation, and (3) not significant. Consideration of the role of compensatory mitigation for projects which are not immediately rejected from further consideration because of the magnitude of the environmental losses must be made on a case-by-case basis.

INTRODUCTION

Compensatory replacement mitigation is the attempted replacement of the functions and values of wetlands proposed for filling through creation of new wetlands or enhancement of existing wetlands.

In order to better understand the ongoing controversy concerning the role of mitigation in evaluating Section 404 permit applications, it is necessary to discuss briefly the history of wetlands mitigation and the role of the Section 404 (b)(1) Guidelines in the permit application review process. This discussion will demonstrate how the inappropriate application of mitigation to projects can transform a straightforward review procedure into a complex and confusing analysis based more upon perceptions than scientific principles.

HISTORY

A review of the legislative and judicial history of the Section 404 program is given by Liebesman (1984, 1986), Want (1984), and Nagle (1985). Section 404 was enacted as part of Public Law 92-500, The Federal Water Pollution Control Act Amendments of 1972 (FWPCA), to control pollution from discharges of dredged or fill material into waters of the United States. Although the Environmental Protection Agency (EPA) is responsible for administration of the Clean Water Act, Congress authorized the Secretary of the Army, acting through the Corps of Engineers, to issue permits under Section 404, since that agency had been regulating dredging and placement of structures in navigable waters under the Rivers and Harbors Act of 1899. However, Congress, in Section 404(b), directed the EPA, in conjunction with the Corps, to develop

[1]The views expressed in this chapter are the author's own and do not necessarily reflect the views or the policies of the Environmental Protection Agency.

the environmental standards, known as the Section 404(b)(1) Guidelines, for the program. Nothing in Section 404 of the FWPCA delineated the role of the Guidelines in the permit review process, but Congress clearly intended that the Guidelines should provide environmental criteria by which to judge the suitability of disposal sites. In addition to the Guidelines, Congress, under Section 404(c), gave EPA the authority to prohibit, withdraw or restrict the specification of a 404 discharge site. This authority, which is known as a 404(c) "veto", can be used by EPA to prevent the unacceptable adverse impact of a 404 project.

On September 5, 1975, EPA, after consultation with the Corps, published in interim final form, the Section 404(b)(1) Guidelines, which established regulatory considerations and objectives to govern decisions concerning issuance of Section 404 permits. These considerations included avoiding discharges that disrupt aquatic food chains and destroy significant wetlands, avoiding degradation of water quality, and protecting fish and shellfish resources. These regulations also set forth a presumption that a permit will not be granted for work in a wetland unless the applicant clearly demonstrates that, for non water dependent projects, there are no less environmentally damaging, practicable alternatives available.

In 1977, Congress amended the FWPCA through passage of the Clean Water Act (CWA). Although sections were added to Section 404 to exempt certain discharges, such as normal agricultural and silvicultural activities, and to establish procedures for transfer of the program to the states, Congress did not change the basic outline of the program which had evolved through Corps Regulations, EPA Guidelines and judicial review.

From 1977 through 1980, there was little conflict between the Corps and EPA in implementing the Program. Although the Corps initially wanted to restrict the extent of its geographical jurisdiction, the Corps complied with a court ruling (NRDC vs. Callaway) and issued revised Regulations on July 19, 1977, which expanded the definition of "waters of the United States" to include wetlands. The Regulations declared that "wetlands are vital areas that constitute a valuable public resource, the unnecessary alteration or destruction of which should be discouraged as contrary to the public interest". The Corps Regulations also reiterated the presumption against filling wetlands for non water dependent projects as stated in the 1975 EPA Guidelines. The Corps proposed that each District Engineer consult with the U.S. Fish and Wildlife Service (FWS), National Marine Fisheries Service (NMFS), Soil Conservation Service, EPA, and state agencies in reaching a decision on whether "the benefits of a proposed alteration outweigh the damage to the wetland resource" and whether "the proposed alteration is necessary to realize those benefits". This has been called the Corps' public interest review. The District Engineer also had to "consider whether the proposed activity is primarily dependent on being located in, or in close proximity to, the aquatic environment and whether feasible alternative sites are available". The Corps Regulations place the burden of proof on the applicant to provide information on the water dependency of a project and evaluation of alternative sites.

The EPA promulgated revised Guidelines on December 24, 1980. These Guidelines reiterated the water dependency tests and presumption against alteration of wetlands found in the 1975 interim Guidelines and the 1977 Corps Regulations. The Guidelines also expanded these presumptions to include special aquatic sites which include sanctuaries and refuges, wetlands, mudflats, vegetated shallows, coral reefs, and riffle and pool complexes. These Guidelines establish a fundamental premise that "the degradation or destruction of special aquatic sites ... may represent an irreversible loss of valuable aquatic resources" and that "dredged or fill material should not be discharged into the aquatic ecosystem, unless it can be demonstrated that such a discharge will not have an unacceptable adverse impact".

The binding, regulatory nature of the Guidelines was emphasized in the 1980 Guidelines because some Corps Districts were issuing Section 404 permits for non water dependent activities when there were less environmentally damaging alternatives. Prior to 1982, when EPA, FWS, or NMFS objected to issuance of a Corps permit, the objecting agency could elevate the permit decision to higher authority. In Region IV during the mid and late 1970's, the threat of elevation of a District Engineer's decisions was usually enough to result in modification, withdrawal, or denial of permit applications for environmentally unacceptable projects. Few permits were elevated each year to Corps Divisions and fewer yet were elevated to the Office of Chief of Engineers, Washington, D.C.

In 1981, the President's Task Force on Regulatory Relief targeted the Section 404 Program for reform. This reform effort seemed to question, among other things, the extent to which the EPA Guidelines should be treated as binding and regulatory. The Corps issued new Regulations as a regulatory relief measure in July 1982 in interim final form. The intent of these Regulations was to expedite the permit issuing process and expand the nationwide permit program.

In July 1982 the Corps revised the memoranda of agreement MOA with EPA, FWS, and NMFS regarding elevation of permit decisions. The new MOA's stated that only specific, higher level, officials of those agencies could request elevation and that the Assistant Secretary of the Army (Civil Works) had the sole discretion to grant such requests. As a result, federal agencies charged with protection of natural resources were less able to influence Corps permitting decisions. Because of shortened processing time, increased workloads, logistics, and interagency politics, compensatory replacement mitigation was frequently used in some parts of the country to resolve differences of opinions between federal agencies concerning the "public interest review". Other Corps Districts issued permits over agency objections without any mitigation. Consideration of compensatory mitigation appearance in permitting decisions was seemingly justified by some promising results in wetland creation projects which had appeared in the scientific literature.

Since the inception of the Program, various factors made compensatory replacement mitigation a popular option in the federal permitting process. The FWS and NMFS did not have any authority similar to the EPA veto authority under Section 404(c). Thus, as early as 1975 agencies would compromise their positions on a permit application as long as there was, at least on paper, no net loss of wetlands. Federal agencies recommended compensatory replacement mitigation, in part, due to EPA's hesitancy to use its 404(c) authority. Also, elevation of Corps decisions was difficult at best, even before regulatory relief measures were adopted. Some agencies may have rationalized that since the Corps, in some cases, would issue permits for projects which were non water dependent and which had practicable alternatives, replacement mitigation was a method of getting some environmental benefit in exchange for filling activities. Agencies within Region IV began writing letters in response to the Corps' public notices which stated that they would not object to permit issuance provided that a similar area of wetlands was created in exchange for the wetlands to be filled. Work in the mid and late 1970's seemed to show that certain wetland systems could be created by man; this was used as a further rationalization to support replacement mitigation. Response of agencies to Corps public notices which included a request for compensatory replacement mitigation became more common after regulatory relief measures were in place and were a clear signal to the Corps that agencies would not elevate permit decisions. A decrease in elevations assured the Corps that it could meet its goal of shortening processing time for permit applications. Commenting federal agencies, particularly after regulatory relief measures were in place, felt that they had "no practicable alternatives" other than to recommend that wetland losses be mitigated through attempted wetland replacement. "Mitigation" came to mean minimize adverse impacts regardless of alternatives, and when that cannot be accomplished, attempt to replace wetlands lost.

On the surface, requiring replacement mitigation seemed to be an equitable solution to a problem. However, the Section 404(b)(1) Guidelines were regularly being ignored in some regions when compensatory mitigation was offered. This was rationalized by concluding that any losses of fish and wildlife habitat and other wetland functions were replaced through attempted wetland creation. Also ignored was the hidden environmental cost of changing or altering existing habitats with values in their present states in hopes of improving these wetland values.

Consultants for applicants quickly adopted mitigation as a means of obtaining permits and undermanned and overworked agency review staffs were soon faced with many difficult decisions. In many cases, because there were little or no data upon which to base decisions, inconsistent recommendations were made concerning acceptable replacement mitigation, including buying and donating lands, mitigation banking, out-of-kind replacement, and off-site replacement. Typically, a permit for filling a wetland, for whatever reason, could be obtained if the applicant was willing to create a similar wetland by scraping down uplands to wetlands elevations and planting the area with appropriate species. There was little or no data available regarding the scientific capability of replicating many kinds of wetlands, particularly during the early years of the program. Also, little or no monitoring of wetland creation projects was required.

Although numbers fluctuated at first, it became standard practice in the southeastern United States to require replacement mitigation at a ratio of 1.5:1 on an acre for acre basis. Agencies rationalized that greater than 1:1 was justified because of the uncertainty of wetland creation and to compensate for the length of time that it would take to replace fully functional systems. Corps Districts in Region IV generally accepted this argument and included 1.5:1 mitigation as a condition to Corps permits.

Wetland creation practices developed concurrently with the regulatory history. Examples of some pioneers in the field on the Atlantic and Gulf Coasts include Savage (1972), Woodhouse, Seneca, and Broome (1972), Eleuterius (l974), Garbisch et al. (1975), and Lewis and Lewis (l978). At this same time, in response to growing environmental awareness and the increasing problem of acceptable disposal of dredged material, Congress authorized the Corps' Dredged Material Research Program in 1973. One of the primary efforts of that program was to assess the feasibility of developing habitats on dredged material substrate. Although research in wetlands creation originally began as attempts to stabilize dredged material and eroding shorelines, it soon developed in the 1970's into the business of planting wetlands in exchange for wetland acres permitted to be filled. A symposium which started modestly in 1974 as a "Conference on the Restoration of Coastal Vegetation in Florida" soon became a major scientific vehicle to demonstrate what kinds of mitigation for dredge and fill projects were available. The conference is now entitled "Conference on the Restoration and Creation of Wetlands" and is international in scope and interest. Initial successes of marsh creation projects in 1975 through 1978 were used as further justification of replacement mitigation for Section 404 permits.

Revised Corps Regulations, published in July 1982, provided little recognition of the regulatory role of the Section 404(b)(1) Guidelines in the review of permit applications for certain types of wetlands and contained Nationwide Permits for all dredge or fill activities in two categories of waters, isolated waters and waters above headwaters. The National Wildlife Federation challenged these Corps Regulations on several counts including the role of the Guidelines in the review process and the cumulative environmental impacts of the nationwide permits for activities in the two categories of waters. This suit was settled in February 1984 and the Corps agreed to promulgate new regulations which acknowledged, among other things, the regulatory nature of the EPA Guidelines. The Corps also agreed to establish acreage limitations for Nationwide Permits for isolated and headwater waterbodies. The Corps published revised regulations in October 1984 and again in November 1986. The primacy of the Guidelines in the Section 404 review process was settled and the Corps agreed that no permit can be issued unless it complies with the requirements of the Section 404(b)(1) Guidelines. A permit that complies with the Guidelines will be issued unless the District Engineer determines that it would be contrary to the public interest. This sequencing clearly and explicitly highlights the priority of the Section 404(b)(1) Guidelines in permitting decisions.

In 1985, the agencies negotiated new memoranda of agreement which reestablished a first stage elevation of permitting decisions to the Corps Division Engineers. In EPA's case a difference in interpretation of the Guidelines is one criterion by which a permit decision may be elevated.

In 1981, the FWS formalized its mitigation policy and included Guideline precepts (FR 456, 15:7644-7663). The policy also established "resource categories" which defines "significant impact" by delineating wetland types which receive different levels of review. Mitigation can be considered by FWS for proposals that:

1. Are ecologically sound.
2. Select the least environmentally damaging alternative.
3. Avoid or minimize loss of fish and wildlife resources.
4. Adopt all measures to compensate for unavoidable loss.
5. Demonstrate a public need and are clearly water dependent.

EPA and the Corps are currently working on a joint mitigation policy. However, as previously mentioned, there remains a difference of opinion between the agencies concerning if and when mitigation should be considered in the permitting process. The recent Attleboro Mall 404(c) case highlighted the differences of opinion between the Corps and EPA on the place of mitigation in the stepwise application of the Guidelines. Although it is beyond the scope of this document to discuss that case fully, it is important to recognize that the Corps position in this case was that if mitigation will theoretically offset the adverse impacts to wetlands with a net result of "zero impact", a permit applicant need not seek a less environmentally damaging alternative. Several other Corps Districts also appear to be taking this position on the application of mitigation in the review process. Conversely, a recent paper by Thompson and Williams-Dawe (1988) presents legal, scientific, and policy grounds to reject this interpretation of the Guidelines in the decision process. They state that the Guidelines support a sequential approach to mitigation and that mitigation cannot substitute for the alternatives test. The presumption that there are alternatives to destruction of a wetland cannot be overcome by the promise to reduce wetland destruction or create new wetlands elsewhere.

Thompson and Williams-Dawe (1988) state further that failure to follow the stepwise approach of review given in the Guidelines creates practical difficulties. For example, it has become commonplace to contemplate, "What acreage of created permanent waterbodies is adequate mitigation for filling of seasonally flooded wetlands for residential development?" Elaborate procedures such as the U.S. Fish and Wildlife Service Habitat Evaluation Procedures and the U.S. Department of Transportation Wetland Functional Assessment Technique, and other methods are available to answer such difficult questions. But proper application of the Guidelines may obviate the need to ask such questions in most cases.

GUIDANCE ON THE APPLICATION OF THE GUIDELINES

The Section 404 Guidelines establish specific restrictions which require that no discharge should be permitted unless:

1. There are no less environmentally damaging practicable alternatives to the proposed plan. These alternatives are presumed for non water dependent activities in special aquatic sites.

2. The discharge will not result in a violation of the water quality standards, toxic effluent standards, jeopardize and endangered species, or violate requirements imposed to protect a marine sanctuary.

3. The discharge will not cause or contribute to significant degradation, either individually or cumulatively, of:

   a. Human health or welfare, water quality supply, fish, plankton, shellfish, wildlife, or special aquatic sites;

   b. Life stages of aquatic life or water dependent wildlife;

   c. Aquatic ecosystem diversity, productivity or stability; or

   d. Recreation, aesthetics or economic values.

4. All practicable steps are taken to minimize adverse impacts.

During the evolution of the Section 404 program, and in accordance with the 404(b)(1) Guidelines, the fourth restriction includes compensatory replacement mitigation as a form of impact minimization. Historically, the role of replacement mitigation in the decision making process has been inconsistent and has resulted in confusion in the application of the Guidelines.

ANALYSIS OF THE APPLICATION OF THE GUIDELINES

Since EPA and the Corps are seeking to develop a joint policy on mitigation, it is therefore premature to give a definitive statement on the role of mitigation in the application of the Guidelines. Thus, the following discussion is an analysis of the author's interpretation of the application of the Guidelines in the review of a permit application.

For many years, EPA has publicly taken the position that mitigation should occur in the sequence of avoidance first, then minimization and, lastly, compensation of unavoidable impacts. EPA considers these specific elements to represent the required sequence of steps in the mitigation planning process as it relates to the requirements set forth in the 404(b)(1) Guidelines. A review of a proposed permit's acceptability under the Guidelines should follow a sequence of events: (1) avoidance [Section 230.10(a)], (2) impact minimization [Section 230.10(d)], and finally, (3) compensation by techniques such as restoration and creation [Subpart H]. The highest level of mitigation appropriate and practicable (as practicable is defined in the Guidelines at Section 230.3) should be achieved at a given step prior to applying techniques in the next step.

In light of the above, compensatory mitigation of wetlands should not be considered in the initial analysis. That analysis should be confined to a consideration of alternative sites or designs, construction methods, or other logistical considerations. If all impacts cannot be avoided, other forms of minimization can be factored into a determination of whether there are less environmentally damaging alternatives. If an applicant fails to demonstrate that there are no practicable, less environmentally damaging alternatives to the proposed action, the applicant fails to meet the test of the first restriction even if the applicant proposes to replace the wetlands intended for filling.

Both minimization of impacts and replacement of wetlands could be factors in determining whether a project passes the second restriction. For example, a project which would result in a violation of a water quality standard, such as turbidity, could be redesigned to reduce the size of the project or treat runoff, and these modifications could result in meeting the standard. It is possible that wetlands created to compensate for wetlands unavoidably lost through filling could also be part of the treatment system.

There is general agreement concerning the relationship between the third and fourth restriction concerning minimization of impacts. It is conceivable that impacts could be minimized to a level which is no longer considered significant. Thus, as proposed impacts of a project are reduced, the significance of the impacts can be reevaluated and, if found acceptable, a project could be determined to comply with the Guidelines. Appropriate and practicable compensatory mitigation will be required for unavoidable adverse impacts which remain after all other appropriate and practicable minimization has been required.

However, there have been different interpretations on the role of compensatory mitigation in the test of significant impacts. The underlying reason for conflicting opinions on this matter is the lack of a standard definition of what constitutes a "significant" impact. There are good reasons for the lack of a standard definition of "significant" since the determination must be made at a local or regional level because of differences in the sensitivity of habitats nationally. Because of the uncertainty regarding the success of compeatory mitigation, a cautious approach should be taken in reaching a finding of no significant degradation based on this type of mitigation. Further, there are some wetland habitats in which any filling would result in significant degradation regardless of the apparent example, it is inconceivable that filling for residential development could be allowed in a vast area of fully functional Everglades wetlands or a pristine intertidal red mangrove swamp. Wetland habitats can be, and indeed have been, ranked locally or regionally and some of these listings provide notice of habitats in which any filling activity would be considered to be significant.

CONCLUSION

In summary, consideration must be made on a case-by-case basis of the role of compensatory mitigation for projects which are not immediately rejected from further consideration, because of the magnitude of the environmental losses they pose. Discharges into wetlands can be significant regardless of mitigation, significant unless offset through mitigation, or not significant. The second category is most complex. For projects in this category, consideration of whether a proposed mitigation plan will actually prevent the significant impacts from occurring must be carefully evaluated. Factors to be considered in making this determination are considered in the next chapter.

LITERATURE CITED

Eleuterius, L.N. 1974. A Study of Plant Establishment on Spoil areas in Mississippi Sound and Adjacent Waters. Contract Report DA l-72-C-000l, U.S. Army Corps of Engineers, Mobile District.

Garbisch, E.W., Jr., P.B. Woller, and R.J. McCallum. 1975. Salt Marsh Establishment and Development. Technical Manual 52, U.S. Army Corps of Engineers, Coastal Engineering Research Center, Fort Belvoir, Virginia.

Lewis, R.R. and C.S. Lewis. 1978. Tidal marsh creation on dredged material in Tampa Bay, Florida, p. 45-67. In Proc. Fourth Annual Conf. Restoration Coastal Vegetation Florida, May 14, 1977. Hillsborough Comm. Coll., Tampa, Florida.

Liebesman, L.R. 1984. The role of EPA's Guidelines in the Clean Water Act Section 404 permit program-Judicial interpretation and administrative application. Environmental Law Reporter, News and Analysis 14: 10272-10278.

Liebesman, L.R. 1986. Recent Developments under the Clean Water Act Section 404 Dredge and Fill Permit Program. American Bar Assoc., Water Qual. Comm. Workshop, January 10, 1986, Washington, D.C.

Nagle, E.W. 1985. Wetlands protection and the neglected child of the Clean Water Act: A proposal for shared custody of Section 404. Virginia Jour. Nat. Res. Law 5: 227-257.

Savage, T. 1972. Florida Mangroves as Shoreline Stabilizers. Fla. Dept. Nat. Res. Prof. Papers Ser. No. 19.

Thompson, D.A. and A.H. Williams-Dawe. 1988. Key 404 Program Issues in Wetland Mitigation, p. 49-53. In J.A. Kusler, M.L. Quammen, and G. Brooks (Eds.), Proceedings of the National Wetland Symposium: Mitigation of Impacts and Losses. Association of State Wetlands Managers, Berne, New York.

Want, W.L. 1984. Federal Wetlands Law: The cases and the problems. Harvard Environ. Law Rev. 8(1): 1-54.

Woodhouse, W.W., E.D. Seneca, and S.W. Broome. 1972. Marsh Building with Dredge Spoil in North Carolina. Bull. 445, Agric. Exper. Sta., North Carolina State University, Raleigh, North Carolina.

OPTIONS TO BE CONSIDERED IN PREPARATION AND EVALUATION OF MITIGATION PLANS

William L. Kruczynski[1]
U.S. Environmental Protection Agency
Region IV

ABSTRACT. Consideration of compensatory mitigation should be confined to projects which comply with the Environmental Protection Agency's Section 404(b)(1) Guidelines. The complexity of designing a successful mitigation plan is due to specific characteristics of many types of wetlands and the many options available at mitigation sites. The types of compensatory mitigation, in order of preference, are: restoration, creation, enhancement, exchange. Preservation should only be considered when the ecological benefits of preservation greatly outweigh the environmental losses of an unavoidable filling activity.

A methodology based upon rating of the options is presented to aid in the selection of an acceptable mitigation plan. In general, on-site, in-kind, up-front mitigation is the preferred option. However, other options may be acceptable based upon availability of sites, plant material, and other variables. The proposed methodology should be used as a guide and not as the only criterion in decision making. Monitoring of mitigation sites is essential to demonstrate creation of functional wetland systems.

INTRODUCTION

This chapter will discuss the advantages and disadvantages of compensatory mitigation options which are available for projects which receive Section 404 permits. This guidance is based upon my experience. It is intended for both preparers and reviewers of mitigation plans. This analysis is proposed for use as part of the analytical framework for evaluating proposals to mitigate the environmental losses of dredge and fill projects. It is not meant to be a step-by-step approach to selecting the most desirable mitigation option. Development of such an approach would be difficult since the site specific characteristics of the wetland community which will be lost and the available mitigation sites and options cannot be anticipated. Discussion is confined to compensatory mitigation and assumes that a project meets the Section 404(b)(1) Guidelines. Examples given in the text to illustrate specific points reflect the author's knowledge of ecosystems in the southeastern United States, but the conclusions and recommendations are intended to be generally applicable to wetland ecosystems. The recommendations in this chapter have not, at this time, been embraced in the form of formal Environmental Protection Agency (EPA) policy or guidance.

The complexity of designing a successful mitigation plan is due to specific characteristics of the many types of wetlands and the many options available for the manipulation of biotic and abiotic factors at mitigation sites. A brief discussion of the common mitigation options is given below. Factors have been arranged from very general to specific. This order reflects the recommended order in the decision process for preparing a mitigation plan.

GOAL OF COMPENSATORY MITIGATION

The goal of compensatory mitigation should be consistent with the goal of the Clean Water Act which is "to restore and maintain the chemical, physical, and biological integrity of our Nation's

[1]The views expressed in this chapter are the author's own and do not necessarily reflect the views or policies of the Environmental Protection Agency.

waters". Replacement wetlands should be designed to replace all the ecological functions provided by the destroyed wetlands such as wildlife habitat, water quality, flood storage, and water quantity functions. Sometimes it is suggested that wetland functions can be provided with the successful regrowth of wetland plant species, but often special project designs, such as slope or channel characteristics or watershed area, are necessary to assure replacement of wetland functions such as flood storage. Monitoring of mitigation sites is also essential to demonstrate creation of fully functional compensatory wetland systems.

PREPARATION AND EVALUATION OF MITIGATION PLANS

Preparation of mitigation plans is an exceedingly complex matter. Federal project managers are often forced to make difficult decisions based upon little or no specific information concerning expected or actual success rates or times necessary to achieve the full functions of created communities. Lack of information is due to either the historical lack of environmental monitoring associated with mitigation efforts or poorly designed monitoring programs for new projects. There are several ongoing efforts to revisit sites where wetlands creation mitigation projects have been attempted as a condition of issued Corps permits. For example, mitigation sites in Florida and in New England are being studied by the Corps of Engineers Waterways Experiment Station. However, there are few published follow-up studies of mitigation sites, and the lack of detailed studies of many community types necessitates a cautious approach concerning decisions on anticipated values of created wetlands.

EPA or U.S. Army Corps of Engineers (Corps) project managers may also not have the broad scientific background or field experience to design mitigation plans for any or all of the many types of wetland systems which exist in each region. In addition, as a matter of policy, the federal agencies are not environmental consultants; design of a project should carry with it assurance of success, and the burden to assure success should be completely on the applicant and his technical consultants. However, while federal project managers should guide applicants through this process, federal agencies should require that applicants prepare and submit detailed mitigation plans for review, rather than actually aiding them in the preparation of plans. Initial input by federal agencies should be limited to generic considerations pertaining to community type, suitable sites, area, source of water, slopes, and watershed size and position.

The complexity of preparing mitigation plans is due to the plethora of options concerning factors such as availability of plant materials, genetic compatibility of stock material with local populations and environmental conditions, handling of plant material, planting schemes, slopes, water depth and periodicity, soils, and fertilization rates. Seasonal timing of planting, flooding and fertilization may be critical to the success of mitigation projects. It would be very unusual for a land owner or developer to have the technical background to personally plan or undertake even a small scale mitigation project. Good intentions alone do not assure mitigation success. Thus, it is recommended that all replacement mitigation be performed by a qualified environmental consultant.

Currently, there are no restrictions on who may call themselves environmental consultants or mitigation specialists. Thus, it is very important that applicants examine the credentials of companies or individuals who may bid on compensatory mitigation projects. Reputable firms which have a long, established record of success in mitigation should be qualified to discuss options which have worked in the past and will be able to give accurate cost estimates. Environmental consultants should be encouraged to publish brochures which list and illustrate their successful and unsuccessful projects. It has been recommended that a national or regional certification process be adopted for environmental consultants specializing in compensatory mitigation. Such a process, modeled after the certification process for professional engineers, has been initiated in Florida. Choice of capable, certified environmental scientists with regional knowledge would reduce the frequency of mitigation projects doomed to failure due to improper planning and design.

TYPES OF COMPENSATORY MITIGATION

There are three basic types of compensatory mitigation which are available as options to replace wetlands lost to dredging and filling activities: restoration, creation, and enhancement (Table 1). The following discussion will demonstrate that restoration and enhancement are part of a continuum which can be extended to include a fourth and least desirable mitigation option, namely wetland exchange.

Wetland restoration refers to the reestablishment of a wetland in an area where it historically existed but which now performs no or few wetland functions. Disturbance of historic wetland functions could be due to human activities, such as filling, channelization, or eutrophication, or due to natural events such as lake level rise, shoreline erosion, sediment deposition, beavers, or decreased flooding. Typically, wetland soils remain at disturbed sites, but they might be drained, oxidized, or buried.

Wetland creation refers to the construction of a wetland in an area which was not a wetland in the recent past. Typically, wetlands are created by removal of upland soils to elevations which will support growth of wetland species. Removal of soils to achieve proper elevation can, by itself, establish proper hydrology for wetland plants, such as along gently sloping shorelines, or may prepare the site to receive necessary inundation from streams or runoff from upland watersheds. Development of the correct elevation and establishment of a proper hydroperiod is the critical factor in the success of created wetlands. In an area where soils have been thoroughly disturbed, such as through surface mining, any replacement of previously existing wetlands must be considered creation. This is particularly true if the soil stratification and the surficial aquifer have been modified.

Enhancement refers to increasing one or more of the functions of an existing wetland, such as increasing the productivity or habitat value by modifying environmental parameters, such as elevation, subsidence rate, or wind fetch. Enhancement sites differ from restoration sites because they already provide some wetland functions. Enhancement implies a net benefit, but a positive change in one wetland function may negatively affect other wetland functions. The net overall result of enhancement depends upon established management goals. For example, the habitat value of a swamp forest can be increased for wood ducks by increasing the amount of open water. Increased flooding will provide more food for ducks and may kill less water tolerant trees and provide more nesting cavities. However, this type of enhancement may lower the value of the wetland for other species, such as the spotted salamander, deer, or marsh rabbit.

Enhancement taken to the extreme merely exchanges wetland types. For example, habitat value of open water may be enhanced for some species by establishing an emergent marsh or swamp forest on fill material placed in open water. This type of enhancement is more properly called exchange since it results in the replacement of one habitat type (submerged) with another (emergent). The net ecological value of this mitigation option depends upon acceptable management objectives. For example, a diverse forested wetland can be clearcut and planted with one tree species. If the planted species is a mast producer, it could be argued that the habitat value for deer or ducks has been enhanced. However, increased food for a few species has been accomplished through elimination of the complex food web of the swamp forest.

The choice of restoration, creation, or enhancement mitigation for any project depends upon the site specific characteristics of available locations. The choice should be based upon an analysis of factors that limit the ecological functioning of the watershed, ecosystem, or region. The first question to ask in reaching this decision is, "Are there degraded wetland communities on-site or nearby which could be restored to full function?" If there are, the first choice of mitigation options should be to restore historic wetland functions of the degraded system.

Restoration of degraded systems should be the first option to be considered since it would reestablish the natural order and ratio of community composition in the regional ecosystem. Moreover, likelihood of success of this type of mitigation is greater than for other options. If a portion of a particular community has been removed from an ecosystem through degradation, it would be ecologically beneficial to restore that same community back into the system. In some cases a sizable wetland area can be restored with little effort. For example, a wetland area which has been diked and drained through ditching to create a pasture may be returned to a functional wetland through removal of all or part of the dike which separates it from flood waters. Because a wetland previously existed on the site, and provided that the soils are still intact, albeit drained, the chances of success of restoring this area to a fully functional wetland are good once the hydrology is reestablished. If the organic component of the soils has substantially oxidized,

Table 1. Compensatory Mitigation Options.

Compensatory Mitigation Options

MITIGATION TYPES			RECOMMENDED ACREAGE
Restoration	–	former wetland, no or few functions	1.5:1; 1:1 upfront
Creation	–	made from different community	2:1; 1:1 upfront
Enhancement	–	increase certain functions	3:1; 1:1 upfront
Exchange	–	enhancement to the extreme	case by case
Preservation	–	purchase and donation	case by case
TIMING OF MITIGATION			
Before	–	most prudent; require if unknowns	
Concurrent	–	encouraged for typical projects	
After	–	discouraged	
LOCATION OF MITIGATION			
On-site	–	same locale in watershed or ecosystem	
Off-site	–	different locale or different ecosystem	
COMMUNITY TYPE			
In-kind	–	same species	
Out-of-kind	–	different species	

resulting in subsidence, reflooding of the area may create an open water lake, rather than an emergent wetland.

Restoration grades into enhancement depending upon how many functions have been removed from the ecosystem. A previously wet area which was historically filled may be restored to full wetland function by grading, planting, and restoring the historic hydrological regime. If the wetland had been partially degraded and had lost one or several of its functions, the area could be enhanced to provide full wetland functions. For example, a wetland which has been impounded, retains wetland vegetation, and is managed for ducks, could be enhanced by removing the impoundment dike. This would reestablish a seasonally flooded wetland which provides pulsed export of organic matter to food chains of receiving waters.

It could be argued that the duck impoundment itself was an enhanced wetland since it produced a significantly greater duck population than unimpounded wetlands. But it can only be considered enhanced for that one function, namely duck habitat. It has lost its function to provide detritus on a timely basis for fishery food chains because of its altered hydroperiod. Thus "enhancement" often reflects little more than preference for certain habitat types or values over others.

Another example of wetland habitat exchange is establishment of an emergent marsh on fill material placed in open water. This has been called marsh creation in the past. It also enhances the primary productivity of an area; but, it is an exchange of one functional habitat type for another. Because it results in a loss of functions of existing aquatic habitats, exchange should often be the last option in the choice of mitigation type. Exchange should only be used when there is ample scientific evidence demonstrating that the functions of an ecosystem or region are limited by the lack of a particular community type. For example, exchange may be the option of choice if data demonstrate that the fisheries productivity and ecological stability of an embayment would be significantly increased by establishing a fringe marsh along an unvegetated shoreline and shallow water habitat.

If restoration of a degraded wetland is the first option, and wetlands exchange is the last option and should only be used when scientifically justified, then enhancement and creation are intermediate options. There are good ecological arguments for consideration of wetland creation before wetland enhancement since the former will add to the total wetland area of a site, while the latter may only provide one or more additional functions to an existing wetland. Thus, a logical and defensible order of consideration of the types of compensatory mitigation is: restoration, creation, enhancement, and exchange. This order of preference may be different for a specific mitigation project in light of regional or site specific circumstances, such as quality of wetlands and availability of mitigation sites.

Preservation of existing wetlands through acquisition should not normally be considered as compensatory mitigation for unavoidable wetland losses since there is a net loss of wetland functions and acreage, and wetlands proposed for preservation are usually already regulated through the Section 404 program and provide ecological functions to the public. However, there could be circumstances of such an unusual character that would justify wetland preservation as a mitigation option. For example, if the environmental effects of the proposed filling are very minimal and the benefits of placement of a large area of wetlands (and/or uplands) into public ownership are great, then preservation may be consistent with the goals of wetland protection although some loss may result. This is particularly true if an area proposed for preservation is unique habitat, subject to general or nationwide permitting, or otherwise vulnerable to development. Any agreement for preservation of existing wetlands should explicitly indicate that the preservation shall be required in perpetuity and shall provide assurance for this requirement through an appropriate method such as fee title conveyance to a well established, responsible conservation organization.

Preservation of a large, mixed community might also be an attractive option if, without such preservation, the full functioning of a system could be destroyed through unregulated development of upland portions of an upland-wetland mosaic, such as a bottomland forest mixed with upland bluffs and stands. Loss of upland habitat corridors, edge, and ecological niches could result in the degradation of the functioning of the entire ecosystem, particularly for larger animals, such as black bear. If the ecological benefits of preservation greatly outweigh the environmental losses which will occur in permitting an unavoidable wetland fill, then preservation through acquisition may be considered.

Some have argued that preservation should be considered as a prime mitigation option since it places wetland areas which might be lost through future permit actions or by the dissolution of the federal permitting program, into public ownership or control. Most indications are that the Section 404 program will be strengthened as permitted losses of wetlands cumulatively result in decreased fisheries production and other losses to the national economy.

CREDIT FOR COMPENSATORY MITIGATION

There is need for guidance to establish and maintain consistency between project managers and between EPA Regions and Corps Districts concerning acceptable ratios between wetland losses and compensation acreages for the different mitigation options. One approach is to require that the ecological functions of the replacement wetland be at least equivalent to those of the wetland proposed for destruction. Attainment of functional equivalency should be the goal of all mitigation activities. However, this approach may result in loss of wetland acreage and requires detailed knowledge of the ecological contribution that the destroyed and replacement wetland systems make to the regional ecosystem. Also, the ecological functions which are considered in the test of equivalency must be carefully chosen. Proposed replacement of a degraded wetland habitat by an improved wetland is another complication which requires thorough analysis. For example, if primary productivity is a major function of a wetland, establishment of less acreage of a very productive wetland may adequately compensate for the loss of more acreage of a low productivity wetland. However, the replacement wetland may provide much less habitat for a particular species than was provided by the destroyed habitat, despite overall improved productivity. Usually there is not enough information available to agencies to formulate scientifically valid, functionally equivalent replacement acreage within the processing period of a permit application. This information must be supplied by the applicant as part of the permit application if he expects the agencies to consider "functional equivalent wetland replacement" as a basis on which to issue a permit.

General ratios between mitigation options can be suggested as flexible guidelines to be considered in each permit decision requiring compensatory mitigation. The following ratios are suggested for on-site, type-for-type (in-kind) replacement mitigation. The analysis becomes more complex when variables such as off-site and out-of-kind mitigation options are considered. The analysis is further complicated by the many types of wetland communities, the varying success rates of community replacement, and the difficulty of justifying ranking and exchange of different wetland values.

RATIOS FOR RESTORATION

In general, the chances of success in restoration of most destroyed herbaceous wetlands (e.g., marshes) is good because this type of wetland generally grows rapidly and because a wetland previously existed at the site. Restoration is a matter of removing the perturbations and reestablishing the soils, plants, and hydrology at the same site where a wetland was created by nature. Restoration (or creation) of forested streams or bottomland hardwood floodplains has been attempted, but is much more difficult. To date, no restoration projects are known which are of a sufficient age to have achieved a fully functional, self-reproducing system; most are under twenty years of age. Restoration (or creation) of submerged seagrass communities seems to be a "hit or miss" proposition with little documentation concerning specific environmental conditions needed for success.

Because of the varying rates of success of restoration of different vegetative communities, it is difficult to justify general criteria for acreage credit for restoration of all types of wetlands. Indeed, if it has not been demonstrated conclusively that restoration of a particular wetland community is possible, then the prudent approach is to reject any proposed replacement mitigation. If the ecological loss through filling of such a wetland is determined to result in significant environmental degradation, then filling of such a wetland is unacceptable. However, if it has been convincingly demonstrated that a particular wetland type can be restored, then 1.5 to 1 mitigation should be required on an acre-for-acre basis; that is, 1.5 acres restored for each acre unavoidably lost. The justification for requiring greater than parity is due to the uncertainty that a particular project will be successful and to compensate partially for the length of time that the restored, planted wetland system takes before becoming fully functional. Planting of the restored system is required unless it can be conclusively demonstrated that a natural colonization will result in the vegetative community of choice.

The ratio of wetlands restored to wetlands lost can be reduced to 1 to 1 if wetland restoration is performed "up front", that is before a filling activity is initiated. Reduction of the ratio to parity assumes that a replacement wetland has been constructed and monitored according to an approved plan and that it has been found to be fully functional. Reduction of the ratio to 1 to 1 might also be justified if data demonstrate that the restored site will provide increased ecological and hydrological value to the area.

RATIOS FOR WETLAND CREATION

Wetland creation involves increased risk since it is an attempt to establish a new wetland at a site where one has never existed, or where the previously existing conditions which supported a wetland community have been greatly modified. Wetlands established on land which has been thoroughly disturbed, such as through surface mining, are created wetlands even if they occur on the same geographical sites as previously existing wetlands. Creation of wetlands from existing uplands is the common form of compensatory creation mitigation associated with dredge and fill permits. If it has been convincingly demonstrated that a particular wetland type can be created, then 1.5 to 1 or 2 to 1 mitigation ratios should be required on an acre-for-acre basis. Increasing the ratio to 2 to 1 can be justified on the basis of the greater risk associated with any particular site. The ratio may also be adjusted depending upon whether planting or natural revegetation is part of the proposal. If successful creation (i.e., similar value between created and natural wetland) is performed upfront of proposed filling, then the ratio can be reduced to 1 to 1.

RATIOS FOR WETLAND ENHANCEMENT

Wetlands proposed for enhancement are performing wetland functions, and it will often be difficult to document net improvements to wetland functions. There is a risk that although some functions will be improved, other currently existing functions could be degraded. Due to this uncertainty, a 3 to 1 mitigation should be required on an acre-for-acre basis. This ratio can be lowered to 2 to 1 if it is performed upfront. It can never be lowered to parity since there was an existing wetland which provided some wetland functions at the site.

Wetland types should not be exchanged except under unusual circumstances. Since exchange is the replacement of one wetland type with another, it is, by definition, on a 1 for 1 basis. Gains in one wetland type cannot be equated with losses of another type since each performs different functions and are unique assemblages of physical, chemical and biological variables. To say that they are equal and that the exchange of one wetland type for another is acceptable is the same as trying to equate apples and oranges; they are judged by different sets of criteria. However, there may be unusual circumstances where one wetland type is particularly rare and one wetland type is particularly abundant; such circumstances could justify exchange of wetland types as compensatory mitigation.

RATIOS FOR WETLAND PRESERVATION

Wetland preservation through acquisition should not be considered as compensatory mitigation except in unusual circumstances because a net overall loss in function and acreage will occur. It can also be argued that preservation through donation, conservation easements, restrictive covenants and the like is tantamount to purchasing a dredge and fill permit, and is limited to developers with sufficient capital to make the offer large enough to be attractive to regulatory agencies. Small landowners seeking an individual permit usually lack the land resources or capital to make such an offer. Thus, formalizing a policy or an exchange ratio justifying such an action is ethically and legally questionable.

ECOLOGICAL COSTS OF COMPENSATORY MITIGATION

There must be a careful analysis concerning the ecological trade-offs associated with conversion of one habitat type to another. This analysis should consider the ecological value of non-wetland as well as wetland sites proposed for compensatory actions. An area proposed to be restored to wetlands or an area proposed for wetland creation may have ecological value in itself. For example, an upland pasture which was historically a wetland that was diked and drained may have become important habitat for terrestrial species, such as doves, quail, raptors, bears, or bobcats.

Unless there are unusual site-specific or regional circumstances, it is not justifiable to scrape down a functional hydric or mesic forest adjacent to an existing marsh to create equal or greater area of marsh in exchange for filling of another area. In a similar vein, an impounded marsh may provide habitat for wading birds or ducks in an area where there is little natural suitable habitat for these species. Restoration of such an impoundment to a seasonally flooded system would disrupt the community which has adapted to the existing conditions. The difficult question which must be answered in analyzing the ecological value of proposals of this sort is, "Do the ecological changes associated with habitat restoration, creation, or enhancement outweigh the overall ecological functions of the 'donor' community?" This question is hard to answer objectively because of the bias which exists for dwindling wetland resources. It also requires gathering and interpreting data for upland or disturbed wetland systems and dredge and fill project managers may have little or no experience in performing these analyses. Comparing existing ecological values with anticipated values of replacement systems is no easy task. This analysis should be performed by a team of experts representing a wide range of disciplines and expertise. The multi-disciplinary team of scientists and engineers at the Corps Waterways Experiment Station is an excellent model for interagency evaluation. They have published many studies evaluating created wetlands and comparing them to natural reference sites; their methods provide an excellent source for standardizing these comparisons.

The use of a "quantitative" methodology, particularly by inexperienced personnel, to solve this problem may only add to the confusion. For example, the Habitat Evaluation Procedures (HEP) of the U.S. Fish and Wildlife Service has often been used to calculate "habitat suitability indices" for sites. These indices are often biased. A HEP index is a numerical expression of the potential use of a site for a particular species of fish or wildlife chosen for evaluation. This measure of potential (quality) of a site for these species, when multiplied by area, yields the number of "habitat units", a numerical expression of the useful habitat within the study area. Habitat units can be used to compare different sites for chosen species and to calculate acreage of replacement habitat as mitigation for habitat lost through regulated filling.

HEP uses models which relate biological needs and tolerances of evaluation species to environmental conditions which occur in their habitats. These conditions are expressed as variables, such as water depth, flooding periodicity, vegetation density, and soil type. Through the use of formal, documented models, HEP provides "standardized" numerical expression of habitat suitability, and thus reduces variability due to subjective differences of opinion.

Because HEP appraises environmental value according to habitat suitability for particular species, the selection of species to be used in the evaluation is one of the most controversial parts of any study. The methodology can be easily misused with improper selection of evaluation species. HEP procedures advise forming an interagency team to select the species list, so that all constituencies can be represented. Those interested in maintaining or enhancing historic conditions commonly select the most sensitive (habitat limited) species in the community; others, wishing to modify or develop the site, select species most consistent with the proposed development or management plan. Moreover, since it is impossible to exactly duplicate a natural system, and since developers generally prefer to replace a wetland with a retention pond or lake, they usually prepare compensatory mitigation plans with different environmental conditions which will support a different species assemblage than exists at the "donor" site.

During a recent application of HEP in a seasonally flooded East Everglades wetland, a consultant chose such a list of species which may have historically existed at the site, but currently occur infrequently due to hydrological modifications to the wetlands. The HEP calculations yielded low habitat units for the chosen species under the existing conditions. The procedure was repeated for the habitat which would result when the site was developed as a residential area; development included several borrow lakes as enhancement (exchange) mitigation. Because of the different, "improved" hydrologic conditions, the HEP analysis concluded that development of the site would be ecologically beneficial (for the selected species).

The underlying premise of this conclusion was that management should optimize habitat for Everglade species which historically were more widespread than they are today, such as large-mouth bass and wading birds. The distribution of these species is limited by suitable hydrological conditions. However, the disruption of historic hydrological conditions in the East Everglades has also resulted in "disturbed" Everglades habitats which are wetlands, and are habitat for species other than wading birds and bass. Thus, the trade of many acres of disturbed Everglades, with the resultant loss of habitat for bobcats, raccoons, red tailed hawks, etc. for a residential community with a few acres of enhanced wetlands (borrow lakes), which would provide habitat for bass and wading birds, is not ecologically equitable. Yet the choice of HEP evaluation species supported that conclusion.

Regulatory agencies should be concerned with habitat restoration, such as restoring historic hydrologic conditions to the disturbed wetlands, instead of enhancement or exchange of a small portion of the wetlands through inappropriate mitigation associated with filling of a majority of the wetland area. Also, regulators should consider all aspects of habitat exchange and realize that advantages to the species of choice may not be balanced by impacts to other displaced species.

Another ecological cost which must be considered in preparing mitigation plans is the availability of appropriate plant material. Plant material may be collected from a donor wetland only if that action does not significantly degrade the ecological functions provided by the donor system. Vegetation plugs may be removed from a herbaceous wetland donor site and planted at a prepared mitigation site. But plugs should be harvested at sufficiently spaced intervals to maintain the functional integrity of the donor site.

If plant material is obtained from a commercial source, care should be taken to assure that the propagated plants are from a stock which is reproductively compatible and has similar ecological requirements as stands which naturally occur in the locale. A method which has recently been shown to be cost-effective in restoring or creating large wetland areas in some locales is mulching of a contoured mitigation site with the upper soil horizon from a donor wetland. This mulch contains viable seeds and rhizomes which usually allows rapid establishment of a diverse plant community. Proper application of this technique has been effective in creating or restoring both herbaceous and wooded wetlands. Appropriate tree species are planted in the mulched area, which provides effective soil, moisture, and shading for young trees. Choice of donor wetlands should be carefully controlled so that existing mature wetland systems are not avoidably lost in order to create other wetlands. It is preferable that this material be obtained from the wetland which is proposed to be impacted, and can be stockpiled for a short period if wetland replacement is not concurrent with alteration of the donor site.

TIMING OF COMPENSATORY MITIGATION

Three options exist in timing of a mitigation project relative to receipt of a Section 404 permit. Mitigation can be performed before the permit is issued, concurrent with project initiation and completion, or after a filling activity is completed. Often an initial Section 404 permit is needed for the work performed as part of the mitigation project itself since dredging or filling in waters of the United States is usually required to restore, create, or connect the mitigation site to a source of water. Upfront mitigation is possible as a separate permit action or as part of a phased permit, with the receipt of the second phase, project construction, contingent upon successful completion of the first phase, demonstration of successful mitigation.

Upfront mitigation is the most prudent of mitigation timing options and should be required for all projects which have considerable ecological risks by virtue of their size, complexity, or uncertainty of community establishment at the mitigation site. For example, a permit was issued by the Corps to a mining company for the connection of two tidal creeks, which were constructed in uplands, with existing waters. These created systems were monitored and only when success criteria were satisfactorily met did the Corps process a permit application to mine across an equal acreage of natural, existing tidal creeks.

If mitigation cannot be completed in advance, it should proceed concurrent with project construction since mitigation becomes an integral part of the proposed project; this discourages viewing mitigation as an "add on" cost of receiving a Corps permit. However, a problem inherent in concurrent mitigation is the association of timing of the receipt of a permit and initiation of a project with the timing of maximum anticipated success rate due to seasonality of biological and/or hydrological factors. In general, early spring is the best time to plant most coastal herbaceous species, whereas optimum transplanting times for scrub-shrub or palustrine forested wetlands is during winter when plant material is senescent. Obviously, the Corps cannot refuse to process a permit application if it is received at a time of year when transplanting is not optimal. However, an issued permit can be conditioned to include a date for initiation of construction and mitigation which is consistent with maximum survival rates of transplanted material. Another option is to allow earth moving associated with project construction and mitigation to proceed simultaneously and delay planting of the site to conform with time of expected maximum survival rates.

Regardless of the time of the year when mitigation is initiated or completed, it is always advantageous to require a guaranteed survival rate of transplanted plant material with every mitigation plan. This is particularly important if the Corps issues a permit for project construction and the applicant performs the required mitigation during a time period when expected transplant survival is less than optimum. Typically in Region IV, a guaranteed survival of 70% of transplants after two years is requested. Replanting should be required until a 70% survival rate is obtained for one year. This requirement necessitates monitoring of the mitigation site.

Mitigation performed after a project is completed should be discouraged since it fragments the project into a construction phase and a mitigation phase. Once the construction phase of a project is completed, there is little incentive to complete the mitigation phase in a timely, satisfactory manner. If post project mitigation is the only practical option, the mitigation plan included in the Corps permit should always contain an initiation date and a completion date. If one or both of these dates are not honored, the Corps should be encouraged and supported to take enforcement action for violation of a permit condition. Post-project mitigation should be restricted to small projects which have a very high probability of success or for situations where project construction must be initiated before the time period when maximum survival of transplants is assured. Even then, the earth moving and other physical amenities necessary for preparation of the mitigation site for planting should be performed concurrent with the remainder of the project. A performance bond should be required from the developer for any mitigation project which has an uncertain chance of success.

LOCATION OF COMPENSATORY MITIGATION

The goal of mitigation is to replace the functions which were provided by the wetland area and which were unavoidably lost through a permitted activity. Since the wetland area provides ecological functions such as food chain support or wildlife habitat to the ecosystem of which it is a part, it is important that ecological values of the replacement wetland be provided to the same ecosystem which was impacted by the filling activity. Thus, both on-site (same locale) and off-site (different locale) mitigation should usually be performed in the same ecosystem and functional watershed as the filled wetland area.

The problem with this rule is the difficulty of defining the limits of a particular ecosystem. Ecosystems may be conceived and studied in various scales. For example, an ocean, an embayment, a lake, a pond, or a small aquarium may be called an ecosystem. For our purposes, it is best to define ecosystem as any area of nature which is part of the same watershed which includes interacting living organisms and nonliving substances, and where there is an exchange of materials and energy.

Restoration, creation or enhancement of wetlands should, in most circumstances, occur on-site, that is within the same ecosystem and in the immediate vicinity of the proposed filling activity. For example, if a permit is issued for fill in wetlands to construct a boat ramp and mitigation is required, a disturbed area along the same reach of stream should be restored to wetland elevation to replace the functions which were lost. (Of course this assumes that it was not practicable to locate the boat ramp in an area of disturbed wetland.) There is adequate ecological justification for this approach since the ecosystem will remain unchanged and the chance of success of the mitigation is maximized since it is close to an area which already supports the vegetative community which is being replaced.

If there are no potential mitigation sites in the immediate area, off-site locations within the same embayment, stream reach, or watershed (ecosystem) should be selected. If a thorough analysis reveals that there are no adequate mitigation sites within these areas, this may be adequate reason to recommend that no permit be issued for the proposed activity.

Only in unusual circumstances should off-site mitigation in a different ecosystem or functional watershed be considered as acceptable mitigation. This is due to the difficulty in equating the impacts of the loss to one ecosystem with the advantages to another system. The burden of proof rests with the applicant to demonstrate that the anticipated advantages to the off-site area greatly outweigh any losses that would result through filling of a wetland site. For example, a proposal to mitigate filling of an intertidal marsh through creation of a forested wetland must contain data which supports the conclusion that the loss of intertidal marsh will not individually or cumulatively result in significant degradation to that portion of the coastline and that the created wetland will greatly improve the ecological functioning of the adjacent, riverine ecosystem.

Wetlands mitigation banking is an off-site compensatory mitigation concept which may be used to aggregate smaller wetland impacts towards restoration, creation, or enhancement of larger wetland mitigation bank sites. However, it also entails considerable legal, scientific, and administrative complexity and has the potential for being seriously misused. Therefore, due to the experimental status of this concept, it is recommended that development and use of a mitigation bank for an individual project be assessed by a thorough case-by-case review.

As with all forms of mitigation, a wetland mitigation bank cannot justify a project not otherwise in compliance with the Section 404(b)(1) Guidelines. Any restoration, creation, or enhancement project should be carefully designed by the applicant and agreed to by all concerned parties through a legally enforceable wetland mitigation banking agreement. The bank should be located in the same geographical area and consist of wetland types similar to the wetland where impacts will eventually occur.

The bank should be operational prior to allowing any project to use the bank's value as compensation for unavoidable impacts. Long-term operational, maintenance, and monitoring plans, and legal guarantees should be included in the mitigation plan which assure that tasks are feasible and will be undertaken by the appropriate parties under the force of law.

COMMUNITY TYPE

Most recommendations provided thus far in this discussion are based upon the assumption that wetlands which are unavoidably lost will be replaced by the same wetland community type. The chances of replacement with the same wetland community can be maximized by planting the site. Natural recolonization of a mitigation site is recommended only when there is an adjacent seed source and the applicant agrees in advance that if the desired density or species composition are not present at the mitigation site after one or two growing seasons, that the site will be planted to achieve the recommended plant community.

In-kind mitigation is desirable since it replaces the same community type which was lost and restores the equilibrium of community types which had developed as a result of natural causes. Out-of-kind mitigation should only be approved under unusual circumstances in which the data demonstrate that replacement of a different vegetative type for the one destroyed would clearly benefit the ecosystem or geographical area being evaluated. For example, if a mining company receives a permit to mine ore which exists under a wetland vegetated predominantly by cattails, a mitigation plan may be approved which includes creating a more diverse herbaceous community at this site. If the created system provides increased diversity of habitat or other ecological functions compared to the monotypic stand of cattails, the replacement ratio may be reduced to 1 to 1.

The decisions and value analyses of out-of-kind mitigation proposals must be made carefully and must include an evaluation of the entire community which exists on the site. For example, it is inappropriate to argue that the loss of a wetland which has a hydroperiod which has been reduced compared to historic levels can be mitigated through creation of a smaller sized lake with a permanent hydroperiod. That conclusion overlooks the ecological values provided to organisms which are currently using the drier wetland site. Such a proposal could be approved only if it could be demonstrated that the acreage of drier wetland habitat was not a limiting factor in the area and the presence of a lake would significantly improve the ecological functioning of the ecosystem. In no case should the replacement ratio of this type of mitigation be reduced to lower than 1 to 1.

Out-of-kind mitigation might also be approved if there is a requirement to rapidly stabilize an area, and there is ample assurance that in-kind species will eventually invade the planted area. For example, on suitable sites, Spartina alterniflora may completely cover a site in one growing season after planting at an appropriate density. Rapid cover may be desirable to stabilize the shoreline at the mitigation site. If there is an adjacent marsh vegetated by the slower growing Juncus roemerianus, it may invade the planted Spartina area and eventually achieve an equilibrium. Thus, planting of an area with Spartina to mitigate the destruction of a Juncus marsh may be justified in this case.

SELECTION OF MITIGATION OPTIONS

Choice of a mitigation option should consider site specific and cumulative impact assessments conducted for the proposed site and should be based on sound ecological principles based upon large-scale, landscape considerations. The best choice of a mitigation plan can only be made when the status of the functions and values that will be affected are known. For example, enhancement of waterfowl habitat by creation of a "green tree" reservoir may be inappropriate for a watershed that already has several such reservoirs, and particularly in a watershed which has water quality problems, which such reservoirs are known to exacerbate. Preservation may be a defensible option if the proposed project is on a site with low functional values and the area proposed for preservation is of high functional value or threatened, and is located in the same watershed. Only when the tradeoffs associated with mitigation options are considered on a scale that is ecologically appropriate (e.g., watershed) can decisions be made which effectively protect or replace wetland functions and values.

The goal of the selection of any mitigation plan is to replace, as near as possible, the ecological functions of the wetlands which will be destroyed. Potential options are summarized in Table 1. One selection method is to rank all potential options by assigned values. Values can be regional or site specific. Options can be evaluated through the development of a matrix with assigned values based on assumptions concerning preference of the options. An example of such a matrix is given in Figure 1 which includes all mitigation options and an example of assigned values.

As discussed above, generally restoration is the preferred option, followed by creation and enhancement. Thus, these options have been assigned values of 3, 2, and 1 respectively. These values are almost exactly the opposite of the recommended acre-for-acre replacement ratios discussed above; that is, the recommended replacement acreage is 1.5:1 for a restoration project, 2:1 for a creation project, and 3:1 for an enhancement project.

Values of 3, 2 and 1 have been assigned to upfront, concurrent, and post project mitigation. This is descriptive of the timing of the initiation and completion of the replacement wetlands. The rationale for this weighting is simply that the faster the mitigation project is completed, the faster the wetland functions are replaced in the ecosystem. Having replacement wetlands in place and functioning before the wetlands permitted for filling are destroyed is most desirable.

Values of 3 and 1 are assigned to in-kind and out-of-kind community composition, respectively. Generally, ecosystems reach an equilibrium of community types which maximizes trophic and nutrient cycling efficiencies. Thus, replacement of the same community as that which is lost to filling may restore the integrity of the ecosystem.

Values of 3 and 1 have been assigned to on-site and off-site replacement, respectively. On-site mitigation fulfills the goal of compensatory mitigation, that is to replace the functions that a filled wetland community provides to the portion of the ecosystem of which it was a part. Thus, on-site mitigation is weighted more than off-site mitigation.

The weighting of upfront, in-kind, and on-site mitigation options are presumed equal since each option represents a similar input toward the success and overall effectiveness of replacement mitigation.

The overall values of the 36 mitigation options given in Figure 1 were calculated by adding the value assigned to each component. Values range from 12 to 4. By this method, on-site, in-kind, upfront restoration is the most desirable option, which is reflected in the high value it received (12). Off-site, out-of-kind, post project enhancement of an existing wetland is the least desirable option (4).

It is recommended that project managers and preparers of mitigation plans strive to achieve the highest value of mitigation type possible for each mitigation project. A limit of acceptable mitigation options can be set. For example, acceptable projects could be confined to mitigation options with values of 9 or higher and only in unusual circumstances would a mitigation plan with a mitigation option lower than 9 be approved. Project managers have more flexibility in the acceptability of mitigation options for community types for which success of replacement has been conclusively demonstrated. For example, the success of creation of a _Spartina alterniflora_ marsh at proper elevations is well documented. Thus, an acceptable mitigation plan might include creation of an _S. alterniflora_ marsh in exchange for filling a similar marsh if it is performed concurrently and on site (10). Mitigation for this marsh might also be possible by concurrently restoring a _Spartina_ marsh in the immediate vicinity (11). However, it would take an unusual set of circumstances to approve a mitigation plan for the loss of this marsh which includes creation of a cypress swamp at some other location (7).

Figure 1 will help simplify the selection of ecologically acceptable mitigation options, and can be used to quickly compare the "values" of different options. However, because of the myriad of site specific possibilities and restrictions, Figure 1 and recommendations made herein should be used as a guide and not as the only criterion used in decision making.

Some reviewers of this method have suggested determining the acreage of mitigation which is required by the value of the mitigation given in such a matrix. For example, if a value of 10 is the acceptable level of mitigation options, but the applicant can only perform mitigation options with a value of 7, the option with the lower value may be acceptable if the acreage was increased by 10/7. It is possible that a refinement of a general technique such as this may be acceptable in some regions.

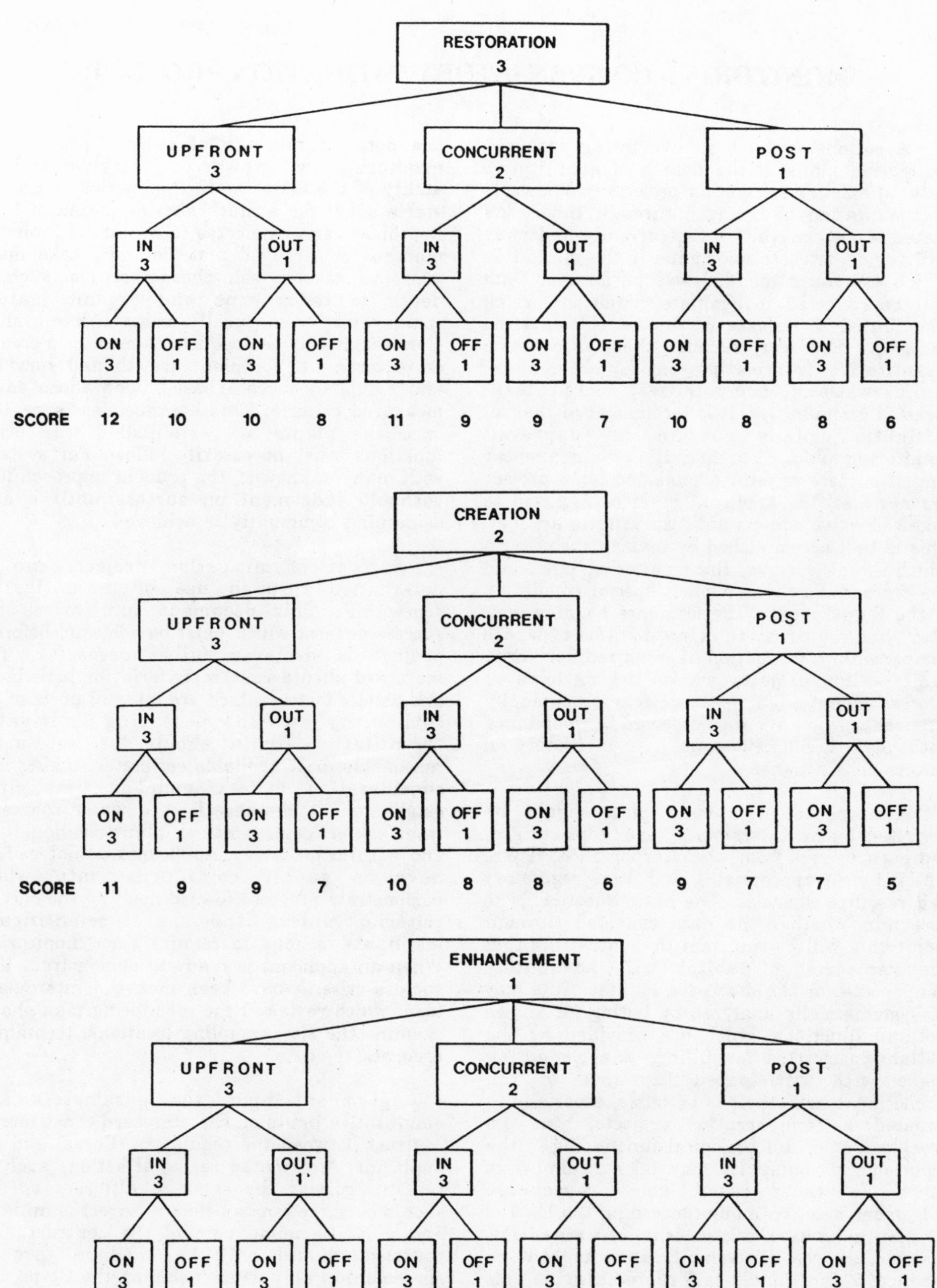

Figure 1. Options to be considered in the preparation and evaluation of mitigation plans. Interpretation of the scores is explained in the text.

MONITORING COMPENSATORY MITIGATION PROJECTS

A serious problem in evaluating proposed mitigation plans is the dearth of quantitative data on existing mitigation projects, particularly documentation of changes through time. As stated above, several studies currently underway will revisit sites to determine if the mitigation which was recommended was performed. Data will be collected to evaluate conditions which contributed to success or failure of completed projects. However, because there is so little quantitative information on replacement of many wetland communities, particularly forested communities, it is recommended that all mitigation plans contain an approved monitoring plan. Further, it is recommended that if success criteria established for a project are not met, the applicant must be required to take corrective actions until the criteria are met. This is best accomplished by making the criteria which define success, the monitoring plan, and the corrective actions explicit special conditions of the Corps permit. Performance bonding may also be required in circumstances where compensatory mitigation is required for large projects, for projects where the anticipated success is not high, and projects proposed by applicants who have a poor record of compliance with permit conditions or have a history of enforcement actions.

A proposed monitoring plan should be reviewed by an interagency team consisting of representatives from the Corps, EPA, FWS, NMFS (when appropriate), and State regulatory and resource agencies. The basic question is to ascertain whether the data collected through monitoring will be sufficient to demonstrate that the replacement habitat will adequately compensate for the destroyed habitat. This may be systematically analyzed by listing all known wetland functions which are provided by the wetland permitted for filling, and comparing these with anticipated functions of the replacement habitat. It is probable, especially in forested wetland creation projects, that the development of full ecological functioning of the replacement community may take a number of years. In these cases, some reasonable judgement must be made concerning the level of function to be used as a measure of success. For example, the goal of swamp creation might be to produce a functional, self reproducing, stable community. This may take 30 years or more to accomplish. It would be reasonable to predict the potential success of such a project by monitoring tree growth and survival and the ability of the faster maturing species to produce viable seeds for a much shorter period of time. Recent evidence indicates that planted intertidal marshes on dredged material may take many years to develop soil characteristics, such as depth to redox zone and organic matter, comparable to naturally occurring marshes. However, it may be adequate to monitor a created or restored site for plant growth and survival, and establish success based upon these easily measured criteria. This approach assumes that once the plants are established, the other functions must necessarily follow. For systems with many unknowns, the prudent approach is to withhold judgement on success until a self-sustaining community is achieved.

Success of mitigation projects can be determined through use of a "mitigation scorecard". This document summarizes the success criteria which must be achieved before a project is declared fully successful. The scorecard should contain criteria for both biotic and abiotic factors which are integral parts of the community which is being mitigated. Quantitative limits should be set using reasonable, best available estimates, taking into consideration factors such as time since establishment, distance from a water source or from donor communities, planting densities, and natural processes. Applicants do not usually have to supply continuous data which demonstrate progress toward meeting the success criteria, unless there are scientifically justifiable reasons to require such monitoring. When an applicant is ready to demonstrate that success criteria have been met, the interagency team which reviewed the monitoring plan should examine the site, sampling locations, techniques used, and the data.

Agreement upon the parameters and quantitative limits on the scorecard constitutes a contract between the regulators (Corps) and the applicant. Anticipated remedial actions such as replanting with the same or different species should be agreed upon before a project is initiated and must be made part of the contract. The contract should only be changed through agreement of all parties including the Corps, the applicant, and the interagency review team.

SUMMARY OF GENERAL RECOMMENDATIONS FOR PREPARATION AND EVALUATION OF MITIGATION PLANS

The following is a list of specific issues which should be explicitly addressed during the permit review process to improve the prospects of successful compensatory mitigation of wetland losses for projects which otherwise comply with the Section 404 Guidelines. A similar listing is found in Reimold and Cobler (1986) and other publications.

SLOPES AND GRADIENTS

A common problem to many unsuccessful sites is steepness of slopes within the mitigation area or in surrounding areas. A gentle slope of 1:5 to 1:15 (vertical:horizontal) is recommended for successful wetland establishment since it provides maximum flooding and minimizes erosion.

SOILS

Plant growth can be facilitated by proper soils. If it is not possible to supply proper wetland soils, the area may be mulched to provide an organic surface horizon, and/or fertilized to stimulate plant growth. Some mitigation sites are slow to become established because of the lack of proper soil microflora. It is possible that such soils must be inoculated with soil microflora during site preparation to assure rapid and healthy plant growth.

PLANT MATERIAL

Transplanting sprigs or other plant stock at mitigation sites is usually preferred over allowing natural colonization, since transplanting promotes favorable community composition and hastens the establishment of a functional wetland. Sites should be planted at appropriate times of the year with stock or seeds that are genetically compatible with vegetation native to the locale. Planting density and survival rates should be specified. Donor sites should be protected from over-harvesting.

HYDROLOGY

Proper water depth and periodicity is the most important element in planning a successful mitigation plan. Long term stage records or tidal data should be used to determine depth and extent of flooding limits, and mitigation sites should be planned at elevations within the flooding tolerance limits of the community type. The methodology of establishing frequency of inundation in tidal areas which are distant from a bench mark is given in Marmer (1951) and Swanson (1974).

MONITORING

Success criteria should be agreed upon before issuance of a permit. Post-project monitoring should be continued by the applicant until success criteria are met. Changes in the mitigation plan or success criteria should be possible only with approval of all members of the interagency review team. However, the Corps has the final authority on permits, including special conditions to permits which might contain mitigation plans and success criteria.

TIMING

Upfront mitigation should be encouraged, particularly when it is determined that risk of failure is high. Concurrent mitigation is acceptable for projects where success is probable.

LOCATION

On site mitigation should be encouraged so that there is no net loss of wetland type or functions to the local ecosystem.

COMMUNITY TYPE

Replacement of the same kind of habitat which was destroyed should be encouraged in order to restore the natural balance of community types in the ecosystem.

LITERATURE CITED

Marmer, H.A. 1951. Tidal Datum Plans. Special Publication 135, Department of Commerce, Coast and Geodetic Survey.

Reimold, R.J. and S.A. Cobler. 1986. Wetland Mitigation Effectiveness. EPA Contract No. 68-04-0015, Metcalf and Eddy, Inc., Wakefield, Massachusetts.

Swanson, R.S. 1974. Variability of Tidal Datums and Accuracy in Determining Datums from Short Series of Observations. National Oceanic and Atmospheric Administration Technical Report NOS 64.

INDEX

References in *italic* refer to illustrations and tables.

ALSO AVAILABLE FROM ISLAND PRESS

Ancient Forests of the Pacific Northwest
By Elliott A. Norse

The Challenge of Global Warming
Edited by Dean Edwin Abrahamson

The Complete Guide to Environmental Careers
The CEIP Fund

Creating Successful Communities: A Guidebook for Growth Management Strategies
By Michael A. Mantell, Stephen F. Harper, and Luther Propst

Crossroads: Environmental Priorities for the Future
Edited by Peter Borrelli

Environmental Agenda for the Future
Edited by Robert Cahn

Environmental Restoration: Science and Strategies for Restoring the Earth
Edited by John J. Berger

The Forest and the Trees: A Guide to Excellent Forestry
By Gordon Robinson

Forests and Forestry in China: Changing Patterns of Resource Development
By S.D. Richardson

From The Land
Edited and compiled by Nancy P. Pittman

Hazardous Waste Management: Reducing the Risk
By Benjamin A. Goldman, James A. Hulme, and Cameron Johnson
for Council on Economic Priorities

Land and Resource Planning in the National Forests
By Charles F. Wilkinson and H. Michael Anderson

Last Stand of the Red Spruce
By Robert A. Mello

Natural Resources for the 21st Century
Edited by R. Neil Sampson and Dwight Hair

The New York Environmental Book
By Eric Goldstein and Mark Izeman

Overtapped Oasis: Reform or Revolution for Western Water
By Marc Reisner and Sarah Bates

Permaculture: A Practical Guide for a Sustainable Future
By Bill Mollison

The Poisoned Well: New Strategies for Groundwater Protection
Edited by Eric Jorgensen

Race to Save the Tropics: Ecology and Economics for a Sustainable Future
Edited by Robert Goodland

Reforming the Forest Service
By Randal O'Toole

Reopening the Western Frontier
From *High Country News*

Research Priorities for Conservation Biology
Edited by Michael E. Soulé and Kathryn Kohm

Resource Guide for Creating Successful Communities
By Michael A. Mantell, Stephen F. Harper, and Luther Propst

Rivers at Risk: The Concerned Citizen's Guide to Hydropower
By John D. Echeverria, Pope Barrow, and Richard Roos-Collins

Rush to Burn: Solving America's Garbage Crisis?
From *Newsday*

Saving the Tropical Forests
By Judith Gradwohl and Russell Greenberg

Shading Our Cities: A Resource Guide for Urban and Community Forests
Edited by Gary Moll and Sara Ebenreck

War On Waste: Can America Win Its Battle with Garbage?
By Louis Blumberg and Robert Gottlieb

Western Water Made Simple
From *High Country News*

Wildlife of the Florida Keys: A Natural History
By William D. Lazell, Jr.

For a complete catalog of Island Press publications, please write:

Island Press
Box 7
Covelo, CA 95428
(1-800-828-1302)